ELECTRONIC STRUCTURE AND PROPERTIES OF HYDROGEN IN METALS

NATO CONFERENCE SERIES

I   Ecology
II   Systems Science
III   Human Factors
IV   Marine Sciences
V   Air–Sea Interactions
VI   Materials Science

VI MATERIALS SCIENCE

ELECTRONIC STRUCTURE AND PROPERTIES OF HYDROGEN IN METALS

Edited by

P. Jena
and
C. B. Satterthwaite

Virginia Commonwealth University
Richmond, Virginia

Published in cooperation with NATO Scientific Affairs Division

PLENUM PRESS · NEW YORK AND LONDON

Library of Congress Cataloging in Publication Data

NATO International Symposium on the Electronic Structure and Properties of Hydrogen
in Metals (1982: Richmond, Va.)
  Electronic structure and properties of hydrogen in metals.

  (NATO conference series. VI, Materials science; v. 6)
  "Proceedings of a NATO International Symposium on the Electronic Structure and
Properties of Hydrogen in Metals, held March 4–6, 1982, in Richmond, Virginia"—
Verso t.p.
  Includes bibliographical references and index.
  1. Metals—Hydrogen content—Congresses. 2. Electronic structure—Congresses. I.
Jena, P. II. Satterthwaite, C. B., 1920–   . III. North Atlantic Treaty Organization.
Scientific Affairs Division. IV. Title. V. Series.
TN689.2.N36 1983   669′.94   82-16569
ISBN 978-1-4684-7632-3   ISBN 978-1-4684-7630-9 (eBook)
DOI 10.1007/978-1-4684-7630-9

Proceedings of a NATO International Symposium on the Electronic Structure and
Properties of Hydrogen in Metals, held March 4–6, 1982, in Richmond, Virginia

© 1983 Plenum Press, New York
Softcover reprint of the hardcover 1st edition 1983

A Division of Plenum Publishing Corporation
233 Spring Street, New York, N.Y. 10013

PREFACE

Hydrogen is the smallest impurity atom that can be implanted in a metallic host. Its small mass and strong interaction with the host electrons and nuclei are responsible for many anomalous and interesting solid state effects. In addition, hydrogen in metals gives rise to a number of technological problems such as hydrogen embrittlement, hydrogen storage, radiation hardening, first wall problems associated with nuclear fusion reactors, and degradation of the fuel cladding in fission reactors. Both the fundamental effects and applied problems have stimulated a great deal of interest in the study of metal hydrogen systems in recent years. This is evident from a growing list of publications as well as several international conferences held in this field during the past decade. It is clear that a fundamental understanding of these problems requires a firm knowledge of the basic interactions between hydrogen, host metal atoms, intrinsic lattice defects and electrons. This understanding is made particularly difficult by hyrogen's small mass and by the large lattice distortions that accompany the hydrogenation process.

The purpose of the "International Symposium on the Electronic Structure and Properties of Hydrogen in Metals" held in Richmond, Virginia, March 4-6, 1982 was to increase our fundamental understanding of hydrogen in metals. Such knowledge is essential in solving technologically important questions. The symposium consisted of twenty-two invited papers and seventy-two contributed poster presentations and attracted nearly 150 participants from thirteen countries. The proceedings of this symposium constitute this book.

The discussions and presentations included a number of topics: phase diagrams, spinodals, order-disorder transformations, thermodynamics, neutron scattering, elastic interactions, electronic structure, magnetic and hyperfine properties, superconductivity, diffusion, interaction with lattice defects, chemisorption, and catalysis. A variety of experimental and theoretical techniques were discussed and the complimentarity of these techniques in probing the electronic structure and properties of metal-hydrogen systems was emphasized.

The structure, phase diagrams, phase transitions, stabilities, stoichiometries and site occupancies were presented from an experimental viewpoint and discussed theoretically in terms of electronic and elastic interactions. The electronic structure and properties were studied with de Haas van Alphen effect, XPS/UPS and energy loss spectroscopy, Mössbauer effect, perturbed angular correlation, muon spin rotation, and electron and nuclear spin resonance techniques. The results were analyzed theoretically using conventional jellium, cluster, and band structure models. The mechanism of trapping of hydrogen by impurities and imperfections and the effects of trapping on diffusion entered the discussion. New work on superconductivity and the status of the theory were presented. Throughout much of the program isotope effects and quantum solid effects associated with the small hydrogen mass played a role. The present understanding of hydrogen chemisorption and catalysis was reviewed. Because of the limited time available for the conference, many important areas of hydrogen in metals had to be left out.

We wish to thank the conferees for the high quality of their participation in the invited and poster presentations and exchange of ideas. We thank the members of the international steering committee: Professors G. Alefeld, B. Baranowski, H. K. Birnbaum, J. P. Burger, J. Friedel, F. D. Manchester, D. T. Peterson, J. M. Rowe, K. S. Singwi, A. M. Stoneham, W. E. Wallace and T. Yamazaki for their advice in the choice of scientific topics and organization of the symposium. We owe special gratitude to the local organizing committee: Professors C. T. Butler, T. D. Doiron, D. W. Hartman, W. J. Kossler, H. E. Schone, J. Ruvalds, C. E. Stronach, and K. R. Squire for their help during the one and half year this symposium was being planned. We are grateful to the faculty and staff of the Physics Department of Virginia Commonwealth University, in particular Professor R. H. Gowdy, Prof. G. B. Taggart, Mr. James Spivey, Ms. Kim Hulce, and Miss Belinda Ashmore for their individual help and advice.

This symposium was made possible by a generous grant from the Special Program Panel on Materials Science of the North Atlantic Treaty Organization. The symposium was hosted by the Virginia Commonwealth University. We also acknowledge with thanks the financial support of the Virginia Commonwealth University, the National Science Foundation, the Exxon Corporation, and Philip Morris, U.S.A.

P. Jena
C. B. Satterthwaite

Richmond, Virginia
May, 1982

CONTENTS

NEUTRON SCATTERING

ELECTRONIC STRUCTURE

SUPERCONDUCTIVITY

DIFFUSION

INTERACTION WITH DEFECTS, CHEMISORPTION

SUMMARY

METAL-HYDROGEN PHASE DIAGRAMS

T. Schober

IFF, KFA Jülich
517 Jülich, W.-Germany

ABSTRACT

   Work on the phase diagrams of the systems NbH(D), TaH(D) and
VH(D) is reviewed. Emphasis is placed on more recent experimental
progress. In the system NbH(D) the occurrence of multiple ordered
low temperature phases and hydrogen density waves is discussed. In
the system TaH(D) a detailed description of the low-temperature
phase relationships is still unavailable. As to the systems VH and
VD (which display a drastic isotope effect) a better characterization of the phase transitions was achieved. Much of the discrepancy between the different phase equilibrium studies of the last
years is now resolved. A new phase transition into an incommensurate
state is described for the system VD. In summary, the three systems
are much more complex than anticipated. They display first and second order transitions, incommensurate structures with hydrogen
concentration waves, possibly metastable phases, miscibility gaps
and pronounced isotope effects. There is strong reason to believe
that a similar variety of effects may also be encountered in other
M-H systems.

INTRODUCTION

   Phase diagrams play a very important rôle in materials science
as they provide "maps" of the temperature and concentration regions in which certain phases or phase mixtures exist. Such information is essential for the preparation of samples, for applications
and for the correct interpretation of experiments. As to metal-hydrogen phase diagrams, there are the fundamental studies of the
reactions of Pd, Nb, Ta and V with H(D,T) and also the investigations

of the technically more important phase diagrams such as the systems
ZrH, TiH, UH, LiH etc. A third group of phase diagrams which are
gaining increasing importance describe the reaction of H(D,T) with
storage materials such as FeTi, LaNi$_5$, Mg$_2$Ni and the Laves phases.
This latter group of diagrams is relatively unexplored and will be
presumably at the focus of further investigations. By "phase diagram" we mean in general a three dimensional p-c-t diagram as
recently discussed[1]. In the case of sealed surfaces (as for Nb, Ta,
V) we consider projections along the p-axis. The present work
is concerned with an overview of phase diagrams and relationships
for the systems NbH(D), TaH(D) and VH(D). It is anticipated that
many of the phenomena encountered in these systems will also be of
importance in the two other groups of diagrams mentionned above.

THE NIOBIUM HYDROGEN SYSTEM

     For purposes of illustration we present in Fig. 1 a recent
NbH diagram[2-5] in which the low-temperature, high concentration
part (see box) has been left out. It will be discussed below. As
to the "terminal solubility for hydrogen" (TSH), most recent studies
as listed in[4,5] are in good agreement. The triple point temperature for H was listed[5] as (88 ± 1)°C and for D[5] (99 ± 1)°C, whereas DTA experiments[4] yielded values about 2° lower. The incoherent
(α,α') miscibility gap was precisely measured in recent work[5].
The temperatures for the $\beta \to \beta + \alpha' \to \alpha$ transition in Fig. 1 are from
DTA and resistivity data[2,3] which are in good agreement with metallographic data[6]. As to the $\varepsilon \to \zeta \to \beta$ transition upon heating
at c ≤ 70 %, a stepwise decay of phase ε was noted[2,3,7]. It first
decays into η[7] which is only stable for about 2°. It then transforms at -66 °C to ζ which in turn changes to β at -45 °C. A different view of these events is[8] that at -66 °C a partial second
order disordering reaction of ε takes place which occurs between
-66 °C and 0 °C, finally leading to β. In the latter view ζ would
not be a distinct phase. (The above reactions apply only to the
hydride phases which are in equilibrium with α at that temperature).

Low Temperature - High Concentration Transitions (see box in Fig. 1)

     Very thorough susceptibility and DTA work[9] recently resulted
in a detailed phase diagram, a portion of which is reproduced in Fig. 2.
It is mainly characterized by low temperature phases at c = 0.73,
0.78, 0.83, 0.85, 0.9 and 1.0. We recognize the more important
phases ε and λ which are well known from previous diagrams.
The phase γ in Fig. 2 is probably identical with Pick's cubic
γ-phase[10]. Birnbaum et al.[8], however, have a different view of
the low-temperature, high concentration region. They report at
least 3 λ-phases with a long-period superlattice in the range
72 - 77 %, 85 - 90 % and 91 - 96 %. These phases consist of a

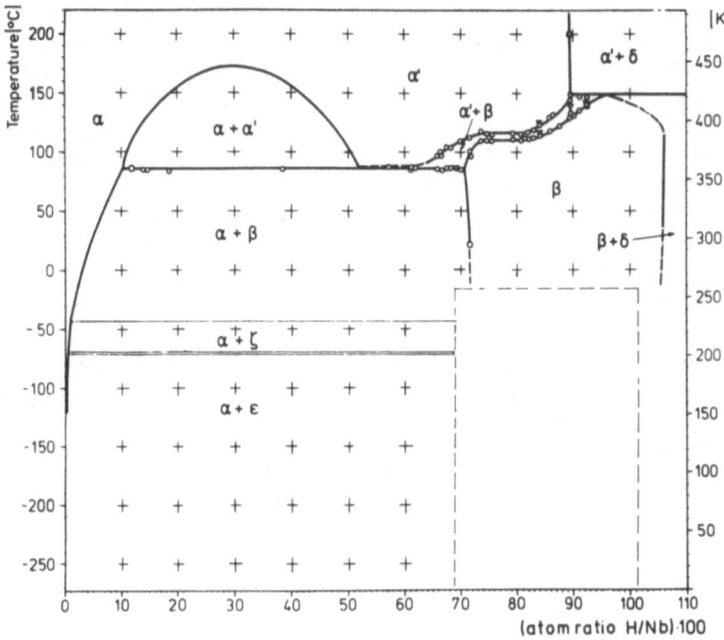

Fig. 1. NbH phase diagram – without low temperature phases. Open circles: DTA work.

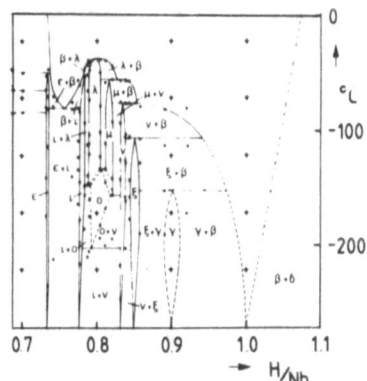

Fig. 2. Partial NbH diagram[9] obtained from susceptibility and DTA measurements. For competing models of this range see Ref. 8,14.

stacking in the $[001]_c$ of β structure unit cells in two variants with superimposed commensurate hydrogen density waves (wavelengths: 14/3, 16/3 and 18/3 a$[100]$). No evidence was obtained in their work for the cubic γ phase. – In different TEM and metallographic work[11] it could be shown that a) λ has an anisotropic crystal structure as witnessed by its domain structure; b) that a transition could be observed at c ≃ 85 % and T ≈ -66 °C from an ordered (high temperature) phase with domain boundaries to a cubic low temperature phase, the phase γ, without domain boundaries. γ was found to display superlattice reflections[11]. Thus, the existence of a cubic γ phase seems well established; further support for it was recently obtained[12,13].

In recent acoustic work[14] several of the above phase transitions were examined. At the composition of 83 % phase transitions were observed at -173 °C and -55 °C. The transition at -173 °C was found to extend from 82% to 93% and to be of second order. This important observation of second order transitions in the Nb-H system suggests that even further transitions in Fig. 2 may be of second order. Obviously, the corresponding two-phase regions in Fig. 2 would have to be omitted.

Furthermore, we wish to mention recent susceptibility data in the range 0 - 70 % and T < -40 °C[9]. Surprisingly, these authors observed anomalies in the temperature dependence of the susceptibility and constructed a "phase diagram" in that range displaying two new phases (ϑ,η) and various phase boundaries. In the present author's opinion these effects do not correspond to true phase transitions; and η are no new hydride phases. It is felt that Fig. 1 is basically correct in that region. There is ample TEM evidence[4,7,8] that the only hydrides which are in equilibrium with α is ε at low T's and ζ at intermediate T's (see Fig. 1). Thus, further work is indicated to clarify this rather serious discrepancy.

Finally, let us consider isotope effects in the systems NbH and NbD. Apart from the above difference in the triple point temperature we note a pronounced isotope effect in the transition temperatures $\beta \to \beta + \alpha' \to \alpha'$ (c > 50 %) when data in Ref. 3 are compared with NbD values in Ref. 15.

THE TANTALUM HYDROGEN (DEUTERIUM) SYSTEM

As a first reference, we present in Fig. 3 a recent TaH diagram[4,19] a portion of which was recently revised[9] (Fig. 4, see discussion below). Starting at low concentrations, the TSH(D) (solvus) has recently been determined again with DTA using a better resolution than previously[4]. The improved solubility limit for H and D is shown in Fig. 5. It is seen that virtually no isotope effect is present. A rather drastic isotope effect between the systems TaH and TaD is evident at elevated concentrations and temperatures[4,16-19]. Two examples may be noted. In the system TaD, the $\varepsilon(=\beta_2)$ phase field is very narrow when compared with TaH[17,18]. The $\delta \to \alpha$ transition at 77 % occurs for TaH between ∿70 and 76 °C, for TaD, however, between ∿96 and 110 °C[16-18]. These isotope effects have been somewhat neglected in the past; a precise determination of the TaD diagram from 40 to 80 % is clearly needed.

As to the low-temperature high-concentration region in Fig. 3 and Fig. 4 we first note that the structure of phase η at 66 % has been determined by neutron diffraction[17,18]. η is pseudo-orthorhombic with A = 4 $\sqrt{2}$ a, B = 3 $\sqrt{2}$ a and C = a. The difference bet-

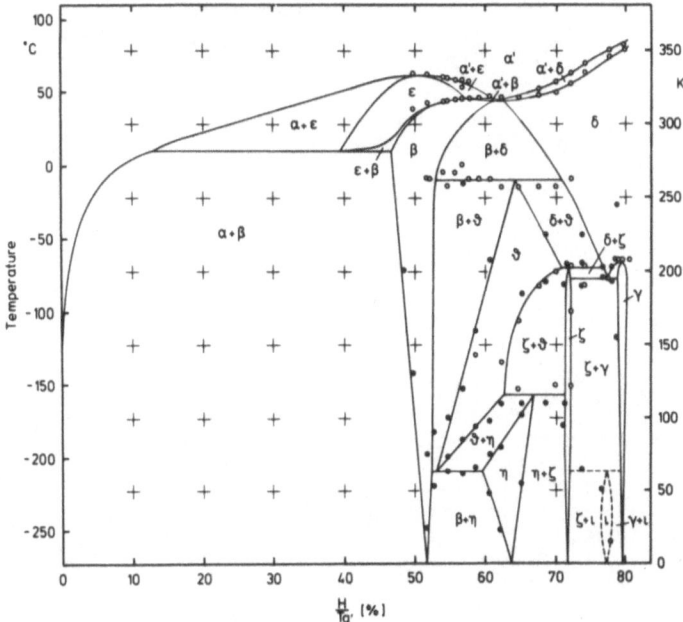

Fig. 3. TaH diagram after Ref. 4,19. It is based
on DTA and susceptiability measurements.

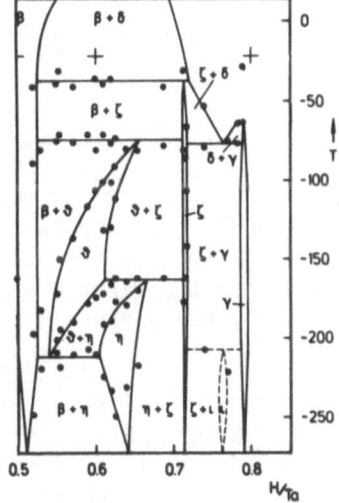

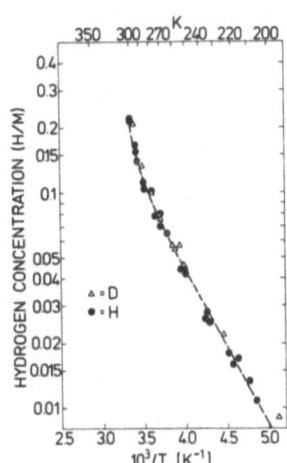

Fig. 4. Partial TaH diagram[9]
containing the revisions
described below.

Fig. 5. DTA determination of the
TaH(D) TSH using rather thick specimens and therefore, improved accuracy.

ween Fig. 3 and Fig. 4 consists in the introduction in Fig. 4 of
triple-point lines at -78 °C and -38 °C between 52 and 72 %, and

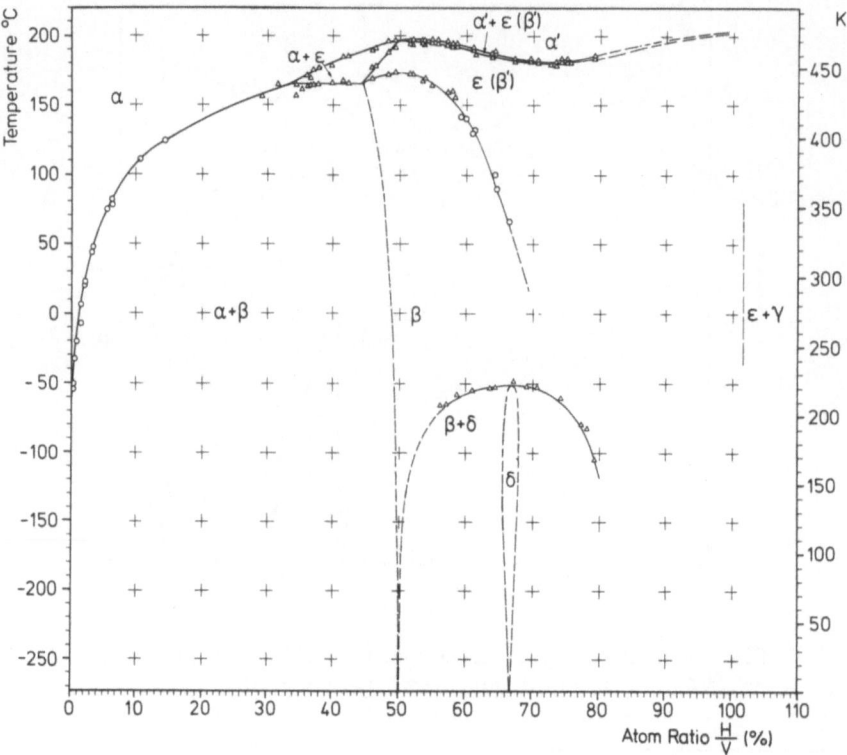

Fig. 6. Updated VH phase diagram containing recent revisions[21].
The data points shown are DTA values[4].

the removal of a triple point line at −13 °C which may have been
due to artifacts[9]. We note however that no DTA signals (but only
χ-anomalies) were observed at the lines of −38 °C and −78 °C. These
transitions should, therefore, be considered with a certain amount
of caution. In weighing all the data, firm evidence seems
to be available for the following TaH phases for the limit T → 0:
the β-phase Ta_2H; the η-phase Ta_3H_2, ζ at ∼72 %, the γ-phase at
80 % and a cubic low temperature phase[11,20]. This latter phase was
observed below −150 °C at concentrations believed to be >85 %. It
was neither assigned a Greek letter nor entered into the phase diagram for lack of further information. We believe this phase to be
similar to Pick's γ-NbH phase[10]. We further assume that additional
TaH phases exist between 80 and 90 %, similar to the system NbH.
We consider 90 % to be the limiting concentration for H in Ta.

THE VANADIUM-HYDROGEN SYSTEM

     A recently updated VH diagram[21,22] is seen in Fig. 6. Its TSH
(solvus) is the same as in earlier work[4]. There is general agree-

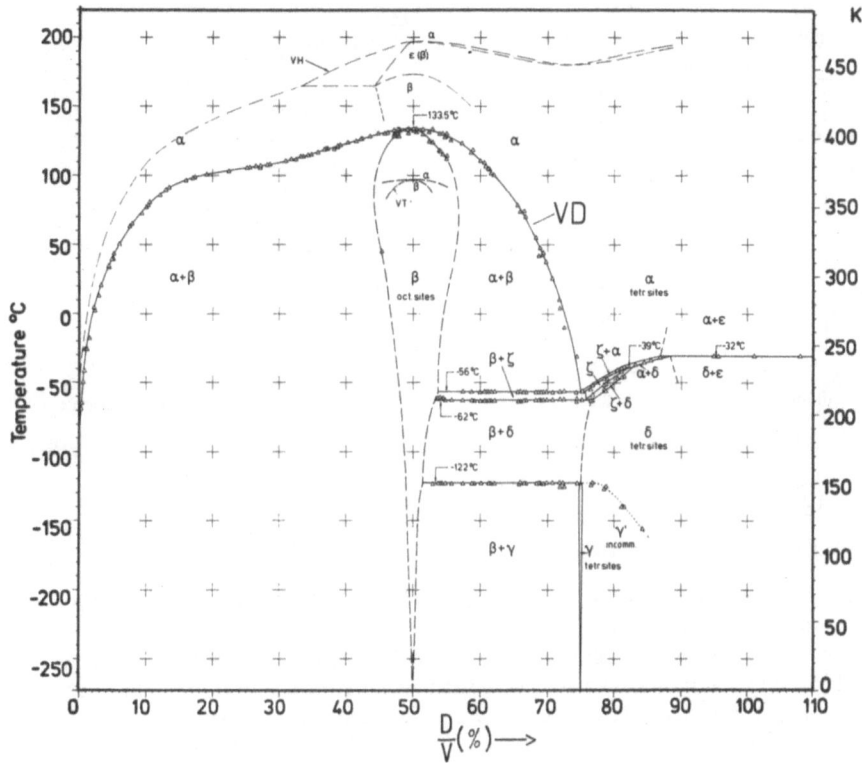

Fig. 7. Updated VD phase diagram with recent revisions[21].
The data points again are DTA values[4]. One VT point is
taken from Ref. 28.

ment now that the β → ε(β') transition is of second order[17,18,23-25].
We have modified our original phase diagram accordingly*). This
transition has become the prototypical second order phase transition in M-H systems. Recent TEM work[21] has shown, however, that
the δ → β transition is of first order since the two phases were
observed in coexistence. An open area in the VH system is still
the region from 80 to 100 % and low temperatures where it is con-
ceivable that further phases with octahedral occupation exist. Recently, a tentative extension of the VH diagram to 200 % was proposed[26].

THE VANADIUM DEUTERIUM SYSTEM

     Again, a recently modified version[21] is shown in Fig. 7. The

_____

*)The old (β+ε) two phase field was dropped. Also, a rounded top
   (Fig. 6) was used to separate (β+δ) and δ from β. This construction is also compatible with the old DTA data[4].

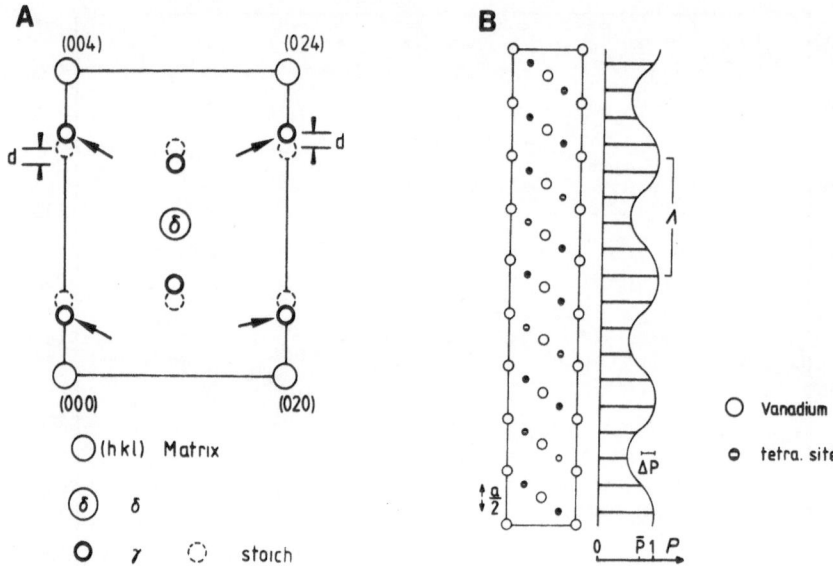

Fig. 8. Incommensurate deuterium density fluctuations in the hyperstoichiometric γ'-VD phase. a) Schematic presentation of a typical electron diffraction pattern, $(110)_c$-orientation. The hydrogen superlattice reflection (arrows) are displaced by d away from the stoichiometric positions. The new positions are incommensurate with the matrix. b) Real space model of an incommensurate deuterium density modulation which would lead to the superlattice reflections in a). Here, Λ = wavelength of the modulation, ΔP its amplitude around the average concentration P̄. Densities are schematically indicated by the white areas within the filled circles.

only changes occur at low temperatures and high concentrations. The γ → δ transition was found to be of second order. Thus, the (γ+δ) two phase field was omitted. If c was raised beyond 75 %, γ transformed to an incommensurate phase (γ') with concentration dependent hydrogen density modulations[21,22]. A typical case is shown in Fig. 8; for details see the caption. The existence of phase ζ could be confirmed[21]. It is uncertain, however, whether a (ζ+δ) two phase field exists. - We note that certain low-temperature deuterides may only be metastable[27] and transform with time to different structures.

THE VANADIUM TRITIUM SYSTEM

    One data point in Fig. 7 refers to VT[28] and shows that the VT diagram is displaced by ∿30 °C to lower temperatures when compared with VD. Work on the VT diagram is in progress in this laboratory[29].

DISCUSSION

The complexity and the isotope effects of the phase diagrams are certainly confusing. We note first and second order phase transitions, incommensurate hydrogen density concentration waves, possibly metastable phases and miscibility gaps. We readily admit that some revisions of the present diagrams may be expected in the future. A theoretical description on the basis of elastic interactions is only available for portions of the NbH(D) diagram[30,31]. Thus, electronic effects seem to determine the general behavior of these systems. As to isotope effects, there are theoretical estimates[32,33] of the direction and the amount of the isotope effect in certain phase transitions. One is far, however, from a quantitative theoretical description which could predict for all the interstitial phases and the three isotopes the range of homogeneity and the transition temperatures.

ACKNOWLEDGEMENT

Helpful discussions with Drs. J.-M. Welter, M. Richardson, U. Köbler and H. Wenzl are acknowledged.

REFERENCES

1. "Gase und Kohlenstoff in Metallen", E. Fromm and E. Gebhardt, ed., Springer, Berlin (1976).
2. H. Wenzl and J.-M. Welter in "Curr. Top. in Mat. Science" Vol. 1, E. Kaldis, ed., North Holland, Amsterdam (1978)
3. J.-M. Welter, M.A. Pick, T. Schober, J. Hauck, H.J. Fenzl and H. Wenzl, Proc. 2nd Int. Congr. H in Metals 1D3, Paris (1977)
4. T. Schober and H. Wenzl in "Hydrogen in Metals II", Top. Appl. Phys. 29, G. Alefeld and J. Völkl, ed., Springer Berlin (1978)
5. H. Zabel and J. Peisl, J. Phys. F: Met. Phys. 9 1461 (1979)
6. T. Schober and H. Wenzl, phys. stat. sol. (a) 33 673 (1976)
7. W. Pesch, T. Schober and H. Wenzl, Scripta Met. 12 815 (1978)
8. B.J. Makenas and H. Birnbaum, Acta Met., to be published (1982)
9. U. Köbler and J.-M. Welter, J. Less-Comm. Met., in press (1982)
10. M.A. Pick and R. Bausch, J. Phys. F: Met. Phys. 6 1751 (1976)
11. T. Schober, Proc. 9th Int. Congr. El. Micr. p. 644 Vol I, Toronto (1978)
12. T. Schober and U. Linke, J. Less-Comm. Met. 44 77 (1976)
13. Y.S. Hwang, D.R. Torgeson and R.G. Barnes, Scripta Met. 12, 507 (1978)

14. V.A. Melik-Shakhnazarov, I.N. Bydlinskaya, I.A. Naskidashvili, N.L. Arabadzhyan and R.V. Chachanidze, Zh. Eksp. Teor. Fiz. (USSR) <u>81</u> 314 (1981)

15. K. Nakamura, Bull. Chem. Soc. Japan <u>46</u> 2028 (1973)

16. F. Reidinger and J.J. Reilly (to be published)

17. H. Asano and M. Hirabayashi,"Metal Hydrides", G. Bambakidis ed. Plenum Publ. Corp. N.Y. 1981

18. M. Hirabayashi and H. Asano,"Metal Hydrides", G. Bambakidis ed. Plenum Publ. Corp. N.Y. 1981

19. U. Köbler and T. Schober, J. Less-Comm. Met. <u>60</u> 101 (1978)

20. T. Schober, Proc. 2nd Int. Congr. H in Metals, 1D2, Paris (1977)

21. W. Pesch, T. Schober and H. Wenzl, Scripta Met. (in press)(1982)

22. W. Pesch, Phil. Mag. (in press) 1982

23. M.W. Pershing, G. Bambakidis, J.F. Thomas, Jr. and R.C. Bowman, Jr., J. Less-Comm. Met. <u>75</u> 207 (1980)

24. H.S.U. Jo and S.C. Moss, Modulated Structures, ed. J.M. Cowley and J.B. Cohen, AIP Conf. Proc. No. 53 (1979)

25. Y. Fukai and S. Kazama, Acta Met. <u>25</u> 59 (1977)

26. T. Schober and W. Pesch, Z. Phys. Chem. NF <u>114</u> 21 (1979)

27. I.R. Entin, V.A. Somenkov and S.Sh. Shil'shtein, Sov. Phys. Sol. State <u>18</u> 1729 (1976)

28. R.C. Bowman, Jr., A. Attala and B.D. Craft, Paper presented at Gen. Meeting Am. Phys. Soc. Chicago (1979)

29. R. Lässer and T. Schober (to be published)

30. G. Alefeld, phys. stat. sol. <u>32</u> 67 (1969)

31. H. Wagner in "Hydrogen in Metals I", Top. Appl. Phys. <u>28</u>, G. Alefeld and J. Völkl, ed., Springer, Berlin (1978)

32. I.R. Entin, V.A. Somenkov and S.Sh. Shil'shtein, Sov. Phys. Solid State <u>16</u> 1569 (1975)

33. L.A. Maksimov, I.Ya. Polishuk and V.A. Somenkov, Sov. Phys. Solid State <u>23</u> 339 (1981)

A REVIEW OF THE STATISTICAL THEORY OF THE PHASE-CHANGE BEHAVIOR OF

HYDROGEN IN METALS

Carol K. Hall

Department of Chemical Engineering
Princeton University
Princeton, New Jersey 08544

ABSTRACT

   Theoretical treatments of the phase-change behavior of hydrogen
in metals based on the lattice-gas model of statistical mechanics are
reviewed. Early model calculations assumed that hydrogen in metals
could be treated as a lattice-gas model in which the hydrogen-hydrogen interactions were either attractive (as in the Lacher model) or
repulsive (as in the blocking models). Models based on attractive
interactions predicted phase separations between the disordered phases
α and α' but did not predict the correct saturation behavior while models based on repulsive interactions predicted the correct saturation
behavior but could not predict phase transitions. In the mid 1970's
work by Hall and Stell showed that both attractions and repulsions
between hydrogen atoms were necessary to obtain the ordered β phase
as well as the phases α and α'. Work by Horner and Wagner, showed
that the elastically-based hydrogen-metal interaction is mathematically equivalent to an effective hydrogen-hydrogen interaction and
that this effective interaction can be calculated in terms of experimentally measurable quantities. Recently, Futran, Coats, Hall and
Welch have developed a model for hydrogen in niobium based on the
Horner-Wagner work which predicts the α, α' and β phases. The model
has been extended to apply to the case where hydrogen is absorbed in
niobium-vanadium alloys.

INTRODUCTION

   Research in metal-hydrogen systems has been motivated in recent
years both by a desire to have a fundamental understanding of the nature of these systems and by a desire to explore and develop techno-

logical applications. Many of the theories developed to explain the
behavior of hydrogen in metals are based on statistical mechanics
which is a mathematical prescription for obtaining the macroscopic
or thermodynamic properties of a substance in terms of its microscopic
or molecular properties. In this paper we will trace the development
of theories of metal-hydrogen systems based on statistical mechanics
and show how increasing levels of sophistication in performing the
calculations over the years have led to a better understanding of the
phase-change behavior of hydrogen in metals.

PHENOMENOLOGY ASSOCIATED WITH THE PHASE-CHANGE BEHAVIOR OF HYDROGEN IN METALS

Hydrogen forms alloys and compounds with many metals*. The hydrogen molecule dissociates upon entering the metal lattice. The
hydrogen sits on tetrahedral interstitial sites in metals with b.c.c.
lattices (such as niobium, tantalum and vanadium) and on octahedral
sites in metals with f.c.c. lattices (such as palladium). Schematic
versions of the phase diagrams of the H-Nb, H-Ta, H-V and H-Pd systems at low to moderate hydrogen concentrations are shown in Figs.
1(a)-(d). It can be seen that the compounds formed are non-stoichiometric and that there are a number of different phases. In all
materials the α and α' phases refer to disordered phases in which
hydrogen atoms occupy the interstitial sites at random. The lattice
parameter in the α' phase is always greater than the lattice parameter in the α phase because the lattice stretches to accommodate the
higher hydrogen concentration. The β phase refers to a high density
phase in which certain of the interstitial sites available to the
hydrogen atom are more likely to be occupied than other sites. These
sites form an ordered array which can be detected by neutron scattering. This ordering is accompanied by a distortion of the host-
metal lattice, i.e. a structural phase change. At even higher concentrations and lower temperatures, these systems exhibit a number
of other ordered phases each of which is accompanied by a change in
the structure of the host-metal lattice.

EARLY MODEL CALCULATIONS

The problem of finding a correct theoretical explanation for
the phase diagram and thermodynamic properties found when hydrogen
is absorbed in metals has challenged scientists for over fifty years.
The usual approach taken is to develop a microscopic model of hydrogen in a metal based on a set of assumptions concerning the nature

*For a review of the phenomenology associated with the behavior of
hydrogen in metals see Ref. 1.

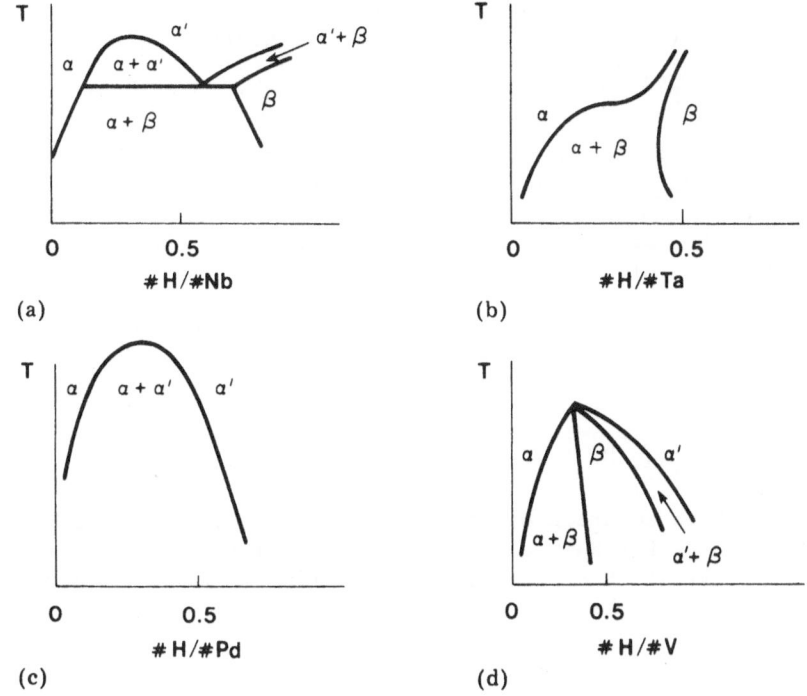

Fig. 1. Schematic versions of the phase diagrams found for metal-hydrogen systems at low to moderate concentrations: (a) H-Nb; (b) H-Ta; (c) H-Pd; (d) H-V.

of the interactions and then to obtain the phase diagram and thermodynamic properties using statistical mechanics. Most microscopic models are of the lattice-gas type[2]. A lattice-gas model is a model in which particles are confined to sites but can jump freely from site to site. Only one particle is allowed per site and interactions between particles are a function of the distances between the sites upon which the particles sit. For the case of hydrogen in metals, the particles are the hydrogen atoms and the lattice is the lattice of interstitial sites within the metal matrix. The term "lattice-gas model" seems to have been coined originally by Lee and Yang[3] who used a model of this type to describe the phase-change behavior of a one-component fluid, however, this type of model had been used much earlier to describe the behavior of solutions[4].

The general procedure in calculating the thermodynamic properties of lattice-gas models according to statistical mechanics[4] is to calculate the grand canonical partition function Ξ (T,μ) which is defined to be

$$\Xi \ (T,\mu) = \sum_{N_H=1}^{N_M} \frac{e^{N_H \mu/kT}}{N_H!} \sum_{States} e^{-E(state)/kT} \tag{1}$$

where T is the temperature, μ is the chemical potential, k is Boltzmann's constant, N_H is the number of hydrogen atoms, N_M is the number of metal atoms, α is the number of interstitial sites available to hydrogen atoms per metal atom, E(state) is the energy associated with a particular state or configuration of N_H hydrogen atoms and the sum is over all possible states or configurations of the hydrogen atoms. Once the grand partition function has been calculated, the equilibrium number of hydrogen atoms, $\bar{N}_H$, can be determined from

$$\bar{N}_H(T,\mu) = kT \left(\frac{\partial}{\partial \mu} \ln \ \Xi \ \right)_T \tag{2}$$

In order to get an equation of state for the model, the expression derived in Eq. (2) for $\bar{N}_H(T,\mu)$ is inverted to give $\mu(T,\bar{N}_H)$. Since hydrogen atoms within the metal are in equilibrium with the hydrogen molecules in the gas phase,

$$\mu(\bar{N}_H,T) = \frac{1}{2} \mu_{H_2,gas} = \frac{1}{2} (\mu^o_{H_2,gas} + RT \ln p) \tag{3}$$

where $\mu^o_{H_2,gas}$ is the standard chemical potential of hydrogen in the gas phase, R is the universal gas constant and p is the pressure. The equation of state is therefore

$$p = \exp(2(\mu(\bar{N}_H,T) - \mu^o_{H_2,gas}))/RT \tag{4}$$

Perhaps the simplest model of hydrogen atoms in a metal is the ideal solution model which was developed to apply to dilute solutions. In the dilute limit, hydrogen atoms are assumed to be so far apart that their mutual interaction may be taken to be zero. The energy of a state may then be taken to be $-N_H W_H$ where W_H is the interaction energy between a hydrogen atom and the site which it occupies. Since all states with N_H hydrogen atoms have the same energy, the sum over all states may be replaced by the number of ways of distributing N_H hydrogen atoms on αN_M sites. The grand partition function becomes,

$$\Xi \ (T,\mu) = \sum_{N_H=1}^{\alpha N_M} \frac{(\alpha N_M)!}{N_H!(\alpha N_M - N_H)!} e^{N_H(\mu-W_H)/kT} \tag{5}$$

$$= (1 - \exp(\mu - W_H)kT)^{\alpha N_M} \qquad (6)$$

where the binomial distribution has been used to perform the sum. The chemical potential is then

$$\mu(T,c) = \ln(c/(1-c)) + W_H/kT \qquad (7)$$

and the equation of state is

$$p = \exp(-2(\mu^0_{H_2,gas} - W_H)/kT)(c/(1-c))^2 \qquad (8)$$

where $c = N_H/\alpha N_M$ is the hydrogen concentration. This equation has the correct behavior in the dilute limit since it obeys Sievert's law, i.e. $c \rightarrow p^{\frac{1}{2}}$. At higher concentrations, however, no phase transitions occur as can be seen by the fact that pressure is a monotonically increasing function of concentration which saturates as $c \rightarrow 1$. It is apparent therefore that regular solution models are not suitable for the explanation of phase-change behavior.

The ideal solution model can be greatly improved upon by considering interactions between hydrogen atoms. In 1937, Lacher[5] developed a model for hydrogen in palladium in which interactions between nearest-neighboring hydrogen atoms were assumed to be attractive. The energy is given by $E(state) = -N_H W_H + N_{HH}W_{HH}$ where W_{HH} is the attractive interaction between nearest-neighboring hydrogen atoms, ($W_{HH} > 0$), and N_{HH} is the number of nearest-neighboring hydrogen atoms. In order to compute the grand partition function, Lacher introduced the Bragg-Williams approximation in which the actual geometrical arrangements of the hydrogen atoms are ignored. Instead one approximates N_{HH} by the number of sites which are occupied, N_H, times the average fraction of the neighbors of these sites which are occupied, $N_H/\alpha N_M$, divided by 2 to prevent double counting. Thus, $E(state) \simeq - N_H W_H + N_H^2 W_{HH}/2\alpha N_M$. The sum over N_H in the grand partition function can then be evaluated by the so-called maximum-term method which involves finding that value of N_H which maximizes the summand and then replacing the sum by the largest term.

The equation of state is

$$p^{\frac{1}{2}} = e^{-(\mu^0_{H_2,gas} - W_H)/kT} \left(\frac{c}{1-c}\right) e^{-W_{HH}c/kT} \qquad (9)$$

A plot of isotherms in the p-c plane shows that below a critical temperature, which is given by $T_c = W_{HH}/4k$, the isotherms exhibit van der Waals loops. By applying Maxwell equal-area constructions, the coexistence curve is obtained. As in the ideal solution model

saturation occurs as $c \to 1$. It can be seen that Lacher's equation
of state is a great improvement over the ideal solution model. By
considering attractions between hydrogen atoms a phase transition
between two disordered phases (the α and α' phases) has been obtained.
Lacher fitted the values of W_{HH} and W_H to the experimental data. He
also fit the value of α to be .58 in order to obtain the correct critical and saturation concentrations thereby assuming that only a
fraction of the interstitial sites were available for hydrogen occupation. Over the years the Lacher model has been refined and extended. Other models based on the assumption of attractive hydrogen-
hydrogen interactions have been developed. For a recent review see
Ref. 6.

Another approach which has been taken in the modeling of hydrogen-metal systems is the so-called blocking approach in which hydrogen-hydrogen interactions are assumed repulsive. The motivation for
this approach is that in many systems saturation occurs at values of
the concentration <1. For example in the H-Nb system, $\alpha = 6$ but saturation appeared to be setting in at $c \simeq 1/6$ indicating that at most
1 in 6 of the sites available to hydrogen could be occupied. One
solution to this is to fit the value of α as Lacher had done, but
another more fundamental approach is to assume that a hydrogen atom
blocks a certain number, say $Z-1$, of its nearest neighbors from being
occupied. Thus the hydrogen atom behaves as if it were a hard-core
atom which occupies Z sites instead of one site. The evaluation of
the grand partition function for hard-core molecules is decidely more
complex than it is for molecules which occupy a single site. One approximation which may be taken is to neglect the overlap of blocked
sites and to replace the combinatorial factor in the regular solution
model by one which incorporates blocking[7]. In this approximation the
chemical potential becomes

$$\mu/kT = \ln(c/(1-Zc)) - W_H/kT \tag{10}$$

A more accurate approximation is to include the overlap of blocked
sites in which case Z becomes a function of concentration, $Z = Z(C)$.
In this case the chemical potential becomes

$$\mu/kT = \ln(c/(1-\overline{Z(c)}c)) - W_H/kT \tag{11}$$

$\overline{Z(C)}$ is the average number of blocked sites. Oates et al.[8] performed
a Monte Carlo calculation of $\overline{Z(c)}$ for hydrogen in niobium and found
that in order to achieve complete filling at $c = 1$, one must choose
$\overline{Z(0)} = 15$ (i.e. the radius of the hard core must be such that the
nearest three shells of neighboring interstitial sites must be excluded from occupation). At concentrations of $c = 1$, $\overline{Z(1)} = 6$. The
blocking models showed that by assuming that the interactions between hydrogen atoms were repulsive, an explanation for the saturation values of the concentration could be obtained. These models
did not however exhibit phase transitions and so could not provide

any explanation for the phase-change behavior of hydrogen in metals.

While the early model calculations provided a good deal of insight into the behavior of hydrogen in metals, they were not capable of providing a comprehensive theory of the phase-change behavior. Those models in which the hydrogen-hydrogen interactions were assumed attractive exhibited phase-changes of the α-α' type but could not (without fitting α) predict the correct saturation behavior or critical point concentration. Those models in which the hydrogen-hydrogen interactions were assumed repulsive, predicted the correct saturation behavior but exhibited no phase changes. Both approaches were incapable of explaining the proliferation of ordered phases (e.g. β, γ, ε etc.) at high concentrations and low temperatures which were discovered in the 1960's. Finally, both approaches ignored the effect of the lattice deformability on the hydrogen-hydrogen interactions. As pointed out by Alefeld[9] and by Wagner[10], the deformability of the metal lattice should strongly influence the interactions between hydrogen atoms and should play an important role in the determination of the phase diagram.

MODEL CALCULATIONS IN THE 1970'S

In the 1970's new research appeared which addressed the problems found in the early model calculations and pointed the way toward the development of a more comprehensive theory of the phase-change behavior in metal-hydrogen systems.

In 1975, Hall and Stell[11] showed that a model of metal-hydrogen systems which contained both long-range attractive and short-range repulsive interactions could explain qualitatively the appearance of disordered phases α and α' and also the appearance of the ordered phase β. This was the first time that any lattice-gas model of metal-hydrogen systems predicted an ordered phase. In addition, as the ratio of the strengths of the attractive to repulsive interactions was varied, the model reproduced qualitatively the phase diagrams found when low to moderate concentrations of hydrogen are absorbed in niobium, palladium, tantalum and vanadium. The results of this research suggested that (a) the direct electronic interactions between hydrogen atoms, which are presumably of short range and primarily repulsive, are responsible for the appearance of ordered phases at high concentrations and (b) the long-range attraction between hydrogen atoms, which is presumably elastic in nature, is responsible for the α-α' transition at low hydrogen concentrations.

Research by Horner and Wagner in 1974[12,13] showed how to take the deformability of the metal lattice into account when modelling hydrogen-metal systems. The deformability of the metal lattice affects the nature of the interactions between hydrogen atoms in the following way: Introducing a hydrogen atom into an interstitial site

causes a strain in the surrounding metal lattice which in turn effects the probability that a neighboring site will be occupied by
another hydrogen atom.

Horner and Wagner developed a lattice-gas model for the hydro-
gen-niobium system. They considered a crystal system of N_M metals on
a b.c.c. lattice loaded with N_H hydrogen atoms which are distributed
on the tetrahedral interstitial sites. A protonic model of the hydrogen atoms is assumed. In addition to a direct electronic interaction between hydrogen atoms (in the form of a screened-Coulomb potential), Horner and Wagner considered elastic interactions between
metal atoms and between hydrogen atoms and metal atoms. Using elasticity theory, Horner and Wagner were able to show that the hydrogen-
metal interaction is mathematically equivalent to an effective hydrogen-hydrogen interaction which contains a weak contribution of
macroscopic range, ΔW_{ab}, and a stronger contribution of microscopic
range, $\tilde{W}_{ab}$ where $\underline{a}$ and $\underline{b}$ are the sites of the interaction hydrogen
atoms. The interaction W oscillates with distance between rep
pulsive and attractive values. The interaction ΔW_{ab}, is a very weak,
long-ranged attraction. Both interactions could be calculated in
terms of experimental quantities such as the phonon spectra, force-
dipole tensor and bulk modulus. The screened-Coulomb interaction between hydrogen atoms, U_{ab}, was approximated by a hard-core repulsion
which extends out to third-nearest neighbors. This was chosen so as
to be consistent with the hydrogen configuration in the β phase in
which hydrogen atoms are never closer than fourth nearest neighbors.

The grand partition function for this model is then given by

$$\Xi \ (T,\mu) = \sum_{N_H=1}^{\alpha N_M} \sum_{\{\tau_a\}} e^{-\frac{1}{kT} \sum_{a,b} (U_{ab} + \tilde{W}_{ab} + \Delta W_{ab}) \tau_a \tau_b} e^{N_H \mu/kT} \qquad (12)$$

where τ_a is the proton occupation number of interstitial site $\underline{a}$,
(i.e. $\tau_a = 1$ if site $\underline{a}$ is occupied by a proton and $\tau_a = 0$ if site
$\underline{a}$ is unoccupied), $\alpha = 6$ and the second sum is a sum over all possible configurations of the hydrogen atoms. Since the ΔW_{ab} term is
very weak and long-ranged it can be treated (probably exactly) using
the Bragg-Williams approximation. In this approximation, the grand
partition function may be conveniently rewritten as

$$\Xi \ (T,\mu) = \sum_{N_H=1}^{\alpha N_M} e^{-F_0/kT} e^{-N_H^2 \sum_a W_{ab}/2kT\alpha N_M + \mu N_H/kT} \qquad (13)$$

where

$$F_0(T,\mu) = -kT\ln \sum_{\{\tau_a\}} e^{-\frac{1}{kT}\sum_{a,b}(U_{ab}+\tilde{W}_{ab})\tau_a\tau_b}$$

is the so-called reference free energy. Since U_{ab} and $\tilde{W}_{ab}$ are rather complicated interaction potentials, Horner and Wagner used Monte Carlo simulation on a lattice of 384 interstitial sites to calculate the reference free energy. Once the reference free energy is determined, the maximum term method may be used to evaluate the grand partition function. The chemical potential then becomes

$$\mu(T,\bar{N}_H)/kT = (\partial F_0(\bar{N}_H,T)/\partial \bar{N}_H)_T - \bar{N}_H \sum_a \Delta W_{ab}/kT\alpha N_M \qquad (14)$$

The coexistence curve may be obtained by plotting isotherms in the μ-c plane and performing Maxwell equal-area constructions whenever van der Waals loops appear. The resulting phase diagram is shown in Fig. 2, which also shows for comparison the experimental results of Walter and Chandler[14], Pryde and Titcomb[15] and Zabel and Peisl[16]. Reasonable agreement with experiment is obtained (especially considering that no adjustable parameters were used) although the Horner and Wagner critical point (T_C = 470K, c_C = .36) is relatively high compared to the recent experimental values and the shape of the coexistence curve is rather flat. Their model calculation was a significant step forward because it showed how the elastic aspects of hydrogen in a metal contribute to its thermodynamic properties.

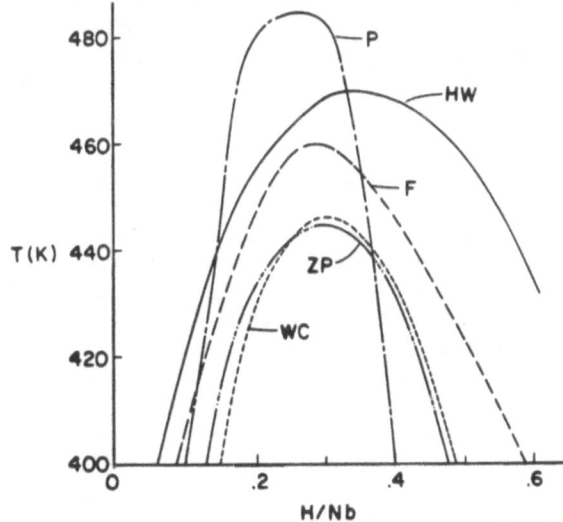

Fig. 2.   The H-Nb phase diagram. The theoretical curves of Horner and Wagner[13] (HW) and Futran et al[17] (F) are compared with the experimental curves of Walter and Chandler[14] (WC), Pryde and Titcomb[15] (P) and Zabel and Peisl[16] (ZP).

RECENT WORK

   In an effort to understand the appearance of ordered phases in
metal-hydrogen systems, Futran, Coats, Hall and Welch[17] have repeated
the model calculations of Horner and Wagner for hydrogen in niobium.
Horner and Wagner did not consider the behavior of their model at
high concentrations and so were unable to ascertain whether their
model predicted the ordered phase β. Futran et al developed a program based on the method of Gilat and Raubenheimer[18] which performs
the numerical integration over the Brillouin zone necessary for the
calculation of the elastic interactions $\tilde{W}_{ab}$. In doing so they found
substantial discrepancies between their values and Horner's and Wagner's values for $\tilde{W}_{ab}$. This could be traced to the numerical integration method used. Horner and Wagner considered 512 points in the
irreducible part of the Brillouin zone and used a uniform mesh size
whereas Futran et al consider 1327 points and used finer mesh sizes
near the origin where rapid variations occur in the integrand. On
this basis Futran et al concluded that their values for $\tilde{W}_{ab}$ were more
nearly correct than those of Horner and Wagner.

   Figure 2 shows the coexistence curve for the α-α' transition
obtained by Futran et al. The critical point is T_C = 460K and c_C =
0.29 as compared with Horner's and Wagner's T_C = 470K and c_C = 0.36.
Since the critical point values obtained by Futran et al and by
Horner and Wagner are within the values obtained experimentally it
is impossible to say on this basis whether the results of Futran et
al are an improvement over those of Horner and Wagner. Of more importance however, is the fact that the shape of the coexistence curve
obtained by Futran et al corresponds more closely to the dome-shaped
experimental curves than does the relatively flat curve of Horner
and Wagner. On this basis, Futran et al concluded that their results
were an improvement over the results of Horner and Wagner.

   Futran et al. investigated the phase-change behavior of their
model at high concentrations. At concentrations $c \geq 0.6$ and moderate
temperatures, $T \sim 350K$, a transition from the α' phase to an ordered
phase occurred. This ordered phase does not have the same structure
as the β phase which occurs experimentally at concentrations of c
≥ 0.7, as shown in Fig. 3(a), but rather has the structure shown in
Fig. 3(b). This result is important because it indicates that the
hard-core approximation for the electronic interactions between hydrogen atoms is too crude. Futran et al performed further investigations to see what changes in the electronic interaction are necessary to obtain the correct ordering. They found that a U_{ab} which had
a hard core out to 3rd neighbors, 4th neighbor repulsions ($\sim$500k),
6th neighbor repulsions ($\sim$100k), 9th shell repulsions ($\sim$100k) and
14th shell attractions ($\sim$-750k) will result in the correct ordering
at high concentrations. Their work underscores the need for having
a first principles calculation of the electronic interactions between hydrogen atoms in a metal.

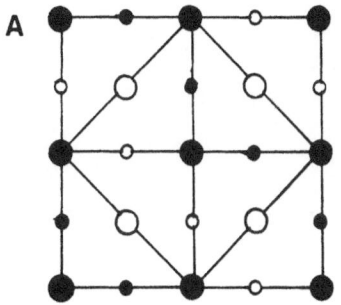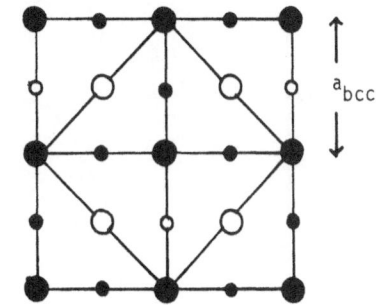

Fig. 3. (a) Structure of the β phase in H-Nb. (b) Structure of
         the ordered phase found by Futran et al at c ≥ 0.6, T ∿
         350K, ● = Nb at Z = 0 and Z = a, ○ = Nb at Z = a/2, • =
         H at Z = 3a/4, o = H at Z = a/4.

Futran et al extended their model of hydrogen in niobium to
the case of hydrogen absorbed in a niobium-vanadium alloy. In modelling the behavior of hydrogen in a binary metal alloy using the formalism of Horner and Wagner a complication arises. This complication is that since there are two types of metal atoms, the effective hydrogen-hydrogen interactions, U_{ab}, W_{ab} and W_{ab}, become
functions of the local configurations of metal atoms surrounding hy-
drogen atoms at sites $\underline{a}$ and $\underline{b}$. Rather than performing a Monte Carlo
simulation of the metal atoms in addition to the simulation of the
hydrogen atoms, Futran et al considered an approximation in which
the hydrogen atoms interact with a single effective metal atom which
has niobium-like or vanadium-like character in proportion to the concentration of niobium or vanadium in the metal alloy. This allowed
them to use the Horner and Wagner theory for hydrogen in a pure metal
but with the interaction parameters, $\tilde{W}_{ab}$, U_{ab} and ΔW_{ab} becoming functions of the alloy composition.

Figure 4 shows the phase diagrams calculated by them for the al-
loys $Nb_{.94}V_{.06}$, $Nb_{.85}V_{.15}$ and $Nb_{.75}V_{.25}$. It can be seen that as vanadium is added to the alloy, the critical point temperature is de-
pressed and the terminal solubility (concentration at the left-hand
side of the coexistence curve) increases. This is in qualitative a-
greement with the experiments of Pick and Welch[19] who found that the
critical temperature of the $Nb_{.94}V_{.06}$ alloy is 418K compared to the
theoretical value calculated by Futran et al of 450K. Thus the model
exhibited the correct qualitative behavior but needed to be improved
before quantitative agreement with experiment could be expected.

Since the effective metal approximation used by Futran et al
did not allow for the possibility of trapping of hydrogen atoms by
vanadium, i.e. for hydrogen atoms to preferentially occupy sites
near vanadium atoms, Futran et al devised a method whereby the effect of trapping could be incorporated into their results. This

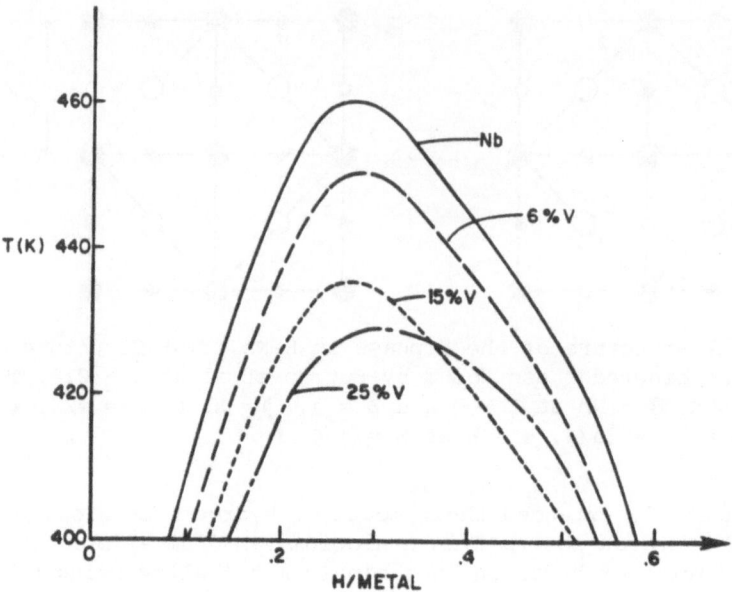

Fig. 4.   Phase diagrams for hydrogen in Nb, $Nb_{.94}V_{.06}$, $Nb_{.85}V_{.15}$ and $Nb_{.75}V_{.25}$ calculated by Futran et al.

involved assuming a value of the hydrogen-vanadium binding energy and then using the quasi-chemical approximation to calculate an effective vanadium concentration of the alloy. This effective vanadium concentration, which is the average concentration of vanadium atoms in the vicinity of the hydrogen atoms, is greater than the nominal vanadium concentration since hydrogen atoms are more likely to sit near vanadium atoms. Using this approximation, and assuming a binding energy of -0.07 eV, (which is the value used by Pick and Welch), the critical temperature for hydrogen in the $Nb_{.94}V_{.06}$ system becomes 428K as compared with Pick's and Welch's experimental critical temperature of 418K. Thus when trapping is incorporated into the model of Futran et al, reasonable quantitative agreement with experiment is obtained. These results also show that trapping of hydrogen by vanadium in niobium-vanadium alloys plays an important role in the determination of the phase diagram.

SUMMARY AND DISCUSSION

     The problem of finding a correct theoretical explanation for the phase-change behavior found when hydrogen is absorbed in metals has challenged scientists for over 50 years and will probably continue to do so in the future. In this paper, we have tried to show how the lattice-gas picture of these systems has evolved with time. Early model calculations assumed that hydrogen in metals could be treated as a lattice-gas model in which hydrogen-hydrogen interactions were either non-existent (as in the ideal solution models) primarily attractive (as in the Lacher model) or primarily repulsive (as in the blocking models). The work of Hall and Stell showed that both attractions and repulsions were necessary in order to obtain both ordered and disordered phases. The work of Horner and Wagner showed that the elastic interaction between hydrogen atoms and metal atoms is equivalent to a rather complicated effective hydrogen-hydrogen interaction. The work of Futran et al points out the fact that in order to obtain phase diagrams which give correct ordered phases at high hydrogen concentrations, it is necessary to have a better understanding of the electronic interactions between hydrogen atoms in a metal.

It is this author's opinion that the next frontier facing workers in this area is the accurate calculation of the electronic interactions between hydrogen atoms in a metal. Such calculations are expected to be rather risky as well as being difficult. Experts in this field point out that these calculations are extremely sensitive to the types of approximations used and often can not be relied upon to even give the correct sign of the interactions[20]. For example, Masuda and Mori[21] have calculated the nearest-neighbor hydrogen-hydrogen interaction in niobium and have concluded that it is attractive. This result does not appear to be consistent however with the structure of the β phase for hydrogen in niobium in which hydrogen atoms are never closer than fourth neighbors apart.

Two attempts at using a first-principles calculation of the electronic interactions between hydrogen atoms in a lattice-gas model have appeared recently in the literature. The first is a calculation of the phase diagram of hydrogen in palladium by Dietrich and Wagner[22]. They used the theory of dielectric screening to calculate the electronic interaction between hydrogen atoms. This electronic interaction was then added to the elastic interaction (calculated in the manner of Horner and Wagner) and the phase diagram was calculated They found that their results were quite sensitive to the types of approximations used and the agreement between theory and experiment was not as good as one would have hoped. Ross and Bond[23] also developed a lattice-gas model for deuterium in palladium in which the electronic interactions were chosen on the basis of the theory of diffuse scattering. They did not include elastic interactions in their model and thus looked only at the high concentration behavior of their model. The calculated phase diagram is quite similar to the experimental phase diagram for deuterium concentrations of up to 68%. Finally, an approach which is promising is that of Demangeat et al[24] who used an extra-orbital model within a generalized "spd" tight-binding approximation to calculate the interaction energy between hydrogen atoms in palladium. The electronic interactions calculated by them have not yet been used in a lattice-gas model calculation of the phase diagram.

In conclusion, although the lattice-gas model of hydrogen atoms
in a metal has seen many advances over the years, a great deal more
work needs to be done before a fully comprehensive theory of the
phase-change behavior of hydrogen in metals can be developed.

ACKNOWLEDGEMENTS

    Acknowledgement is made to the National Science Foundation
(Grant Number CHE 7909735) for support of this work.

REFERENCES

1. G. Alefeld and J. Volkl, Eds., "Hydrogen in Metals", Springer-
    Verlag, Berlin (1978).
2. F. D. Manchester, J. Less-Common Metals 49:1 (1976).
3. T. D. Lee and C. N. Yang, Phys. Rev. 87:410 (1952).
4. R. A. Fowler and E. A. Guggenheim, "Statistical Thermodynamics",
    Cambridge University Press, London (1939).
5. J. R. Lacher, Proc. Roy. Soc. (London), Ser. A 161:525 (1937).
6. W. A. Oates and T. B. Flanagan, Prog. Sol. St. Chem. 13:193
    (1981).
7. R. Speiser and J. W. Spretnak, Trans. Amer. Soc. Met. 47:493
    (1955).
8. W. A. Oates, J. A. Lambert and P. T. Gallagher, Trans. Met.
    Soc. AIME 245:47 (1969).
9. G. Alefeld, Ber. Bunsenges, Phys. Chem. 76:746 (1972).
10. C. Wagner, Acta. Met. 19:843 (1971).
11. C. K. Hall and G. Stell, Phys. Rev. B 11:224 (1975); C. K. Hall
    and M. Futran, J. Less-Common Metals 74:237 (1980).
12. H. Wagner and H. Horner, Adv. Phys. 23:587 (1974).
13. H. Horner and H. Wagner, J. Phys. C 7:3305 (1974).
14. R. J. Walter and W. T. Chandler, Trans. AIME 233:762 (1965).
15. J. A. Pryde and C. G. Titcomb, Trans. Faraday Soc. 65:2758
    (1969).
16. H. Zabel and H. Peisl, Phys. Stat. Sol. (a) 37:K67 (1976).
17. M. Futran, S. G. Coats, C. K. Hall and D. O. Welch, J. Chem.
    Phys. (to be published).
18. L. J. Raubenheimer and G. Gilat, Oak Ridge Nat'l. Lab. Internal
    Report No. ORNL-TM-1425 (1966).
19. M. A. Pick and D. O. Welch, Z. Physik Chem. (N.F.) 114:37 (1979).
20. T. L. Einstein (private communication).
21. K. Masuda and T. Mori, J. Physique 37:569 (1976).
22. S. Dietrich and H. Wagner, Z. Physik B 36:121 (1979).
23. R. A. Bond and D. K. Ross, J. Phys. F (to be published).
24. C. Demangeat, M. A. Khan, G. Moraitis and J. C. Parlebas, J.
    Physique 41:1001 (1980).

SPINODALS AND SPINODAL MECHANISMS IN M-H SYSTEMS[†]

F.D. Manchester

Department of Physics
University of Toronto
Toronto, Ontario, M5S 1A7, Canada

ABSTRACT

A brief, descriptive, review is given of the physics of spinodal curves and of spinodal mechanisms in general as they occur in
various physical systems, e.g., binary alloys, glasses and liquids.
The unique features of spinodals in M-H systems are discussed from
the point of view of the special nature of the density fluctuations,
in the critical point region, for coherent M-H systems. Methods of
locating spinodals and detecting spinodal decomposition are reviewed
for M-H systems.

INTRODUCTION

The study of spinodals and spinodal mechanisms in M-H systems
is, apparently, in a very different situation from the study of the
critical point in these systems, primarily because the theoretical
predictions of mean-field-theory for spinodals seem less clear than
they are for critical point behaviour, and also because, the predictions of theories depicting more 'realistic' systems appear to call
into question[1] the existence of a spinodal (even for mean field
systems), or at least, the existence of one with a well defined
meaning for dimensionality where real experimental measurements (i.e.
with d = 2 or 3) can be made, as distinct from "computer experiments"[2].
There is a further feature concerning spinodals in M-H systems[3-5]
which has to do with the fact that M-H systems provide an almost
unique situation where the class of spinodals connected with macroscopic density modes in a coherent M-H crystal can be observed. For
these reasons alone, the study of spinodals and related matters in
M-H systems, has been, and continues to be, a fruitful one.

[†]Work supported in part by a grant from the Natural Sciences and
Engineering Research Council of Canada.

SIMPLE CLASSICAL PICTURE OF A SPINODAL

The idea of a spinodal can be traced back to the early work of
Gibbs, Maxwell and Van der Waals, on metastability in one component
systems. For such a system, which we could here think of as an M-H
system[*], the principal thermodynamic parameters required to give a
simple meaning[**] to the idea of a spinodal curve are shown in
Fig. 1. The original difficulty with the Van der Waals isotherm,
was that for part of that isotherm, the existence of non-stable
states was implied, i.e. that the pressure decreased with increasing
density (Fig. 1(a)). That this region was bounded by the locus of
the turning points of the Van der Waals isotherm and related to the
limits of stability discussed in terms of the energy-volume-entropy
surface of Gibbs gave a formal specification to the idea of
instability[6] in these systems. It can be simply shown[7] and related
to a wider discussion of stability[8], that the stability limit can
be expressed as

$$\frac{\partial^2 F}{\partial \rho^2} = 0$$

and this condition is shown as giving the points of inflection on
an $F(\rho)$ plot (for $T < Tc$)[***] and the turning points in a plot of
$\mu(\rho)$, the chemical potential. In Fig. 1(a) then, we have three
regions: (a) the region of stable, equilibrium states outside the
coexistence curve; (b) the meta-stable region between the coexistence curve and the spinodal; and (c) the unstable region inside
the spinodal curve. The purpose of Fig. (1) is not only to point
out the simple features of a spinodal curve for a one-component
system but also to indicate the different experimental approaches
which are used to investigate spinodals. For those concerned with 26
spinodals in metallic alloys, glasses and liquids, and particularly
with spinodal decomposition, the experimental approach emphasizes
quenching along lines of constant ρ from the stable into the metastable or unstable regions. In these quenching experiments, the

---

[*]We represent a system of hydrogen gas inside a metal as a one
component system - in the sense of the Gibbs phase rule. This is
clearly so because the effect of pressure and temperature on the
metal lattice is clearly inferior to that on the H_2. Also, we regard
the hydrogen-metal system as a lattice gas.

[**]Descriptions of the principal features of spinodals have been given
elsewhere[9,10], the only reason for a certain amount of repetition
here is to set up the discussion for the particular context of M-H
systems.

[***]Here Tc denotes the critical temperature as commonly understood
in physics for a one component fluid system and the consolute temperature as used in connection with binary alloy systems in
metallurgy.

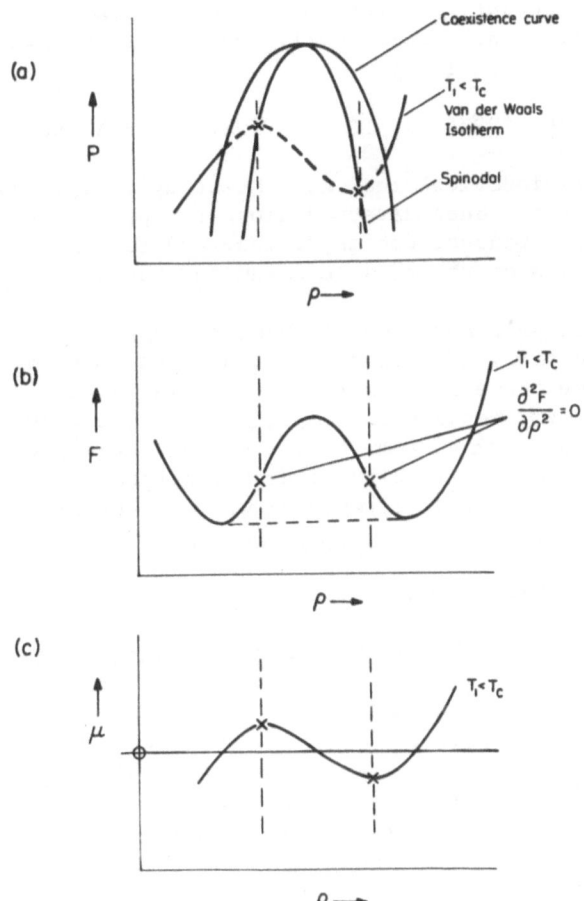

Fig. 1. (a) Coexistence curve, v de W isotherm and spinodal for a simple, mean field fluid system; (b) Helmholtz free energy and (c) chemical potential for same system.

emphasis is on entering the region inside the spinodal curve while
for some liquid-vapour systems an approach along an isotherm (for
T < Tc) toward the spinodal is appropriate, especially when the
study of metastability and the limit of stability is emphasized.
For M-H systems, spinodals can be located by extrapolation from
measurements sampling the lattice gas compressibility in the stable
region outside the coexistence curve[11,12] and in one case[7] both
approach along isotherms and extrapolation of compressibility were
used. These different, operational, approaches to spinodals will be
discussed in more detail below.

THE TWO PRINCIPAL APPROACHES TO THE STUDY OF SPINODALS

    Partly for historical reasons, the study of spinodals has, up
to the present time, been directed along two principal lines:
(a) the study of spinodal decomposition, and (b) the location of
spinodals in terms of the phase diagram for the system involved.

    The requirements and aims of these two approaches really come
together in the study of spinodals and spinodal mechanisms in M-H
systems, because these systems readily display a wide range of
spinodal behaviour amenable to sampling by a wide range of techniques,
which are very different from those used up to the present time in
the study for instance of spinodal decomposition in alloys. To
clarify this statement, we first review, the general features of
spinodal decomposition and spinodal location.

(a) SPINODAL DECOMPOSITION

    Understanding of this mechanism came from the concepts of
Cahn[9] and co-workers[13,14] and provided a framework for various
observations which had accumulated as a consequence of quenching
experiments with metallic-alloys, e.g. the occurrence of G.P.
zones[15,16] when the (binary) alloy was quenched from above Tc into
the mixed phase or miscibility gap region below Tc. Quenching in
such a fashion results in one of two processes coming into play.
In the one case, for which the quenching produces a final state in
the meta-stable regions of the miscibility gap the final state is
a meta-stable equilibrium with a local minimum in the free energy
and the fluctuations in the free energy have to be above a threshold
size to succeed in producing phase separation; there has to be an
activation energy. In the other case, which corresponds to quenching
into the region inside the spinodal, the activation energy goes to
zero at the spinodal line and fluctuations without the threshold
size restriction are able to generate the final state arrangements and
the system is unstable. In this case the final state is a regular
and modulated structure, (i.e. A-rich, B-rich, A-rich... etc. in an
A-B alloy) with its wave vector directed along one of the elastically
softer crystal directions (e.g. <100> or <111> in a cubic crystal)
and with a wavelength of the order of 100 Å - what is now well

recognized as the structure of spinodal decomposition, an example of which is given in Fig. 2 for a metallic alloy, $(Fe_{0.4}Al_{0.1})Ti_{0.5}$.

The structures arising from these two processes in an alloy can grow; in the spinodal decomposition case, to a coarsening of the modulated structure and in the nucleation and growth case, more or less isolated islands of one phase in the other are formed.

This distinction in activation energy requirements can be explained in terms of Fig. (3) which shows the Helmholtz free energy for a simple (e.g. mean field fluid) system for T < Tc. The densities for the two equilibrium phases are represented by ρ_a and ρ_b and we first focus our attention on the case where the mean density of the fluid has the value ρ_1, which lies between the equilibrium value ρ_a, of one phase, and a spinodal point represented by a cross

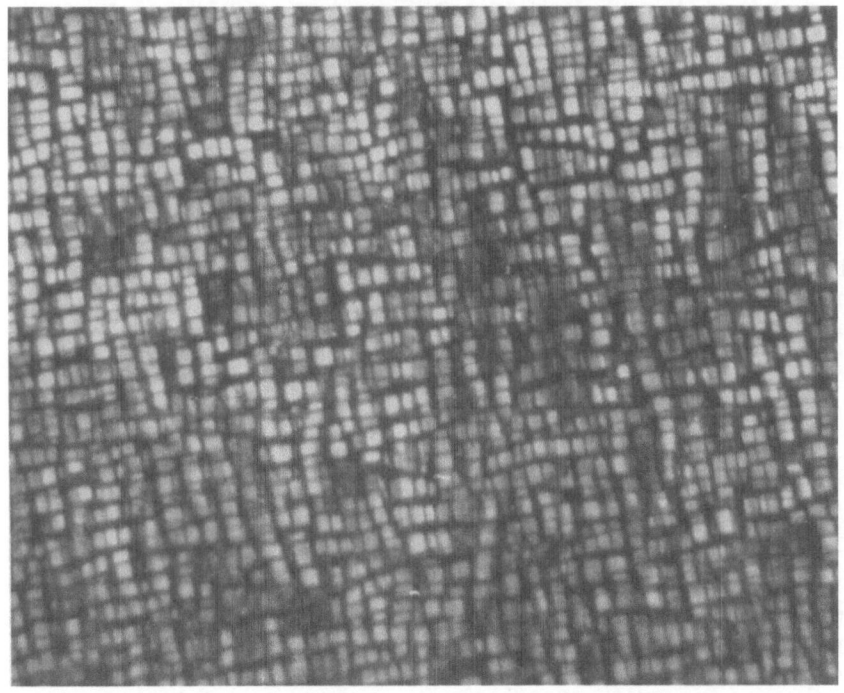

Fig. 2. Spinodal decomposition observed in a $(Fe_{0.4}Al_{0.1})Ti_{0.5}$ alloy after cooling from the melt. This (T.E.M.) photograph shows an advanced (coarsened) stage in the process, but the principal features of the periodic variation in alloy content can be seen. The evident principal directions in the photograph are <100> and the wavelength of the concentration variations is $\sim$ 200 Å, at this stage. Photograph courtesy of Prof. G.C. Weatherly.

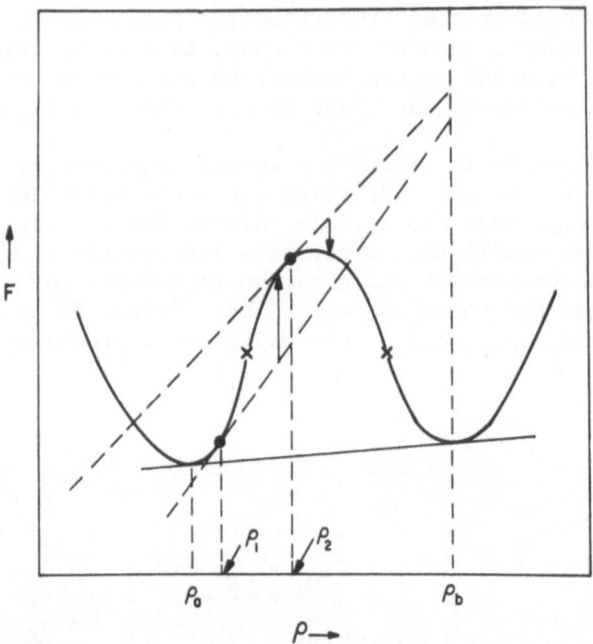

Fig. 3. F(ρ) plot for a simple, mean field, fluid system, for
T < Tc (solid curve). The solid line joining the F(ρ) curve at the
ρ_a and ρ_b points is the Maxwell construction. For further details
see text.

(the point of inflection, where $\partial^2 F/\partial \rho^2 = 0$) i.e. ρ_1 corresponds to
a density in the metastable region. Now we consider fluctuations
in density about the value ρ_1 i.e. that some small volume in the
fluid could have a value of the density $\rho \neq \rho_1$. To do this we draw
a tangent to the F(ρ) curve at the point F(ρ_1) and from this
construction[17] the free energy, per mole, required to produce a
fluctuation in density is given by the vertical distance between
this tangent and the F(ρ) curve. If, for instance, the density of
the small volume were to undergo a fluctuation giving it a density
$\rho = \rho_1 + \Delta\rho$ then the free energy required would be such as that
indicated by the upward directed arrow to the right of ρ_1. This

represents a free energy change which has to be supplied to the liquid to provide this change, $\Delta\rho$. A more extreme case of a density fluctuation would be for ρ to increase to ρ_b and in that case the free energy change would have opposite sign, and some small volume would change to the density of the equilibrium phase farthest from ρ_1, and the energy corresponding to the difference between the tangent at $F(\rho_1)$ and the point $F(\rho_b)$ would be released. It can be seen from the relation of the tangent at $F(\rho_1)$ to the $F(\rho)$ curve that there is a wide range of $\rho \neq \rho_1$, in the coexistence region for which a free energy increase is required and only for very large (and correspondingly improbable) ρ changes is there a free energy decrease involved. Similar considerations apply in the case of other points in the metastable region, and the free energy increase required for the system represents a thermodynamic barrier to be overcome in order for phase separation to proceed - an activation energy is required. This is not the case for points such as ρ_2, which are inside the spinodal region. For such points, the free energy difference for an excursion of the density from a value such as ρ_2, is represented by the downward pointing arrow and represents a release of energy, i.e. there is no thermodynamic barrier for any density value in the region between the spinodal points - no activation energy is required and a system quenched from above Tc into this region is unstable.

As has been discussed by Cahn[9], the spinodal decomposition can be treated using an effective diffusion coefficient which is negative, and the growth of the concentrations of one alloy component with respect to the other in the modulated structure for a binary mixture is produced by uphill diffusion.

Table 1 lists several features of Spinodal Decomposition and nucleation and growth which can readily be contrasted. It should be mentioned here that the contrast with nucleation and growth processes is to compare the more novel spinodal decomposition with a process that in some respects, at least, is more familiar and also because these are the only processes which apply, each to its distinct region of validity, in the coexistence region.

COHERENT STRAIN ENERGY AND "COHERENT-SYSTEMS"

Apart from the mechanism of spinodal decomposition, the concepts which Cahn introduced included one which is of central concern for the discussion of spinodals in M-H systems - that of coherent strain energy. Coherent strains occur in a crystal when the two coexisting phases (of Fig. 1 for example) share the same lattice with the lattice planes undergoing continuous deformation without relief from the formation of dislocations[10]. This strain energy results in an additional contribution to the free energy of the system[17], which can be included to give

Table I. Some points of comparison between spinodal decomposition
and nucleation and growth. $\hat{D}$ may be taken as the intercomponent
diffusion coefficient for a binary alloy. The concentration profiles
are meant to contrast the gradual accumulation of a concentration
increase with a sudden, nucleated, change, and 'downhill' diffusion
into a surrounding depleted region. (after Cahn[9]).

Feature / Mechanism	Activation Energy	Sign of $\partial^2 F/\partial \rho^2$	Sign of $\tilde{D} = \dfrac{M \partial^2 F}{\partial \rho^2}$	Concentration Profile	Fluctuation Threshold
Spinodal Decomposition	zero	negative	negative		zero
Nucleation and Growth	finite	positive	positive		finite

$$F = \int_V \left\{ f(\bar{\rho}) + \frac{\eta^2 E}{1-\nu} (\rho - \bar{\rho})^2 + K(\nabla \rho)^2 \right\} dV$$

where $f(\bar{\rho})$ is the free energy density for a uniform system of density
$\bar{\rho}$, (as in Fig. 1), the second term in the integrand gives the
coherent strain energy, and the third term, the 'gradient energy'
term gives the extra energy which arises because of deviations from
uniformity. In the coherent strain energy term; E is Young's modulus,
ν is Poisson's ratio, $(\rho - \bar{\rho})$ is the local departure from the uniform
density, $\bar{\rho}$, and, $\eta = 1/a \, da/d\rho$, is the linear expansion per unit
composition change (a = lattice spacing of unstrained solid), a
familiar quantity in work with M-H systems.

With these extra terms present in the free energy the apex of
the spinodal curve (here, as is usual, considered as a curve in the
T-ρ plane) is lowered below the usual coexistence curve, so that
the spinodal apex temperature, T_s, is less than the critical temperature Tc. This "spinodal suppression" is directly proportional to
the coherent strain energy term and this proportionality has been
observed experimentally[9]. The strain field always increases the
free energy and this acts to suppress the temperature of the phase
separation. This can be appreciated from the following simple
considerations.

If we consider a simple, mean field, fluid system for which
the free energy density may be written in the form

$$f(\rho) = k_B T \, f_o(\rho) - \tfrac{1}{2} W\rho^2$$

where W is the interaction energy parameter and $f_0(\rho)$ includes the configurational contribution to the free energy and k_B is the Boltzmann constant. Then applying the stability condition to give the spinodal curve we get for the spinodal apex temperature:

$$T_s = W/k_B f_2(\rho_c) \qquad\qquad f_2 = \partial^2 f/\partial\rho^2$$

which is also equal to Tc for this system. If we now look at the system with just coherent strain energy (the effect of gradient energy term need not be included for this discussion[17]) then the spinodal apex temperature would be obtained in a similar way, and we would have

$$T_s^{inc} - T_s^{coh} = \frac{2\eta^2 E}{(1-\nu)k_B f_2(\rho_c)}$$

where the superscript "inc" has been used to label the system without coherency strain energy as incoherent[9].

COHERENT PHASE DIAGRAM

Table 2 gives a very brief summary of the types of physical system in which spinodals and spinodal mechanisms have been studied. Note: that liquids have a very small typical spinodal quench (suppresion) listed. As they cannot support coherency strain energy, their spinodal decomposition region can be easily entered. For the crystal which can support coherency strains (this can usefully be thought of as one which is fairly free of dislocations) the phase diagram is altered from the incoherent case and it has been shown by Cahn[18] that such a crystal has a coherent coexistence curve (miscibility gap) associated with the coherent spinodal and depressed by the same amount for the same elastic conditions. This gives a further meaning to the concept of a coherent system, it can undergo coherent nucleation when the coherent coexistence curve (coherent solvus) is crossed and also spinodal decomposition when the associated spinodal region is entered. An example of a coherent solvus, placed in relation to the incoherent phase diagram for the same system, Al-Zn, is shown in Fig. 4. The coherent strain energy is not large in this system, so the spinodal suppression is not very pronounced. In Fig. 5, examples of coherent nucleation for the system Pd-Hx, are shown, together with some incoherent nucleation showing very high dislocation densities.

It is important to clarify the difference between the incoherent and coherent phase diagrams, particularly in the field of M-H systems, where there has been something of a tradition of measurement of phase diagrams by physical chemists. By a number of different techniques the coexistence curves and critical points have been established using samples in a wide range of physical conditions but almost all, until very recently, having one thing in

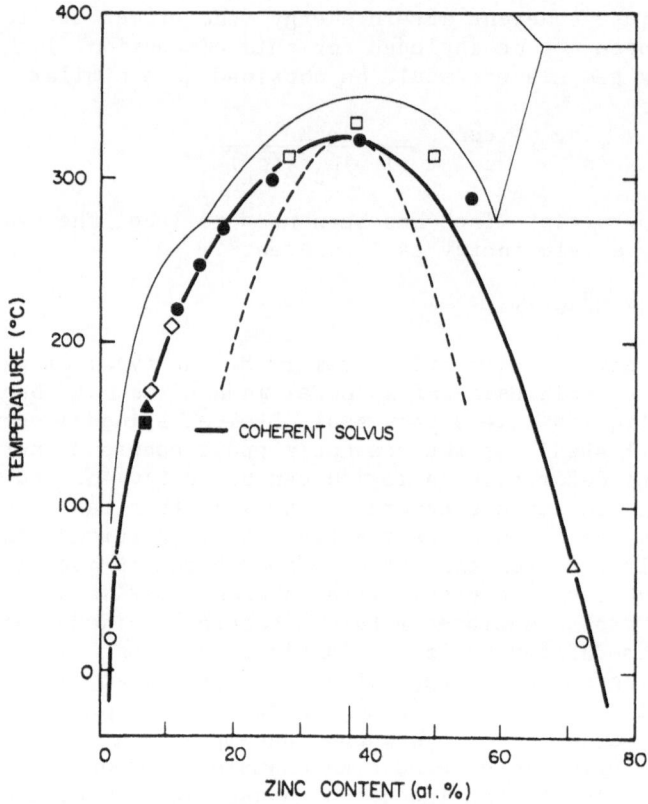

Fig. 4. The coherent solvus for the alloy system Al-Zn, taken
from reference 10, in which are given the details of the various
sources of information on the solvus location. Diagram reproduced
courtexy of Prof. D. de Fontaine and Plenum Publishing Corp.

TABLE II. Some simple features of spinodal occurrences and processes in several physical systems. For a review of spinodal decomposition in liquids see Goldburg et al[19] and in glasses see James[20].

Class of System	Spinodal Types	Examples	Typical Experiment Method	Typical Quench ΔT
Alloys	Bulk Modes Coherent	Al - Zn	X-ray Scatt Microscope	$10-10^2$ °C
Liquids	Incoherent	Methanol - cyclohexane	Laser Light Scatt.	$\pm 2 \times 10^{-3}$ °C
Glasses	Incoherent or Bulk Modes Coherent	Boro-silicate glass	Microscope Light scatt.	$0-10$ °C
M-H Systems	Incoherent, Microscopic and Bulk Coherent	Nb-Hx Pd-Hx	Gorsky effect	$0-10^2$ °C

common - a high dislocation density. Such measurements[7,21,22], give the incoherent phase diagram, also regarded, implicitly, as the equilibrium phase diagram. The addition of the idea of the coherent phase diagram has broadened the concept of what phase separation in the coexistence region means for alloys, glasses and M-H systems.

WIDE SPECTRUM DENSITY FLUCTUATIONS FOR COHERENT SYSTEMS

Cahn used periodic boundary conditions in finding solutions to the diffusion equation which governed the growth of the spinodally decomposing structure and thus his analysis did not take into account any effect of boundary conditions on the coherent strain energy of the crystal, which, for the long range particle-particle interaction in M-H systems, can be very significant[24]. Thus the Cahn solution deals with fluctuations in the density, ρ (as referred to in discussion of Fig. 3) typical of bulk material, and on a scale for which the typical wavelength turns out to be of the order of 100 A. Any fluid system, including the lattice gas of an M-H system, displays thermodynamic fluctuations of the density which have a wide range of wavelengths and these are related to the isothermal compressibility, K_T, of the system through.[25]

$$K_T \simeq \int G(r) dr$$

where $G(r)$ is the density-density correlation function for the
fluid system. The compressibility diverges at the critical point
and also along the spinodal line (for a mean field system), so there
is a corresponding increase in the range of the correlation function.
This relation expresses the concept that a response function of the
system, K_T, is dependent on a wide range of density fluctuation
wavelengths in a general system. Horner, Wagner and Bausch,[3-5]

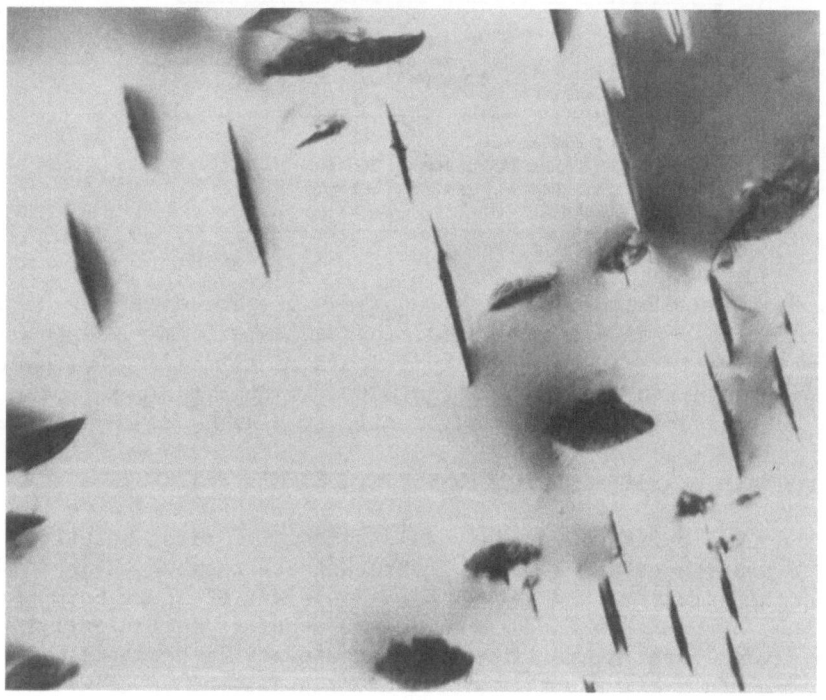

Fig. 5. A transmission electron microscope photograph of coherent
precipitation in Pd-Hx at a temperature of approximately 90K.
The coherent precipitates (of α-phase in β-phase) are the lenti-
cular shapes with almost vertical direction. Some of these, and
almost all the thicker precipitates, directed more or less
horizontally,have become incoherent, with very high dislocation
densities ($\sim 10^{12}/cm^2$). The precipitates (average length $\sim$ 1 micron)
have their long axes in < 100 > directions. For more details, see
Ho et al.[21]

examined the occurrence of spinodals in M-H systems, and, in terms
of this picture of fluctuations in the hydrogen density, looked
at all possible wavelengths,* from those comparable to, or even
longer than, the dimensions of the crystal, down to those of short
wavelength considered by Cahn, which are, unlike the long wavelength fluctuations, independent of boundary conditions.

Horner et al[3-5] found a series of spinodals with apex
temperatures lying in between T_s^{inc} and T_s^{coh} and having a relative
position which depends on the particular density fluctuation
allowed by the experimental sample geometry.[5] While there are
some variations within each of the spinodal groups (e.g. two dimensional group, three dimensional group) there is a denumerable,
discrete, set of spinodals, corresponding to a discrete set of
density fluctuations, which are allowed, for instance, by the
boundary conditions on the M-H system. Thus, in contrast to the
case of an incoherent M-H system, there is, for a coherent M-H
system, a compressibility response which is related to a given,
allowed, fluctuation wavelength, determined by a given set of
boundary conditions, which can, in principle, be highly selective.
The compressibility response of the M-H system (lattice gas) can be
sampled using the Gorsky effect, which gives a measure of Δ_E, the
anelastic relaxation strength, (elastic susceptibility) and as[26]

$$\Delta_E \alpha K_T \alpha \left(\frac{\partial \mu}{\partial \rho}\right)^{-1}$$

a plot of $(\Delta_E)^{-1}$ vs T, for a fixed concentration, can be extrapolated
to give a spinodal temperature appropriate for the boundary conditions which apply in a chosen situation - the anelastic response in
these systems obeys a Curie-Weiss type law in the region of interest[26].
Such a plot is shown in Fig. 6 for a (210) sheet of a coherent Pd-H_x
single crystal, with the sheet being mechanically loaded parallel
to a <100> direction in a plane stress bending mode[27]. The location
of the incoherent spinodal temperature for the same $\rho (= H/Pd)$ value is
shown on the abscissa, indicating that spinodal supression is being
observed. The salient feature of Fig. 6, as far as the present
discussion is concerned, is that the boundary conditions and the
manner of stressing put restrictions on the density fluctuation
modes which can respond, and if one can be sure enough of the restrictions and absence of coupling to other density fluctuation
modes in the crystal, one could claim that just one mode was respond-
ing. Thus this M-H system in the form of a coherent, single crystal,
sheet, shows the possibility of a very specialized response in the
Ornstein-Zernike sense, which would make cases like this very unique
in the study of critical point fluctuations[28].

---

*Fluctuation wavelengths is used here in the sense of Fourier
components of the size of fluctuations being considered.

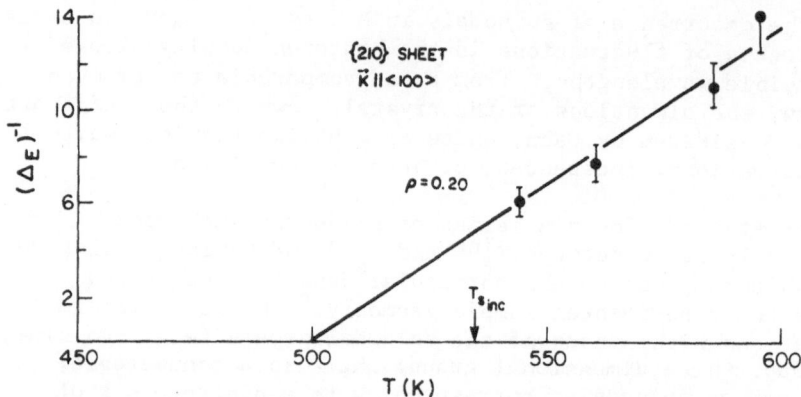

Fig. 6. A plot of reciprocal anelastic relaxation strength (reciprocal elastic susceptibility) against temperature for a coherent
Pd-H_x crystal sheet with H/Pd = 0.20. The sheet surface is in the
{210} plane and the plane stress loading of the sheet is along the
<100> direction. The location of the incoherent spinodal tempera-
ture, T_s, is indicated[27].

SUMMARY OF SPINODAL EXPERIMENTS WITH M-H SYSTEMS

        Fluctuations of the hydrogen density in the short wavelength
(Cahn) region have been examined by using coherent neutron scattering[12,29] to sample the elastic susceptibility. By extrapolation
of scattered neutron intensity data from outside the coexistence
region of Nb-D_x, an estimate of the position of the Cahn type
spinodal (in a T-ρ plane) was obtained. The spinodal suppression
was ΔT ∿ 270K (Conrad et al[12]) and ∿ 200K (Munzing and Stump[29]).

        Spinodals in crystals displaying a large degree of coherency
with sampling of density fluctuation wavelengths comparable with the
sample size have been investigated by Tretkowski et al[11] for the
Nb-H_x and Ta-H_x systems using the Gorsky effect. Spinodal suppression with ΔT ∿ 100K was found for both systems, for samples in the
form of foils. These foils had a 'preferred orientation' metallurgical structure, and were not single crystals, so that spinodal suppression could be demonstrated but not the excitation of a precisely
specified density mode as discussed above. A detailed treatment of
the possible density modes for two dimensional coherent M-H crystals
has been given by Goldberg[30,31].

        Incoherent spinodals have been determined using the Gorsky
effect for Pd-H_x by de Ribaupierre and Manchester[7], for Nb-H_x, Ta-H_x
and V-H_x by Bauer et al[32]. In the case of Pd-H_x and Nb-H_x which have
phase diagrams with clearly defined critical points, corresponding
at least in part to a fairly simple lattice gas situation, the apex

temperature for the observed incoherent spinodal coincides with the
critical temperature in the expected way. In the case of Ta-H_x and
V-H_x, the location of the critical point is less clear and the spinodal apex temperature is located well below the nearest phase boundary.
No comment about the spinodal temperature suppression or lack of it
can be made until there is some independent and reliable information
on the location of the critical point and the miscibility gap. For
Ta-H_x Bauer et al[32] show points at the higher concentration end of
the incoherent spinodal as occurring at negative absolute temperatures*, an indication that the assumptions used in making the extrapolation of the elastic susceptibility measurements no longer hold,
and that a more detailed examination of how to locate the spinodal
at higher temperatures is required.

There has been one series of observations of spinodal decomposition in a coherent M-H system capable of supporting macroscopic
density modes. Zabel and Peisl[33] cooled Nb-H_x single crystals from
well above Tc into the coexistence region and followed the behaviour
of an X-ray Bragg reflection as the cooling proceeded. Typically,
the X-ray line showed some broadening (due to strain) as the temperature fell below Tc but did not split into lines corresponding to the
lattice parameter values for the 'wings' of the coexistence curve.
At $\Delta T \sim$ 15K ($\sim$ 3% of Tc = 444K) the X-ray line showed significant
changes in lattice parameter value on the side of the crystal being
examined. Cooling to $\Delta T \sim$ 50K produced some evidence of the onset
of an incoherent state in the crystal. Companion experiments looked
at the appearance, as revealed by X-ray lattice parameter shifts, of
macroscopic distributions of hydrogen density across the thickness
of single crystal discs of Nb-H_x, consistent with the occurrence of
macroscopic density modes as the crystal was cooled below Tc. Thus
the spinodal decomposition which appears to be encountered in these
experiments is for density fluctuations of large wavelength compared
to crystal dimensions - at the opposite end of the density fluctuation 'spectrum' from Cahn type spinodal decomposition. Some experimental observations on coherent nucleation in Pd-H_x have been made
by Ho et al[21] in which the coherent solvus was crossed and re-crossed
without entering the spinodal region. Examples of this coherent
nucleation are shown in Fig. 5.

The wavelengths of Cahn-type density fluctuations in coherent
crystals are sufficiently small, that microscope pictures give
direct and convenient evidence of coherent nucleation and spinodal
decomposition, whereas for longer wavelength density fluctuations

---

*A negative temperature is a well defined physical concept (e.g. for
a magnetic spin system) and without discussion of a suitable physical
picture, applicable in the present instance, it seems hard to justify
bringing in negative temperatures.

the evidence for spinodal decomposition will have to be more indirect and more varied as in the work of Zabel and Peisl[33].

In most of the experiments on spinodals in M-H systems, the
emphasis has been on location of the spinodals, whereas for other
physical systems, particularly alloys, the emphasis has been on
observing spinodal decomposition, with the result that determination
of the location of the coherent solvus is, as far as metallurgists
have gone in determining the coherent phase diagram. On the other
hand, working from outside the coexistence region, it has been
possible to determine spinodal curves for M-H systems fairly readily
and, in time, it should be possible to have, microscopic information
on precipitation fitted together with the thermodynamic data on
curves derived from elastic measurements.

OUTLOOK

There is still a great deal to be done on completing the location and delineation of spinodal curves in M-H systems and the
examination of the morphology of spinodal decomposition in these
systems has only really just begun. Likewise, the dynamics of
spinodal decomposition for M-H systems is almost untouched, experimentally.

The operational situation that experimental evidence for
locating spinodals is obtained from outside the coexistence region
and on spinodal decomposition, from well inside the spinodal region,
has meant that there has been very little experimental work done
close to, or passing through, the boundary between metastable and
unstable regions. It has been suggested[34] that, in general, there
is some breadth to this boundary region, and it has also been
suggested[1], that it is not at all clear that the notion of a point
on a spinodal curve has a well-defined meaning - and, that the
evidence for even mean field system type spinodals in two and three
dimensions is abmiguous[1]. Some of these inconsistencies, can
probably be removed by a clarification of what realities are shared
by model system concepts, and experimental evidence of spinodals,
and, at least for the mean field system case, M-H systems should
provide suitable means for doing this. The investigation of just
how sharp the spinodal boundary region can be for the good mean
field system that an M-H system can provide, should also be of some
help in the general clarification of spinodal concepts.

REFERENCES

1. W. Klein, Phys. Rev. Lett. 47:1569 (1981).
2. F.F. Abraham, M.R. Mruzik and G.M. Pound, J. Chem. Phys. 70:2577
       (1979).
3. H. Wagner and H. Horner, Adv. Phys. 33:587 (1974).
4. H. Horner and H. Wagner, J. Phys. C: Solid State Phys. 7:3305
       (1974).

5.   R. Bausch, H. Horner and H. Wagner, J. Phys. C: Solid State
       Phys. 8:2559 (1975).
6.   J.C. Maxwell, "Theory of Heat" 2nd Ed., Longmans Green, London
       (1899) p. 195.
7.   Y. de Ribaupierre and F.D. Manchester, J. Phys. C: Solid State
       Phys. 8:1339 (1975).
8.   H.B. Callen, "Thermodynamics", Wiley, New York (1961).
9.   J.W. Cahn, Trans. Met. Soc. A.I.M.E. 242:166 (1968).
10. D. de Fontaine, in "Treatise on Solid State Chemistry", Vol. 5
       edited by N.B. Hannay, Plenum, New York, (1975), p. 129.
11. J. Tretkowski, J. Volkl and G. Alefeld, Z. f. Phys. B28:259
       (1977).
12. H. Conrad, G. Bauer, G. Alefeld, T. Springer and W. Schmatz,
       Z. f. Phys. 266:239 (1974).
13. M. Hillert, Acta. Met. 9:525 (1961).
14. J.W. Cahn and J.E. Hilliard, J. Chem. Phys. 28:258 (1958).
15. A. Guinier, Nature, 142:569 (1938).
16. G.D. Preston, Proc. Roy. Soc. (London) A167:526 (1938).
17. J.E. Hilliard, in "Phase Transformations" edited by
       H.I. Aronson, Am. Soc. for Metals, Metals Park, Ohio (1970),
       p. 497.
18. J.W. Cahn, Acta Met. 10:907 (1962).
19. W.I. Goldburg, A.J. Schwartz and M.W. Kim, Prog. Theor. Phys.
       (Japan) Suppl. 64:477 (1978).
20. P.F. James, J. Mat. Sci. 10:1802 (1975).
21. E.C. Ho, H.A. Goldberg, G.C. Weatherly and F.D. Manchester,
       Acta. Met. 27:841 (1979).
22. H. Frieske and E. Wicke, Ber. Bunsenges. Phys. Chem. 77:48 (1973).
23. T. Schober and H. Wenzl in "Hydrogen in Metals" II, Topics in
       Applied Physics, Vol. 29, edited by G. Alefeld and J. Volkl
       Springer-Verlag, Berlin (1978).
24. J.D. Eshelby in "Solid State Physics" Vol. 3, edited by F. Seitz
       and D. Turnbull, Academic Press, New York, (1956), p. 79.
25. H.E. Stanley, "Introduction to phase transitions and critical
       phenomena" Oxford University Press, London (1971).
26. G. Alefeld, G. Shauman, J. Tretkowski and J. Volkl, Phys. Rev.
       Lett. 22:697 (1969).
27. M. Fairlie and F.D. Manchester, Proceedings of International
       Conference on Solid-Solid Phase Transformations, Pittsburgh,
       U.S.A. (Aug. 1981) in press.
28. H. Wagner in "Topics in Applied Physics" Vol. 28 edited by
       G. Alefeld and J. Volkl, Springer-Verlag, Berling (1978) p. 5.
29. W. Munzing and N. Stump, J. Appl. Cryst. 11:588 (1978).
30. H.A. Goldberg, J. Phys. C: Solid State Phys. 10:2059 (1977).
31. H.A. Goldberg, J. Phys. C: Solid State Phys. 11:3147 (1978).
32. H.C. Bauer, J. Tretkowsi, J. Volkl, U. Freundenberg and
       G. Alefeld Z. f. Phys. B28:255 (1977).
33. H. Zabel and H. Peisl Acta. Met. 28:589 (1980) and references
       contained therein.
34. J.S. Langer, Physica 73:61 (1974).

WHAT IS THE RELEVANCE OF THERMAL FLUCTUATIONS FOR THE SPINODAL DECOMPOSITION IN A COHERENT METAL-HYDROGEN SYSTEM?

Joseph W. Haus and Harald King[*]

GHS Essen, FB Phys., 4300 Essen 1, FRG
[*]P. T. B., 3300 Braunschweig, FRG

ABSTRACT

In order to provide an answer to the title question, we examined the dynamics of unstable hydrogen density modes within the framework of a stochastic theory which describes the coupling of the hydrogen density to their atomic environment. We find that the thermal fluctuations could be a dominant mechanism inducing the spinodal decomposition. Point defects could play the role of a inhomogeneous decomposition mechanism; however, we find their role to be minimal in the decomposition process. Experiments are proposed to determine the validity of our calculations and explicit results are presented based upon the well studied niobium-hydrogen system.

I. INTRODUCTION

We study the spinodal decomposition of the coherent metal-hydrogen systems using a model for macroscopic density modes proposed by Wagner and Horner.[1] Our main interest is in the hitherto neglected analysis of the role which thermal fluctuations play in initiating the decay of the labile state.

The macroscopic hydrogen density modes owe their existence to the important long-range elastic interaction between point defects in a coherent (i. e. single crystal) medium.[1-3] Experiments have been able to identify

43

several of these modes[4] and their existence has been
further supported by an examination of fully decomposed
crystals.[5] An x-ray experiment to observe the time evolution of the macroscopic modes has been suggested by
Burkhardt and Wöger.[6]

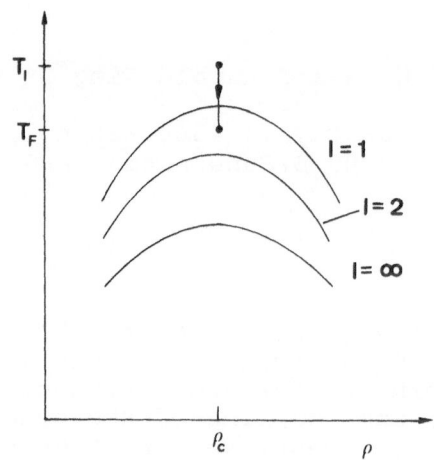

Fig. 1. Schematic of the stability boundaries. The quench
is shown for the critical hydrogen concentration, ρ_C ,
from a initial temperature T_I to a final temperature T_F.

We restrict our analysis to the case of isotropic elastic crystals and we consider only the case where a few
modes are unstable, ie. the so-called shallow quench
(fig. 1). In fig. 1 we have shown only modes labeled
with l=1 as unstable, but deeper quenches with some finite number of unstable modes could, in principle, be
analyzed by this method[7].

II. INCLUDING THERMAL FLUCTUATIONS

Local fluctuations in the hydrogen density are
caused by the interaction of the hydrogen atoms with
the lattice, which is in thermal equilibrium at a temperature T. The evolution of the hydrogen density $\vec{\rho}(\vec{r})$,
from some initial state $\rho_0(\vec{r})$ is governed by an equation
of motion involving the chemical potential and additive

terms which act as a source of noise. This noise arises
from the coupling of the hydrogen atoms to the thermal
phonons and it is, of course, proportional to $K_B T$. The
local chemical potential, $\mu(\vec{r})$, is a functional derivative of the total free energy F:

$$\mu(\vec{r}) = \delta F / \delta \rho(\vec{r}) \tag{1}$$

The free energy, on the other hand, contains two additive contributions. One contribution is from short-
ranged interactions and mixing effects (entropy):

$$F_0 = \int d^3 \left\{ (\frac{a}{2}) \rho(\vec{r})^2 + \frac{b}{4} \rho(\vec{r})^4 \right\} \tag{2}$$

and the second, and most decisive, contribution from
the elastic properties of the medium in which the hydrogen atoms are immersed:

$$H_{el} = -\frac{1}{2} \int d^3 r \int d^3 r' \rho(\vec{r}) W(\vec{r},\vec{r}') \rho(\vec{r}'), \tag{3}$$

where the kernel $W(\vec{r},\vec{r}')$ is related to the elastic
Green's function and the force-dipole constants[.] The
constants a and b appearing in eq(2) have been determined for the niobium-hydrogen system[3], we restrict our
analysis to this example. The density is expanded in a
set of eigenfunctions (surface modes) of the kernel
$W(\vec{r},\vec{r}')$[2,7,8]

$$\rho(\vec{r}) = \sum_{\vec{\ell}} \rho_{\vec{\ell}} \xi_{\vec{\ell}}(\vec{r}). \tag{4}$$

As mentioned earlier, the modes can become unstable at
a given temperature $T_{\vec{\ell}}$ (fig. 1).

The amplitudes $\rho_{\vec{\ell}}(\tau)$ are random variables and the
surface modes, $\xi_{\perp}^{\ell}$, vary over length scales of the
order of a linear dimension of the sample. Thus, each
sample supports, at any given point in time, a single
realization of the stochastic amplitudes $\rho_{\vec{\ell}}(\tau)$.

It is important to keep in mind since we eventually
make statements about averages with respect to an ensemble of trajectories, that an experiment must be repeated many times and the system must be prepared in
the same initial state (in fig. 1, $T_I \rightarrow T_F$) over and over
again. We shall concentrate on the spinodal decomposition process when the fluctuations are greatly amplified. The effect of thermal fluctuations is to initiate
the decomposition process[7-10], after this initial period, the mode amplitudes can be described by deter-

ministic dynamics. Our analysis in this later time regime shows that the equations of these modes are just
those considered by Jannsen[11] in his description of the
dynamics of the density amplitudes. We give, in addition, the distribution of initial values of these amplitudes due to the thermal fluctuations.

The applicability of this theory depends on the system
not being too close to the critical point ($Tc, \rho c$) (the
maximum of the l=1 curve in fig. 1), since we have made
an assumption that the thermal noise be small as compared to the saturation value of the amplitudes. When
we use the modes derived from a spherical elastically
isotropic sample and the niobiom data for the constants
of the free energy, we must require that

$$Tc - T_F \gtrsim \frac{10^{-7}}{R^{3/2}} \ (^{\circ}K) \tag{5}$$

R, the radius of the sphere, is in units of millimeters.
The analysis shows further that for the temperature
difference between the spinodal temperature of the unstable and stable density amplitudes, is 15°K. This interval is sufficient to perform experiments for which
the treatment of a few unstable and stable modes is
justified. In fact, Burkhardt[12] has presented an analysis of the minimum allowed temperature interval below which the coherency of the metal is endangered by
stresses exceeding the yield stress material; this interval is only a few degrees.

Finally, we have considered the effects of point defects and temperature inhomogeneities. These effects
do not put stringent demands on the purity of the sample or upon the temperature control. We conclude that,
indeed the effects of thermal fluctuations should be
observable and could be the dominating mechanism initiating the spinodal decomposition process. Possible
experimental schemes to observe these effects will be
discussed in the final section.

III. Prospects for Observation

As on illustration of the results, we consider
here only the unstable surface modes. There are three
such amplitudes in a elastically isotropic spherical
sample and in suitably scaled units[7], their equations
of motion are:

$$\frac{d\vec{x}}{dt} = \vec{x} \ (1 - \frac{15}{7}|\vec{x}|^2) \tag{6}$$

These equations can be easily solved. The initial conditions are taken from realizations of the initial
Gaussian distribution for the amplitudes:

$$Q \ (\vec{x},o) = (2\pi N)^{-3/2} \ exp \ (-|\vec{x}|^2/2N) \tag{7}$$

where N is proportional to the thermal energy, $k_B T$, and
is scaled by terms inversely proportional to $(T_c - T_F)^2$.

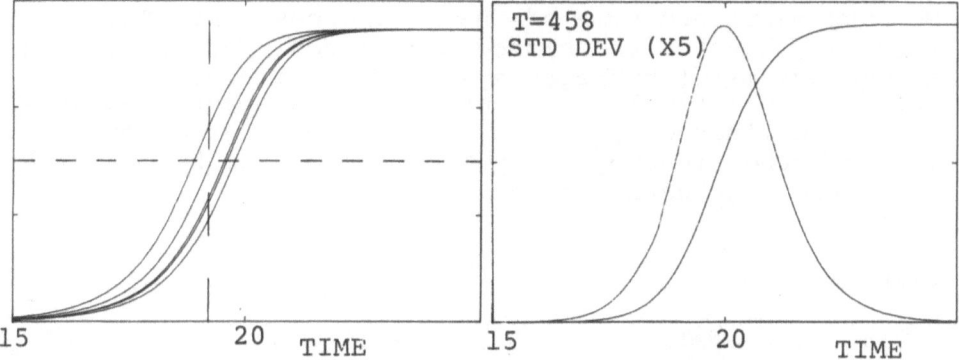

Fig. 2 Sample trajectories.   Fig. 3 Second Moment and
                                      standard deviation.

The width of the Gaussian is much smaller than unity.
The amplitude of the order parameter $|\vec{x}|$ is initially
small and remains so for a very long time, while being
amplified through the dynamics of the order parameter
(fig. 2). Each line in fig. 2 represents the evolution
of a single sample after a quench of the temperature
into the spinodal region $T=458^0 K$, $T=460^0 K$. Consider
the second moment, $<|x(t)|^2 >$, this[1] is the result of
summing all the trajectories created from the distribution of initial values eq.(7). For the temperature
$T=458^0 K$, fig. 3 presents the second moment and its
standard deviation, $\sigma=\sqrt{<(|\vec{x}|^2-<|\vec{x}|^2>)^2>}$, the fluctua-
tions of the order parameter amplitude are greatest at
the time where the rate of change of the amplitude is
largest. Repeated experiments could be performed and
the standard deviation of the measured quantity should
be analyzed after a preset time. The vertical line in
fig. 2 represents one possible point in time where the
fluctuations of the order parameter are nearly maximal
$(\sigma/<|\vec{x}|^2> \approx 35\%!)$.

Another type of data analysis for which we have also
determined the statistics is represented by the horizontal line in fig. 2. A preset value of the order parameter amplitude is chosen, and the fluctuations in
the times at which each trajectory reaches this value
are measured[7], the distribution of these times is the
so-called passage time distribution[10]. For a threshold
value $|\vec{x}|^2 = 7/30$, the relative standard deviation is:

$$\sigma_T = 1/|\ln(15N/7)|. \tag{8}$$

This is about 3% at $T = 458^\circ K$.

We have mentioned some of the influences which thermal
noise may have on the spinodal decomposition process in
a coherent metal-hydrogen system; experiments can distinguish in a simple way whether the decomposition process is dominated by intrinsic noise or whether there
is some other inhomogeneity, not considered here, which
spoil the fluctuation effects.

REFERENCES

1. G. Alefeld, Phys. Stat., Solid. 32, 67 (1969);
      Ber. Bandes ges. Phys. Chem. 76, 746 (1972)
2. H. Wagner and H. Horner, Adv. Phys. 23, 587
      (1974); H. Wagner, in Hydrogen in Metals I,
      edited by G. Alefeld and J. Völkl (Sprin-
      ger, Berlin, 1978).
3. H. Horner and H. Wagner, J Phys. C7, 3305 (1974)
4. J. Tretkowski, J. Völkl and G. Alefeld, Z. Phy-
      sik B28, 259 (1977).
5. H. Zabel and H. Peisl, Phys. Rev. Lett. 42, 511
      (1979); Acta Metall 28, 589 (1980).
6. T. Burkhardt and W. Wöger, Z. Physik B21, 89
      (1975).
7. J. W. Haus and H. King, Phys. Rev. B, to appear
      (1982).
8. W. Kappus and H. Horner, Z. Physik B27, 215
      (1977).
9. F. Haake, Phys. Rev. Lett. 41, 1685 (1978); M.
      Suzuki Prog. Theor. Phys. (suppl) 64, 402
      (1978)
10. F. Haake, J. W. Haus and R. Glauber, Phys. Rev.
      A23, 3255 (1981).
11. H. K. Janssen, Z. Physik B23, 245 (1976).
12. T. W. Burkhardt, Z. Physik 269, 237 (1974).

ENTHALPY OF METAL-HYDROGEN SYSTEMS: A PAIR-BOND MODEL

P. S. Rudman

Inco Research & Development Center, Inc.
Suffern, NY 10901

A model is developed that takes into account the variation in pair-bond energy as the interatomic distances vary with hydrogen content in metal-hydrogen (M-H) systems. It is observed that the general behavior of the relative partial molar enthalpy of hydrogen, ΔH_H, is that ΔH_H vs x (where x = H/M) tends to exhibit a minimum. This is accounted for by the model as due to the generality of the existence of a repulsive MH interaction and an anti-bonding HH interaction.

INTRODUCTION

Fig 1 presents representative ΔH_H vs x behavior: FeTi-H[1], Pd-H[2], V-H[3], and βTi-H[4]. All of these systems exhibit negative deviations from ideality, that is, ΔH_H decreases with increasing x, at least initially. A negative deviation implies a tendency toward miscibility gap formation[5]. The behavior of the Pd-H system is not atypical and thus the unique protonic, d-band filling models that have been applied to this system[6] appear questionable as band theory calculations have shown[7].

We shall investigate the ability of a model, in which the interaction energies vary due to the expansion of the lattice by the dissolved hydrogen, to account for this thermodynamic behavior.

The relative integral molar enthalpy, in the random interstitial solution approximation can be written as

$$\Delta H = [W_{MM} + xW_{MH} + x^2 W_{HH}]/(1 + x), \qquad (1)$$

where: W_{MM} = difference between MM interaction energy in solution and in pure metal M; W_{MH} = difference between MH interaction energy in solution and H_2 bonding energy; and W_{HH} = HH interaction energy in solution. We determine the relative partial molar enthalpy from Eq. (1) by

$$\Delta H_H = \Delta H + (1 + x)\partial \Delta H/\partial x, \qquad (2)$$

which yields

$$\Delta H_H = (W_{MH} + W'_{MM}) + x(2W_{HH} + W'_{MH}) + x^2 W'_{HH}, \qquad (3)$$

where $W' = \partial W/\partial x$.

Eq. (3), with invariant interactions assumed, yields

$$\Delta H_H = W^O_{MH} + 2W^O_{HH}x. \qquad (4)$$

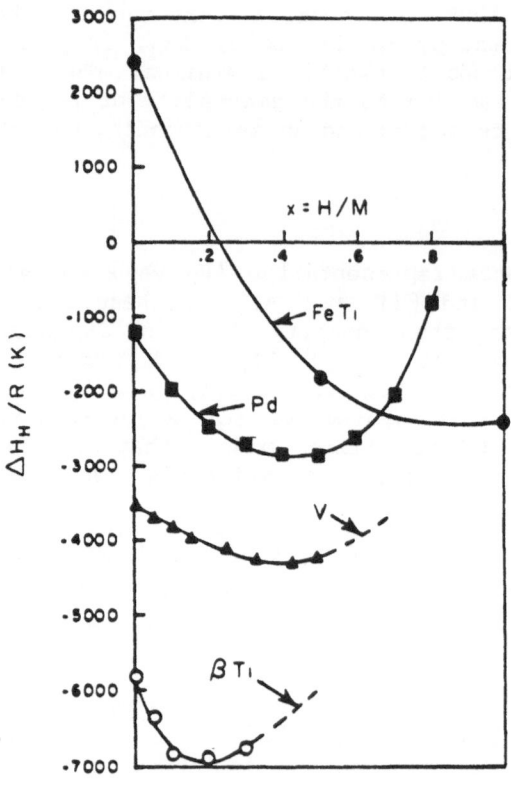

FIG 1 Representative behavior of the relative partial molar enthalpy of hydrogen in metal-hydrogen systems.

This leads to the concept that a negative deviation from ideality is due to an HH interaction, $W_{HH}^{o} < 0$. Eq. (4) accounts for the low composition behavior in the Pd-H and many other systems and its employment to calculate the HH interaction energy is common practice[3,5,8]. By comparing constant volume vs constant pressure thermodynamics, Wagner[9] showed that the bonding energy change due to lattice expansion is sufficient to account for the composition variation in ΔH_H. This paper should have been fatal to the invariant interaction model. In fact the model is alive, but W_{HH} is now an "apparent HH interaction energy"[8].

COMPOSITION DEPENDENT INTERACTION ENERGIES

We assume that the composition dependence of the interaction energies is determined only by the variation in interatomic distance as depicted in Fig 2. The pure metal MM interaction is minimized when the lattice parameter is a_{MM}^{o}. At a_{MM}^{o} the MH interaction must exert the dilational force that results in the generally observed expansion.

The role of the HH interaction which is so central to invariant interaction theory, is here relegated to a more modest role. As noted by Friedel[10], the proton screening radius is much shorter than the HH interatomic radius so the HH interaction must be small. Two alternate HH interactions are drawn: one a bonding interaction with $W_{HH} < 0$, and the other an anti-bonding interaction with $W_{HH} > 0$, over the

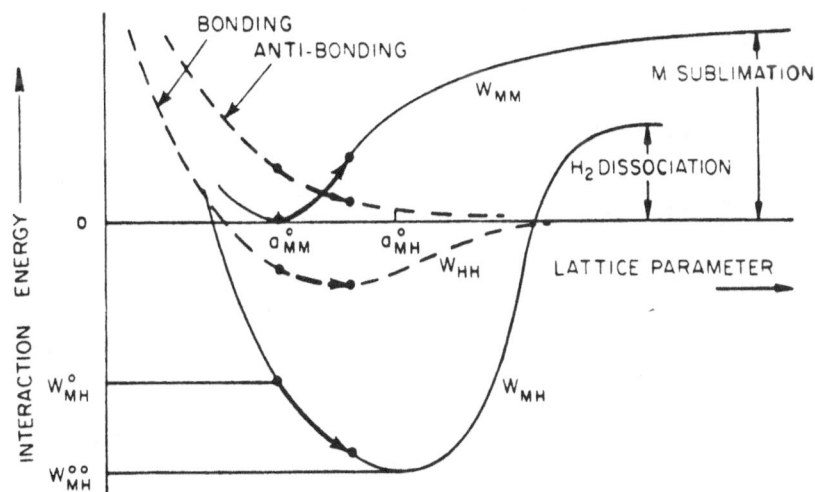

FIG 2 Schematic dependence of interaction energies on lattice parameter in metal-hydrogen systems.

observable lattice parameter range. Many, if not all, M–H systems exhibit ordering of H atoms wherein the H atoms tend to avoid each other. This implies that the HH interaction is generally repulsive as illustrated in Fig 2.

The lattice expansion is determined by the condition for mechanical stability: the sum of all internal forces is zero. Differentiating Eq. (1) we have

$$\partial W_{MM}/\partial a + x \partial W_{MH}/\partial a = 0. \tag{5}$$

We have assumed that $\partial W_{HH}/\partial a = 0$, which is probably a generally reasonable approximation. We express the lattice parameter dependence by parabolic potential wells

$$W_{MM} = k_{MM} \ (a - a_{MM}^{o})^{2}/2, \tag{6a}$$

$$W_{MH} = W_{MH}^{oo} + k_{MH} \ (a - a_{MH}^{o})^{2}/2, \tag{6b}$$

but for the HH interaction we simply let $W_{HH} = W_{HH}^{o}$.

Introducing Eqs. (6) into the mechanical stability criterion, we obtain the lattice parameter dependence

$$a - a_{MM}^{o} = x(a_{MH}^{o} - a_{MM}^{o})/(k_{MM}/k_{MH} + x), \tag{7}$$

The mechanical stability criterion reduces Eq. (3) to

$$\Delta H_{H} = W_{MH} + 2W_{HH}^{o} x. \tag{8}$$

We obtain the composition dependence of W_{MH} by employing the lattice parameter dependence in Eq. (6b) which yields

$$W_{MH} = W_{MH}^{o} - k_{MH}(a_{MH}^{o} - a_{MM}^{o})^{2} \ \frac{[2x(k_{MH}/k_{MM}) + x^{2}(k_{MH}/k_{MM})^{2}]}{2[1 + x(k_{MH}/k_{MM})]^{2}}, \tag{9}$$

where $W_{MH}^{o} = W_{MH}^{oo} + k_{MH} \ (a_{MH}^{o} - a_{MM}^{o})^{2}/2$.

SUMMARY DISCUSSION

Eq. (9) can be understood by reference to Fig 3. When H atoms are introduced into an M lattice the MH bond cannot form at its energy minimum of W_{MH}^{oo} which would require a lattice parameter of a_{MH}^{o}. Rather, the MH bond is compressed and the elastic energy of this compression raises the MH interaction energy to W_{MH}^{o}. However, as more H atoms are added, the lattice expands and some of the elastic energy ($W_{MH}^{o} - W_{MH}^{oo}$) is recovered. The higher the ratio of the elastic constant of the MH bond to that of the MM bond, the more rapidly the elastic energy is recovered.

If we compare Fig 3 with the experimental curves of Fig 1, we note that our model is not capable of accounting for the upturn in ΔH_H at higher compositions. Thus, in real systems some other factor(s) must be acting. Switendick's rule[7], that hydrogen concentration upper limits are accounted for by the assumption that ~ 0.2 nm marks the onset of a strong HH repulsion, certainly points strongly to an anti-bonding HH interaction, that is $W_{HH}^{o} > 0$, as the generally operating other factor.

Fig 4 presents a fit of Eq. (8) to the Pd-H data which yields $W_{MH}^{o} - W_{MH}^{oo} = 8840R$ and $W_{HH}^{o} = 3230R$ with $k_{MH}/k_{MM} = 1$. Taking from experiment that $W_{MH}^{o} = -1250R$, we derive that $W_{MH}^{oo} = -10,090R$. Considering the approximations employed, the fit is not unreasonable. The great depth of the uncompressed MH potential well W_{MH}^{oo} is noteworthy. In many, if not all M-H systems, there is an anomalously low ΔH_H, $\lim x \to 0$ which has been attributed to MH bonding at defect and surface sites[11], sites that do not require compression of the MH bond as a normal interstitial site requires.

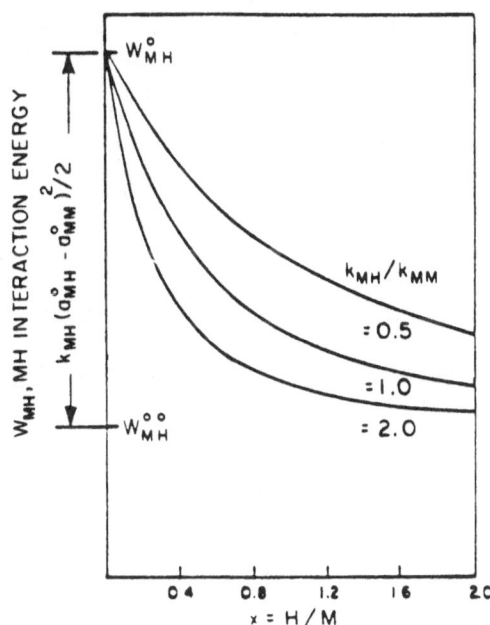

FIG 3 Dependence of MH interaction energy W_{MH} on composition and ratio of elastic constants of MH and MM bonds.

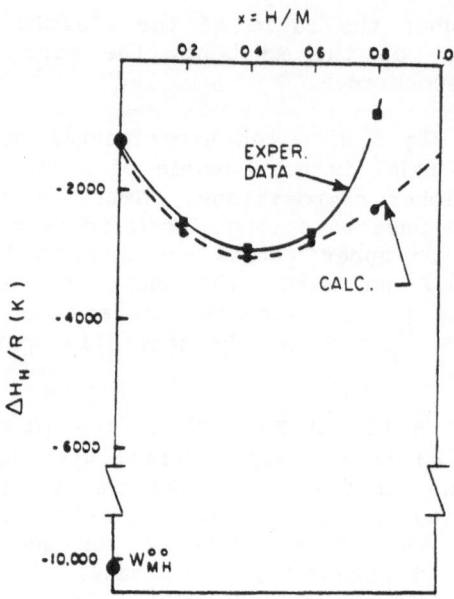

FIG 4 Fit of composition-dependent-interaction-energy-model to Pd-H system relative partial molar enthalpy of hydrogen data.

We further emphasize the radical change in conclusions when we compare data analysis by the present model with analysis by the invariant interaction model. The latter model leads to an "apparent" HH interaction, of exactly the opposite sign to the HH interaction obtained using the present model.

REFERENCES

1 H. Wenzl and E. Lebsanft, J. Phys. $\underline{10}$, 2147 (1980).
2 T.B. Flanagan and J.F. Lynch, J. Phys. Chem., $\underline{79}$, 444 (1975).
3 R.K. Edwards and E. Veleckis, as quoted in Metal Hydrides, W.M. Mueller, et al Eds., Academic Press, NY, 1968, p. 609.
4 J.F. Lynch and J. Tanaka, Acta Met., $\underline{29}$, 537 (1981).
5 P.S. Rudman, Int. J. Hydrogen Energy, $\underline{3}$, 431 (1978).
6 H. Brodowsky, Ber. Bunsenges., $\underline{76}$, 740 (1972).
7 A.C. Switendick, Z. Physik. Chem. Neue Folge, $\underline{117}$ 89 (1979).
8 T.B. Flanagan and W.A. Oates, Ber. Bunsenges., $\underline{76}$, 706 (1972).
9 C. Wagner, Acta Met., $\underline{19}$, 843 (1971).
10 J. Friedel, Ber. Bunsenges., $\underline{76}$, 829 (1972).
11 S. Tanaka and T.B. Flanagan, J. Less-Common Metals, $\underline{51}$, 79 (1977).

THERMAL DESORPTION SPECTRA OF PdH$_x$ SYSTEM (0<x<0.9) IN DIFFERENT

SAMPLES: A POWDER, A FOIL AND A WIRE

A. Stern, A. Resnik, D. Shaltiel and S.R. Kreitzman

Center of Energy and the Racah Institute of Physics
Hebrew University of Jerusalem, Jerusalem 91904, Israel

ABSTRACT

A preliminary report on the thermal desorption spectra of hydrogen in the PdH$_x$ system is presented for various initial hydrogen concentrations (0<x$_o$<0.9) and for various samples: a powder, a thin foil and a wire. The results show a two-peak structure for the powder and the foil, and a structure that consists of two peaks and a shoulder for the wire. It is shown that in the powder and the foil the desorption rate is surface limited and the bulk protons are in equilibrium, whereas for the wire, bulk diffusion limits the desorption rate. Therefore, while a correlation between the spectra of the powder and the foil to the Pd-H phase diagram can be made, such a correlation for the wire is not readily attainable.

INTRODUCTION

Thermal Desorption Spectroscopy (TDS) is a technique where the rate of desorption, -dx/dt (x ≡ [H]/[Metal]), into a pressure lower than the equilibrium pressure is being measured while the temperature is increased linearly with time. As a result, one obtains a spectrum composed of peaks of the desorption rate versus temperature, T. This technique has already been applied to investigate the desorption of hydrogen from the bulk in various hydrides[1,2]. We have used here the TDS technique to investigate the desorption kinetics of hydrogen from palladium of various forms; a powder, a foil and a wire. Due to the limited space we shall concentrate only on some important features. The results indicate that the desorption rate is surface limited in the power and foil samples due to their small diffusion length within the bulk. TDS are partially diffu-

sion limited in the wire sample. We analyse the spectra with respect
to the Pd-H (x,T) phase diagram and find that while the peaks in the
powder spectra are correlated to various regions in the phase diagram,
small deviations occur in the foil case and larger deviations, including the appearance of a new uncorrelated peak, occur in the wire
spectra. A more detailed report is now being prepared.

EXPERIMENTAL

     The experimental set-up, as well as an additional improvement
whereby the thermocouple was placed in direct contact with the sample, has been described elsewhere.[2]

     The wire sample was 140 mm long x 0.5 mm diameter (purity 5N),
the foil dimensions were 300x1.2x0.07 mm (5N) and the powder particles size were 2-4 μm (99.8% purity).

RESULTS: THE SPECTRA

     Figures 1-3 show the powder, foil and wire spectra for various
initial concentrations x_0. The heating rates were approximately
1.7, 0.8, 0.6 ^{o}C/sec for the wire, foil and powder respectively.

     The powder desorption spectra shown in Fig. 1 exhibit a single
peak curve (peak A) at low initial concentrations. Its intensity
grows gradually with x_0. For $x_0 \gtrsim 0.6$ a shoulder is developed at the
low temperature side of peak A.   This shoulder grows into a second
peak (peak B) for $x_0 > 0.8$. We notice that while the onset temperature
of the desorption (the lowest temperature where a given desorption is
observed) increases gradually when x_0 decreases for $x_0 > 0.6$, it is approximately constant, independent of x_0, for $x_0 < 0.6$. The temperature
range of the spectra is from -100 C to +100 C.

     The foil absorption spectra is given in fig. 2. They are simi-
lar to the powder spectra and include a peak A and a shoulder B.
However, the desorption occurs at higher temperatures compared to
the powder (from -30 C to +200 C).

     The wire desorption spection spectra are shown in fig. 3. Their
behavior is not as regular as the powder spectra, and in effect, they
were found to be dependent on the cycling. At low initial concentration the spectra consist of peak A and a shoulder (appearing
on the high temperature side of peak A). The intensity of the peak
increases while the shoulder develops into a second peak (peak C)
as x_0 increases. For $x_0 > 0.6$ a new shoulder appears on the low temperature side of peak A (shoulder B). Similar to the case of the
powder, the onset temperatures of desorption are about the same
(from 20 to 50 C) for different initial concentrations below $x_0 = 0.6$

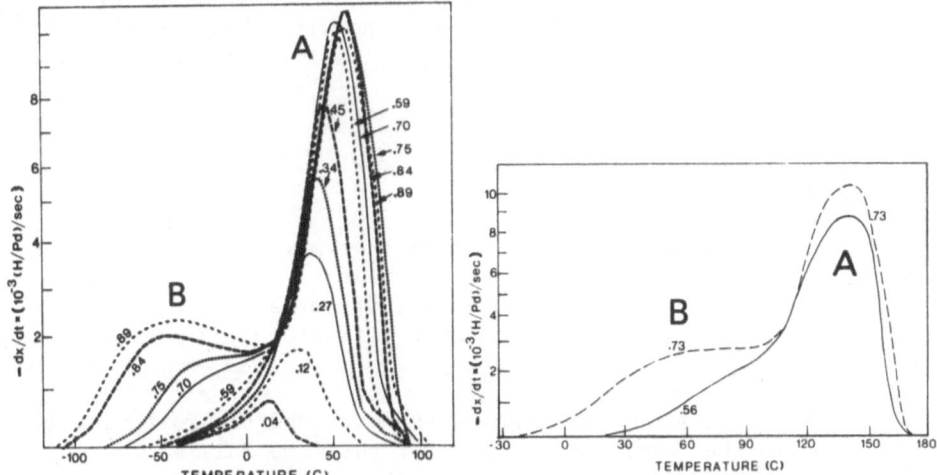

Figure 1:
The powder thermal desportion
spectra for various initial concentrations, denoted by the
members beside each graph.

Figure 2:
The foil thermal desorption
spectra, for two initial
concentrations.

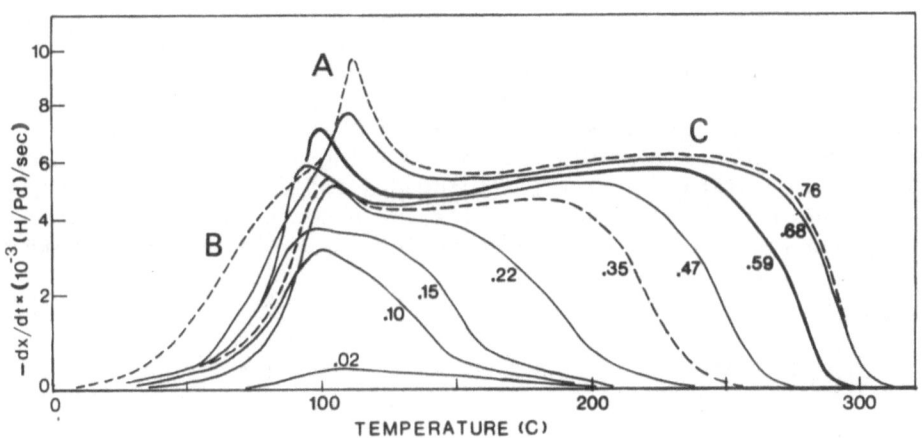

Figure 3: The wire thermal desorption spectra. The numbers
besides each graph denote the initial concentrations.

Comparing the temperature of desorption for the various samples, we note that the initial desorptions from the powder are at lower temperatures than the foil and the wire while the last two have closer onset temperatures (for the same initial concentrations).

DISCUSSION

a. The Rate Limiting Step

There are two important parameters that may cause the differences between the spectra of the wire, the foil and the powder: 1. The ratio, M_S/M_B, between the number of surface sites to the number of bulk sites. 2. The shortest diffusion distance to the surface.

Since M_S/M_B is about the same in the wire and the foil, but is mush higher in the powder, we expect for the same initial concentration (see ref. 2), similar onset temperatures of desorption in the wire and in the foil spectra, but lower onset temperatures in the powder - in agreement with our results. On the other hand, the similarity between the powder and the foil spectra (fig. 1 and 2) is due to their shorter diffusion distance that allows for a quasi-equilibrium state inside the bulk (see below). The C peak in the wire spectra is therefore attributed to concentration gradients, probably formed within the α phase in the outer part of the sample, and perhaps in the β phase in the interior.

The kinetics of desorption is composed of at least two steps:
1. Diffusion from the bulk to the surface,
2. Passage through the surface, including the recombination of protons to form a single H_2 molecule.

In cases where the diffusion flow is fast compared to the passage through the surface, the rate determining-step will be step 2. Otherwise, the diffusion will limit the desorption rate. To determine whether the bulk diffusion limits the desorption, we estimate the maximum variation of the concentration Δx in the bulk by assuming that a formed in a homogeneous system:

$$\Delta x = (-\frac{dx}{dt}) . \tau_{diff} . \tag{1}$$

Here $\tau_{diff} \equiv L^2/g^D$ is a characteristic decay time for the diffusion[3,4], D is the diffusion constant, L is a diffusion distance and g is a geometrical factor. For the cases of a sphere, a cylinder and a plate, L is taken as the radius of the sphere, the radius of the cylinder and the thickness of the plate, and g = π^2, 2.4^2 and π^2 respectively[4]. Δx is compared to $\bar{x}$, the average concentration in the bulk as follows: When $\Delta x << \bar{x}$ we have a quasi-equilibrium state of hydrogen in the bulk. Otherwise, concentration gradients are formed and the diffusion limits the desorption.

Diffusion in a two phase system with a moving boundary is a more complicated problem[3,4]. However, in order to obtain a crude estimate for Δx, we calculate (1) using for D either its value in the α phase or in the β phase ($D_\alpha = 2.0 \times 10^{-3} \exp(-2670/T)$, $D_\beta = 0.9 \times 10^{-3} \exp(-2650/T)$, D is measured in cm^2/sec and T in degrees K[5,6]). When the Δx values estimated as above are small compared to the hydrogen concentration in the α phase there is an equilibrium state of hydrogen in the bulk. Otherwise, concentration gradients may be formed either in the α or in the β phase.

We have calculated Δx for various temperatures during the desorption, and for various samples. The results indicate that:
1. The desorption from the powder is a surface limited process ($\Delta x \sim 10^{-3}$–10^{-6}), except, perhaps, at the high temperature region of the decreasing part of peak A.
2. The wire desorption is limited by diffusion ($\Delta x \sim 0.01$–1) at all temperatures, except at the onset of the desorption.
3. The foil is an intermediate case ($\Delta x \sim 10^{-3}$) being closer to the powder case (which is surface limited).

These conclusions are further supported by attempting to correlate the desorption spectra to the equilibrium phase diagram (see below).

b. Correlating the Desorption to the Phase-Diagram

By integrating the spectrum with respect to time we obtain a path $\bar{x}(T) = x_0 - \int_0^{t=T/b} (-dx/dt)dt$, ($\bar{x}$ is an average hydrogen concentration and b is the heating rate) in the (x,T) plane, along which the system proceeds during the desorption. Plotting the path $\bar{x}(T)$ together with the phase diagram[7] in the (x,T) plane enables us to relate the various parts of the spectra to different regions in the phase diagram. It is found that in the powder spectra, peak B is related to the β region, peak A is related almost entirely to the $\alpha+\beta$ region and the minimum between the two peaks corresponds to a crossing of the boundary line between the β and the $\alpha+\beta$ region. However, such an attempt to correlate peaks A and B in the wire case to the phase diagram was not successful. For example, in the wire spectrum with $x_0=0.76$, the shoulder B, peak A, as well as part of the curve up to 150 C, are already inside the β region. This again, is an indication that there is a formation of concentration gradients which cause a low chemical potential - compared to the value determined by $\bar{x}$ - near the surface. (On the role of the chemical potential in the desorption process, see ref. 2).

The foil is an intermediate case, closer to that of the powder where peak B is correlated to the β region and peak A is almost entirely correlated to the $\alpha+\beta$ region.

Other experimental results (effects of different heat rates) as well as extraction of activation energies and (Arrhenius) pre-exponential factors from various parts of the spectra (such as the onset of the desorption) and a further discussion on the desorption kinetics will be published elsewhere.

ACKNOWLEDGEMENTS

We wish to thank N. Kaplan and V. Zevin for most helpful discussions. This research was partially supported by The Hebrew University Research Center for Hydrogen and Redox Fuels Synthesis, Utilization and Storage.

REFERENCES

1. M.H. Mendelsohn and D.M. Gruen, Mat. Res. Bull., 16, 1027 (1981)
2. A. Stern, S.R. Kreitzman, A. Resnik, D. Shaltiel and V. Zevin,
        Solid State Commun., 40, 837 (1981).
3. J. Crank, "The Mathematics of Diffusion", Clarendon Press,
        Oxford (1975),
4. W. Jost, "Diffusion in Solids, Liquids, Gases", Academic Press
        Inc., Publishers, New York (1952).
5. J. Völkl and G. Alefeld, Diffusion of Hydrogen in Metals, in
        "Topics in Applied Physics", Vol. 28, "Hydrogen in Metals I",
        G. Alefeld and J. Völkl ed., Springer-Verlag, Berlin
        Heidelberg, New York (1978).
6. E.F.W. Seymour, R.M. Cotts and W.D. Williams, Phys. Rev. Lett.
        35, 165 (1975).
7. E. Wicke and H. Brodowsky, Hydrogen in Palladium and Palladium
        Alloys, in "Topics in Applied Physics", Vol. 29, "Hydrogen
        in Metals II", G. Alefeld and J. Volkl ed., Springer-Verlag,
        Berlin Heidelberg, New York (1978).

FUNDAMENTAL ASPECTS OF THERMODESORPTION SPECTRA IN BULK

METAL-HYDRIDE SYSTEMS

S. R. Kreitzman, A. Stern, D. Shaltiel

Center of Energy and the Racah Institute of Physics
Hebrew University of Jerusalem
Jerusalem 91904, Israel

ABSTRACT

A basic model is introduced to describe the thermodesorption
process in bulk metal-hydride systems. The model is then applied
to progressively more complex and realistic situations in order to
illustrate the physical processes which are important. Specifically,
we compare and contrast the situation to be found in non-interacting
multisite hydrides with those exhibiting a phase structure. Furthermore, it is shown that by representing the thermodesorption spectra
in normalized variables one obtains a straightforward appraisal of the
nature of the surface recombination, as well as the stoichiometries
and energies associated with the various states in the bulk.

INTRODUCTION

Recent experimental studies[1][2][3] on thermodesorption of
hydrogen from bulk metal-hydride (MH) systems yield spectra (TDS)
which show a rich and varied structure. This work presents a
simple model which describes the basic physics. The resulting
differential equation has been numerically solved for various
typical situations likely to be found in MH systems.

The physical situation at the surface is illustrated in Figure 1.
We take the equation governing the desorption process to be

$$\frac{dx}{dt} = -c' \cdot \beta^{\delta} \cdot [\sum_{ij} \chi_{\delta}^{s} e^{-(\epsilon_i^* - \epsilon_j^s)\beta}]^{\eta} = -c' \cdot \beta^{\delta} \cdot [\frac{Q^* \lambda}{1 + Q^s \lambda}]^{\eta}. \qquad (i)$$

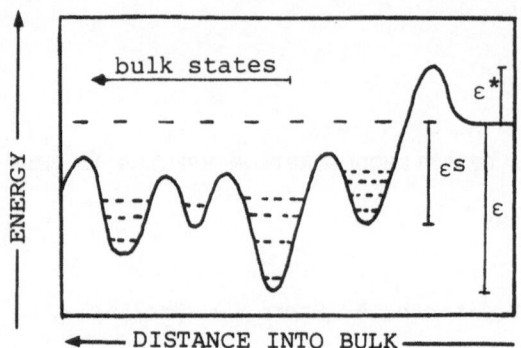

Figure 1. Schematic representation near the surface. Energies are taken with respect to the hypothetical noninteracting H_2 molecule.

Here, $\lambda = e^{\mu\beta}$ is the activity. μ is the chemical potential. $Q^*\eta$ is the partition function of the excited state excluding the contribution from the spatial reaction coordinate. Q^s is the surface well partition function, which can in principle include interactions with other bulk and surface states. The independent variable here is τ, the time. In this work the temperature, $T \equiv 1/\beta$, is taken to be linear in τ. The constant c' depends on the heating rate, the physical geometry (i.e. the surface to bulk ratio), the order of the surface reaction (η), and a quantum mechanical constant determining basically the amount of semiclassical phase space available for the hydrogen at hte surface. δ is a small integral or half integral number. The transition to the right side of (i) was effected by utilization of lattice gas formula for the probability that only one of the jth levels at a surface site can be occupied;

$$\chi_j^s = \frac{\lambda e^{-\varepsilon_j^s}}{1 + \lambda Q^s} \qquad \text{(ii)}$$

The denominator is simply the grand partition function of the surface site subject to the restriction of single occupancy. Assuming that the occupancy of the lowest energy levels yields the funcamental equation,

$$\frac{d\chi}{dt} = -c \cdot \beta^\alpha \cdot \left[\frac{e^{\varepsilon^*\beta}}{1/\lambda + e^{-\varepsilon^s\beta}} \right]^\eta . \qquad \text{(iii)}$$

MODELS

Non-interacting, Single Site

The model is defined by the ideal lattice gas formulation found in standard texts[4]. The activity for a bulk site of energy ε is

$$\lambda = e^3\beta \, \frac{x}{1-x} \, , \qquad\qquad\qquad\qquad (iv)$$

which yields a desorption equation of the form

$$\frac{dx}{dt} = -c \cdot \beta^\alpha \cdot [\frac{e^{-\varepsilon^*\beta}x}{e^{-\varepsilon\beta}(1-x)+e^{-\varepsilon s}\beta x}]^\eta = -c \cdot \beta^\alpha \cdot [\frac{e^{-\varepsilon^*\beta}}{e^{-\mu\beta}+e^{-\varepsilon s\beta}}]^\eta . \quad (v)$$

Desorption from such a system yields three typical patterns of behaviour depending on whether $(\varepsilon - \varepsilon^s)/<4kT>$ is positive, zero or negative, corresponding to a deep surface site, a bulk site at the surface, and a shallow surface well. TDS for these cases are illustrated in Figure 2 for a second order interaction and $\alpha = -z$. The development of these curves with increasing concentration shows two basic patterns. In Figure 2a, the onset of desorption is constant and as the initial concentration increases the desorption continues to higher temperature. In Figures 2b and c, however, the change in μ as a function of x dominates the TDS. Specifically, as the initial concentration increases so does the chemical potential and therefore the probability that a surface site is occupied. The onset of desorption then decreases with increasing initial concentration.

The variables x and T are not the most useful quantities for extracting information about the overall nature of the desorption TDS displayed as $d(^x/x_0)/d\beta$ vs $(^x/x_0)$ yield direct information with regard to the energies and the order of the surface step, at least if one does not have a very deep surface well. This is most easily

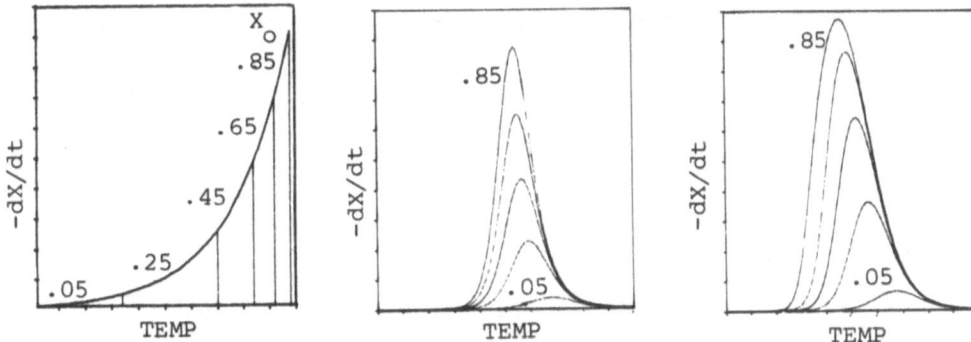

Fig. 2. Single site desorption when $(\varepsilon - \varepsilon^\delta)/<4kt>$ is a) <0, b) $=0$, c) >0.

shown by using a low temperature approximation integration of (v),

$$\int_{\beta i}^{\beta f} f(\beta) e^{-\epsilon\beta} d\beta \approx \frac{-1}{\epsilon} [f(\beta_f) e^{-\epsilon\beta_f} - f(\beta i) e^{-\epsilon\beta i}] , \qquad (vi)$$

where $f(\beta)$ is a slowly varying function. The resulting equations for the TDS are,

$$\frac{1}{\eta(\epsilon^*-\epsilon)} \frac{d(^x/x_o)}{d\beta} = -\frac{g^\eta(x)}{x_o} \int_{x_o}^{x} \frac{1}{g^\eta(x')} dx', \qquad (vii)$$

with

$$g(x) = x \qquad\qquad (\epsilon-\epsilon^s)/<4kt>>0,$$

or

$$g(x) = x/(1-x) \qquad (\epsilon-\epsilon^s)/<4kt><0.$$

In the first case, with $g(x)$ a homogeneous function, it can immediately be seen that the RHS of (vii) is a function of $(^x/x_o)$ only. This implies that the shape of the TDS in the normalized variables is independent of x_o and a function of η only, with a 'height' proportional to $\eta(\epsilon^*-\epsilon)$. For this case the TDS are given by

$$\frac{d(x/x_o)}{d\beta} = (\epsilon^*-\epsilon)\frac{x}{x_o} \ln(\frac{x}{x_o}) \qquad \eta = 1, \qquad (viiia)$$

$$\frac{d(x/x_o)}{d\beta} = (\epsilon^*-\epsilon)\frac{x}{x_o}(1-\frac{x}{x_o}) \qquad \eta = 2. \qquad (viiib)$$

When $(\epsilon-\epsilon^s)/<4kt><0$, the TDS in natural variables (NTDS) show a chatacteristic shape, and more prominently, amplitude dependence on x_o. It is this feature which allows one to quickly distinguish, from the NTDS the order and the nature of the desorption for the non-interacting systems to be considered shortly. The form of the equations similar to (viii) also allow an easy approximation to the energy of a given peak directly from the NDTS. Using the fact that the maximum of the TDS is approximately twice the value of the desorption at the half width we find the following relation in terms of widths, positions and energies of fully developed peaks in a TDS,

$$(\epsilon^*-\epsilon) \approx 10\eta T_{max}^2 /\Delta T \qquad\qquad (ix)$$

where ΔT is the temperature half width of a peak in the TDS.

Independent Site Model

    We focus attention on the simplest nontrivial situation. The inhomogeneities of the hydrogen-metal interaction is larger than any hydrogen-hydrogen interaction and the energy of a particular proton is then determined only by its position in the host lattice. The equation of state $\lambda=\lambda(x,\beta)$ is determined by the solution of

$$x = \sum_{\alpha} \left(\frac{g_{\alpha}\lambda e^{-\varepsilon\alpha\beta}}{1+\lambda e^{-\varepsilon\alpha\beta}} \right), \qquad\qquad (x)$$

Where g_{α} are the stoichiometries for the various sites in the metal[4]. Figure 3a illustrates the resulting behaviour for $\mu=T\ln\lambda$ for a three site case. We see that for $\varepsilon^{\alpha'+1}-\varepsilon^{\alpha'}>4T$, μ displays a step-function like behaviour as x reaches $\Sigma_{\alpha=1g\alpha}^{\alpha'}$. The behaviour is analogous to the filling of non-interacting electron states and the chemical potential μ is basically the energy of the highest filled state. The TDS and NTDS shown in Figure 3b and c reflect the fact that each 'plateau' area is similar to the equation of state for a single site MH system, except that it develops in the concentration range appropriate to its place in the multisite energy spectrum. This stems fundamentally from the quasi-Fermi statistics, Eq. (x) which governs x and from the result that μ, on a global scale, is almost solwy a function of x. The multipeaked structure of TDS is basically a reflection of $d\mu/dx|_{T}$.

In this model, all of the basic information can be easily extracted from the NTDS[*]. The maxima of the NTDS peaks, (proportional to $(\varepsilon^{*}-\varepsilon^{\alpha})$), occur at approximately half the site stoichiometry. Distances between successive minima yield the values of g_{α}. Use of relation (ix) allows an approximate determination of ε^{*} as well as ε^{α}. Correlation of the g_{α} with site stoichiometries in intermetallic compounds should provide fundamental data with regard to the hydride structure.

Two-phase Model

In attempting to calculate TDS from multiphase systems it was first necessary to develop a reasonable and calculable model for typical MH phase diagrams. We have used an N particle cluster equation including terms due to concentration fluctuations. A discussion of the nature of the model, its solutions and the methods of obtaining a numerical solution will be the subject for another work. The model has been chosen so that the phase structure resembles the morphological nature of the PdH isotherms above 150 K[5]. Results are shown in Figures 4a and b.

The desorption model is a natural extension of equation (v), but explicitly accounts for the possibility of desorption from both the alpha (a) and beta (b) phases.

$$\frac{dx}{dt} = -c\cdot\beta^{\alpha} \cdot \left\{ X_a \left[\frac{e^{-\varepsilon_a^{*}\beta}}{e^{-\mu\beta}+e^{-\varepsilon_a^{s}\beta}} \right]^{\eta} X_b \left[\frac{e^{-\varepsilon_b^{*}\beta}}{e^{-\mu\beta}+e^{-\varepsilon_b^{s}\beta}} \right]^{\eta} \right\} \qquad (xi)$$

---

[*]For this situation the NTDS taken as $dx/d\beta$ with respect to x.

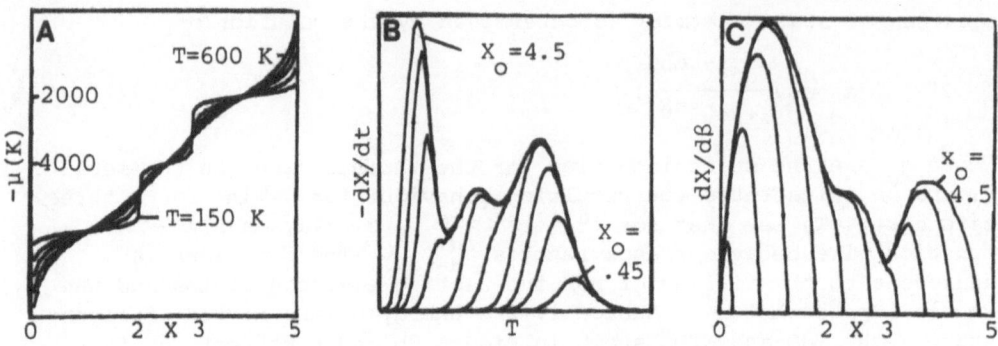

Fig. 3. a) Chemical potential, b) TDS, c) and NTDS - three site system.

Here X_a and X_b are the amounts of a and b phases, determined by the lever rule in conjunction with the position in the phase diagram. Surface energy parameters must be consistent so that the equation is valid in the fluid as well as in the phase separated region. Thus the surface energy terms must have identical functional dependence, i.e. $\varepsilon_{a,b}^{*,s} = \varepsilon_o^* + \Delta\varepsilon_o^{*,s} \cdot X$. Various choices of $\varepsilon_o^{*,s}$ and $\Delta\varepsilon_o^{*,s}$ yield a wide variety of TDS. Figure 4c shows a TDS, using the phase diagram in 4a, quite typical of that found in PdH[3]. A detailed discussion of this spectrum must be published elsewhere.

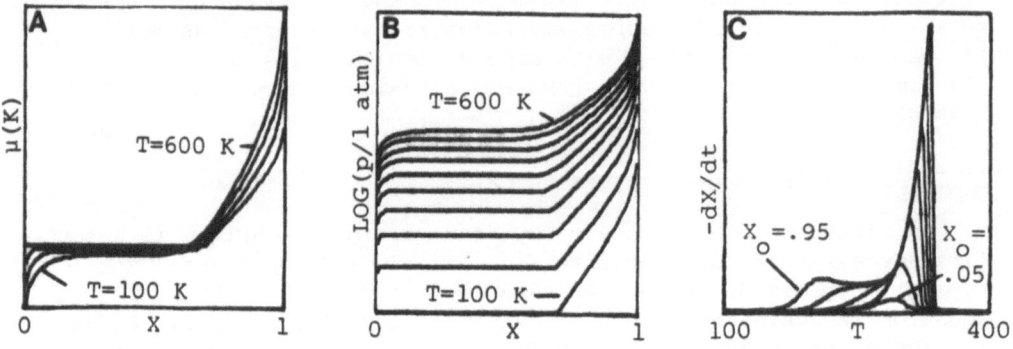

Fig. 4. a) Chemical potential, b) Pressure, c) TDS of PdH-like system.

In conlusion, we hope that concepts presented will allow a more meaningful appreciation of the TDS experiment.

REFERENCES

1 M.H. Mendelsohn, D.M. Gruen, Mat. Res. Bull. 16, 1027 (1981)
2 A. Stern, S.R. Kreitzman, A. Resnik, D. Shaltiel, V. Zevin,
     Solid State Commun. 40, 837 (1981)
3 A. Stern, A. Resnik, D. Shaltiel, S.R. Kreitzman
     To appear in this journal
4 T.L. Hill, Introduction to Statistical Thermodynamics
     Addison Wesley, Reading, Mass. (1962)
5 H. Frieske, E. Wicke, Ber. Bunsenges. Phys. Chem. 76, 847 (1972)

THERMAL DESORPTION SPECTROSCOPY (TDS) OF HYDROGEN FROM NIOBIUM*

V.J. Ghosh, M.A. Pick, D.O. Welch, and G.J. Dienes

Brookhaven National Laboratory
Upton, New York 11973

INTRODUCTION

Thermal desorption spectroscopy is often used to determine the activation energy of desorption of an adsorbate on a substrate as well as the order of the kinetics. In the case of flash desorption of an adsorbate from a surface obeying second order kinetics, a plot of $Log(\theta_0 T_P^2)$ vs. $1/T_P$ (where θ_0 is the initial surface coverage and T_P is the temperature of the maximum desorption rate, i.e., the "desorption temperature") will yield a straight line, the slope of which is determined by the chemisorption energy. However, Davenport et al.[1] argued on the basis of a simple theoretical model that in the case of a system with hydrogen absorbed into the bulk in quasiequilibrium with adsorption sites on the surface, a similar plot, where the coverage θ_0 must now be replaced by the initial concentration x_0, will yield a slope determined by the heat of solution. In the following experiments we have tried to verify the functional dependence of the desorption temperature on hydrogen loading as outlined above. Deviations from these theoretical predictions were observed, and numerical kinetic simulations were made to aid in understanding them.

EXPERIMENTAL METHOD

Samples consisted of thin foils (50 μm, MARZ grade) of niobium, 5 mm wide by 20 mm long. They were spot welded onto 2 mm diameter tungsten rods as supports in an ultrahigh vacuum system

---
*Research supported by the U.S. Department of Energy under contract No. DE-AC02-76CH00016.

with a base pressure in the low 10^{-10} Torr range. The samples
were cleaned by resistive heating to over 2300 K in the ultrahigh
vacuum after baking them in 10^{-6} Torr oxygen at 2000 K. This
treatment removes the interstitial impurities.[2] The frequent
heating and cooling of the sample caused recrystallization and
grain growth. Using an attached low-energy electron diffraction
apparatus we observed that the grains were all aligned with their
(110) planes parallel to the surface. The experimental results
can therefore be taken as characteristic of the (110) plane of
Nb. For the desorption experiments the samples were heated by a
dc current which was programmed to vary in such a way as to provide a linear time dependence of the sample temperature for heating rates up to ~200°/sec. The hydrogen desorption was monitored
using the output of a UTI mass analyzer 100C tuned to the H_2
peak.

The sample was allowed to accumulate hydrogen at pressures
which varied from 5×10^{-10} Torr to 1×10^{-4} Torr for periods of time
usually shorter than 2 minutes. The pressure drop during absorp-
tion and the area under the resulting desorption peaks correlated
well and, knowing the volume of the vacuum system, could be used
to estimate the absolute amount of hydrogen absorbed or adsorbed
by the sample.

EXPERIMENTAL RESULTS

The data obtained in this manner are shown in Figure 1.
The data are plotted so that, according to the discussion of
Davenport et al.[1], they should fall on a straight line with a
slope equal to $2E_s/k_B$ where E_s is the heat of solution per
hydrogen atom. There are two features of the desorption data
which are in disagreement with the predicted behavior. First,
the slope is such that if it is given by $2E_s/k_B$ then the value
of E_s is found to be 12.8 rather than the accepted value[3] of 8.6
Kcal/mole of hydrogen atoms. Second, it may be seen (especially
from the 200 Ks^{-1} data) that at high hydrogen loadings the de-
sorption-peak temperature T_p, falls with increasing loading to a
minimum value and then rises again. As we shall see below, the
occurrence of such a minimum is a feature of the kinetic model
upon which Davenport et al. based their discussion although the
analytic approximations they used caused them to overlook its
existence.

DISCUSSION

To help understand the origins of the discrepancies mentioned above we numerically solved the kinetic equations of the model
used by Davenport et al.[1] (albeit with various approximations):

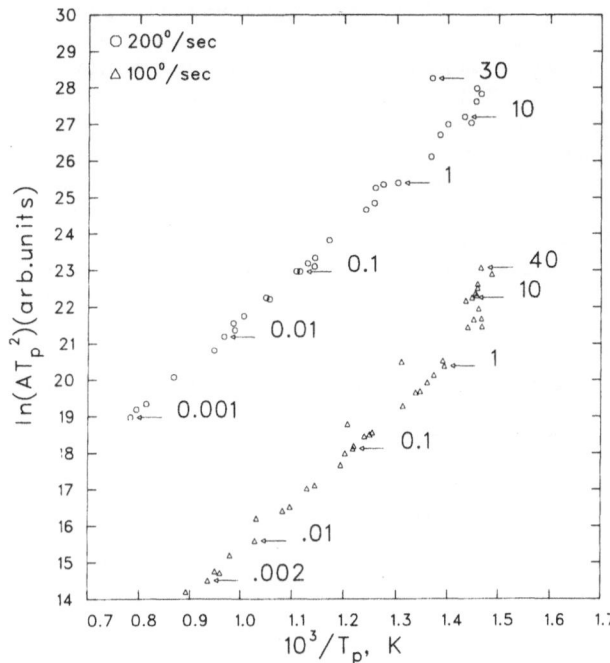

Fig. 1. The variation of the temperature of the maximum hydrogen desorption rate, T_p, with the amount of hydrogen loading for a (110) niobium surface for two rates of heating. The amount of hydrogen loading is characterized by the integrated desorption rate, A. The numbers indicate the approximate loading in monolayers of hydrogen atoms.

bulk hydrogen sites "communicate" with a set of surface sites which "communicate" with the vacuum (the single-site model). We also numerically calculated the properties of models with two types of surface sites (two-site models). The kinetic equations of the single-site model for desorption into the vacuum are:

$$d\theta/dt = -K\theta^2 - \nu\theta(1-x) + \beta x(1-\theta),$$ (1)

$$N_\ell dx/dt = \nu\theta(1-x) - \beta x(1-\theta),$$ (2)

where x and θ are bulk and surface concentrations and K, β, and ν are temperature-dependent rate constants, N_ℓ is the number of layers of sites in the bulk. Using rate constants consistent with experimental data on hydrogen uptake kinetics[4] and heat of solution[3] E_s, the numerical calculations yield the results in Figure 2, which are qualitatively like the experimental data. The slope of the linear region is $\simeq 2E_s/k_B$ in agreement with the

results of Davenport et al.[1], and quasiequilibrium between bulk
and surface sites was found to occur over the entire range of
loading as they had assumed. The calculated minimum desorption
peak temperature, $\simeq$570 K, is less than the experimental value,
$\simeq$680 K, but agreement is achieved by reducing the pre-experimen-
tal factor of the rate constant K from 1×10^{14} s^{-1} to 5×10^{12} s^{-1}.
(Such a minimum was not predicted by Davenport et al. because of
their assumptions that $\beta x \ll 1$.) It was not possible to choose
parameters consistent with the experimental uptake kinetics and
the heat of the solution which would yield quantitative agreement
between the value of the scope calculated with the single-site
model and the experimental value. Numerical simulations based on
two different two-site models also failed to match the experimental value. The origin of the discrepancy remains unknown.

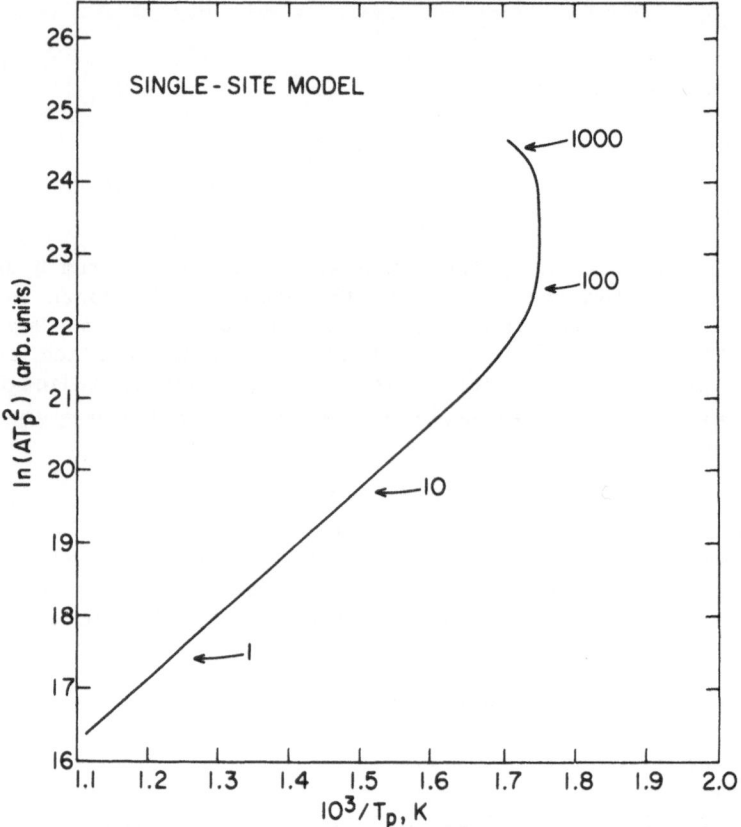

Fig. 2.   The dependence of the temperature of the maximum hydro-
           gen desorption rate, T_p, upon hydrogen loading (char-
           acterized by the integrated hydrogen desorption rate,
           A) as calculated with the single-site model. The num-
           bers indicate the approximate loading, in monolayers of
           hydrogen atoms.

REFERENCES

1. J. W. Davenport, G. J. Dienes, and R. A. Johnson, Phys. Rev. B, March 15 (1982).
2. H. Wenzl and J.-M. Welter in: <u>Current Topics in Material Science</u>, Vol. 1, Ed. E. Kaldis (North Holland, Amsterdam, 1978).
3. J. A. Pryde and C. G. Titcomb, J. Phys. C <u>5</u>, 1293 (1972).
4. M. A. Pick, Phys. Rev. B <u>24</u>, 4287 (1981).

REFERENCES

1.

2.

3.

4.

THE CONCENTRATION PROFILE OF HYDROGEN NEAR THE SURFACE OF NIOBIUM

M. A. Pick, A. Hanson, K. W. Jones, and A. N. Goland

Brookhaven National Laboratory
Upton, New York 11973

INTRODUCTION

Hydrogen in niobium has been studied for many years for reasons motivated both from a basic as well as an applied point of view. On the one hand it has generated interest within the theoretical physics community because it is a very close physical realization of the theoretical lattice-gas model. On the other hand, it is also studied as a model substance by those interested in the practical aspects of hydrogen storage and other applications of metal-hydrogen systems.

The numerous experimental and theoretical investigations of the Nb-H system have provided us with detailed information concerning many of the more important physical properties of the system, e.g., phase diagram, structures, diffusion coefficients, heat of solution, etc.[1] More recently work has been published dealing with the hydrogen absorption-desorption kinetics of this system and its related surface properties.[2,3] From this work it is apparent that the rate at which hydrogen can be inserted into and extracted from Nb, and also other metal-hydrogen systems more suitable for practical applications, is strongly influenced by the hydrogen concentration on and near the surface of the material.

Zabel, et al.,[4] used different x-ray wavelengths to measure the lattice parameters near the surface of a Nb-single crystal containing hydrogen at the critical concentration, 31 at.%. They found that a layer approximately 1 μm thick at the surface contained appreciably less hydrogen than the bulk. They interpreted their findings as evidence for critical wetting. On

73

the other hand, it can be assumed that the energy it costs to expand the metal lattice by the insertion of a hydrogen interstitial is minimized at or near the surface. This would result in the "effective" heat of solution being more negative at the surface and thus leading to a higher hydrogen concentration at the surface,[5] contrary to the findings of Zabel et al. It is therefore important, from both a basic and an applied point of view, to measure the depth-concentration profile of hydrogen in niobium by an independent technique. This can be conveniently accomplished by the use of nuclear reaction analysis using the $^1H(^{15}N,\alpha\gamma)^{12}C$ reaction.

EXPERIMENTAL

Samples hydrided in two different ways were investigated. The materials used were thin foils (50 μm, MARZ grade) of niobium, 10 x 25 mm in size. The two methods of hydriding produced different surface conditions which were the subject of the present investigation. One method consisted of the following procedures. The samples were outgassed and cleaned in UHV to remove virtually all the interstitial impurities from the bulk and surface.[6] A small spot of Pd was then deposited onto the clean surface of the niobium in situ in the UHV-system by evaporation from a hot Pd filament. Such a layer allows hydrogen to subsequently be absorbed at low temperatures, 100-250 C, in a diffusion-pumped and cold-trapped hydriding unit.[2,7] Finally, the palladium is removed by immersion in mercury.

The second method of hydriding was similar to that used by most investigators of metal-hydrogen systems. A sample is placed in the hydriding unit and allowed to absorb a premeasured amount of hydrogen while cooling from ~ 800 C where the surface oxide is dissolved into the bulk. In order to study the effect of a poor vacuum in the hydriding unit, samples were heated to 800 C in different base pressures prior to the admission of hydrogen. One sample was heated in 1 x 10^{-5} Torr oxygen.

The hydrogen concentration depth profiles were measured utilizing the 6.385-MeV (lab) resonance of the $^1H(^{15}N,\alpha\gamma)^{12}C$ nuclear reaction (Q = 4.966 MeV). The hydrogen-containing target is bombarded with a ^{15}N beam, and the number of 4.43-MeV gamma rays from the deexcitation of the residual carbon-12 nuclei is counted.[8] As an ion beam passes through a target, it loses energy continuously within the target. Therefore, when the incident beam energy, E_B, is greater than the resonance energy, E_R, the beam slows through the resonance at a depth D_x from the surface where:

$$D_x \approx \frac{(E_B-E_R)}{dE/dx}$$

This expression is valid when the stopping power, dE/dx, does not change appreciably over the energies of interest. In our case the stopping power (calculated for $NbH_{0.55}$ from the Northcliffe and Schilling tables[9]) can indeed be assumed to be constant, 2.9 keV/nm, between the beam energies 6.385 and 10.0 MeV. The depth resolution varies from approximately 3 nm at the resonance energy to 40 nm at 10 MeV.

RESULTS

Two typical results of measurements made simultaneously on several samples prepared by the two methods described above are shown in Fig. 1. The spectra were taken at a temperature of 120 C. Plotted on the ordinate is the number of 4.43-MeV photons detected as a function of the incident energy of the ^{15}N ions. The incident energy scale is indicated on the top abscissa of the graph for curve A. Curve B is shifted in energy by 0.2 MeV so as to facilitate comparisons. The lower abscissa shows depth scales for both curves. The gamma count can be taken to be proportional to the local hydrogen concentration.

Both samples contained 55 at.% H (H/Nb = 0.55). We chose this concentration for most of the samples because at this value the NbH phase diagram indicates that the samples will be in the α'-single phase region at temperatures above ~ 90 C. Sample A was cleaned in UHV, Pd-plated in situ, and hydrided at approximately 230 C. Sample B was cleaned in UHV and then charged in the diffusion-pumped hydriding unit after heating to 800 C in 1 x 10^{-6} Torr.

DISCUSSION

The experimental curves show several distinguishable features. The first feature to appear as a function of energy is the sharp surface peak at the resonance energy (6.385 MeV). This peak appears in all samples with roughly the same intensity. It is due to hydrogen-containing gases (water, hydrocarbons) adsorbed to the surface of the samples. This layer cannot be eliminated with the present experimental set-up. Under the vacuum condition in the sample chamber of the Van de Graaff accelerator, an overlayer of the residual gas will develop in seconds.

Below the surface peak the samples generally show an intensity minimum. This minimum is more pronounced in those samples heated to 800 C prior to hydriding, and is associated with an oxide layer at the surface. A study of the low temperature oxidation of niobium by Halbritter[10] has shown that the oxidation in this temperature range consists mainly in the growth of a 4 to

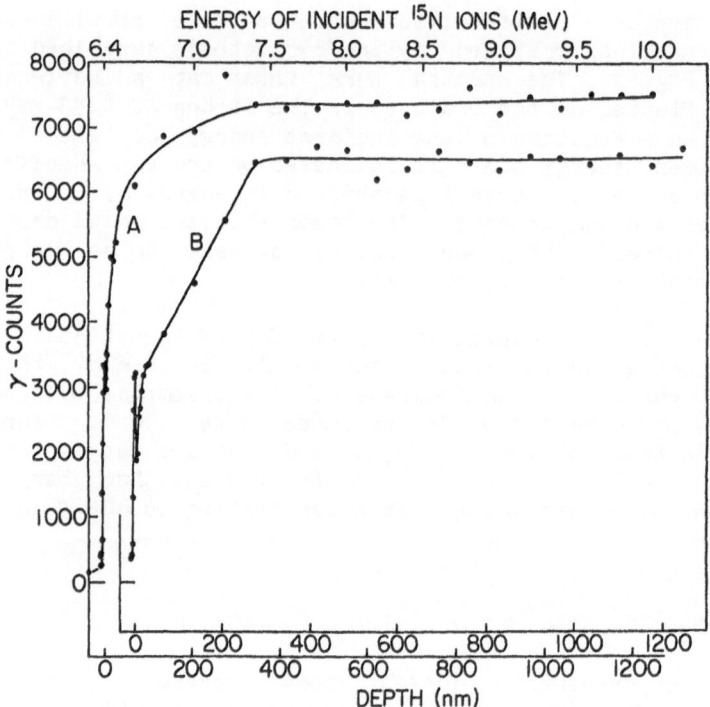

Fig. 1. Depth concentration profile of hydrogen in two representative samples of niobium containing 55 at.% (H/M).
See text for the differences in sample preparation.

10 nm thick layer of compact and amorphous Nb_2O_5. The minimum in the hydrogen concentration, caused by the oxide layer, is followed by an increase in hydrogen concentration, which is more or less rapid and depends on the sample, to a value which reflects the bulk concentration.

The results shown as curve A represent material which was outgassed in UHV and then Pd-plated, therefore containing virtually no interstitial impurities such as oxygen, carbon or nitrogen. It was heated to 230 C for one hour prior to hydriding. We can explain the hydrogen profile by the fact that oxygen, carbon, and maybe nitrogen, which are invariably present on the surface of the sample and in the residual gas of the hydriding unit, can diffuse into the sample at this temperature. Taking into account the diffusion lengths, $\ell = \sqrt{2Dt}$ using literature values for the diffusion coefficients,[11] we conclude that a penetration of the Nb surface by an appreciable amount of oxygen is consistent with the experimental findings. The penetration depth of both carbon and nitrogen is insufficient to explain the hydrogen concentration gradient near the surface. This is in agreement with the study by Halbritter mentioned above.[10] It remains, however, to be shown that high concentrations of oxygen in niobium tend to displace hydrogen. It is known that very low concentrations (< 1 at.%) of oxygen and nitrogen in niobium can act as hydrogen traps.[12,13]

Sample B was heated to approximately 800 C for one hour prior to hydriding. The diffusion length ℓ for all three interstitials is so great at this temperature that the concentration of dissolved gases should be virtually uniform throughout the sample. The fact that the hydrogen concentration is not constant throughout the sample indicates that a different effect must predominate, namely the precipitation of oxides in the near-surface region. Electron microscopic investigations by Hurlen[14] have shown that there is a nucleation-type oxide formation stage with the formation of dispersed oxide particles under conditions similar to those occurring in our experiments. In this stage the oxide particles increase first in number and later in size. The oxide being formed in this stage is also the pentoxide, Nb_2O_5. Hydrogen is almost certainly insoluble in this Nb oxide because it is a very thin layer of this oxide on the surface which inhibits the desorption of hydrogen from NbH_x samples. The hydrogen concentration increases therefore as the volume fraction of the Nb oxides decreases. This region is then followed by a region from 400 to well over 1200 nm in which the hydrogen concentration is depressed below the bulk value due to the presence of oxygen in the Nb. The effect was more pronounced in the case where oxygen was deliberately added to the residual gas in the vacuum system at 800 C. In this case, due to the availability of sufficient oxygen, the oxides grew much further into the material.

CONCLUSIONS

The results show that experiments on samples of niobium-hydrogen, or similar materials hydrided in the manner described must take into account the possible existence of an appreciable hydrogen concentration gradient towards the surface. Experiments which are especially affected by this result are x-ray diffraction experiments and other techniques which probe the near-surface regions of a sample. In x-ray measurements the sampling depth is in the range of the hydrogen-depleted and oxygen-enriched zone, and this may very well lead to misleading results. For example, this may be the reason for the finding of Zabel, Schönfeld, and Moss[4] that there was a ~ 1 μm thick hydrogen-depleted layer on the surface of a crystal which was essentially in the α'-phase.

ACKNOWLEDGEMENTS

This research was supported by the U. S. Department of Energy, Division of Basic Energy Sciences, under Contract No. DE-AC02-76CH00016.

REFERENCES

1.   For a recent review see: Hydrogen in Metals, Vols. I and II, ed. G. Alefeld and J. Völkl (Springer Verlag, Berlin, 1978).
2.   M. A. Pick, F. W. Davenport, Myron Strongin, and G. J. Dienes, Phys. Rev. Lett. 43, 286 (1979).
3.   M. A. Pick, Phys. Rev. B 24, 4287 (1981).
4.   H. Zabel, B. Schönfeld, and S. C. Moss, J. Phys. Chem. Solids 42, 897 (1981).
5.   D. O. Welch and M. A. Pick, to be published.
6.   H. Wenzl and J.-M. Welter in: Current Topics in Material Science, Vol. 1, Ed. E. Kaldis (North-Holland, Amsterdam, 1978).
7.   M. A. Pick, M. G. Greene, and Myron Strongin, J. of the Less-Common Met. 73, 89 (1980).
8.   W. A. Lanford, Nucl. Instrum. Methods 149, 1 (1978).
9.   L. C. Northcliffe and R. F. Schilling, Nucl. Data Tables 7, 233 (1970).
10.   J. Halbritter, to be published in Surf. Sci.
11.   E. Fromm and G. Gebhardt, Gase und Kohlenstoff in Metallen, (Springer Verlag, Berlin, 1976).
12.   G. Pfeiffer and H. Wipf, J. Phys. F: Metal Phys. 6, 167 (1976).
13.   P. E. Zapp and H. K. Birnbaum, Acta Met. 28, 1523 (1980).
14.   T. Hurlen, J. Inst. of Met. 89, 273 (1960-61).

THEORY OF HYDROGEN ABSORPTION IN METAL HYDRIDES

Tomoyasu Tanaka and Daniel E. Azofeifa

Department of Physics
Ohio University
Athens, OH 45701

INTRODUCTION

The mechanism of hydrogen absorption by transition metals and alloys has been a subject of intensive investigations in recent years because of its prospective energy storage applications[1]. The amount of hydrogen absorbed by all metals, at low pressures, is proportional to the square root of the pressure of the external hydrogen gas. This is a clear indication that the hydrogen molecule of the gas is dissociated into atoms in the metal phase. Many transition metals and alloys, such as Pd, Ti, La, Th, and FeTi, absorb a large quantity of hydrogen at relatively low temperatures and pressures. The concentration of saturation absorption of hydrogen in these metals and alloys goes as high as 100 to 200 % of metalic atoms, and the metals are regarded as forming hydrides represented conveniently by chemical formulas such as PdH and TiH_2. Mott and Jones[2] argued that the hydrogen atoms inside metals are ionized into electrons and protons. These protons go into some interstitial positions in the matrix of the host metal while electrons join the main body of the d-band electrons. This is most remarkably demonstrated by the paramagnetic susceptibility of Pd. Pd is a strong paramagnetic metal due to electron holes in the 4d band. When Pd metal is loaded with hydrogen, the paramagnetism decreases linearly with the concentration of hydrogen and this metal hydride becomes diamagnetic at and above 55 at% of hydrogen concentration.

   The high concentrations of hydrogen in many metal hydrides are interpreted as due to a cooperative condensation of absorbed hydrogen from a gas-like state to a liquid-like state. Lacher[3] was the first to formulate a statistical mechanical theory of hydrogen absorption by palladium. The condensation of hydrogen in Pd is due to some kind of attractive interaction between hydrogen atoms. Lacher assumed this interaction to be a nearest neighbor interaction, applied the Bragg-Williams (the mean field) approximation, and derived absorption isotherms. Lacher's theory explained qualitatively the observed features of the absorption isotherms of the Pd-H system.

   In the Pd-H system, protons go into interstitial positions which are octahedrally surrounded by Pd atoms and the metal hydride forms a NaCl structure. It is, therefore, quite natural to assume that the number of interstitial positions available for protons is equal to that of Pd atoms. One of the consequences of the mean field approximation, as applied to hydrogen absorption, is that the critical point of absorption occurs at 50 at% hydrogen concentration, while the experimentally known value in Pd-H system is somewhere near 27 at%. In the traditional theories of hydrogen absorption, including Lacher's formulation, this discrepancy is explained by a somewhat ad hoc assumption, i.e., the total number of interstitial positions available for protons is equal to 60% of Pd atoms. This assumption may seem to be acceptable at first sight because all the isotherms, which have a plateau portion in the neighborhood of the critical pressure, appear to approach the saturation concentration of 60 at%, after taking off sharply upwards. For the sake of definiteness let this concentration be called the incipient saturation concentration, which should be distinguished from the true saturation concentration defined below. If one looks at the general profile of absorption isotherms more closely (see Fig.1 in reference 4), however, one notices that all of the isotherms converge to the 100 at% concentration limit at sufficiently high pressures. This 100 at%concentration must be the true saturation concentration. It is, therefore, more reasonable to assume that the total number of hydrogen sites available is equal to the number of Pd atoms, which is consistent with crystallographic considerations. The present authors[4] pointed out that the incipient saturation can be regarded as arising from the nonavailability of electron energy levels in the d-band. The energy of extra electrons increases almost discontinuously as soon as the concentration exceeds the 60 at% value. The correct enumeration of the number of available sites for protons should be directly reflected in the calculation of the entropy, and similarly, the knowledge of the electronic energy band structure is reflected in the incipient saturation. Both of these features were taken into account in our previous paper. In the present paper it will be shown that the asymmetry of the isotherms with respect to the critical point can arise not only from the energy band effects but also from the correlations of proton distribution in the metal phase.

EFFECT OF PROTON CORRELATIONS BY MEANS OF THE PAIR APPROXIMATION

Since the concentration of protons in the metal phase can go as high as that of the liquid hydrogen, the mean field approximation may not be appropriate to deal with the cooperative condensation phenomenon. Furthermore, in an approximation in which the particle correlation is taken into account the critical concentration of absorption is expected to occur at a lower density than the 50 at% predicted by the mean field approximation. For this reason, let us work with the two-body approximation of the absorption isotherms.

The equation for the absorption isotherm is obtained by setting the chemical potential per atomic hydrogen in the gas phase equal to the corresponding quantity in the metal phase. The chemical potential per hydrogen atom in the gas phase is given by

$$\mu_{gas} = -\tfrac{1}{2} kT \left(\ln C + \frac{7}{2} \ln T - \ln p \right) - \left(\chi - \frac{h\nu}{2} \right)$$

where

$$C = e \left(2\pi M \right)^{3/2} k^{5/2} / \left(2 h^3 \Theta_{rot} \right) ,$$

where M, Θ_{rot}, χ, ν are, respectively, the mass, the rotational temperature, the dissociation energy, and the vibrational frequency of hydrogen molecule. In the pair approximation, the Helmholtz free energy per proton site in the metal phase is given by

$$F/N = -E_a \theta - \tfrac{1}{2} z w \sigma + kT (1 - z)[\theta \ln \theta +$$
$$+ (1 - \theta) \ln (1 - \theta)] + \tfrac{1}{2} z kT [\sigma \ln \sigma +$$
$$+ 2 (\theta - \sigma) \ln (\theta - \sigma) + (1 - 2\theta + \sigma)$$
$$* \ln (1 - 2\theta + \sigma)] ,$$

Where $-E_a$, w, σ, z and θ are, respectively, the energy of hydrogen, the proton-proton interaction energy, the proton-proton pair correlation function, the proton coordination number and the proton concentration. The proton correlation function is determined by minimizing the free energy with respect to σ. The equation for this quantity will, then, be given by

$$\sigma (1 - 2\theta + \sigma) = (\theta - \sigma)^2 \exp(w/kT).$$

The chemical potential per hydrogen atom in the metal phase is, thus, given by

$$\mu_{metal} = -E_a + kT (1 - z) [\ln \theta - \ln(1 - \theta)]$$
$$+ zkT [\ln(\theta - \sigma) - \ln(1 - 2\theta + \sigma)]$$

Now equating the two chemical potentials, one finds an equation for the absorption isotherm:

$$\ln p^{\frac{1}{2}} = \frac{1}{2} \ln C + \frac{7}{4} \ln T - (E_a - \frac{X}{2})/kT$$

$$+ (1-z) [\ln\theta - \ln(1-\theta)] + z [\ln(\theta - \sigma) -$$

$$- \ln(1 - 2\theta + \sigma)] .$$

In the above formulation, the effect of the electronic energy band, which changes drastically as the amount of the absorbed hydrogen increases, is not explicit, except for the fact that the energy E_a represents the energy of proton and the accompanying electron relative to the energy of the hydrogen atom outside the metal. To some extent, the proton-proton pair interaction energy may also depend upon the electron density, however, this will not be discussed in the present paper. In the previous paper, we took into account the effect of the change of the electronic energy band due to the change in the amount of hydrogen. According to recent calculations[5] based on the extended coherent potential approximation of the density of states in Pd-H system, it is found that an S-like orbit is formed around each absorbed proton, while the density of states near the Fermi surface rapidly changes as the hydrogen concentration increases. One may represent this situation by a formula

$$- E_a = - E_s + \Delta E \exp \beta (\theta - B),$$

where $-E_s$ is the energy of an S-like orbit, which is sufficiently lower than the Fermi energy. The second term represents a rapid increase of the Fermi energy as the amount of absorbed hydrogen exceeds a concentration given by B. These parameters are determined by fitting the theoretical isotherm with the experimentally observed isotherms.[6] The interaction constant w, the energy E_s, and β are determined from the critical data, while other parameters are determined from one point along the critical isotherm and another point on a different isotherm, both at rather high pressures. These values are:

$$w = 0.618, \quad E_s = 52.040, \quad \Delta E = 1.735 \ k \ cal/mole$$

$$\beta = 6.994, \quad B = 0.595.$$

The quality of the agreement between the theoretical formula and the experimental data may be judged, for instance, by looking at the densities of coexisting phases along individual isotherms.   The agreement seems to be rather satisfactory. With these constants, the pressure-concentration diagram for the Pd-H system is obtained and is shown on the following page.

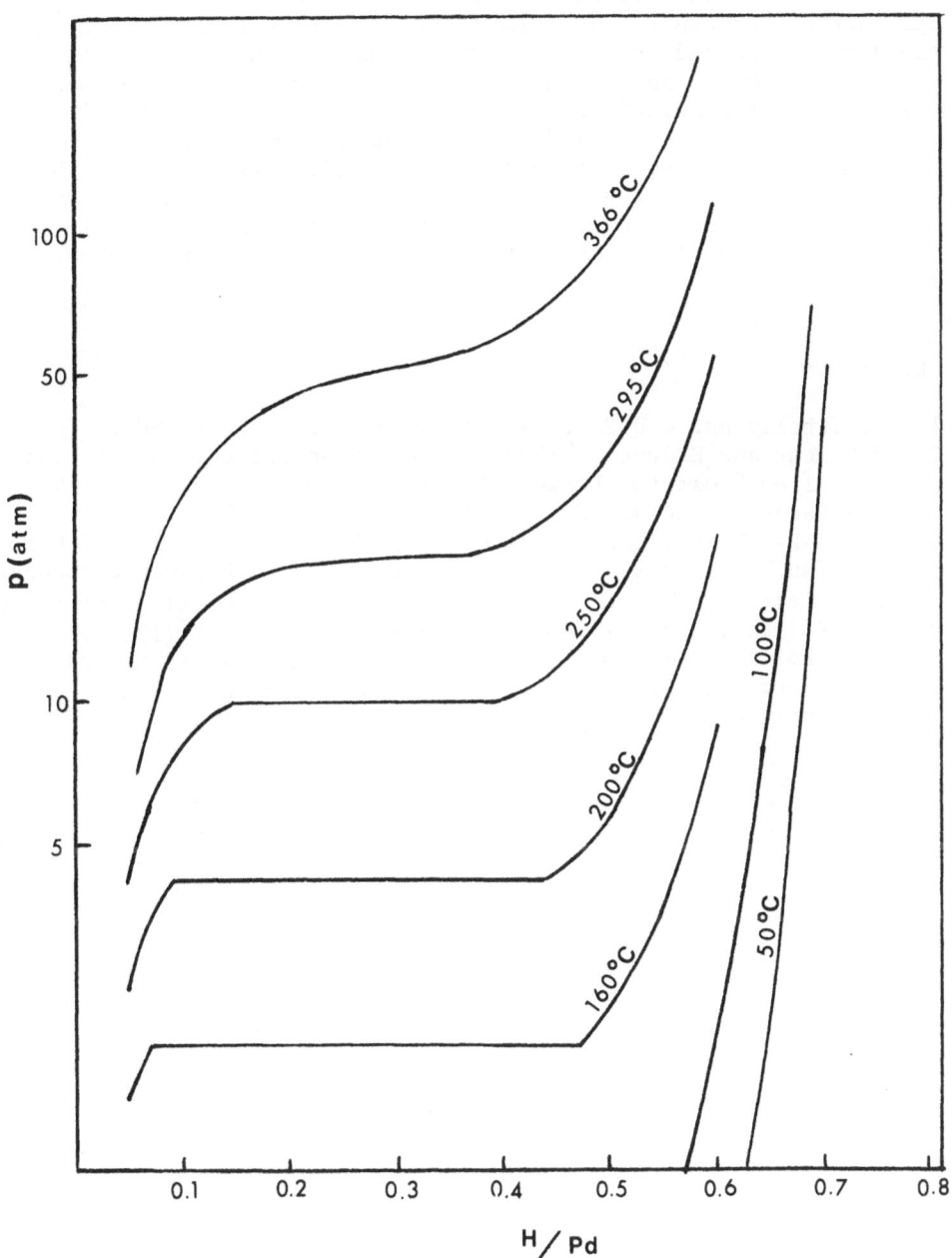

Fig. 1. Pressure–concentration diagram for the Pd–H system
(pair approximation).

CONCLUSION

    The observed asymmetry of the hydrogen absorption isotherms
with respect to the critical point in the Pd-H system and the
distinction between the incipient and the true saturation concentrations are explained in terms of the proton pair correlation function and the change in the energy density of states of electrons in
the process of absorption. Agreement between the theoretical isotherm and the experimental data is satisfactory. No attempt is made
to formulate the theory of hydrogen absorption starting with the
Hamiltonian for both the electrons in the disordered lattice and
the protons distributed over interstitial sites and then applying
statistical mechanics. This will be a subject for future investigations.

REFERENCES

1. J.J.Reilly and G.D.Sandrok, Sci.Am.$\underline{240}$ (2), 118 (1980).
2. N.F.Mott and H.Jones, "Theory of the Properties of Metals and
       Alloys" Oxford, London (1937).
3. J.R.Lacher, Proc.Roy.Soc. London $\underline{161A}$, 525 (1937).
4. T.Tanaka, M.Keita and D.E.Azofeifa, Phys.Rev. $\underline{B24}$, 1771 (1981).
5. J.S.Faulkner, Phys.Rev. $\underline{B13}$, 2391 (1976), D.A.Papaconstantopoulos,
    B.M.Klein,E.N.Economou and L.L.Boyer, Phys. Rev. $\underline{B17}$, 141 (1978).
6. L.J.Gillespie and L.S.Galstaun, J.A.C.S. $\underline{58}$, 2565 (1936).
    E.Wicke and H.Brodowsky, in "Hydrogen in Metals II" Vol.29 of
       "Topics in Applied Physics" ed. G.Alfeld and J.Vökl,
       Springer, Berlin (1978).

STABILITIES, STOICHIOMETRIES AND SITE OCCUPANCIES IN HYDRIDES

OF INTERMETALLIC COMPOUNDS

D. G. Westlake

Materials Science Division
Argonne National Laboratory
Argonne, Illinois 60439 USA

ABSTRACT

In the literature, one can find numerous attempts to explain the observed stabilities, stoichiometries and site occupancies in hydrides of the various families of intermetallic compounds. Some of the approaches to these problems are critically reviewed here. For some, but not all such hydrides, the stabilities have been shown by different researchers to correlate with the enthalpy for formation of the intermetallic compound, itself, or with cell size, or electronic properties, or elastic properties. It appears, therefore, that all of these effects may play a role, but none is dominant in all cases. The development of the procedure for qualitative and quantitative determinations of H-site occupancy from calculations of enthalpies for the formation of imaginary binary hydrides was reviewed. Such inspection raises the question of possible fortuitous agreement between experimental observations and predictions arising from the technique. The concepts of minimum hole size for H occupation and minimum H-H distance in stable hydrides of metals or intermetallic compounds have been discussed in terms of their importance to preferred H sites and to stoichiometry, and considerations necessary to a geometric model have been outlined. The model is used to rationalize observed H sites and stoichiometry of $LaNi_5H_x$. The review points up the need for theoretical treatment leading to fundamental understanding of such systems.

INTRODUCTION

Some intermetallic compounds absorb almost no hydrogen, even under very high H_2 pressure, while others absorb as much as, or more

than, two H atoms per metal atom under only moderate H_2 pressure,
and still others absorb intermediate amounts of hydrogen. Numerous
studies have been conducted to gain an understanding of the factors
that influence the hydriding behavior of the various intermetallic
compounds in terms of their stability, their stoichiometry and the
preferred interstitial sites for occupation by H.

In a pioneering report, Beck[1] stated, "One of the purposes of
this investigation was to amass a wealth of information concerning
the hydriding characteristics of intermetallic compounds as a means
for discovering some of the factors which govern hydride formation". From his compilations, he concluded that several of the
structure types seemed to have critical lattice parameters, i.e.,
minimum cell sizes, for hydride formation. "An attempt was made to
relate these apparent critical parameters to various factors such as
electronegativity, valence, electron transfer and the mechanism and
temperature of formation. However, the only variable which appeared
to be of significance was the size factor itself." Because it was
assumed that the relative size of any interstice within a given
structure type would be proportional to the overall size of the unit
cell, it was concluded that there must be a critical hole size for
each structure type below which hydrogen occlusion would not occur.

In his review, Oesterreicher[2] concluded that the influence of
size is beyond doubt, but, often, it is not possible to distinguish
clearly between size and electronic effects as both vary periodically with atomic number. An early attempt to explain the stoichiometry of metal hydrides was made by Robins[3], who assumed that H
contributes electrons to the metal (protonic theory). Energy band
calculations[4] show, however, that the addition of hydrogen to a
metal introduces new states, below the Fermi level of the metal,
that can be filled with the added electrons. Switendick[5] has
treated the case of simultaneous occupation of tetrahedral and octahedral interstices in YH_x with $2 < x < 3$. Addition of H's to octahedral interstices changes the band structure importantly. Each
hydrogen added lowers one band that was already partially filled.
There is little difference in charge on the octahedral and tetrahedral interstices, but that charge is significantly greater than
that of a comparable atomic sphere; thus, the protonic model is
simply not applicable.

Maeland et al.[6] have demonstrated the importance of the availability of appropriate sites to H-absorption capacity by studying
both amorphous and crystalline hydrides of TiCu. Amorphous TiCu
absorbs H to a composition of $TiCuH_{1.35}$ at room temperature and one
atmosphere H_2 pressure compared with a composition of TiCuH for
crystalline TiCu. They suggest that, possibly, the maximum hydrogen
capacity is determined by electronic structure, which probably would
not be greatly different for the amorphous and crystalline structures, but the number, type and size of the available interstices

may determine the amount of H <u>actually</u> absorbed. They contrast this with the behavior of the Pd-H system. The structurally limited composition would be PdH, but compositions of PdH_x with $x > 0.6$ are difficult to achieve. They proposed that this is probably because there are only 0.6 empty states below the Fermi level in palladium hydride that are available for filling by electrons from the H's.[4] Thus, electronic structure, rather than crystal structure, limits the H content of Pd.

The possible importance of cell size and hole size were mentioned earlier. Another significant "geometric" factor may be hydrogen spacing. Switendick[4] has stated that, when the addition of hydrogen to a metal introduces new states below the Fermi level, these states are very sensitive to H-H distance, and the position of the states depends on the close approach of the H's to each other. From a survey of the stable hydrides of seven transition metals, it seems that there may be a minimum H-H distance of about 2.1 Å. This value has been used by the present author in the development of a geometric model of the hydrides of AB_2 and AB_5 compounds to be discussed later.

Carter[7] has tried to explain the behaviors of various intermetallic compounds toward hydrogen with a proposal that appears to encompass certain aspects of both size and electronic effects. He generalizes that intermetallic compounds that involve transition metals or rare earth metals which undergo large contractions during compound formation will absorb significant quantities of hydrogen. The H's will occupy those sites that will weaken the metal-metal bonds that necessitated the aforementioned contraction, thereby allowing the contracted atom to expand toward its initial volume. Carter's proposal[7] seems to be inherently related to one by Takeshita et al.[8] who claim that one can classify the several theories and semiempirical rules to account for the stability of metal hydrides in two broad categories: 1) electronic and 2) elastic. They conclude, however, that because hydrogen absorption definitely changes the electronic properties of a metal, the theory based on elastic interactions between neighboring H's may need to be modified to include electronic effects. From low temperature heat capacity measurements[8], they conclude that the density of states at the Fermi surface is probably not an important factor in determining hydrogenation characteristics, but they did report a correlation between H-absorption capacity and the elastic properties.

In a series of papers[9-14], Miedema and his colleagues have presented semiempirical equations for calculating the heats of formation of ternary hydrides. These take into account electronegativities and electron densities of the elements involved. The authors[12] forewarn their readers that, because only nearest neighbor interactions are considered, and because their assumed distribution between A-H and B-H contacts is arbitrary, the calculations of ΔH are, necessarily, crude approximations. Nevertheless, it was from their formulation that Shaltiel and his co-workers[15-19] evolved their concept of "imaginary binary hydrides" formed by the H atom and its nearest neighbor metal atoms in ternary hydrides. This latter group calculates an enthalpy of hydride formation for each type of site in the structure, and from the relative "stabilities" they attempt to rationalize the observed preferential occupation of particular sites. A section of this review is devoted to their procedure.

In this contribution, we have attempted to provide a compendium of the research on some of the factors affecting H-absorption behavior of intermetallic compounds and a personal assessment of the current status. The literature on the subject has burgeoned, in recent years, to such a degree that it has been impossible to make this review comprehensive in the space available.

STABILITY

Hole Size vs. Cell Size vs. Electronic Effects

The free energies of formation ΔF have been determined for the hydrides of many intermetallic compounds from the observed plateau H_2 pressures at various temperatures. For hexagonal AB_5 compounds, Lundin et al.[20] found that these free energies correlated with the radii they calculated for one type (AB_3, or 12o in Wyckoff notation) of tetrahedral interstice; as hole size increased, $-\Delta F$ increased. It seems clear, now, that the correlation is probably not basically meaningful, because later studies[21] show that only a fraction of the H's occupy 12o sites. Gruen et al.[22] argued that the stability correlates with cell volume and that this happens, because the enthalpy of formation of the compound, itself, depends on one term that is related to the volume contraction on going from the pure metals to the alloy and on another term that is related to molar volume[9]. The same group[23] concluded that electronic factors, manifested in the volume, are more important than structural differences in determining ΔF. For the dihydrides of the rare earth metals and Group III metals, Magee[24] reported a correlation between ΔF and the shortest M-H distance, with $-\Delta F$ reaching a maximum at a M-H distance of about 2.24Å. In his treatment, he compares calculated cohesive energies with thermodynamic results. Magee suggests that the correlation between hole size and stability of the hydrides of intermetallic compounds might be explained in the same way.

Bechman et al[25], also, doubt that variations in stability are attributable to hole size, per se. In RT_x systems, where R = rare earth, T = transition metal and x approximates 3, the affinity for hydrogen decreases as the atomic number of R or T increases and as the value of x increases. All three of these changes cause reductions in the lattice parameters and, therefore, in the sizes of the

interstitial holes available for occupancy by hydrogen atoms. They
suggest, however, that the variations probably affect the band
structure, thereby resulting in the observed systematic trends.

Compressibility

Takeshita et al.[8] have pointed out that certain intermetallic
compounds are anomalous with respect to the reported[22] correlation
between ΔF and cell volume. For the isoelectronic compounds YNi_5,
$LaNi_5$ and $GdNi_5$, stable hydrides form only when the unit-cell volume
exceeds 82 Å^3. According to this criterion, $ThNi_5$, with a unit-cell
volume of 83.3 Å^3, would be expected to form a hydride, but it
doesn't. Thus, size alone cannot account for hydrogenation
behavior. Further, they stated[26,27] that electronic specific heat
constants, γ, which are proportional to densities of states at the
Fermi surface, are not related to hydride stabilities, at all.
They[8] suggested that the high value of the e/a ratio for $ThNi_5$ might
account for the anomaly. Because the compressibility depends on the
electronic wave functions of the valence electrons, it embodies both
the e/a ratio and the size factor. The authors[8] proposed, therefore, that the compressibility might be accountable, by itself, for
the hydrogenation characteristics of RM_5 intermetallic compounds.
They reported[26] later, however, that there are some anomalies and
that the valency of the A atom in AB_5 compounds, may be still
another factor contributing to the relative stability of the
hydride. Scrutinizing the work of Takeshita et al[8,26-28], one has
to conclude that, while compressibility may play some role in
determining hydrogenation behavior of AB_5 compounds, it cannot be
the only important factor, any more than unit cell volume can be the
only important factor.

The Rule of Reversed Stability

It has been suggested[10] that the more stable the binary intermetallic compound, the less stable is its hydride. While the rule
has been shown to hold for numerous systems, its validity is not
universal[15,29]. Shinar et al.[15] claim that the calculation of the
enthalpy of formation should take into account the possible interstitial sites to be occupied by the hydrogen atoms.

PREFERRED INTERSTICES FOR HYDROGEN OCCUPATION

Enthalpies of Formation for Imaginary Binary Hydrides

Attempts have been made[15-19] to rationalize observed pref-
erences for H occupation of particular interstices in intermetallic
compounds from calculated enthalpies of formation $\Delta H'$ of "imaginary
binary hydrides" at each type of interstice. A hydrogen atom in a
particular site is considered to be equally associated with each of
its nearest-neighbor, metal atoms, and the imaginary hydride formed

at the site is made up of one H atom and fractional parts of A and B atoms. The enthalpies are calculated from the semiempirical equations due to Miedema[9].

In 1975, Miedema et al.[11] improved upon their empirical approach to the calculation of enthalpies of formation for solid alloys, and Bouten and Miedema[14] claimed further improvement, specifically for hydrides, in 1980. Jacob and Shaltiel[17] applied their technique to the hydrides of several AB_2 cubic C15 compounds. In these calculations, however, they used Miedema's earliest equations, rather than the later, allegedly improved ones. From their results, they drew two very qualitative conclusions: 1) A_2B_2 sites should be occupied preferentially at low hydrogen concentrations because $\Delta H'$ is more negative for these than for AB_3 sites, and 2) AB_3 sites should be partially occupied at higher hydrogen concentrations.

Didisheim et al.[18] attempted to make somewhat more quantitative predictions for D-site occupancy, as a function of x, in ZrV_2D_x. Because of the license taken with Miedema's equations, however, one may be justified in considering the reported agreement with experimental observations with some caution. Fruchart et al.[30] suspected, and Didisheim et al.[31] later agreed, "that the close fit[17] between our observed and calculated occupancy factors could be fortuitous." This latter group,[16] also, reported an apparent limitation of their own model. The calculation[16] of $\Delta H'$ correctly predicted the preference for A_2B_2 sites in $ZrMn_2D_3$ but could not predict the observed preferential occupation of <u>certain</u> A_2B_2 sites.

In a different approach to the problem of simultaneous occupation, Jacob et al.[19] claimed that use of the Boltzmann distribution function allowed them to explain, quantitatively, the observed occupancy of A_2B_2 and AB_3 sites in $TiCr_2$, ZrV_2, $ZrCr_2$ and $ZrMn_2$. Miedema's[9] earliest formulation was used to calculate $\Delta H'$. We calculated $\Delta H'$, however, for the A_2B_2 and AB_3 sites in ZrV_2H_x, $ZrCr_2H_x$ and $ZrTi_2H_x$ using the improved equations of Bouten and Miedema[14], and found that the model[19] failed.

A Phenomenological Model

Jacob et al.[32,33] have presented a phenomenological model, not based on calculations of $\Delta H'$, for the hydrogen absorption capacity in pseudobinary Laves phase compounds. Compounds of $Zr(A_xB_{1-x})_2$ have been considered, with A = V, Cr and Mn and with B = Fe and Co. They concluded that the ratio H/Zr is a step function of x, and, therefore, that the occupancy of interstices around a given Zr atom is critically dependent on the number of B atoms coordinating that Zr atom. On the other hand, Oesterreicher[34] has presented a convincing argument that the step functions are merely manifestations of the $\alpha \rightarrow \beta$ phase transitions for the various hydrogenated pseudobinary alloys.

Minimum Hole Size

As mentioned in the INTRODUCTION, the idea that "hole size" might be important to the hydrogenation behavior of intermetallic compounds is not new[1]. The present author has surveyed reported experimental results for the hydrides of numerous metals and intermetallic compounds in order to ascertain whether there appears to be a minimum hole size for occupation by H in a stable hydride. The compilation will be reported elsewhere, but the conclusion was that, in stable hydrides, occupation by H has been reported only for interstices that could accommodate a sphere of radius $r_h \geqslant 0.40$ Å. This criterion and another regarding minimum H-H distance in stable hydrides have been used to rationalize[35] and even predict[36] experimentally observed behavior for the hydrides of ZrNi. More recently, we developed a geometric model for rationalizing the hydrogenation behavior of AB_2 and AB_5 compounds. More will be said about the model, in the next section.

STOICHIOMETRIES

H-H Distance

Our own survey of the hydrides of metals and intermetallic compounds (to be published elsewhere) is in agreement with that of Switendick[4], who reported that the H-H distance appears to be never less than 2.1 Å. The repulsive interaction of near-neighbor H atoms is widely accepted and has been used by many researchers[21,37-42] to help explain why H atoms do not fill all seemingly available interstices. Shoemaker and Shoemaker[37] adopted an "exclusion rule" that two interstices sharing a trigonal face cannot be simultaneously occupied by H's. If, in fact, there is a minimum H-H distance in stable hydrides of intermetallic hydrides, it is not only first, second and third distances that are important, however. We have considered, some hydrides of cubic AB_2 compounds in which even the sixth neighbor $A_2B_2-A_2B_2$ distance is less than the value of 2.1 Å.

A Geometric Model

Using criteria of 0.4 Å for minimum hole radius and 2.1 Å for minimum H-H distance, we have been able to rationalize the observed stoichiometries for hydrides of ZrNi[35,36] and of $LaNi_5$ and numerous AB_2 compounds (to be published elsewhere). Here, we shall present only a summary of the rationale for $LaNi_5$. There has been controversy[21,43-45] over the structure of $LaNi_5D_{6.5}$, but because of the refined techniques used in the later research[21], we have chosen the reported[21] P6/mmm space group for our model. Schematic diagrams of

the hexagonal structure are shown in Figs. 1 and 2. The lattice
parameters are a_o = 5.399 Å and c_o = 4.290 Å. There are 37 interstices potentially available for occupation by H(D) in each unit
cell containing one La atom and five Ni atoms. In Wyckoff notation,
a) 3f's are octahedral, A_2B_4 sites; b) 4h's are B_4 sites; c) 12o's
are AB_3 sites; d) 12n's are AB_3 sites; and e) 6m's are A_2B_2 sites.
Examples of these can be found in Fig. 1: a) lying midway between
atoms La_1, and La_3; b) coordinated by (I), (II), III and B'; c)
coordinated by La_4', (I), (II) and B'; d) coordinated by La_3, A, B,
and III; and e) La_2, La_2', I and II. Published[46] metal-atom radii
were used to calculate the respective hole sizes (in Å): a) 0.313;
b) 0.393; c) 0.433; d) 0.448; and e) 0.555.

In the compounds we have considered, to date, there appears to
be a preference for H occupation of the larger sites. In the model,
therefore, we choose to have H occupy 6m sites first. First, second
and third 6m-6m distances are 1.146 Å, 1.985 Å and 2.292 Å. Thus,
only two of the 6m sites (Fig. 2) can be filled simultaneously without violating the 2.10 Å criterion. The second largest sites are
12n. Around each 3f site, there are four 12n sites. These five
sites are all so close together (<1.329 Å) that only one can be
occupied at a time. The shortest distance between 12n sites situated around a given La atom is 2.2694 Å, and the shortest distance
for a 6m-12n pair is 2.155 Å, so it should be possible to put three
H's in 12n sites. In an analysis that is too detailed to be
included in a review paper, we have concluded that only 1.5 of the
12o sites are not blocked by the previously occupied sites. All 37
sites in the unit cell have now been filled with H atoms or excluded
from occupancy by the distance criterion. The sum of the filled
sites is 2 + 3 + 1.5 = 6.5, exactly the same as the observed H/La
ratio. The predicted occupancy, however, differs slightly from the
experiment. Percheron-Guégan et al.[21] reported observing the
following numbers of occupied sites in a unit cell: 3f, 0.64; 4h,
0.52; 12o, 1.29; 12n, 2.14; and 6m, 1.91. The lower value for 12n
and the higher one for 3f are expected ramifications of 12n-3f-12n
diffusion. A similar explanation may account for the fractional
occupancy of 4h sites, because the shortest distance between two 12o
sites is through a 4h site. Partial occupancy of 3f and 4h sites is
not expected to increase the hydrogen capacity above H/La = 6.5,
because of the large number of 6m, 12n and 12o sites blocked by
filled 3f and 4h sites.

As stated above, the geometric model has demonstrated considerable usefulness in predicting site occupation and stoichiometries in
hydrides of AB_2 and AB_5 intermetallic compounds. Its primary limitation is that it does not predict, or even rationalize, relative
stabilities. It seems quite obvious that the successes of such a
simple model must be explicable in terms of more fundamental
physical principles, but, as yet, such an explanation is lacking.

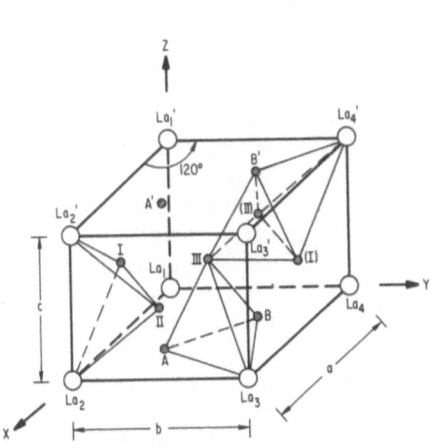

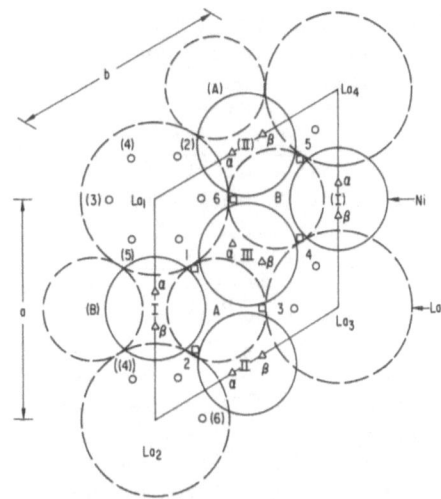

Fig. 1. Schematic diagram of the hexagonal structure for LaNi$_5$ showing the four types of tetrahedral interstices.

Fig. 2. Projection of the LaNi$_5$ structure on the basal plane. o , 6m; Δ , 12n; □ , 12o. 4h sites project onto the letters A and B. 3f sites project onto Roman numerals.

CONCLUSION

    For the hydrides of intermetallic compounds, whether the stabilities are correlated with hole size, cell volume, enthalpy of formation for the intermetallic compound, or compressibility, one can always find exceptions to the rule. It appears safe to say that stability is influenced by many factors, none of which predominates in all cases. Early attempts to predict maximum stoichiometries of hydrides were probably not restrictive enough regarding H-H distances; there is considerable evidence for a minimum distance of 2.10 Å. Critical review of the model based on enthalpies of imaginary binary hydrides and used for rationalizing observed H-site occupancies has led us to question whether its successes might have been fortuitous. Agreement between observations for the hydrides of AB$_2$ and AB$_5$ compounds and the predictions of a simple geometric model point up the need for an explanation based on the underlying fundamental principles.

This work supported by the U.S. Department of Energy.
References:
1. R. L. Beck, Investigation of hydriding characteristics of intermetallic compounds, Report Number LAR-55, Nov. 1961.
2. H. Oesterreicher, Hydrides of intermetallic compounds, Appl. Phys. 24:169 (1981).

3. D. A. Robins, An interpretation of some of the properties of the transition metals and their alloys, J. Less-Common Metals 1:396 (1959).

4. A. C. Switendick, Band structure calculations for metal hydrogen systems, Zeit. für Phys. Chem. 117:89 (1979).

5. A. C. Switendick, Electronic structure of non-stoichiometric cubic hydrides, J. Less-Common Met. 74:199 (1980).

6. A. J. Maeland, L. E. Tanner and G. G. Libowitz, Hydrides of metallic glass alloys, J. Less-Common Met. 74:279 (1980).

7. F. L. Carter, Atomic volume contraction in intermetallic hydride formers: A valuable new clue, J. Less-Common Met. 74:245 (1980).

8. T. Takeshita, K. A. Gschneidner, Jr., D. K. Thome and O. D. McMasters, Low-temperature heat-capacity study of Haucke compounds $CaNi_5$, YNi_5, $LaNi_5$ and $ThNi_5$, Phys. Rev. B 21:5636 (1980).

9. A. R. Miedema, The electronegativity parameter for transition metals: Heat of formation and charge transfer in alloys, J. Less-Common Met. 32:117 (1973).

10. H. H. Van Mal, K. H. J. Buschow and A. R. Miedema, Hydrogen absorption in $LaNi_5$ and related compounds: Experimental observations and their explanation, J. Less-Common Met. 35:65 (1974).

11. A. R. Miedema, R. Boom and F. R. deBoer, On the heat of Formation of solid alloys, J. Less-Common Met. 41:283 (1975).

12. K. H. J. Buschow and A. R. Miedema, Hydrogen absorption in rare earth intermetallic compounds, in: "Proc. Int. Symp. Hydrides for Energy Storage", A. F. Andresen and A. J. Maeland, eds., Pergamon, New York (1978).

13. P. C. P. Bouten and A. R. Miedema, On the stable compositions in transition metal-nitrogen phase diagrams, J. Less-Common Met. 65:217 (1979).

14. P. C. P. Bouten and A. R. Miedema, On the heats of formation of the binary hydrides of transition metals, J. Less-Common Met. 71:147 (1980).

15. J. Shinar, I. Jacob, D. Davidov and D. Shaltiel, Hydrogen sorption properties in binary and pseudobinary intermetallic compounds, in: "Proc. Int. Symp. Hydrides for Energy Storage", A. F. Andresen and A. J. Maeland, eds. Pergamon, New York (1978).

16. J.-J. Didisheim, K. Yvon, D. Shaltiel and P. Fischer, The distribution of the deuterium atoms in the deuterated hexagonal Laves-phase $ZrMn_2D_3$, Solid State Comm. 31:47 (1979).

17. I. Jacob and D. Shaltiel, Hydrogen sorption properties of some AB_2 Laves-phase compounds, J. Less-Common Met. 65:117 (1979).

18. J.-J. Didisheim, K. Yvon, P. Fischer and D. Shaltiel, The deuterium site occupation in ZrV_2D_x as a function of the deuterium concentration, J. Less-Common Met. 73:355 (1980).

19. I. Jacob, J. M. Bloch, D. Shaltiel and D. Davidov, On the occupation of interstitial sites by hydrogen atoms in intermetallic hydrides: A quantitative model, Solid State Comm. 35:155 (1980).

20. C. E. Lundin, F. E. Lynch and C. B. Magee, A correlation between the interstitial hole sizes in intermetallic compounds and the thermodynamic properties of the hydrides formed from these compounds, J. Less Common-Met. 56:19 (1977).

21. A. Percheron-Guégan, C. Lartigue, J. C. Achard, P. Germi, and F. Tasset, Neutron and x-ray diffraction profile analyses and structure of $LaNi_5$, $LaNi_{5-x}Al_x$ and $LaNi_{5-x}Mn_x$ intermetallics and their hydrides (deuterides), J. Less-Common Met. 74:1 (1980).

22. D. M. Gruen, M. H. Mendelsohn and I. Sheft, Absorption of hydrogen by the intermetallics $NdNi_5$ and $LaNi_4Cu$ and a correlation of cell volumes and desorption pressures, in: "Proc. Symp. Electrode Materials and Processes for Energy Conversion and Storage", The Electrochemical Society, 1977, p. 482.

23. M. H. Mendelsohn and D. M. Gruen, The pseudo-binary system $Zr(V_{1-x}Cr_x)_2$: Hydrogen absorption and stability considerations, J. Less-Common Met. 78:275 (1981).

24. Charles B. Magee, Structures and stabilities of the group IIIa dihydrides, J. Less-Common Met. 72:273 (1980).

25. C. A. Bechman, A. Goudy, T. Takeshita, W. E. Wallace and R. S. Craig, Solubility of hydrogen in intermetallics containing rare earth and 3d transition metals, Inorganic Chemistry 15:2184 (1976).

26. T. Takeshita, K. A. Gschneidner, Jr. and J. F. Lakner, High pressure hydrogen absorption study on YNi_5, $LaPt_5$ and $ThNi_5$, J. Less-Common Met. 78:P43 (1981).

27. T. Takeshita, G. Dublon, O. D. McMasters and K. A. Gschneidner, Jr., Low temperature heat capacity studies on hydrogen absorbing intermetallic compounds, in: "The Rare Earths in Modern Science and Technology", Vol. 2, G. J. McCarthy, J. J. Rhyne and H. B. Silber, eds., Plenum, New York (1980).

28. Y. Chung, T. Takeshita, O. D. McMasters and K. A. Gschneidner, Jr., Influence of the lattice and electronic factors on the hydrogenation properties of the RNi_5-base (R is a rare earth) Haucke comounds: Results of low temperature heat capacity measurements, J. Less-Common Met. 74: 217 (1980).

29. M. H. Mendelsohn, D. M. Gruen and A. E. Dwight, The effect of aluminum additions on the structural and hydrogen absorption properties of AB_5 alloys with particular reference to the $LaNi_{5-x}Al_x$ ternary alloy system, J. Less-Common Met. 63:193 (1979).

30. D. Fruchart, A. Rouault, C. B. Shoemaker and D. P. Shoemaker, Neutron diffraction studies of the cubic $ZrCr_2D_x$ and $ZrV_2D_x(H_x)$ phases, J. Less-Common Met. 73:363 (1980).

31. J.-J. Didisheim, K. Yvon, P. Fischer and P. Tissot, Order-disorder phase transition in $ZrV_2D_{3.6}$ Solid State Comm. 38:637 (1981).

32. I. Jacob, D. Shaltiel, D. Davidov and I. Miloslavski, A phenomenological model for the hydrogen absorption capacity in pseudo-binary Laves phase compounds, Solid State Comm. 23:669 (1977).

33. I. Jacob, A. Stern, A. Moran, D. Shaltiel and D. Davidov, Hydrogen absorption in $(Zr_xTi_{1-x})B_2$ (B ≡ Cr, Mn) and the phenomenological model for the absorption capacity in pseudo-binary Laves-phase compounds, J. Less-Common Met. 73:369 (1980).

34. H. Oesterreicher, Queries concerning local models for hydrogen uptake in metal hydrides, J. Phys. Chem. 85:2319 (1981).

35. D. G. Westlake, Stoichiometrics and interstitial site occupation in the hydrides of ZrNi and other isostructural intermetallic compounds, J. Less-Common Met. 75:177 (1980).

36. D. G. Westlake, H. Shaked, P. R. Mason, B. R. McCart, M. H. Mueller, T. Matsumoto and M. Amano, Interstitial site occupation in ZrNiH, to be published in J. Less-Common Met., "Proc. Int. Symp. on the Properties and Applications of Metal Hydrides-II", Toba, Japan, 1982.

37. D. P. Shoemaker and C. B. Shoemaker, Concerning atomic sites and capacities for hydrogen absorption in the AB_2 Friauf-Laves phases, J. Less-Common Met. 68:43 (1979).

38. D. P. Shoemaker, C. B. Shoemaker and D. Fruchart, Predictions and observations concerning hydrogen occupancy of tetrahedral interstices in certain binary alloys, in: "Program and Abstracts of the Summer Meeting, American Crystallographic Association", Calgary, Canada, 1980.

39 C. B. Magee, James Liu and C. E. Lundin, Relationships between intermetallic compound structure and hydride formation, J. Less-Common Met. 78:119 (1981).

40. A. V. Irodova, V. P. Glazkov, V. A. Somenkov and S. Sh. Shil'shtein, Hydrogen ordering in the cubic Laves phase HfV_2, J. Less-Common Met. 77:89 (1981).

41. G. Boureau, A simple method of calculation of the configur-ational entropy for interstitial solutions with short range repulsive interactions, J. Phys. Chem. Solids 42:743 (1981).

42. J.-J. Didisheim, K. Yvon, D. Shaltiel, P. Fischer, P. Bujard and E. Walker, The distribution of the deuterium atoms in the deuterated cubic Laves-phase $ZrV_2D_{4.5}$, Solid State Comm. 32:1087 (1979).

43. W. E. Wallace, H. E. Flotow and D. Ohlendorf, Configurational entropy and structure of β-LaNi$_5$ hydride, J. Less-Common Met. 79:157 (1981).

44. G. Busch, L. Schlapbach, W. Thoeni, Th. v. Waldkirch, P. Fischer, A. Furrer, and W. Haelg, Hydrogen in La-Ni compounds: Localization and diffusion, in: "Proc. of the 2nd Int. Congress on Hydrogen in Metals", Paris, Vol. 1, Pergamon Press, New York (1978).

45. J. C. Achard, C. Lartigue, A. Percheron-Guégan, J. C. Mathieu, A. Pasturel and F. Tasset, Reply to "Configurational entropy and structure of β-LaNi$_5$ hydride", J. Less-Common Met. 79:161 (1981)

46. E. J. Teatum, K. A. Gschneidner, Jr., and J. T. Waber, Compilation of calculated data useful in predicting metallurgical behavior of the elements in binary alloy systems, Report No. LA-4003, 1968.

Octahedral Site Occupation in Lanthanum Dihydride[*]

E. L. Venturini

Sandia National Laboratories
Albuquerque, New Mexico 87185

ABSTRACT

The distribution of hydrogen ions among the octahedral and tetrahedral sites in lanthanum hydride has been determined at liquid helium temperatures for hydrogen/metal ratios from 1.90 to 2.04. Electron spin resonance spectra of dilute erbium ions substituted for host metal atoms establish the hydrogen site occupation adjacent to the erbium versus hydrogen loading. These data together with a lattice-gas calculation yield the site occupation versus hydrogen concentration in bulk lanthanum hydride, which is such that 2% of the octahedral sites are filled in $LaH_{2.00}$.

INTRODUCTION

The various binary metallic hydrides offer very simple systems to study the behavior of hydrogen in solids. One of the numerous interesting problems in these materials is the distribution of hydrogen among the available sites as a function of the hydrogen/ metal ratio. One approach[1] to measuring this distribution at low temperatures involves recording electron spin resonance (ESR) spectra of dilute erbium magnetic ions substituted for nonmagnetic host metal atoms in samples with different hydrogen content. These spectra are fitted to a lattice-gas model[2] which in turn yields the hydrogen distribution in the bulk host hydride. This method has been applied previously to yttrium and scandium hydrides and deuterides,[3] and here it is applied to lanthanum hydride to obtain

---

[*]This work performed at Sandia National Laboratories supported by the U.S. Department of Energy under contract #DE-AC04-76DP00789.

2% O-site occupation in $LaH_{2.00}$ at liquid helium temperatures.

LaH_x has an unusually large single-phase region, retaining a face-centered-cubic (fcc) metal lattice for H/metal ratios x from 1.9 to 3.[4] There are two distinct sites available to the H ions in this fcc structure, one octahedral (O) and two tetrahedral (T) sites per metal atom. The T-site has lower energy, and the ideal dihydride has all T-sites filled and all O-sites vacant.

The actual distribution of H among O-sites and T-sites in LaH_x has been studied by proton nuclear magnetic resonance[5] and inelastic neutron scattering[6]. The NMR measurement involves a comparison of experimental proton second moments with the values calculated from dipolar broadening assuming random partial O-site occupation by H predominantly located on T-sites. The second moment data[5] measured at 76 K suggest that the T-site filling stops when the H/metal ratio reaches 1.95, followed by random occupation of O-sites. This implies 5% O-site occupation in $LaH_{2.00}$. Inelastic neutron scattering spectra[6] for several light rare earth hydrides show separate peaks for T-site and O-site H vibrations. The relative area under the two peaks suggests an O-site occupation of 5 to 10% near the dihydride composition at room temperature.

ERBIUM ESR SPECTRA IN LANTHANUM HYDRIDE

The samples used in this work were prepared by first arc-melting 0.1 atomic percent Er into purified La metal using an inert atmosphere (argon) and water-cooled copper hearth. The resulting ingot was placed in a modified Sieverts' apparatus, heated to 600 C under a dynamic vacuum of 10^{-6} torr, and exposed to H_2 gas purified by diffusion through a Pd-Ag tube. The H/metal ratio was determined volumetrically and gravimetrically, the loading from the two independent measurements agreeing to better than 1%. Due to the large volume expansion[4] of hydrided lanthanum, the samples consisted of small polycrystalline pieces following loading. Very light grinding with a mortar and pestle produced a coarse powder suitable for ESR studies. The powder was sealed in quartz capillary tubes to prevent excessive oxidation of the hydride.

Fig. 1 compares derivative ESR absorption spectra for Er in two LaH_x powders, the upper trace for a sample with a H/metal ratio x = 2.04, the lower trace for x = 1.96. Both spectra were recorded at 2 K and 9.8 GHz in a homodyne ESR spectrometer using magnetic field modulation. The spectra have a large isotropic absorption near 1035 Oe which is asymmetric due to the metallic conductivity of the samples. There are small hyperfine lines associated with this main absorption due to the 23% natural abundance of ^{167}Er with nuclear spin 7/2.

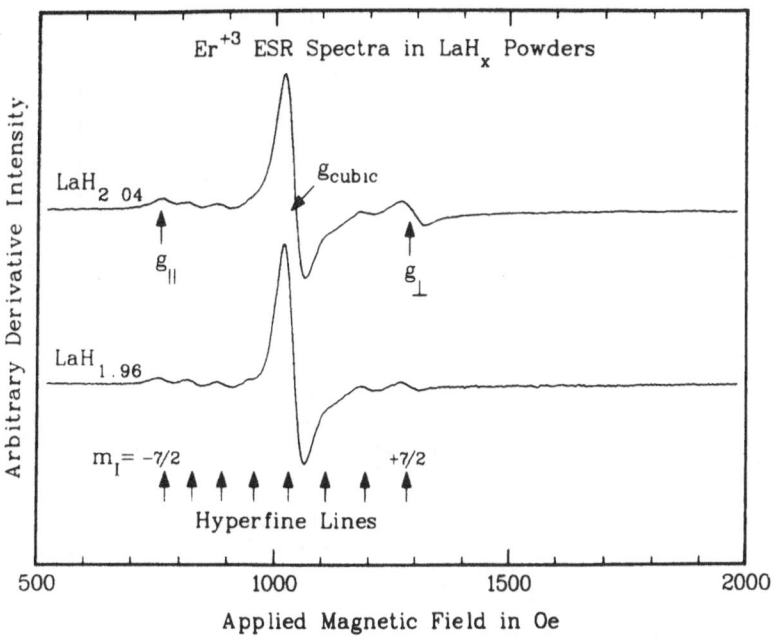

Fig. 1. Er ESR spectra in two LaH_x powders at 2 K and 9.8 GHz.

The two spectra in Fig. 1 are identical except for a small additional resonance in the upper trace with features indicated by $g_{\parallel}$ and $g_{\perp}$. Based on our previous work[1-3] with Er ESR spectra in Sc and Y hydrides and deuterides, the dominant signal in both spectra in Fig. 1 is attributed to Er in a cubic site, i.e., an Er ion surrounded by a simple cube of 8 nearest-neighbor (nn) T-site H ions. The cubic g-factor is 6.778(3). The additional resonance in the upper trace arises from Er in an axial site, i.e., an Er ion surrounded by the same cube of 8 T-site H ions plus one H on an O-site. This latter next-nearest-neighbor (nnn) H ion lowers the Er site symmetry to axial, causing a splitting of the ESR g-factor into $g_{\parallel}$ = 9.2(1) when the applied magnetic field is parallel to the direction from the Er to the nnn O-site H ion and $g_{\perp}$ = 5.38(2) when the applied field is normal to this direction. Since the ESR spectra are recorded with powder samples, signals from all magnetic field orientations are observed.

There has been one prior report[7] of Er ESR in LaH_x, a sample containing 1% Er with x = 2.02. Although the authors do not show an ESR spectrum, they describe a single asymmetric isotropic absorption with a linewidth of 50 Oe at 2 K and no noticeable hyperfine structure. The g-factor for this resonance is 6.68(5) in fair agreement with our value of 6.778(3). Since this report[7] does not

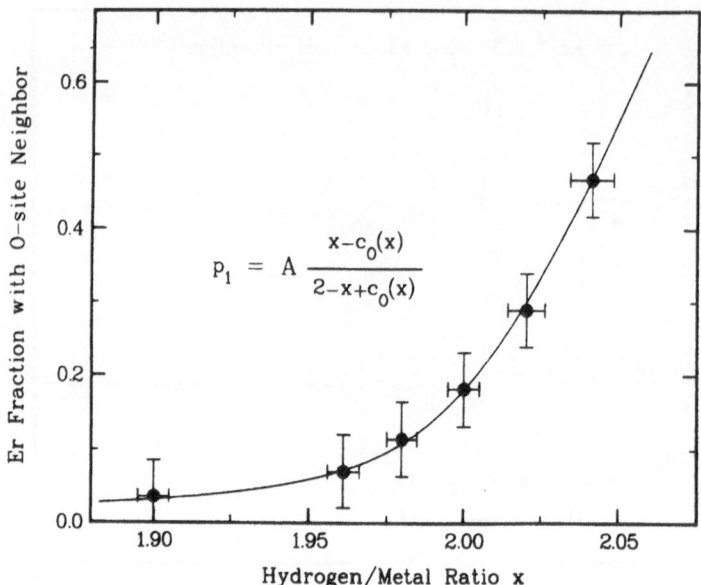

Fig. 2.   Fraction of Er ions with one nnn O-site occupied by H.

define the ESR linewidth used, we assume they are employing a
Lorentzian lineshape with a half-width at half-maximum of 50 Oe.
For comparison our samples with 1/10 the Er content have a
Lorentzian half-width at 2 K which increases with H/metal ratio from
27 Oe for x = 1.90 to 38 Oe for x = 2.04.

HYDROGEN SITE DISTRIBUTION

        The ESR spectra in Fig. 1 have been analyzed numerically[1] using
derivative Lorentzian lineshapes for a cubic and an axial signal in
a powder sample. The least-squares fit yields directly the number
of Er ions with one adjacent O-site occupied (axial site) relative
to those ions with all nnn O-sites vacant (cubic site). Using this
analysis on the spectra for six samples, we obtain the occupation
probability for an O-site adjacent to an Er impurity in LaH_x versus
H/metal ratio x as shown in Fig. 2.
        The solid line in Fig. 2 is the "best" fit of a lattice-gas
model[2] to the ESR data. The probability p_1 that one and only one
adjacent O-site is occupied is given by[2]

$$p_1 = A[x - c_0(x)]/[2 - x + c_0(x)]$$

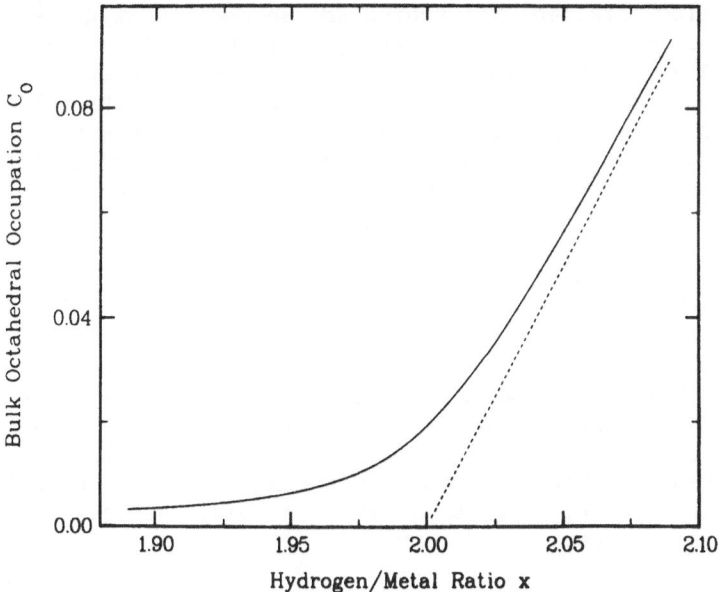

Fig. 3. Fraction of O-sites occupied in bulk LaH_x from lattice-gas model (solid line), and for ideal fcc metal hydride (dashed line).

where A is an adjustable parameter related to the site energies of a bulk T-site and an O-site adjacent to the Er impurity, and $c_0(x)$ is the bulk O-site occupation probability. $c_0(x)$ is given by[2]

$$c_0(x) = x - 4x/[2 + x + B(1 - x) + \sqrt{8B + \{2 - x - B(1 - x)\}^2}]$$

where B is an adjustable parameter related to the bulk T-site and bulk O-site energies. Parameters A and B are assumed independent of H/metal ratio. Fitting the lattice-gas model to the ESR data in Fig. 2 gives $A = 1.8 \times 10^{-3}$ and $B = 1.9 \times 10^{-4}$.

      This value of B yields the bulk O-site occupation in LaH_x versus H/metal ratio x shown as a solid line in Fig. 3. From this curve we obtain an O-site occupation of 2% in $LaH_{2.00}$ at 2 K. For comparison the dashed line is the expected O-site filling in the ideal hydride, where the site energy difference is sufficiently large to prevent significant H on O-sites below x = 2. In this latter case the O-sites fill linearly with x above x = 2, reaching complete occupation at x = 3.

DISCUSSION

      This result of 2% O-site occupation in LaH_2 is approximately half the value obtained from proton NMR second moment data[5]. However, the NMR data suggest that 2.5% of the T-sites remain vacant for all x > 1.95, which is not consistent with our results. It is unlikely that the O-site occupation changes between the ESR measurement temperature of 2 K and the NMR temperature of 76 K, since the hydrogen diffusion times[5] are extremely long at these temperatures. Inelastic neutron scattering or neutron diffraction data versus temperature might provide an explanation.

In previous papers[1-3] we have applied this impurity ESR plus lattice-gas model to both ScH_x and YH_x. In the case of ScH_x less than 0.5% of the O-sites are occupied in the bulk lattice at the highest H/metal ratio of 1.99. For YH_x 8% of the O-sites are filled at 2 K when x = 2. A simple explanation for the increase between ScH_x and YH_x is that the larger YH_x lattice leads to a lower energy difference between T-sites and O-sites and hence greater disorder. The data presented here for LaH_x indicating 2% O-site occupation contradict this simple argument, since the LaH_x lattice is the largest of the three.

In summary we have prepared samples of dilute Er in LaH_x for several H/metal ratios x between 1.90 and 2.04. The Er ESR spectra show two distinct sites for the Er impurities, one having all six nnn O-sites vacant, and one having one such O-site occupied by H. The dependence of this adjacent O-site occupation on H content is explained by a lattice-gas model with two adjustable parameters. This model in turn yields the distribution of H among the sites in bulk LaH_x over the loading range investigated.

ACKNOWLEDGEMENT

Special thanks to P. M. Richards for applying the lattice-gas model to this problem and encouraging the experimental work.

REFERENCES

1. E. L. Venturini and P. M. Richards, Solid State Commun. 32:1185 (1979).
2. E. L. Venturini and P. M. Richards, Phys. Lett. 76A:344 (1980).
3. E. L. Venturini, J. Less-Common Met. 74:45 (1980).
4. W. M. Mueller in: "Metal Hydrides", W. M. Mueller, J. P. Blackledge and G. G. Libowitz, eds., Academic Press, New York (1968), Chapter 9.
5. D. S. Schreiber and R. M. Cotts, Phys. Rev. 131:1118 (1963).
6. P. P. Parshin, M. G. Zemlyanov, M. E. Kost, A. Yu. Rumyantsev and N. A. Chernoplekov, Inorganic Mater. 14:1288 (1978).
7. H. Drulis, K. P. Hoffman and B. Stalinski, Solid State Commun. 36:973 (1980)

THERMODYNAMIC AND STRUCTURAL PROPERTIES OF THE

RARE-EARTH Co_3 HYDRIDES

Henry A. Kierstead

Argonne National Laboratory
Argonne, Illinois   60439

ABSTRACT

The thermodynamic and structural properties of the hydrides of the intermetallic compounds $LnCo_3$ (Ln = Nd, Gd, Tb, Dy, Ho, Er, Tm, and Y) are presented. Systematic changes in the absorption isotherms; the heat, entropy, and free energy of absorption; and the x-ray lattice parameters are discussed and interpreted.

INTRODUCTION

We have previously reported[1] the preparation and thermodynamic properties of the hydrides of the compounds $LnCo_3$, where Ln stands for Nd, Gd, Tb, Dy, Ho, Er, and Y. The purpose of this paper is to summarize the properties of these hydrides, to point out the similarities between them, and to examine the changes in the properties as a function of the atomic number of the rare-earth component. In addition to the above-mentioned compounds we have made preliminary measurements on $TmCo_3$ hydrides, which are reported for the first time here.

EXPERIMENTAL PROCEDURE

The $TmCo_3$ sample was prepared from 99.9% pure thulium and cobalt by repeated arc melting in an argon atmosphere. It was annealed in vacuum at 950 °C for 14 days. Debye-Scherrer X-ray patterns showed single-phase hexagonal $PuNi_3$-type crystals. Preparation of the hydride and measurement of desorption isotherms used standard techniques.[2] Pressures were measured using Mensor Corporation digital quartz pressure gauges. Temperatures were measured with a platinum resistance thermometer.

Preparation of the other hydrides has been reported previously.[1] The techniques were similar.

THERMODYNAMIC PROPERTIES

Desorption isotherms of the hydrides are shown in Fig. 1. All were measured at 20 °C except for $NdCo_3$ and $TmCo_3$, which were measured at 80 °C and 40 °C, respectively. The isotherms are displaced along the pressure axis for clarity, and are placed in order of the atomic number of the rare-earth component. The symbols are experimental measurements; the lines are calculated by the generalized multi-plateau theory.[3]

The isotherms can be separated into three groups. Those of $GdCo_3$ and $NdCo_3$, which are representative of the early members of the rare-earth series, have two plateaus each two atoms/mole wide with very narrow one-phase regions. The isotherms of $TbCo_3$ through $ErCo_3$ and of YCo_3 have a first plateau which is about one atom/mole wide and a second plateau which is markedly less than two atoms/mole wide. There are broad one-phase regions above and below the second plateau. $TmCo_3$, the highest member of the series, has plateaus about 1.25 and 2 atoms/mole wide with a narrow one-phase region between them.

In the middle group the shape of the isotherm in the one-phase region between the two plateaus suggests a phase transition whose critical temperature is below the measurement temperature. The pressure at the inflection point moves upward with increasing atomic number, from just above the first plateau in $TbCo_3$ to just below the second plateau in $ErCo_3$. This is shown graphically in Fig. 2, where we plot the free energy of absorption

$$\Delta F^\circ = RT \ln P \qquad\qquad\qquad (1)$$

for each of the plateaus and for the inflecton point of the one-phase region. It is apparent that in $GdCo_3$ and $NdCo_3$ the free energy of absorption on the sites responsible for the one-phase region falls at or below that of the first plateau, while in $TmCo_3$ it falls above that of the second plateau. Hence the absorption on those sites is included in the first plateau in $NdCo_3$ and $GdCo_3$ and in the second plateau in $TmCo_3$.

Thermodynamic properties of the compounds, calculated from isotherms at different temperatures, are summarized in Table 1. Heats of absorption were calculated from the equation

$$\Delta H^\circ = RT^2 (\partial \ln P / \partial T)_x \qquad\qquad (2)$$

and entropies of absorption were calculated from

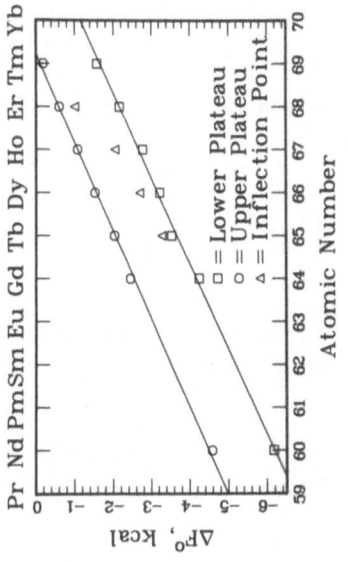

Fig. 2. Free Energy of Absorption of LnCo₃ Hydrides

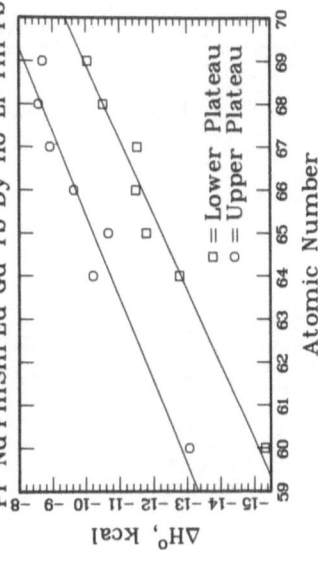

Fig. 3. Heat of Absorption of LnCo₃ Hydrides

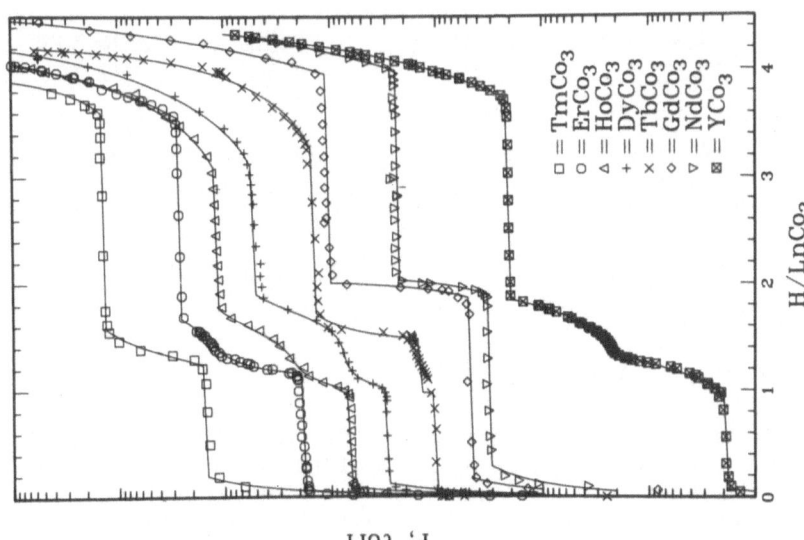

Fig 1. Isotherms of LnCo₃ Hydrides

Table 1.   Thermodynamic Properties at 20 °C in the 2-Phase Regions

Phases	Plateau Pressure torr	$\Delta F°$ kcal mol^{-1}	$\Delta H°$ kcal mol^{-1}	$\Delta S°$ cal K^{-1} mol^{-1}
YCo$_3$				
α–β	.198	−4.807	−13.41	−29.35
β–γ	20.9	−2.093	−10.48	−28.61
NdCo$_3$				
α–β	.0189	−6.176	−15.33	−31.23
β–γ	.292	−4.581	−13.07	−28.96
GdCo$_3$				
α–β	.518	−4.248	−12.78	−29.11
β–γ	11.4	−2.446	−10.22	−26.51
TbCo$_3$				
α–β	1.74	−3.542	−11.79	−28.14
β–γ	23.1	−2.035	−10.67	−29.46
DyCo$_3$				
α–β	2.98	−3.227	−11.48	−28.15
β–γ	54.8	−1.532	−9.64	−27.66
HoCo$_3$				
α–β	6.50	−2.774	−11.52	−29.83
β–γ	119	−1.081	− 8.92	−26.74
ErCo$_3$				
α–β	18.7	−2.160	−10.52	−28.52
β–γ	269	−0.604	− 8.58	−27.19
TmCo$_3$				
α–β	50.5	−1.580	−10.05	−28.90
β–γ	551	−0.188	− 8.70	−29.05

$$\Delta S° = (\Delta H° - \Delta F°)/T \tag{3}$$

The heats of absorption are plotted in Fig. 3.   They show more scatter than the free energies because of the differentiation with respect to temperature.   The entropies show little consistent variation with atomic number, but are all close to −28 cal K^{-1}(mol H$_2$)$^{-1}$.

Table 2. Lattice Parameters of LnCo₃ Compounds and Hydrides

	$a(\text{Å})$	$c(\text{Å})$	$\Delta a/a$	$\Delta c/c$	$\Delta v/v$
YCo_3	5.004	24.36			
$YCo_3H_{1.06}$	5.012	25.86	.002	.062	.065
$YCo_3H_{1.75}$	5.012	26.57	.002	.091	.094
$YCo_3H_{3.73}$	5.257	26.30	.051	.080	.192
$NdCo_3$	5.067	24.76			
$NdCo_3H_2$	5.70	27.37	.001	.106	.107
$NdCo_3H_{4.5}$	5.392	27.30	.064	.103	.249
$GdCo_3$	5.031	24.50			
$GdCo_3H_2$	5.011	27.20	−.004	.110	.101
$GdCo_3H_{4.5}$	5.284	26.98	.050	.101	.215
$TbCo_3$	5.012	24.39			
$TbCo_3H_{1.57}$	5.004	26.94	−.002	.105	.101
$TbCo_3H_{3.32}$	5.250	26.8	.047	.099	.206
$DyCo_3$	4.999	24.36			
$DyCo_3H_{1.06}$	5.001	26.00	.000	.067	.068
$DyCo_3H_{1.75}$	4.988	26.80	−.002	.100	.095
$DyCo_3H_4$	5.256	26.77	.051	.099	.215
$HoCo_3$	4.984	24.29			
$HoCo_3H_{1.07}$	4.988	25.79	.001	.062	.064
$HoCo_3H_4$	5.240	26.40	.051	.087	.201
$ErCo_3$	4.978	24.25			
$ErCo_3H_{1.2}$	4.980	25.53	.000	.053	.054

LATTICE PARAMETERS

Lattice parameters of the compounds and some of their hydrides, measured from Debye-Scherrer X-ray patterns, are tabulated in Table 2. They all have hexagonal PuNi₃-type lattices. The hydride compositions were in the one-phase region between the two plateaus and in the one-phase region above the second plateau. In the case of YCo₃ and DyCo₃ hydrides samples were taken just above the first plateau and also just below the second plateau.

In all the hydrides the volume of the unit cell increases by 5 or 6% per hydrogen atom absorbed, but the expansion is not isotropic. Up to the beginning of the second plateau the expansion is entirely in the c direction; across the second plateau there is an expansion in the a direction and a small decrease in the c direction.

DISCUSSION

In a discussion of the structural relations in rare earth-transition metal hydrides, Dunlap et al.[4] point out that the crystal structure of the AB_3 ($PuNi_3$-type) compounds consists of alternating layers of the AB_5 ($CaCu_5$-type) and AB_2 ($MgZn_2$-type) structures, stacked along the hexagonal (c) axis. Thus one third of the A atoms would have AB_5 local symmetry, and two thirds of them would have AB_2 local symmetry. They propose a model in which hydrogen first enters the AB_5 layers. When all the available sites in this layer are filled, hydrogen enters the AB_2 layer.

Since only a third of the A atoms are in the AB_5 layer, we would expect the first plateau to be one atom/mole long, as in $SmCo_5H_3$, or two atoms/mole, as in $LaNi_5H_6$. $NdCo_3$ and $GdCo_3$ have this second type of isotherm, while all the other compounds are of the first type. In any event, in this model, all the AB_5 sites are filled when x, the number of hydrogen atoms absorbed per mole of $LnCo_3$, is equal to 2; subsequent absorption is on AB_2 sites. In $TmCo_3$ the second half of the AB_5 sites and all of the AB_2 sites fill within the long second plateau and in the one-phase region above it.

The behavior of the lattice parameters supports this interpretation. While the sites in the AB_5 layer are filling, this layer cannot expand laterally because it is sandwiched between unfilled AB_2 layers; hence it expands in the c direction to provide space for the hydrogen atoms. However, when both layers are filled, the lattice can expand in the a direction, and the c parameter relaxes somewhat.

ACKNOWLEDGEMENTS

We wish to express our gratitude to P. J. Viccaro and D. G. Niarchos for preparing the intermetallic compounds and making the x-ray measurements.

This work was supported by the U.S. Department of Energy.

REFERENCES

1.   H. A. Kierstead, J. Less-Common Met.; 73, 61 (1980); 78, 29
       (1981); 78, 61 (1981); 80, 115 (1981); and 81, 221 (1981).
2.   H. A. Kierstead, J. Less-Common Met., 71, 311 (1980).
3.   H. A. Kierstead, J. Less-Common Met., in press.
4.   B. D. Dunlap, P. J. Viccaro, and G. K. Shenoy, J. Less-Common
       Met., 74, 75 (1980).

MEASUREMENT OF THREE DIMENSIONAL DISTRIBUTIONS OF HYDROGEN IN THIN SAMPLES USING THE ^{15}N HYDROGEN PROFILING METHOD

W.A. Lanford[+] and C. Burman

Department of Physics
SUNY/Albany
Albany, New York 12222

ABSTRACT

The use of the $^{15}N+^{1}H\rightarrow^{12}C+^{4}He+\gamma$-ray resonant nuclear reaction, with ^{4}He particles recorded in plastic track detectors, is described as a method of measuring full three dimensional distributions of H in thin samples. To use this reaction to measure H concentrations, the sample is bombarded with ^{15}N ions at an energy greater than the resonance energy at which the reaction occurs. As these ions penetrate the sample, they lose energy and reach the resonance energy at some depth. The yield of ^{4}He particles from the reaction is proportional to the concentration of hydrogen at that depth. By recording the He particles in plastic track detectors, an autoradiograph of the lateral hydrogen distribution at the resonance depth is obtained. Limitations and potentials of this method are discussed along with some measurements of distributions of H ion implanted into silicon.

INTRODUCTION

Analysis for hydrogen is notoriously difficult. It's atomic charge is too small for Auger or x-ray methods, and it is mobile and so common a contaminant in vacuum systems that many ion beam techniques are unsatisfactory. The one quantitative technique which has become widely used over the past five years is nuclear reaction analysis (NRA).

+An Alfred P. Sloan Foundation Fellow

Nuclear reaction analysis makes use of a nuclear reaction between an energetic heavy ion and ^{1}H as a probe for the hydrogen in the sample being analyzed. While there are many reactions which can and have been utilized[1] the most generally satisfactory reaction is: ^{15}N+^{1}H→^{12}C+^{4}He+4.4 MeV gamma-ray.[2] Because this is a resonant reaction, by varying the beam energy, various depths within the sample can be independently analyzed for hydrogen. Because this method relies on a nuclear reaction which is insensitive to the chemical environment of the hydrogen, the method is easily made quantitative. Conversely, the method tells nothing about how the hydrogen is bound within the solid.

While nuclear reaction analysis has been successfully applied to a wide range of problems in physics, chemistry, material science and archaeology where one needs to know the hydrogen concentration vs depth in planar samples, there are a large number of intrinsically interesting and technologically important problems where one needs to know the lateral distribution of hydrogen on a surface or at a given depth within a sample. One example is the study of hydrogen (or water) enhanced crack propagation in the fracture of metals (or glasses). Perhaps equally interesting would be the study of hydrogen distributions and transport along surfaces or on grain boundaries.

Here we report for the first time the extension of the well established ^{15}N hydrogen profiling technique to give not only the depth distribution of hydrogen but also the lateral distribution at any depth. The depth resolution is of order 10 nanometers and the lateral resolution is of order 10 micrometers.

PROCEDURE

Hydrogen Lateral Distribution Measurements

The ^{15}N hydrogen profiling technique[2] makes use of the nuclear reaction: ^{15}N+^{1}H→^{12}C+^{4}He+4.4 MeV gamma-ray to measure the hydrogen concentration vs depth in any solid. The usual way to utilize this reaction is to bombard the sample being analyzed with ^{15}N ions from an accelerator and to measure the yield of the characteristic 4.4 MeV gamma-rays. Because there is a strong isolated resonance in this reaction, the probability for the reaction occuring is neglible except when the ^{15}N is at the resonance energy (E_r=6.405 MeV). If a sample is bombarded with ^{15}N beam at $E=E_r$, the yield of characteristic gamma-rays is proportional to the concentration of hydrogen on the surface. If the sample is bombarded with ^{15}N at an energy greater than E_r, there is negligible yield from reactions with surface hydrogen, but as the ^{15}N ions lose energy penetrating the solid, they reach E_r at some depth, and now the gamma-ray yield is proportional to the hydrogen content at this depth. Hence, by measuring the gamma-ray yield vs ^{15}N beam energy, the concentration of hydrogen vs depth is determined.

A natural way to extend this procedure to measure the full three dimensional distribution of hydrogen in a solid would be to obtain a microbeam of ^{15}N and raster this beam across the sample, recording the gamma-ray yield as a function of beam spot location. The difficulty with this procedure is that even with only a few nanoamperes of ^{15}N beam and a beam spot 10 microns across, the energy density being deposited in the samples is of order of 10 kilowatts/cm^2. Such a high energy density is very likely to drive the hydrogen away from the analyzing beam.

An alternative method[3] which avoids these problems is to use a very large ^{15}N beam spot and determine the lateral distribution of hydrogen at a given depth by recording the lateral distribution of the characteristic radiations created by reaction between the ^{15}N ions and the hydrogen in the sample. The ^{4}He (alpha particles) produced in the ^{15}N+H→^{12}C+^{4}He+4.4 MeV gamma-ray nuclear reaction are ideal for this purpose.

A schematic representation of this autoradiographic technique is shown in Figure 1. The ^{15}N beam bombards the sample at an energy E>E$_r$. The ^{15}N ions slow down reaching the resonance energy at some depth creating the "H resonance window". Reactions are induced between the ^{15}N ions and hydrogen within the resonance window emitting both a gamma-ray and a He particle. He particles emitted forward are very energetic and have a long range. The ^{15}N ions, on the other hand, have a residual range which is much less. Hence, by choosing a sample of appropriate thickness, the ^{15}N ions will stop in the sample and He particles will emerge from the back of the sample where they can be recorded in a track detector. If the track detector is mounted in contact with the back of the thin sample, the density of latent tracks created by the He particles will be proportional to the lateral hydrogen concentration at the depth of the hydrogen detection window.

We do not wish to discuss track detectors here because there is extensive literature concerning them[4]. However, we need to make use of the following special properties of our detector (plastic CR-39): Any energetic charged particle with Z≥2 which penetrates this track detector will create a latent track along its path. Less densely ionizing particles (such as protons, electrons, gamma-rays) will not produce latent tracks. The latent tracks can be made visible by etching the track detector. In our case, we etched in 6N KOH at 65° for 100 minutes. The etch pits created by such a procedure can then be counted or photographed under a microscope.

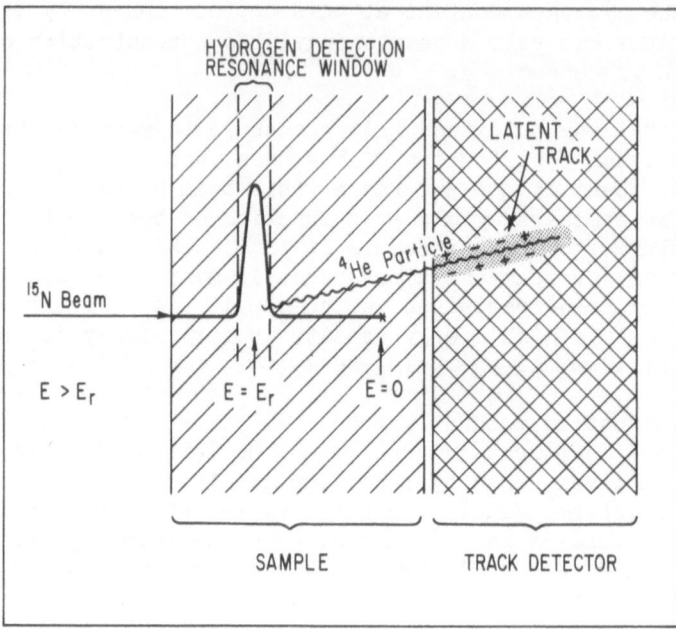

Figure 1. The schematic representation of the autoradiograph
H profiling technique.

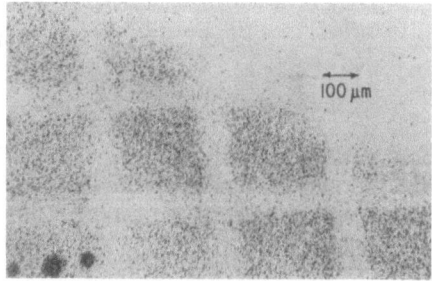

Figure 2. An autoradiograph of H ion implanted into silicon
through a wire grid.

Preparation of a Test Sample

To test the feasibility of the procedure outlined above a test sample was needed which had a known, but not homogenous, hydrogen distribution. We prepared such a sample by ion implanting hydrogen (30 KeV) through a mask (a wire mesh grid) into silicon. The range distribution and the immobility of such implanted hydrogen had been previously established by a number of measurements.[1]

Because the autoradiographic technique requires thin samples, this implant of 10^{17} H/cm^2 was made into a 5 micron thick silicon crystal.

RESULTS

A typical autoradiograph obtained is shown in Figure 2. The result was obtained with a ^{15}N beam at E=6.75 MeV, which means we are measuring the lateral hydrogen distribution near the peak in the ion implant distribution.

The square "grid" pattern resulting from the ion implantation through the wire grid is clearly evident. There are a few features which bear comment. First, while the track density within the shadowed region is small, it is not zero. The principal source of these tracks is from nuclear reactions between the ^{15}N ions and hydrogen on the surface of the sample. The reaction is ^{15}N+^{1}H→^{12}C+^{4}He with the ^{12}C nucleus in its ground state (and no nuclear gamma-ray emitted). The cross-section for this reaction varies smoothly with energy and, hence, the yield of these He tracks varies smoothly with beam energy. Their contribution can be subtracted by measuring at various beam energies or by measuring the amount of surface hydrogen on the sample. A better solution would be to work in a higher vacuum (the present data were taken at 10^{-5} Torr) and with cleaner samples.

The lateral resolution can be estimated, from the data of Figure 2, to be of order 10 microns. This experimental estimate is only approximate but it is consistent with the a priori estimate based on the geometry and kinematics of the reaction.

CONCLUSION AND DISCUSSION

We have demonstrated for the first time the utilization of a nuclear reaction with a narrow isolated resonance to measure the lateral distribution of H within a plane inside a thin sample. The depth resolution of this method is of order of 10 nanometers and the lateral resolution is roughly of order 10 micrometers.

This technique can be made quantitative by counting the track density and by evaluating the reaction cross section and detector efficiency.

While the use of track detectors is attractive because of the ease of data taking, this method may also be carried out using position sensitive electronic detectors. This procedure would allow for an immediate live-time display of the lateral hydrogen distribution as the data were recorded. We are presently working on developing this procedure.

ACKNOWLEDGEMENTS

We wish to thank C. Buel for preparing the thin silicon crystal, R. Fleischer for advice on track detectors, K. Davis and W. Means for help on making micrographs, and the Office of Naval Research for supporting this research.

REFERENCES

1. J. Ziegler et al Nucl. Inst. and Methods 185 (1981) 505.
2. W.A. Lanford, H.P. Trautvetter, J. Ziegler and J. Keller
   Applied Physics Letters 28 (1976) 566.
3. R. Ilic and C. Altstetter, Nucl. Inst. and Methods 185 (1981) 505.
4. R.L. Fleischer, P.B. Price, R.M. Walker, Nuclear Tracks in Solids
   Principles and Applications, University of California Press,
   Berkeley (1975).

STRAIN EFFECTS IN THE X-RAY DETERMINATION

OF LATTICE PARAMETERS IN Pd-H$_x$

P.A. Hardy, I.S. Balbaa and F.D. Manchester

Department of Physics, University of Toronto
Toronto, Ontario, Canada M5S 1A7

ABSTRACT

    X-ray lattice parameter determinations for the β-phase in Pd-H$_x$ have revealed a variation in a(x), the lattice parameter, that depends on the method of preparation. Samples loaded to a concentration of x > 0.6, by avoiding the mixed phase region, have a lattice parameter slightly lower than samples loaded to the same concentration by passing through the mixed phase region. The difference is thought to result from unrelieved strain produced as a result of the α → β phase change. These results point to a significant correction of the a(x) data of Schirber and Morosin[1], which have become widely used for determining hydrogen concentrations from lattice parameter measurements. A full account of these experiments is being published elsewhere.

REFERENCES

1. J.E. Schirber and B. Morosin, Phys. Rev. B12, 117 (1975).

LATTICE DISTORTIONS AND PHASE TRANSITIONS OF HYDROGEN IN NIOBIUM

H. Peisl

Sektion Physik der Universität München
W-Germany

ABSTRACT

   Lattice distortion due to hydrogen in niobium have been studied by various X-ray and neutron scattering methods. All the important quantities describing the lattice distortions have been determined: The defect volume, the long-range displacement field described by the force dipole tensor, the local displacements of the hydrogen neighbors and the Kanzaki forces necessary to create the observed lattice distortions.

   Lattice distoritions are closely related to the high temperature phase transitions (α-α', α'-β, α'-δ). The before mentioned quantities have been studied at various temperatures and H concentrations. At high concentrations some novel and unexpected effects have been observed. Pretransitional fluctuations connected with various phase transitions occur: Partial ordering of H above T_c, precipitation of coherent microscopic H density modes and an abnormal change of the properties of the host lattice.

FORCES, DIPOLE FORCE TENSOR AND ELASTIC BINDING ENERGY IN

α-PALLADIUM HYDRIDES

J. Khalifeh, G. Moraitis and C. Demangeat

L.M.S.E.S. (LA CNRS 306) - Université Louis Pasteur
4, rue Blaise Pascal 67070 STRASBOURG CEDEX, France

ABSTRACT

The forces resulting from the presence of one interstitial Hydrogen in Palladium are deduced from the variation of the total energy of the alloy up to first order in the atomic displacement. This variation includes the band term together with the electron-electron and ion - ion terms but, unfortunately, neglects the zero point motion of the Hydrogen. This tight-binding calculation is based on a rigidly moving wave function basis where the electronic structure of the unrelaxed alloy is determined by taking into account a perturbing potential up to the nearest neighbours of the Hydrogen atom. Expressions of the distribution of the forces and of the dipole force tensor P are derived in terms of the variation of the hopping integrals and of the energy levels. Estimations of P and of the forces up to third nearest neighbour shell of the Hydrogen atom are presented. Once this has been done the elastic binding energy can be obtained if we know the lattice Green function of the alloy. In the present model we replace this unknown exact Green function by a phenomenological expression of the host lattice Green function.

INTRODUCTION

The relaxation effects around impurities are essential to study the thermodynamic properties of solid solutions and for the understanding of local order. The importance of the relaxation can be qualitatively deduced from the experimental value of the relative variation of the average alloy atomic volume per added impurity. Diffuse neutrons (or X rays) scattering and EXAFS experiments are directly related to the displacement.

From a theoretical point of view a few authors have discussed
the physical origin of the relaxation and its relation to the electronic structure of solid solutions. A first attempt to calculate
the size effect was done using the elastic model[1],[2]. Such a macroscopic theory is only able to determine the long range part of the
displacements. In the lattice statics theories[3],[4] the discrete nature of the lattice was considered. The lattice displacement $u(\lambda)$
are deduced from the force constants of the alloy and from the
forces $F(\lambda)$ introduced by the impurity on the host ion λ. These
quantities are defined in terms of the derivative of the variation
of the total energy induced by the relaxation. Usually the force
constants are approximated by the host force constants deduced
from phonon spectra, whereas the forces $F(\lambda)$ are assumed to be localized on the nearest neighbours of the impurity atom. These forces $F(\lambda)$ are deduced by the relation between the size effect, the
host elastic constants and the dipole force tensor P associated to
the distribution of the forces. These approximations are dubious
from a quantitative point of view : (i) the force constants are
certainly strongly modified by the impurities when the size effect
is large[5] ; (ii) the distribution of forces $F(\lambda)$ decreases slowly
and presents Friedel oscillations in dilute alloys[6].

A simple self-consistent expression for the field of forces
induced by substitutional impurities has been obtained recently in
the case where the band structure is approximated by a tight-binding
(TB) scheme[7]. It is assumed that the basis of the TB representation
is built from orbitals which are rigidly shifted by the relaxation.
The total energy change δE due to the relaxation induced by the
presence of one impurity arises from the one electron energies
(band structure contribution E_{bs}), from the electron-electron interaction E_{ee} (which has been counted twice in E_{bs}) and the direct
ion-ion interaction E_{ii}. The band structure energy is modified by
the explicit contribution of the displacement of the TB orbitals :
the forces $F(\lambda)$ contain a new term which is expressed in terms of
the variation with distance of the transfer integrals, the crystalline potential remaining constant and equal to its value in the
frozen (unrelaxed) lattice. This expression presents the advantage
to eliminate the screening potential induced by the relaxation in
the alloy, a quantity which is difficult to determine self-consistently[8]. However, this expression is not well suited for numerical
calculations[9],[10] because large cancellation occurs between the
attractive part (δE_{bs}) and the repulsive part ($\delta E_r = \delta E_{ee} + \delta E_{ii}$)
of the forces. This leads to an expression which contains two
terms : (i) the first one comes from the screening of the impurity
excess charge : it is expressed in terms of the screening charge
and of the variation of the host transfer integrals (and energy
levels) with distance ; (ii) the second is directly related to
the difference of size via the difference of the impurity and host
atoms. A general expression of the dipole force tensor P is then
obtained in terms of the impurity t matrix and the change of host

energy levels and transfer integrals with distance[11]. Numerical estimation of P in terms of the filling of the band and impurity excess charge is shown to reproduce qualitatively the physical predictions of the Vegard's law for transition metals[11].

The present objective is to study the forces, the dipole force tensor of Hydrogen in Palladium in the very dilute limit.

THE FORCES AND THE DIPOLE FORCE TENSOR IN α-PALLADIUM HYDRIDES

The aim of this chapter is to relate the forces $F(\lambda)$ on Palladium site λ and resulting from the presence of a Hydrogen impurity at interstitial site (either tetrahedral or octahedral) to the electronic structure of the alloy. The one electron Hamiltonian, per spin, of the alloy is written as[12,13] :

$$H = \sum_{\mu m} |\mu m> \varepsilon_\mu^m <\mu m| + \sum_{\substack{\nu \mu m m' \\ \nu \neq \mu}} |\nu m> \beta_{\nu\mu}^{mm'} <\mu m'| \tag{1}$$

where $|\mu m>$ represents the orbital of symmetry m centred either on one of the N metallic sites (in this case m = s,p,d) or on interstitial site I (in this case m = s). For simplicity we assume that these functions define a complete set of orthogonal functions. $\beta_{\nu\mu}^{mm'}$ is the transfer integral and ε^m is the energy level. The tight-binding orbital $\phi_\lambda^m(r) = <r|\lambda m>^\mu$ (for m = d) follows rigidly the core displacement $u(\lambda)$ without undergoing any deformation. The displaced tight-binding basis $\overline{\phi}_\lambda^m(r)$ is then related to the undisplaced one $\phi_\lambda^m(r)$ by :

$$\overline{\phi}_\lambda^m(r) = \phi_\lambda^m(r - u(\lambda)) \tag{2}$$

Up to first order in $u(\lambda)$, $\overline{\phi}_\lambda^m(r)$ is given by :

$$\overline{\phi}_\lambda^m(r) = \phi_\lambda^m(r) - \vec{u}_\lambda \cdot \vec{\nabla}_\lambda \phi_\lambda^m(r) \tag{3}$$

The effect of the relaxation is to replace the Hamiltonian H by :

$$\overline{H} = H + \delta V \tag{4}$$

where δV is the variation of the potential induced by the relaxation. The resulting total change in energy δE arises from the sum of the one electron energies (band structure contribution E_{bs}), from electron-electron interactions E_{el-el} (which have been counted twice in E_{bs}) and the direct ion-ion interactions $E_{ion-ion}$. Once δE is known, the force $F(\lambda)$, on site λ, induced by the defect is given by (α = x, y, z) :

$$F_\alpha(\lambda) = - \frac{\partial \delta E}{\partial u_\alpha(\lambda)} |u_\alpha(\lambda) = 0 \tag{5}$$

As discussed extensively[7],[14] the three terms δE_{bs}, δE_{el-el} and $\delta E_{ion-ion}$ combine in order to give the following expression :

$$\delta E = \text{Tr} \left[\delta W \, \Delta N \right] \qquad (6)$$

where δW is a pseudopotential associated with the displacement of quasineutral pseudoatoms and ΔN, in the atomic orbital basis is given by :

$$\Delta N_{\lambda \mu}^{mm'} = - \frac{\text{Im}}{\pi} \int^{E_F} <\lambda m | \, G - G^o \, | \mu m'> \, dE \qquad (7)$$

where G is the Green function related to the Hamiltonian H given by equation (1) : $G = |E - H|^{-1}$ and G^o is related to the Hamiltonian H_o of the pure Palladium. The matrix elements of δW in the basis of the undisplaced orbitals are given by[7] :

$$<\lambda m | \, \delta W \, | \mu m'> = (1 - \delta_{\lambda \mu}) \, \delta \beta_{\lambda \mu}^{mm'} + \delta mm' \, \delta_{\lambda \mu} \, \delta \varepsilon_{\lambda}^{m} \qquad (8)$$

In Hamiltonian H, the summation is over $N+1$. This can be split in terms of a summation over the N host sites, a term containing only the interstitial site and a coupling term between the Hydrogen atom and the neighbouring metallic sites :

$$H = H_o + \sum_{Rm}' |Rm> v^d(R) <Rm| + |Is> \varepsilon_I^s <Is|$$
$$+ \sum_{\mu m}' \{ |\mu m> \beta_{\mu I}^{ms} <Is| + c.c. \} \qquad (9)$$

where the prime in the summation means that the summation is restricted to lattice sites only. ε_I^s is the Hydrogen level and $|Is>$ its s orbital centred on interstitial site I. The diagonal disorder term $v^d(R)$ in the d bands is introduced to be in accord with Friedel's rule. With evident notations we have :

$$H = H^I + v^d + v^{nd} \qquad (10)$$

H^I is typically a Hamiltonian of $N+1$ particles but without interaction between the Hydrogen atom and the metal. The scattering matrix T is defined by :

$$T = (v^d + v^{nd}) + (v^d + v^{nd}) \, G^I T \qquad (11)$$

The component α ($\alpha = x,y,z$) of the force acting on the matrix atom sitting at the λ site is given by[15] :

$$F_\alpha(\lambda) = F_\alpha^o(\lambda) + F_\alpha^I(\lambda) \qquad (12)$$

where $F_\alpha^o(\lambda) = \sum_{\rho m}' \Delta N_{\rho \rho}^{mm} D_{\alpha \lambda} \, \delta \varepsilon_\rho^m + \sum_{\substack{\rho \mu mm' \\ \rho \neq \mu}}' \Delta N_{\rho \mu}^{mm'} D_{\alpha \lambda} \, \delta \beta_{\mu \rho}^{m'm} \qquad (13)$

Table 1. Forces from a Hydrogen interstitial at the octahedral site in Palladium for extended $(v^d_{\lambda 1} \neq 0)$ and
non extended perturbing potential $(v^d_{\lambda 1} = 0)$. The α
component $(\alpha = x, y, z)$ of the forces $F_\alpha(\lambda_i)$ are reported up to third nearest neighbour λ_3 to the Hydrogen
impurity. Lattice parameter a_d; forces $(10^{-4}$ dyn) ;
perturbing potential energy $v^d_{\lambda 1}$ (from reference 15).

$v^d_{\lambda 1}$ (eV)	$\lambda_1 = \frac{a}{2}(0,0,1)$			$\lambda_2 = \frac{a}{2}(1,1,1)$			$\lambda_3 = \frac{a}{2}(0,2,1)$		
	$F_x(\lambda_1)$	$F_y(\lambda_1)$	$F_z(\lambda_1)$	$F_x(\lambda_2)$	$F_y(\lambda_2)$	$F_z(\lambda_2)$	$F_x(\lambda_3)$	$F_y(\lambda_3)$	$F_z(\lambda_3)$
0	0	0	1.037	-.012	-.012	-.012	0	.045	.025
-.042	0	0	1.112	-.020	-.020	-.020	0	.039	.021

$$F^I_\alpha(\lambda) = \Delta N^{ss}_{II} D_{\alpha\lambda}\, \delta\varepsilon^s_I + \sum_{\mu \neq I} (\Delta N^{sm}_{I\mu} D_{\alpha\lambda}\, \delta\beta^{ms}_{\mu I} + \Delta N^{ms}_{\mu I} D_{\alpha\lambda}\, \delta\beta^{sm}_{I\mu}) \qquad (14)$$

with the notations m

$$D_{\alpha\lambda}\, \delta\varepsilon^m_\rho = - \frac{\partial}{\partial u_\alpha(\lambda)}\, \delta\varepsilon^m_\rho \qquad (15)$$

$$D_{\alpha\lambda}\, \delta\beta^{m'm}_{\mu\rho} = - \frac{\partial}{\partial u_\alpha(\lambda)}\, \delta\beta^{m'm}_{\mu\rho} \qquad (16)$$

A component $P_{\alpha\alpha}$ of the dipole force tensor is given by :

$$P_{\alpha\alpha} = \sum_\lambda \lambda_\alpha\, F_\alpha(\lambda) \qquad (17)$$

where λ_α is the α component of the vector of the λ-th lattice site
relative to interstitial site I.

The values of the forces on first, second and third nearest
neighbours of the Hydrogen interstitial are than deduced[15,16]
(Table 1).

CONCLUSION AND OUTLOOK

In the framework of the tight-binding approximation we have
estimated the dipole force tensor P and the forces distribution
around the defect, in α-PdH, by taking into account a perturbing
potential up to nearest neighbours of the Hydrogen atom. We have
shown that P is mainly determined by the variation of the hopping
integrals introduced by the relaxation whereas the forces are sensitive to δv^d_λ which describes the extension of the perturbation.
It is found that the forces at nearest and third nearest neighbours
are directed outward whereas the forces on second nearest neighbours are directed inward.

In connection with the elastic binding energy Dederichs and Deutz[17,18] have discussed the range of validity of the continuum approximation in the case of substitutional point defects. Extension to the case of hydrides has been done by Kramer[19]. Following these lines we are presently investigating the elastic binding energy of a pair of Hydrogen atoms in α-Palladium hydrides by using our results on the forces and on P. Numerical estimations is less trivial than it was thought at the beginning and results will be probably appear in the latter part of 1982[16].

REFERENCES

1. J. Friedel, Adv. in Phys. 3 : 446 (1954).
2. D. Eshelby, Acta Met. 3 : 487 (1955).
3. H. Kanzaki, Phys. Chem. Sol. 2 : 24 (1957).
4. V.K. Tewary, Adv. Phys. 22 : 757 (1973).
5. H.R.Schober and V. Lottner, Zeit. Phys. Chem. Wiesbaden 114 : 203 (1979).
6. A. Blandin and J.L. Deplanté, J. Phys. Rad. 23 : 609 (1962).
7. G. Moraitis, B. Stupfel and F. Gautier, J. Phys. F 11 : L 79 (1981).
8. G. Moraitis and F. Gautier, J. Phys. F 9 : 2025 (1979).
9. G. Moraitis, D. Sc. Thesis, Strasbourg, 1978.
10. B. Stupfel, Thèse de 3e cycle, Strasbourg, 1980.
11. G. Moraitis, B. Stupfel and F. Gautier, to be published.
12. M.A. Khan, J.C. Parlebas and C. Demangeat, Phil. Mag. B 42 : 111 (1980).
13. M.A. Khan, G. Moraitis, J.C. Parlebas and C. Demangeat in : 11th Int. Symp. on Elec. Struc. of Metals and Alloys, P. Ziesche ed., Gaussig, 50 (1981).
14. J. Khalifeh, G. Moraitis and C. Demangeat, J. Physique 43 : 165 (1982).
15. J. Khalifeh, G. Moraitis and C. Demangeat, J. Less Common Metals (1982) ; under press.
16. J. Khalifeh, D. Sc. Thesis, Strasbourg (1982).
17. P.H. Dederichs and J. Deutz, in Continuum Models of Discrete Systems, University of Waterloo Press, 329 (1980).
18. P.H. Dederichs and J. Deutz, Berichte der K.F.A. Jülich 16 00 (1979).
19. I. Kramer, Diplomarbeit in Physik, Aachen (1980).

RELAXATION, ELECTRONIC AND VOLUME EXPANSION CONTRIBUTIONS TO THE ELASTIC CONSTANT CHANGES INDUCED BY HYDROGEN AND DEUTERIUM IN Nb

F.M. Mazzolai* and H.K. Birnbaum

University of Illinois at Urbana-Champaign
Urbana, IL 61801

ABSTRACT

The high frequency moduli and acoustic attenuation have been measured in Nb-H(D) alloys over a wide range of H/Nb and temperatures. An anelastic relaxation has been investigated, whose strength depends on T according to a Curie-Weiss type of relationship. This relaxation effect only partly accounts for the softening of the elastic constants C' and C induced by hydrogen or deuterium. The moduli are also affected by a positive electronic contribution and a negative contribution due to the volume expansion.

INTRODUCTION

Acoustic techniques have been extensively used to investigate the elastic constants of the Nb-, V-, and Ta-H (D) systems (1-8). It has been found that the rigidity modulus G as well as the shear constant C' decrease, while C and the bulk modulus B increase as a consequence of H(D) absorption. The decrease in C' has been interpreted by some authors (2,5,7) as evidence for a Snoek relaxation effect due to H. However, the modulus defects ΔG and $\Delta C'$ do not show (1,3) the temperature dependence expected from the theory of an anelastic relaxation effect. This observation has led a second group of authors to the conclusion (1,3,4) that a Snoek effect due to H would have a relaxation strength at least

---

*Permanent address: Istituto di Acustica O.M. Corbino V. Cassia I216, Rome-Italy.

two orders of magnitude smaller than that associated with heavier interstitials (O,N and C), and cannot account for the observed decrease in C' and G.

In the present note we discuss the changes in elastic constants of Nb caused by H(D) and show that these can be divided into contributions associated with an anelastic relaxation, the volume change caused by solution of H(D) and changes of electronic nature, which have been proven to be of great relevance in the case of Nb-Zr (9) and Nb-Mo (10) alloys.

EXPERIMENTAL

A Nb single crystal having pairs of faces parallel to (100) or (110) planes was outgassed in UHV $\approx 10^{-10}$ torr, (at $\approx$ 2100 C for $\approx$ 10 h). Two x-cut and one y-cut 10 MHz transducers were bonded to the crystal faces to allow simultaneous determination of C', C_L and C_{11}, by a pulse echo overlap technique. In the calculations use has been made of the following relationships;

$$C'=1/2(C_{11}-C_{12})=(\rho v_T^2) \begin{array}{l} [1\bar{1}0] \\ [110] \end{array} \tag{1}$$

$$C_L=1/2(C_{11}+C_{12}+2C_{44})=(\rho v_L^2)[110] \tag{2}$$

$$C_{11}=(\rho v_L^2)[100] \tag{3}$$

where the upper and lower indexes indicate the polarization and the propagation directions, respectively. The density ρ, was measured by a hydrostatic balance technique and was equal to 8.588 (g/cm^3). In the calculations, dimensional changes due to H absorption or temperature variation were taken into account. The specimen surfaces were activated to allow in situ absorption of H(D) and the transducer bonds allowed unconstrained expansion of the specimen due to H(D) absorption or temperature changes.

RESULTS

Figures 1, 2 and 3 show the temperature dependence of the as measured moduli (and as will be discussed shortly, correspond to fully relaxed moduli) at various H contents. The moduli C'_R and C_{11}^R decrease with increasing atomic ratio, n=H/Nb, and show a gradually increasing softening as the temperature is reduced. A similar effect at the composition n=0.78 has previously been reported (11).

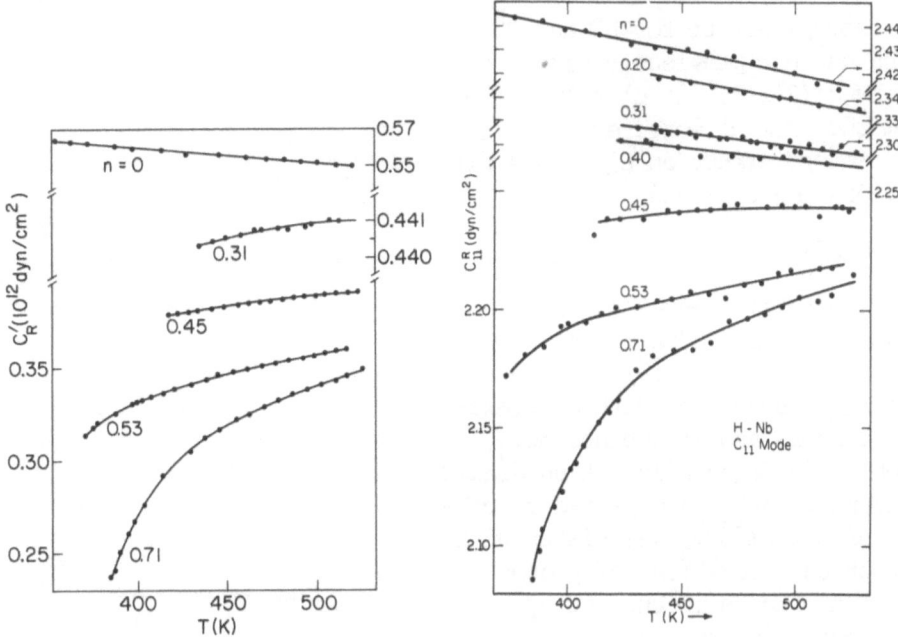

Figures 1 and 2 - Temperature dependence of the relaxed elastic
constants C' and C_{11} at various H contents
(n=H/Nb).

In a contrast to this behavior a marked increase with n is shown by C_L^R at each temperature. C_L^R linearly decreases with T over the entire temperature range of the present measurements for n ≤ 0.45. At the higher values of n softening of the C_L^R modulus is observed at the lower temperatures. Similar measurements carried out on Nb-D alloys show a clear isotopic effect as seen for C_R', B^R and C_{44}^R in Fig. 4, where values measured at 528K and normalized to the values for pure Nb are reported. B^R and C_{44}^R have been calculated from the measured quantities according to the relationships

$$B^R = C_{11}^R - (4/3)C_R' \qquad (4)$$

$$C_{44}^R = C_L^R + C_R' - C_{11}^R \qquad (5)$$

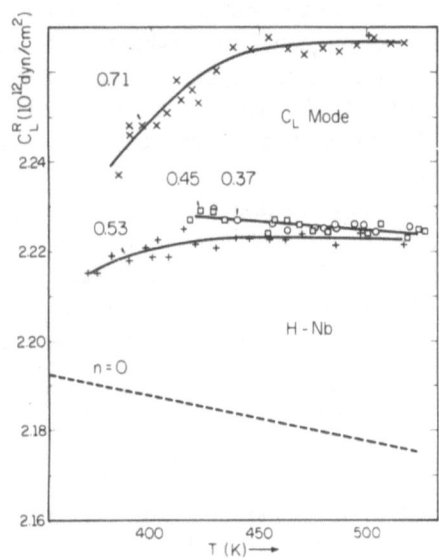

Figure 3 - Temperature dependence
of the relaxed elastic
constant C_L at various
H contents (n=H/Nb).

The isotope effect for B^R and C^R
which has not previously been
observed (2,3,7), is clearly evident
from the present measurements while
the isotope effect on C'_R agrees with
previous measurements.

DISCUSSION

The total changes in the elastic
constants caused by alloying Nb with
H(D), $\Delta C_{R_{ij}}$, can be divided into a
part ΔC_{ij}^R associated with a relaxation process and in another ΔC_{ij}^U,
which is the change in the unrelaxed
constants. ΔC_{ij}^U is expected to consist of a term ΔC_{ij}^E associated with
the optical vibrations of protons
and with changes in the energy of
the electron states and a term ΔC_{ij}^V
associated with changes in the volume
and in the density of states of
acoustic phonons. The electronic
contribution to ΔC_{ij}^U is not expected
to be strongly dependent on T (12)
in the range of the present measurements (390-530K). The optical
phonon contributions are expected
to be relatively insensitive to
temperature and to exhibit only small
changes on alloying with H(D) in view

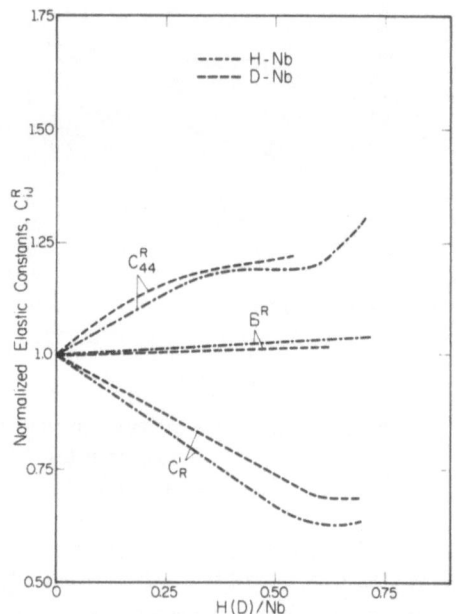

Figure 4 - Isotopic effect due
to H and D at 528K.

of the slight changes in the phonon spectrum induced by H. Thus the
temperature dependence of ΔC_{ij} is expected to essentially derive
from ΔC_{ij}^R according to the following relation

$$\Delta C_{ij}(n,T) = C_{ij}^R(n,T) - C_{ij}^R(0,T) = \Delta C_{ij}^R(n,T) + \Delta C_{ij}^U(n,T) =$$

$$\Delta_o/(T-T_{C_{ij}}) + \Delta C_{ij}^E - (\Delta V/V)B_T^R \, \partial C_{ij}/\partial P|_T \qquad (6)$$

where ΔC_{ij}^R and ΔC_{ij}^V are given in an explicit form. $C_{ij}^R(n,T)$
and $C_{ij}^R(o,T)$ are the measured elastic constants at H/Nb=n and of pure
Nb, respectively. B_T^R is the isothermal bulk modulus, $\Delta V/V$
the relative volume change associated with solution of H, $T_{C_{ij}}$ is a
critical temperature for self ordering, which depends on
the acoustic mode and $\partial C_{ij}/\partial P|_T$ is the pressure derivative of the
moduli.

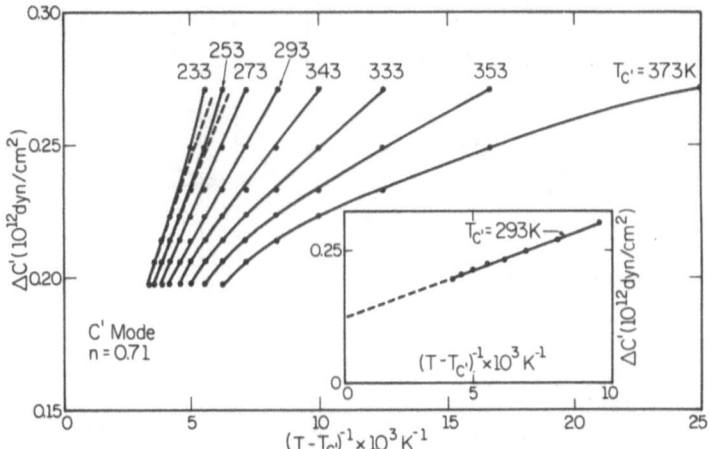

Figure 5 —Total change in the relaxed elastic constant C'_R as a function of $(T-T_{C'})^{-1}$ for various $T_{C'}$ values.

In Fig. 5 the quantity $\Delta C'$ has been plotted against $(T-T_{C'})^{-1}$, where $T_{C'}$ has been used as a fitting parameter. A linear relation is obtained for $T_{C'}=293K$ with the intercept equal to ΔC_{ij} in agreement with Eqn. 6.

The present analysis contains the implicit assumption that the temperature dependence of $C_{ij}(n)$ would be the same as that of $C_{ij}(o)$ in the absence of the relaxation process. While this assumption is only an approximation in view of the concentration dependence of the thermal expansion coefficient (13) and of possible contributions to the temperature dependence by optical phonons associated with H(D), it is more adequate than that used in a previous note (8), where C'_R was linearized, rather than $\Delta C'_R$.

While the source of the relaxation process cannot be discussed at this time, the concentration and temperature dependences of ΔC^R_{ij} have been determined from plots analogous to that of Fig. 5 for C', C_L, C_{44} and B. These results strongly suggest that the relaxation process is of the Zener type, such as occurs in the β phase of the Pd hydride (14) rather than a Snoek relaxation. It was shown (8) that under the conditions of the present experiments $\omega\tau \ll 1$ and the measured moduli correspond to the fully relaxed moduli. A full discussion of these data will be given elsewhere.

Fig. 6 shows ΔC^U_{ij} as calculated using Eqn. 6. For the C' mode preliminary results obtained with deuterium are also

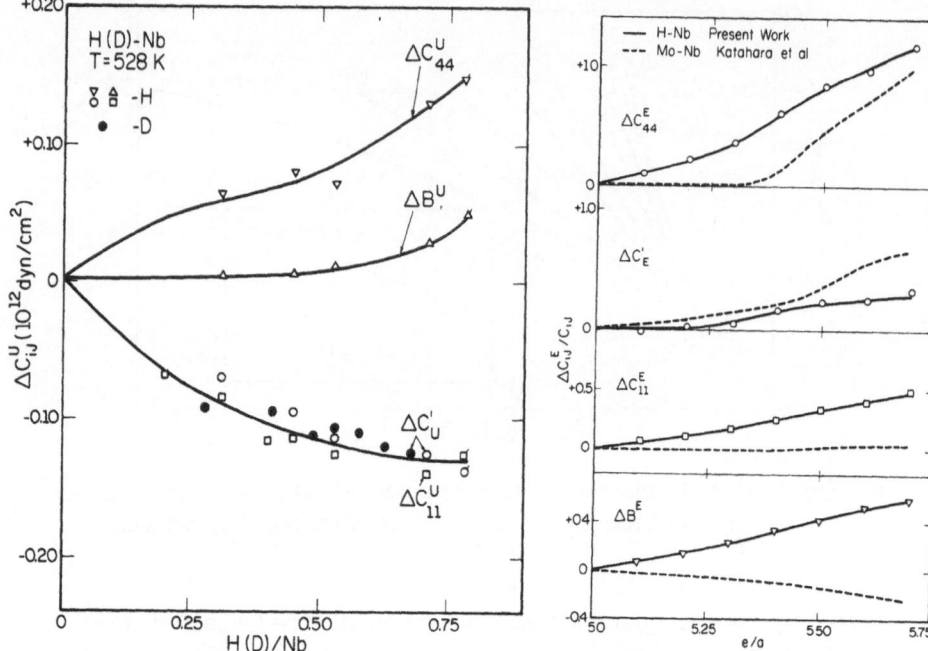

Figure 6 - Change in the relaxed elastic constants as a function of the H content: $\Delta C_{ij}^U(n) = C_{ij}^U(n) - C_{ij}^U(0)$.

Figure 7 - Change in the electronic contribution to the moduli with the electron to atom ratio. Data of Ref. (10) have been corrected for the effect of volume changes.

reported. The most interesting features are the isotope independence of $\Delta C_U'$ and the large negative values of $\Delta C_U'$ and ΔC_{11}^U. This later result shows that a large fraction of the mode softening on alloying with H or D is not relaxational in nature, contrary to previous suggestions (1,5,7).

The values of ΔC_{ij}^U derived above can be further analyzed using the relations given in Eqn. 6, $\Delta V/V$ from x-ray data (13) and $\partial C_{ij}/\partial P|_T$ from the values measured in the Nb-Mo alloys (10). The last assumption is justified by the fact that $\partial C_{ij}/\partial P|_T$ seems to be affected primarily by the number (e/a) of conduction electrons/atom (10). In our case e/a is the ratio between the cumulative number of electrons of Nb and H atoms and the number of Nb atoms. The results of such calculations are plotted in Fig. 7 where data for Nb-Mo, taken from reference (10) and corrected for volume contraction which occurs on alloying Nb with Mo, are also plotted for comparison. It can be seen that the electronic plus protonic contribution associated with H shows trends similar to those of the electronic contribution associated with Mo, at least for C', C_{44} and C_{11}. A relatively larger effect is observed for H in Nb than for alloying with Mo.

Concluding, the main results can be summarized as follows: 1) The decrease in C' and C_{11} due to alloying with H(D) is partly associated with a Zener relaxation process; 2) isotope effects exist for C_{44} and B as well as for C' and C_{11}; 3) significant changes in the moduli can be identified as due to increases in the volume and to changes in the electronic structure on alloying with H(D).

ACKNOWLEDGEMENTS

This work was supported by the Department of Energy contract DE-AC02-76ER01108.

References

1) H.C. Buchholz, J. Völkl and J. Alefeld, Phys. Rev. Lett. 30: 318 (1973).
2) E.S. Fisher, D.G. Westlake and S.T. Ockers, Phys. Stat. Sol. (a). 28: 591 (1975).
3) A. Magerl, B. Berre and G. Alefeld, Phys. Stat. Sol. (1). 36: 161 (1976).
4) W.L. Stewart III, J.M. Roberts and N.G. Alexandropolous, J. Appl. Physics. 48: 75 (1977).
5) E.S. Fisher, J.F. Miller, H.L. Alberts and D.G. Westlake, J. Phys. F: Metal Phys. 11: 1557 (1981).
6) H.L. Alberts, E.S. Fisher, K.W. Katahara and M.H. Manghnani, J. of Physics: Metal Phys. 9: L209 (1979).
7) E.S. Fisher and J.F. Remark, J. Appl. Phys. 51: 927 (1980).
8) F.M. Mazzolai and H.K. Birnbaum, Seventh International Conference on Internal Friction and Ultrasonic Attenuation in Solids, Les E'ditions de Physique, Lausanne (1981).
9) J. Ashkenazi, M. Dacorogna, M. Peter, Y. Talmor and E. Walker, Inst. of Phys. Conf. Ser. 39: 695 (1978).
10) K.W. Katahara, M.H. Manghnani and E.S. Fisher, J. Phys. F: Metal Phys. 9: 773 (1979).
11) M. Amano and H.K. Birnbaum, Internal friction and ultrasonic attenuation in solids, University of Tokyo Press (1977.
12) E. Walker and M. Peter, J. Appl. Phys. 48: 2820 (1977).
13) H. Zabel and J. Peisl, J. Phys. F: Metal Phys. 9: 1467 (1979).
14) F.M. Mazzolai, P.G. Bordoni and F.A. Lewis, J. Phys. F: Metal Phys. 11: 337 (1981).

LOW TEMPERATURE ULTRASONIC ATTENUATION IN RAPIDLY COOLED NIOBIUM

CONTAINING OXYGEN AND HYDROGEN*

K.F. Huang[+], A.V. Granato[+], and H.K. Birnbaum[≠]

Physics Department[+], Metallurgy Department[≠] and
Materials Research Laboratory
University of Illinois, Urbana, Illinois 61801

ABSTRACT

Measurements of ultrasonic attenuation and velocity in dilute Nb-O-H alloys were made as a function of temperature, frequency, polarization, annealing temperature, isotope, and defect concentration. In addition to the stable 2.4 K relaxation peak found earlier by Poker, et. al., an additional peak at 6.3 K at 10 MHz was found when the specimen was rapidly cooled to He temperature. The peak correlated with the resistivity recovery found by Hanada for quenched specimens and corresponds to a complex with tetragonal symmetry.

INTRODUCTION

Measurements by Sellers, Anderson and Birnbaum[1,2] of a large isotope effect below 1 K in the specific heat of niobium containing hydrogen showed the existence of tunnelling effects for H in Nb. A model proposed by Birnbaum and Flynn[3] showed that this behavior could be described by a delocalized hydrogen defect undergoing tunnelling between interstitial sites. The low temperature solubility limit of hydrogen in niobium is quite small, but ultrasonic measurements at 150 K and 10 MHz[4], by Mattas and Birnbaum[5], as well as resistance measurements (Pfeiffer and Wipf[6]) show that interstitial oxygen or nitrogen traps hydrogen. The latter authors found that only one hydrogen atom traps at any one nitrogen interstitial, with any excess hydrogen precipitating. A

133

binding energy of 0.12 eV was found, consistent with the result of Baker and Birnbaum.

The specific heat measurements were redone by Morkel, Wipf and Neumaier[7], who found that no excess heat capacity was present in nitrogen free samples. They concluded that the excess heat capacity could be accounted for by assuming that the hydrogen traps into a localized site near a nitrogen interstitial without undergoing delocalized tunnelling. They estimated that a concentration of about 3000 ppm of nitrogen, which was used as the trapping defect, could cause splittings as large as 10^{-3}eV, more than the amount needed to lift the degeneracy of different orientations of the N-H complex enough to account for the heat capacity effect.

Ultrasonic measurements on Nb-O-H alloys by Poker, Setser, Granato and Birnbaum[8] showed an attenuation peak near 2.5 K at 10 MHz with a large isotope effect. The defect responsible for the peak was identified as hydrogen trapped at an oxygen interstitial in a complex with tetragonal symmetry. The location of the peak was not sensitive to the oxygen concentration, suggesting that the strain field effects of neighboring oxygen atoms do not play a dominant role. An elaboration of the Birnbaum-Flynn model to take account of the trapping interstitial was proposed to account for the relaxation. In this bound delocalized model, only the lower states were supposed to be occupied at the measurement temperature.

Quenching experiments on Nb with dilute concentrations of O and H by Hanada[9] showed a resistivity increase at low temperature, indicating that hydrogen was frozen into the lattice. The annealing data also showed that most of the quenched in resistivity annealed out between 4.2 K and 100 K. Two major recovery stages were found, one centered at 40 K and the other at 80 K. The 80 K stage was observed only in the specimen with high oxygen content. It was interpreted as the detrapping of hydrogen freed from oxygen interstitials. However, since Pfeiffer and Wipf had shown that the O-H pair was rather stable at 80 K, it was suggested by Hanada that the quenched in defects were O-H complexes with n = 2 ~ 4.

To explain their inelastic neutron scattering results, Wipf, et al.[10] gave a model for the effect of strain by neighboring oxygen atoms which was more specific than that[7] given earlier. They supposed that a hydrogen atom trapped near an oxygen atom would tunnel between two equivalent interstitial sites whose energies differ randomly by a shift resulting from strain-induced interaction effects between different O-H pairs. While such a model can be fit to the neutron scattering and specific heat results, it would predict no elastic constant relaxation for low concentrations of oxygen, contrary to the results of Poker, et al.

RESULTS

In the present work, measurements of ultrasonic attenuation and velocity in dilute Nb-O-H alloys were made as a function of temperature, frequency, polarization, annealing temperature, isotope, and defect concentration. In addition to the stable 2.5 K relaxation peak (peak 1) found by Poker, et. al., an additional low temperature isotope sensitive peak (peak 2) was found when the specimen was rapidly cooled to He temperature. Peak 2 is larger than the peak at 2.5 K, and anneals out at the same temperature, both for H and D, as those found by Hanada for quenched specimens. As with the 2.5 K peak, the defect causing peak 2 has tetragonal symmetry, and its position is not sensitive to the 0 concentration at low concentrations.

The annealing experiments consisted of a series of linear temperature ramps (3 K/min) from He temperature to a chosen anneal point, followed by a linear ramp back to He temperature. The elastic wave mode which measures the $C' = (C_{11} - C_{12})/2$ elastic constant (E_{2g} symmetry strain) decrement and annealing behavior as a function of temperature is shown in Fig. 1.

The decrement is plotted versus temperature for a 10 MHz C' mode in a niobium crystal containing about 100 ppm 0 and 700 ppm H after a rapid cool down from 240 K, followed by annealing at 70, 80, 90, 100 and 120 K. The decrement is made up of a temperature independent background due to bonding and diffraction losses, an electronic contribution porportional to the electrical conductivity which decreases below the superconducting transition temperature, and two peaks due to hydrogen. The peak at 6.3 K anneals out near 80 K, while the peak at 2.5 K increases only a small amount. Both peaks were observed in the C' mode, but not to the C_{44} mode (T_{2g} symmetry) indicating a tetragonal symmetry. For deuterium peak 1 was buried in peak 2, which appeared near 10.5 K with a high temperature shoulder.

Figure 2 shows the decrement versus temperature for three niobium specimens with differing oxygen concentrations after rapid cooling from 240 K. The hydrogen concentration is much greater than the 0 concentration. The background has been subtracted in this figure. The peak heights scale with the 0 concentration while the electronic contribution decreases with increasing impurity content. The background attenuation at the Nb superconducting critical temperature shows a near-discontinuity which is sensitive to the 0 content. In the purest sample (1 week UHV anneal at 2500 K), the discontinuity drop was large but peak 2 was small. However, in the sample with about 200 ppm 0, the drop was hardly observed and peak 2 was large. After unloading the hydrogen in the specimen, a small discontinuity was found. The superconducting effect in the background attentuation serves as a monitor for the impurity content.

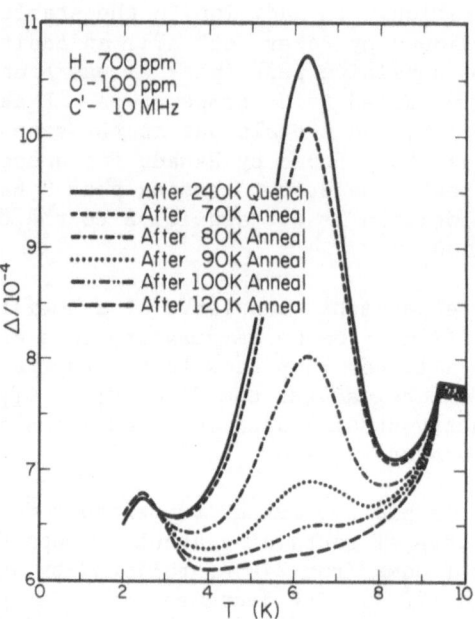

Fig. 1. Decrement Δ versus temperature T for a 10 MHz C'
mode in a niobium crystal containing about 100 ppm 0 and
700 ppm H after a rapid cool down from 240 K, followed by
annealing at 70, 80, 90, 100 and 120 K.

With the previous resistivity experiments mentioned above
and the ultrasonic results described here, it can be concluded
that peak 2 at 6.3 K arises from an OH complex. The simplest
interpretation is that it represents an OH_2 complex. In a slow
cooling process, one oxygen traps only one hydrogen and this OH
pair is rather stable. In a rapid cooling process, some of the
oxygen traps more than one hydrogen, forming OH_2. However, the
second hydrogen is less strongly bound than the first. During
annealing, detrapping of one of the two hydrogens occurs and an OH_2
becomes an OH with the freed hydrogen going into precipitates.

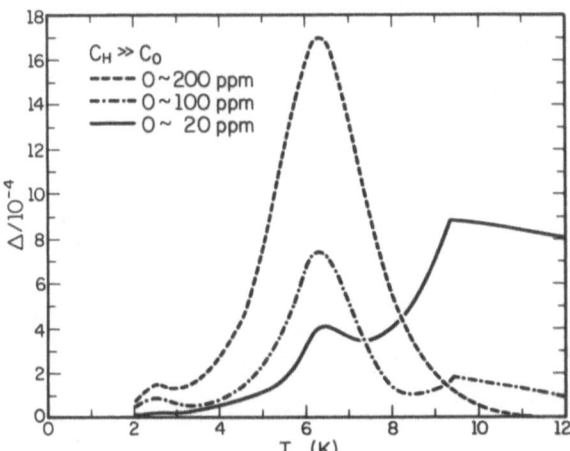

Fig. 2. Decrement versus temperature for three niobium specimens with differing oxygen concentrations after rapid cooling from 240 K.

ACKNOWLEDGEMENT

    This work is supported by the Department of Energy under contract DE-AC02-76ER01198.

REFERENCES

1.   G. J. Sellers, A. C. Anderson, and H, K. Birnbaum (1973),
     Phys. Lett. 44A, 173.
2.   G. J. Sellers, A. C. Anderson, and H. K. Birnbaum (1974),
     Phys. Rev. B10, 2771.
3.   H. K. Birnbaum and C. P. Flynn (1976), Phys. Rev. Lett. 37,
     25.
4.   C. Baker and H. K. Birnbaum (1973), Act. Metall. 21, 865.
5.   R. F. Mattas and H. K, Birnbaum (1975), Acta. Metall. 23, 973.
6.   G. Pfeiffer and H. Wipf (1976), J. Phys. F6, 167.
7.   C. Morkel, H. Wipf and K. Neumaier (1978), Phys. Rev. Lett.
     40, 947.
8.   D. B. Poker, G. G. Setser, A. V. Granato and H. K, Birnbaum
     (1979) Zeitschrift fur Physikalische Chemie, 260, 636.

9.   R. Hanada (1977), Proc. 2nd Internat. Congress on Hydrogen
     in Metals, Paris (Oxford: Pergammon), Vol. 3, p. 1136.
10. H. Wipf, A. Magerl, S. M. Shapiro, S. K. Satija and W. Thom-
     linson (1981), Phys. Rev. Lett. 46, 947.

ULTRASONIC ATTENUATION MEASUREMENTS IN SINGLE CRYSTAL PdH$_{.65}$

T. Kanashiro,[*] D. K. Hsu, R. G. Leisure and P. C. Riedi[†]

Department of Physics
Colorado State University
Fort Collins, Colorado 80523

ABSTRACT

Ultrasonic attenuation measurements are reported for single-crystal PdH$_{.65}$ over the temperature range 80-300 K and the frequency range 10-150 MHz. At 10 MHz the C_L and C' modes show peaks at approximately 190 K while the C_{44} mode shows a peak at approximately 180 K. These peaks are interpreted as relaxational attenuation due to stress induced ordering of vacancies on the fcc H lattice. The peaks are too broad to be accounted for by a single relaxation process.

INTRODUCTION

There have been several studies[1] of elastic energy dissipation (internal friction) in palladium hydrogen alloys. The internal friction Q^{-1} is usually interpreted in terms of a Debye-like response,

$$Q^{-1} = \Delta\omega\tau/(1 + \omega^2\tau^2) \tag{1}$$

where $\omega/2\pi$ is the vibration frequency, τ is a relaxation time associated with the movement of hydrogen in the lattice and Δ characterizes the magnitude of the effect. Δ is usually taken to depend on temperature, in the simplest case as T^{-1} and in other cases as $(T - T_c)^{-1}$ where T_c is a self-ordering temperature. It is usually assumed that the relaxation is due to a thermally activated process so that $\tau = \tau_0\exp(W/kT)$ where W is the activation energy. Due to the site symmetry, an isolated H interstitial will not give rise to relaxation effects. The relaxation effects

139

occurring in the bulk, undamaged lattice are then attributed to H pairs or higher order clusters (Zener relaxation), or vacancy pairs[2] and clusters.

While low frequency internal friction methods have received extensive use for the study of atomic motion in hydrides and other alloys, high frequency pulse propagation methods have received relatively little use. For alloys other than hydrides the high frequencies involved often mean that the relaxation peaks occur at high, inconvenient temperatures. The situation is different with the hydrides. The high mobility of hydrogen means that the peaks occur at convenient temperatures, even at MHz frequencies. With the ultrasonic method there is the interesting possibility of making measurements with different acoustic modes on the <u>same</u> sample. For example, with a cubic single crystal oriented along the [110] axis three independent acoustic modes may be propagated. In general both Δ and τ will be different for different modes. The various values of τ and Δ are associated with various relaxation normal modes in the crystal.[3]

We present here ultrasonic attenuation measurements on a single crystal of $PdH_{.65}$ over the frequency range of 10–150 MHz and the temperature range of 80–300 K. The hydride single crystal was prepared from a single crystal of palladium (obtained from Materials Research Corporation) by charging from the gas phase. The temperature and pressure were controlled so as to avoid the mixed $\alpha+\beta$ phase region and thus avoid the damage associated with passage through this region. The axis of the cylindrical sample was oriented along the [110] crystalline direction which was also the direction of propagation of the ultrasonic waves. The longitudinal, C_L, and two transverse modes, C_{44} and $C' = 1/2 (C_{11} - C_{12})$, were studied. The attenuation was measured by monitoring the heights of two echoes in the usual way. The attenuation coefficient is related to the internal friction by

$$\alpha = (\omega/2v)Q^{-1} \tag{2}$$

where α is the amplitude attenuation in Np-cm^{-1}, and ω and v are, respectively, the angular frequency and velocity of the ultrasonic waves.

RESULTS

Experimental data for the three independent modes are given in Figs. 1–3 for a frequency of 10.8 MHz. The longitudinal and slow transverse (C') modes have somewhat similar shapes and each shows a maximum at about 190 K. The fast transverse (C_{44}) has a somewhat different shape and peaks at about 180 K. The magnitudes of the attenuation are quite different in the three cases.

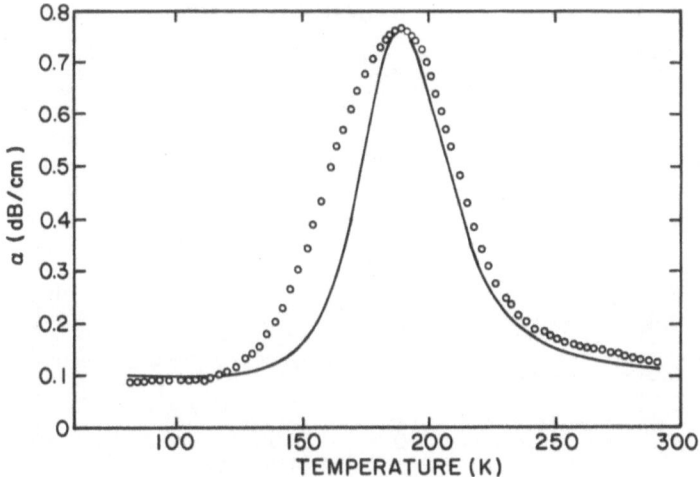

Fig. 1. Attenuation of the C_L mode at 10.8 MHz.

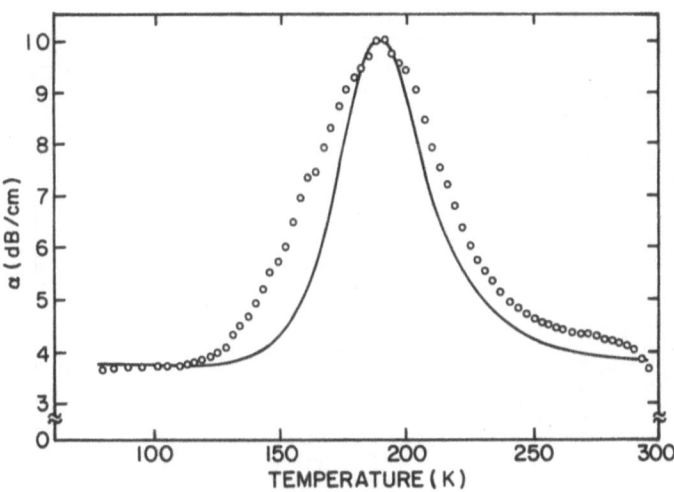

Fig. 2. Attenuation of the C' mode at 10.8 MHz.

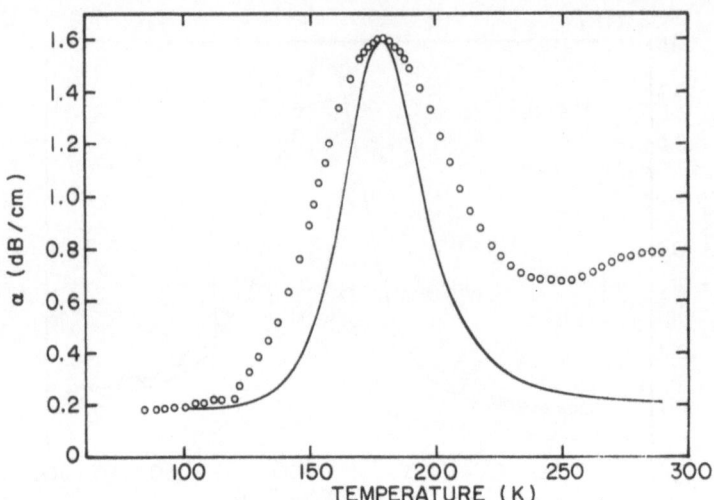

Fig. 3.   Attenuation of the C_{44} mode at 10.8 MHz.

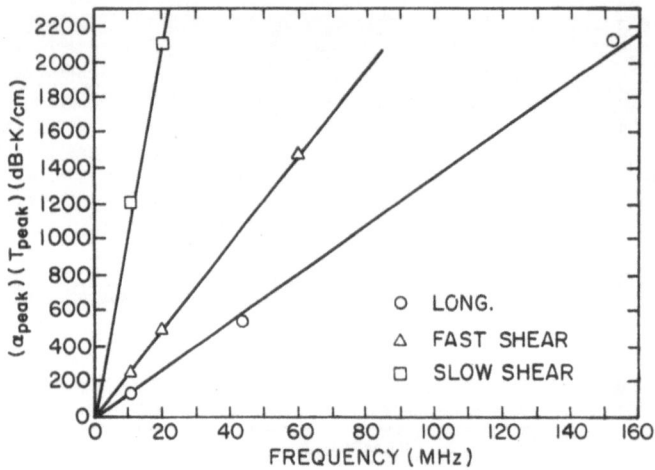

Fig. 4.   Peak attenuation times peak temperature versus frequency.

Measurements were also made at higher frequencies. The maximum
in the attenuation shifted to higher temperatures with increasing
frequency, as expected for a thermally activated process. Some
of the higher frequency data are presented in Fig. 4. The peak
attenuation values have been multiplied by the temperature at
which the peak occurred for reasons which will be discussed below.

DISCUSSION

Using the data in Figs. 1-3, Eqs. (1) and (2), and the known
sound velocities,[4] we can calculate the parameter Δ which corresponds to Q^{-1}_{max}. The results for the three modes are $\Delta_{C_L} = 2.2 \times$
10^{-3}, $\Delta_{C_{44}} = 2.4 \times 10^{-3}$, and $\Delta_{C'} = 6.8 \times 10^{-3}$. From the internal
friction data of Mazzolai[1] et. al. we have $2Q^{-1}_{max} = \Delta = 10.6 \times$
10^{-3} for H concentrations similar to ours. If we assume a T^{-1}
dependence to correct their results (at $T = 114$ K) to 190 K we
find $\Delta = 6.4 \times 10^{-3}$, essentially the same as our results for $\Delta_{C'}$.
If instead of a T^{-1} dependence we use a $(T-T_c)^{-1}$ dependence with
$T_c = 50$ K we find $\Delta = 4.8 \times 10^{-3}$. As their measurements were for
polycrystalline bars vibrated flexurally, it seems likely that
their results will reflect contributions from all modes. The
most heavily damped mode would probably dominate, consistent with
the above results.

Arrhenius plots of our results at various frequencies yield
activation energies of approximately 0.20 eV for each mode. Using
this value for W we calculate the attenuation expected for each
mode for a single relaxation process. The results are shown as
the solid lines in Figs. 1-3. It can be seen that the experimental peaks are in every case wider than the calculated ones.
Such a situation is usually taken as evidence for a spectrum of
relaxation times.

From Eqs. (1) and (2), we expect the peak attenuation to
vary as

$$\alpha_p = \omega\Delta/4v \ . \tag{3}$$

Because Δ is expected to vary as T^{-1}, or possibly $(T - T_c)^{-1}$, we
checked Eq. (3) by plotting the peak attenuation times the peak
temperature versus frequency. The results are shown in Fig. 4.
In all three cases the data are well-described by a straight
line through the origin as expected for a relaxation process.
(Because of the rather narrow temperature range covered, the
results would not be significantly changed if we used a $T_c \simeq 50$ K
for this analysis.)

One of the most interesting features of the data is the fact
that the peak for C_{44} occurs at a lower temperature than the peaks

for the other modes. If exactly the same activation energy is
involved for all modes, then the implication is that the attempt
frequency for the C_{44} relaxation is <u>higher</u> than for the other
modes. Actually a consideration of the reorientation rates of an
isolated pair of interstitial defects in the fcc lattice leads
to the opposite conclusion. The C_{44} mode should couple to relaxation modes having a <u>lower</u> attempt frequency than the relaxation
modes associated with C'. This result, while based on an over-
simplified model, suggests that the C_{44} relaxation may occur at a
lower temperature because of a lower activation energy. There is
some indication from our data that such is indeed the case,
although more accurate measurements are needed to decide the
issue.

The authors wish to acknowledge helpful discussions with
Dr. J.-Y. Prieur. This research was supported by the National
Science Foundation under Grant DMR-8012689.

*Permanent Address: Department of Physics, Tokushima University,
        Tokushima, JAPAN.
†Permanent Address: Department of Physics, St. Andrews University,
        St. Andrews, SCOTLAND.

REFERENCES

1. See F. M. Mazzolai, P. G. Bordoni, and F. A. Lewis, J. Phys.
        F <u>11</u>, 337 (1981), and references therein.
2. F. M. Mazzolai, M. Nuovo, and F. A. Lewis, Il Nuovo Cimento
        <u>33B</u>, 242 (1976).
3. A. D. Franklin, J. Res. Nat. Bur. Standards <u>67A</u>, 291 (1963).
4. D. K. Hsu and R. G. Leisure, Phys. Rev. B <u>20</u>, 1339 (1979).

CONTRIBUTION OF OPTICAL PHONONS TO THE ELASTIC MODULI OF PdH$_x$ AND PdD$_x$

B.M. Geerken[o], R. Griessen[o], L.M. Huisman[o] and E. Walker[+]

[o] Vrije Universiteit Amsterdam, The Netherlands
[+] Université de Genève, Switzerland

ABSTRACT

Sound velocity measurements carried out between 10 and 300K on α'-phase PdH$_x$ and PdD$_x$ show that the temperature coefficients dC_{ij}/dT of the elastic moduli C_{ij} are significantly more negative in PdD$_x$ than in PdH$_x$. This isotope effect is well described by a quasi-harmonic model in which the transverse and longitudinal optical phonons are treated as Einstein oscillators with different Grüneisen parameters γ_t and γ_ℓ. The Einstein temperatures are $\theta_t^H = 650K$, $\theta_\ell^H = 910K$ in PdH$_{0.66}$, $\theta_t^D = 450K$ and $\theta_\ell^D = 640K$ in PdD$_{0.652}$. Our analysis implies that $|\gamma_\ell - \gamma_t|$ is large compared to the average optical phonon Grüneisen parameter $\bar{\gamma} = (\gamma_\ell + 2\gamma_t)/3$ determined from thermal expansion measurements. For C_L, C' and C_{44} a maximum in ultrasonic absorption is found around 220K. It is interpreted as arising from a reorientation of pairs (or clusters) of vacancies in the hydrogen sublattice under uniaxial or shear strains.

INTRODUCTION

Around roomtemperature, a significant isotope dependent contribution to the electrical resistivity[1,2] and the thermal expansion[3] of PdH$_x$(D$_x$) has been observed that arises from the excitation of optical phonons. In order to obtain information on the average frequency of optical phonons and their volume dependence we have measured the temperature dependence of the elastic moduli C_{ij} of PdH$_x$(D$_x$) using ultrasonic techniques.

EXPERIMENTAL PROCEDURE

Two ultrasonic methods are used in this work. The *absolute* value of the elastic moduli is determined at room- and/or liquid nitrogen temperature by means of a continuous wave technique. The frequency of the ultrasonic wave is varied around 30MHz and the elastic moduli are determined with a precision of typically 1% from the frequency intervals between consecutive resonances of standing waves. The *relative* changes of the elastic moduli with temperature are measured by comparing the phase of a short pulse which has been reflected several times at the sample boundaries with that of the excitation signal. The accuracy is typically 0.02%. The different waves are generated by 10MHz X- and Y-cut quartz transducers with glycerol used as bonding agent.

The alloying of single crystal palladium cylinders with hydrogen (deuterium) into the α'-phase is carried out by absorption from the gas phase in a high pressure cell at temperatures $\sim 25°C$ above the critical temperature in order to avoid segregation in α- and α'-phases.

EXPERIMENTAL RESULTS

Figs. 1, 2 and 3 show the temperature dependence of C_L, the tetragonal shear modulus C' and the angular shear modulus C_{44} for pure palladium and various PdH_x and PdD_x alloys. The temperature dependence of the bulk modulus $B = C_L - C_{44} - \frac{1}{3} C'$ is shown in Fig. 4. All elastic moduli of Pd and the C_L and B of $PdH_{0.66}$ agree within about 1% with those of Hsu and Leisure[4] . The deviations of the C_{44} and C' of $PdH_{0.66}$ are slightly larger. The temperature coefficients dC_{ij}/dT of the elastic moduli C_{ij} of $PdH_x(D_x)$ are isotope dependent and significantly more negative than that of Pd.

For the *amplitude* of the ultrasonic signal a weak minimum around 220K is found for the longitudinal mode both in PdH_x and PdD_x. For the shear modes a reduction of the signal is observed around 150K and the amplitude remains very low up to room temperature.

DISCUSSION OF EXPERIMENTAL RESULTS

The temperature dependence of the amplitude of the ultrasonic signal mentioned above can be explained by assuming that the ultrasonic strains induce a reorientation of pairs (or clusters) of vacancies in the hydrogen (deuterium) lattice. According to standard theory of anelasticity the attenuation of the sound wave should be a maximum when $\omega\tau \simeq 1$ where ω is the angular frequency of the wave and τ^{-1} the jump frequency of the interstitials. Since τ^{-1} is equal to $6D_r/b^2$ (where b is the jump length and D_r the reduced diffusion constant) the temperature T_m for which maximum attenuation is observed is determined by the conditions $D_r(T_m) = \frac{b^2\omega}{6}$.

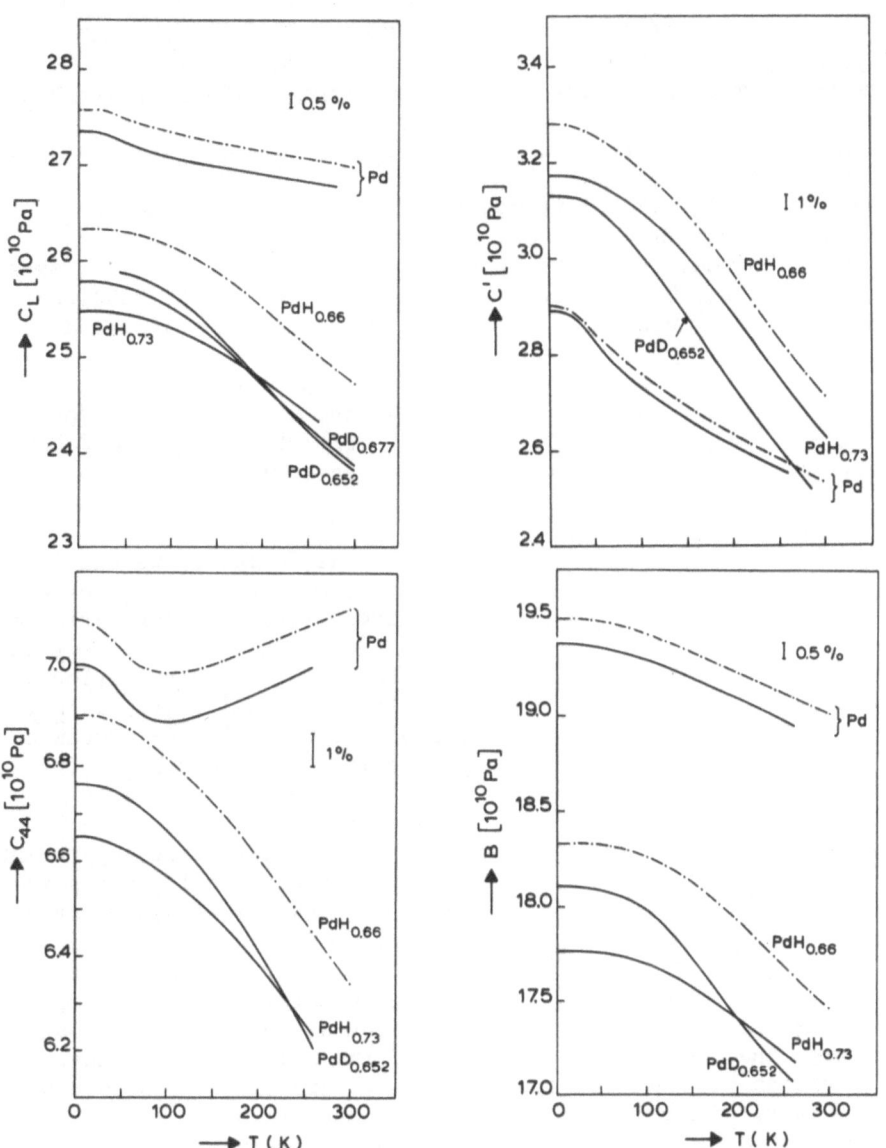

Fig. 1-4. Temperature dependence of C_L, C', C_{44} and B in Pd, Pd$_{0.73}$ and PdD$_{0.652}$.
The estimated absolute error is indicated in the upper right corner. The relative error for the temperature dependence itself is at least one order smaller.

Using the diffusion measurements of ref. 5 for D(T), b = 2,87Å and ω = 30MHz one obtains T$_m$ = 220K in excellent agreement with our experimental data.

Since internal friction peaks[6] are often associated with elastic moduli defects it would be natural at this point to attribute the large differences between dB/dT in Pd and in $PdH_{0.66}$ (and $PdD_{0.652}$) to some relaxation process in the sample. The corresponding modulus defect would then approximately be given by

$$\delta B_{H(D)}(T) = B_{Pd}(T) - B_{PdH_x(D_x)}(T) - \left[B_{Pd}(0) - B_{PdH_x(D_x)}(0) \right] \qquad (1)$$

and from $T_m(PdH_x) \simeq T_m(PdD_x)$ it would follow that $\delta B_H(T)$ should be proportional to $\delta B_D(T)$. However, as shown in Fig. 5 quite a different behaviour is observed, the PdD_x curve being *shifted* towards lower temperatures by a factor of ~ 1.45 relative to the PdH_x curve. Such a large *shift* cannot be understood within a relaxation model.

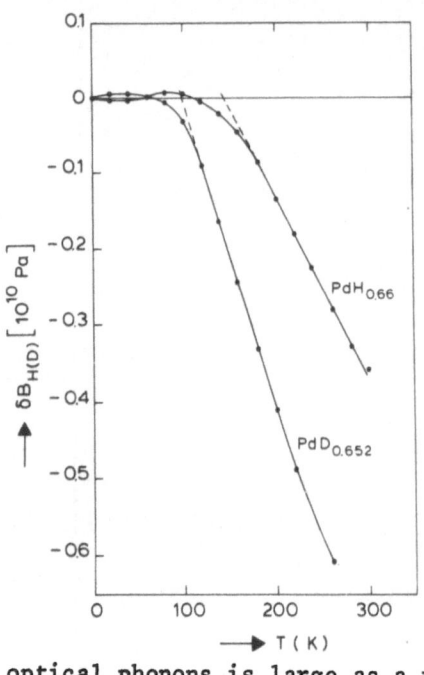

Fig. 5   Temperature variation of the bulk moduli differences δB for $PdH_{0.66}$ and $PdD_{0.652}$ according to Eq.1.

In the following we shall therefore assume that, in spite of the large ultrasonic absorption in C_{44} and C' (and to some extent in C_L), the temperature variation of the elastic moduli $C_{ij}(T)$ is <u>not</u> influenced by a stress-induced rearrangement of H ions but originates from the excitation of optical phonons.

Neutron scattering experiments show that the optical phonons in PdH_x and PdD_x are well separated in energy from the top of the acoustic phonon branches. The dispersion of the longitudinal optical phonons is large as a result of the strong proton-proton interaction and the non-stoichiometry of the alloys. The transverse optical phonon branches, on the other hand, are rather narrow. The phonon density of states has therefore a peak at 650K in PdH_x (450K in PdD_x) with a high energy shoulder whose top reaches about 1100K in PdH_x (750K in PdD_x). The position of the density of states peak is approximately independent of the concentration of interstitials.

THEORETICAL MODEL

In ultrasonic experiments such as those carried out in this work one measures the adiabatic elastic moduli C_{ij}, defined by

$$C_{ij} = \frac{1}{V} \left(\frac{\partial^2 U}{\partial \varepsilon_i \, \partial \varepsilon_j} \right)_S , \qquad (2)$$

where $\{\varepsilon_i\}$ is the sixcomponent strainvector, U the internal energy given by $U=U_0(\varepsilon_i)+\Sigma_q(n_q + \frac{1}{2})\hbar\omega_q$ with $n_q=(\exp(\beta\hbar\omega_q)-1)^{-1}$ and $\beta=(k_BT)^{-1}$ and S the entropy. The summations are over all the phonon states and $\hbar\omega_q$ is the energy of a phonon with wavevector q. The frozen lattice energy U_0 depends only on the strainvector $\{\varepsilon_i\}$.

We assume that the ω_q are functions of the strain vector $\{\varepsilon_i\}$, with $\partial\ln\hbar\omega_q/\partial\varepsilon_i = -\gamma_q^i$ where γ_q^i is the Grüneisen parameter associated with the C_{ij}.

Under adiabatic conditions, the elastic moduli C_{ij} can be written as

$$C_{ij} = \frac{1}{V}\,\partial^2 U_0/\partial\varepsilon_i\partial\varepsilon_j + \frac{1}{V}\,\Sigma_q(n_q+\tfrac{1}{2})\partial^2\hbar\omega_q/\partial\varepsilon_i\partial\varepsilon_j$$

$$-1/V\beta\,\Sigma_q(\beta\hbar\omega_q)^2 n_q(n_q+1)\,(\gamma_q^i-\gamma^i)\,(\gamma_q^j-\gamma^j) \qquad (3)$$

where γ^i is a weighted average of the γ^i's. The first term on the right hand side of this equation depends implicitly on the phonons through the zero point motion and thermal expansion[7]: $\frac{1}{V}\,\partial^2 U_0/\partial\varepsilon_i\partial\varepsilon_j = C_{ij}^0 + (V(T)-(V_0)\,\partial/\partial V\,C_{ij}^0 = C_{ij}^0 + C_{ij}^0/B_0V_0\partial\ln C_{ij}^0/\partial\ln V\ (n_q+\tfrac{1}{2})\hbar\omega_q\gamma_q\gamma(4)$ where V_0 and B_0 are the volume and the bulk modules of the crystal at T = OK and in the absence of zero point motion and γ_q is the volume Grüneisen constant.

One often makes the additional assumption[8] that the occupation numbers n_q themselves do not depend on the strain state of the crystal. This is equivalent to assuming that all the Grüneisen constants γ_q^i or γ_q are equal to their average (γ^i or γ respectively), and consequently vanish for pure shears. Also, the terms proportional to $n_q(n_q+1)$ vanish identically when the γ_q are equal. These terms describe the redistribution in the occupancy of the phonon states[9] due to the relative shifts of the phonon energies, as is shown by the fact that these terms depend only on the *deviation* of the Grüneisen constants from their average value.

We now separate the elastic moduli in contributions from the frozen lattice (C_{ij}^0), the acoustic phonons and the optical phonons. For the latter contribution, C_{ij}^{opt}, we assume that the optical phonon dispersion curves can be described by two Einstein oscillators with frequencies ω_t and ω_ℓ. We cannot assume that there are only two optical shear Grüneisen constants, one for the longitudinal modes and one for the transverse modes, because the sum of γ_q^i over the star of q has to vanish for a pure shear[10]. We can assume, however, that there are only two *volume* Grüneisen constants, denoted by γ_ℓ and γ_t.

FIT TO THE EXPERIMENTAL DATA

As the phonon dispersion curves of Pd are not uniformly deformed on alloying with H(D) we consider the difference $\Delta C_{ij}(T)=C_{ij}(T,PdH_x)-C_{ij}(T,PdD_x)-(C_{ij}(0,PdH_x)-C_{ij}(0,PdD_x))$ which are independent of the acoustic modes. The ΔC_{ij} obtained from the experimental data in Figs.2,3 and 4 are shown in Fig. 6. For all three moduli ΔC_{ij} is very small at low temperature although $\Delta C'(T)$ increases much more rapidly with temperature than ΔC_{44} and reaches a maximum around 200K. ΔC_{44} on the other hand is a smoothly increasing function of T.

As shown in Fig. 6 the experimental moduli differences can accurately be fitted by setting $\Delta C_{ij}(T)=C_{ij}^{opt}(PdH_x)-C_{ij}^{opt}(PdD_x)$ (5) and by using the following values for the optical phonon energies in PdH_x ($\hbar\omega_t^H$ = 56 meV, $\hbar\omega_\ell^H$ = 78 meV) and in PdD_x ($\hbar\omega_t^D$ = 39 meV, $\hbar\omega_\ell^D$ = 55 meV).

In the case of the bulk modulus, $\Delta B(T)$ depends essentially on $(\gamma_\ell-\gamma_t)^2$. From a fit to the experimental curve one obtains $|\gamma_\ell-\gamma_t|$ = 12.3. On the basis of existing data it is not possible to decide whether $\gamma_\ell-\gamma_t$ is positive or negative. We are presently carrying out thermal expansion experiments using high sensitivity capacitance dilatometric techniques to determine the temperature variation of the average Grüneisen $\bar{\gamma}$ = $\frac{1}{3}(\gamma_\ell+2\gamma_t)$. According to the quasiharmonic model described above, $\bar{\gamma}$ should *decrease (increase)* with temperature when $\gamma_\ell-\gamma_t<0$ (>0).

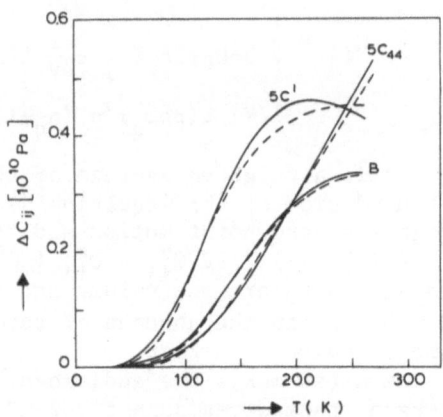

Fig. 6. Elastic moduli differences $C_{ij}(T)$ according to Eq.5. The experimental data are represented by full lines and the fits are given by dashed lines.

ACKNOWLEDGEMENT

We are grateful to the "Stichting voor Fundamenteel Onderzoek der Materie" for financial support of this work.

REFERENCES

1.   D.S. MacLachlan, R. Mailfert, J.P. Burger and B. Souffaché, Solid State Commun. 17, 281 (1975).
2.   A. Gorska, A.M. Gorski, J. Igalson, A.J. Pindor and L. Sniadower, Proc. H in metals conference, Paris 1977.
3.   R. Abbenseth and H. Wipf, J. Phys. F10, 353 (1980).
4.   D.K. Hsu and R.G. Leisure, Phys. Rev. B20, 1339 (1979).
5.   F.M. Mazzolai and H. Züchner, Z. Phys. Chem. N.F. 124, 59 (1981).
6.   G.J. Zimmermann, J. Less Common Metals 49, 49 (1976).
7.   J.A. Garber and A.V. Granato, Phys. Rev. B11, 3990 (1975).
8.   G. Leibfried and W. Ludwig, Solid State Phys. 12, 275 (1961).
9.   T.H.K. Barron, Phys. Rev. 137A, 487 (1965).
10. A.M. Gray, D.M. Gray and E. Brown, Phys. Rev. B11, 1475 (1975).

NEUTRON SPECTROSCOPIC STUDIES OF DILUTE HYDROGEN IN PURE AND

DEFECTED METALS

J. J. Rush

National Measurement Laboratory
National Bureau of Standards

ABSTRACT

A review is presented of recent work at NBS on the study of vibration spectra of hydrogen isotopes bound at low concentration ($\gtrsim 1\%$) in both pure metals and metals containing interstitial and substitutional traps. Results will be discussed for neutron spectra of Nb, Ta, V, and Pd-hydrogen systems, including preliminary work on palladium and titanium alloys and studies of isotope dependence. Spectroscopic results for hydrogen trapped by O, N, and Ti impurities will be discussed in detail and compared to other systems which do not exhibit trapped states at low temperature. The implications of these results in terms of models for the local environment of trapping sites will be considered. Very recent neutron data on the concentration dependence of local modes in α- and β-phase palladium hydrides and deuterides will also be presented.

STRUCTURE AND PHASE TRANSITIONS IN

V_2D AND V_2H

S. C. Moss

Physics Department
University of Houston
Houston, TX   77004

ABSTRACT

The structure and phase transitions in V_2D and V_2H are reviewed. Prior neutron studies of site occupancy and phase transitions are compared with more recent x-ray studies in which the
interstitial ordering is accompanied by modulated host atom displacements. This has permitted x-ray scattering to be used as a
sensitive probe of both the ordered and disordered states. We suggest that the transfer from disordered tetrahedral to ordered octahedral occupancy occurs through a [1/2 1/2 0] modulation wave
which slides, on cooling, from local tetrahedral to local octahedral modulation. New results on the second order phase transition
in V_2H suggest that it is an excellent example of an Ising-like
transition modified by the presence of a nearby tricritical point.

INTRODUCTION

The hydrides and deuterides of vanadium are especially attractive for combined x-ray and neutron scattering studies. This is
due to the fact that x-rays scatter neglibibly from the interstitial hydrogen or deuterium while neutrons see an appreciable coherent cross section for deuterium and hydrogen (as well as a very
large incoherent hydrogen cross section) but scatter coherently to
an essentially neglible degree from vanadium. X-rays (or electrons)
see only the vanadium atoms while neutrons see mainly the interstitial hydrogen or deuterium atoms in a <u>coherent</u> scattering experiment. The structure(s) of the hydrides βV_2D and βV_2H have therefore been extensively studied by neutrons on powder samples by
Somenkov and co-workers[1], Westlake, Mueller and Knott[2], Asano and
Hirabayashi[3,4] and Asano, Abe and Hirabayashi[5]. These studies

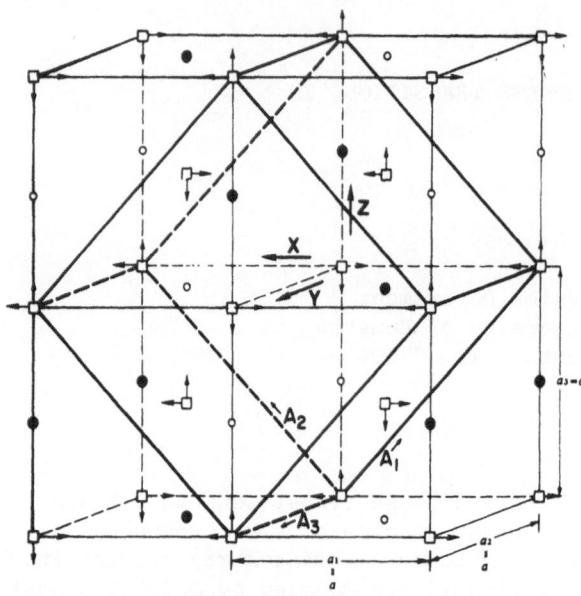

Fig. 1.   Structure of ordered $\beta V_2 D$ ($V_2 H$) with z axis (μ_z) and
          x axis (μ_x) displacements: $\mu_z = 0.15 \pm .02$ Å;
          $\mu_x = 0.03 \pm .01$ Å; □ V; ● filled O_{z1} sites; o empty
          O_{z2} sites.   (from Ref. 12).

have shown that the low temperature ordered phases of both $V_2 D$ and
$V_2 H$ are monoclinic with the D(H) atoms occupying one set of octahedral sites (O_{z1}) along the z axis of the pseudo-tetragonal cell
whose c/a = 1.1.[21] The second set of z-axis octahedral sites,
labelled O_{z2} by Asano et al.[3-5], is ideally empty at T = 0° K and
exact stoichiometry.

As the phase diagrams of both Asano and Hirabayashi[4] and
Schober and Wenzl[6] show, there is a large isotope effect in the
phase transitions of vanadium hydride and deuteride attributed by
Entin et al.[7] to zero point vibrations. At $V_2 D$ the phase diagram
indicates a first order transition from a disordered (α) bcc interstitial solid solution to the ordered β phase at T_c = 406°K. $V_2 H$
has an upper first order transition from the disordered α phase
to a tetragonal β' or ε phase at T_{c2} = 471°K. This ε phase has

random occupancy of the two sets of O_z sites. Below T_c = $441^{\circ}K$
(the precise assignment of critical temperatures depends on concentration), the tetragonal ε phase orders into the monoclinic β
phase with selected occupation of O_{z1} sites. The neutron studies
on this sequence of phase transitions has most recently been reviewed by Hirabayashi[8] who discusses both powder and single crystal
results.

The question of site occupancy in these ordered and disordered
phases is of great interest. At high temperature in the α phase,
early neutron work of Somenkov et al.[1] and Reidinger[9] indicated
through an analysis of Bragg peaks from ~ VD_5 (Reidinger analyzed
several bcc V-D alloys) that the nuclear density peaked strongly
at the tetrahedral sites. In the ordered βV_2H Rush and Flotow[10]
observed, via incoherent neutron scattering, characteristic octahedral symmetry and strength for the hydrogen modes. This octahedral assignment was confirmed by crystallographic studies[1-5].
More recently Hirabayashi[8], as noted above, reported on the various
site occupancies determined by neutron scattering as a function of
temperature in V_2D and V_2H. The former shows mostly O_{z1}, in the
ordered state which disorders to random tetrahedral (T) and slight
O_{z2} occupation. At T_c the site occupancy appears to switch to
mostly tetrahedral and slight octahedral in the bcc α phase. In
V_2H, O_{z1} and O_{z2} come together at T_{c1} with a small random T occupancy. In the ε phase, average or random O_z and random T switch
slightly and at T_{c2} the occupancy, as in V_2D, is said to jump to
mainly T sites (~90%). This neutron assignment however is not entirely unambiguous and we will return to this question in analyzing
the available x-ray data.

X-RAY STUDIES OF V_2D

Single crystal x-ray work on these phase transitions was first
initiated by Metzger et al.[11] who demonstrated that the ordering
in V_2D was accompanied by host-atom (vanadium) displacements.
These ordered displacements produced a separate set of x-ray modulation or superlattice peaks and permitted a detailed study of the
phase transition in which the thermodynamic order was more clearly
established as first-order. Jo et al.[12] refined this study verify-
ing the first-order character of the phase transition and determining the precise values of the vanadium displacements about an octa-
hedral D atom. These are μ_z = 0.15 $\pm$ .02 $\overset{\circ}{A}$ and μ_x = 0.03 $\pm$ .01 $\overset{\circ}{A}$.
Fig. 1 shows the resultant unit cell in both monoclinic and (pseudo)
tetragonal representations. This unit cell is appropriate, of
course, for both V_2D and V_2H in the low temperature β phase. In
as much as there is no isotope effect on the lattice parameters[4],
the displacements can be considered equal for the two compounds.

Fig. 2 from Metzer, Jo and Moss[13] compares the displacements
in V_2D (V_2H) with the displaements in Ta_2H in which the hydrogen

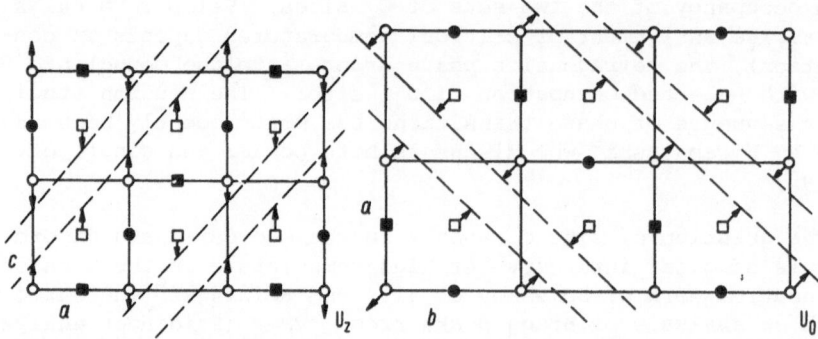

Fig. 2. Lattice structure in the β-phase a) V lattice projuected in
the (010) plane; O,□V at (000), (1/2 1/2 1/2); ●,■D at (001/2), (1/2
1/2 0). b) Ta lattice projected in the (001) plane; O,□Ta at (000),
(1/2 1/2 1/2); ●,■H at (0 1/2 3/4); (1/2 0 1/4) $u_0=\sqrt{2}u_x$ (Ref. 13).

atoms occupy tetrahedral sites. This comparison is interesting
because an ordering of both octahedral and tetrahedral sitings of
the H(D) atoms can give rise to the 110 modulation noted in Fig. 2.
However, the magnitude of the metal atom displacements is considerably smaller in Ta_2H where μ_0 in Fig. 2 is estimated by Petrunin et
al.[14] to be about .04 Å and by Asano et al.[15] to be ~.05 Å. This
is to be compared with the projected value of μ[110] = 0.13 Å for
V_2D. In addition $c/a(V_2D) = 1.1$ while $c/a(Ta_2D) = 1.01$. Clearly
the tetrahedral occupancy produces a similar ordered structure to

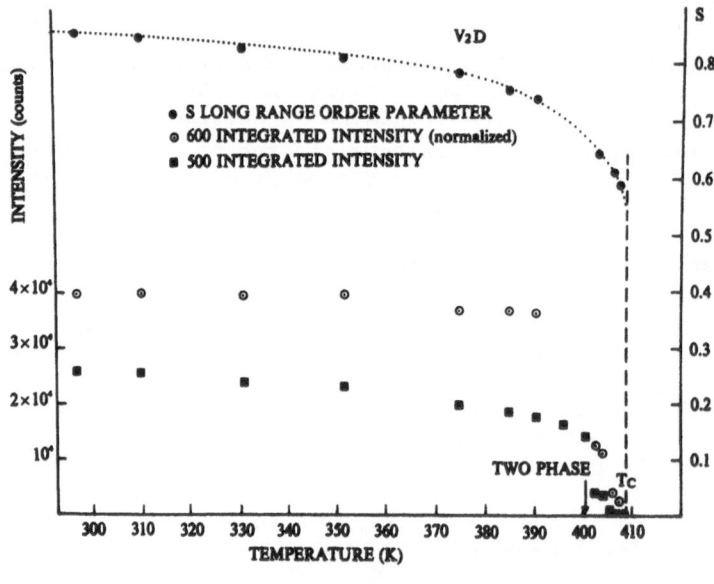

Fig. 3. Tem-
perature depen-
dence of the
600(O) funda-
mental and
500(■) super-
lattice inten-
sities, and the
long-range-
order parameter
S(●) in the
V_2D β phase.
(from Ref. 12).

the octahedral occupancy, including the [110] modulation, at an
apparently much smaller cost in overall elastic deformation.

Fig. 3 and Fig. 4 are from Jo, Moss and Westlake[12] and are
presented to indicate both the first order nature of the phase transition in V_2D and the success with which the lattice parameters can
be fit. The only inputs are the order parameter, the distortion
per tetrahedral site and the thermal expansion, all of which are
measured quantities. The calculated lattice parameters fail only
within about $10°K$ of the phase transition, T_c. This clearly can
be attributed to partial disordering on octahedral sites which
will not produce the same collapse of c/a as random tetrahedral
occupancy.

The order parameter, S, is defined[8,12] as:

$$P_1 - P_2 = S$$

where P_1 = fraction of O_{z1} sites filled

P_2 = fraction of O_{z2} sites filled

Since $P_T + P_1 + P_2 = 1.0$, where P_T is the fraction of randomly occupied tetrahedral sites, we can __assume__ in V_2D:

$$P_1 + P_T = 1 \quad P_2 = 0,$$

and this will produce the fit in Fig. 4 which is excellent up to
$T_c - 10°K$. At that point the disordering into alternate O_{z2} sites
becomes appreciable (about 10% or $P_2 = 0.1$) and the lattice parameters fall off more slowly.

We may then follow this tendency with x-rays into the α phase
in Fig. 5. It is clear from Jo and Moss[16] that, at $T_c + 7°K$ =
$413°K$ ($140°C$), the superstructure modulation remains above T_c. It
persists as a short range fluctuation above the first order transition but it is clearly there with, roughly, a full width at half
maximum corresponding to a correlation range of about 15-20 Å. As
is often the case above first order transitions, these correlated
fluctuations persist well above T_c and in this case remain at $200°C$
($473°K$). This observation seems particularly interesting in light
of the reasonably well-documented fact that above T_c in V_2D, or in
bcc V-D(H) alloys generally, the tetrahedral site is preferred[1,8,9].
It seems clear therefore that in these alloys there is an inclina-
tion towards a [110] zone boundary softness--that there is a tendency to deform with the $\vec{k}_2$(1/2 1/2 0) wave structure as defined
by Khachaturyan and Shatalov.[17] It is our interpretation[18] of
these results that the 1/2 1/2 0 instability, observed here in
V_2D (or V_2H), switches from a modulation of tetrahedral occupancy
to a modulation of octahedral occupancy as the temperature is

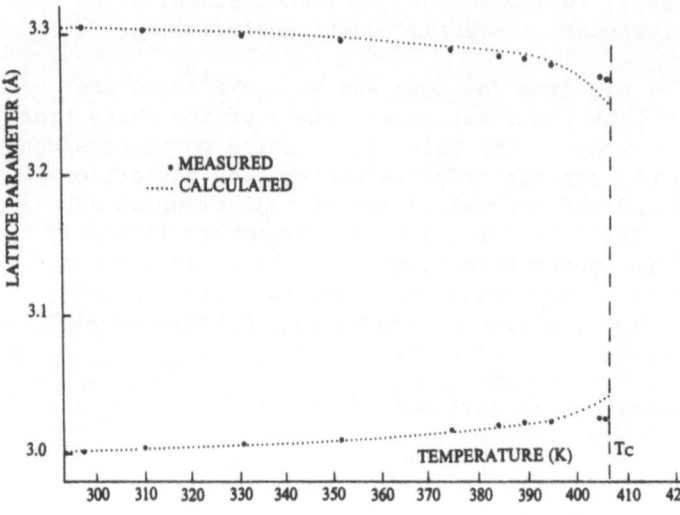

Fig. 4. Lattice parameters c and a in the pseudo-tetragonal $\beta V_2 D$ below T_c. Calculated values assume random T occupancy for the disordered D atoms. (see text). (from Ref. 12).

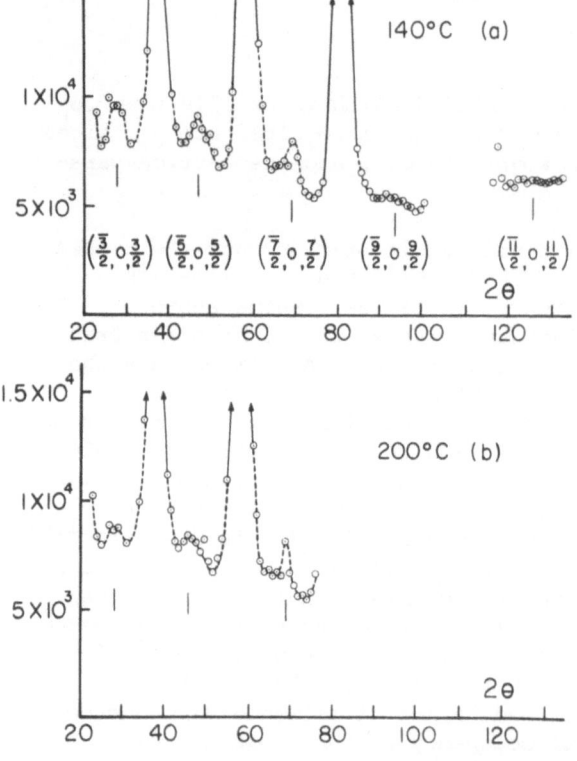

Fig. 5. Short-range order diffuse intensity in α phase V_2D. The measured counts are per 2000 sec. The data are uncorrected (raw) at (a) 140°C and (b) 200°C. The indices $(\bar{h}_3 o h_3)$ refer to the cubic superlattice positions where, for example, $\bar{3}/2$, 0, 3/2 is equivalent to 300 in β phase monoclinic notation. (from Ref. 16).

lowered. Another way of stating this is to note that in the
Fourier transform of the hydrogen-metal interaction energy
$V(k (1/2 \, 1/2 \, 0))$ is a minimum. This interaction energy arises in
the ordering case from short range elastic and electronic contributions into which enters the short wavelength deformation energy
of the bcc alloy. A general softness at $1/2 \, 1/2 \, 0$ for the hydrogen
solutions of the group Vb metals (V, Nb, Ta) may also be responsible
for the unusual ordering fluctuations reported by Burkel et al.[19]
over a wide range of concentration and temperature in H-Nb alloys.
The reason for the switch in vanadium from tetrahedral to octahedral still remains uncertain.

X-RAY STUDIES OF V₂H

In V₂H the sequence of phase transitions permits more elaborate
x-ray studies to be performed. We shall concentrate here on the
second order transition at T_{c1} = 441°K. Fig. 6 from Jo, Moss and
Westlake[20] shows the temperature variation of a superlattice modulation peak (the 300 monoclinic reflection = 3/2 0 3/2 tetragonal
reflection) compared with a 600 fundamental (monoclinic) reflection.
It is clear that the disordering is continuous and that there is
rounding of the phase transition at T_{c1} due to pronounced critical
scattering. Plotted also are the c and a parameters of the pseudo-
tetragonal cell. At T_{c2} there is a discontinuous jump in the lattice parameters while below T_{c1} there is only a continuous, albeit
rapid, change in ordering. It is clear that, in the ε (β') phase,
$P_2 = P_1$ and the modulation superstructure vanishes. According to
Hirabayashi[8], there is also a measureable occupancy of random T
sites which we are attempting to verify with x-rays. The change in
Fig. 6 is due both to the vanishing modulation structure and this
presumed T site population.

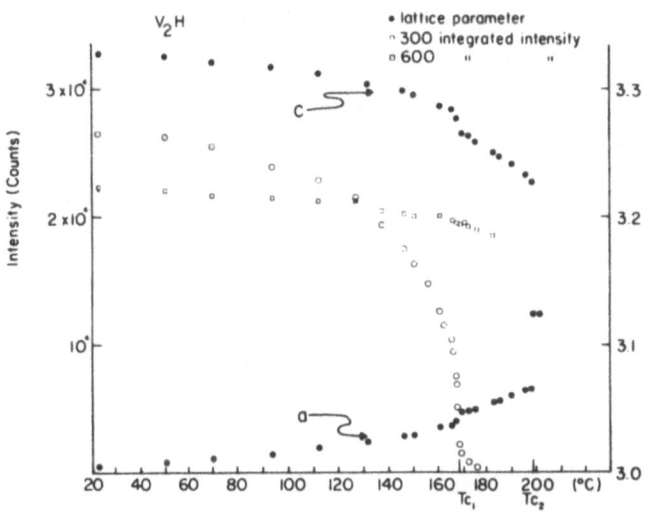

Fig. 6. Tempera-
ture dependence of
the BCT cell para-
meters and mono-
clinic 300 super-
lattice and 600
fundamental reflec-
tions in V₂H through
the β → ε
transition.
(from Ref. 20).

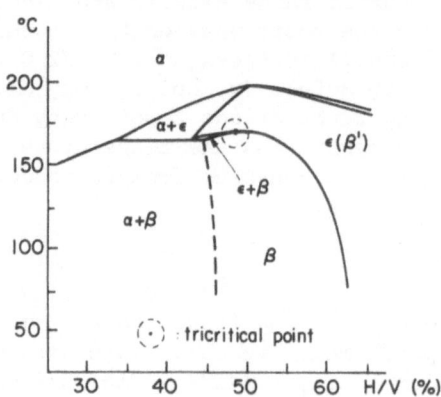

Fig. 7. A portion of the phase diagram of Pesch
et al.[21] reproduced to illustrate the phase transition region including the postulated tricritical
point.

In Fig. 7 is sketched a schematic view of the relevant portion of the up-dated phase diagram of Pesch, Schober and Wenzl[21]
drawn to emphasize the nearby tricritical point. Exactly at V_2H
there is the β-ε transition while at lower H concentrations the β
phase clearly must enter a two phase β+ε field on heating. This
proximity of a tricritical point is important in analyzing the
temperature dependence of the order parameter near $T_{cl} = T_c$.
Fig. 8 shows an expanded plot due to Jo[22], of the variation of the
300 monoclinic superlattice intensity versus thermocouple voltage
in mv. Included in this figure is a plot of the background scattering at 2° removed from the 300 reciprocal lattice point. The
peaking of the background critical scattering occurs roughtly at,
but not exactly at, the critical thermocouple voltage V_c (T_c). It
is clear that the rounding of the plot of the 300 peak count rate
versus temperature is due to intense critical scattering which
also peaks at the 300 position.

In the vicinity of T_c, it is well known[23] that the order
parameter in a second order phase transition varies as $S = A\frac{(Tc-T)}{Tc}^\beta$.

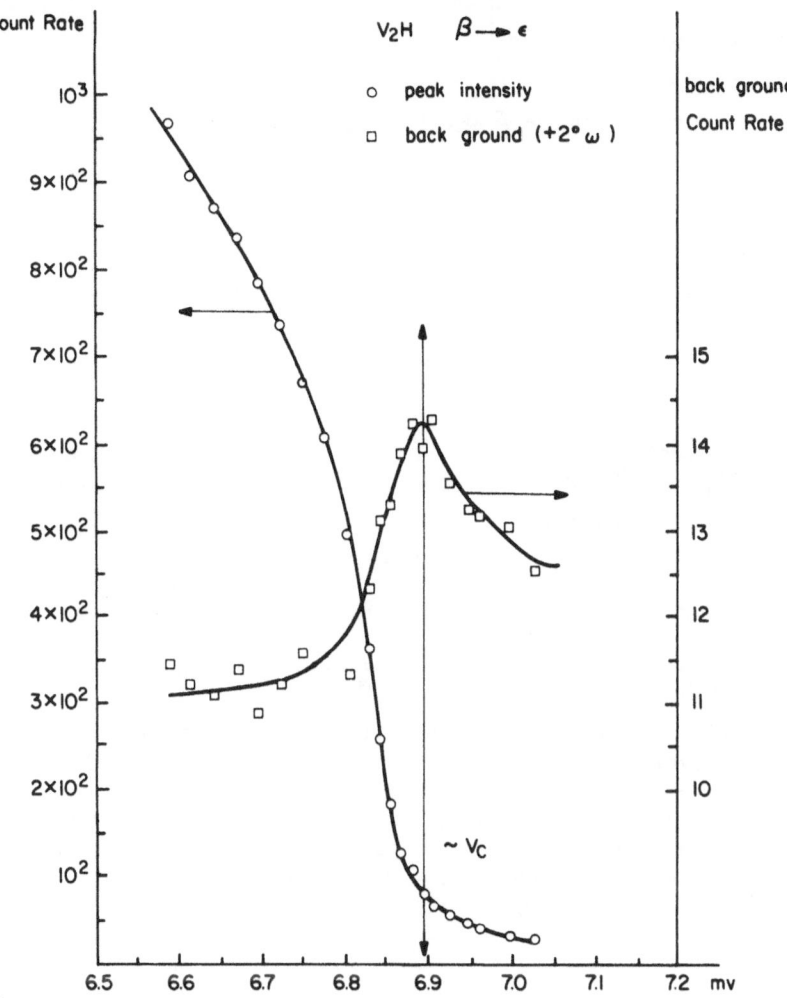

Fig. 8. The peak intensity of the 300 monoclinic superlattice reflection versus T through the β→ε transition. Included also is the background critical scattering 2° away from the 300 position. The critical thermocouple voltage, V_c, is indicated. (from H.S. U. Jo, unpublished results).

The range of validity for this expression is thought to be inside, say, 10^{-2} in $\Delta = T_c - T$. In Fig. 9 we show a log-log plot[22] of peak intensity over background vs. Δ in a least square fit (LSF) for β and T_c where I (peak) $\propto S^2$. If we take 10 points in the range indicated we have a good fit with β = 0.20 ± .01. Both the values of β are preliminary and more recent and detailed data on both the 300 and 500 integrated reflections by Schönfeld[24] suggest that β is closer to 0.16! What we are observing in this con-

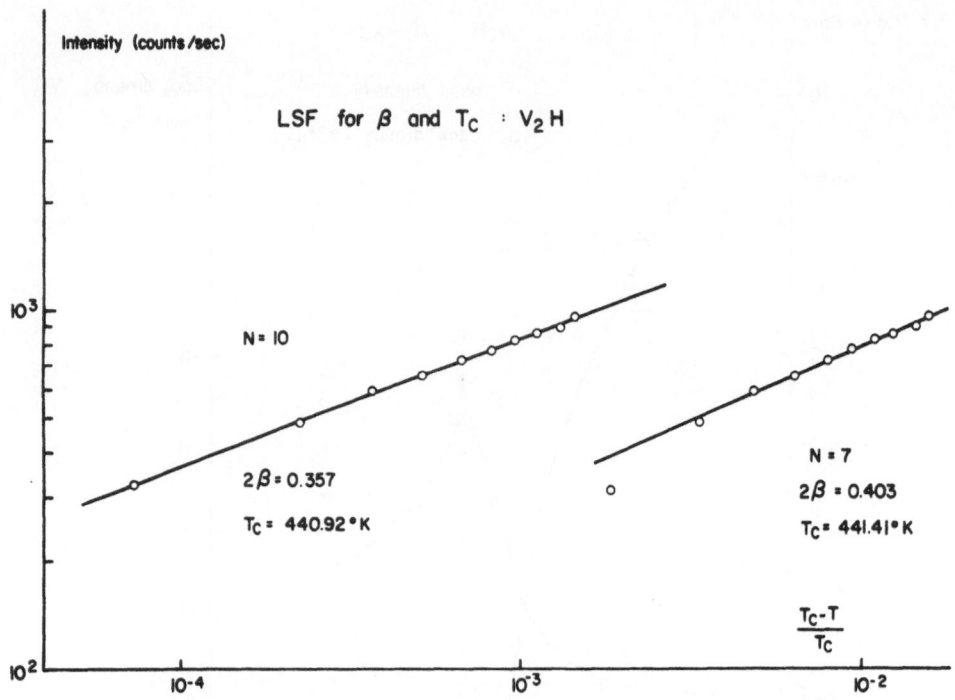

Fig. 9. Log-log plot of the superlattice intensity in
Fig. 8 versus reduced temperature $\Delta = T_c - T/T_c$. For
N = 10, the regime of critical scattering is included.
For N = 7, critical scattering is not so crucial but
the true critical regime is also excluded which yields
a somewhat larger exponent, β. (from H. S. U. Jo,
unpublished results).

tinuous disordering of the β phase V_2H is an excellent example of
a second order phase transition whose critical behavior is modified by a nearby tricritical point. In such cases one must be
especially careful to cross the second order line and not pass
through the 2-phase first order region. Checks on this include
evaluating T_c from above and below and looking carefully for
hysteresis. We have done both[24] and are convinced of the continuous nature of the transition in this sample. What is not clear is
over what range of T the tricritical regime leaves off and the exponents become representative of a true second order transition.
In the case of V_2H we are within 10^{-4} of T_c. The mean field value
of $\beta = 0.25$. This is most likely to be an upper limit and a $\beta =$
0.16 - .20 is not unreasonable especially when we consider the
effect of lattice contraction which usually lowers the observed
β[25].

A consistent set of exponents for this transition is currently being collected and evaluated[24]. It is clear that V_2H provides us with one of the most elegant examples of a second order (Ising-like) phase transition among alloys and the first example to our knowledge in an interstitial system.

ACKNOWLEDGEMENTS

The author acknowledges with pleasure the contribution and collaboration of H. Metzger, H. S. U. Jo, and D. G. Westlake. He also wishes to thank H. Zabel, M. Mori, H. Peisl, and K. Ohshima for many helpful discussions. Finally, he wishes to acknowledge the contribution of B. Schönfeld, who has currently been collecting and analyzing new V_2H data to extract a proper set of exponents from both above and below T_{cl}. This program of research has been supported entirely by the U. S. DOE, Division of Basic Energy Sciences, under contract no. DE-AS05-76ER05111.

REFERENCES

1. V. A. Somenkov, I. R. Entin, A. Yu Chervyakov, S. Sh. Shil'shtein, and A. A. Cherykov, Sov. Phys. Solid State 13, 2178 (1972).

2. D. G. Westlake, M. H. Mueller and H. W. Knott, J. Appl. Cryst. 6, 206 (1973).

3. H. Asano and M. Hirabayashi, Phys. Stat. Solidi A 15, 269 (1973).

4. H. Asano and M. Hirabayashi, Proc. Second Int'l. Cong. on Hydrogen in Metals, Paris (1977), paper 1D6.

5. H. Asano, Y. Abe and M. Hirabayashi, Acta Met. 24, 95 (1976); J. Phys. Soc. Japan 41, 974 (1976).

6. T. Schober and H. Wenzl in: "Hydrogen in Metals-II, ed. G. Alefeld and J. Völke, Topics in Appl. Phys. 28 (Springer-Verlag, Berlin, 1978) p. 77.

7. I. R. Entin, V. A. Somenkov and S. Sh. Shil'shetein, Sov. Phys. Solid State 16, 1569 (1975).

8. M. Hirabayashi, Proc. Jimis-2, Hydrogen in Metals, 1980 p. 49.

9. F. Reidinger, Brookhaven National Lab, unpublished results (1978).

10. J. J. Rush and H. E. Flotow, J. Chem. Phys. 48, 3795 (1968).

11. H. Metzger, H. Jo, S. C. Moss and D. G. Westlake, Phys. Stat. Solidi (a) 47, 631 (1978).

12. H. S. U. Jo, S. C. Moss and D. G. Westlake, J. Appl. Cryst. 13, 486 (1980).

13. H. Metzger, H. Jo, and S. C. Moss, Proc. of the Int'l. Meeting on Hydrogen in Metals, Münster 1979, Zeit Phys. Chem. 116, 87 (1979).

14. V. F. Petrunin, V. A. Somenkov, S. Sh. Shil'shtein and A. A. Chertkov, Sov. Phys. Crystallography 15, 137 (1970).

15. H. Asano, K. Kishi and M. Hirabayashi, Proc. Jimis-2, Hydrogen in Metals, 1980 p. 93.

16. H. S. U. Jo and S. C. Moss, Sol. St. Commun. <u>30</u>, 365 (1979).
17. A. G. Khachaturyan and G. A. Shatalov, Acta Met. <u>23</u>, 1089 (1975).
18. H. S. U. Jo and S. C. Moss, AIP Conf. Proc. No. 53 (Modulated Structures) p. 400 (1979).
19. E. Burkel, H. Behr, H. Metzger and J. Peisl, Phys. Rev. Lett. <u>46</u>, 1078 (1981); see also J. Peisl, this conference.
20. H. S. U. Jo, S. C. Moss and D. G. Westlake, AIP Conf. Proc. No. 53 (Modulated Structures) p. 403 (1979).
21. W. Pesch, T. Schober and H. Wenzl, Scripta Met. (1982) in press; T. Schober, this conference.
22. H. S. U. Jo, unpublished results (1979). (The publication of this data was postponed until a proper assessment of the order of the transition could be made; see ref. 24).
23. H. E. Stanley, <u>Introduction to Phase Transitions and Critical Phenomena</u>, New York, Oxford University Press (1971).
24. B. Schönfeld, research in progress; B. Schönfeld and S. C. Moss, Bull. Am. Phys. Soc. <u>27</u>, 162 (1982).
25. M. D. Rechtin, S. C. Moss and B. L. Averbach, Phys. Rev. Lett. <u>24</u>, 1485 (1970).

HYDROGEN VIBRATIONS IN NICKEL HYDRIDE

J. Eckert[*], C. F. Majkrzak and L. Passell[**] W. B. Daniels[+], and T. A. Kitchens[++]

Los Alamos National Laboratory[*], Brookhaven National Laboratory[**] University of Delaware[+], and Los Alamos National Laboratory[++]

ABSTRACT

Nickel hydride was prepared in a BeCu high pressure cell at room temperature by applying a hydrogen gas pressure of 7kbar to pellets pressed from fine nickel powder. The rate and degree of completion of the hydride formation was first checked by neutron diffraction using deuterium gas with the result that a deuterium to metal ratio of approximately 0.75 was reached. After desorption of the deuterium, the nickel sample was charged with hydrogen for the inelastic incoherent neutron scattering measurements. Measurements were performed with the Be filter technique at the Brookhaven High Flux Beam Reactor. The optic modes appear as a broad band in the phonon density of states from about 70 to 110 meV with peaks at approximately 88 and 108 meV. Included in the discussion of these results will also be the effect of hydride formation on the acoustic phonon density of states.

*Work supported by the U.S. Department of Energy, at Brookhaven under contract no. DE-AC02-76CH00016, and at the University of Delaware by the National Science Foundation.

165

LOCALIZED VIBRATIONS OF H IN Nb AND Ta AT HIGH

CONCENTRATION AND LOW TEMPERATURES

J. Eckert and J. A. Goldstone*, and D. Richter**

Los Alamos National Laboratory*, and
Kёrnforschungsanlage Jülich**

ABSTRACT

Extensive measurements have been performed of the optic mode
frequencies in the hydrides of Nb and Ta at concentrations between
70% and 95% using inelastic incoherent neutron scattering techniques
at the WNR pulsed neutron source of the Los Alamos National Laboratory. The aim of this study was to gain new information on anhar-
monicities of the H potential as well as to relate observations of
broading or splitting of the local mode peaks to details in the
high concentration, low temperature portion of the NbH_x phase diagram. The higher harmonic of the singlet excitation was observed
in all cases at energies about 3-5% less than the harmonic value.
For some cases a simultaneous excitation of the singlet and doublet
is apparent, which provides additional information on the hydrogen
potential when these vibrations are coupled by anharmonic interactions. A previously reported[1] splitting in the optic mode peaks
of $NbH_{0.87}$, which occurs at temperatures below about 210K, as well
as broadening of the optic mode peaks of $NbH_{0.78}$ and $NbD_{0.8}$ can
be discussed using new information on the NbH_x phase diagram in
terms of coexisting phases with differing hydrogen sites.

---

*Work supported in part by the Division of Basic Energy Sciences,
U.S. Department of Energy.
[1]J. Eckert, J. A. Goldstone, and D. Richter, Miami International
Symposium on Metal Hydrogen Systems, 13-15 April 1981.

TETRAHEDRAL AND OCTAHEDRAL SITE OCCUPANCY AND OPTIC MODE VIBRATIONS

MODE VIBRATIONS OF HYDROGEN IN YTTRIUM

J. A. Goldstone and J. Eckert,[*] E. L. Venturini and
P. M. Richards[**]

Los Alamos National Laboratory[*] and Sandia National
Laboratory[**]

ABSTRACT

The yttrium–hydrogen system exhibits both tetrahedral and octahedral site occupancy prior to reaching the dihydride concentration. Using the WNR pulsed neutron source at the Los Alamos National Laboratory, temperature dependence studies from 15K to 275K of the optic modes of $YH_{1.99}$ were made by inelastic incoherent neutron scattering. Room temperature neutron powder diffraction patterns were also taken. The powder data yielded a cell parameter for $YH_{1.99}$ of 5.2021 ± 0.0004A and the occupation numbers of tetrahedral and octahedral sites were 1.95 ± 0.08 and 0.05 ± 0.02, respectively. The inelastic scattering data clearly showed the vibrational components due to tetrahedral and octahedral protons and a two peaked structure for the fundamental tetrahedral vibration. Several higher harmonics of the tetrahedral peak and one harmonic of the octahedral were evident. The observed energies are 117 and 127 meV for the tetrahedral site split fundamental and 81 meV for the octahedral site. The higher harmonics were a few percent lower than the expected harmonic value. The occupation fractions obtained from the elastic and inelastic scattering were compared to a model which predicts occupation factors as a function of temperature.

---

*Work supported by the U. S. Department of Energy.

INFLUENCE OF SUBSTITUTIONAL AND INTERSTITIAL

IMPURITIES ON LOCAL HYDROGEN MODES IN Nb

A. Magerl[1,2], J.J. Rush[2], J.M. Rowe[2], D. Richter[3] and H. Wipf[4]

[1] Institut Laue-Langevin, 156X Grenoble Cedex, France

[2] National Bureau of Standards, Div. 566, Washington DC 20234, U.S.A.

[3] Kernforschungsanlage Jülich, 517 Jülich, W. Germany

[4] Physik Department der TU München, 8046 Garching W. Germany

ABSTRACT

Neutron vibrational spectroscopy has been used to study the association of H with interstitial (O) and substitutional (V) defects in a Nb host. From these results we propose a model for the site occupancy of H bound by O at low temperatures.

INTRODUCTION

Local modes of H dissolved in metals have found scientific interest since the beginning of neutron spectroscopy [1]. Until now this is still the only technique enabling us to study H spectra. The H modes are of interest to us because they reflect directly interatomic potentials. With improved instrumentation very detailed information of the vibrational spectra became available; e.g. large intrinsic linewidths have been found in disordered phases [2]. These have been associated with the statistical occupation of tetrahedral sites by H [3]. Also higher order transitions [3,4] and the excitation spectra for the three isotopes ^1H, ^2H and ^3H with Nb as a host [5] have been measured, yielding information about the anharmonicity of the potential.

171

Fig. 1 shows with TaH$_{.18}$ as an example the general behaviour of neutron spectra measured in the ordered and disordered phases. At 295K (α phase) we observe two broad peaks in the density of states centered at 113 meV and 163 meV. In the ordered phase at 150K (β phase) the peak widths are considerably reduced. They are measured only slightly larger than the experimental resolution. In general a change in the peak positions at the order disorder rearrangement of H is found. In Fig. 1 we observe a shift in the position of the first peak to 120 meV, whereas the position of the second peak remains the same. The different shifts found for both excitations upon ordering demonstrates how sensitively the H modes reflect changes in the local environment of the H.

In the past impurities were found to shift the solus curve for metal-H systems to lower temperatures [6,7]. Meanwhile the defect-H bonding and the H diffusion in the presence of impurities have been studied by a variety of techniques. Perhaps the most interesting phenomena found so far in these metallic systems is translational tunneling of H bound by an interstitial defect. This was observed both by specific heat [8] and by inelastic neutron scattering experiments [9]. We have repeated these measurements at the NBS research reactor with a single crystalline sample of NbO$_{.011}$H$_{.010}$ and with improved instrumental resolution of 92 μeV. An energy scan for a wave vector transfer along the [110] direction is shown in Fig. 2 at an intermediate temperature. Clearly, we observe inelastic scattering mainly on the positive energy side (neutron energy loss) originating from the H tunneling motion.

Several proposals have been made for the H-defect geometry based on particular experimental aspects [10,11]. However, these models may be doubted considering all the experimental information available [12], so that the site of a H trapped by a defect and moreover the spatial arrangement of the H defect pair is not well understood. The knowledge of the sites occupied by H is vital for a quantitative understanding of the metal-H-defect systems, e.g. when interpreting the measured tunneling matrix element [9]. We have used neutron vibrational spectroscopy of H modes to study the association of H with impurities and to gain directly from the local mode frequencies information about the H positions.

EXPERIMENTAL

Nb crystals were purified at 2300°C in an ultra high vacuum and subsequently loaded in a controlled manner by either interstitial (O) or substitutional (V) impurities. The concentrations were about 1 atomic % or less so that single defects could be studied. The H concentration was lower than the impurity concentration in order to avoid saturation of the traps which could

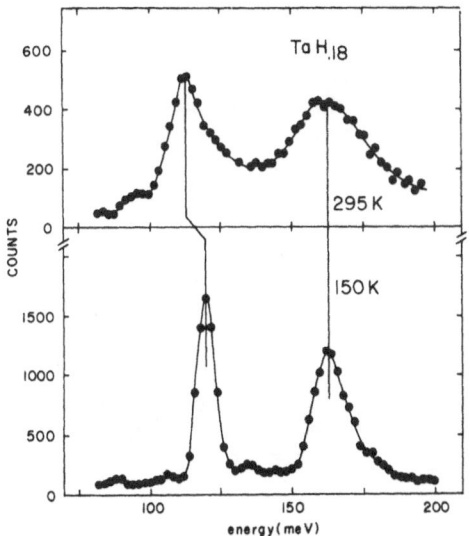

Fig. 1. Vibrational spectra of TaH.18 at 295K (α phase) and at 150K (β phase).

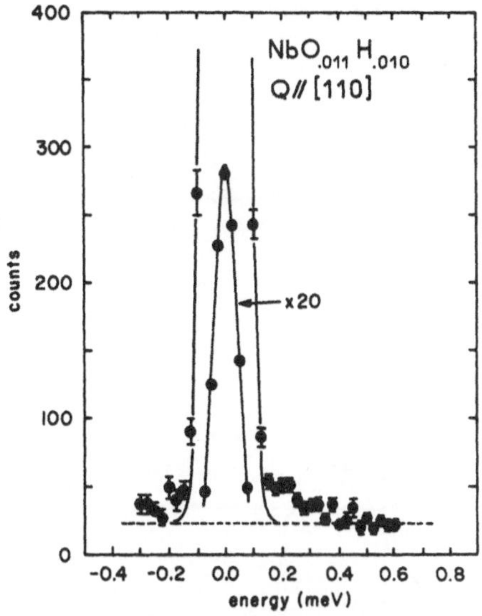

Fig. 2. Energy scan with wave-vector parallel to the [110] direction on a single crystal of NbO.011H.010 at about 2K. The dashed and the solid line represent background and resolution, respectively. The resolution has been carefully measured with a V sample.

result in a precipitation of remaining H in an ordered phase. In all cases particular care was taken to avoid clustering of the impurities during the various steps of the sample preparation. Homogeneity and sample purities were checked by a variety of techniques. For a more detailed description of the sample preparation and characterization we refer to Ref. 9 and 12.

RESULTS AND DISCUSSION

Non Trapping Defect V

Fig. 3a shows neutron spectra from NbV.008H.005 measured between 295K and 78K. At 295K two very broad density of states peaks at 106 meV and 164 meV are observed. As the temperature is decreased the peaks in the inelastic scattering become better determined : the widths of both peaks are reduced and consequently their heights increase. At the same time we observe an energy shift for both excitations. A vertical line in Fig. 3 demonstrates for

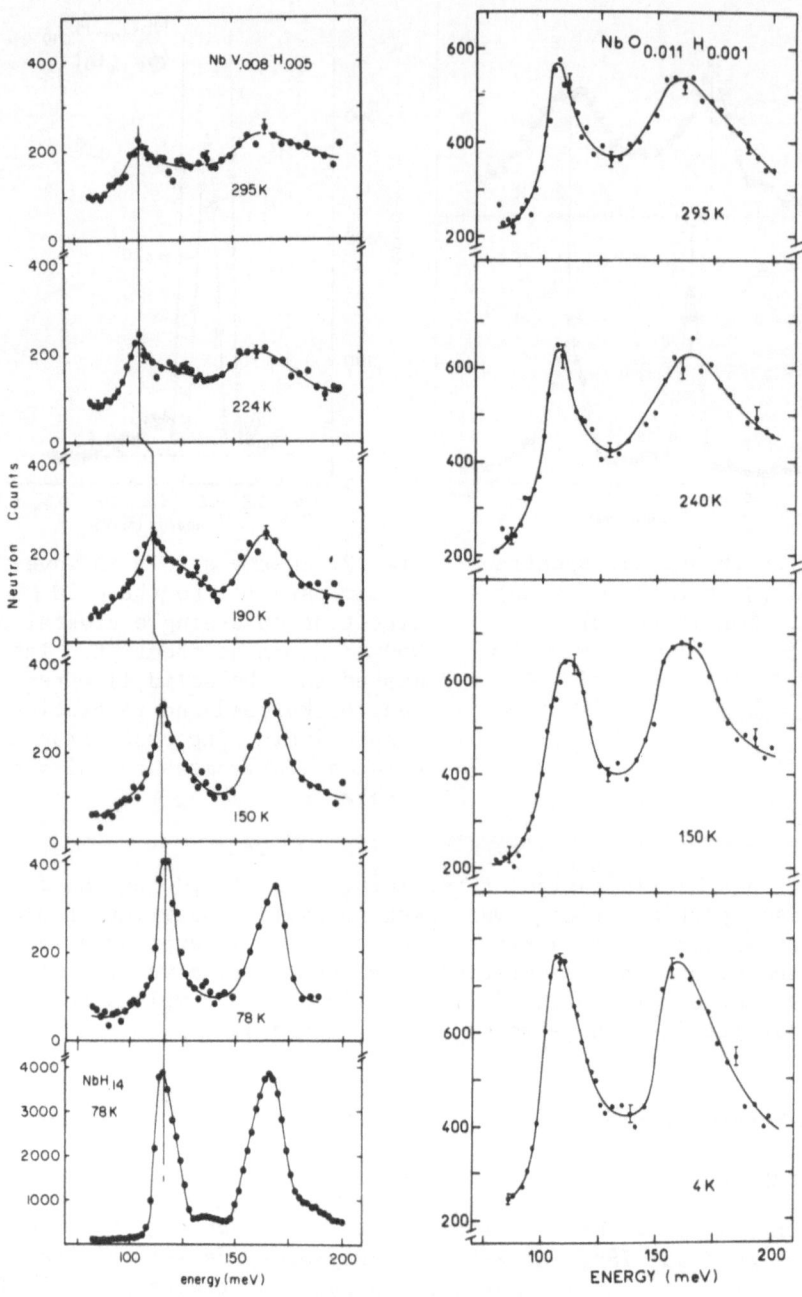

Fig. 3.   Neutron spectra from NbV.008H.005 and Nb.011H.010 measured
at various temperatures. A spectra from NbH.14 at 78K is
added for easy comparison.

the first excitation the gradual increase from 106 meV at 295K to
116 meV at 78K. At this temperature peak positions and widths
are the same as for ordered $NbH_{.14}$, also shown in Fig. 3. These
data show that substitutional V cannot trap H at low temperatures.
It precipitates into an ordered hydrid phase which is easily identified by its characteristic H modes.

However, from the spectra at the intermediate temperatures we
find a remarkable shift of the solvus line. At 150K literally all
H in defect free $NbH_{.005}$ is precipitated. The neuron spectrum,
however, reveals a broad component - particularly easy to see for
the first excitation - which is characteristic for nonprecipitated
H. In addition, we observe a narrow component on top of it,
noticeable by the increase in the height of the peak between 190K
and 150K.

To summarize, we find an attractive interaction between H and V,
which suppresses H precipitation in the temperature range known for
pure Nb ($\sim$ 190K). However, this interaction is not strong enough
to avoid the formation of an ordered hydride phase at a lower temperature ($\sim$ 150K).

Trapping Defect O

Neutron spectra for $NbO_{.011}H_{.010}$ are displayed in Fig. 3b.
Again at 295K we observe two broad excitations, however not quite
as smeared out as for the defect V. Although there are subtle
changes in peak widths and positions at the intermediate temperatures
here we want to point out the similarity of the spectra at 295K and
at 4K. In particular, the positions for the two peaks at 107 meV and
at 160 meV are the same at both temperatures. This shows immediately
that the interaction between O and H is strong enough to suppress
any H precipitation down to low temperatures. Moreover the identity
of the peak positions directly suggests that the sites occupied by
free H at 295K and by H associated with O must have a very close
resemblance. The vibrational spectra provide direct evidence that H
is trapped in the neighborhood of an O at a tetrahedral site whose
environment is only weakly distorted.

An O located at an octahedral site (black dot in Fig. 4) will
heavily displace the two nearest Nb (indicated by arrows). As our
results allow only weakly distorted tetrahedral sites we discard all
sites which involve one of these two Nb as their nearest neighbor.
These sites are marked a in Fig. 4. To account for the tunnel split
ground state mentioned before the H must in addition be localized
between (at least) two equivalent sites with respect to the O defect.
The positions b or c in Fig. 4 fulfill all the above requirements
However, the distance between the positions b is too large for a
reasonable overlap in the calculation of the tunneling matrix element.

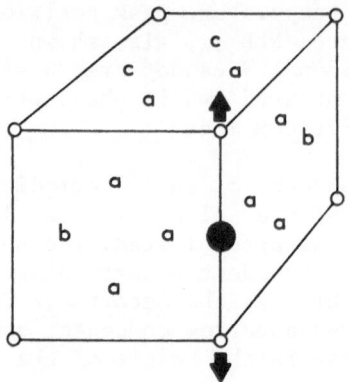

Fig. 4.   Tetrahedral sites (a,b,c) around an octahedral site (•)
          in a bcc lattice. For further explanation see text.

Therefore we propose the positions   c   as possible sites for H
bound by O at low temperatures.

REFERENCES

[1]   For a review see T. Springer in "Hydrogen in Metals, 1",
      Topics in Applied Physics, edited by G. Alefeld and J.Völkl,
      Springer Verlag, N.Y. 1978.
[2]   A. Magerl, V. Wagner and N. Stump, Solid State Comm. 33, 627
      (1980).
[3]   D. Richter and S.M. Shapiro, Phys. Rev. B 22, 599 (1980).
[4]   R. Hempelmann, D. Richter and A. Kollmar, Z. Phys. B -
      Condensed Matter 44, 159 (1981).
[5]   J.J. Rush; A. Magerl, J.M. Rowe, J.M. Harris and J.L. Provo,
      Phys. Rev. B 24, 4903 (1981).
[6]   P. Kofstadt, W.E. Wallace and L.J. Hyvönen, J. Am. Chem. Soc.
      81, 5015 (1959).
[7]   P. Kofstadt and W.E. Wallace, J. Am. Chem. Soc. 81, 5019
      (1959).
[8]   C. Morkel, H. Wipf and K. Neumaier, Phys. Rev. Lett. 40, 947
      (1978).
[9]   H. Wipf, A. Magerl, S.M. Shapiro, S.K. Satija and W.Thomlinson,
      Phys. Rev. Lett. 46, 947 (1981).
[10] P.E. Zapp and H.K. Birnbaum, Acta Metall. 28, 1275 (1980).
[11] H.D. Carstanjen, Phys. Stat. Sol. (a) 59, 11 (1980).
[12] A. Magerl, J.J. Rush, J.M. Rowe, D. Richter and H. Wipf, to
      be published.

QUASIELASTIC NEUTRON SCATTERING IN METAL HYDRIDES: EFFECTS OF THE

QUANTUM MECHANICAL MOTION OF INTERSTITIAL HYDROGEN ATOMS*

D. W. Brown and V. M. Kenkre

Department of Physics and Astronomy
University of Rochester
Rochester, New York 14627

The traditional theme of metal hydride transport analysis has employed an equation for the probability $P_\ell(t)$ that a hydrogen atom occupies an interstitial site ℓ in the host metal at a time t[1]

$$\frac{d}{dt} P_\ell(t) = \Sigma_{\ell'} [\Gamma_{\ell\ell'} P_{\ell'}(t) - \Gamma_{\ell'\ell} P_\ell(t)] \tag{1}$$

where $\Gamma_{\ell\ell'}$ is the hopping rate from site ℓ' to site ℓ. The inadequacy of such treatments has led more recent investigations to consider correlations between jumps and correlations in continuous time.[2] We will discuss here a transport instrument particularly well suited to the analysis of such memory effects, and we will illustrate the nature of these effects on the quasielastic scattering function.

The hopping picture of hydrogen motion can be understood as arising through a process of coarse-graining from a microscopic, "flowing" picture, wherein motion is described by the Schrödinger equation. Under certain initial conditions (including the completely (de)localized ones) the Schrödinger equation has been shown to lead, without approximation, to the generalized master equation[3]

$$\frac{d}{dt} P_\ell(t) = \int_o^t dt' \Sigma_{\ell'} \{W_{\ell\ell'}(t-t')P_{\ell'}(t') - W_{\ell'\ell}(t-t')P_\ell(t')\}. \tag{2}$$

As a direct consequence of quantum mechanics, (2) includes effects not accessible to (1), which is recovered under the Markoffian approximation, in which the "memory functions" $W_{\ell\ell'}(t)$ are replaced by time delta functions $\delta(t)\Gamma_{\ell\ell'}$. In the case of translationally invariant systems one of us[4] has calculated the memory functions of

*Supported by NSF(DMR-7919539) and ONR(N00014-76-0001).

(2), which may be simply written in the Fourier-Laplace domain as

$$\tilde{W}^q(\epsilon) = - \{\frac{1}{N} \Sigma_k [\epsilon + i(V^{k+q} - V^k)]^{-1}\}^{-1} \tag{3}$$

where N is the number of sites, $V_{\ell\ell'}$ the intersite Hamiltonian matrix element (The discrete Fourier space (q) is conjugate to the lattice.). Dimensionality does not restrict (3), so (2) and (3) together give a purely quantal description of motion in perfect crystalline matter.[4]

A familiar description of motion with site-to-site interaction V and a randomizing parameter α is provided by stochastic Liouville equations for the density matrix ρ[5]

$$\frac{\partial}{\partial t} \rho_{\ell m} = \frac{1}{i\hbar} [V,\rho]_{\ell m} - \alpha(1-\delta_{\ell m})\rho_{\ell m} + \delta_{\ell m} \Sigma_{\ell'}(\Gamma_{\ell\ell'}\rho_{\ell'\ell'} - \Gamma_{\ell'\ell}\rho_{\ell\ell}) \tag{4}$$

shown here without a term in $\rho_{m\ell}$, often neglected for simplicity. It has further been shown[6] that (4) is equivalent to (2) with

$$W_{\ell\ell'}(t) = \Gamma_{\ell\ell'}\delta(t) + W^c_{\ell\ell'}(t)e^{-\alpha t} \tag{5}$$

where $W^c_{\ell\ell'}(t)$ is the memory function whose Fourier-Laplace transform is (3). The two-channel form of (5) is directly due to the natural separation of transport events into a class <u>assisted</u> by a change in the bath state, and a class in which no such change occurs. Furthermore, the last term in (5), even by itself, provides a simple, continuous bridge between the "hopping" and "flowing" extremes of transport. The two primary aspects of microscopic motion being thus accessible to the generalized master equation (2) with (5), we take it as the instrument for our calculation.

In the usual vein we write the scattering function S(q,w) as a product of a Debye-Waller factor involving only the mean square displacement $<u^2>$ and the scattering function $S^D(q,w)$ arising from the hydrogen transport alone[1]

$$S(q,w) = e^{-q^2<u^2>}S^D(q,w) \tag{6}$$

The quantity of interest is $S^D(q,w)$, which is related to the van Hove self-correlation function $G_s(\ell,t)$ through $\tilde{G}(q,\epsilon)$ as

$$S^D(q,w) = \text{Re}\tilde{G}_s(q,iw) \tag{7}$$

Defining $A_{\ell\ell'}(t) \equiv -W_{\ell\ell'}(t)$, $A_{\ell\ell}(t) \equiv \Sigma_{\ell'}W_{\ell'\ell}(t)$, and invoking the translational invariance of the system, we see that the propagator $\psi_\ell(t)$ of (2) has the Fourier-Laplace transform

$$\tilde{\psi}^q(\epsilon) = \{\epsilon + \tilde{A}^q(\epsilon)\}^{-1} = \{\epsilon + \tilde{W}^o(\epsilon) - \tilde{W}^q(\epsilon)\}^{-1} \tag{8}$$

and is identical to $G_s(\ell,t)$ at high enough temperatures. Therefore, within the approximations discussed, we have the result

$$S^D(q,w) = Re\{\tilde{W}^q_c(\alpha+iw)-\alpha+(\Gamma^o-\Gamma^q)\}^{-1} \qquad (9)$$

The restriction to high temperatures (narrow bands) is made only because this is the normal regime of hydrogen transport; the extension
of (9) to arbitrary temperature has been developed,[7] but is not
presented here.

For simplicity we shall consider a one-dimensional crystal with
nearest-neighbor interactions V and nearest-neighbor hopping rates
Γ, whereupon (9) results in

$$S^D(q,w) = Re\{\sqrt{(\alpha+iw)^2 + 16V^2 \sin^2(q/2)}-\alpha+4\Gamma\sin^2(q/2)\}^{-1} \qquad (10)$$

The "flowing" limit is obtained from (10) by allowing α and Γ to
vanish: defining $V_q \equiv 4V \sin(q/2)$ and $\Gamma_q \equiv 4\Xi\sin^2(q/2)$, we have

$$S^D(q,w) = \{\sqrt{V_q^2-w^2} -\Gamma_q\}^{-1} \qquad (11)$$

The "hopping" limit is extracted by keeping $2V^2/\alpha$ finite as both α
and V are allowed to diverge, resulting in the Lorentzian

$$S^D(q,w) = \frac{4(\Gamma+2V^2/\alpha) \sin^2(q/2)}{w^2 + [4(\Gamma+2V^2/\alpha)\sin^2(q/2)]^2} \qquad (12)$$

The intermediate regime forsakes such simplicity, requiring the
more complicated form

$$\begin{aligned}
S^D(q,w) = &\{[V_q^2-w^2-\Gamma_q(\alpha-\Gamma_q)][\tfrac{1}{2}((V_q^2+\alpha^2-w^2)^2+4\alpha^2w^2)^{\frac{1}{2}}+\tfrac{1}{2}(V_q^2+\alpha^2-w^2)]^{\frac{1}{2}} \\
&+ \alpha w[\tfrac{1}{2}((V_q^2+\alpha^2-w^2)^2+4\alpha^2w^2)^{\frac{1}{2}} - \tfrac{1}{2}(V_2^2+\alpha^2-w^2)]^{\frac{1}{2}} \\
&+ (\alpha-\Gamma_q)[V_q^2-w^2-\Gamma_q)]\}/\{[V_q^2-\Gamma_q(\alpha-\Gamma_q)-w^2]^2+4\alpha^2w^2\}
\end{aligned} \qquad (13)$$

Figure 1 displays the general features of our result for the
case of negligible Γ_q and fixed V_q. The parameter V_q/α represents
the degree of persistence of quantum mechanical phase. The decrease
in linewidth (after an initial increase) with increasing randomization
of phase testifies to the occurrence of motional narrowing. The diffusion constant varies among the curves of Fig. 1; therefore, if
known (e.g., from measurements of the mean square displacement at
long times) the ratio $2V^2/\alpha$ may be used to present our results in
another meaningful way. Fig. 2 displays $S^D(q,w)$ for fixed $2V^2/\alpha$ and
negligible Γ_q, which requires both α and V_q to vary. At fixed q, the
content of Fig. 2 is essentially the same as Fig. 1; however, in the
experimentally reasonable case that α and V are fixed (and probably

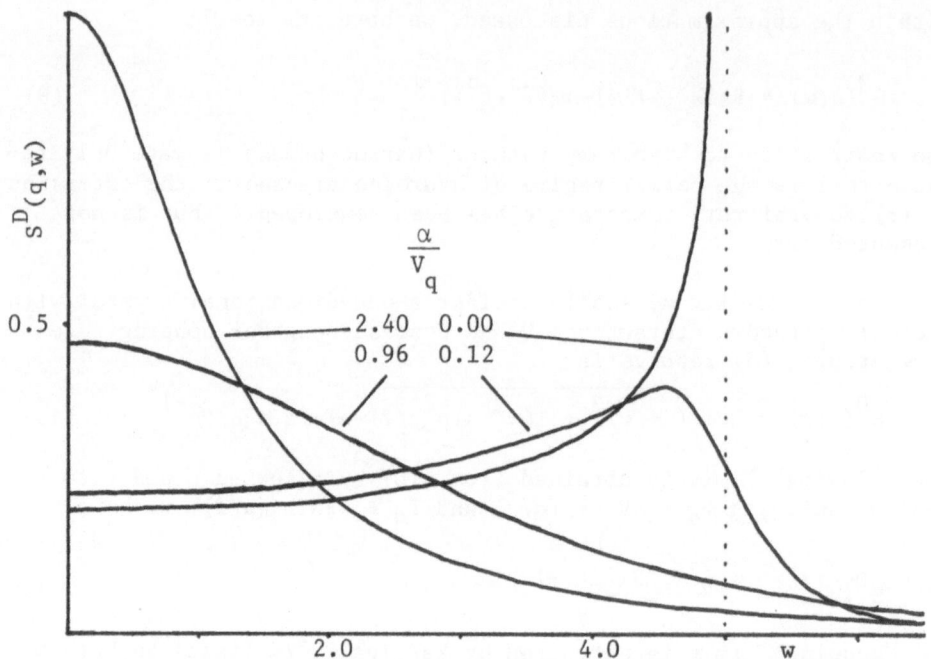

Fig. 1. Scattering function $S^D(q,w)$ plotted against frequency w for several values of α/V_q for $\Gamma_q=0$. V_q has been set arbitrarily 5.

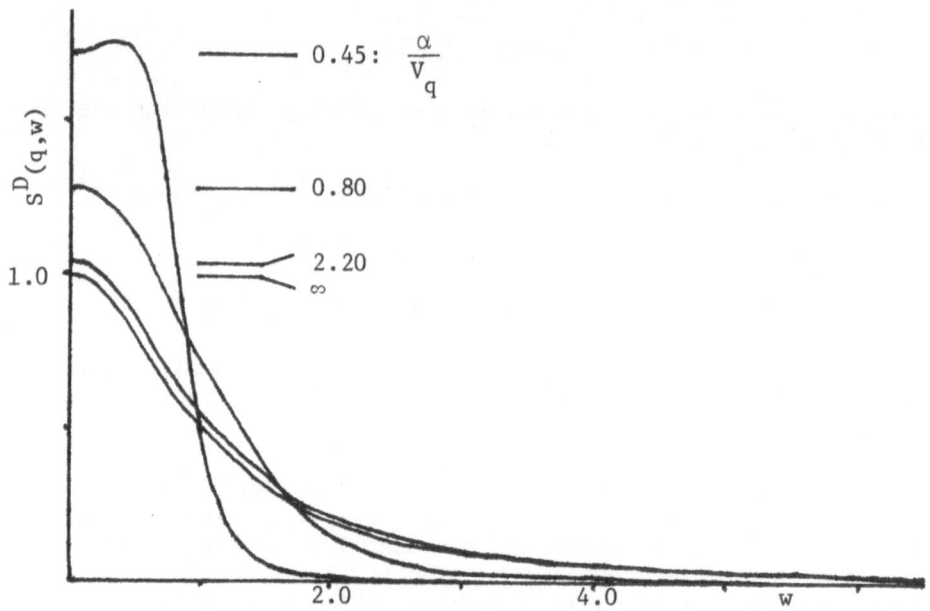

Fig. 2. Scattering function $S^D(q,w)$ in which $\Gamma_q=0$ and $2V^2/\alpha=4$. When $\alpha/V_q\gg1$, $2V^2/\alpha$ represents the diffusion constant.

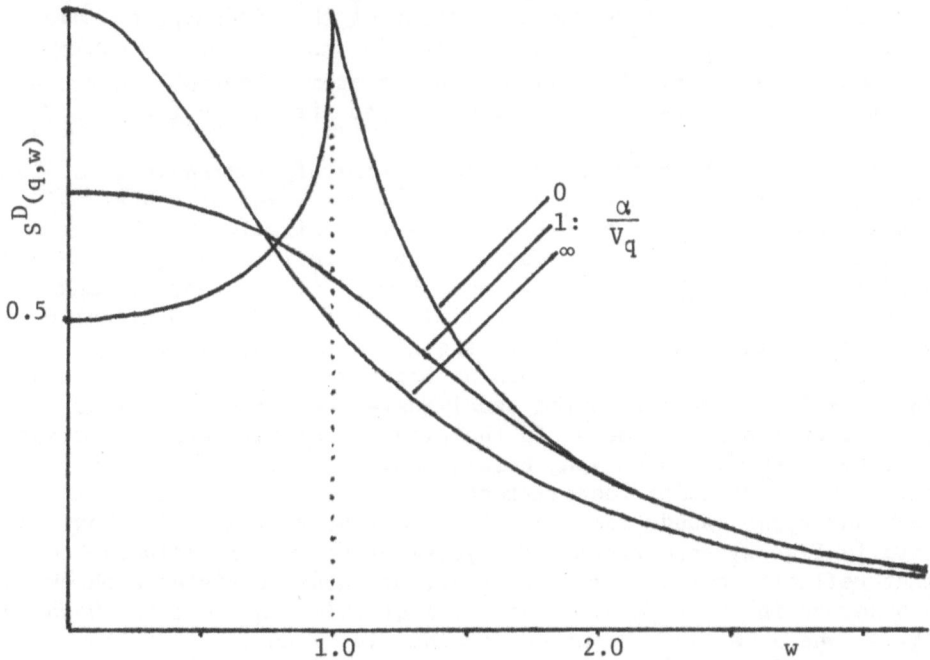

Fig. 3. Scattering function $S^D(q,w)$ in the presence of bath-assisted transport. Γ_q and V_q have been set arbitrarily to 1.

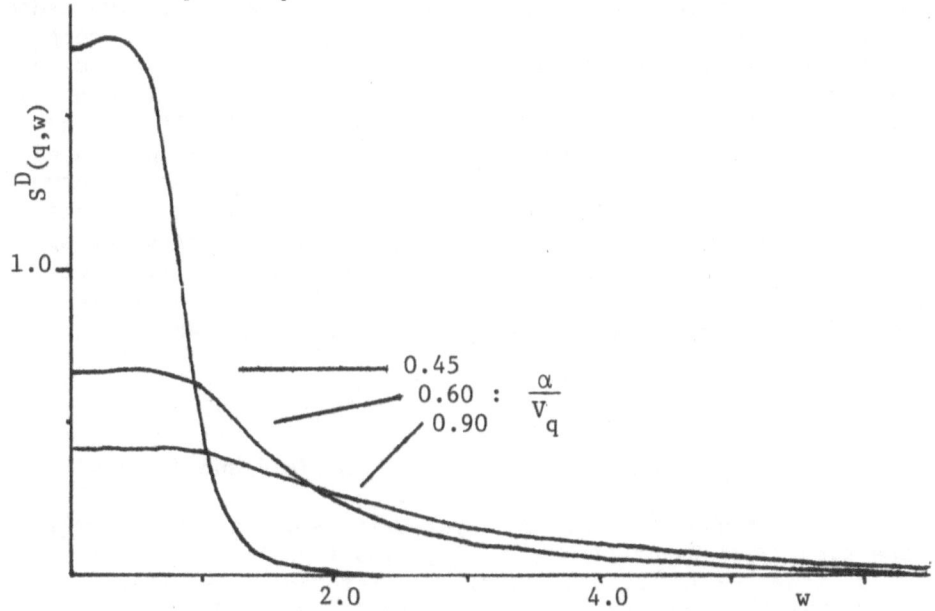

Fig. 4. Scattering function in the presence of bath-assisted transport. For comparison with Fig. 3, $(2V^2/\alpha+\Gamma)=4$ and $V=.9$. When $\alpha/V_q \gg 1$, $(2V^2/\alpha+\Gamma)$ represents the diffusion constant.

unknown), Fig. 2 displays the dependence of the lineshape on momentum transfer $[\alpha/_q = \alpha/4V \sin(q/2)]$. It must be noted that whatever values of α and V may be available in a system, observable effects become completely washed out at low momentum transfers (large α/V_q).

Figs. 3 and 4 relax the restriction that Γ_q vanish in Figs. 1 and 2 respectively. In Fig. 3, V_q and Γ_q are fixed at nontrivial values, showing now the effect of phase randomization in the presence of bath-assisted transport. Characteristic features are that $S^D(q,0)$ is pinned between the values Γ_q^{-1} and $(V_q+\Gamma_q)^{-1}$, and that the maximum value attained by $S^D(q,w)$ in both limits of Γ is Γ_q^{-1}. Motional narrowing is again evident; however, the lines now narrow to the Lorentzian (12) whereas there was no Lorentzian in Fig. 1. The hopping rate $1/\tau$ appearing in the usual Master equation treatments[1] of hydrogen motion corresponds, in the presence of bath-assisted transport, to $2V^2/\alpha+\Gamma$. Specifying this quantity fixes the diffusion constant, and conversely, the existence of a diffusion constant fixes lower and upper bounds for α and Γ respectively (fixed V). Thus the curve in Fig. 2 appropriate to a system with a known diffusion constant reflects the most extreme effect of phase persistence which can be observed in that system. The sense of this statement is shown in Fig. 4, where we have chosen one of the Fig. 2 curves and allowed Γ to increase from zero.

We have illustrated the manifestations in the scattering observable of two distinct features of the microscopic motion of the hydrogen atom: (i) quantum mechanical phase persistence and (ii) competition of bath-assisted and bath-hindered transport. Such manifestations include generally non-Lorentzian lines, motional narrowing, and a characteristic distension in the half-widths of near-Lorentzian lines. Limitations on the possible appearance of these have been noted, importantly the experimental demand of significant momentum transfer. Our transport instrument in this analysis has been the well-known stochastic Liouville equation in a high-temperature approximation, or equivalently, the corresponding generalized master equation. The detailed analysis of these effects for realistic lattices and arbitrary temperatures will be reported elsewhere.

REFERENCES

1. K. Sköld in Hydrogen in Metals, eds. G. Alefeld and J. Völkl (Springer-Verlag, New York, 1978).
2. J. W. Haus and K. W. Kehr, J. Phys. Chem. Sol. 40, 1019 (1979).
3. See e.g. R. W. Zwanzig, Physica 30, 1109 (1964); V. M. Kenkre, J. Stat. Phys. 19, 333 (1978), and references therein.
4. V. M. Kenkre, Phys. Lett. 63A, 367 (1977); Phys. Rev. B18, 4064 (1978).
5. H. Haken and P. Reineker, Z. Phys. 249, 253 (1972)
6. V. M. Kenkre, Phys. Lett. 65A, 391 (1978).
7. V. M. Kenkre, University of Rochester preprint.

A STUDY OF HYDROGEN SITE OCCUPATION IN THE YTTRIUM-HYDROGEN SOLID

SOLUTION PHASE USING INELASTIC NEUTRON SCATTERING

J.E. Bonnet [†], S.K.P. Wilson [*] and D.K. Ross [*]

[*]Department of Physics, University of Birmingham,
P.O. Box 363, Birmingham B15 2TT, U.K.
[†]Defauts dans les Métaux, Bâtiment 350, Université Paris
Sud, 91405 Orsay, France

ABSTRACT

The solid solution phase of YH_x systems has a hcp structure
and can exist with H/M ratios up to 24% at temperatures below 500 K.
In this phase the hydrogen can occupy either tetrahedral (T) or
Octahedral (O) sites. Existing neutron diffraction and NMR
measurements however offer conflicting evidence as to the amount of
hydrogen occupying each type of site. Neutron energy loss experiments
have been successfully applied to the problem of locating hydrogen
in solution in Zr and Ti. We therefore made further measurements
on samples of $YH_{0.1}$ and $YH_{0.2}$ at 297 K and 104 K using a Be filter
spectrometer. In all cases two peaks were observed centred around
137 and 102 meV. By comparison with previous solid solution and
hydride phase measurements we attribute the higher energy peak to
the T site and the lower peak to the O site.

INTRODUCTION

Hydrogen in solid solution in transition metals with hcp
structures such as Ti and Zr occupy tetrahedral (T) interstices
(1,2,3). In some rare earths with hcp structures (Y, Er, Lu and Tm)
the solid solution phase (α phase) extends to quite high hydrogen
contents; in the case of yttrium it extends to 24%, and is
independent of temperature below 500 K. Above this the maximum
extent of the α phase increases with temperature (fig. 1). NMR
and neutron diffraction measurements have shown both tetrahedral (T)
and octahedral (O) sites to be occupied in yttrium (4,5) and the

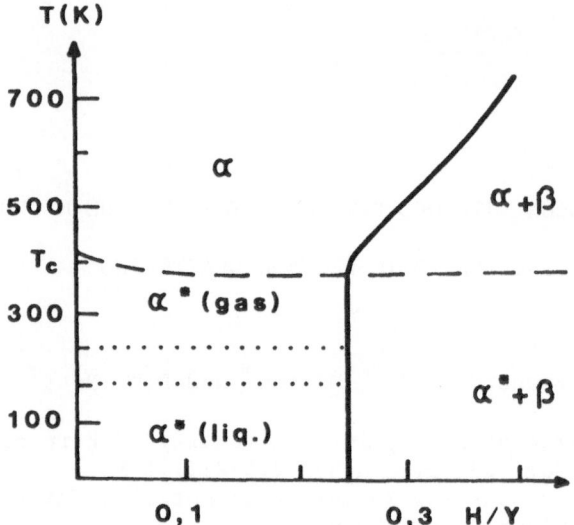

Fig. 1.   Phase diagram of Yttrium Hydride.

same result was obtained on La(6) but this is in conflict with
recent channeling measurements(7). Moreover there is an observed
discontinuity in the resistivity of these solid solution phases in
the temperature range 170-240 K(8). One possible explanation(8)
is that the hydrogen atoms are independent at high temperatures but
form pairs or some other short range ordered arrangement in the
above temperature range - the "liquid-gas" transition model. An
alternative explanation is that the atoms move from O sites to
T sites.

Incoherent inelastic neutron scattering is a useful tool to
determine site occupation as the area of each observed optical mode
peak is proportional to the number of hydrogen atoms in a given site
times the degeneracy of the mode(9). $YH_{0.2}$ has a c/a ratio of
1.57 compared with a c/a ratio of 1.63 required for its T and O sites
to have true cubic symmetry. This slight distortion would result in
some splitting of the peak. However calculations indicate that this
will not be significant. Thus the area of each peak can be thought
of as being proportional to the occupation of each site.

In order to interpret such results empirical formulae have been
produced to predict site frequencies. In the case of regular T sites
in fcc dihydrides the correlation $E = AR^{-n}$ (n = 3/2 or 1) is found
to work well(10,11). Attemps have been made to fit metal-hydrogen
potentials to evaluate the modes of vibration(14). If we assume a
Born-Mayer form for the metal-hydrogen potential $f = \Gamma \exp - \gamma R$ where
Γ and γ are constants and R is the metal hydrogen distance and if we
also assume that the metal atoms are fixed and the hydrogen interstitials
behave like simple harmonic oscillators, it may be shown(14), that for

regular T sites the frequency is given by

$$\omega^2 = \Gamma\gamma^2 e^{-R_T\gamma} \frac{4}{3M} \left(1 - \frac{1}{\gamma R_T}\right) \tag{1}$$

or for a regular octahedral site

$$\omega^2 = \Gamma\gamma^2 e^{-R_0\gamma} \frac{2}{M} \left(1 - \frac{1}{\gamma R_0}\right) \tag{2}$$

where R_T and R_0 are the M-H distances for tetrahedral and octahedral sites respectively. This kind of approach is also readily applied to distorted sites.

EXPERIMENTAL DETAILS

A sample of Y metal with better than 99.9% purity was used for this experiment. It was outgassed and annealed under vacuum ($<10^{-6}$ Torr) and 1000°C. The metal was then hydrogenated to $YH_{0.2}$ as previously described (12). After hydrogenation the sample was kept at 600°C to permit the hydrogen to be distributed uniformly through the metal, and then cooled slowly to 200°C where it stayed for 5 hours. This was done to let thermodynamic equilibrium be established in the α phase before entering the metastable α^* phase.

Inelastic neutron scattering measurements were performed using the Be filter spectrometer on the DIDO reactor at AERE Harwell (U.K.) The sample was mounted in a liquid nitrogen cryostat and measurements were performed well above and below the region of resistivity discontinuity. At both temperatures (104 ± 5 K and 297 ± 1 K) three scans of energy transfer in the range 80-200 meV were made with 1 meV steps in the incident neutron beam energy. Single scans were also performed from 20-80 meV and 170-220 meV with larger steps. The incident energy was selected by Bragg scattering from an Al monochromator. The scattered neutrons which have transferred most of their energy to the hydrogen atoms pass through two liquid nitrogen cooled Be filters and are detected by arrays of BF_3 counters. The mean energy of the detected neutrons is thus 3.5 meV.

The background was measured in the same manner using an empty cryostat and the results of each scan were averaged and the mean background subtracted. The same sample was later outgassed and rehydrogenated to form $YH_{0.1}$ and the experiment was repeated.

RESULTS AND DISCUSSION

Figures 2 and 3 show the energy transfer at 104 K and 297 K for $YH_{0.2}$. Similar results were obtained for $YH_{0.1}$ but with larger statistical errors in the counts per channel. Two distinct peaks

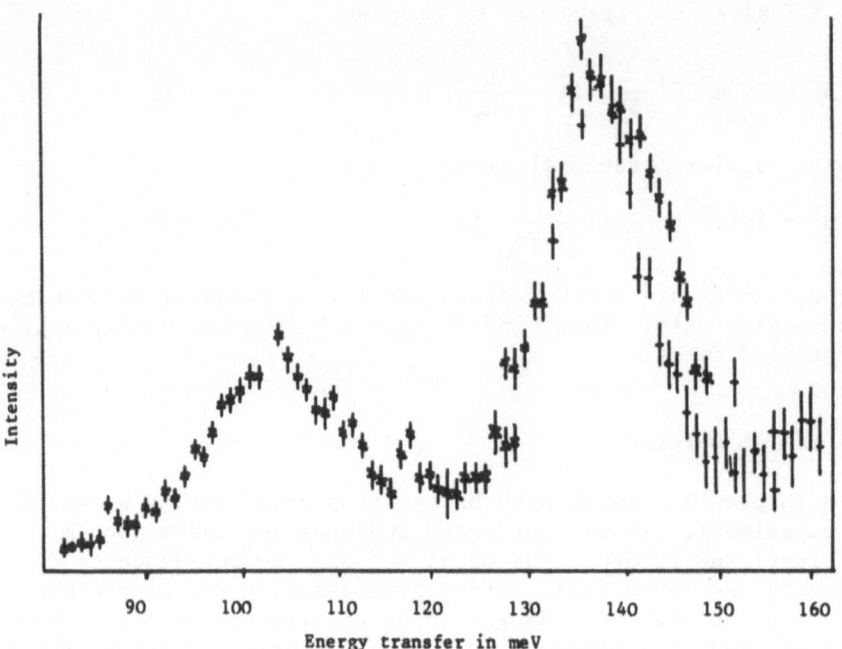

Fig. 2. Energy transfer measurements for $YH_{0.2}$ at 104 K.

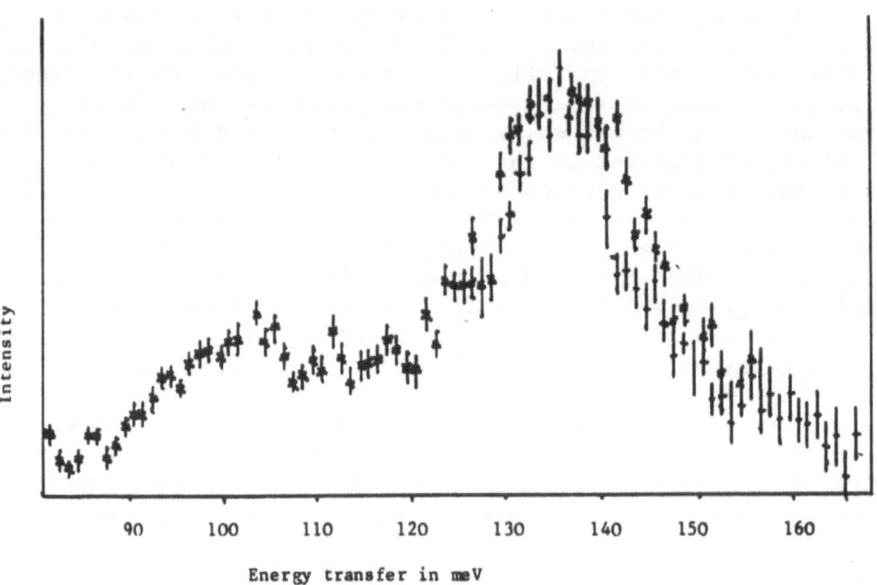

Fig. 3. Energy transfer measurements on $YH_{0.2}$ at 297 K
measurements using 200 monochromator plane
measurements using 220 monochromator plane

are observed supporting the dual site location of hydrogen. Gaussian
profiles were fitted to these data and the resulting parameters for
$YH_{0.2}$ are given in the table. The values show a slight fall in site
energies as T increases and a small change in the ratio of peaks
areas. There is also a slight splitting in the high energy peak at
104 K. This may be present at 297 K but masked by peak broadening.

Previous measurements on the dihydride phase of yttrium show a
peak at 127 mm meV(10). This result is 7 meV higher than would be
predicted by the $E = AR^{-3/2}$ correlation. In changing from the solid
dihydride phase to the solid solution the lattice shrinks and the
possibilities for lattice distortion increase. It has been observed(2)
that the T site frequencies rise or fall slightly in this transformation. Thus we believe the higher energy peak in the α phase around
137 meV is due to to tetrahedral site occupation.

There is no evidence of a peak around 70 meV as might be
expected from measurements of regular O sites in PdH_x, PrH_{2+x}(9) and
CeH_{2+x}(15). However, the presence of T site hydrogens could well be
responsible for an increase in frequency and we have therefore identified the lower frequencies as the O site. The NMR and diffraction results
(4,5) show different O and T occupations . This may be due to differences
in sample preparation, e.g. how the samples were cooled across the
$\alpha-\alpha*$ transition.

The small change in the O:T ratio with temperature eliminates
$O \rightarrow T$ transitions as a possible explanation of the low temperature
resistance anomaly and also indicates that protons in the ground
states of O and T potential wells have about the same total energy.

Table 1   Optical peak energies E and widths (Gaussian fitting)
          for $YH_{0.2}$

	O site	T site	
E (meV) at 104 K	103.5	138.5	139
σ (meV	9.3	5.9	9.8
H concentration(%)	6.5	3.15	10.25
E (meV) at 297 K	101.9	136.5	
σ (meV)	10.3	12.2	
H concentration (%)	5.1	14.5	

REFERENCES

1. P.P. Narang, G.L. Paul, K.N.R. Taylor, J. Less Comm. Metals, 56:125 (1977).
2. R. Khoda Bakhsh and D.K. Ross, J. Phys. F. Met. Phys. 12:15 (1982)
3. H. Pinto, C. Korn, S. Goren and H. Shaked, Solid State Commun. 32:397 (1979).
4. D.L. Anderson, R.G. Barnes, S.O. Nelson and D.R. Torgeson, Phys. Letters 74A:427 (1979).
5. D. Khatamian, C. Stassis, B.J. Beaudry, Phys. Rev. B23:624 (1981).
6. J.E. Bonnet, J. Less Common Metals 49:451 (1976).
7. R. Danielou, J.N. Daou, E. Ligion and P. Vajda. Phys. Stat. Sol. (a) 67:453 (1981).
8. J.E. Bonnet, C. Juckum and A. Lucasson, J. Phys. F., 12:699 (1982).
9. D.G. Hunt and D.K. Ross, J. Less Common Metals 49 (1976).
10. D.K. Ross, P.F. Martin, W.A. Oates and R. Khoda Bakhsh Z. Phys. Chem., 114 (1979).
11. Y. Fukai and H. Sugimoto, J. Phys. F. 11:L137 (1981).
12. J.E. Bonnet and J.N. Daou, J. Phys. Chem. Solids 40:421 (1979).
13. R. Khoda Bakhsh, Ph.D. Thesis, University of Birmingham (1981).
14. J.E. Bonnet, D.K. Ross and S.K.P. Wilson, to be published.
15. W. Wagener, P. Vorderwisch and S. Hauteiler. Phys. Stat. Sol. (b) 98:K171 (1980).

NEUTRON DIFFRACTION STUDIES OF THE LOW

TEMPERATURE TRANSITION IN α'PdD

R.A. Bond[4], I.S. Anderson[2], B.S. Bowerman[1],
C.J. Carlile[3], D.J. Picton[1], D.K. Ross[1], D.G. Witchell[1]
and J.K. Kjems[5]

[1]Department of Physics, University of Birmingham
P.O. Box 363, Birmingham B15 2TT, U.K.
[2]Institut Laue-Langevin 156X Centre de Tri, 38042
Grenoble, Cedex, France
[3]Rutherford Appleton Laboratory, Didcot, Oxon OX11 OQX, U.K.
[4]Present address: Culham Laboratory, Abingdon, Oxon
OX14 3DB, U.K.
[5]Risø National Laboratory, Post Box 49, DK-4000 Roskilde
Denmark

ABSTRACT

Neutron diffraction measurements of the equilibrium intensity
of the $(1 \frac{1}{2} 0)$ peak in the region of the low temperature transition
for PdD_x, $x = 0.665$ and $x = 0.640$ are presented. An accurate
determination of the transition temperature and critical exponent
are made, $T_C = 42.5 \pm 1.0$ K and $\gamma = 1.27 \pm 0.06$ for $x = 0.665$,
$T_C = 53.0 \pm 0.5$ K and $\gamma = 1.31 \pm 0.06$ for $x = 0.640$. The relaxation
of the peak intensity following a temperature change shows fast and
slow time components in agreement with heat release measurements.
The neutron slow time constants exhibit critical slowing down in
the region of T_C.

INTRODUCTION

In Palladium Deuteride in the concentration range $0.6 < x =$
$D/Pd < 1.0$ at low temperatures there are anomalies in several
physical parameters[1] such as specific heat, resistivity, internal
friction and Hall co-efficient, which are known collectively as the
"50 K Anomaly". Anderson et al[2] first showed that this anomaly
was due to an order-disorder transformation of the deuterons on the
octahedral interstitial sites to form a superlattice with the $I4_1/amd$

189

space group. Later it was discovered that at higher concentrations
(x > 0.72) the deuterons formed a new superlattice with the I4/m
space group[3,4].

A complete series of measurements of the position of the
resistance anomaly (the peak in the temperature dependence of the
resistance) with concentration was carried out by Ellis[5] for both
PdD and PdH, although he noted that the exact position of the
anomaly depended on the heating rate at which the 50 K region was
traversed. Ellis found that the anomaly temperature was roughly
constant at around 53 K in the concentration range $0.5 < x < 0.63$,
above this concentration it rose slowly at first but then more
steeply until flattening out at around 80 K above $x = 0.8$.

Other measurements of the anomaly position by internal
friction[6] and specific heat[7] also found that the magnitude and
position of the anomaly depended on the cooling or heating rate;
no precise value for the transition temperature can be drawn from
these measurements. The use of neutron scattering which looks
directly at the structure of the system is obviously the best
method for determining exactly where the phase transition occurs.
However, because the $I4_1/amd$ superlattice gives a reflection at
$(1\ \frac{1}{2}\ 0)$ which is coincident with the peak in the diffuse scattering
intensity, it is difficult to distinguish exactly where long range
order begins and the transition occurs. This distinction is not a
problem for the higher concentration I4/m structure as the
superlattice reflection occurs at the (0.8 0.4 0) point.

Previous work on spontaneous heat release in $PdH_{0.63}$[7] had shown
that after quenching to temperatures in the range 40-52 K the thermal
relaxation had both a fast and a slow component except at the two
highest temperatures 50 K and 52 K at which only a fast time
constant could be measured. Both sets of time constants obeyed an
Arrhenius relationship.

In order to make an estimate of the transition temperature
therefore a detailed study of the growth of the intensity at the
$(1\ \frac{1}{2}\ 0)$ position was made.

EXPERIMENTAL MEASUREMENTS

The experiments described here were carried out on single
crystal samples of Palladium Deuteride of two concentrations
$x = 0.665 \pm 0.007$ and $x = 0.640 \pm 0.003$. The samples were made by
charging with deuterium gas, care being taken to avoid the two-
phase region so as to minimise damage to the crystal; the
concentration was determined by weight gain. Both samples were
cylindrical in shape with a diameter of ∿6 mm, the $x = 0.665$ sample
had a height of 12 mm and the $x = 0.64$ a height of 5 mm.

The measurements on the PdD$_{0.665}$ crystal were carried out on the D7 spectrometer at the ILL in Grenoble. A series of rapid (∿10 minutes) quenches from 100 K to temperatures around 50 K were made, and the (1 ½ 0) region repeatedly scanned using the bank of eight detectors in eight different positions enabling both the peak intensity and width to be studied. The measured intensities were normalised using a vanadium single crystal.

The measurement on the PdD$_{0.64}$ crystal were made on the four circle diffractometer D10 at the ILL which allowed scans to be made through the (1 ½ 0) point along the three principle axes using the single detector. Again a series of rapid quenches to temperatures around 50 K were made and narrow scans in the (100) (010) and (001) directions performed after about 10 hours at each temperature. In addition after each quench the temperature was raised in steps back to 100 K and the subsequent development of the intensity of the (1 ½ 0) position studied. At one temperature well above the anomaly temperature a series of scans along the principle directions was made.

RESULTS AND DISCUSSION

The time dependence of the intensity at (1 ½ 0) for the PdD$_{0.665}$ crystal following each of the quenches showed a sharp initial increase followed by a slower approach to an equilibrium value. At the highest temperature to which the sample was quenched, 52.8 K, the intensity could be fitted by a single exponential of the form

$$I(t) = (I(0) - I(\infty)) \exp(-t/\tau) + I(\infty) \qquad (1)$$

$I(\infty)$ is the equilibrium intensity at (1 ½ 0), $I(0)$ the initial

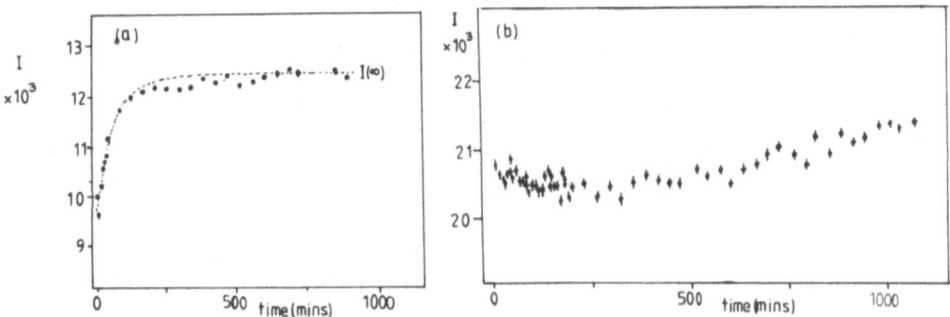

Fig. 1.   Time dependence of scattered intensity at (1 ½ 0) in Pd$_{0.665}$ a) following a quench to 52.8 K b) following an increase in temperature from 47.8 K to 48.8 K.

Table 1.   Relaxation time constants τ_1 and τ_2 for PdD$_x$ for
a) x = 0.665 and b) x = 0.640. All results are for
heating from lower temperatures except those starred
which are from quenches.

	(a)			(b)	
Temp (K)	τ_1(mins)	τ_2(mins)	Temp (K)	τ_1(mins)	τ_2(mins)
52.8*	65 ± 7	-	63.7	4 ± 1	-
52.7	68 ± 7	-	59.9	7 ± 2	-
50.9*	190 ± 19	17 ± 2	58.0	13 ± 2	3 ± 1
49.2*	530 ± 53	64 ± 4	56.2	24 ± 2	7 ± 1
47.8*	880 ± 88	90 ± 9	55.2	59 ± 6	8 ± 1
			54.3	62 ± 6	10 ± 1
			53.8	54 ± 4	13 ± 2
			53.4*	765 ± 80	5 ± 1
			52.7*	883 ± 80	

intensity and τ the relaxation time constant, as shown in figure 1a).
The equilibrium intensity and time constant at a particular
temperature were found to be independent of the direction in which
the temperature is approached.

The quenches to lower temperatures could be better fitted by
a double exponential of the form

$$I(t) = (I_1(0)-I(\infty))\exp(-t/\tau_1)+(I_2(0)-I(\infty))\exp(-t/\tau_2)+I(\infty) \quad (2)$$

The short and long time constants extracted from this fit are given
in Table 1a.

The fast relaxation process as exhibited by the short time
constant reacts immediately to a change in temperature. When the
temperature was raised by 1 K from 47.8 K to 48.8 K the intensity
first dropped before slowly rising again (fig. 1b); this rise was
because the crystal had not reached equilibrium at 47.8 K.

Measurements of the equilibrium intensity were also made at
temperatures between 55 K and 80 K; in this range the intensity
reached equilibrium very quickly, and these are shown in figure 2a.
Above a phase transition the critical scattering should obey the
relationship[8]

$$I(T) = \text{Const} \times \left| \frac{T - T_c}{T} \right|^{-\gamma} \quad (3)$$

where T_c is the transition temperature and γ is the critical exponent.
Using equation 3 values for T_c and γ where extracted. These
values are

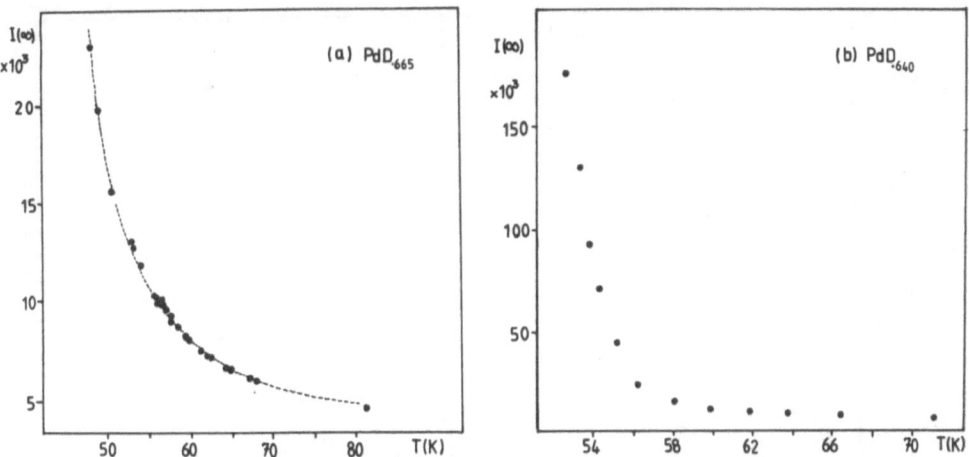

Fig. 2.   Equilibrium intensity at $(1\ \frac{1}{2}\ 0)$ as a function of temperature
a) in $PdD_{0.665}$ b) in $PdD_{0.64}$

$T_C = 42.5\ \pm 1.0$ K
$\gamma = 1.27 \pm 0.06$

The value of γ compares with the nearest neighbour Ising model
value of 1.25.

As the equilibrium intensity and the time constants are the
same at a particular temperature whether one quenches to that
temperature or approach from below, for the $PdD_{0.64}$ crystal
information could be obtained at more temperatures without needing
to quench to every temperature. When the temperature was raised
following a quench the intensity decreased very rapidly before
reaching an equilibrium value; it was possible to fit the data to
equation 3 and extract two time constants except at higher
temperatures where the relaxation was very rapid and only one time
constant could be obtained. The time constants are shown in Table 1b.

The equilibrium intensity rose very rapidly around 54 K as shown
in figure 2b. Values of T_C and γ were determined using equation 3.
The values obtained were

$T_C = 53.0\ \pm 0.5$
$\gamma = 1.31 \pm 0.06$

The value of γ is comparable with that obtained for $PdD_{0.67}$
($\gamma = 1.27 \pm 0.05$), but T_C is 10 K higher.

A comparison of the time constants for $PdD_{0.64}$ with those
obtained for the heat release measurements of Jacobs and Manchester[7]
are shown in an Arrhenius Plot in figure 3. The heat release data

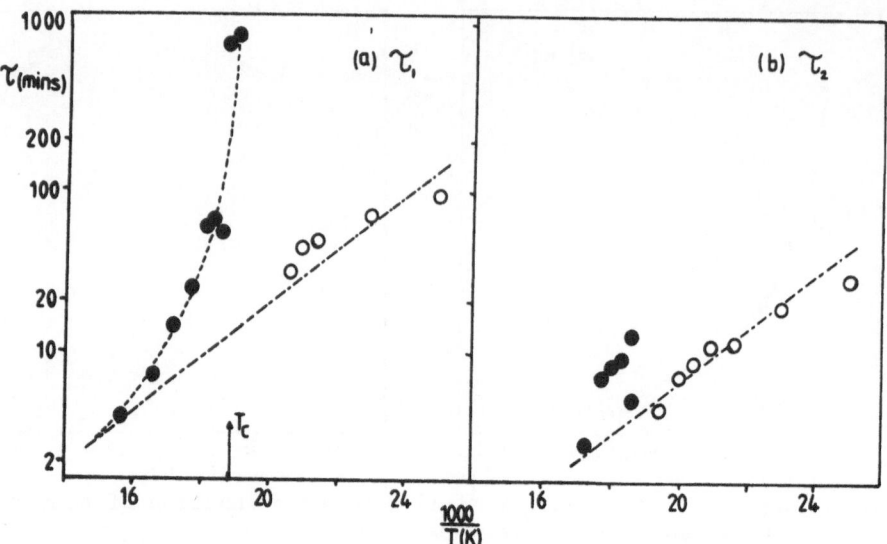

Fig. 3.   Arrhenius plots of the two time constants τ_1 (a) and τ_2
(b) from present neutron diffraction measurement (●) and
from the heat release measurement (o) (ref. 7).

was obtained for $PdH_{0.63}$ and has been corrected to the relaxation
times for the deuteride by using $\tau_D = 1.4\ \tau_H$[9].   As can be seen
the short time constant seems to fit the Arrhenius relationship
observed by Jacobs and Manchester. The long time constant fits the
Arrhenius relationship at 60 K but below this temperature it diverges
from this relationship exhibiting critical slowing down similar to
that observed in Cu_3Au[10].

        Preliminary measurements made more recently at Risø on the
$PdD_{0.64}$ crystal indicate that when the temperature is dropped slowly
in stages through the transition region the diffuse peak gradually
narrows and grows and then a sharp peak grows out of this diffuse
peak.   The sharp peak grows in height with time but the diffuse
peak remains unchanging. The scans on D10 were unfortunately not
wide enough to show whether there were two components present in
the peak. The Risø peak is much narrower; this may be due to there
being a narrow range of temperatures around T_c in which the ordered
crystallites can grow rapidly. By quenching rapidly through this
region as in the D10 experiment one has only small crystallites which
grow slowly during the anneal at the quench temperature. This would
lead to a gradual narrowing of the peak as it grows, unfortunately
we did not wait long enough or do sufficient scans to observe if
this was so. At Risø, however, the crystal passed through this
region much more slowly and so in consequence the crystallites were
larger giving rise to a narrower peak. Further investigation of
the difference between quenched and slowly cooled peak shapes is
proceeding.

CONCLUSION

In the concentration range $0.64 < x < 0.665$ palladium deuteride seems to show the characteristics of a second order phase transition just above the critical temperature with a lower transition temperature at the higher concentration. This is in agreement with the Monte Carlo calculations of Bond and Ross[11], who using interaction parameters determined by the position and shape of the diffuse scattering intensity in palladium deuteride predict a first order transition to the $I4_1$/amd structure for $x < 0.64$ and a second order transition in the range $0.64 < x < 0.69$ with a lower transition temperature at the higher concentrations.

ACKNOWLEDGEMENTS

We wish to thank Dr. W. Just, Dr. N. Lehner and the staff at the ILL and Risø for their help. Also Dr. R. Jordan of the Materials Science Centre, Birmingham University for preparing the palladium single crystals. Work supported in part by the U.K. Science and Engineering Research Council.

REFERENCES

1. J.K. Jacobs and F.D. Manchester, J. Less Common Metals 49:67 (1976)
2. I.S. Anderson, D.K. Ross and C.J. Carlile, Phys. Lett. 68A: 249 (1978).
3. T.E. Ellis, C.B. Satterthwaite, M.H. Mueller and T.O. Brun, Phys. Rev. Lett. 42:456 (1978).
4. O. Blaschko, R. Klemencic and P. Weinzierl, Acta Cryst. A36: 605 (1980).
5. C.B. Satterthwaite, T.E. Ellis and J.R. Miller, Physics of Transition Metals (Toronto 1977) IOP Phys. Conf. Ser. 39: 501 (1978).
6. J.K. Jacobs, C.R. Brown, V.S. Pavlov and F.D. Manchester, J. Phys. F. 6:2219 (1976).
7. J.K. Jacobs and F.D. Manchester, J. Phys. F. 7:23 (1977).
8. J. Als Nielsen, 'Phase Transitions and Critical Phenomena', C. Domb and M.S. Green, eds., vol 5a, p. 87, Academic Press, London (1976).
9. Y. de Ribaupiere and F.D. Manchester, J. Phys. C. 6:L390 (1973).
10. T. Hashimoto, T. Miyoshi and H. Ohtsuka, Phys. Rev. B13:1119 (1976).
11. R.A. Bond and D.K. Ross, J. Phys. F. to be published.

PROPERTIES OF VANADIUM-ALLOY DIHYDRIDES

H. G. Severin, E. Wicke

Institut für Physikalische Chemie
D-4400 Münster, Germany (FRG)

Abstract

Dihydrides of binary V-alloys with Ti and Cr have been investigated by X-ray diffraction, inelastic neutron scattering, measurements of the magnetic susceptibility and low temperature calorime-
try. A strong H induced bulk segregation in V/Ti-alloys required
special care in sample preparation. All dihydrides exhibit f.c.c.
structure, the lattice constant decreases with decreasing Ti content, but remains almost unaffected by the Cr concentration. Inelastic neutron scattering shows only little dependence of the
vibrational band modes on the alloy composition. Because the frequency range of the band modes and the local vibrations of H in the
tetrahedra are widely separated, no contribution of the optical
modes to the electron-phonon coupling and thus no superconductivity
is possible. The local H frequencies (mean peak energies) of all
the alloy dihydrides are lower than those of VH_2. The neutron
scattering spectra of the V-alloy dihydrides show a structure in
the peak of the optical modes (splitting). The steep rise of the
dihydride electronic heat capacity as well as of the susceptibility
is discussed in terms of band-structure calculations and reduced
screening charge at the H site.

Introduction

The hydrides of Vanadium and of Palladium are probably the metal-
hydrogen systems studied most frequently and in greatest detail.
Much of the understanding of hydride formation by Pd has been
derived from the examination of Pd-alloy hydrides /1/. Addition of
alloying elements permits a systematic variation of properties of the

host metal, and thereby the separation of factors determining the
hydride forming capacity. Much less however is known about the influence of alloying on the properties of V-hydrides. Therefore a
systematic investigation of V-alloy dihydrides with Ti and Cr has
been undertaken. Ti and Cr have been chosen because of their complete
miscibility with V what enables a wide range of alloy compositions.
Beyond this, Ti and Cr as nearest neighbors of V will not affect the
electronic structure of the host metal essentially, so that the basic
features of V remain unchanged.

Sample preparation and characterization

The samples were prepared from M.R.C. material (Orangeburg N.Y.) by
arc melting in an atmosphere of high purity Ar and remelted several
times. Subsequently they were annealed for 2 h at 1500 °C on a water
cooled Cu-finger by inductive heating. Homogeneity was examined by
electron-microprobe analysis, and by low-temperature heat capacity
measurements. These are suitable for homogeneity tests, because the
transition to superconductivity is sensitive to inhomogeneous distribution of the alloy components. The samples were charged with hydrogen from the gas phase after heating them in a vacuum better than
10^{-5}mbar for 2 h. The temperature of the heat treatment depended on
the thermal stability of the hydrides to be formed. The V/Ti-samples
were prepared at H_2 pressures below 1 bar in a quartz vessel, where
temperatures of 1000 °C could be applied. Most of the V/Ti-samples
immediately absorbed H after exposure to H_2 atmosphere and cooling
down. Contact with H at elevated temperatures had to be minimized
for them to avoid segregation /2/. The V/Cr-samples were charged
under H_2 pressures up to 160 bar in a different device limiting the
temperature treatment to 700 °C. They had to be reheated several
times in the H atmosphere in order to activate them sufficiently.

During the charging process the samples became brittle and disinte-
grated to small grains. Therefore the copper tabletting method had
to be applied for the low-temperature heat capacity investigations
as described elsewhere /2/.

X-Ray diffraction revealed f.c.c. structure for all the dihydrides
investigated. As shown in fig. 1 the lattice constant decreases
with decreasing Ti content, but remains almost constant for the
V/Cr-dihydrides.

Low-temperature heat-capacity and magnetic properties

The low-temperature heat capacity of V-alloy dihydrides exhibits
straight lines in the familiar Cp/T versus T^2 plot. This indicates
that there is no other contribution to the heat capacity than the
lattice and the electronic one. Especially no evidence for superconductivity can be found down to 1.4K. This is in agreement with
the results for VH_2 /7/. Thus the V/Ti/H_2 samples do not reflect the

high superconducting transition temperatures of the V/Ti-system.
Some dihydride samples have been investigated by susceptibility
measurements in the temperature range from 5K to room temperature
/8/.

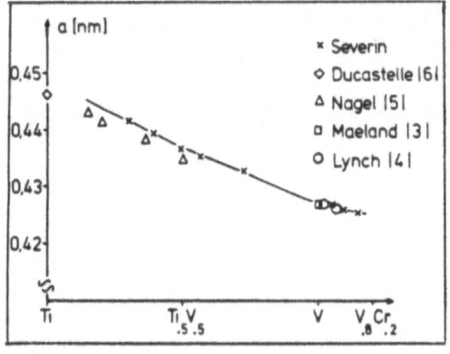

Fig. 1. Lattice constants of V-
alloy dihydrides

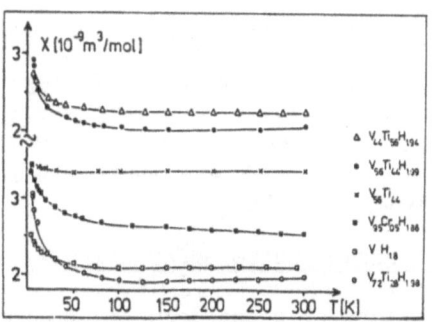

Fig. 2. Susceptibility as a
function of temperature

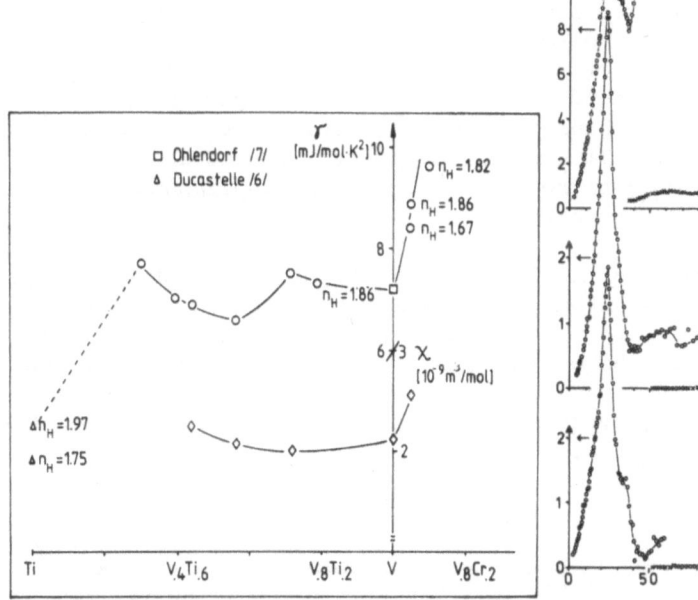

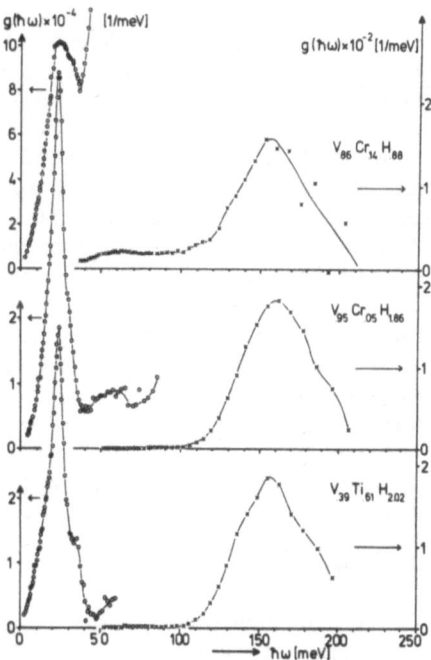

Fig. 3. Comparison of room-temperature susceptibilities
and coefficients γ of the
electronic heat capacity

Fig. 4. Spectra of lattice vibrations in V-alloy dihydrides

The susceptibility increases towards lower temperatures dependent on
the alloy composition (fig. 2). The increase grows on with lower Ti
contents, passes through a minimum for VH_2, and rises again for the
sample with 5at % Cr. The susceptibility of an alloy, wherefrom the
H has been removed, shows no increase at low temperatures (fig. 2);
this proves that impurity effects can be excluded. The coefficients
γ of the electronic heat capacity and the room temperature susceptibility are compared in fig. 3. As can be seen, they follow the
same trend as predicted by simple theoretical models. The deviation
in the γ value of the sample $V._{72}Ti_{28}H_{1.98}$ may be attributed to the
large low-temperature increase of the susceptibility in this sample
(fig. 2). So the low-temperature heat capacity may be enhanced by
a magnetic contribution.

Inelastic neutron-scattering

The lack of superconductivity in the V-alloy dihydrides can be understood readily from the spectra of the lattice vibrations as revealed
by inelastic neutron scattering (fig. 4) /9/. The band modes below
50 meV show high similarity for the V/Ti and the V/Cr dihydrides.This
is   consistent with nearly identical Debye-temperatures as obtained
from heat-capacity measurements. The maximum in the density-of-states
appears at 24 meV, in agreement with the Debye-temperatures and with
the results for VH_2 /10/. Band modes and local H modes are separated
by more than 100 meV due to the fact that the H atoms occupy tetrahedral sites (in contrast to PdH). This large energy difference
prevents any contribution of the optical modes to the electron-phonon
coupling. This has been found to be essential for superconductivity
in PdH /11/. On the other hand the phonon spectra of the acoustic
part of the V-alloys with Ti and Cr /12/ and the corresponding alloy
dihydrides are too much different to maintain the superconducting
properties of V-alloys themselves.

Investigations of the local H vibrations with the Be-filter technique
show clear evidence for a structure in the peak with two distinct
maxima for the samples with the lowest Ti and Cr content /13/(fig.5).
The high energy part of the peak nearly coincides in energy with the
local H modes in VH_2/14/, but as the peaks are significantly broader
(36 meV at half maximum compared with 18 meV for VH_2), the mean peak
energies are lowered by about 7 meV.

Discussion

In VH_2 the hydrogen density reaches an appreciable value, and there
are indications from theory /15, 16/ that this value is nearby the
maximum possible, corresponding to the minimum H-H distance. These
ideas are supported by the fact that the lattice constant of V is
reduced markedly by alloying of Cr, whereas the lattice constant of
the dihydride decreases only little compared with VH_2. The local
mode frequencies in VH_2 are the highest of all dihydrides with CaF_2

structure /14/. This points to a narrow and deep potential well,
originating from high M-H binding energies and also high repulsive
forces due to the smallness of the tetrahedral site. Admixtures
of Ti and of Cr both diminish the local mode frequencies of the
dihydrides, as mentioned above, but for different reasons. Cr
admixtures reduces the average M-H binding energy (as will be shown
in detail later), whereas admixture of Ti primarily increases the
lattice constant and thereby decreases the frequencies, in agreement with the relationship of Ross et al. /14/. As shown in Fig. 5,
the optical phonon peaks exhibit a splitting into two distinct
maxima. No structure in the density-of-states of the optical modes
has been observed for Me-H-systems with CaF_2-structure before. These
results cast some doubt on the purely local character of the optical
H-vibrations in the dihydrides investigated.

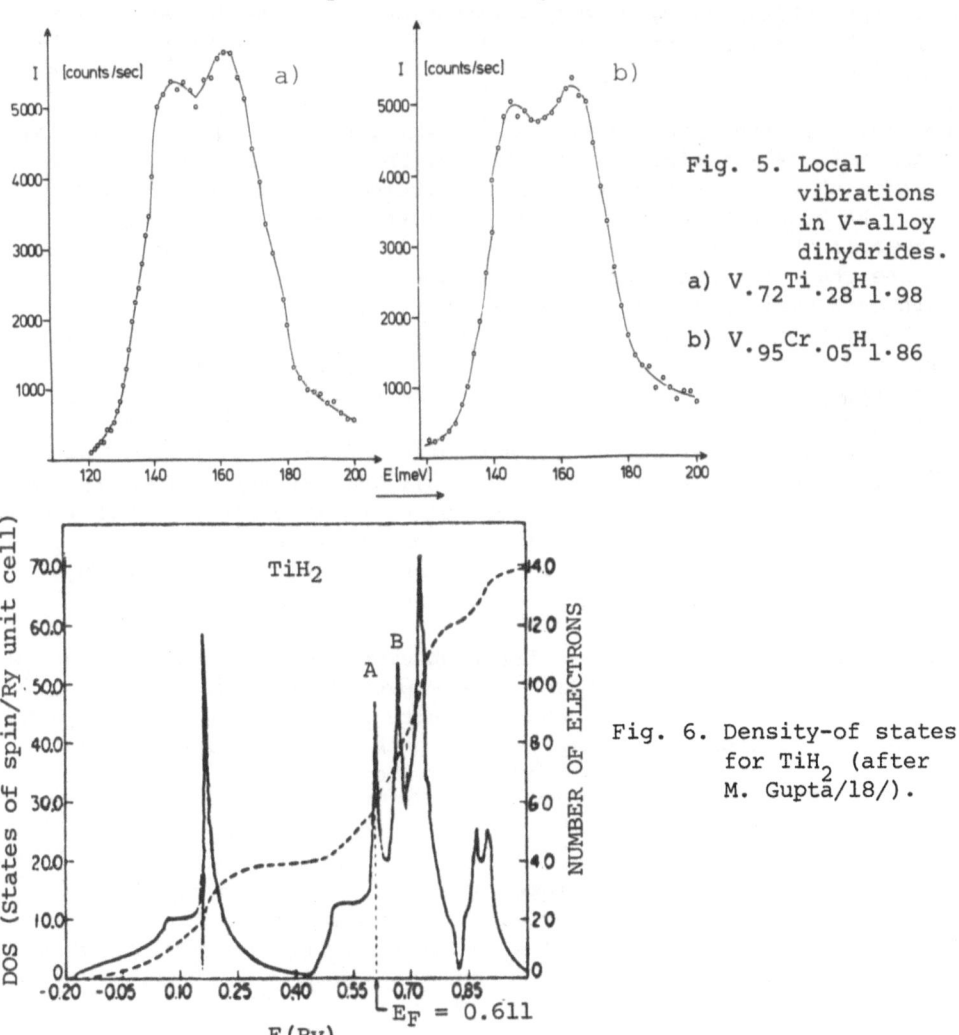

Fig. 5. Local
vibrations
in V-alloy
dihydrides.

a) $V_{.72}Ti_{.28}H_{1.98}$

b) $V_{.95}Cr_{.05}H_{1.86}$

Fig. 6. Density-of states
for TiH_2 (after
M. Gupta/18/).

The dependence of the γ-values and the room-temperature susceptibilities on the alloy composition can be deduced from band structure calculations for TiH_2 (Switendick /16, 17/, Gupta /18/, which are in basic agreement with one another and with UPS measurements /19/. Following Switendick /16/ we assume that the general features of the TiH_2 band-structure will not change very much upon alloying Ti with V. From calculations of Gupta /18/ (fig. 6) we can then follow the density-of-states values at the Fermi level, passing through a minimum and rising up again to a sharp maximum (marked B in fig. 6), when V is added to Ti and finally Cr to V. The question arises, whether the degenerate d-states leading to peak B will cause a Jahn-Teller distortion analogous to the d-states represented by peak A in TiH_2 /6/. However, low-temperature X-ray diffraction of a sample $V._{95}Cr._{05}H_{1.86}$, as well as the temperature dependence of the magnetic susceptibilities, gave no indication of a Jahn-Teller effect down to 5 K. May be the effect appears at higher Cr contents in the dihydrides.

The course of the γ-values in fig. 3 exhibits an unusually steep increase for small admixtures of Cr to V. This gives rise to the speculation that more electrons are contributed to the metal d-bands than just by the Cr atoms, that means more extra electrons from H-atoms in the interstitial sites. In a simple consideration one may imagine that as a consequence of the decrease of the lattice constant - although small - upon Cr addition a critical M-H distance is reached, where accomodation of the screening charge at the H-site becomes difficult. This would enhance the fraction of the electronic charge that is raised up to and above the Fermi level, and thereby no longer available for the M-H bonding. Therefrom may originate the lowering of the local mode frequencies observed with the V/Cr dihydrides in comparison to VH_2 (fig. 5). This effect is expected to be different for the H isotopes because of their different vibrational amplitudes. In fact a large enrichment of deuterium in the solid phase has been observed in isotope exchange measurements with VH_2, indicating the higher stability of the deuteride /20/.

The difference in the isotope effect observed between the dideuteride and tritium, however, is smaller than expected /20/. With regard to the foregoing considerations these results can be interpreted as due to the smaller vibrational amplitudes the heavier isotopes do not reach the critical distances mentionned above.

ACKNOWLEDGEMENTS

We are indebted to Dr. J. Ihringer (Inst. f. Kristallographie, Tübingen) for the performance of the low-temperature X-ray investigation. Cooperation of K. Frölich in the magnetic investigations is gratefully acknowledged.

REFERENCES

1/ E. Wicke, H. Brodowsky; Hydrogen in Metals II, eds.: G. Alefeld,
    J. Völkl, Springer Verlag, Berlin 1978

2/ D. Ohlendorf, H. G. Severin, E. Wicke; Z. Physik. Chem. NF
    in press
    D. Ohlendorf, H. G. Severin, E. Wicke; Thermochimica Acta 49
    (1981) 11

3/ A. J. Maeland; J. Phys. Chem. 68 (1964) 2197

4/ J. F. Lynch, J. J. Reilly, F. Millot; J. Phys. Chem. Sol. 39
    (1978) 883

5/ H. Nagel, R. S. Perkins; Z. Metallkde 66 (1975) 362

6/ F. Ducastelle, R. Caudron, P. Costa; J. Phys. (Paris)
    31 (1970) 57

7/ D. Ohlendorf, E. Wicke; J. Phys. Chem. Solids 40 (1979) 721

8/ H. G. Severin, K. Frölich, E. Wicke; to be published

9/ H. G. Severin, P. Schweiß; unpublished

10/ J. J. Rush, H. E. Flotow; J. Chem. Phys. 48 (1968) 3795

11/ B. N. Ganguly; Z. Phys. 265 (1973) 433

12/ G. F. Syrykh, M. G. Zemlyanov, N. A. Chernoplekov,
     B. I. Saveliev; Zh. Eksp. Teor. Fiz. 81 (1981) 308

13/ H. G. Severin, E. Wicke, S. Wilson, D. K. Ross; to be
     published

14/ D. K. Ross, P. F. Martin, W. A. Oates, R. Khoda Bakhsh;
     Z. Phys. Chem. NF 114 (1979) 341

15/ P. C. P. Bouten, A. R. Miedema; J. Less-Common Met. 71
     (1980) 147

16/ A. C. Switendick; Z. Phys. Chem. NF 114 (1979) 221

17/ A. C. Switendick; J. Less-Common Met. 49 (1976) 283

18/ M. Gupta; Solid State Commun. 29 (1979) 47

19/ J. H. Weaver, D. J. Peterman, D. T. Peterson, A. Franciosi;
     Phys. Rev. B, 23 (1981) 1692

20/ R. H. Wiswall, J. J. Reilly; Inorg. Chem. 2 (1972) 1691

TEMPERATURE DEPENDENCE OF INTRINSIC DISORDER IN LaH_x and YH_x

Peter M. Richards

Sandia National Laboratories
Albuquerque, New Mexico 87185

ABSTRACT

Intrinsic disorder in fcc metal hydrides MH_x, defined as $x_o(2)$, the fraction of octahedral (0) sites occupied at stoichiometry $x=2$, has been measured by both spectroscopic[1] and thermodynamic[2] techniques for M=La and Y. The spectroscopic (NMR, ESR and neutrons) data are all for temperature t < 300K while the thermodynamic (pressure vs. x isotherms) results have t > 850K. A most surprising feature is that $x_o(2)$ at 300K is greater than or equal to its value at high temperature. This is in gross disagreement with any theory which attributes $x_o(2)$ solely to the finite difference between 0-site and tetrahedral (T) site energies, including vibrational states.

My proposed explanation is based on a model in which large amplitude hydrogen vibrations combined with repulsive interactions prevent occupation of both a T and neighbor 0-site at high temperature. Thus $x_o(2)$ goes through a maximum vs. temperature as disorder is established more via vibrational entropy than by 0-site occupation.

A quantitative calculation is performed by assuming an energy level structure of bound ($\varepsilon < 0$) harmonic oscillator states and extended ($\varepsilon > 0$) particle-in-a-box states. Because of repulsion, a T(0) site $\varepsilon > 0$ state is possible only if the neighbor 0(T) sites are vacant. The vibrational energy is known from neutron data, and the box size for $\varepsilon > 0$ states is taken to agree with the available volume per site assuming cubic close packing of the metal ion spheres. The difference in well depths $U_0 - U_T$ is given by the low temperature data[1]; so the only really free parameter is the depth of the T-site well-U_T. The value-U_T=0.02eV gives good agreement and corresponds to the low temperature activation energy[3] for NMR in YH_2. Sensitivity of the model to U_T and implications for NMR relaxation are discussed.

ACKNOWLEDGEMENTS

This work was performed at Sadia National Laboratories and supported by the U.S. Department of Energy under contract number DE-AC04-76DP00789.

REFERENCES

1.   E. L. Venturini and P. M. Richards, Phys. Lett. <u>76</u> A, 344 (1980).
2.   P. Dantzer and O. J. Kleppa, J. Solid State Chem. <u>35</u>, 34 (1980).
3.   D. L. Anderson et al, J. Less Comm. Metals <u>73</u>, 243 (1980).

ELECTRONIC STRUCTURE OF METAL HYDRIDES: A REVIEW

OF EXPERIMENTAL AND THEORETICAL PROGRESS

J.H. Weaver,* D.J. Peterman* and D.T. Peterson**

*Synchroton Radiation Center
University of Wisconsin–Madison
Stoughton, Wisconsin 53589 USA

**Ames Laboratory–USDOE and Department of
Materials Science and Engineering
Iowa State University, Ames, Iowa 50011 USA

ABSTRACT

In this paper we discuss metal-hydrogen electronic interactions in bulk hydrides by reviewing recent theoretical and experimental results for typical monohydrides (VH, NbH, and TaH), dihydrides (LaH_2, PrH_2, and NdH_2), and trihydrides (LaH_3).

INTRODUCTION

At this conference there have been many stimulating papers in which metal hydrogen interactions have been discussed. From the interest of the conferees, it is clear that the field of "hydrogen in metals" is alive and active. Its interdisciplinary character is also evident and this allows, or should allow, each of us to gain insight into the global problems of hydrogen in metals and hydrides. It is the purpose of this paper to review recent studies relating to one aspect of hydrides: their electronic structure. To do so, we will consider experimental results obtained using photoemission spectroscopy studies and will compare them to calculations of the electronic structure, $E(\vec{k})$, of ordered hydrides. A review of non-stoichiometric properties will be given by Bansil elsewhere in these proceedings.

Until recently the understanding of how the hydrogen atom
interacts electronically when introduced into a metal lattice has
been limited. Freidel[1] first studied the behavior of a proton in
a simple metal in 1953 but it wasn't until ∿1970 that the details
of metal-hydrogen interactions were considered in the context of
band structure. At that time Switendick[2] performed insightful band
calculations for a representative transition metal, a monohydride,
a dihydride and a trihydride, $YH_0-YH_1-YH_2-YH_3$, and predicted how
the addition of a proton with nuclear charge Z = 1 might alter the
electronic properties. It was not until several years later (1977)
that experiments[3] were performed which unambiguously demonstrated
that Switendick's model was correct and that his treatment of
composition-dependent M-H interactions was valid.

Within the last several years, there have been a series of
sophisticated calculations which have made the understanding of the
electronic structure more quantitative for a variety of hydrides of
various compositions,[4-35] stoichiometric and non-stoichiometric.
In addition, Gelatt et al.[32,33] established important systematics
for 3d-4d metal monohydrides and displayed wave functions for Pd and
PdH which showed a clear and intuitive picture of charge rearrangement and transfer due to hydrogen perturbation. Deviations from
stoichiometry have been examined for several hydrides through
ATA,[33-35] CPA,[13,14] VCA,[15] and supercell[2,4,5,9,28-31] calculations.
Calculations of the effects of a hydrogen impurity in metal clusters
and treatments of H as an isolated impurity have also been
reported.[36,37] In this rich arena, there have been relatively few
experimental studies of the band structure using, for example,
photoemission or optical spectroscopy.[38-55]

The clear message that comes from these calculations and
experiments is that metal-hydrogen interactions have a profound
impact on the electronic structure of the host, that any model which
treats a hydride as a "slightly-modified" metal is limited, and that
changes in the electronic structure must be appreciated if related
physical and chemical properties of hydrides are to be understood,
e.g. hybriding[53-55] in $LaNi_5-LaNi_5H_{6.7}$.

METAL MONOHYDRIDES

Particular insight into M-H interactions in hydrides can be
gained by examining, for example, the band structure systematics
demonstrated by Switendick's PdH_x and YH_x calculations[2,4,7] and the
wavefunction calculations for Pd-PdH by Gelatt et al.[33] The former
showed (and his calculations have been substantiated by others) that
in monohydrides the hydrogen potential influences, in a particularly
strong way, the lower states in the band structure and, more
generally, alters states which have substantial real-space charge

distributions near the proton site. It is reasonable to expect that extended states with the proper symmetry can mix with the H 1s-derived states and can be lowered in energy. Indeed, occupied states at the bottom of the metal sp-band (furthest from E_F) pull away from the higher states. Furthermore, some empty states of the metal can be lowered below E_F by the M-H interaction and become occupied. Finally, to account for the number of electrons added in excess of the number of states lowered, there must be an upward shift of Fermi level. Change is transferred toward the proton, screening it from its environment.

This intuitive picture of screening can be extended to the dilute limit where hydrogen acts as an impurity. Bansil[35] has reviewed this important problem in his paper at this conference. In addition to the results coming from band calculations, important insight into local effects are gained by considering the spatial distribution of charge around a proton in a Ni cluster, as has been reported recently by Rudolf and Chaney.[37] They showed the charge distribution around the proton to be greater than that of a single electron and that the radial distribution was non-hydrogenic.

From the various calculations for PdH, we would expect to see a band (two electrons) separated from other occupied states of the solid. These predictions were more or less substantiated in early photoemission experiments[51] with PdH_x although there remains some ambiguity concerning sample characterization. Recent optical spectroscopy studies[52] of PdH_x also support the above picture. In the future, a critical study of the electronic structure of PdH using synchrotron radiation photoemission might be appropriate since there is already a wealth of theoretical information for comparison.

We recently undertook a study of the electronic properties of the hydrides of V, Nb, and Ta[49] as part of our on-going program involving experimental and theoretical studies of hydrides. Of particular interest was the formation of the lowered band since it had not been well-characterized in the monohydrides. These refractory metal hydrides display many different ordered structures in which the metal lattice is bcc or slightly distorted and, in most cases, the hydrogen occupies the tetrahedral sites.[56] Unfortunately, the lower crystal symmetry of these phases, compared to the simple NaCl structure of PdH, increases the difficulty and cost of band calculations. Furthermore, since the hydrogen mobility is very high at room temperature, our experiments had to be performed at low temperature lest the samples evolve too much hydrogen. In our synchrotron radiation photoemission studies, then, we cleaved bulk hydride samples _in situ_ at $\sim 5 \times 10^{-11}$ Torr at $\sim 50K$. Although the hydrogen partial pressure increased, as measured with a quadrupole mass spectrometer, the experimental results showed no time dependences which could not be related to contamination from non-

hydrogen residual gases. Analogous measurements of NbD_x at room
temperature showed no evidence of deuterium release and the photoemission results for the deuterides agreed with those for the
hydrides.

In Fig. 1 we show representative background-subtracted energy
distribution curves (EDC's) measured at a photon energy of 21 eV
using fractured polycrystalline samples of VH, NbH, and TaH. Also
shown are the results for $NbD_{.75}$. The two low-lying features for
these hydrides, falling approximately 5.6 and 7.6 eV below E_F reflect
the drawn-off band due to M-H interactions. In the metals themselves,
there are no such H-induced features below the sp-d bands.[57] The
presence of these low-lying features, of course, indicates that the
rigid band model fails to describe hydride formation in these monohydrides and we look to band theory for a detailed interpretation.

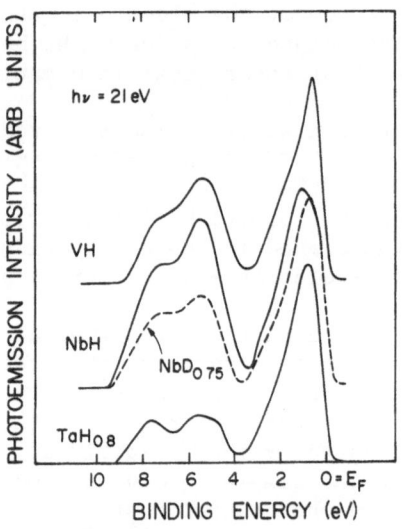

Fig. 1   Photoemission spectra for VH, NbH, and $TaH_{0.8}$ showing the
         d-derived states within ~3 eV of E_F and the H-induced band
         centered at ~7 eV.

Several monohydride energy band calculations based on simplified crystal structures can be found in the literature. Switendick's[5] calculation of VH, based on a bcc metal lattice and a hydrogen atom above every vanadium site at $(0,0,a/2)$, indicated a split-off band with two prominent peaks in the density of states. These were located 2-3 eV below where they are observed experimentally (Fig. 1) but the full width of that band is in reasonable agreement with experiment. However, the remaining "metal" bands were calculated to have a width of $\sim$5 eV, compared to an observed value of $\sim$3 eV.

A different crystal structure was employed in the calculation of NbH by Gupta and Burger.[21,22] The metal atoms were placed on an fcc lattice with hydrogen occupying octahedral sites $(a/2,a/2,a/2)$. This work, undertaken to study trends in superconductivity, predicted a single feature in the density of states related to the split-off band but also predicted two distinct structures in the metal bands near E_F. As expected, there appeared even less agreement with the experimental results for NbH shown in Fig. 1 because of the assumption of an fcc lattice.

Motivated by this lack of detailed agreement, we undertook a theoretical study using a more realistic crystal structure. The pseudo-cubic γ-phase of NbH was chosen for this calculation because it occurs in both NbH and TaH and its relatively high crystal symmetry (D_{2d}) reduces computation time compared to the less symmetric β-phase. In γ-NbH, the metal atoms fall on a bcc lattice with hydrogen in tetrahedral sites $(a/2,0,a/4)$ relative to the metal.[56] While calculational details and results will be presented elsewhere,[49] we discuss here preliminary results of non-self-consistent (NSC) energy band calculations using the Korringa-Kohn-Rostoker (KKR) method.

In Fig. 2 we show the NSC bands and density of states (DOS) calculated for γ-phase NbH. The symmetry labels are for the bcc lattice and we include two inequivalent N points given by $(\pi/a, \pi/a, 0)$ and $(\pi/a, 0, \pi/a)$. The DOS shows a prominent hydrogen-induced peak near -8 eV and gives a full width of the occupied metal band of about 5 eV. The presence of only one hydrogen peak in Fig. 2, compared to two found by Switendick for VH,[5] is related to the different site symmetries and nearest-neighbor-hydrogen coordination numbers. Comparison of the experimental results of Fig. 1 and the calculations of Fig. 2 still shows a dismaying lack of agreement; this shortcoming is significantly greater than expected based on our previous success with analogous comparisons for the dihydrides of the group III and IV transition metals. Most puzzling is that the predicted metal band-width for NbH, which is about the same as that found by Switendick for VH, is too large ($\sim$5 eV predicted, $\sim$3 eV measured). Noteworthy, but less surprising, is the apparent discrepancy in the calculated width of the split-off band (2-3 eV

predicted, 4-5 eV observed). These H-induced features are particularly sensitive to the details of the calculation but the results
for the d-bands should be considerably less sensitive.

We suggest two possibilities for further theoretical work.
First, self-consistent calculations allowing for charge transfer may
be important. For the dihydrides, iteration to self-consistency
shifted the hydrogen-induced bands toward E_F and narrowed the d-bands.
Second, a calculation for the β-phase would show the effects of
reduced symmetry and would probably have a noticeable effect on the
appearance of the split-off H-induced band. It may also show substantial changes in the d-bands because of increased M-H interaction
and lowering of empty states, thereby requiring less shift in E_F and
giving a narrower predicted d-band width. In the meantime, we
conclude that band calculations, which are beginning to be applied
to complex structures, have yet to provide the necessary details to
understand the experimental results.

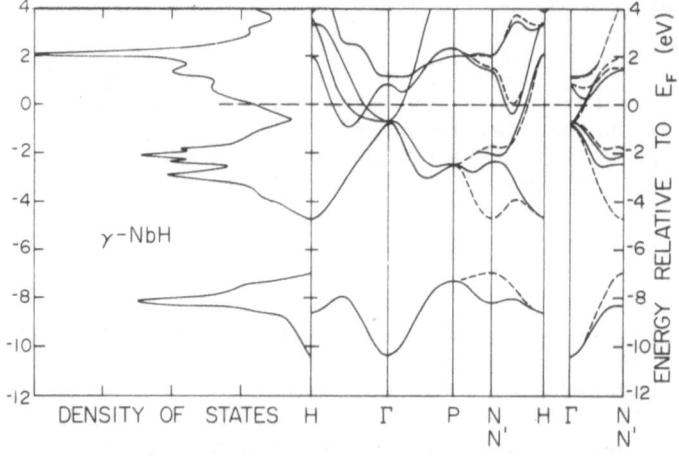

Fig. 2   Calculated NSC energy bands and DOS for γ-NbH. The dashed
         lines represent bands connecting to the non-equivalent
         N-N´ points in the Brillouin zone (see text).

METAL DIHYDRIDES

The electronic structure of metal dihydrides, as revealed by experiment and calculation, appears at first glance to be very different from those of the monohydrides, but the interaction mechanisms have a great deal in common. As with monohydrides, we must consider those states of the crystal which are sensitive to the potential energy perturbation introduced by the proton. Intuitively, one should expect that the extended states will again be most affected. In the dihydride, however, there are now two protons and this difference, compared to the monohydride, makes itself felt in a way which is more difficult to visualize but is related to the symmetries of the H-1s states and the lattice itself. With two protons, the H-derived charge distribution can be written as $\psi_h + \psi_h$ (bonding) and $\psi_h - \psi_h$ (antibonding). Now the correct question to ask is which states of the solid have the appropriate symmetry to hybridize with these bonding and antibonding states.

Switendick was the first to discuss the formation of dihydrides and he elegantly described this bonding-antibonding combination of H-1s states in their interaction with the metal.[2] He showed that in metal dihydrides with CaF_2 structure (H at tetrahedral sites) these hydrogen combinations result in a hybridization and lowering in energy of certain metal states. For example, the antibonding state $\psi_h - \psi_h$ mixes with Γ_2' and lowers it by $\sim$10 eV. Localized states such as the purely d-derived Γ_{12} and Γ_{25}' states are virtually unchanged. Others are more difficult to visualize because of their inherently-mixed angular momentum character, but it is reasonable to return to the picture where delocalized states are those which are most likely to be influenced and then ask about their symmetry properties.

There have been several calculations of the electronic structure of different dihydrides but they all demonstrate the influence of the bonding and antibonding symmetries and agree as to the overall picture. Gupta and Burger[19] also reported an interesting result for Al, AlH, and AlH_2 in which the hydrogen influence was examined in the context of a classic nearly-free-electron metal. Their results showed the pulling off of the extended states in the monohydride and the lowering of empty states in the dihydride, analogous to the transition metal hydrides in that the same states were affected because of their symmetry but the contribution of d-character to bonding was absent. The results suggest how one should approach the alkali-earth dihydrides, including MgH_2.

We have presented detailed comparisons between experimental spectra (obtained through optical and photoemission studies of a variety of hydrides) and theoretical predictions in a series of papers[40-48] which have focussed on the dihydrides. The reader is referred to those papers for details. In this brief overview, we offer in Fig. 3 new photoemission spectra for LaH_2, PrH_2, and NdH_2

for comparison to theory. The results show partially occupied,
largely-d-derived bands within about 2 eV of the Fermi level and
hybridized M-H bands at greater binding energy (centered near -5 eV).
Some states comprising the bonding bands were formerly occupied in
the metal, others were drawn from above E_F, but all have mixed with
the hydrogen charge to produce hybridized s-p-d states in a region
where nothing is observed in the elemental metals. The overall width

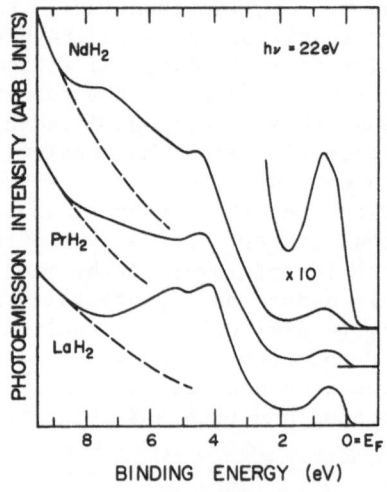

Fig. 3 Photoemission spectra for LaH_2, PrH_2, and NdH_2 showing
        metal-hydrogen bands around -5 eV; d-derived states within
        ∿2 eV of E_F; 4f emission overlapping the bonding bands.

of these bands varies with the H-H separation, a correlation which
Switendick predicted based on the separation of the hydrogen atoms
via the resulting bonding and antibonding combinations of H-1s
orbitals. In the spectra of Fig. 3, there is an additional contribution due to emission from the highly localized, core-like 4f
electrons which accounts for a broadening of the bonding band
features in PrH_2 compared to LaH_2 and in the structure near -7.5 eV.

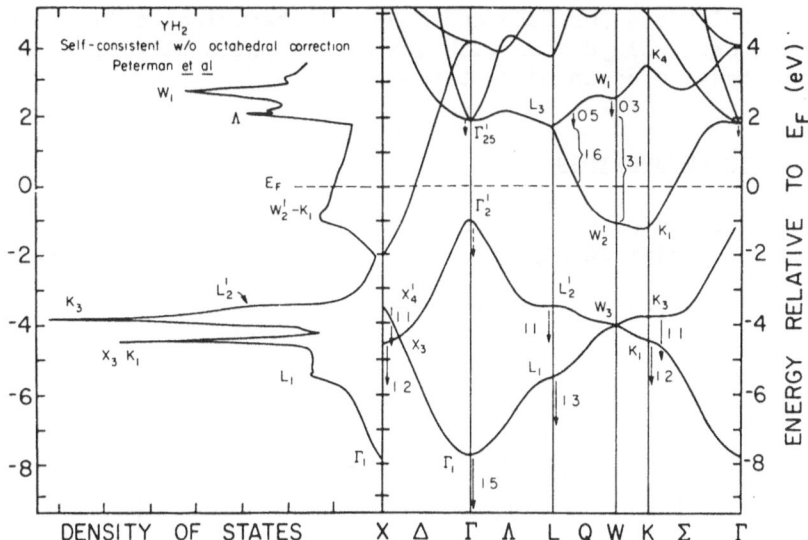

Fig. 4   Calculated self-consistent band structure and DOS for YH_2.
The arrows indicate changes suggested by photoemission,
optical reflectance, and thermoreflectance studies, as
discussed in Ref. 48.

The results shown in Fig. 3 and those obtained for the dihydrides
($x \simeq 2$) of Sc, Y, La, Ce, Er, Lu, Ti, Zr, and Hf demonstrate[40-48]
that the basic modeling of dihydride interaction is correct. In
Fig. 4 we reproduce the calculated bands for YH_2 and the density of
states for comparison to the experimental results of Fig. 3. The
arrows were drawn after detailed analysis of photoemission results[43]
(deeper features) and optical reflectance or thermoreflectance
results[41] (states within about 2 eV of E_F) and indicate how the
calculations should be shifted to agree with experiment. Subsequent
refinements in the calculations[23-24] for ScH_2 led to excellent agreement with optical measurements of band separations within the d-band
complex but the calculated position of the hydrogen-derived states
remained less accurate. As part of our optical and photoemission
study of the dihydrides, we also showed that octahedral site occupancy
occurs for $x < 2$ and that new low-energy optical transitions occur.[41]
Calculations by Switendick[9] and Peterman and Harmon[24] provided
plausible explanations which would explain these x-dependent features.

Further calculations that would be of interest include CrH and
CuH. While these are monohydrides in concentration, they contain
2 formula units per unit cell. Switendick has postulated[2] that the
energetics of these compounds is such that the "dihydride-like"
antibonding-bonding combination of H orbitals is necessary to lower
more states below E_F than occurs with only one formula unit per unit

cell. Calculations for CaH_2 and MgH_2 will be valuable in comparing transition metal dihydrides with the alkali earth dihydrides which, formally, have no occupied d-states.

TRI- AND HIGHER HYDRIDES

The addition of yet more hydrogen to the unit cell continues the mixing (hybridization) of metal- and hydrogen-derived states

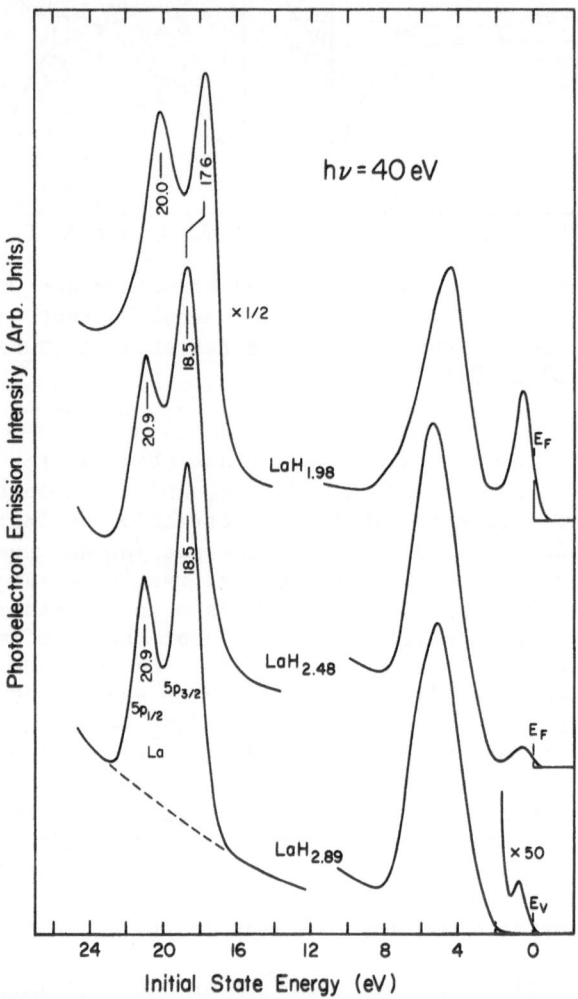

Fig. 5 Photoemission spectra for $LaH_{1.98}$, $LaH_{2.48}$, and $LaH_{2.89}$ showing the disappearance of states near E_F (drawn down by H-interaction), the increase of emission from La-H hybridized bands, and the shift of the La 5p core levels due to charge transfer from the metal site.

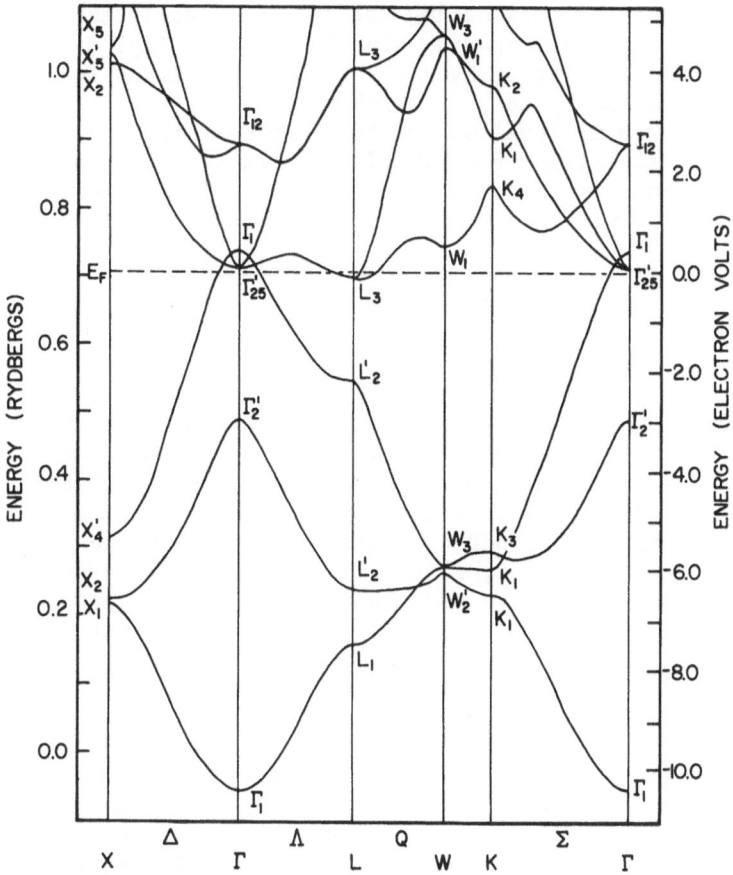

Fig. 6   Calculated NSC bands for YH_3. The Γ_1 state falls above E_F
in YH_3 but below E_F in LaH_3, giving rise to a semiconducting
gap, consistent with the results of Fig. 5. Comparison of
Figs. 4 and 6 shows how the M-H interactions lower states
from near or above E_F.

because there are now additional symmetries of the H-1s states which
can mix with and lower additional states from near or above E_F.
Switendick[2] first theoretically addressed this problem studying
YH_2-YH_3. Experimental studies have been performed by our group for
the LaH_2-LaH_3, CeH_2-CeH_3, and ThH_2-Th_4H_{15} systems using optical
spectroscopy for the first[48] and photoemission[3,48] for all three.

The photoemission spectra for $LaH_{1.98}$-$LaH_{2.48}$-$LaH_{2.89}$ shown
in Fig. 5 indicate diminished emission from states near E_F and
increased emission from the bonding bands with increased hydrogen
content. In $LaH_{2.89}$, the metal band at E_F has disappeared and the
results suggest that $LaH_{2.89}$ is a semiconductor. As yet unpublished
results for $CeH_{2.2}$ and $CeH_{2.6}$ support the observed trends. These

can be understood by considering the non-self-consistent bands shown in Fig. 6 for YH_3 and comparing them to those of Fig. 4 for YH_2. The additional coupling of hydrogen 1s orbitals leads to a mixing with such states as Γ_2 and L_2', which were well above E_F in YH_2, and a shift of the Fermi level upward to accomodate the 6 electrons per unit cell. Calculations for LaH_3 (which has a considerably larger lattice constant) indicate that the Γ_1 state falls below E_F and the trihydride is a small gap semiconductor, consistent with the results shown in Fig. 5.

SUMMARY

In this brief review we have tried to highlight the basic physics behind the metal-hydrogen interaction in a variety of hydrides, seeking to be intuitive in our description wherever possible. Comparison of experimental results with theoretical band structures and densities of states has shown generally good agreement, with exceptions as noted. With the understanding of these relatively simple hydrides, it should be possible to examine the role of the electronic structure in the more complicated systems such as $LaNi_5H_{6.7}$. Such studies are currently underway.[53-55]

ACKNOWLEDGEMENTS

The continuous support of the personnel of the Ames Laboratory and the Synchrotron Radiation Center is gratefully acknowledged. We are grateful for stimulating discussions with many colleagues over the years. The work at the University of Wisconsin is supported by USDOE OBES-DE-ACO2-80ER10584. The Ames Laboratory is operated for the USDOE by ISU under contract W-7405-ENG-82 (WPAS-KC-02-02). The Synchrotron Radiation Center is supported under NSF DMR 7821888.

REFERENCES

1. J. Friedel, Philos. Mag. _43_, 153 (1952).
2. A. Switendick, Electronic Band Structures of Metal Hydrides, Solid State Commun. _8_, 1463 (1970); A. C. Switendick, Metal Hydrides - Structure and Band Structure, Int. J. Quantum Chem. _5_, 459 (1971).
3. J. H. Weaver, J. A. Knapp, D. E. Eastman, D. T. Peterson, and C. B. Satterthwaite, Electronic Structure of the Thorium Hydrides ThH_2 and Th_4H_{15}, Phys. Rev. Lett. _39_, 639 (1977).
4. A.C. Switendick, Electronic Energy Bands of Metal Hydrides - Pd and Ni Hydrides, Ber. Bunsenges Physik. Chem. _76_, 535 (1972).

5. A. C. Switendick, Hydrogen in Metals - A New Theoretical
    Model, in: "Hydrogen Energy, Part B," T. N. Veziroglu, ed.,
    Plenum Press (1975) p. 1029.
6. A. C. Switendick, Bandstructure Calculations for Metal Hydrogen
    Systems, Z. Physik Chemie $\underline{117}$, 89 (1979).
7. A. C. Switendick, The Change in Electronic Properties on Hydrogen
    Alloying and Hydride Formation, in: "Topics in Applied Physics,
    Vol. 28, Hydrogen in Metals I: Basic Properties," G. Alefeld
    and J. Volkl, eds., Springer Verlag (1978) p. 101.
8. A. C. Switendick, Influence of the Electronic Structure on the
    Titanium-Vanadium-Hydrogen Phase Diagram," J. Less-Common
    Metals $\underline{49}$, 283 (1976).
9. A. C. Switendick, Electronic Structure of Non-Stoichiometric
    Cubic Hydrides, J. Less-Common Metals $\underline{74}$, 199 (1980).
10. D. A. Papaconstantopoulos, Electronic Structure of Metal
    Hydrides, "Proceedings of the NATO Advanced Study Institute
    on Metal Hydrides," Plenum Press (1981) and detailed
    references therein.
11. D. A. Papaconstantopoulos and B.M. Klein, Superconductivity in
    the Pd-H System, Phys. Rev. Lett. $\underline{35}$, 110 (1975).
12. D. A. Papaconstantopoulos, B. M. Klein, E. N. Economou, and
    L. L. Boyer, Band Structure and Superconductivity of PdD_x
    and PdH_x, Phys. Rev. B $\underline{17}$, 141 (1978).
13. D. A. Papaconstantopoulos, B. M. Klein, J. S. Faulkner, and
    L. L. Boyer, Coherent-Potential-Approximation Calculations
    for PdH_x, Phys. Rev. B $\underline{18}$, 2784 (1978).
14. J. S. Faulkner, Electronic States of Substoichiometric Compounds
    and Application to Pd-H, Phys. Rev. B $\underline{13}$, 2391 (1976).
15. J. Zbasnik and M. Mahnig, Electronic Structure of β-Phase PdH,
    Z. Phys. B $\underline{23}$, 15 (1976).
16. M. Gupta and A. J. Freeman, Electronic Structure and Proton
    Spin-Lattice Relaxation in PdH, Phys. Rev. B $\underline{17}$, 3029 (1978).
17. M. Gupta, Electronic Structure of ErH_2, Solid State Commun. $\underline{27}$,
    1355 (1978).
18. M. Gupta, Electronically Driven Tetragonal Distortion in TiH_2,
    Solid State Commun. $\underline{29}$, 47 (1979).
19. M. Gupta and J. P. Burger, Electronic Structure and Electron-
    Phonon Interaction in Aluminum Hydrides, J. Physique $\underline{41}$,
    1009 (1980).
20. M. Gupta and J. P. Burger, Electronic Structure of Rare-Earth
    Hydrides: LaH_2 and LaH_3," Phys. Rev. B $\underline{22}$, 6074 (1980);
    M. Gupta and J. P. Burger, Electronic Structure and Its
    Relationship to Superconductivity in NiH, J. Phys. F $\underline{10}$,
    2649 (1980).
21. M. Gupta and J. P. Burger, Trends in the Electronic-Phonon
    Coupling Parameter in Some Metallic Hydrides, Phys. Rev. B
    $\underline{24}$, 7099 (1981).
22. M. Gupta, Electronic Properties and Electron-Phonon Coupling
    in Zirconium and Niobium Hydrides, private communication.

23. D. J. Peterman and B. N. Harmon, Electronic Structure of ScH_2,
     Phys. Rev. B 20, 5313 (1979).
24. D. J. Peterman, B. N. Harmon, D. L. Johnson, and J. Marchiando,
     Electronic Structure of Trivalent Metal Dihydrides: Theory,
     Z. Phy. Chem. 116, 47 (1979).
25. N. I. Kulikov, V. N. Bobzunov, and A. D. Zvonkov, The Electronic
     Band Structure and Interatomic Bond in Nickel and Titanium
     Hydrides, Phys. Stat. Sol. B 86, 83 (1978).
26. N. I. Kulikov and V. V. Tugeshev, An Electronic Band Structure
     Model for the Metal-Semiconductor Transition in Cerium-Group
     Hydrides, J. Less-Common Metals 74, 227 (1980).
27. N. I. Kulikov and A. D. Zvonkov, Band Structure and Metal-to-
     Semiconductor Transition in the Cubic Hydrides of 3B-Subgroup
     Elements, Z. Physik. Chemie 117, 113 (1979).
28. A. Fujimori and N. Tsuda, Electronic States in Non-Stoichiometric
     Rare-Earth Hydrides, J. Phys. C 14, 1427 (1981).
29. A. Fujimori, F. Minami, and N. Tsuda, Electronic Structure of
     Cerium Hydrides: Augmented-Plane-Wave-LCAO Energy Bands,
     Phys. Rev. B 22, 3573 (1980).
30. A. Fujimori and N. Tsuda, Electron-Phonon Interaction and
     Composition-Dependent Phonon Anomaly in CeH_x, J. Phys. C
     14, L69 (1981).
31. A. Fujimori and N. Tsuda, Electronic Structure of TiH_2, Solid
     State Commun. in press; A. Fujimori and N. Tsuda, Electronic
     Structure of Nonstoichiometric Titanium Hydrides, private
     communication.
32. M. Methfessel and J. Kubler, Bond Analysis of Heats of Formation:
     Application to Some Group VIII and IB Hydrides, J. Phys. F
     12, 141 (1982).
33. C. D. Gelatt, J. A. Weiss, and H. Ehrenreich, Heats of Formation
     of 3d and 4d Transition Metal Hydrides, Solid State Commun.
     17, 663 (1975); C. D. Gelatt, H. Ehrenreich, and J. A. Weiss,
     Transition Metal Hydrides: Electronic Structure and the Heats
     of Formation, Phys. Rev. B 17, 1940 (1978).
34. A. Bansil, R. Prasad, S. Bessendorf, L. Schwartz, W. J. Venema,
     R. Feenstra, F. Blom and R. Griessen, Electronic States and
     Fermi Surface Properties of α-Phase PdH_x, Solid State Commun.
     32, 1115 (1979).
35. A. Bansil, Proceedings of this conference and references therein.
36. M. I. Darby, G. R. Evans, and M. N. Read, Self-Consistent
     Screening of Hydrogen in Zirconium, J. Phys. F 11, 1023 (1981).
37. P. G. Rudolf and R. C. Chaney, Electronic Structure of Hydrogen
     Impurity in Nickel Using the Linear-Combination-of-Atomic-
     Orbitals Method, Phys. Rev. B in press.
38. D. L. Westlake, C. B. Satterthwaite, and J. H. Weaver, Hydrogen
     in Metals, Physics Today 31, 32 (1978).
39. J. H. Weaver and D. T. Peterson, The Influence of Interstitial
     Hydrogen on the Band Structure of Nb and Ta: An Optical
     Study of $NbH_{0.453}$ and $TaH_{0.257}$, Phys. Lett. A 62, 433 (1977).

40. J. H. Weaver, R. Rosei, and D. T. Peterson, Optical Interband
    Structure of the Low Energy Plasmon in ScH_2, Solid State
    Commun. <u>25</u>, 201 (1978).

41. J. H. Weaver, R. Rosei, and D. T. Peterson, Electronic Structure
    of Metal Hydrides I: Optical Studies of ScH_2, YH_2, and LuH_2,
    Phys. Rev. B <u>19</u>, 4855 (1979).

42. D. J. Peterman, B. N. Harmon, J. Marchiando, and J. H. Weaver,
    "Electronic Structure of Metal Hydrides II: Band Theory of
    ScH_2 and YH_2, Phys. Rev. B <u>19</u>, 4867 (1979).

43. J. H. Weaver, D. T. Peterson, and R. L. Benbow, Electronic
    Structure of Metal Hydrides III: Photoelectron Studies of
    ScH_2, YH_2, and LuH_2, Phys. Rev. B <u>20</u>, 5301 (1979).

44. J. H. Weaver and D. T. Peterson, Photoelectron Spectroscopy
    of Metal Dihydrides, Z. Physik. Chemie <u>116</u>, 501 (1979).

45. D. J. Peterman, D. T. Peterson, and J. H. Weaver, Optical and
    Photoemission Studies of Lanthanum Hydrides, J. Less-Common
    Metals <u>74</u>, 167 (1980).

46. J. H. Weaver and D. T. Peterson, Electronic Structure of Metal
    Hydrides, J. Less-Common Metals <u>74</u>, 207 (1980).

47. J. H. Weaver, D. J. Peterman, D. T. Peterson, and A. Franciosi,
    Electronic Structure of Metal Hydrides IV: TiH_x, ZrH_x, HfH_x,
    and the fcc-fct Lattice Distortion, Phys. Rev. B <u>23</u>, 1692
    (1981).

48. D. J. Peterman, J. H. Weaver, and D. T. Peterson, Electronic
    Structure of Metal Hydrides V: X-Dependent Properties of
    LaH_x ($1.9 \leq x \leq 2.9$) and NdH_x ($2.01 \leq x \leq 2.27$), Phys. Rev.
    B <u>23</u>, 3903 (1981).

49. D. J. Peterman, D. Misemer, and J. H. Weaver, Electronic
    Structure of Metal Hydrides VI: Photoemission Studies of VH,
    NbH, and TaH and Band Calculations of NbH," manuscript in
    preparation.

50. B. W. Veal, D. J. Lam, and D. G. Westlake, X-Ray Photoemission
    Spectroscopy Study of Zirconium Hydride, Phys. Rev. B <u>19</u>,
    2856 (1979).

51. D. E. Eastman, J. K. Cashion, and A. C. Switendick, Photoemission
    Studies of Energy Bands in the Pd-H System, Phys. Rev. Lett.
    <u>27</u>, 35 (1971).

52. G. A. Frazier and R. Glosser, Hydrogen-Induced Changes in the
    Electronic Structure of Pd-H Measured by Thermoreflectance,
    Solid State Commun. <u>41</u>, 245 (1982).

53. J. H. Weaver, A. Franciosi, W. E. Wallace, and H. Kevin Smith,
    Bulk Electronic Structure and Surface Oxidation of $LaNi_5$,
    Er_6Mn_{23} and Related Systems, J. Appl. Phys. <u>51</u>, 5847 (1980).

54. J. H. Weaver, D. J. Peterman, A. Franciosi, T. Takeshita, and
    K. A. Gschneidner, Electronic Structure and Surface Oxidation
    of the Haucke Compounds $CaNi_5$, YNi_5, $LaNi_5$, and $ThNi_5$,
    J. Less-Common Metals (1982).

55. W. E. Wallace, J. H. Weaver, D. J. Peterman, and F. Pourarian,
    Photoemission Studies of $LaNi_{5-x}Cu_x$ Alloys and Relation to

Hydride Formation, J. Phys. Chem. (1982).

56. T. Schober and H. Wenzl, The Systems NbH(D), TaH(D), VH(D): Structures, Phase Diagrams, Morphologies, Methods of Preparation, in "Topics in Applied Physics, Vol. 29, Hydrogen in Metals," G. Alefeld and J. Volkl, eds., Springer Verlag (1978) chapter 2 and references therein.

57. For Nb see, for example, R. J. Smith, Photoemission Studies of Hydrogen Chemisorption on Nb, Phys. Rev. B $\underline{21}$, 3131 (1980).

THE ELECTRONIC STRUCTURE OF PdH$_x$ STUDIED WITH PHOTOEMISSION AND ELECTRON ENERGY LOSS SPECTROSCOPY

P. Bennett and J. C. Fuggle

Bell Laboratories, Murray Hill, NJ 07974
and
KFA-Jülich

ABSTRACT

A 10μ thick Pd foil was charged to a concentration of H$_x$∿.75 by cooling slowly from 400 K to 80 K under 2000 Torr H$_2$ pressure in a preparation chamber and then transferred under vacuum into the spectrometer. Upon hydrogenation the following changes in the spectra are observed: in the valence band, a broad peak appears at ∿8 eV BE, the total d-band width is reduced ∿10% and the density of states at E$_f$ decreases to nearly zero. In the 3d core lineshape, a satellite structure ∿7 eV below the main line disappears, and the asymmetry of the line is reduced. In the energy loss spectrum, a prominent peak at 7 eV is reduced in intensity and shifted to 4.5 eV.

INTRODUCTION

The photoemission technique has contributed greatly to the understanding of the electronic structure of metallic hydrides, however, extreme surface sensitivity and UHV requirements have hindered its use for technologically important "high-pressure" materials like palladium hydride. Previous experiments have reported "hydrogen-induced" levels at 5.4 eV in the UPS spectrum.[1] Another study has inferred from XPS core lineshapes a filling of the d-bands for hydrogen concentrations near 0.7 atomic fraction.[2,3] These results have been questioned due to poor sample characterization.[4] Here we report XPS and EELS data on a well characterized sample prepared by in-situ cleaning and loading, and measured at 80 K to retain the hydrogen.

SAMPLE PREPARATION

The sample was a 1cm. × 1cm. × 10μ thick palladium foil of 4N purity from W. Hereaus. After an initial ion bombardment cleaning, the sample was hydrided by exposure to hydrogen of 5N purity at 2000 Torr and 400 K. It was then cooled at ∿10°C/minute to 80 K. As the sample cools under pressure, the hydrogen concentration increases, following the 2000 Torr isobar until the slowing kinetics prevent further uptake. The resulting bulk concentration is estimated to be $H_x \approx$ .75 atomic fraction.

The fully loaded sample was then introduced into the measuring chamber where a background pressure of 2×10^{-10} Torr was maintained. The XPS spectra of the loaded, cold sample showed oxygen and carbon contamination corresponding to 0 = .16 and C = .14 atomic fractions relative to palladium. These figures are calculated from core-line intensities using atomic cross-sections from Ref. 5. Attempts to remove these contaminants by ion sputtering or scraping gave levels of 0 = .13 and C = .20, or .20 and .10, respectively. It was found, however, that briefly warming the sample to 100 K reduced these to $0 \leq .03$ and C = .14, which were the same as obtained for the ion-sputtered pure palladium sample. The preferred treatment was therefore no cleaning except for a series of "flashings" to remove weakly adsorbed contaminants. The presence of these contaminants produced no noticeable change in the valence band, Pd core-level or EELS spectra. We also measured these spectra after exposing the cleaned, cold (120 K) sample to ∿1000L H_2 or 1000L of atmospheric gases and again found no noticeable change. The spectra shown for the "drhydrided" sample were measured at 220 K, a temperature just sufficient to desorb hydrogen from the bulk after several hours waiting.

RESULTS AND DISCUSSION

The valence band spectra are shown in Fig. 1 for the fully hydrided and degassed samples. The following changes are observed upon hydrogenation: New states appear at ∿8 eV BE, the d-bands "sink" such that the half-maximum near the top of the valence band is ∿0.2 ± .1 eV below E_f (determined from a gold reference), the total width of the spectrum is decreased ∿10% (as indicated in the figure) and a small peak appears near the top of the band. These features are in good qualitative agreement with several recent bulk band-structure calculations for the palladium-hydrogen system.[6,7,8] The new states at 8 eV are strongly hybridized bonding orbitals between the palladium and hydrogen. The fact that they are visible in the XPS spectrum demonstrates that considerable d-character remains in these states, since the photoemission cross-section for the H 1s atomic orbital is 10^{-5} that of the palladium 4d orbitals.[5] The narrowing of total

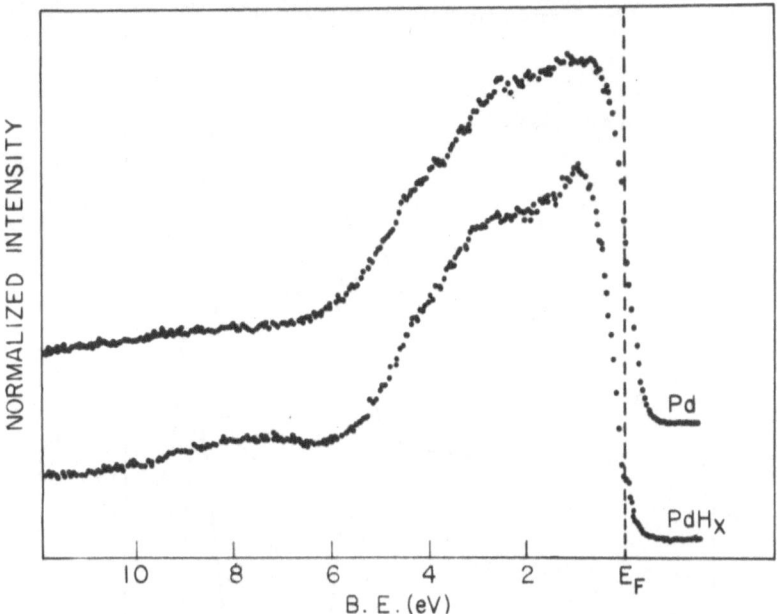

Fig. 1 XPS valence band spectra of the pure and fully loaded
        sample measured at 220 K and 80 K, respectively. H$_x$ ∿ .75
        (see text). The Fermi level is determined with a gold
        reference.

bandwidth is understandable as a result of the 3% increase of the
lattice parameter upon hydrogenation. The calculated "rise" of the
Fermi level on a total energy scale cannot be measured experimentally
without a fixed reference, however, the observed change with respect
to the valence band is consistent with the substoichiometric
calculations of Ref. 7. The increased intensity near the top of the
bands is not present in the calculations, but is presumed to result
from an instrumental broadening which obscures the peak in pure
palladium and enhances the peak in the hydride where it is further
below the Fermi cutoff.

     These spectra appear different from UPS spectra, which show new
states near 6 eV BE. The discrepancy may result from sensitivity
to possible contamination or true surface states,[9] or from structure
in the final DOS and matrix element variations across the valence
band. These effects are presumably much reduced in our XPS
measurements.

     The spectra of the palladium 3d core levels are shown in Fig. 2.
The intensity scale is arbitrary and the energy scale has been
adjusted to align the peaks. Upon hydrogenation, the satellite
feature present a 6 eV below each of the spin-orbit split lines is
seen to essentially disappear, and the lines become less asymmetric,

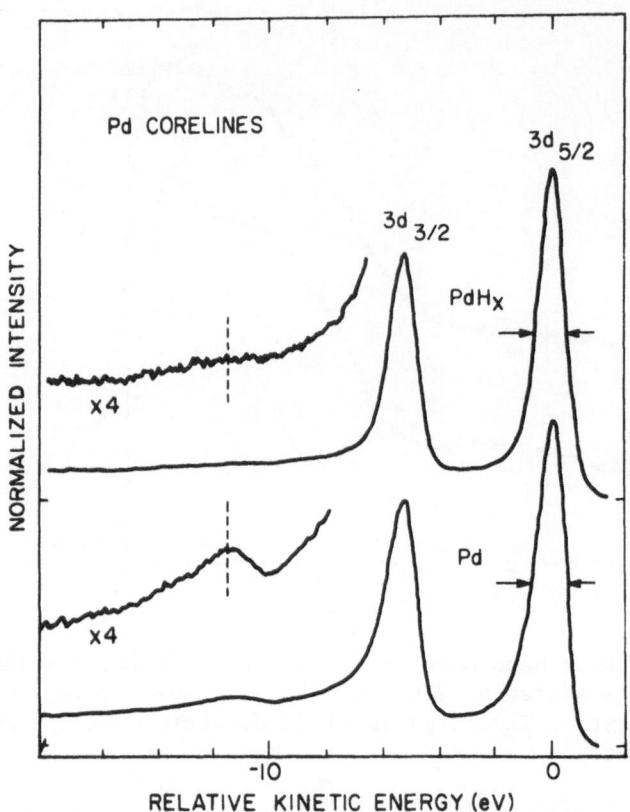

Fig. 2   Pd 3d core-lines of the pure and fully loaded sample. The
         intensity has been normalized and the energy scale shifted
         to align peak maxima. Instrumental width ∼0.6 eV FWHM.

which reduces their measured FWHM from 1.45 eV to 1.20 eV. Not
shown on the drawing is a 16% decrease of the peak maximum
corresponding to a 20% reduction in integrated intensity. Also,
the energy of the half-maximum on the high KE side of the line
decreases $0.2 \pm .1$ eV with respect to the gold reference lines.

    We interpret the changes in lineshape to result from a filling
of the d-bands upon hydrogenation and the subsequent decrease to
nearly zero of the density of states at E_f. This follows since the
satellite and shake-up tail are understood to result from the
presence of empty d-states just above the Fermi level, analogous to
the case of nickel and its alloys.*[10,11,12] The -0.2 eV
shift of the line is apparently smaller than expected on the basis of

*There is a possibility in the case of palladium that some intensity
 in the region of the satellite arises through inelastic
 scattering.

initial state calculations of Ref. 6.

The electron energy loss spectra (EELS) are shown in Fig. 3 for several different primary beam energies. The vertical scale has been normalized to the elastic peak intensity for all traces. In pure palladium, a prominent peak is present at 7 eV energy loss, followed by a weaker and much broader peak around 25 eV. This structure does not change much at lower primary energies, except that a shoulder appears on the smaller loss side of the 7 eV peak. In the hydrided spectra the 7 eV peak is shifted to $\sim$4.5 eV and is reduced to 1/2 its intensity, and the broad structure near 25 eV appears somewhat larger. Again, little variation is seen with decreasing primary energy. The lack of sensitivity to primary beam energy indicates that surface states or possible surface contamination contribute little to the spectra. The similarity of these spectra to the reported energy loss function for bulk palladium prompt us to assign the 7 eV feature to a plasmon loss ($\varepsilon_1 = 0$, $\varepsilon_2 \sim 2$) and the 25 eV feature to interband transitions, principally into bands, 8, 9 and 10.[13,14] The shoulder on the plasmon peak may be a surface plasmon, which should occur at energies

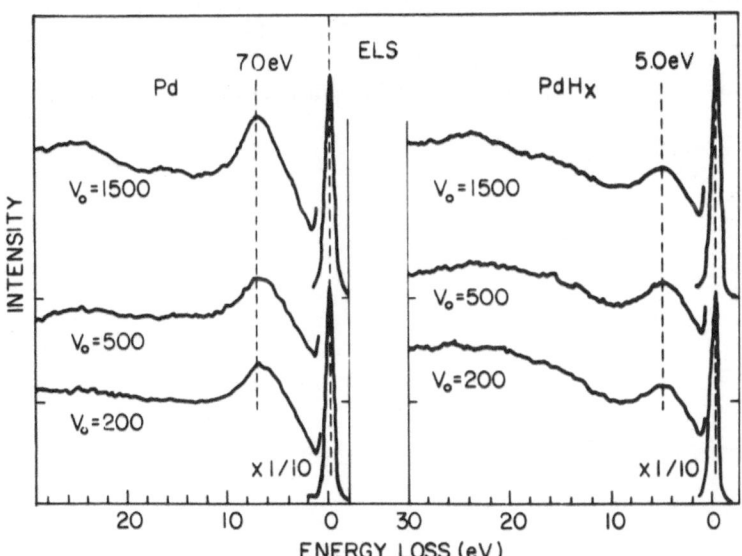

Fig. 3   EELS spectra, angle-integrated in reflection, shown for the pure and loaded sample and for three primary energies: 1500, 500 and 200 eV. Intensity is normalized to the elastic peak, shown × 1/10.

very near the bulk plasmon loss because ε_1 is changing rapidly here. The shift of the plasmon to lower energy in the hydride presumably results from enhanced interband transitions of energy $\gtrsim 8$ eV, which one may expect due to a large "anti-bonding" density of states ~ 8 eV above E_f.

REFERENCES

1. D. Eastman, J. Cashion and A. Switendick, Phys. Rev. Lett. 27: 35 (1971).
2. F. Antonangelli, A. Balzarotti, A. Bianconi, P. Perfetti, P. Ascarelli and N. Nistico, H. Nuovo Cimento 39B:720 (1977).
3. F. Antonangelli, A. Balzarotti, A. Bianconi, E. Burattini, P. Perfetti and N. Nistico, Phys. Lett. 55A:309 (1975).
4. P. Légaré, L. Hilaire, G. Maire, G. Krill and A. Amamou, Surf. Sci. 107:533 (1981).
5. J. H. Scofield, J. Elec. Spectros. 8:129 (1976).
6. P. Jena, F. Y. Fradin and D. E. Ellis, Phys. Rev. B20:3543 (1979).
7. J. S. Faulkner, Phys. Rev. B13:2391 (1976).
8. C. D. Gelatt, H. Ehrenreich and J. Weiss, Phys. Rev. B17:1940 (1978).
9. S. Louie, Phys. Rev. Lett. 42:476 (1979).
10. U. Hillebrecht, to be published.
11. J. Fuggle and Z. Zolnierek, Solid State Comm. 38:799 (1981).
12. L. A. Feldkamp and L. C. Davis, Phys. Rev. B22:3644 (1980).
13. J. H. Weaver, Phys. Rev. B11:1416 (1975).
14. R. Lässer and N. V. Smith, to be published.

XPS/UPS STUDY OF THE ELECTRONIC STRUCTURE OF $PdH_{0.6}$

L. Schlapbach and J.P. Burger*

Lab. für Festkorperphysik ETH, CH-8093 Zürich

*Lab. de Physiques des Solides, Université de Paris-Sud
F-91405 Orsay

ABSTRACT

We have studied the electronic structure of polycrystalline $PdH_{0.6}$ at 100K by means of photoelectron spectroscopy (XPS and UPS). In $PdH_{0.6}$ as compared to Pd the 4d-band is shifted away from E_F and $N(E_F)$ is considerably decreased. In contradiction to earlier investigations we do not observe a band at 5eV. However, we see weak emission at 8eV probably related with the hydrogen induced states or with chemisorbed hydrogen. The chemical shift of the $3d_{5/2}$ core level is very small (+0.15 ± 0.10eV). A weak additional peak at 5.5eV is found on Pd samples which have been exposed to O_2 at 600K.

INTRODUCTION

Pd dissolves at room temperature some hydrogen in the α-phase $(PdH_{0.01})$ and forms at the equilibrium pressure of ≈10mbar the β-phase $PdH_{0.6}$. At higher pressure the hydrogen content of the β-phase increases towards stoichiometric PdH [1].

Many bandstructure calculations were performed for PdH_x [2-4]. An overview on the different methods and the results was given recently by Switendick [2]. All the calculations have the following features in common, which can be tested experimentally:

- The 4d-band of Pd is modified and shifted relatively to E_F. The density of states $N(E_F)$ decreases from 1.1 states/eV unit cell for Pd to 0.25 states /eV unit cell in PdH [3].
- A new band is formed about 7eV below E_F. It is mainly derived

from s- and d-states of Pd. Its position shifts with the hydrogen concentration.

- A small charge transfer seems to occur.

The decrease of $N(E_F)$ is in good agreement with results from specific heat (cf. ref[3]) and magnetic susceptibility measurements[5].

The electronic structure of PdH_x and of hydrogen chemisorbed on Pd was investigated by several groups by means of photoelectron spectroscopy using X-rays (XPS) or UV radiation (UPS):

Eastman et al.[6] evaporated Pd at $5 \cdot 10^{-6}$ Torr H_2 onto a substrate at 350°C and exposed the film subsequently at room temperature to 170 Torr H_2 for 30 min and analyzed it with UPS. An additional peak was detected at 5.4eV and interpreted as the hydrogen induced band of β-phase Pd hydride. No decrease of $N(E_F)$ was observed. The analysis was probably made at room temperature in ultrahigh vacuum.

Antonangeli et al.[7] analyzed electrolytically charged $PdH_{0.6}$ with XPS. They observed additional states from 5.5 to 8eV which they related to the hydrogen induced states. No variation of $N(E_F)$ was observed. The Pd $3d_{5/2}$ core level is shifted from 334.8eV in Pd to 334.9eV in $PdH_{0.6}$. Again, sample temperature and hydrogen pressure during the analysis are not indicated.

Veal et al.[8] and Gilberg et al.[9] failed to see the hydrogen induced states of PdH_x by means of XPS and X-ray emission spectroscopy because the hydrogen escaped through the clean surface.

Demuth[10] found a hydrogen induced peak at 6.5eV for $1 \cdot 10^{-6}$ Torr-sec H_2 chemisorbed on Pd(111) at 80K, in agreement with theory[11], whereas Eberhardt et al.[12] observed a similar peak at 7.9eV on Pd(111) which was exposed to $2 \cdot 10^{-6}$ Torr sec H_2 at 100K.

The discrapancies between the experimental results of different groups about the hydrogen induced states of PdH_x and between theory and experiment about $N(E_F)$ led us to the reinvestigation of PdH_x by means of XPS and UPS.

EXPERIMENTAL

The analysis was performed on a VG Escalab spectrometer using Mg Kα (1253eV), He I (21.2eV) and He II (40.8eV) radiation. The analyzer was set to give a resolution of ≈1.0eV for Mg Kα and 0.2eV for He I,II. E_F and the binding energy scale were calibrated on Au. Apart from the valence band we also investigated the Pd $3d_{5/2,3/2}$ core levels. The photoemission was measured normal to the surface.

The working pressure was $6 \cdot 10^{-11}$mbar before the inlet of He and before the desorption of hydrogen started.

Polycrystalline Pd foils (0.125 mm thick, 99.99%, Goodfellow), mounted on Cu sample holders, were cleaned in the spectrometer by Ar^+ bombardement and subsequently annealed at 600K for 1 hour. After the analysis of Pd metal the sample was transferred into the high pressure autoclave of the spectrometer and kept for 30 min at 15 bar H_2 at room temperature. Before releasing the high pressure the sample was cooled to 100K and then transferred on a cooled support to the precooled sample manipulator and analyzed at 100K.

The pressure in the analyzing chamber rose to $5 \cdot 10^{-9}$ mbar H_2 and dropped to $5 \cdot 10^{-10}$mbar H_2 within 10 min. A pressure increase was also observed when the power of the X-ray source was set to 200 Watt or more. Thus the XPS-spectra were measured at 50 and 100 Watt. The residual gas was analyzed with a mass spectrometer. All pressure variations were due to H_2 or He (He lamp on). The peaks at the masses 16, 18 and 28 (O_2, H_2O, N_2, CO) showed the same intensity as at $6 \cdot 10^{-11}$mbar total pressure.

The contamination of the sample was checked by the analysis of the 1s levels of carbon and oxygen before and after the hydrogenation and again after desorption. No carbon could be detected. The oxygen 1s level overlaps with the Pd $3p_{3/2}$ peak. However, from the comparison of the peak areas divided by theoretical cross[13] section of the $3d_{5/2}$ and $3p_{3/2}$ peaks we conclude that within the accuracy of the background subtraction and of the cross sections no oxygen was present.

From the thermodynamic properties of the Pd-H system [1] we estimate that we charged our samples to the concentration slightly above PdH$_{0.6}$. At 100K the equilibrium pressure should be of the order of 10^{-10}mbar H_2 and the kinetics very slow. Thus we estimate that our samples, after some initial desorption during the first 10 minutes, correspond to PdH$_{0.6}$.

At temperatures above 150K and ultrahigh vacuum all our PdH$_{0.6}$ samples with clean surface desorbed very quickly their hydrogen at least within the escape depth of XPS ($\simeq$ 25Å for the VB and Pd $3d_{5/2}$. Electrolytically charged samples kept the hydrogen under similar conditions, but their surface was contaminated.

RESULTS

In Fig.1 the He I and He II spectra are shown for PdH$_{0.6}$ at 100K and $5 \cdot 10^{-10}$mbar H_2 (curves A), for PdH$_x$ after the beginning of

the desorption ($\approx$ 200K, $1 \cdot 10^{-6}$mbar H_2, curve B) and for Pd at the
end of the desorption (250K, $5 \cdot 10^{-7}$mbar, curve C). In PdH$_{0.6}$
the d band is displaced relative to E_F and accordingly $N(E_F)$ is
rather small. No peak appears between 5 and 6eV. With increasing
desorption (B) the displacement becomes smaller and at the end of the
desorption (C) the d band begins at E_F and $N(E_F)$ is large. The He I

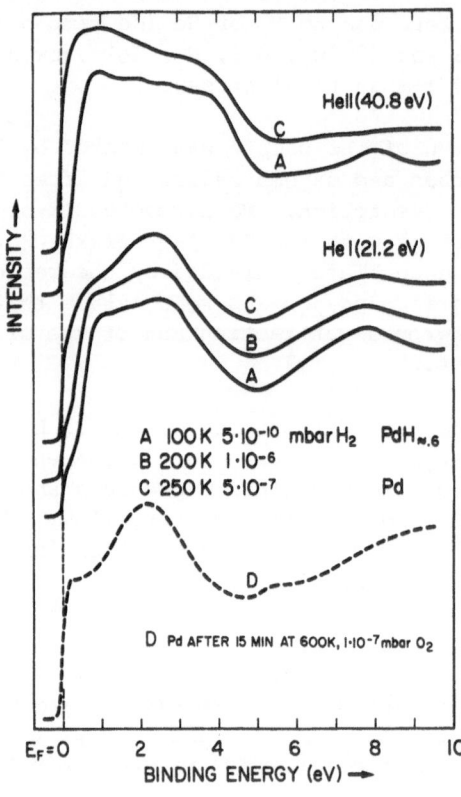

Fig. 1. UPS spectra of PdH$_x$ before (A), during (B) and at
        the end (C) of hydrogen desorption. Curve D was
        measured on Pd which was exposed to $1 \cdot 10^{-7}$ mbar O_2
        at 600 K for 15 min.

and He II spectra of PdH$_{0.6}$(A) as compared to those at the end of
the desorption (C) exhibit a weak additional emission around 8eV.
The He I spectrum of Pd which was exposed to 10^{-7}mbar oxygen at
600K for 15min shows a weak emission at 5.5eV (Fig.1, D).

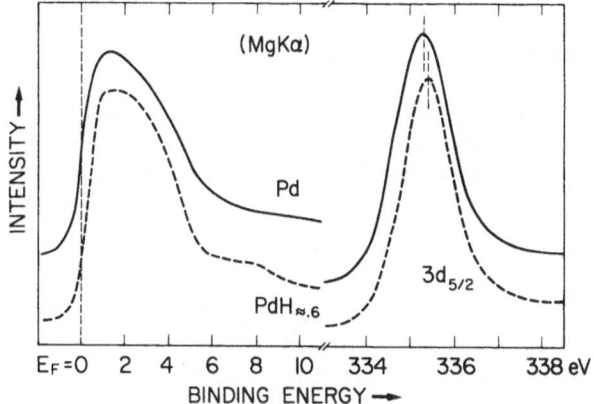

Fig. 2. XPS spectra of the valence band and of the 3d$_{5/2}$ core
level of PdH$_{0.6}$ and Pd at 100 K.

The XPS valence band spectrum of PdH$_{0.6}$ (Fig.2) as compared
to that of Pd shows the same features as the He spectra show: A
small displacement of the d-band relative to E$_F$ together with the
decrease of N(E$_F$), no peak between 5 and 6eV, but some additional
emission around 8eV. The d-band intensity of PdH$_{0.6}$ at 3eV is
slightly larger than in Pd. The chemical shift of the 3d$_{5/2}$ core
level (Fig.2) to higher binding energy is very small (0.15 $\pm$ 0.10eV)
and compares fairly well with the shift of 0.3eV calculated by
Jena et al.[4].

CONCLUSIONS

Hydrogen desorbes from PdH_x with clean surface within a few minutes at temperatures above $\simeq$ 150K under ultrahigh vacuum conditions. We have shown that the XPS and UPS spectra of $PdH_{0.6}$ at 100K differ from those of Pd in the following points: The d-band of $PdH_{0.6}$ is displaced relative to E_F and $N(E_F)$ is considerably reduced in agreement with theory. A weak emission around 8eV is probably due to the hydrogen induced states but could also originate from chemisorbed hydrogen. We do not observe, however, hydrogen induced states betweeen 5 and 6eV. The $3d_{5/2}$ core level is shifted not more than 0.15 $\pm$ 0.10eV.

ACKNOWLEDGEMENTS

We should like to thank P. Brack for technical assistance, H.C. Siegmann for continuous interest and support and the Swiss National Science Foudation for financial support.

REFERENCES

1. E.Wicke, H.Brodowsky and H.Zürcher in Topics in Appl.Physics, Vol.29, Hydrogen in Metals, G.Alefeld, ed., Springer 1978
2. A.C.Switendick, Zeitschrift für Physik. Chemie N.F. 117, 89 (1979). and ref. cited therein.
3. M.Gupta and A.J.Freeman, Phys.Rev.B 17, 3029,(1978)
4. P.Jena, F.Y.Fradin and D.E.Ellis, Phys.Rev.B 20,3543,(1979)
5. R.J.Miller, T.O.Brun and C.B.Satterthwaite, Phys.Rev.B 18, 5054 (1978)
6. D.E.Eastman, J.K.Cashion and A.C.Switendick, Phys.Rev.Letters 27, 35 (1971)
7. F.Antonangeli, A.Balzarotti, A.Bianconi, E.Burattini, P.Perfetti and N.Nistico, Phys. Letters 55A, 309 (1975) and Solid State Commun. 21, 201 (1977)
8. B.W.Veal, D.J.Lam and D.G.Westlake, Phys.Rev. B 19, 2856 (1979)
9. E.Gilberg, Extended Abstracts, Int.Conf. on the Physics of X-Ray Spectra, National Bureau of Standards, Gaithersburg, Maryland (1976)
10.J.E.Demuth, Surf. Sci. 65,369 (1977)
11.S.G.Louie, Phys.Rev.Letters 42,476 (1979)
12.W.Eberhardt, F.Greuter and W.E.Plummer, Phys.Rev.Letters 46, 1085 (1981)
13.J.H.Scofield, J.Electron Spectr. 8, 129 (1976)

ELECTRONIC STRUCTURE OF METAL HYDRIDES AND DEUTERIDES FROM DE HAAS-VAN ALPHEN MEASUREMENTS

R. Griessen and L.M. Huisman

Natuurkundig Laboratorium
Vrije Universiteit, Amsterdam, The Netherlands

I. INTRODUCTION

The electronic structure of metal-hydrogen systems has tradition-
ally been investigated by means of low temperature specific heat and
magnetic susceptibility measurements. It was for example on the basis
of susceptibility measurements that Mott proposed the proton model[1]
for the palladium-hydrogen system. According to this model the extra
electron brought in by dissolving a hydrogen atom in palladium fills
empty states of the host metal at the Fermi energy.

On the basis of supercell calculations on Pd_4H, Pd_4H_3 and PdH
Switendick[2] showed that quite independently of the precise form of
the proton potential, i) the filled palladium states with s-
character relative to the octahedral interstitial site where the
hydrogen is located are strongly lowered at a rate of approximately
0.2 Ryd as a result of the attractive potential of the proton,
ii) additional empty palladium states are lowered below the Fermi
energy E_F, iii) approximately 0.5 electrons per added hydrogen fill
empty states at E_F. The lowering of s-states mentioned under i) has
been observed by Eastman et al.[3], Antonangeli et al.[4] and Schlapbach
and Burger[5] in $PdH_{\sim 0.6}$, by Eastman et al.[6] in TiH_x, by Fukai et al.[7]
in VH_x and by Gilberg[8] in NbH_x.

Experimental evidence for the lowering of empty palladium states
below E_F is that ~ 0.6 H atoms per Pd atom[9] must be added to
palladium to fill the 0.36 holes of its d-band[10-12]. From caculations
on metal-hydrogen compounds Switendick[13] concluded that similarly
to the case of PdH 0.4 electrons per added hydrogen in TiH_2, 1.0 in
CrH and $\gtrsim 0.5$ in VH fill empty states at E_F.

The influence of *small* amounts of hydrogen on the electronic
structure of a metal is not as well known as that of stoichiometric
hydrides. In the only supercell calculation carried out so far for a

dilute alloy Switendick[14] found that in $Pd_{32}H$ only few unoccupied palladium states were lowered below the Fermi energy. He concluded from this, that at low concentrations not 0.6 electrons but ~ 1.0 electron per H atom fills the empty states of Pd. This is exactly what is expected on the basis of the proton model[1].

Quite a different situation is found in the average t-matrix calculations of Gelatt et al.[15] and Bansil et al.[16]. In contrast to supercell calculations[14] where PdH_x behaves in a virtual-crystal manner (i.e. where Pd bands are gradually lowered in energy[17]), in the averaged t-matrix approximation[15,16] PdH_x is already in the split-band regime[18]. For dilute disordered PdH_x alloys a low-lying band containing 2x electrons is formed just below the muffin-tin zero. These states which are suggestive of an H^--ion are not new states but have been extracted from the palladium states. As a result of this loss of spectral weight (which is especially strong for states with s-symmetry at the proton site) the Fermi energy *increases* although *two* electrons per added H-atom are put in the low-lying band. From calculations on disordered $PdH_{0.05}$ and $PdH_{0.10}$ Bansil et al.[16] found that at E_F the palladium s-p band is lowered at a rate roughly twice as large as that of the d-states.

From existing magnetic susceptibility[9,19] and specific heat data[20] it is not possible to extract information about the filling of electronic bands in dilute hydrides. In the case of magnetic susceptibility the difficulty arises from the strong Stoner enhancement whose hydrogen concentration dependence is superimposed on band filling effects. Similarly, the conversion of electronic specific heat data to density of states information is complicated by the hydrogen concentration dependence of the electron-phonon and paramagnon enhancement factor. Furthermore none of the electronic specific heat data obtained so far pertain to a homogeneous dilute PdH_x α-phase[21].

One method, if not the only one, which is very well suited to measure small changes in the electronic structure of a dilute alloy is based on the de Haas-van Alphen effect. In Section II we indicate briefly how this effect can be used to measure the Fermi surface and the electron scattering in an alloy. In Section III we present experimental results for copper and palladium hydrides and in Section IV we discuss among others the remarkable isotope effects found in PdH_x and PdD_x.

II. THE DE-HAAS VAN ALPHEN EFFECT

The de Haas-van Alphen effect[22] (i.e. the oscillatory part of the magnetization of a metal in a magnetic field) as well as all other quantum oscillatory effects in, for example, magnetostriction, sound velocity and electrical resistivity (Shubnikov-de Haas effect) have their origin in an oscillatory contribution Ω_{osc} to the Gibbs free energy of an electron gas in a homogeneous magnetic field B. As shown by Lifshitz and Kosevich[23] the corresponding oscillatory contribution M_{osc} to the magnetization is given by

$$M_{osc} = \sum_{r=1}^{\infty} Z_r \frac{T \exp(-\alpha r m^* T_D/B)}{\sinh(\alpha r m^* T/B)} \sin(2\pi r F/B + \gamma) \tag{1}$$

where T is the temperature, m^* the effective mass of the electron and γ is a phase which is equal to $\frac{1}{2}$ for free electrons. The amplitude Z_r depends on the local curvature of the Fermi surface and on the effective g-factor of the electron. The de Haas-van Alphen frequency F is related to the area A of an extremal-cross section of the Fermi surface perpendicular to the applied field direction by means of $A = 2\pi eF/(\hbar c)$. The Dingle temperature T_D is a measure of the broadening ΔE of electronic energy levels by dislocations and impurities and is given by $k_B T_D = \Delta E/\pi$. The constant α in the argument of the exponential and sinh-functions in Eq.1 is $\alpha = 2\pi^2 k_B m/(e\hbar) = 14.6925 T/K$.

From Eq(1) follows that i) the dimensions of the Fermi surface of a metal can be determined by measuring the *frequency* F of the de Haas-van Alphen oscillations and ii) the scattering of electrons by impurities can be determined from the field dependence of the *amplitude* of the magnetization oscillations.

De Haas-van Alphen oscillations are usually determined by means of the field-modulation technique or the torque technique. In the field modulation technique one exploits the non-linearity of the M(1/B) function to apply phase sensitive techniques to the detection of second (or higher) harmonics of the frequency at which the magnetic field is modulated. The reader is refered to ref. 24 for a detailed description of this method.

Instead of measuring the magnetization directly one can also measure the torque $\tau = V\vec{M}_{osc} x \vec{H}$ acting on a sample of magnetic moment $V\vec{M}_{osc}$ (V is the volume of the sample) in a homogeneous field $\vec{H}$. In contrast to the field modulation technique where eddy currents in the single crystalline samples have to be minimized, the torque method is a static method which can be used to measure de Haas-van Alphen oscillations of relatively large samples. For metal hydrides this is a nice feature as it makes it possible to use crystals which are large enough for a gravimetric determination of their hydrogen content. A description of a compact torque balance is given in ref. 25.

III. EXPERIMENTAL RESULTS

a) The CuH_x-system

The first succesful de Haas-van Alphen measurements on a metal-hydride were those of Wampler and Lengeler[26] on dilute CuH_x alloys. Using the field modulation technique they determined the hydrogen concentration dependence of the Dingle temperature of several extremal orbits on the Fermi surface of copper. Experimental values for dT_D/dx corresponding to high symmetry orbits are compared to theoretical dT_D/dx's obtained from an average-t-matrix calculation[27] in Fig. 1.

The highest value for dT_D/dx is observed for the neck orbit. This is at first sight surprising as in several copper-based alloys (CuAu, CuNi, see ref. 28) dT_D/dx is lowest for the neck. The large value of dT_D/dx on the neck orbit of CuH_x is however easily understood if one assumes that the scattering of an electron out of a Bloch state is roughly proportional to the electron density near the scattering potential. At the octahedral interstitial site where the proton is

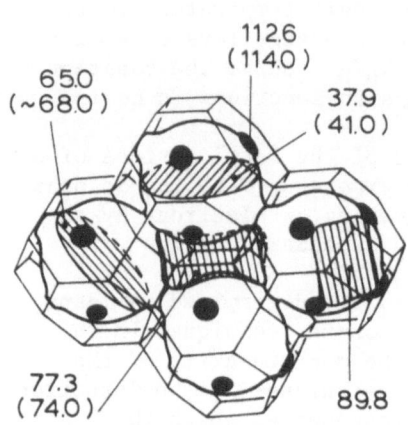

Fig. 1. Hydrogen concentration dependence dT_D/dx of the Dingle temperature of extremal orbits of the Fermi surface of dilute CuH_x alloys[26]. The values are given in units of K per at.% hydrogen. The theoretical values[27] in brackets have been corrected for electron-phonon enhancement effects.

localised, the density of electrons is highest for states close to the L_2' state. (The same holds for palladium see Fig. 3) The L_2' state itself has s-symmetry with respect to the octahedral site, although it is p-like with respect to the Cu-lattice site. The Bloch states on the belly orbit have also a strong s-like character but the electron density at the proton site is about two times smaller than for the L_2' states and thus $(dT_D/dx)_{belly} \approx 0.5 \, (dT_D/dx)_{neck}$ as observed experimentally.

The excellent agreement between experimental and theoretical values for dT_D/dx is remarkable for the following reason. In all band structure calculations carried out so far for metal hydrogen systems it is assumed that the protons are *frozen-in* at interstitial sites. For the case of hydrogen in copper this means that one neglects a zero-point-motion amplitude of the order of a tenth of the lattice spacing. From the good agreement between experiment and theory in Fig. 1 one could therefore conclude that zero-point-motion effects do not affect the electronic structure of metalhydrides. As discussed in the next section this is only true for states with s-symmetry with respect to the proton site.

b) The PdH_x-system

PdH_x is probably the most extensively studied metal hydrogen system. PdH was also one of the first hydrides for which band structure calculations became available[2]. Until now several calculations have been carried out, both for compounds [13,14,30-32] and substoichiometric hydrides[15,16,31].

One of the major reasons for the continued interest in the electronic structure of PdH_x is without doubt the superconductivity of concentrated PdH_x (x $\gtrsim$ 0.8) alloys discovered by Skoskiewicz[33] in 1972 and the inverse isotope effect in the superconducting transition temperature T_c reported by Stritzker and Buckel[34] in the same year.

The increase in T_c with hydrogen concentration for 0.8 $\gtrsim$ x $\leq$ 1 has been shown by Papaconstantopoulos et al.[32] to be related to the increase with concentration of the density of states with s-character relative to the octahedral site. As more and more hydrogen is added to palladium the Fermi energy moves away from the s-d hybridization gap and the coupling of electrons to the (hydrogen) optical phonons increases.

For the inverse isotope effect in T_c (T_c(PdH) $\simeq$ 9K, T_c(PdD) $\simeq$ 11K) two different types of mechanisms have been proposed so far. Ganguly[35] proposed that the anharmonicity of the proton-palladium potential is responsible for a smaller electron-phonon enhancement parameter in PdH than in PdD. As shown in ref. 32, the electron-phonon enhancement factor is essentially given by the contribution $\lambda_{H(D)}$ resulting from the coupling of electrons to the optical phonons. Following McMillan[36] $\lambda_{H(D)}$ can be written as

$$\lambda_{H(D)} = \frac{\eta_{H(D)}}{M_{H(D)} <\omega^2_{H(D)}>} \qquad (2)$$

where $\eta_{H(D)}$ depends only on electronic properties at the Fermi energy, $M_{H(D)}$ is the mass of the proton (deuteron) and $<\omega^2_{H(D)}>$ is a weighted average over the optical phonons. Papaconstantopoulos et al. and Ganguly assume that the electronic term $\eta_H(D)$ is independent of the isotope mass (i.e. $\eta_H = \eta_D$). Consequently $\lambda_H/\lambda_D = 2\omega^2_D/\omega^2_H$. The ratio of optical frequencies in PdH_x and PdD_x has been determined from a comparison of inelastic neutron scattering[37] in $PdD_{0.63}$ and time-of-flight measurement[38] in $PdH_{0.63}$ to be $\omega_D/\omega_H \simeq$ 0.64. This leads to $\lambda_H \simeq$ 0.83 λ_D. However, significantly smaller anharmonicity has been derived from measurements of the temperature dependence of the electrical resistivity[39], the thermal expansion[40] and the elastic moduli[41] and from spectroscopic methods such as superconducting tunnelling[42] and Raman scattering[43]. Taking the average of all these experiments one obtains $\lambda_H \simeq$ 0.9 λ_D. Also as pointed out by Ginodman and Zherikhina[44] the isotope effect in T_c predicted by the theory[32] is approximately two times smaller than that observed even if λ_H = 0.83 λ_D is used for the evaluation of the transition temperatures of PdH_x and PdD_x.

An other mechanism, which should also lead to an inverse isotope effect in T_c, was proposed by Miller and Satterthwaite[45]. According to these authors the difference in the zero-point amplitude of a proton and a deuteron influences *directly* the band structures of PdH_x and PdD_x. Miller and Satterthwaite did not give a quantitative theory of these zero point effects on T_c. Quite recently, however, Morozov[46] considered the influence of electron-optical phonon interactions on

the electronic band structure of a metal-hydrogen system. Besides
the well-known effects of mass renormalization and electron density
of states enhancement, the energy band of s-electrons is lowered by
an amount Σ_s by the electron-phonon interaction. As the chemical
potential is lowered by the same amount the latter effect is generally ignored as it leaves the Fermi momentum unchanged. For a metal
such as palladium, where s-states hybridize strongly with d-states,
the situation is however quite different. In contrast to s-states,
the d-states are, according to Morozov, not lowered ($\Sigma_d = 0$). As
$\Sigma_s \neq 0$ the s-states are shifted relatively to the d-states. Since
Σ_s is of order $\lambda_{H(D)}\hbar\omega_{H(D)}$ it is larger in PdH_x than in PdD_x and the
Fermi surface of PdH_x is slightly different from that of PdD_x.

According to de Haas-van Alphen studies[10] and band structure
calculations[11,12] the Fermi surface of pure palladium consists of a
sixth-band electron sheet centered at point Γ of the Brillouin zone,
six equivalent fourth-band hole ellipsoids at X, eight equivalent
fifth-band hole ellipsoids at point L and a complex hole surface,
the so-called fifth-band jungle-gym. The extremal cross-sections
investigated by Venema[47] and Bakker et al.[48] are shown in Fig. 2.

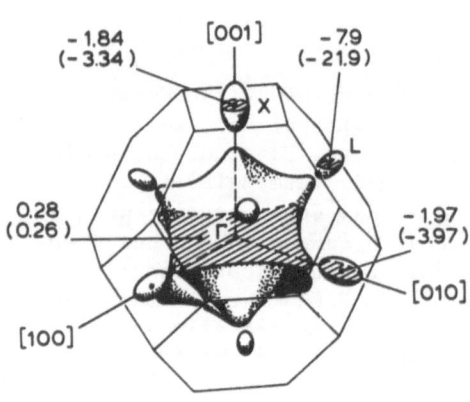

Fig. 2. Hydrogen (deuterium)
concentration dependence dlnA/dx
of the area of extremal cross-
sections of the Fermi surface of
palladium as obtained from de
Haas-van Alphen experiments[47,48].
The derivatives dlnA/dx are
corrected for volume expansion
effects.

By using a dual de Haas-van
Alphen spectrometer they were
able to determine the H(D)-
concentration dependence
$dlnA/dx_{H(D)}$ of various extremal
cross-sectional areas A of the
Fermi surface of palladium. For
both hydrides and deuterides they
found that the size of the electron sheet at Γ increases with increasing interstitial concentration while the hole ellipsoids at X
and L decrease in size. As shall be shown in Section IV this is in
qualitative agreement with the predictions of a rigid-band model.

The data in Fig. 2 show for the first time that *important isotope
effects may be present in the electronic structure of metals containing light interstitials*. For the hole ellipsoids at X, $dlnA/dx_D$ is
approximately twice as large as $dlnA/dx_H$ in PdH_x. For the hole
ellipsoids at L $dlnA/dx_D$ is almost three times larger than $dlnA/dx_H$.
On the other hand no isotope effect has been found for the hydrogen
(deuterium) concentration dependence of the large Γ-electron sheet.

A similar behaviour is also found for the concentration dependence $dT_D/dx_{H(D)}$ of the Dingle temperature associated with the various extremal orbits shown in Fig. 2. For the X- and L-hole ellipsoids dT_D/dx_D is about two times larger than dT_D/dx_H while no isotope effect is again observed for the electron sheet at Γ.

IV. DISCUSSION OF EXPERIMENTAL RESULTS

In this section we compare the experimental results on PdH_x, PdD_x (and CuH_x) with the predictions of various band structure models.
a) Experimental energy shifts
Shifts of energy levels E_k with respect to the Fermi energy can be deduced from the measured derivatives $d\ln A/dx$ by means of the following relation

$$\frac{d\ln A}{dx} = - \frac{\pi m_b^*}{A} < \frac{d(E_{\vec{k}} - E_F)}{dx} >$$ (3)

where m_b^* is the band mass defined as $m_b^* = (\partial A/\partial E)/\pi$ and the brackets $< >$ denote an average taken over the orbit corresponding to the Fermi surface cross-section of area A. As shown by Friedel[49,50], $dE_F/dx=0$ in the dilute limit.

The average energy shifts calculated by means of Eq. 3 can not be directly compared to theoretical predictions as they also contain a contribution due to the lattice expansion produced by the dissolution of H or D in the host metal. The area of an extremal cross-section of the Fermi surface depends thus not only explicitly on x but also implicitly on x via the volume of the sample. Writing $A = A(x,V(x))$ we obtain

$$\frac{d\ln A}{dx} = \frac{\partial \ln A}{\partial x}\bigg|_{V=const} + \frac{\partial \ln A}{\partial \ln V} \frac{d\ln V}{dx}\bigg|_{matrix}$$ (4)

where $(d\ln V/dx)_{matrix}$ is the dilation of the palladium matrix which is related to the total macroscopic dilation of the sample by means of the relation

$$\frac{d\ln V}{dx}\bigg|_{matrix} = \gamma \frac{d\ln V}{dx}\bigg|_{total}$$ (5)

The Eshelby factor $\gamma = 2(1+2\nu)/(3-3\nu)$ is small ($\gamma = 0.153$) since the Poisson ratio of palladium at 0K ($\nu = 0.435$ determined from the elastic moduli of Geerken et al.[41]) is close to $\frac{1}{2}$. Taking $(d\ln V/dx)_{total} = 0.19$ we find that $(d\ln V/dx)_{matrix} = 0.0291$. For most orbits $(\partial \ln A/\partial x)_V$ at constant volume is thus only slightly different from the measured $\partial \ln A/\partial x$ since the volume derivatives $d\ln A/d\ln V$ measured by Skriver et al.[51] are of order 1.

The energy shifts $<d(E_{\vec{k}} - E_F)/dx>_V$ at constant volume calculated by means of Eq.3 with $\partial \ln A/\partial x$ replaced by $(\partial \ln A/\partial x)_V$ are given in Table I. For both electron- and hole-orbits the energy shifts are negative.
These shifts can be compared with the rigid-band-model shift given

by

$$\frac{d(E_{\vec{k}} - E_F)}{dx}\bigg|_{RBM} = -\frac{\zeta}{N_{Pd}(E_F)} \qquad (6)$$

where ζ is the effective number of electrons per added H(D) that is put at the Fermi surface. In the proton model[1] and according to the $Pd_{32}H$ supercell calculation[13] it is equal to 1.0, while average-t-matrix calculations[16] indicate a value of roughly 0.5. Recent magnetic susceptibility measurements[52] indicate a value of roughly 0.6 for *both* PdH_x and PdD_x. ζ need not be equal to 1.0 because some states that are above E_F in Pd may disappear when H is added and reappear in the low lying H-band[15,16]. In Eq.6 $N_{Pd}(E_F)$ is the density of states of pure Pd at the Fermi energy. From bandstructure calculations[11,12] it follows that $[d(E_{\vec{k}} - E_F)/dx]_{RBM} \approx -32\zeta$ mRy. As the density of states of Pd is dominated by the contribution from the jungle gym, this rigid band result indicates the rate at which that sheet of the Fermi surface is pulled down in energy.

b) <u>Calculation of energy shifts</u>

To first order, the shift ΔE of a state $\Psi_{\vec{k}}$ in PdH_x or PdD_x is given by

$$\frac{d(E_{\vec{k}} - E_F)}{dx} = \int \psi_{\vec{k}}^* V(\vec{r}) \psi_{\vec{k}} d^3r \qquad (7)$$

where $V(F)$ is the screened proton (deuteron) potential.

From non-linear screening theories[53,54] one knows that the charge density $\Delta\rho$ around a proton embedded in jellium resembles closely that of a free hydrogen atom, i.e. $\Delta\rho \sim e^{-r/\lambda}$. The corresponding potential $V(\vec{r})$ is then given by

$$V(\vec{r}) = -\frac{2}{r}\left[1 + \frac{r}{2\lambda}\right]e^{-r/\lambda} \qquad (8)$$

As shown in Fig. 3 the states on the Γ electron sheet (designed by 1 along ΓX and by 2 along ΓL) have a strong s-character at the interstitial site. The electron density averaged over the three directions [100], [010] and [001] depends only weakly on the distance from the octahedral site. The integral in Eq.5 reduces thus to an integral over the potential alone. This potential leads to a shift

$$d(E_{\vec{k}} - E_F)/dx = -16\pi\lambda^2 \ \overline{\Psi^*\Psi} \qquad (9)$$

Electron states representative for the Γ-[001] orbit have an average density $\overline{\Psi^*\Psi}$ 0.0025/a.u.^{3} around the interstitial site. Using $<d(E_{\vec{k}} - E_F)/dx>_\Gamma = -43$mRy we find $\lambda = 0.58$ a.u. which is quite close to $\lambda_{atom} = 0.5$ a.u. obtained for the time averaged potential of a neutral hydrogen atom. The good agreement between λ and λ_{atom} can be viewed as an a posteriori justification of the procedure used to calculate the shifts of energy levels on the Γ-sheet of the Fermi surface.

Orbit	$\langle d(E_{\underset{\tilde{k}}{}}-E_F)/dx\rangle_{orbit}$ [mRy]	
	PdH_x	PdD_x
Γ	-47	-43
X[001]	-24	-43
X[010]	-23	-47
L	-23	-64
jungle gym	-10/-30	-10/-30

Table I. Average energy shifts in dilute PdH_x and PdD_x alloys. The shifts for the Γ, X and L orbits were obtained from de Haas-van Alphen measurements. The shifts for the jungle gym are rigid band model estimates.

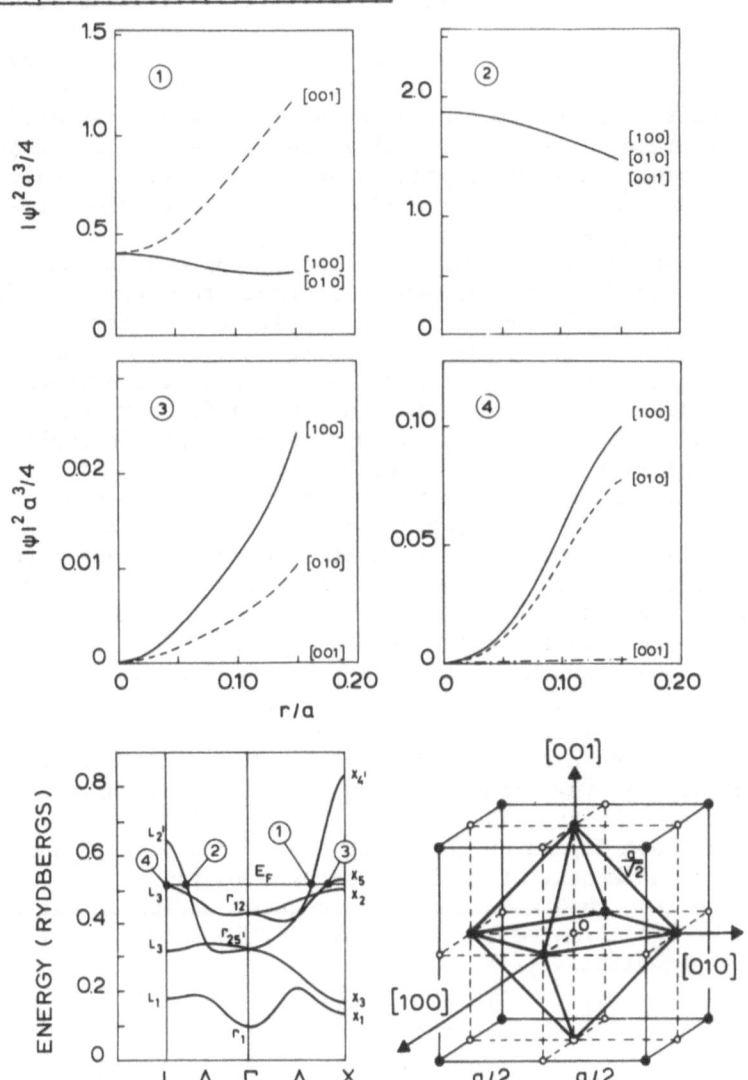

Fig. 3. Electron densities near the octahedral interstitial site in Pd (a = 7.353 a.u.) for various Bloch states at E_F.

c) The isotope effect in dlnA/dx

The most striking result of the de Haas-van Alphen measurements in
PdH$_x$ and PdD$_x$ is the isotope dependence of the energy shifts
corresponding to the X- and L-hole pockets.

It is perhaps important to point out first that in the de Haas-van
Alphen investigations of the Fermi surface of PdH$_x$ and PdD$_x$, the H(D)
concentration dependence of the Γ-orbit is determined in the same run
as that of the X- and L-orbit. The observed isotope effect for the X-
and L-orbit can thus not be due to a systematic experimental error as
in the *same* run, on the *same* sample, no isotope effect is measured
for the Γ orbit. The isotope effects have been observed both in field
modulation and in torque runs on samples of various shapes and sizes.

The dinstinguishing features of this isotope effect are i) that
the energy shifts of the states on the small hole pockets differ by
amounts that are as large as the shifts themselves and ii) that in
PdH$_x$ all d-states (small hole pockets and jungle gym) are shifted at
roughly the same rate (-20 to -30 mRy) while in PdD$_x$ the rates depend
strongly on the character of the state.

The states on the X-ellipsoid (3 in Fig. 3) and on the L-
ellipsoid (4 in fig. 3) have densities that are identically zero
at the interstitial site and increase rapidly with increasing
distance from the interstitial site. In the majority of existing band
structure calculations the proton potential is confined to the
muffin-tin radius. For such a potential the shifts for the X- and L-
pockets should be approximately two orders of magnitude smaller than
that of typical states on the Γ-sheet as $|\psi|^2 a^3/4 \approx 1$ for Γ-states
compared to a few hundreths for the X- and L-states. Since
$<d(E_{\vec{k}} - E_F)/dx>$ is of the same order of magnitude for all sheets of
the Fermi surface (see Table I) it must be concluded that a potential
confined to the hydrogen (deuterium) muffin-tin sphere is not
applicable to PdH$_x$ and PdD$_x$.

The simplest way to simulate long-range effects of the proton
potential is to adjust the muffin-tin potential so as to satisfy
Friedel's self-consistency condition[49,50] $dE_F/dx = 0$ in the dilute
limit. This amounts in fact simply to shifting rigidly the energy
band structure. Such a rigid shift is obviously not sufficient to
explain the de Haas-van Alphen data which lead to strongly $\vec{k}$-
dependent energy shifts. Also, a homogeneous shift of the metal
potential does not depend on the exact location of the protons and is
consequently independent of their zero-point motion.

To first order, the influence of zero-point motion on the energy
shifts can be approximated by replacing $V(\vec{r})$ in Eq.7 by $<V(\vec{r})>$.
The average potential $<V(\vec{r})>$ is given by

$$<V(\vec{r})> = \left(\frac{3}{2\pi<\rho^2>}\right)^{3/2} \int e^{-\frac{3}{2<\rho^2>}\rho^2} V(\vec{r}+\vec{\rho})\,d\vec{\rho} \qquad (10)$$

where $\vec{\rho}$ is the instantaneous position of the interstitial and $<\rho^2>$
the mean squared amplitude of its zero point motion. It is equal to
0.32 a.u.^{2} for the proton and 0.23 a.u.^{2} for the deuteron.

The influence of zero-point motion on the potential V(r) given in

Eq.8 is to increase its strength. This is in disagreement with the de Haas-van Alphen data which show that the energy shifts are smaller in the hydride than in the deuteride.

Venema et al.[55] have proposed that the long range Friedel oscillations of the screened impurity potential have to be taken into account in order to explain the observed isotope effect. At large distances from the proton the screening potential is given by

$$V(\vec{r}) \sim C \frac{\cos(2k_F r + \varphi)}{(k_F r)^3} \qquad r \to \infty \qquad (11)$$

where the amplitude C, the phase φ and the Fermi wave vector k_F depend in a complicated way on the electronic structure of the host and on the scattering properties of the impurity. In the simplest possible model, the screening is done predominantly by the Γ-sheet electrons. k_F is then of the order of 0.5 a.u.$^{-1}$ and C should roughly be equal to $\frac{1}{2\pi}$ [56]. After averaging we find:

$$\langle V(\vec{r}) \rangle \sim \frac{C}{(k_F r)^3} \left(1 - \frac{1}{6} \langle \rho^2 \rangle (2k_F)^2\right) (\cos(2k_F r + \varphi)$$

$$+ \frac{1}{3} \langle \rho^2 \rangle (2k_F)^2 \frac{\sin(2k_F r + \varphi)}{(k_F r)}$$

$$+ O\left(\frac{1}{(k_F r)^2}\right)) \qquad (12)$$

With a k_F of order 0.5 a.u.$^{-1}$, the factor $1 - \frac{1}{6} \langle \rho^2 \rangle (2k_F)^2$ is within a few per cent equal to 1.0 and can therefore be neglected. The term proportional to $\sin(2k_F r + \varphi)/k_F r$ can have a large effect however. This term is alternately attractive and repulsive with a wavelength $\pi/k_F \sim 6$ a.u. By choosing φ properly, it can be made repulsive in the region where the d-charge closest to the impurity is concentrated. As the strength of this term falls off as r^{-4} subsequent attractive regions can be neglected.

The extra term in $\langle V(\vec{r}) \rangle$ can explain to some extent the observed isotope effect. As k_F is small, the region in which this term is repulsive is fairly large. With C of the order of 1.5, k_F equal to 0.5 and φ roughly equal to $\frac{1}{2}\pi$, the energy shifts at the X pocket can be reproduced. There are however a number of problems.

In the first place, it is not clear why φ should have the particular value necessary to make the additional term repulsive in the region where the d charge is concentrated. Furthermore, the amplitude C is considerably larger than one would estimate from simple calculations. Finally, the explanation based on averaged Friedel oscillations is qualitatively very unsatisfactory. We have mentioned above that the d-states in PdH_x are all shifted downwards at roughly the same rate but that in PdD_x they are shifted downwards at rates that depend very strongly on the character of the state. This strong $\vec{k}$ dependence is very difficult to explain with the averaged Friedel potential because the added term oscillates only slowly in space (wavelength of the order of 6 a.u.).

Another mechanism that takes explicitly into account the strong $\vec{k}$ dependence of the energy shifts in PdD$_x$ seems to be called for.

V. CONCLUSIONS

The de Haas-van Alphen experiments on dilute PdH$_x$ and PdD$_x$ have confirmed the general features of the proton model[1]. States at the Fermi level are lowered in energy and some of the H(D) electrons are accomodated at the Fermi level. As no de Haas-van Alphen data are available on the behaviour of the jungle-gym when H or D is intro-duced in Pd, the exact number of H(D) electrons put at the Fermi surface can not yet be measured.

The de Haas-van Alphen experiments have also shown that large isotope effects exist for the hydrogen (deuterium) concentration dependence of energy levels with a vanishing electron density at the octahedral interstitial site. This raises the question whether marked isotope effects should be observable or not in macroscopic physical quantities such as specific heat or magnetic susceptibility.

In the dilute limit this question can be answered by considering that isotope effects have only been observed for small pockets of the Fermi surface. The number of states accomodated in the X- and L-ellipsoids is less than 1% of the 0.36 electrons contained in the large Γ-sheet. Within experimental errors no isotope effect has been observed for this sheet of the Fermi surface. This implies that within 1% the same number of electrons must be accomodated in the jungle-gym of PdH$_x$ and PdD$_x$. For dilute alloys the difference in the density of states of the hydrides and deuterides is thus expected to be less than one percent.

ACKNOWLEDGEMENTS

"This work is part of the research program of the Stichting voor Fundamenteel Onderzoek der Materie (Foundation for Fundamental Research on Matter) and was made possible by financial support from the Nederlandse Organisatie voor Zuiver-Wetenschappelijk Onderzoek (Netherlands Organization for the Advancement of Pure Research)."

REFERENCES

1. N.F. Mott and H. Jones, "The Theory of the Properties of Metals and Alloys", Clarendon, Oxford 1936
2. A.C. Switendick, Ber. Bunsenges. Physik. Chem. <u>76</u> (1972) 535
3. D.E. Eastman, J.K. Cashion and A.C. Switendick, Phys. Rev. Lett. <u>27</u> (1971) 35
4. F. Antonangeli, A. Balzarotti, A. Bianconi, E. Buranttini, P. Perfetti and N. Nistico, Phys. Lett. <u>55A</u> (1975) 309
5. L. Schlapbach and J.P. Burger, J. Physique, to be published
6. D.E. Eastman, Solid State Commun. <u>10</u> (1972) 933
7. Y. Fukai, S. Kazama, K. Tanaka and M. Matsumoto, Solid State

Commun. 19 (1976) 507

8. E. Gilberg, Phys. Stat. Sol. (b) 69 (1975) 477
9. E. Wicke and J. Blaurock, Ber. Bunsenges. Phys. Chem. 85 (1981) 1091
10. J.J. Vuillemin, Phys. Rev. 144 (1966) 396
11. F.M. Mueller, A.J. Freeman, J.O. Dimmock and A.M. Furdyna, Phys. Rev. B1 (1970) 4617
12. O.K. Andersen, Phys. Rev. B2 (1970) 883
13. A.C. Switendick in Topics in Applied Physics 28 edited by G. Alefeld and J. Völkl, (Springer Verlag 1978) p. 101
14. A.C. Switendick, Z. Phys. Chem. NF117 (1979) 447
15. C.D. Gelatt, H. Ehrenreich and J.A. Weiss, Phys. Rev. B17 (1978) 1940
16. A. Bansil, R. Prasad, S. Bessendorf, L. Schwartz, W.J. Venema, R. Feenstra, F. Blom and R. Griessen, Solid State Commun. 32 (1979) 1115
17. J. Zbasnik and M. Mahnig, Z. Physik B23 (1976) 15
18. H. Ehrenreich and L. Schwartz, Solid State Phys. 31 (Academic Press, New York 1976, edited by H. Ehrenreich, F. Seitz and D. Turnbull) p. 149
19. H.C. Jamieson and F.D. Manchester, J. Phys. F:Metal Phys. 2 (1972) 323
20. U. Mizutani, T.B. Massalski and J. Bevk, J. Phys. F:Metal Phys. 6 (1976) 1
21. As Shown by W.J. Venema (PhD-Thesis, Vrije Universiteit, Amsterdam 1980) homogeneous α-phase PdH$_x$ alloys can only be obtained by cooling samples from room temperature to 4.2K at a rate faster than 30K/min. For specific heat measurements the cooling rates are ~ 1K/min (see ref. 20)
22. A.V. Gold in Solid State Physics 1: "Electrons in Metals", edited by J.F. Cochran and R.R. Haering, Gordon and Breach, New York 1968
23. I.M. Lifshitz and A.M. Kosevich, Zh. Eksperim. i Teor. Fiz. 29 (1955) 730; Sov. Phys. JETP 2 (1956) 636
24. R.W. Stark and L.R. Windmiller, Cryogenics 8 (1968) 272
25. R. Griessen, M.J.G. Lee and D.J. Stanley, Phys. Rev. B16 (1977) 4385
26. W. Wampler and B. Lengeler, Phys. Rev. B15 (1977) 4614
27. L. Huisman and J.A. Weiss, Solid State Commun. 16 (1975) 983
28. M. Springford in "Electrons at the Fermi Surface" edited by M. Springford (Cambridge University Press, 1980) p. 362
29. H. Teichler, Hyperfine Interactions 6 (1979) 251
30. M. Gupta and A.J. Freeman, Phys. Rev. B17 (1978) 3029
31. J.S. Faulkner, Phys. Rev. B13 (1976) 2391
32. D.A. Papaconstantopoulos, B.M. Klein, E.N. Economou and L.L. Boyer, Phys. Rev. B17 (1978) 141
33. T. Skoskiewicz, Ber. Bunsenges. Physik. Chem. 76 (1972) 847
34. B. Stritzker and W. Buckel, Z. Physik. 257 (1972) 1
35. B.N. Ganguly, Z. Physik 265 (1973) 433
36. W.L. McMillan, Phys. Rev. 167 (1968) 331

37. J.M. Rowe, J.J. Rush, H.G. Smith, M. Mostoller and H.E. Flotow, Phys. Rev. Lett. 33 (1974) 1297
38. A.N. Rahman, K. Sköld, C. Pelizzari and S.K. Sinha, Phys. Rev. B14 (1976) 3630
39. J.P. Burger and D.S. MacLachlan, J. Physiqe 37 (1976) 1227
40. R. Abbenseth and H. Wipf, J. Phys. F:Metal Phys. 10 (1980) 353
41. B.M. Geerken, R. Griessen and L.M. Huisman, to be published
42. A. Eichler, W. Wühl and B. Stritzker, Solid State Commun. 17 (1975) 213
43. R. Sherman, H.K. Birnbaum, J.A. Holy and M.V. Klein, Phys. Lett. 62A (1977) 353
44. V.B. Ginodman and L.N. Zherikhina, Sov. J. Low Temp. Phys. 6 (1980) 278
45. R.J. Miller and C.B. Satterthwaite, Phys. Rev. Lett. 34 (1955)144
46. A.I. Morozov, Sov. Phys. Solid State 20 (1978) 1918
47. W.J. Venema, PhD-Thesis, Vrije Universiteit, Amsterdam 1980
48. H.L.M. Bakker, R. Griessen, L.M. Huisman and W.J. Venema, to be published
49. J. Friedel, Adv. in Phys. 3 (1954) 446
50. E.A. Stern, Phys. Rev. B5 (1972) 366
51. H. Skriver, W.J. Venema, E.Walker and R. Griessen, J. Phys. F: Metal Phys. 8 (1978) 2313
52. E. Wicke, private communication
53. Z.D. Popovicz, M.J. Stott, J.P. Carbotte and G.R. Piercy, Phys. Rev. B13 (1976) 590
54. P. Jena, K.S. Singwi and R.M. Nieminen, Phys. Rev. B17 (1978) 579
55. W.J. Venema, R. Griessen, R.S. Sorbello, H.L.M. Bakker and P.E. Mijnarends, Inst. Phys. Conf. Ser. 55 (1981) 579
56. J.S. Langer and S.H. Vosko, J. Phys. Chem. Solids 12 (1959) 196

ELECTRONIC STRUCTURE OF NON-STOICHIOMETRIC TRANSITION
METAL HYDRIDES

A. Bansil, R. Prasad[1] and L. Schwarz[*2]

Physics Department, Northeastern University
Boston, Massachusetts 02115
*Physics Department, Brandeis University
Waltham, Massachusetts 02154

ABSTRACT

We review the theoretical work concerning the effects of non-stoichiometry on the electronic spectra of transition metal hydrides. The treatment of these effects generally falls outside the scope of the conventional Bloch band theory. However, the recently developed techniques of band theory of random alloys can be applied to this problems by modelling the metal-hydrogen system as an "alloy" of hydrogen atoms and vacancies. This approach involves the use of the average t-matrix (ATA) and coherent potential (CPA) approximations within the Korringa-Kohn-Rostoker framework. We discuss the premises underlying such a scheme and the progress that has been possible on this basis in understanding the metal-hydrogen system.

INTRODUCTION

Application of band theory to the metal-hydrogen system, begun in the late sixties, is widely acknowledged to have yielded a good overall understanding of the electronic structure of the stoichio-

---

[1]Supported, in part, by the United States Department of Energy
[2]Supported, in part, by the United States National Science Foundation

metric hydrides.[1-3] These techniques, however, are not appropriate
for discussing the non-stoichiometric phases, which form over wide
composition ranges. During the last decade, especially the late
seventies, the elements of what may be called a band theory of
random substitutional alloys have emerged by the application of
multiple scattering techniques to the muffin-tin Hamiltonian and
the use of average t-matrix (ATA) and coherent potential (CPA)
approximations to incorporate disorder effects.[4-9] Since hydrogen
enters the metal lattics <u>interstitially</u>, it is possible to model
the system as a random "alloy" of hydrogen atoms and vacancies;[10-13] the methods of alloy theory then become applicable and
a band description follows. This approach for treating the effects
of non-stoichiometry constitutes the main subject of this review.
We refer the reader to the literature for a discussion of a variety
of other important aspects of the electronic spectra of metal
hydrides.[1-3,14-16]

Concerning some of the other relevant studies, we note that
Switendick[17] has considered the band structures of ordered phases
(including hypothetical ones) to infer, by interpolation, the
properties for intermediate compositions. A number of authors
have used the CPA on an appropriately chosen tight-binding model
Hamiltonian to discuss the hydrides.[3,18,19] Insights into the
nature of the hydrogen potential in metals have been obtained by
considering the problem of a single H impurity in the free electron gas.[20-22] Molecular cluster calculations have also been
carried out in connection with the hydrides.[23-24]

An outline of this article is as follows. Section II provides
an overview of the quasi-particle spectrum in a substitutional
alloy. We discuss how the conventional Bloch bands generalize to
complex energy bands[25] in the disordered case. Among the electronic properties of random muffin-tin alloys, which have been
investigated on the basis of the ATA and CPA in the recent years
are: momentum densities,[26-28] charge densities,[5-7] band spectroscopies,[29,30] relativistic effects,[31] itinerant magnetism,[32]
transport,[33,34] etc. Extensions of the theory to allow the possibility of short range ordering and clustering in the system[35] and
to go beyond the single site framework of ATA and CPA[36] are at
various stages of development. Although some of these advances
in the alloy theory will undoubtedly be brought to bear on the
hydride problem in the future, their discussion is not considered
germane to this review.

Section III describes the model for treating interstitial
impurities as an "alloy" of impurities and vacancies. The intrinsic limitations of the present ATA and CPA type framework are
clarified. Although we discuss the example of sub-stoichiometric
PdH$_x$ in Section III, this approach can be extended easily to model
related systems such as the ternary or polyhydrides.

Section IV presents a comprehensive picture of the electronic spectrum of α-phase PdH_x within the muffin-tin ATA. Our results concerning the disorder smearing of states have not been published before. While the level shifts in Pd on H uptake are described satisfactorily by the theory, the disorder smearing of states is not. The question of isotope effects in α-PdH_x[37] is not addressed; these delicate effects may not be amenable to a quantitative treatment within the band theory framework.

BAND THEORY OF SUBSTITUTIONAL ALLOYS

As noted above, the application of ATA and CPA to the muffin-tin Hamiltonian yields a band theory of random alloys. In briefly summarizing the basic elements of such a theory, we note that these approximations amount, in a certain sense, to replacing the disordered system by an equivalent perfect crystal of effective atoms. The associated quasi-particle spectrum is then given by the determinantal equation (in angular momentum space):[5]

$$\| \ \tau_{eff}^{-1}(E) \ - \ B(\vec{k},E) \ \| \ = 0 \ , \tag{2.1}$$

where τ_{eff} is the matrix of the on-the-energy-shell elements of the effective atom t-matrix (either ATA or CPA), and $B(\vec{k},E)$ is the matrix of the usual KKR structure functions. In a perfect A (or B) crystal $\tau_{eff} \rightarrow \tau_A$ (or τ_B) and (2.1) reduces to the KKR equation, well-known in the band theory of perfect crystals. Thus, Eq. (2.1) provides a simple conceptual basis for constructing a band theory of random alloys. It is convenient to fix the value of the crystal momentum $\vec{k}$ and solve (2.1) for the bands $E(\vec{k})$. The solutions are real in a perfect crystal, but become complex in the alloy, with the imaginary part representing the disorder smearing of states. From a formal viewpoint, the real and imaginary parts of the complex bands yield respectively the positions (in energy) and half-widths of the corresponding peaks in the spectral density function. The usefulness of the complex energy bands in connection with disordered alloys would appear by now to be well-established.[29,38-40]

As emphasized frequently in connection with the CPA and more recently the ATA, the disordered system also involves the properties of a single A or B impurity embedded in the effective medium.[41] These contributions to the quasi-particle spectrum are given by the secular equation:[5]

$$\| \ 1 - T^{eff}(\tau_{eff}^{-1} \ - \ \tau_{A(B)}^{-1}) \ \| \ = 0, \tag{2.2}$$

where the matrix T^{eff} is defined by

$$T^{eff}(E) = \frac{1}{N} \sum_{\vec{k}} \frac{1}{\tau_{eff}^{-1} - B(\vec{k},E)} \cdot \tag{2.3}$$

as a Brillouin-zone summation. In the limit of x (or y) $\rightarrow$ 0, Eq. (2.2) can be shown to yield real impurity levels for an A (or B) impurity in an otherwise perfect B (or A) crystal. More generally, therefore, (2.2) represents the complex levels arising when an A (or B) atom is placed in the effective medium.

In the use of Eqs. (2.1) and (2.2) the distinction between the ATA and the CPA arises only from the choice of the effective atom. In the ATA, $t_{eff} \equiv t^{ATA}$ corresponds to the simple choice

$$t^{ATA}(p,p') = \langle t(p,p') \rangle \equiv x t^{A}(p,p') + (1-x) t^{B}(p,p') \; , \tag{2.4}$$

whereas the CPA t-matrix is given by the equation[6]

$$t^{CPA}(p,p') = \langle t(p,p') \rangle - [t^{A}(p,\kappa) - t^{CPA}(p,\kappa)] \; F^{CPA}(E)$$

$$x \; [t^{B}(\kappa,p') - t^{CPA}(\kappa,p')] \; . \tag{2.5}$$

Here, F^{CPA}, related to T^{eff} of Eq. (2.3), is defined by

$$F^{CPA}(E) \equiv \frac{1}{N} \sum_{\vec{k}} B_{\vec{k}} \; (1 - \tau_{CPA} B_{\vec{k}})^{-1} \; . \tag{2.6}$$

CPA Eq. (2.5) is a complicated self-consistency condition for the quantities τ_{CPA}. Its solution requires repeated Brillouin zone summations (2.6).

Eqs. (2.1) and (2.2) determine the nature of the quasi-particle spectrum in the alloy. The solutions of (2.1) are Bloch-type in that these states evolve from Bloch levels in a perfect A or B crystal. In contrast, the levels given by (2.2) are non-Bloch-type, and their appearance in the alloy is a characteristic effect of disorder. Note also that the quasi-particle levels in the alloy do not, in general, possess uniform spectral weights. This difference between the complex bands and their perfect crystal counterparts is fundamental and is the basic reason why the calculation of density of states in the disordered system involves a level of intricacy well beyond the perfect crystal case.[42] In particular, the density of states cannot be obtained by constructing a histogram of the complex energy levels.

Figures 1 and 2 provide illustrative examples of the ATA and CPA complex bands and densities of states in the CuNi system.

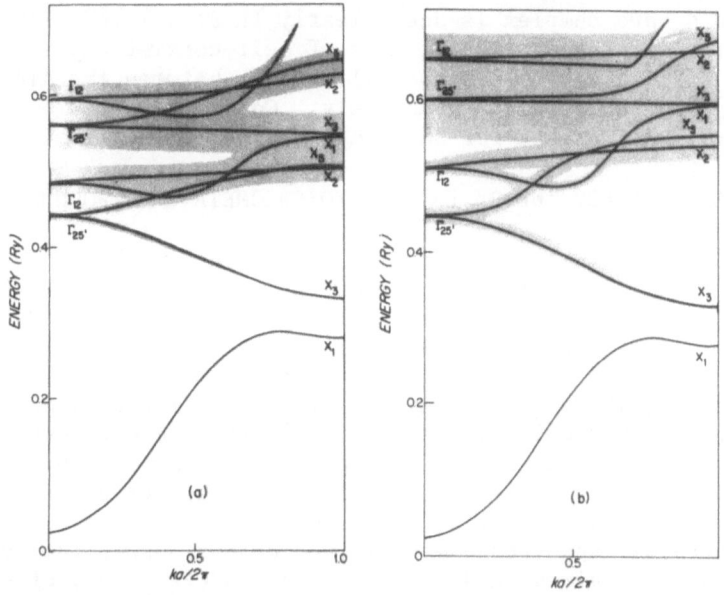

Fig. 1 Complex energy bands (i.e., solutions of Eq. (2.1)) in
$Cu_{0.75}Ni_{0.25}$ along the direction Δ in the Brillouin zone for
(a) the ATA and (b) the CPA. The vertical length of shading
around the bands equals two times the imaginary part of the
energy. (After Bansil[5]).

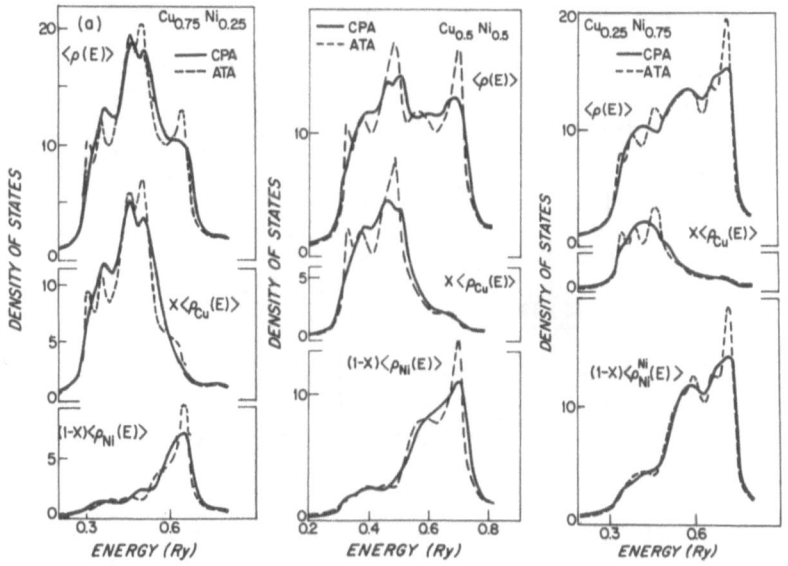

Fig. 2 The average density of states $\langle\rho(E)\rangle$, in CuNi solid solutions.
The weighted Cu and Ni component densities $x \langle\rho_{Cu}(E)\rangle$ and
$(1-x)\langle\rho_{Ni}(E)\rangle$ are also shown. (After Bansil[5]).

Ni-derived d band complex is seen clearly in Fig. 1 and also in
Fig. 2(a), as expected. The effects of self-consistency in the
treatment of disorder (i.e., the differences between the ATA and
the CPA) have been studied extensively. Generally speaking, such
effects are small except for detailed spectral properties.

APPLICATION OF ALLOY THEORY TO NON-STOICHIOMETRIC COMPOUNDS

To illustrate how the preceding ideas can be applied to non-
stoichiometric compounds, we consider the example of the Pd-H
system. The stoichiometric phase, PdH, possesses the FCC NaCl
structure. It is generally believed that at off-stoichiometric
compositions, the H atoms continue to enter into the octahedral
sites, and thus in PdH_x a fraction x of these sites will be occupied (shown schematically in Fig. 3).[43,44] It is clear that the substoichiometric PdH_x may be modelled as a two atoms per unit cell
system in which every site of one sublattice is occupied by Pd
atoms, while the second sublattice consists of a "binary alloy"
of hydrogen atoms and vacancies. If we further assume the hydrogen to occupy the sites randomly, the ATA and CPA theory of substitutional alloys can, in principle, be extended rather straightforwardly by interpreting the various matrices in the formalism
(such as $B(\vec{k},E)$, τ_A, τ_B etc.) as (2x2) supermatrices. For example,
the secular equation (2.1) would now correspond to an ordered
system of two atoms per unit cell; one of the atoms being Pd and
the other an appropriate effective atom. In particular, the ATA
Eq. (2.4) generalizes to the equation,

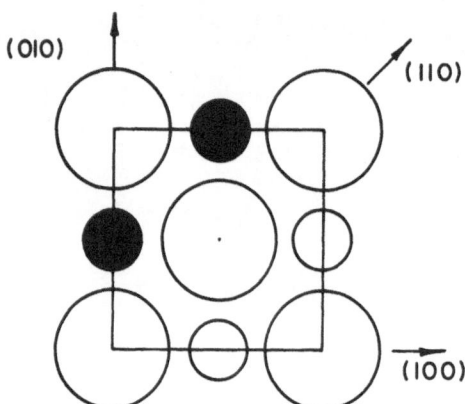

Fig. 3.   Schematic representation of a cube face in the NaCl
          lattice. In $PdH_{0.5}$, two of the four octahedral sites
          will be occupied by hydrogen atoms (filled circles),
          while the other two (small unfilled circles) will be
          empty.

$$t^{ATA} = xt_H + (1-x) \; t_{vac} \tag{3.1}$$

As a first approximation, the vacancy potential may be taken to be equal to the average potential in the interstitial region. Since this constant value implicitly defines the zero of energy in the muffin-tin formalism, it follows that

$$t_{vac} = 0 \; . \tag{3.2}$$

To date, all reported computations[10-13] have employed approximation (3.2). Effects of using a more realistic vacancy potential should be examined.

The preceding model for Pd-H can be extended to consider ternary and poly hydrides and related systems. For example, Pd-Ag-H would correspond to a situation where both the metal and the hydrogen sublattices are disordered. To model a polyhydride (i.e., MH_x for x > 1), more than two atoms per unit cell will, in general, be necessary, because at least two different types of hydrogen sites are involved.

Finally, we note that, by using a composition dependent lattice constant, the present approach can take account of lattice distortions around impurities in an average way. No experience exists at this time in treating non-uniform distortions even in substitutional muffin-tin alloys. In this vein, we further note that to incorporate the effects of non-randomness (i.e., short range ordering and clustering) would also require a considerable increase in the complexity of the formalism.

RESULTS ON α-PHASE PdH_x

A typical plot of the ATA complex energy bands in $PdH_{0.05}$ (i.e., using Eqs. (3.1) and (3.2)) is shown in Fig. 4; the density of states is given in Fig. 5. We will use these results to infer the characteristic effects of H on the spectrum of Pd. Most notably, H induces a bonding Pd-H band approximately 2 eV below the bottom of the Pd bands. Thus the host spectrum is modified drastically in this low energy regime. With increasing H concentration, this band, in fact, develops continuously into the lowest valence band of the ordered compound PdH. This is evident from Switendick's band structures for odered Pd-H phases[45] and also in the ATA complex bands[10] of PdH_x for 0 < x < 1. On the basis of these computations and those on hydrides of other transition metals (e.g., V, Nb, Ta, Y),[1,2,17] the presence of a metal-hydrogen bonding band, well below the Fermi energy is a well established prediction of the band theory.[46]

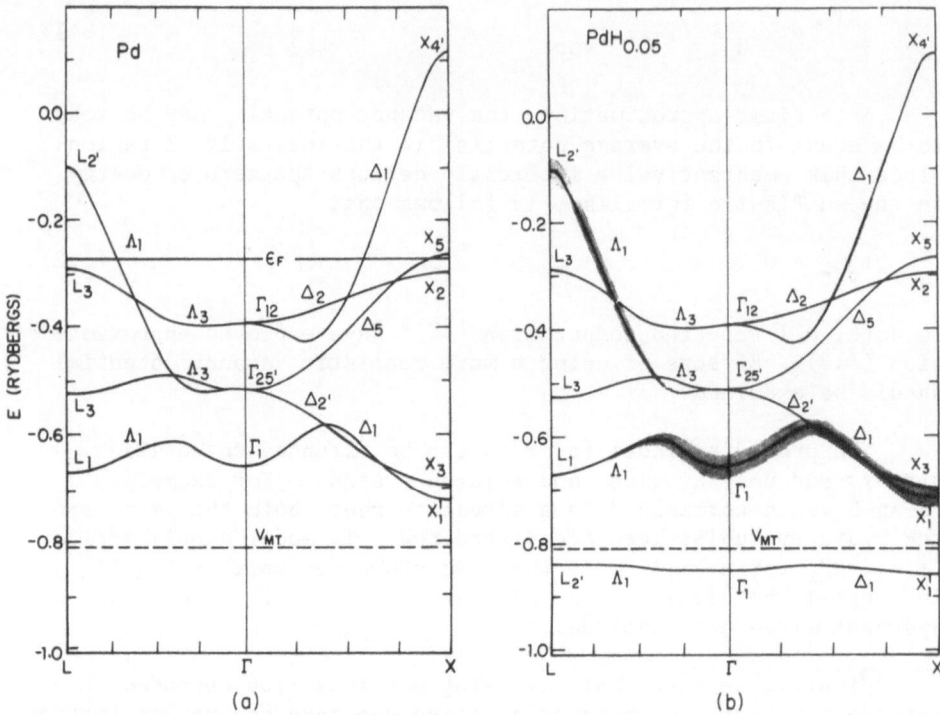

Fig. 4   ATA Complex energy bands in (a) Pd and (b) PdH$_{0.05}$. (After
        Gelatt et. al.[10]).

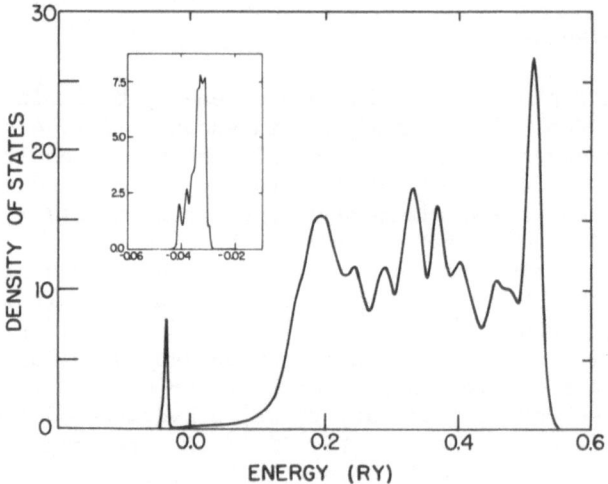

Fig. 5   ATA density of states in PdH$_{0.05}$.   (After Bansil et. al.[12])

The bonding Pd-H band in Figs. 4 and 5 is only 0.1 to 0.2 eV wide (see inset in Fig. 5). This unusually small width is a consequence of the strict lack of disorder smearing in the ATA theory for states lying below the alloy muffin-tin zero, and is probably not physical. However, the ATA is expected to give the correct total weight of this band. Density of states calculations of Bansil et. al.[12] show that this band accommodates a total of 2x electrons per unit cell in dilute PdH_x. We emphasize that, despite this fact it does not necessarily follow that the Pd states at the Fermi energy will be depleted on adding H. Indeed, when the modifications in the host spectrum (i.e., bonding with H, disorder smearing, changes in the spectral weights) are fully taken into account via a density of states computation, we find that the Fermi energy, $E_F(x)$, _increases_ at approximately half the rigid band rate corresponding to the various Pd and H potentials employed in Refs. 11 and 12. In other words, half of the extra electrons are added at the Fermi energy;[47,48] neither the simple anionic nor the protonic model is applicable.

Extensive calculations on ordered compounds have shown that the interstitial atoms act to lower the energies of metal states possessing s-p symmetry around the impurity sites. Furthermore, the rather localized d states such as X_5 or L_3, which possess little weight in the interstitial region, are hardly influenced by impurities. These effects are found also in the ATA results on PdH_x. In addition, the complex bands of Fig. 1 depict the disorder smearing of states (shown by shading). In fact, the level shifts and the smearing of states go together in this case, because both effects derive here from the same physical mechanism, i.e., the presence of H impurities.

We turn now to a discussion of the changes in the Fermi surface of Pd on H uptake.[37] These changes are directly accessible via dHvA experiments.[37] Fig. 6 shows a plot of the dHvA frequency F_{dHvA} against H concentration for the X[100] hole orbit.[49] For x < 0.003, F_{dHvA} decreases relatively rapidly, while for x > 0.007, the x dependence is weaker; in this regard, the results of Fig. 6 are typical of measurements on other extremal orbits.[50] The data for x > 0.007 can be interpreted to correspond to the formation of the mixed α - α' phase. The slope of the curve for x < 0.003, on the other hand, is representative of the α-phase; this regime can be extended to values of x > 0.003 by using faster quenching rates on the (high temperature) α-phase samples.[49,50] The conversion to the mixed phase was not recognized in the early experiments, and those results are, therefore, not relevant in discussing the properties of the solid solutions.[51]

Table 1 lists results for the α-phase (i.e., for x < 0.003), together with the ATA and rigid band predictions for the Γ-centered electron sheet and for several X-centered hole orbits. The avail-

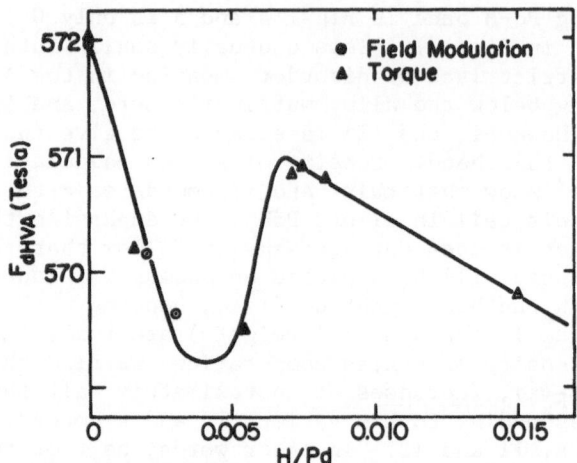

Fig. 6   Concentration dependence of the dHvA frequency, F_{dHvA}, for
the X[100] hole orbit in PdH$_x$. A smooth curve has been
drawn through the experimental points. (After Venema et.
al.[49]).

able data on the L-centered hole pockets is not considered,
because the appearance of these pockets in the computations
depends sensitively on the choice of the Pd crystal potential.
Calculations based on a number of different sets of Pd and H
potentials (see legend for details) are considered.[52] The ATA
predictions in column c are in an excellent absolute agreement
with the measurements. Differences between the theoretical
columns c, d, and e are due almost entirely to the differences
in the densities of states in the corresponding pure Pd. Thus,
these differences are representative of the uncertainties inherent
in the first principles band theory of _perfect_ transition metals
and ordered hydrides; they should _not_ be taken to reflect difficulties in the present _ab. initio_ treatment of non-stoichiometry.

By scaling the ATA results with the corresponding rigid band
values, a striking pattern is seen to emerge in Table 1. Columns
c', d' and e' show that the Γ-centered electron orbit (row one)
grows at approximately the rigid band rate, while the X-centered
hole pockets (rows two, three and four) shrink at about _half_ the
corresponding rigid band rate. This non-uniform variation of the
Fermi surface results from a combination of several effects (e.g.,
the insensitivity of the d-like hole states and the lowering of
the s-p-like electron states of Pd on adding H, the bonding
between Pd and H levels and the consequent variation of $E_F(x)$, etc.)
It is probably not very useful to describe these complicated changes
in terms of simple model parameters.

Table 1. Concentration dependence of the extremal areas of the Γ-centered electron sheet (in [100] field direction) and of the X-centered hole ellipsoids (for several field directions) on the Fermi surface of α-PdH$_x$. The measured values of the parameter d ℓn A/dx, where A denotes the orbit area, are compared with the ATA and rigid band (RB) predictions. Theoretical calculations employ a concentration independent lattice constant. Details of the various columns are listed in the legend. (After Bansil et. al.[12]).

Orbit	$\left(\dfrac{d\ \ell n\ A}{dx}\right)_{Expt.}$		$\left(\dfrac{d\ \ell n\ A}{dx}\right)_{ATA}$			$\left(\dfrac{d\ \ell n\ A}{dx}\right)_{ATA} \Big/ \left(\dfrac{d\ \ell n\ A}{dx}\right)_{RB}$		
	a	b	c	d	e	c'	d'	e'
Γ[100]	0.22		0.23	0.19	0.15	1.03	1.24	0.98
X[100]	-1.5	-1.7	-1.53	-1.04	-0.80	0.56	0.67	0.49
X[001]	-1.9		-1.64	-1.10	-0.84	0.56	0.66	0.48
X[110]		-1.7	-1.48	-0.98	-0.75	0.55	0.66	0.48

a.   Field modulation measurements.

b.   Torque measurements.

c and c'.   Renormalized atom Pd potential, touching sphere radii.

d and d'.   Overlapping atoms Pd potential, touching sphere radii.

e and e'.   Overlapping atoms Pd potential, 65/35 sphere radii.

The PdH$_x$ calculations in Table 1 employ a lattice constant independent of H concentration. The lattice expansion effect so neglected can be estimated by writing[49-51]

$$\left(\frac{d\ \ell n\ A}{dx}\right)_{expt} = \left(\frac{\partial\ \ell n\ A}{\partial x}\right)_{\Omega} + \left(\frac{\partial\ \ell n\ A}{\partial\ \ell n\ \Omega}\right)_{Pd} \left(\frac{\partial\ \ell n\ \Omega}{\partial x}\right) . \qquad (4.1)$$

The value of the quantity ($\partial\ \ell n\ A/\partial\ \ell n\ \Omega$) can be determined from dHvA pressure measurements on various orbits in Pd. Note that, in estimating the ($\partial\ \ell n\ \Omega/\partial x$) term, one should only consider the uniform part of the dilation of the Pd lattice, since the dHvA experiment samples Pd-like "coherent" states; local (non-uniform) distortions around the H sites are presumably less important. An upper limit of 0.2 for the size of ($\partial\ \ell n\ \Omega/\partial x$) can be obtained by associating the total volume change on hydrogenation with a uniform expansion of the Pd lattice. However, Venema et. al.[49]

argue, based on a theory of Eshelby,[53] for a substantially smaller
value of 0.03 for the quantity $(\partial \ln \Omega/\partial x)$, and thus estimate the
size of the correction term $C \equiv (\partial \ln A/\partial \ln \Omega)$ $(\partial \ln \Omega/\partial x)$ to be
less than 10% of the experimental values in Table 1. We emphasize
that the comparison of the ATA and rigid band predictions in
columns c', d' and e' is insensitive to how the corrections C are
calculated, since both results will require the same correction.

In considering the disorder smearing of states at the Fermi
energy, Table 2 presents the theoretical and experimental Dingle
temperatures.[54] As another (but related) measure of this smearing,
the values of the disorder induced width $\Delta\vec{k}$ of the Fermi surface
dimension $\vec{k}$ are also listed for the (100) direction in the orbit
plane.[55] The quantity $\Delta\vec{k}$ will be relevant in a measurement of the
Fermi surface radius via, for example, the angular correlation of
annihilation radiation.[56] No direct measurement of $\Delta\vec{k}$ exists at
this time.

Table 2 shows that the measured Dingle temperatures for the
hole orbits are considerably larger than the calculations. Note
that the measurements, in general, involve scattering mechanisms
other than disorder. Since the latter effect is computed here to
be negligibly small, the measured values may be representative of
scattering due to lattice strain fields, imperfections, etc.
Venema et. al.[57] have argued that a proper treatment of the long
range part of the H potential (i.e., beyond the H muffin-tin sphere
radius), neglected in the present KKR framework, may explain this
discrepancy. Indeed, the hole Dingle temperature in column e,
which corresponds to a larger H sphere radius, is larger than
column d; but, its magnitude is still quite small. Whether further

Table 2. Disorder smearing of states on the Fermi surface of
α-PdH$_x$. Dingle temperatures for the Γ-centered
electron orbit and for a typical orbit on the
X-centered hole ellipsoid are listed. The disorder
induced width $\Delta k(100)$ of the orbit radius $k(100)$ in
(100) direction, is also listed as $\Delta k/k$ (in percent).
The Pd and H potentials employed for the theoretical
columns c, d and e (also c', d', and e') are as indicated in the legend to Table 1. Experimental results
are after Venema.[56]

| Orbit | Dingle Temperatures (K per % H) | | | | $(\Delta k(100)/k(100)) \times 10^2$ (per % H) | | |
	Expt.	c	d	e	c'	d'	e'
Γ[100]	7.5 ± 1.5	38	24	25	0.37	0.26	0.27
X[100]	2.1 ± 0.4	0.01	0.02	0.2	0.001	0.002	0.03

increases in the hydrogen sphere radius can explain the observations is a matter of speculation.

Turning to the electron orbit, the measured Dingle temperature is 3 to 5 times smaller than the calculated values. This is surprising because generally the measured values are expected to be greater than the predicted disorder smearing, as in the case of the hole orbits above. Also, this level of discrepancy (i.e., a factor of 3 to 5) would appear to be substantially larger than that observed in calculations using the alloy theory on a number of other solid solutions such as $CuZn$,[38,39] $CuGe$,[40] $CuA\ell$,[29] $AgPd$,[34] and CuH[13]. It has been suggested that the partial ordering of hydrogen may be responsible for the reduction of the Dingle temperature from the (random) ATA predictions.[58]

CONCLUDING REMARKS

Given the success of the muffin-tin Hamiltonian in describing the ordered hydrides, and of the ATA and CPA in describing the influence of disorder, it is reasonable to expect that the present framework would generally provide a satisfactory picture of the effects of non-stoichiometry. [The aforementioned discrepancies in α-PdH$_x$ may not be typical in this regard.] The uncertainties inherent in the band theory of perfect systems (as, for example, the choice of the H muffin-tin potential) are, of course, unavoidable in the present first principles treatment. The muffin-tin ATA and CPA formalism should be used to carry out more extensive studies of interstitial impurities over wider composition ranges.

REFERENCES

1. A.C. Switendick, Topics in Applied Physics, Vol. 28: Hydrogen in Metals I: Basic Properties, G. Alefeld and J. Völkl, eds., Springer Verlag, Berlin 1978, p. 101.
2. A.C. Switendick, Zeits. Physik. Chem. Neue Folge, Bd. 117, S89 (1979).
3. D.A. Papaconstantopoulos, Proceedings of the NATO Advanced Study Institute on Metal Hydrides . (Plenum, 1981).
4. For recent theoretical discussions of the muffin-tin alloys see Refs. 5 to 9 below. A review up to 1975 is found in Section IX of H. Ehrenreich and L. Schwartz in Solid State Physics, edited by H. Ehrenreich, F. Seitz and D. Turnbull (Academic, New York, 1976), Vol. 31.
5. A. Bansil, Phys. Rev. Letters 41, 1670 (1978); A. Bansil, Phys. Rev. B20, 4025 (1979); ibid, B20, 4035 (1979).
6. A. Bansil, R.S. Rao, P.E. Mijnarends and L. Schwartz, Phys. Rev. B23, 3608 (1981).
7. J.S. Faulkner and G.M. Stocks, Phys. Rev. B21, 3222 (1980).
8. B.E.A. Gordon, W.E. Temmerman and B.L. Gyorffy, J. Phys. F. 11, 821 (1981).

9. See contributions by L. Schwartz, by A. Bansil and by R.
    Prasad, R.S. Rao and A. Bansil in the Proceeding of NATO
    Advanced Study Institute on Excitations in Disordered Systems,
    Michigan State University, M.F. Thorpe, ed. (Plenum, 1982).
10. C.D. Gelatt, Jr., H. Ehrenreich and J.A. Weiss, Phys. Rev. B17,
    1940 (1978).
11. A. Bansil, S. Bessendorf and L. Schwartz, Inst. Phys. Conf.
    Ser. 39, 493 (1978).
12. A. Bansil, R. Prasad, S. Bessendorf, L. Schwartz, W.J. Venema,
    B. Feenstra, F. Blom and R. Griessen, Solid State Commun. 32,
    1115 (1979).
13. L. Huisman and J.A. Weiss, Solid State Commun., 16, 983 (1975).
14. J.H. Weaver, D.J. Peterman and D.T. Peterson, Proceedings of
    this conference.
15. M. Gupta, Proceedings of this conference.
16. M.A. Khan, J.C. Parlebas and C. Demangeat, Proceedings of this
    conference.
17. A.C. Switendick, J. Less-Common Metals 74, 199 (1980).
18. J.S. Faulkner, Phys. Rev. B13, 2391 (1976).
19. D.A. Papaconstantopoulos, B.M. Klein, J.S. Faulkner, and L.L.
    Boyer, Phys. Rev. B18, 2784 (1978).
20. P. Jena and K.S. Singwi, Phys. Rev. B17, 3518 (1978).
21. P. Jena, F.Y. Fradin and D.E. Ellis, Phys. Rev. B20, 3543
    (1979).
22. M.I. Darby, G.R. Evans and M.N. Read, J. Phys. F 11, 1023 (1981).
23. H. Adachi, S. Imoto, J. Phys. Soc. Japan, 46, 1194 (1979).
24. H. Adachi, S. Imoto, T. Tanabe and M. Tsukada, J. Phys. Soc.
    Japan 44, 1039 (1978).
25. Here, "complex bands" refers to bands prossessing an imaginary
    part. This terminology has often been used in the literature
    in conncection with energy bands in perfect crystals with
    complicated structures.
26. For a recent review, see P.E. Mijnarends in Positrons in Solids,
    ed. P. Hautojärvi (Springer Verlag, Berlin, 1979), p. 25.
27. P.E. Mijnarends and A. Bansil, Phys. Rev. B19, 2919 (1979).
28. B.L. Gyorffy and G.M. Stocks, J. Phys. F10, L321 (1980).
29. M. Pessa, H. Asonen, R.S. Rao, R. Prasad and A. Bansil, Phys.
    Rev. Letters 47, 1223 (1981).
30. M. Pessa, H. Asonen, M. Lindroos, A.J. Pindor, B.L. Gyorffy
    and W.M. Temmerman, J. Phys. F11, L33 (1981).
31. J. Staunton, B.L. Gyorffy and P. Weinberger, J. Phys. F (1981).
32. S. Kaprzyk and A. Bansil, Phys. Rev. B (1982).
33. L. Schwartz, Phys. Rev. B23, 3608 (1981).
34. G.M. Stocks and W. Butler, Phys. Rev. Letters 48, 55 (1982).
35. L.J. Gray and T. Kaplan, Phys. Rev. B24, 1872 (1981).
36. R. Mills and P. Ratnavararaska, Phys. Rev. B18, 5291 (1978);
    H.W. Diehl and P.L. Leath, Phys. Rev. B19, 587 (1979); T.
    Kaplan, P.L. Leath, L.J. Gray, and H.W. Diehl, Phys. Rev. B21,
    4230 (1980).
37. R. Griessen and L.M. Huisman, Proceedings of this conference.

38. A. Bansil, H. Ehrenreich, L. Schwartz and R.E. Watson, Phys. Rev. B9, 445 (1974).
39. R. Prasad, S.C. Papadopoulos and A. Bansil, Phys. Rev. B23, 2607 (1981).
40. R. Prasad and A. Bansil, Phys. Rev. Letters 48, 113 (1982).
41. The exact density of states in the disordered alloy will, of course, have contributions from impurity levels associated with clusters of more than one atom. However, within the framework of ATA and CPA only the properties of the single A or B impurities occur in a transparent manner.
42. In fact, different versions of the muffin-tin ATA for densities of states and other properties can be obtained, depending upon how the impurity contributions, given by Eq. (2.2), are treated. See Ref. 5 for a discussion of this point.
43. For structural information see, for example, G.G. Libowitz, The Solid State Chemistry of Binary Metal Hydrides, (Benjamin, New York, 1965).
44. Of course, there are many delicate questions concerning the way H enters metals, e.g., the formation of hydrogen density waves. [See review by D.G. Westlake and by T. Schober, in Proceedings of this Conference]. However, the average electronic spectrum will likely be insenstive to some of these fine structural details.
45. A.C. Switendick, Ber. Bunsenges. Physik. Chem. 76, 535 (1972).
46. A hydrogen induced structure in PdH_x was seen in the early photoemission data of D.E. Eastman, J.K. Cashion and A.C. Switendick [Phys. Rev. Letters 27, 35 (1971)]. More recent XPS/UPS studies have been presented by L. Schlapbach and J.P. Burger [Proceedings of this conference].
47. A. similar conclusion has also been reached by Switendick (to be published) by supercell calculations on $Pd_{32}H$ and by Klein and Pickett (Ref. 48), who consider a single H impurity in the Pd lattice.
48. B.M. Klein and W.E. Pickett, Proceedings of this conference.
49. W.J. Venema, R. Feenstra, F. Blom and R. Griessen, Zeits. Physik. Chem. Neue Folge, Bd. 116, S125 (1979).
50. W.J. Venema, Ph.D. Thesis (Free University of Amsterdam, 1980).
51. R. Griessen, W.J. Venema, J.K. Jacobs, F.D. Manchester and Y. de Ribaupierre, J. Phys. F7, L133 (1977).
52. All calculations in Table 1 are based on an unscreened $(1/r)$ H Potential, which is cut-off at the H sphere radius. However, we carried out a number of additional computations, which show that our conclusions are insensitive to reasonable variations in the H potential.
53. J.D. Eshelby, J. Appl. Phys. 25, 255 (1954).
54. Dingle temperatures corresponding to column c were also inferred in Ref. 10.
55. $\Delta k/k$ values for other directions in the orbit are not presented; the results listed for the (100) direction are typical in this regard.

56. See Ref. 39 for a detailed discussion of this point.
57. W.J. Venema, R. Griessen, R.S. Sorbello, H.L.M. Bakker and
    P.E. Mijnarends, Inst. Phys. Conf. Ser. 55, 579 (1981). See,
    also, Ref. 37.
58. R. Griessen (private communication).

A SIMPLE TIGHT BINDING DESCRIPTION OF THE ELECTRONIC STRUCTURE OF A SINGLE AND A PAIR OF HYDROGEN ATOMS IN fcc TRANSITION METALS

M.A. Khan, J.C. Parlebas and C. Demangeat

L.M.S.E.S. (LA CNRS 306) - Université Louis Pasteur
4, rue Blaise Pascal 67070 STRASBOURG CEDEX, France

ABSTRACT

The electronic structure of an Hydrogen impurity embedded in a face centred cubic transition metal is obtained by a generalized tight binding Slater-Koster fit to the first principles band structure for the host, combined with one s orbital for the isolated impurity atom. Matrix elements of the perturbing potential up to the first nearest neighbours of the interstitial are explicitely taken into account. The hopping integrals between the Hydrogen and the metallic sites are deduced from ab initio calculation of the corresponding hydride. The intrasite matrix elements are given in terms of the local Coulomb correlation U on the Hydrogen site and are adjusted with the help of Friedel's screening rule. Nearest neighbours and next nearest neighbors pairs of Hydrogen atoms in α-Palladium hydrides are shown to be forbidden. Elements of an estimation of elastic binding energy are also presented.

INTRODUCTION

Since the pioneering work of Switendick[1] a considerable number of papers dealing with the electronic structure of stoichiometric and off-stoichiometric hydrides have appeared in the literature. Let us recall that besides the usual stoichiometric hydrides TH, TH_2 (where T is a transition metal) Switendick[2] and others[3] have discussed hydrides with periodic arrangement of Hydrogen atoms (like $Pd_{32}H$, Pd_6H...) by the classical APW or KKR methods. On the other hand disordered off-stoichiometric hydrides have been discussed in the coherent potential approximation in the framework of tight binding[4] or KKR method[5].

The present paper is devoted to a description of one and two interstitials in a face centred transition metal. In the case of two Hydrogen interstitials in a transition metal we discuss the stability of the pair in terms of the distance between the two Hydrogen atoms. In this case the presence of impurities breaks the translational symmetry and classical APW or KKR methods are useless. Different models presented recently have tried to overcome this difficulty. On one hand Sholl and Smith[6] using a screened proton model have essentially shown that the energy of the tetrahedral position of Hydrogen in Palladium is lower than the octahedral position which is in contradiction with channelling experiments[7]. On the other hand, KKR Green's function method developed by Beeby[8] for substitutional defects has been recently extended to the case of interstitial defects[9,10,11]. More precisely Kanamori's group[9,10] proposed a theory for the description of Hydrogen in Nickel and α-Iron within the restriction of a single muffin-tin for the impurity potential. Also this type of calculation has been made self-consistent[11]. Nevertheless, it has been shown[12] that,within the concept of one muffin-tin for the description of Hydrogen in transition metals, it is impossible to satisfy Friedel's screening rule. Extension of the range of the impurity potential has been discussed by Inglesfield[13] in the case of Hydrogen in Copper. Besides these ab initio calculations, more empirical methods have been used for the description of Hydrogen in transition metals[14,15,16]. In these calculations a generalized Slater-Koster fit to ab initio band structure calculation for the host together with one s extra orbital for the interstitial have been used. This kind of calculation was performed along the lines of previous studies of Carbon interstitial in Nickel and α -Iron[17,18] and other light interstitials in α-Iron (B, N, O)[19].

Dietrich and Wagner[20] developed a model for the determination of the binding energy of a pair of Hydrogen in incoherent Palladium-Hydrogen system. They estimated in a crude model the electronic binding energy of a pair of Hydrogen. Besides this electronic term, they included an elastic term deduced from the experimental dipole force tensor P[21]. Moreover they used the asymptotic form of the binding energy which is not valid for the nearest neighbouring distances[22]. Determination of P in terms of the electronic structure of an interstitial Hydrogen in Palladium has been described recently[23]. As far as the chemical (or electronic) binding energy of a pair of Hydrogen atoms is concerned an estimation has been given by Masuda and Mori[24] who used a rather schematic representation for the pure metal band structure. More recently Demangeat et al[25,26,27] have estimated the stability of a pair of Hydrogen atoms in α-Palladium hydrides in terms of the electronic structure[15].

The outline of the paper is as follows. Section 2 presents a formal derivation of a tight-binding model used for the description of Hydrogen in metals containing strongly bound d electrons. We will

discuss explicitely two types of approximations which are based on
i) the local neutrality criterion[14],[15],[16], and ii) on an explicit
Hartree-Fock model with different values of the Coulomb correlation term[28]. Section 3 presents the results obtained for Hydrogen
at octahedral position in Palladium. It is shown that a virtual
bound state appears at the bottom of the sp bands when the local
neutrality condition is used. On the other hand, for realistic
values of the Coulomb term a bound state falls below the bottom
of the sp bands together with a charge transfer from metal to metalloid. Section 4 gives a formal calculation of the electronic
binding energy of two Hydrogen atoms. Then it shows that Hydrogen
atoms, in α-Palladium hydrides repel each other at nearest and
next nearest interstitial positions. Also, some elements of the determination of the dipole force tensor is briefly reported. Our
results are summarized in section 5.

THE ELECTRONIC STRUCTURE OF HYDROGEN AS A SINGLE IMPURITY
IN A fcc TRANSITION METAL

The band structure of pure transition metal is described by an
spd Slater-Koster fit to a first principle calculation. The metallic
spd orbitals are labelled $|Rm\sigma>$, where R is a metallic site, m the
orbital symmetry and σ the spin. Let λ be the position of an inters-
titial Hydrogen and $|\lambda s\sigma>$ the corresponding s extra orbital of ener-
gy $E_\lambda^{s\sigma}$. The perturbed Hamiltonian for a spin σ is given by[15] :

$$H = H_o + |\lambda s\sigma> E_\lambda^{s\sigma} <\lambda s\sigma| + \sum_{R,d} |Rd\sigma> v_d^\sigma(R) <Rd\sigma|$$
$$+ \sum_{Rm} \{ |Rm\sigma> \beta_{R\lambda}^{ms} <\lambda s\sigma| + c.c. | \tag{1}$$

where H_o is the pure metal Hamiltonian and c.c. $\equiv$ complex conjugate.
$\beta_{R\lambda}^{ms}$ is the hopping integral between $|Rm\sigma>$ and $|\lambda s\sigma>$ with m =
(s, x, y, z, xy, yz, zx, x^2-y^2, $3z^2-r^2$). The diagonal disorder term
$v_d^\sigma(R)$ in the d bands is introduced to be in accord with Friedel's
rule: that the number of external electrons brought by the Hydrogen
atom is equal to the total number of displaced states $\sum_\sigma Z^\sigma(E)$ up to
the Fermi level E_F :

$$\sum_\sigma Z^\sigma(E_F) = 1 \tag{2}$$

where $Z^\sigma(E)$ is given by[17],[28] :

$$Z^\sigma(E) = -\frac{1}{\pi} \{ arg \Sigma_\lambda^{s\sigma}(E) - \pi \} - \sum_{Rd} v_d^\sigma(R) \, n_d^{o\sigma}(E) \tag{3}$$

$$\Sigma_\lambda^{s\sigma}(E) = E - E_\lambda^{s\sigma} - \Delta_\lambda^{s\sigma}(E) - \Gamma_\lambda^{s\sigma}(E) \tag{4}$$

In equation (3), $v_d^\sigma(R)$ is treated in perturbation, $n_d^{o\sigma}(E)$ is the d
density of states of pure metal. $\Delta_\lambda^{s\sigma}(E)$ and $\Gamma_\lambda^{s\sigma}(E)$ are given in terms
of β, G^o and v_d.

In the present paper we use a very simple tight binding scheme where the overlap integrals $<\lambda s\sigma|Rm\sigma>$ are supposed to be zero. Therefore, the local density of states (LDOS) of spin σ, at the Hydrogen site λ, is given by :

$$n_\lambda^{s\sigma}(E) = - \frac{Im}{\pi} G_{\lambda\lambda}^{ss\sigma}(E) \quad ; \quad G_{\lambda\lambda}^{ss\sigma}(E) = [\Sigma_\lambda^{s\sigma}(E)]^{-1} \qquad (5)$$

The variation of the LDOS on a metallic site R at a nearest neighbouring position to the Hydrogen atom is equal to :

$$\delta n_R^\sigma(E) = - \frac{Im}{\pi} \sum_m (G_{RR}^{mm\sigma}(E) - G_{RR}^{omm\sigma}(E)) \qquad (6)$$

where $G_{RR}^{mm\sigma}(E)$ is a complicated expression[28],[29] given in terms of β, G^o and v_d. The quantities $n_\lambda^{s\sigma}(E)$ and $\delta n_R^\sigma(E)$ are given by expressions which contain the same denominator. This means that a bound state located at energy E_b below the sp conduction bands and defined by $\Sigma_\lambda^{s\sigma}(E_b) = 0$ will be present at the interstitial site and at its nearest neighbouring metallic sites. Let $N^{s\sigma}(\lambda, E_b)$ and $\delta N^\sigma(R, E_b)$ be the local filling of these bound states : the total number of s electrons, of spin σ, in the Hydrogen orbital at site λ is then given by :

$$N^{s\sigma}(\lambda) = N^{s\sigma}(\lambda, E_b) + \int^{E_F} n_\lambda^{s\sigma}(E) \ dE \qquad (7)$$

and the variation of the number of electrons on a site R by :

$$\delta N^\sigma(R) = \delta N^\sigma(R, E_b) + \int^{E_F} \delta n_R^\sigma(E) \ dE \qquad (8)$$

As usual, $E_\lambda^{s\sigma}$ is related to $N^{s\sigma}(\lambda)$ by the following equation :

$$E_\lambda^{s\sigma} = E_\lambda^{at} + \alpha_\lambda^\sigma + U_\lambda^{ss} N^{s-\sigma}(\lambda) \qquad (9)$$

where E_λ^{at} is the atomic level of Hydrogen, α_λ^σ the crystal field and U_λ^{ss} the local Coulomb correlation on the Hydrogen site λ.

NUMERICAL RESULTS FOR HYDROGEN IN PALLADIUM

We will discuss here the results obtained in the case of Hydrogen at octahedral position in Palladium. The band structure of pure Palladium is given by the fitting scheme of Papaconstantopoulos et al[4] whereas the hopping integrals between Hydrogen and Palladium atoms are those of Faulkner[30]. We restrict these metal-metalloid hopping integrals in terms of nearest neighbours two center integrals $ss\sigma$, $sp\sigma$ and $sd\sigma$. In fact the values of these hopping integrals may not be exactly the same as for stoichiometric hydrides because the distance Hydrogen - metal is probably sensitive to the environment. However, in the present calculation we will neglect this difference.

The determination of the position of the Hydrogen level at

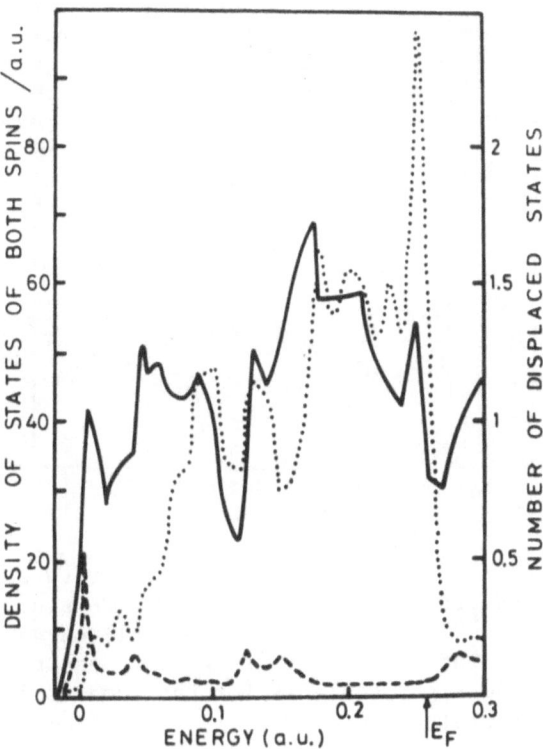

Fig. 1.   Electronic structure of Hydrogen at octahedral position in
          Palladium in the charge neutrality approximation. The dashed
          curve is the s LDOS at the Hydrogen site whereas the full
          curve is the total number of displaced states. The dotted
          curve shows the total density of states per atom of pure
          Palladium [Ref. 15].

octahedral position in α-Palladium hydrides requires the knowledge
of the energies $E_\lambda^{s\sigma}$, α_λ^σ and U_λ^{ss} and the position of the Palladium
bands relative to E_λ^{at}. It is convenient to choose the vacuum as
the zero of energy. With that convention we have $E_\lambda^{at} = -0.5$ a.u.
(atomic units). The position of the bands is fixed so as to give
the experimental value (0.19 a.u.) for the work function of pure
Palladium[31]. As usual in this kind of calculation we neglect α_λ^σ so
that the value of U_λ^{ss} is still a remaining problem. Following

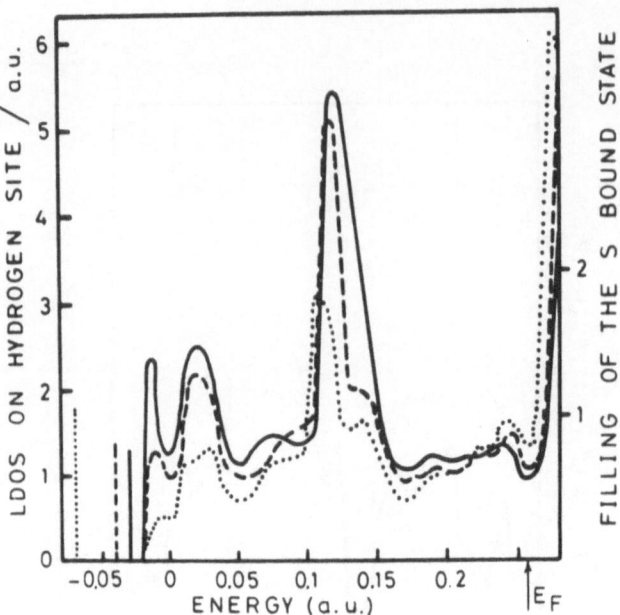

Fig. 2. The s LDOS on the Hydrogen impurity for three values of U_λ^{ss}. The dotted curve is for U_λ^{ss} = 0.26 a.u., the dashed curve for U_λ^{ss} = 0.40 a.u. and the full curve for U_λ^{ss} = 0.475 a.u. The left scale is for the density of states whereas the right scale is for the number of electrons in the bound states [Ref. 28].

Gelatt et al[32] and Darby et al[33] we take U_λ^{ss} = 0.26 a.u., 0.475 a.u. and a value in between (U_λ^{ss} = 0.4 a.u.) to see the influence of U_λ^{ss}.

We are now left with two unknown parameters $E_\lambda^{s\sigma}$ and $v_d^\sigma(R)$. They do not depend on the spin σ in the present case. It is then rather trivial to solve equations (3) and (9) if the inter-site matrix elements of G^o are neglected. If local charge neutrality condition is taken into account[15,16] a virtual bound state appears at the bottom of the sp bands. This is shown in Fig. 1 where the s DOS on the Hydrogen site together with the total number of displaced states Z(E) are reported. However if we give[28,29] realistic values to U_λ^{ss} a bound state appears at the bottom of the sp bands of Palladium. The local filling of this Hydrogen bound state is decreased when U_λ^{ss} is increased. A bound state appears not only

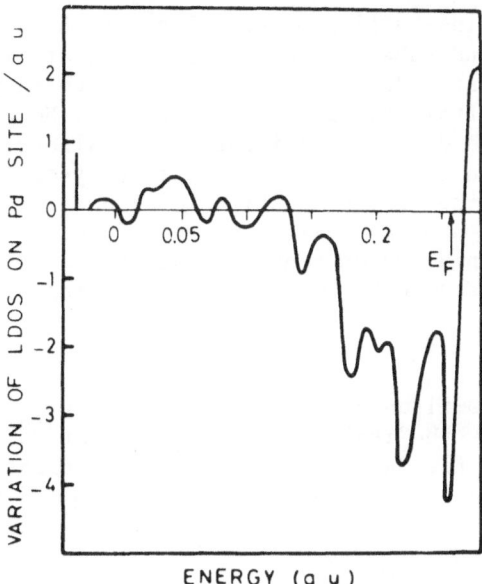

Fig. 3.   The variation of the LDOS on a Palladium site, the near-
          est neighbour to the Hydrogen impurity for U_λ^{ss} = 0.475 a.u.
          [Ref. 28].

on the Hydrogen site but also on the nearest neighbouring Palladium
atoms. On Fig. 2 we report the s DOS on the Hydrogen site whereas
on Fig. 3 the variation of the total DOS on a nearest neighbouring
site is reported.

    The conclusion of this calculation is that the most probable
solution of our problem contains a Hydrogen bound state near the
bottom of the sp bands of Palladium together with a charge transfer from metal to the metalloid. The total variation of the density of states $\Delta n(E)$ = $dZ(E)/dE$ is negative at E_F ; this is in agreement with an experimental decrease of the specific heat[34] and magnetic suscpetibility[35].

CHEMICAL BINDING ENERGY OF A PAIR OF HYDROGEN ATOMS IN
α-PALLADIUM HYDRIDES

    A calculation of the chemical binding energy between two de-
fects has to take into account the electron-electron modification
resulting from the alloying. It has been shown that the term pro-

portional to the interacting potential $v_{\lambda\mu}^{el}(r)$ between the two screenings clouds around substitutional impurities at site λ and μ cancels, in the lowest order of $R_{\lambda\mu}$ (distance between the impurities), the term arising from the band contribution[36]. However, this proof cannot be extended directly to interstitial alloys because it uses the Dyson equation which is valid only in a system with a fixed number of sites like in the substitutional case. This is the reason of the introduction of an auxiliary Green function $G'(E)$ relative to the Hamiltonian H' which concerns the system with two interstitials but at distance $R_{\lambda\mu} \to \infty$ (non interacting impurities) :

$$H' = H_o + v_\lambda + v_\mu \tag{10}$$

where H_o is the Hamiltonian of pure metal and $v_{\lambda(\mu)}$ the perturbation introduced by the isolated interstitial at site $\lambda(\mu)$. With these definitions the Hamiltonian H of the interacting system is given by :

$$H = H' + v_{\lambda\mu}^{el} \tag{11}$$

The chemical binding energy between the pair of interstitials can be formally written as :

$$\Delta E(\lambda,\mu) = \Delta E_{bs}(\lambda,\mu) - \Delta E_{elel}(\lambda,\mu) \tag{12}$$

In this expression $\Delta E_{bs}(\lambda,\mu)$ is the sum of one electron band energies :

$$\Delta E_{bs}(\lambda,\mu) = \int^{E_F} E[\delta n_{\lambda\mu}^{(2)}(E) - \delta n_\lambda^{(1)}(E) - \delta n_\mu^{(1)}(E)]dE \tag{13}$$

where $\delta n_{\lambda\mu}^{(2)}(E)$ is the variation of the density of states due to the pair ; $\delta n_{\lambda(\mu)}^{(1)}(E)$ is the variation of the density of states due to an interstitial atom at site $\lambda(\mu)$. Integrating equation (13) by parts :

$$\Delta E_{bs}(\lambda,\mu) = - \int^{E_F} [Z_{\lambda\mu}(E) - Z_\lambda(E) - Z_\mu(E)]dE \tag{14}$$

with :

$$Z_{\lambda\mu}(E) = - \frac{Im}{\pi} \int^E Tr[G_{\lambda\mu}(E') - G^o(E')]dE' \tag{15}$$

$$Z_{\lambda(\mu)}(E) = - \frac{Im}{\pi} \int^E Tr[G_{\lambda(\mu)}(E') - G^o(E')]dE' \tag{16}$$

$G_{\lambda\mu}(E)$ and $G_{\lambda(\mu)}(E)$ are the Green functions of the system containing a pair of defects at sites λ and μ and the Green function of the system containing one isolated interstitial at site $\lambda(\mu)$ respectively. $G_{\lambda\mu}(E)$ is associated to the Hamiltonian H whereas $G_{\lambda(\mu)}(E)$ is relative to the Hamitonain $H_{\lambda(\mu)} = H_o + v_{\lambda(\mu)}$.
The electron-electron term is given after straightforward extension

of reference 36 by (to the lowest order in $1/R_{\lambda\mu}$) :

$$\Delta E_{elel}(\lambda,\mu) = \Delta E_{elel}^{(1)}(\lambda,\mu)$$

$$+ \int v_{\lambda\mu}^{el}(R)[n^{o\ el}(r) + \Delta n_{\lambda}^{el}(r) + \Delta n_{\mu}^{el}(r)]d^3r \qquad (17)$$

where $n^{o\ el}(r)$ represents the electronic density of the pure Hamiltonian H_o and $\Delta n_{\lambda(\mu)}^{el}(r)$ is the variation of the electronic density resulting from the presence of isolated impurities :

$$\Delta n_{\lambda(\mu)}^{el}(r) = - \frac{Im}{\pi} \int^{E_F} <r|G_{\lambda(\mu)}(E) - G^o(E)|r>dE \qquad (18)$$

The Green function $G_{\lambda\mu}(E)$ for the interacting pair is related to the Green function $G'(E)$ of the non-interacting pair through the Dyson equation. Also $G'(E)$ can be expressed in terms of $G_{\lambda(\mu)}(E)$ and $G^o(E)$. Finally, after straighforward manipulations where -for example, $Z_{\lambda\mu}(E)$ can be expressed in terms of $v_{\lambda\mu}^{el}(r)$- the chemical binding energy can be written as[27] :

$$\Delta E(\lambda,\mu) = - \int^{E_F} [Z_{\lambda\mu}^{(1)}(E) - Z_{\lambda}(E) - Z_{\mu}(E)]dE - \Delta E_{elel}^{(1)}(\lambda,\mu) \qquad (19)$$

where (1) means that only non-interacting potentials enter in the definition of these quantities. Estimation of $\Delta E(\lambda,\mu)$[25,26] in the case where local neutrality condition is used have shown Hydrogen-Hydrogen repulsion for nearest and next nearest position of the interstitial atoms. This is in partial agreement with the so called blocking model[37,38,39]. However it has been shown in the preceeding section that charge transfer may be present in α-Palladium hydrides so that a repulsive short range electrostatic term[40] has to be taken into account in a more quantitative calculation.

Besides this chemical term, elastic binding energy is always present[20]. This elastic interaction energy of two Hydrogen atoms at a distance $R_{\lambda\mu}$ is calculated by means of lattice statics[41] for a lattice Green function of the alloy approximated by the lattice Green function G^{oL} of pure Palladium[23]. For large distance between the two defects, asymptotic expansion of G^{oL} leads to an expression containing the dipole force tensor P of isolated Hydrogen which has been estimated recently[42,43,44]. P is determined essentially from a tight binding electronic structure of a single Hydrogen impurity in Palladium. At short distance, this kind of expansion is no more valid and the binding energy is given in terms of G^{oL} and the forces acting on neighbouring Palladium atoms to the Hydrogen impurity.

CONCLUDING REMARKS

In this paper we have presented a simple tight binding method for the description of the electronic structure of Hydrogen in fcc transition metals. An application to Hydrogen at octahedral position in Palladium has shown that a bound state is present at the

bottom of the sp conduction bands ; this is in accord with other
theoretical calculations. Moreover, the decrease in the density
of states at Fermi level is in qualitative agreement with specific heat and susceptibility measurements.

An estimation of the binding energy of a pair of Hydrogen
atoms is presented and shown to be in agreement with the blocking
model.

ACKNOWLEGEMENTS

The authors are grateful to R. Riedinger, J. Khalifeh and
G. Moraitis for their many pertinent remarks and advices.

REFERENCES

1. A.C. Switendick, Solid State Comm. 8 : 1463 (1970).
2. A.C. Switendick, Z. Phys. Chem. Wiesbaden 117 : 89 (1979).
3. P. Jena, F.Y. Fradin, D.E. Ellis, Phys. Rev. B 20 : 3543 (1979).
4. D.A. Papaconstantopoulos, B.M. Klein, J.S. Faulkner and L.L.
        Boyer, Phys. Rev. B 18 : 2784 (1978).
5. A.J. Pindor and W. Temmerman in Workshop on Hydrogen in Me-
        tals : problems related to the impurity and to non-stoichio-
        metric compounds, CECAM, Orsay (1981).
6. C.A. Sholl and P.V. Smith, J. Phys. F : Metal Phys. 10 : 11
        (1980).
7. J.P. Bugeat and E. Ligeon, Phys. Lett. A 71 : 93 (1979).
8. J.L. Beeby, Proc. Roy. Soc. London A 302 : 113 (1967).
9. H. Katayama, K. Terakura and J.Kanamori, Solid State Comm. 29 :
        31 (1979).
10. J. Kanamori, Proceedings JIMIS 2, Hydrogen in Metals 33 (1980).
11. M. Yussouff and R. Zeller in : Recent Developments in Condensed
        Matter Physics, J.T. Devreese, L.F. Lemmens, V.E. Van Doren
        and J. Van Royen eds, Plenum, New-York 3 : 135 (1981).
12. P.H. Dederichs in Electronic Structure and Phonon Properties in
        Dilute and Concentrated Metal Hydrides, Orsay (1980).
13. J.E. Inglesfield, J. Phys. F : Metal Phys. 11 : L 287 (1981).
14. C. Demangeat, M.A. Khan and J.C. Parlebas, J.M.M.M. 15-18 :
        1275 (1980).
15. M.A. Khan, J.C. Parlebas and C. Demangeat, Phil. Mag. B 42 :
        111 (1980).
16. M.A. Khan, G. Moraitis, J.C. Parlebas and C. Demangeat in :
        11th Int. Symp. on Elec. Struc. of Metals and Alloys,
        P. Ziesche ed., Gaussig, 50 (1981).
17. C. Demangeat and J.C. Parlebas in : 9th Int. Symp. on Elec.
        Struc. and Alloys, P. Ziesche ed., Gaussig, 43 (1979).
18. J.C. Parlebas, C. Demangeat and M.C. Cadeville, J. Appl. Phys.
        50 : 7545 (1979).
19. C. Demangeat, M.A. Khan and J.C. Parlebas, J.M.M.M. 15-18 :
        885 (1980).

20. S. Dietrich and H. Wagner, Z. Phys. B36 : 121 (1979).
21. J. Völkl, G. Wollenweber, K.H. Klatt and G. Alefeld, Z. Natur-forsch. A 26 : 922 (1971).
22. P.H. Dederichs and J. Deutz in : Continuum Models of Discrete Systems, University of Waterloo Press, 329 (1980).
23. J. Khalifeh, G. Moraitis and C. Demangeat, poster session of this conference.
24. K. Masuda and J. Mori, J. Physique 37 : 569 (1976).
25. C. Demangeat, M.A. Khan, G. Moraitis and J.C. Parlebas, J. Physique 41 : 1001 (1980).
26. C. Demangeat, M.A. Khan, G. Moraitis and J.C. Parlebas in : Recent Developments in Condensed Matter Physics, J.T.Devreese, L.F. Lemmens, V.E. Van Doren and J. Van Royen eds., Plenum, New-York 4 : 365 (1981).
27. G. Moraitis and C. Demangeat, Phys. Lett. A 83 : 460 (1981).
28. J. Khalifeh and C. Demangeat, to be published.
29. J. Khalifeh, D. Sc. Thesis, Strasbourg (1982) unpublished.
30. J.S. Faulkner, Phys. Rev. B 13 : 2391 (1976).
31. H.B. Michaelson, J. Appl. Phys. 48 : 4729 (1977).
32. C.D. Gelatt Jr, H. Ehrenreich and J.A. Weiss, Phys. Rev. B 17 : 1940 (1978).
33. M.I. Darby, G.R. Evans and M.N. Read, J. Phys. F : Metal Phys. 11 : 1023 (1981).
34. U. Mizutani, T.B. Massalski and J. Bevk, J. Phys. F : Metal Phys. 6 : 1 (1976).
35. H. Frieske and E. Wicke, Ber. Bunsenges, Phys. Chem. 77 : 48 (1973).
36. M. Natta and G. Toulouse, Phys. Lett. A 24 : 205 (1967).
37. R. Speiser and J.W. Spretnak, Trans. Am. Soc. Metals 47 : 493 (1955).
38. G. Boureau and J. Campserveux, Phil. Mag. 36 : 9 (1977).
39. G. Boureau, J. Phys. Chem. Sol. 42 : 743 (1981).
40. V. Heine in : Solid State Phys., Academic Press 35 : 92 (1980).
41. V.K. Tewary, Adv. Phys. 22 : 757 (1973).
42. G. Moraitis and C. Demangeat in Proceedings of Physics of Transition Metals 1980, Inst. Phys. Conf. Ser. 55 : 583 (1981).
43. J. Khalifeh, G. Moraitis and C. Demangeat, J. Physique : 43 : 165 (1982).
44. J. Khalifeh, G. Moraitis and C. Demangeat, J. Less Common Metals, under press.

THEORETICAL STUDIES OF DILUTE HYDROGEN

IMPURITIES IN PALLADIUM

Barry M. Klein and Warren E. Pickett

Naval Research Laboratory
Washington, D.C. 20375

INTRODUCTION

There have been numerous theoretical investigations of hydride systems in recent years due in large part to their technological importance as energy storage materials[1] and their possible potential as high transition temperature (T_c) superconductors.[2,3] Palladium hydride is one of the more studied materials primarily due to the relatively high $T_c \sim 10$ K for the stoichiometric β-phase material, especially since bulk fcc Pd metal does not superconduct. Conventional band structure methods[3] together with sophisticated techniques such as the coherent potential approximation have been successful in studying PdH_x close to stoichiometry $(x \gtrsim 1.0)$[4], while the dilute H limit $(x \gtrsim 0.0)$ has been studied via jellium models,[5] molecular cluster methods[6] and the average t-matrix approximation (ATA).[7,8] Here we present results of a study of an isolated hydrogen impurity at an octahedral site in an fcc Pd host lattice, using a self-consistent muffin-tin Green's function (MTGF) technique that has recently been developed for studying the electronic structure of defects.[9,10] This method is particularly well-suited for studying the palladium hydride system because effects due to lattice strains and transformations, neglected in the present calculations, are probably not of prime importance in this system since the Pd lattice maintains its integrity with a rather small lattice constant increase even for large H concentrations. After giving some details regarding some important modifications that we have made to the MTGF method, we will present results for the local density of states (LDOS) and charge density (ρ) of the H defect in Pd, compare our results with previous theoretical work and experiments, and indicate the future directions of our research on this system.

THEORETICAL FORMULATION

The MTGF method has been reviewed recently by Podloucky et al.[10] In this method a self-consistent solution for the Green's function in the defect MT is formulated in terms of the host Green's function which is readily determined via a standard band structure calculation (e.g., APW or KKR methods). In principle, electronic and/or nuclear relaxation beyond the defect MT may be included, but the calculations become considerably more complicated and these extensions have not as yet been implemented. We have made two modifications to the MTGF method as used by other workers: (1) our calculations are semi-relativistic with only spin-orbit coupling neglected and, more importantly, (2) we have developed an approach for accounting for the defect charge density <u>outside</u> the defect MT which is particularly important for properly determining the defect Coulomb potential relative to the MT zero of the host lattice. The latter point can be considered as the Green's function equivalent of the boundary problem one encounters in cluster calculations. The problem was recognized early on by Terakura and Kanamori[11] who carried out pioneering Green's function calculations of Al and Cu impurities in Pd using non-self-consistent defect potentials which were empirically adjusted to achieve approximate charge neutrality in the Wigner-Seitz cell of the defect. The approach that we have adopted can be considered a logical extension of the earlier models for the screening of a defect. Consider the difference $\Delta\rho(r)$ between the defect and host charge densities. Since the defect is neutral, the corresponding difference $\Delta V_c(r)$ in Coulomb potentials vanishes as $r \to \infty$, and therefore it can be determined <u>uniquely</u> from $\Delta\rho$, so that the total defect Coulomb potential V_c^{def},

$$V_c^{\text{def}}(r) = V_c^{\text{host}}(r) + \Delta V_c(r) \tag{1}$$

is determined relative to the host Coulomb potential $V_c^{\text{host}}(r)$. It is straightforward to account for the local density exchange-correlation potentials in the host and defect systems.

Up to this point the development is exact. We make the approximation of treating the potentials and charge densities as spherical, which should be quite adequate for the purpose of determining the self-consistent spherical potential within the MT region at high symmetry sites. Direct calculation using the MTGF method gives $\rho^{\text{def}}(r)$, and hence $\Delta\rho(r)$, for $r < r_{MT}$. For $r > r_{MT}$ we propose the model charge perturbation

$$\Delta\rho(r) = (A + Br + C\,r^2)\,\exp(-r/\lambda). \tag{2}$$

For λ we take the Thomas-Fermi screening length corresponding to the host (constant) charge density in the interstitial region ($\lambda = 0.612$ a.u. in this case). This exponential decay of $\Delta\rho$ for $r > r_{MT}$ builds in the metallic screening of the defect. The parameters A, B and C are adjusted to satisfy three physical requirements: charge neutrality, and continuity of $\Delta\rho$ and its derivative at $r = r_{MT}$.

RESULTS AND DISCUSSION

To begin the MTGF calculation one needs to have the host MT potential to determine the host phase shifts, LDOSs, wavefunctions and Green's function. In the present case, the host MT is an <u>empty</u> octahedral interstitial site in an fcc Pd lattice. We have used a Pd potential,[3] determined self-consistently using the APW method, to redo the Pd band structure as a binary compound with the octahedral sites treated as "empty spheres," having zero total potential. These calculations were done on a uniform 20 $\overline{k}$ point mesh followed by an interpolation onto an 89 $\overline{k}$ point mesh[12] and tetrahedral integration to obtain the host LDOSs. The resulting host octahedral site potential and charge density were nearly constant, as expected, yielding host phase shifts $\delta_l \lesssim 0.05$ at and below the Fermi energy E_F. We emphasize that in the present case H is considered to be a substitutional defect at an empty octahedral site. The MT radius of the defect was taken as 1.076 a.u., the maximum value for touching defect and Pd MTs, where the Pd MT radius was 2.599 a.u. for touching Pd-Pd MTs. The local density form of exchange-correlation potential of Hedin and Lundqvist[13] was used for the defect calculation. The real part of the host GF was calculated by Kramers-Kronig transforming the imaginary part, with an energy cut-off of 1.85 Ry above the host MT zero. An energy grid of 2 mRy was used throughout, and self consistency in the H defect potential to better than 0.02 Ry r.m.s. was achieved.

Figure 1 shows the resulting H defect charge density $\rho(r)$ and volume integrated charge density $N(r)$, while Fig. 2 shows the H LDOS both with and without including the $\Delta\rho$ correction in Eq. 2. We see first from Fig. 1 that ρ for the H defect in Pd is substantially greater than for a hydrogen atom indicating that the proton is well-screened at the octahedral site. In addition we see that setting $A = B = C = 0$ and thereby neglecting the $\Delta\rho$ correction in Eq. 2 leads to a grossly inaccurate H defect charge density even <u>inside</u> the defect MT. From Fig. 2 we see that the LDOS for the H defect has developed a strong bonding peak at the bottom of the Pd valence bands where the occupied Pd s states reside, and furthermore there is an H defect *bound state* indicated by a vertical solid line, just below the bottom of the Pd valence bands. The bonding and bound states are shifted downward relative to the host lattice peak by 2 eV, roughly the value found in band structure calculations[3] for stoichiometric PdH. This remarkable occurrence strongly indicates that the isolated H defect, without including defect-defect interactions or other lattice effects, already qualitatively accounts for the strong bonding of H in Pd. The location of the low-energy bonding structure relative to E_F of Pd is somewhat lower than was found in photoemission measurements of Eastman, et al.;[14] however their measurements were done on $PdH_{0.6}$. New photoemission measurements for dilute H concentrations would be particularly interesting to compare with our work.

The results shown in Figs. 1 and 2 show many similarities to the molecular cluster results of Jena, et al.,[6] the ATA results of Bansil, et al.,[8] and the MTGF results of Yussouff and Zeller.[15] The latter results are particularly interesting to compare with ours since we have used similar techniques except for relativistic effects included here (which probably accounts for our host valence bandwidth being 1 eV larger than theirs), and our inclusion of a correction for the charge outside the H defect MT, neglected by Yussouff and Zeller.[15] They find an occupied H defect bonding peak around 1 eV higher in energy than ours, and an H-induced bound state approximately 4.5 eV below the bottom of their host valence bands. Our bound state falls at 0.2 eV below our valence bands. We believe that the bound state location that we have found is much more physical, as this state will undoubtedly broaden and merge with the bonding band as the H concentration increases, but only small bound state energy shifts (i.e., much less than 3-4 eV) would be expected. We further note that our bound state contains 0.334 electrons within the MT sphere versus 0.8 electrons found by Yussouff and Zeller[15] (who may have employed a Wigner-Seitz sphere, however). This quantitative discrepancy between our results and theirs indicates the importance of accounting for $\Delta\rho$ for $r > r_{MT}$ in defect calculations. In this regard, it is a bit surprising that our $A = B = C = 0$ results for the bound state energy shown in Fig. 2b are so similar to the results in Fig. 2a even though the charge density differences shown in Fig. 1 are quite large. We believe that these similarities in Fig. 2 are somewhat "accidental"; in

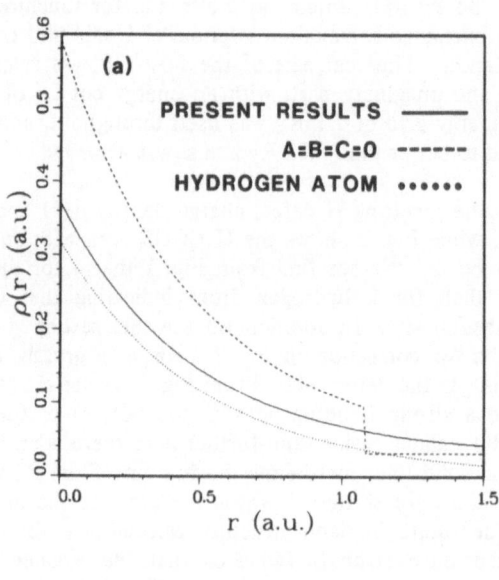

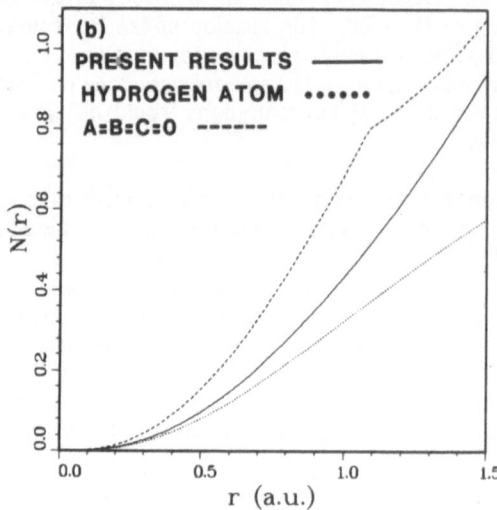

Fig. 1.   (a) Charge densities $\rho(r)$ for the self-consistent H octahedral interstitial defect
and for the free hydrogen atom. The dashed curve refers to a calculation
neglecting the defect charge density outside the defect muffin-tin
($A = B = C = 0$ in Eq. 2 of the text). (b) Volume integral $N(r)$ of $\rho(r)$
in (a).

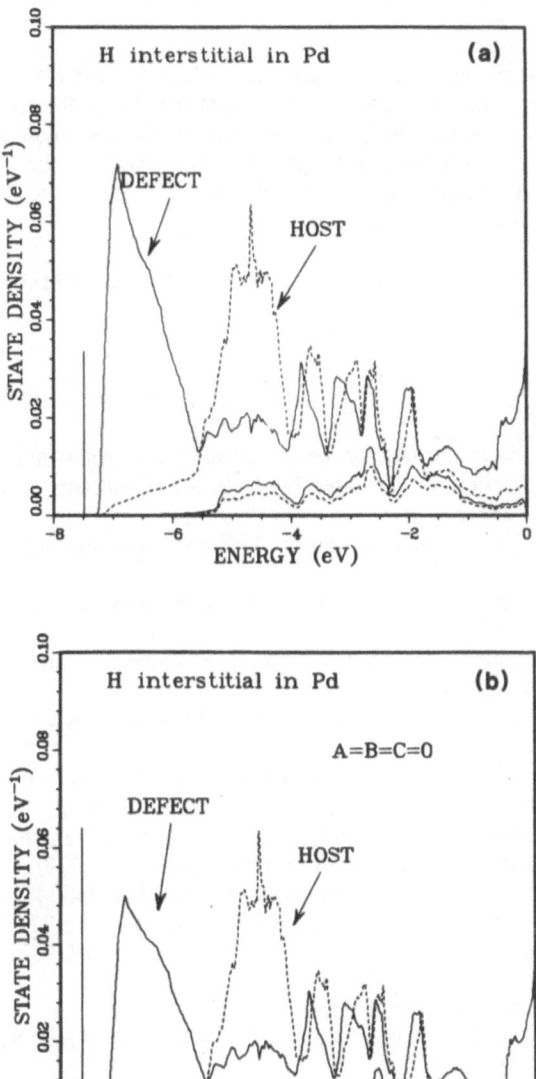

Fig. 2. (a) Upper solid and dashed curves refer to the defect and host total $(s + p)$ local density of states, respectively. The lower solid and dashed curves are the p-like contributions only. The defect induced bound state is shown by a vertical solid line just below the host and defect band states. (b) Defect results obtained by neglecting the defect charge density outside the defect muffin-tin $(A = B = C = 0$ in Eq. 2 of the text). Same notation as in (a).

another situation we have studied, the vacancy in Al, the differences are much more drastic.[16]

It is of interest to repeat the present calculations for the tetrahedral H defect in the Pd lattice to see if we can theoretically account for the preferred occupation of the octahedral site. The results of the MTGF calculations will also allow us to perform calculations of several experimentally accessible quantities such as the Knight shift, spin-lattice relaxation rate, residual resistivity and Dingle temperature. Calculational estimates of these quantities will be presented elsewhere.

We are indebted to R. Zeller for communication on the MTGF method and calling our attention to Ref. 15.

REFERENCES

1. J. J. Reilly, Jr., in *Proceedings of the Internatonal Symposium on Hydrides for Energy Storage, Geilo, Norway, 1977*, edited by A. F. Anderson and A. J. Maelland (Pergamon, Oxford, 1978).

2. B. M. Klein, E. N. Economou, and D. A. Papaconstantopoulos, Phys. Rev. Lett. 39, 574 (1977).

3. D. A. Papaconstantopoulos, B. M. Klein, E. N. Economou, and L. L. Boyer, Phys. Rev. B 17, 141 (1978).

4. D. A. Papaconstantopoulos, B. M. Klein, J. S. Faulkner, and L. L. Boyer, Phys. Rev. B 18, 2784 (1978).

5. See for instance, P. Jena, and K. S. Singwi, Phys. Rev. B 17, 3518 (1978), and references therein.

6. P. Jena, F. Y. Fradin, and D. E. Ellis, Phys. Rev. B 20, 3543 (1979).

7. C. D. Gelatt, Jr., H. Ehrenreich, and J. A. Weiss, Phys. Rev. 17, 1940 (1978).

8. A. Bansil, R. Prasad, S. Bessendorf, L. Schwartz, W. J. Venema, R. Feenstra, F. Blom, and R. Griessen, Solid State Commun. 32, 1115 (1979).

9. M. Hamazaki, S. Asano, and J. Yamashita, J. Phys. Soc. Jpn. 41, 378 (1976).

10. R. Podloucky, R. Zeller, and P. H. Dederichs, Phys. Rev. B 22, 5777 (1980).

11. K. Terakura and J. Kanamori, J. Phys. Soc. Jpn. 34, 1520 (1973).

12. L. L. Boyer, Phys. Rev. B 19, 2824 (1979).

13. L. Hedin and B. I. Lundqvist, J. Phys. C 4, 2064 (1971).

14. D. E. Eastman, J. K. Cashion, and A. C. Switendick, Phys. Rev. Lett. 27, 35 (1971).

15. M. Yussouff and R. Zeller, in *Recent Developments in Condensed Matter Physics, Vol. 3*, edited by J. T. Devreese, L. F. Lemmens, V. E. Van Doren, and J. Van Royen (Plenum, 1981), p. 135.

16. W. E. Pickett and B. M. Klein (unpublished).

ELECTRONIC STRUCTURE OF INTERSTITIAL

HYDROGEN IN ALUMINIUM

Raju P. Gupta

Centre d'Etudes Nucléaires de Saclay
Section de Recherches de Métallurgie Physique
91191 Gif sur Yvette Cedex, France

ABSTRACT

A supercell calculation for the electronic structure of in-terstitial hydrogen in Al has been performed. The metal states are pertuebed by the H and the lowest band is split-off resulting in the formation of a metal-hydrogen bonding state. The formation of an additional state (bound state) is not observed.

INTRODUCTION

A detailed knowledge of the electronic structure and the local perturbation created by the hydrogen is important in the understan-ding of many physical phenomena, including the hydrogen embrittle-ment and the hydrogen diffusion. Several methods have been used to treat the problem of hydrogen as an impurity in metals :
(1) the pseudopotential approach with linear screening approxima-tion. In this approach the metallic lattice is represented by a pseudopotential but the potential perturbation due to the hydrogen atom is taken to be the full unscreened coulombic potential $(-e^2/r)$ screened linearly. Since the $-e^2/r$ potential is an extremely strong perturbation and <u>not</u> a weak perturbation appropriate for li-near screening, this approach has been shown to be invalid and un-trustworthy for treating the problem of hydrogen in metals.[1]
(2) the jellium model with non-linear screening of hydrogen [2]. In this model the ionic lattice is replaced by a uniform jelly of po-sitive backgrond and the discrete nature of the ions is thus ne-glected. The screening problem of a point positive charge can then

be solved easily without further approximations. This model, by
its very nature, is limited to the case of simple metals. A bound
state $\sim$ 0.01 eV below the bottom of the conduction band has thus
been predicted for H in Al. The neglect of the discrete nature of
the ionic lattice, however, creates serious uncertainties about
the predictions from this model. The position or the existence of
the bound state in particular, will be significantly affected by
the inclusion of the lattice, something which will be important
for the local screening of the hydrogen and the perturbation thus
created. This is also an important consideration for the question
of energetics, needed to study diffusion paths and activation
energies. A recent variation of the jellium model is the spherical
solid model [3] where the ionic lattice is represented by positive
spherical shells. This model is again far from a realistic representation of the ionic lattice.
(3) the cluster models [4]. Here, one studies the local electronic
structure in terms of a molecular agregate. A hydrogen atom is
surrounded by a small number of metal atoms. The results evidently
depend upon the size of the cluster and the boundary conditions
imposed.
(4) The tight binding model [5]. This could be an interesting approach
for transition metals if properly applied. The existing calculations, however, are too crude and make numerous approximations [5].
As a result the numerical output from these calculations are
very sensitive to the various input parameters which cannot
always be predicted on physical grounds.
(5) the Green's function method using scattering theory for the
perturbed states [6]. This would ideally be the best method for
treating the electronic perturbation due to an impurity. However,
in its existing calculations the perturbation due to the hydrogen
has been restricted to the muffin-tin (MT) sphere of hydrogen alone and the electronic redistribution in neither the interstitial
region nor the neighboring metal atom MT spheres is taken into account. A bound state for interstitial H in some transition metals
has been obtained but the result is presumably quite sensitive to
the value of the constant potential in the interstitial region,
the radius of the MT sphere of the H atom, and to the method of
construction of the potential. This immediately implies that the
inclusion of neighboring metal atoms in the t-matrix treatment is
absolutely essential, something which is difficult at this time.
Further, the calculation of the total energy of a system is not a
simple task in the Green's function approach. These calculations
are nevertheless necessary if the intention is to calculate the
diffusion paths and the activation energy of diffusion.

We propose here a different approach to investigate the electronic structure of H in Al; it is based on the supercell method [7]
and has been used successfully in previous defect calculations.
This method is simple and intuitively appealing provided the supercell is large enough, and in addition permits straight forward
total energy calculations. The method also permits the inclusion

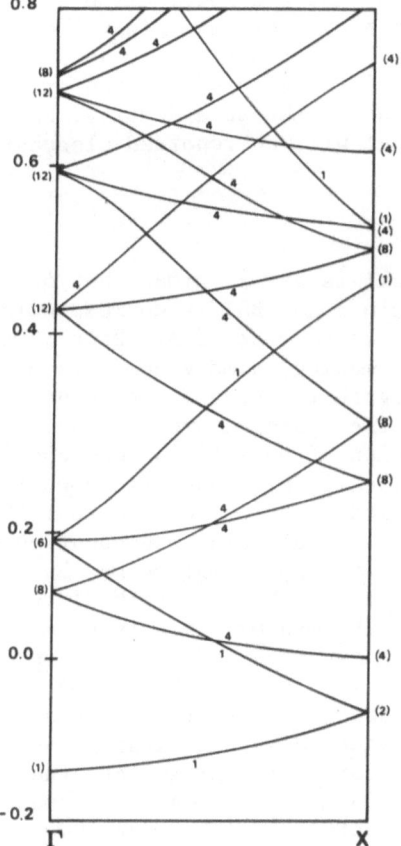

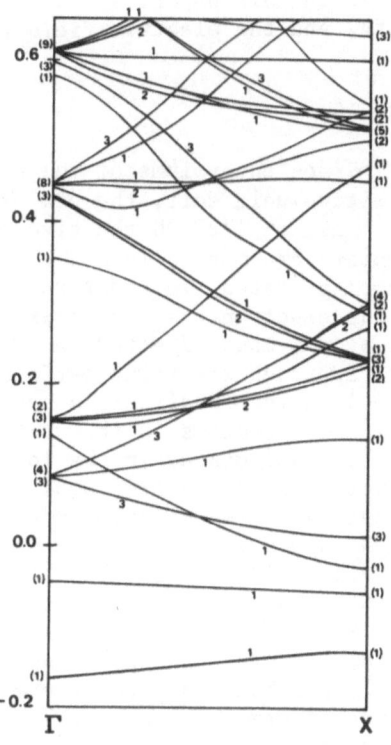

Fig.1. Energy bands of Al_{27}
All energies are in
Rydbergs

Fig.2. Energy bands of $Al_{27}H$.

of the relaxation around the defect and a full treatment of the electronic perturbation created by the defect. The principle of the method is simple; the defect, here a hydrogen atom, is surrounded by a cluster of metal atoms and the whole unit is repeated periodically to form an infinite crystal. In our present calculation of H in Al we have chosen the H to occupy the octahedral position as is the case in the face-centered-cubic (fcc) metals, and we have chosen an fcc supercell with a lattice parameter three times that of the primitive unit cell of Al. This gives a H-H distance of $\sim$ 8.6 A° which is large enough that the H-H interaction can be ignored. The supercell thus contains a hydrogen atom in the octahedral position and is surrounded by 27 Al atoms. The results on the electronic structure and the total charge distribution are reported in this paper. The calculations on the spin density, knight shift, and the electric field gradient will be reported elsewhere.

RESULTS

Since the volume of the unit cell is now 27 times that of the primitive unit cell, the new Brillouin zone (BZ) is correspondingly smaller, 1/27 th the size of the regular BZ of Al. Hence, many k-points are mapped into the new BZ, making it more complicated to identify states. In order to investigate the general perturbation in the electronic states created by the hydrogen and to pinpoint the states most affected, a calculation of pure Al in the supercell approach was first performed. The energy bands are shown in Fig. 1 in the ΓX direction and labelled Al_{27} to indicate that there are 27 Al atoms in the unit cell. The potential was constructed in the usual manner in the local density approximation by the superposition of the free atom selfconsistent Hartree-Fock-Slater charge densities calculated using the Herman-Skillman program and X_α approximation with α = 1.0 for exchange and correlation. The eigenvalues were fully converged. The integer numbers on different bands in Fig. 1 indicate the degeneracies of these states (not including of course, the spin degeneracy). At first glance, the electronic structure looks complicated; in particular, the nearly parabolic nature of the bands in Al is no longer apparent. A careful examination of the band structure does however show this behavior.

In Fig. 2 is shown the band structure, again in the ΓX direction, with the hydrogen in the octahedral position. Comparison with Fig. 1 shows that practically all the states have been somewhat perturbed by the introduction of hydrogen and some degeneracies lifted. For the slightly perturbed states, a direct parentage with the host metal states can be made. The largest perturbation is seen by those metal states which have an s type symmetry at the H site in the unperturbed lattice. This is the case of the first band which has been split-off the band complex and lowered by $\sim$ 0.48 eV at Γ and by $\sim$ 1.05 eV at X, the second band which is de-

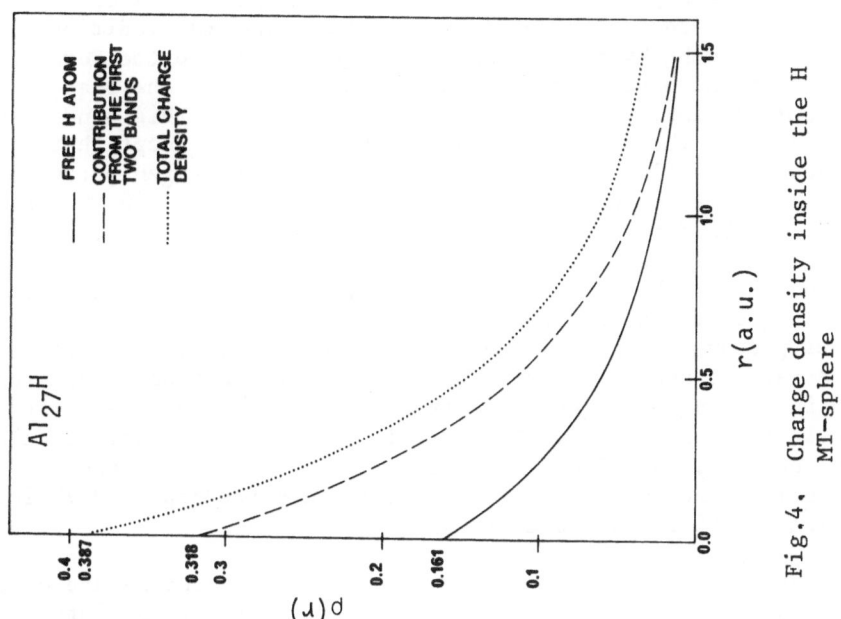

Fig.4. Charge density inside the H
MT-sphere

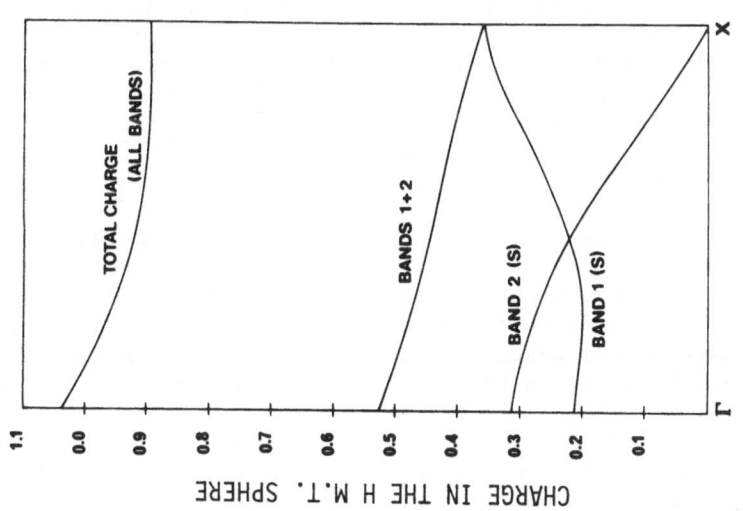

Fig.3. The integrated charge inside
the H MT-sphere

rived from the eightfold complex at Γ and states in the region ~ 0.6 Ry in energy. It is important to note that the addition of the hydrogen does not create any new states and the additional electron brought by the hydrogen thus goes to raise the Fermi level. The first band in Fig. 2, although split from the rest of the band complex, is not an additional band since this band already existed in the pure Al_{27} case shown in Fig. 1 except that it has now been pulled down substantially by the hydrogen potential. A similar behavior was found earlier by Switendick[9] in the case of Pd. Our band structure calculations thus show it to be a metal-hydrogen bonding state. This is in contrast to the previous calculations where a bound state has been predicted, meaning an additional state and a consequent lowering of the Fermi energy. Thus in our picture the addition of hydrogen only deforms the metal states and no new states are created, the deformation increasing with increasing hydrogen concentration. This is consistent with calculations on stoichiometric metal hydrides[10]. In our calculation the first band lies ~ 0.72 eV below the bottom of the Al_{27} conduction bands which is also approximately the gap (~ 0.78 eV) between the first two bands in $Al_{27}H$.

Further evidence for the metal-hydrogen bonding nature of the first band in $Al_{27}H$ comes from an analysis of the total integrated charge inside the hydrogen MT sphere (of radius 1.50 au) at various k-points inside the BZ. In Fig. 3 we have plotted this charge in the ΓX direction (spin degeneracy has been included). We see that the band 1 charge strongly depends on the k-vector, varying approximately by a factor of 2 in the ΓX direction (from 0.21 at Γ to 0.38 electron at X). The charge due to a genuine bound state is, on the contrary, independent of k. While the first band charge increases, the one from the second band decreases but the total of the two remains almost constant (within $\sim 10\%$) at 0.45 electron in the ΓX direction. The total charge in the H sphere from all bands upto the Fermi energy is also shown in Fig. 3 in the ΓX direction and is approximately constant (~ 0.95 electron).

In Fig. 4 we have shown the spherically averaged charge distribution inside the H muffin-tin sphere. The full curve shows the contribution from the first two bands and the dotted curve the total from all the bands upto the Fermi energy; the dashed curve shows the free H atom charge density. This calculation was performed using 256 k-points in the supercell BZ which corresponds to approximately 8000 points in the regular BZ of Al. The present calculation predicts a value of $\rho(o) = 0.387$ au^{-3} for the charge density at the proton site compared to a value of $\rho(o) = 0.318$ au^{-3} in the free hydrogen atom. There is thus $\sim 20\%$ increase in the charge density at the proton site in our calculation. This is to be compared to an increase of $\sim 50\%$ found in the jellium[2] and cluster model[4] calculations. The jellium and cluster models thus predict an overscreening of the proton, something which is expected since these models neglect the discrete nature of the ionic lattice. The total integrated charge in the H MT-sphere turns out to be 0.95 electron of which 0.38 is contributed by the first two bands. A free hydrogen atom contains 0.61 electron in the MT-sphere. Thus the proton is well screened and is almost fully neutralized at a distance of $\sim$ 1.6 au. The total charge contained in our MT sphere is essentially the same as is obtained after the superposition of atomic charge densities indicating that our calculation is reasonably selfconsistent.

CONCLUSION

We have performed a supercell calculation for an interstitial hydrogen in Al, and have shown that the H does not induce a bound state since there is no additional state created by the introduction of H. The metal bands are perturbed by the H potential and the lowest band is split-off from the rest of the band complex to form a metal-hydrogen bonding state. The increasing H concentration only creates more perturbation in the metal states. This picture thus allows a better and consistent understanding of the metal-hydrogen interaction. Our calculation also shows that the jellium and the cluster models overestimate the screening of the proton.

The author is indebted to Dr. Y. Adda for his encouragement and fruitful discussions on various aspects of this work. This work was supported by the French Commissariat à l'Energie Atomique.

REFERENCES

1. R. Evans and M.W. Finnis, J. Phys. F 6, 483 (1976).
2. C.O. Almbladh, U. Von Barth, Z.D. Popovic and M.J. Stott, Phys. Rev. B 14, 2250 (1976); P. Jena and K.S. Singwi, Phys. Rev. B 17, 3518 (1978).
3. F. Perrot and M. Rasolt, Phys. Rev. B 23, 6534 (1981).
4. R.W. Simpson, N.F. Lane and R.C. Chaney, J. Nucl. Mater. 69-70, 581 (1978).
5. C. Demangeat, this conference.
6. M. Yussouff and R. Zeller, in "Recent Developments in Condensed Matter Physics", Vol. 3, edited by J.T. Devreese, L.F. Lemmens, V.E. Van Doren and J. Van Royen (Plenum, 1981), p. 135; B.M. Klein and W.E. Pickett, this conference.
7. R.P. Gupta and R.W. Siegel, Phys. Rev. Lett. 39, 1212 (1977); Phys. Rev. B 22, 4572 (1980).
8. F. Herman and S. Skillman, "Atomic Structure Calculations" (Prentice-Hall, Englewood cliffs, N.J. 1963).
9. A.C. Switendick, in "Hydrogen in Metals", Proc. of the Second Japan Institute of Metals (JIM) Symposium, November

1979), Suppl. to Trans. JIM Vol. 21 (1980), p. 57.
10. M. Gupta, J. de Phys. (Paris) $\underline{41}$, 1009 (1980).

CALCULATION OF THE GROUND STATE ENERGY OF HYDROGEN AT INTERSTITAL

SITES IN A LITHIUM CLUSTER

D.D. Shillady*, T. Nguyen*, and P. Jena**

*Department of Chemistry, **Department of Physics
Virginia Commonwealth University
Richmond, Virginia 23284

ABSTRACT

A small, extended basis set of Gaussian lobe functions has
been used to map the potential of a hydrogen atom in lithium using
molecular cluster approximation. The relative energies of several
interstitial sites for hydrogen were calculated using a basis of
(3G1s, 1G2s, 2G2p) on lithium and (4G1s, 1G2s, 1G2p) on hydrogen
in closed-shell Hartree-Fock-Roothaan calculations. At 0°K, hydrogen was found to be preferentially located at the tetrahedral site.
A three dimensional mapping of the hydrogen potential reveals that
the most probable diffusion of hydrogen would take place from te-
trahedral-octahedral-tetrahedral sites. The sensitivity of the results to the cluster size and its implications on studying hydrogen
defect interaction are discussed.

INTRODUCTION

In this paper, we have used self-consistent-field (SCF) Hartree-Fock-Roothaan procedures to study the interaction of hydrogen
with host metal atoms. This scheme has been successfully developed for large molecules in chemical studies of electronic structure.
Recently, the interaction of impurities with host atoms has been
studied by approximating the infinite defected lattice by a molecular cluster consisting of the impurity atom and near neighbor
host atoms. Such a model is reasonable in view of the short
screening length of the impurity and host atom charge by the conduction electrons of the medium. Thus, molecular clusters mimic
the total environment and can provide information on the electron
charge distribution, and lattice distortion around impurities.

.We have initiated a program for a systematic study of the interaction of hydrogen with defect as well as host sites in metals.
Here we present our results on $Li_{11}H$ and Li_9H clusters. By placing
the hydrogen atom at different locations inside a lithium cluster,
we have calculated the ground state energies of the metal-hydrogen
system. The results enable us to study the equilibrium site of
hydrogen, the activation barrier, and the most likely path a diffusing hydrogen atom in lithium would take. The limitation of minimum cluster size is explored. The extension of our technique to
study defect-hydrogen interaction is outlined.

COMPUTATIONAL PROCEDURE

In this work, we have used the closed shell Hartree-Fock-Roothaan[1] method for neutral clusters. In brief, this involves a
self-consistent solution of the Hamiltonian equation,

$$H_{ij} + \Sigma P_{k\ell} \left((ij \mid k\ell) - \tfrac{1}{2} (ik \mid j\ell) \right) = F_{ij}$$

$$FC = ESC \tag{1}$$

where F_{ij} the matrix-element of the Hartree-Fock-Roothaan Hamiltonian and S_{ij} is the overlap matrix,

$$S_{ij} = \int \phi_i^+ \phi_j \, dt . \tag{2}$$

$\phi_i(\vec{r})$ is given in terms of a linear combination of orbitals centered on each atom, namely,

$$\phi_i = \sum_n a_{in} X_n , \tag{3}$$

where X_n's are gaussian orbitals[2]

$$X_n = \exp [-\alpha_n (r-R_n)^2] . \tag{4}$$

The wave function, Φ_i is given by,

$$\Phi_i = \sum_m C_{im} \phi_m . \tag{5}$$

The remaining terms in (1) are defined as,

$$(ij \mid k\ell) = \int\int \phi_i^+ (\vec{r}_1) \, \phi_j (\vec{r}_1) \, \frac{1}{|\vec{r}_1 - \vec{r}_2|} \, \phi_k^+ (\vec{r}_2) \, \phi_\ell (\vec{r}_2) \, d\tau_1 d\tau_2. \tag{6}$$

and

$$P_{k\ell} = \sum_m 2 C_{km} C_{\ell m} . \tag{7}$$

The gaussian-lobe basis[3] expansion in Eq. (4) has been used

in the past for calculations of molecular electronic structure.
Its usage stems from the realization that spherical harmonic angular functions can be built from appropriate linear combinations of
gaussian spheres. Furthermore, many difficult operators can be
easily evaluated in a gaussian-lobe basis and s, p, d, and f orbitals[4] can be easily treated, provided one sums over a sufficient
number of expansion spheres. We have made use of the more efficient closed-shell case by keeping the total number of electrons
"even". This method requires only about 40 iterative improvements
of the Fock matrix.

RESULTS AND DISCUSSION

     In Fig. 1, we plot the total energy of LiH diatomic molecule
as a function of interatomic distance. We have used four spheres
for 1s orbital (4G1s), one sphere for 2s (1G2s) and two pairs of
spheres for 2p orbitals (2G2p) on Li with scaling of (2.690:1s,
0.634:2s, 0.761:2p) and a similar basis on H except for only using
one pair of spheres for the 2p set (1G2p) scaled as (1.37:1s,
0.96:2s, 0.92:2p). We find a minimum energy of -7.94672 au at
1.655A°. This has to be compared to the experimental value[5] of
-8.0705 au at 1.595A° and the configuration interaction result[5] of
-7.98984 au at 1.606A°. While it is easy to improve the accuracy
of the LiH results by adding more spheres and including configuration interaction, these steps are presently too costly for the
cluster calculations.

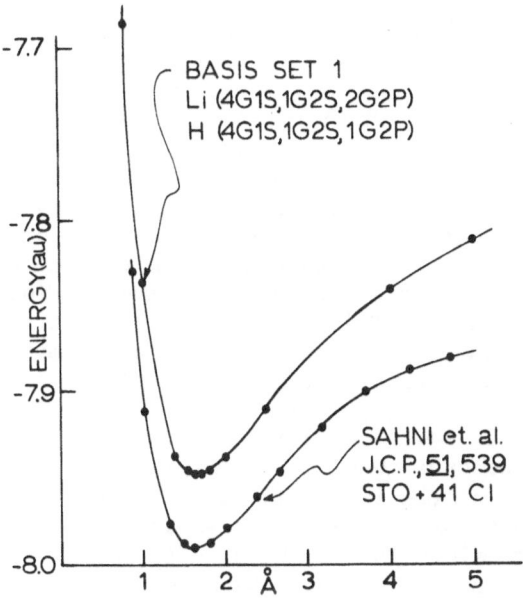

Fig. 1. Li-H Potential

For all our cluster calculations involving Li and H, the body center of the cube was taken as the origin of the coordinate system. Initial studies were carried out on Li_9H cluster keeping a 4G1s set on Li in the center of the unit cell and the H-atom but reducing the 1s orbital on the 8 corner Li atoms to a 3G1s. Ground state energies were calculated for various locations of hydrogen along the (1, 1, 1) direction (joining the body centered Li to that at a corner) and along the line joining (0, 0, ½) and (½, 0, ½) (from face center to edge center). The results are plotted in Fig. 2. The dark circles represent actual calculations and the line passing through them is the result of a least-squares fit.

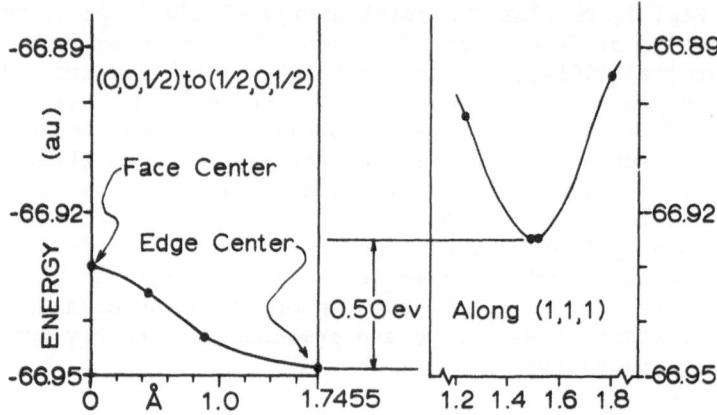

Fig. 2.   Calculated H-Site Energy in Li_9H

Two interesting points are to be noted. First, a deep minimum exists at a point midway between the corner and body-centered Li atoms. This minimum, however, is about 0.5eV above the configuration in which hydrogen occupied the edge-center (octahedral site). Second, the energies for the hydrogen atom at face center and edge centered in Li_9H are different and no minimum is found along (0, 0, ½) - (½, 0, ½) line. This result clearly is a reflection of the cluster size and originates from lack of symmetry of the face center (Oh) and edge center (Oh') sites in Li_9H.

To remedy this shortcoming as well as to test the sensitivity of our results, we have considered a $Li_{11}H$ cluster shown in Fig. 3. The configuration of the 11 lithium atoms in Fig. 3, held fixed during our calculations, were chosen so that a hydrogen atom at various locations along the (0, 0, ½) - (½, 0, ½) line is "properly dressed" with nearest neighbor Li atoms. A Li atom at (-1, 0, 1) was added to the cluster to keep the number of electrons even in the closed shell treatment outlined previously.

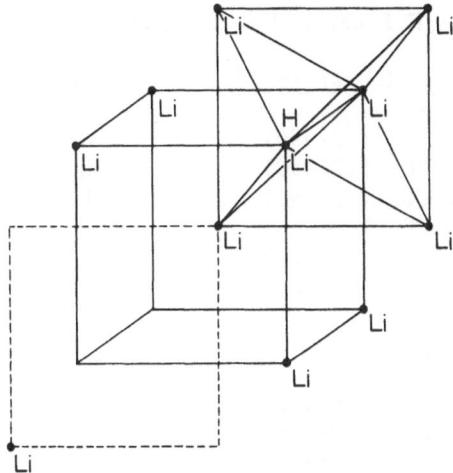

Fig. 3. Li$_{11}$H Cluster Geometry

The results of the ground state energies for various locations of H in Li$_{11}$H cluster is shown in Fig. 4 and again rounded to 10^{-5} au in Table I. Note that the shape of the energy curve along (1, 1, 1) direction in Li$_{11}$H cluster is very similar to that in Li$_9$H cluster. This result indicates that the ground state properties can be reasonably described by taking the local environment into account. On the other hand, there is a qualitative difference in the energy curves along the (0, 0, ½) - (½, 0, ½) line. In the Li$_{11}$H cluster, a minimum exists at the tetrahedral site (T$_d$). This minimum, interestingly, is about 0.5eV below the minimum along (1, 1, 1) direction (see Fig. 2). The energies of Oh and Oh' sites are not yet identical. This is due to the fact that the hydrogen atom at the Oh' site still does not have all its nearest neighbor Li atoms present as does the hydrogen atom at Oh site. From the symmetry of the Li$_{11}$H cluster, we believe that the barrier height for hydrogen diffusion from T$_d$-Oh site is about 0.2eV. This is, however, significantly larger than the corresponding barrier heights in transition metal-hydrogen systems.

Our results of a tetrahedral minimum for Li$_{11}$H cluster is in agreement with an earlier calculation of Hinterman et. al.[6] However, these authors did not include 2p orbitals on Li or 2s and 2p orbitals on the positive muon. In this regard, we believe the present basis to be more realistic. It should also be emphasized that the present work as well as that of Hinterman et. al. includes exact exchange and inhomogeneities in the entire electron charge cloud.

Table I. $Li_{11}H$ Interaction Energies (bcc, a = 3.491 A°)

Across Top Face H(X,Y,Z) in A°	Site	Energy (au)
(0.00000, 0.87275, 1.74550)	Td	−81.61686
(0.00000, 1.74550, 1.74550)	Oh	−81.60812
(0.00000, 0.00000, 1.74550)	Oh'	−81.59875
(0.00000, 1.30910, 1.74550)		−81.61416
(0.00000, 0.43640, 1.74550)		−81.60981
(0.00000, 0.97013, 1.74550)		−81.61706

Along (1,1,1) Direction

(0.87275, 0.87275, 0.87275)		−81.59679
(0.70985, 0.70985, 0.70985)		−81.56941
(1.03565, 1.03565, 1.03565)		−81.56818

Midpoint between (1,1,1) and (T_d-O_h) Minima

(0.43640, 0.91750, 1.30410)		−81.60722

All Li: (3G1s, 1G2s, 2G2p)
Roving H: (4G1s, 1G2s, 1G2p)

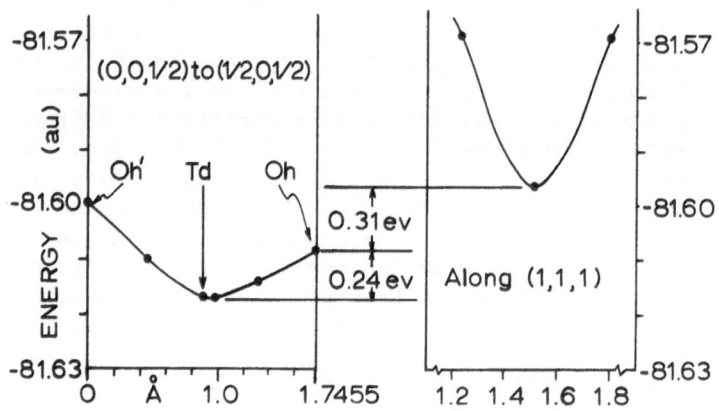

Fig. 4.   Calculated Diffusion Barrier in $Li_{11}H$

In the present calculation, we have used the Born-Oppenheimer approximation[7] and decoupled the electrons' motion from that of either Li or hydrogen. While non-Born-Oppenheimer vibronic coupling may eventually be invoked in order to explain the reverse isotope effect in superconductivity in metal hydrogen systems quantitatively, our use of a static potential can still be developed

further to compute the vibrational levels of hydrogen and its isotopes. The cluster size and the symmetry of the impurity atom under focus are found to have important effects in locating the preferential site of the impurity. Hydrogen is found to occupy the
tetrahedral site in lithium. The larger barrier height signifies
low diffusitivity of hydrogen in Li and the tendency of lithium to
form a hydride.[8] We are presently studying the interaction of hydrogen with defects by replacing one of the host atoms by the defect. These results will be published in due course.

ACKNOWLEDGEMENT

We wish to thank Dr. D. Hartford and the Academic Services
Staff at the VCU Computer Center for their assistance and generous
allocations of machine time on the twin-IBM 370/168 system.

REFERENCES

1. J.C. Slater, "Quantum Theory of Molecules and Solids, Vol. 1",
   McGraw-Hill Book Co., New York (1963), Appendix 7.
2. J.L. Whitten, Gaussian Expansion of Hydrogen-Atom Wavefunctions, J. Chem. Phys. 39:349 (1963).
3. D.D. Shillady and S. Baldwin-Boisclair, Dipole Optimized Gaussian Orbitals for Rapid Computation of Electrostatic Potential
   Contour Maps, Int. J. Quantum Chem., Quantum Biology Symp. No.
   6:105 (1979).
4. D.D. Shillady and D.B. Talley, Spherically Symmetric Axial
   Gaussian Lobe 3d and 4f Orbitals, J. Computational Chemistry,
   in press (1982).
5. R.C. Sahni, B.C. Sawhney and M.J. Hanley, Electronic States of
   Molecules II. Calculation of the Potential-Energy Curve and
   Molecular Constants of LiH ($X^1\Sigma^+$), J. Chem. Phys. 51:539 (1969).
6. A. Hinterman, A.M. Stoneham and A.H. Harker, Cluster Calculations of Hyperfine Fields and Knight Shifts for Muons in Li
   and Be, J. Hyperfine Interactions 8:475 (1981).
7. Ref. 1, Appendix 2.
8. R.C. Werner and T.A. Ciarlariello, Metal Hydride Fuel Cells as
   Energy Storage Devices, in "Proceedings, United Nations Conferences for New Sources of Energy", Rome (1961), pp. 213-218.

PARAMETRIZATION OF THE SELF-CONSISTENTLY CALCULATED ELECTRONIC

DENSITIES AROUND A PROTON IN JELLIUM

S. Estreicher and P.F. Meier[*]

Institut für Theoretische Physik
*Physik-Institut, Universität Zürich
CH-8001 Zürich

ABSTRACT

The self-consistent density-functional method has been
used to calculate electronic densities around impurities in jellium. The resulting charge density around a
proton has been fitted to a simple analytic expression
and the fit parameters determined as a function of r_s
for all metallic densities. The parametric function
reproduces with high precision all the characteristic
features of the self-consistently calculated density.
The results are compared to previous calculations.

INTRODUCTION

In recent years, the density functional formalism introduced
by Hohenberg, Kohn and Sham[1,2,3] has been used extensively to calculate the electron distribution around impurities in metals. The
first self-consistent calculations of the non-linear screening of
a proton in jellium which include exchange and correlation through
the local-density approximation were made by Popovic and Stott[4].
Since then various authors have published similar results and extended the calculations to systems where the hydrogen is trapped
at a vacancy. The charge distribution around other impurities have
also been treated within the jellium model.

The results of the induced charge distribution around an

impurity in jellium can be used to study the influence of the
lattice ions on the total energy by first-order perturbation
theory. Using a localized pseudo-potential, the interaction potential between the screened impurity and the lattice ions can
be determined as a function of the impurity site. In this way,
energy profiles for protons in Al and Mg have been calculated in
Ref.4. Other authors however, have obtained quite different results for Al, indicating that the interaction energy depends critically on the form of the induced charge distribution and the
details of the pseudo-potential.

As part of a general theoretical program to investigate the
behavior of positive muons in simple metals, we have recalculated
the charge distribution induced by a positive charge in jellium
and have studied the interaction energy with the lattice ions for
a number of simple metals. In this publication we report on the
results for the electron density distribution. The densities are
given in parametrized form in order to allow easy use for further
applications.

METHOD OF CALCULATION

In the density-functional formalism of Hohenberg, Kohn and
Sham, the following system of one-particle equations is to be
solved self-consistently:

$$[- \frac{1}{2}\Delta + V_{eff}(\vec{r})] \psi_i(\vec{r}) = E_i \psi_i(\vec{r}),$$

$$V_{eff}(\vec{r}) = V_{ext}(\vec{r}) + \int d\vec{r}' \frac{n(\vec{r}')}{|\vec{r}-\vec{r}'|} + \mu_{xc}[n(\vec{r})], \qquad (1)$$

$$n(\vec{r}) = \sum_{E_i < \mu} |\psi_i(\vec{r})|^2.$$

Here the effective potential consists of the external potential
($- 1/r$ in the case considered here), the average electrostatic
potential and μ_{xc}, which describes the effects of exchange and
correlation. The calculations were performed within the local
density approximation[5] for μ_{xc}, and we used the parametrized form
of the result of Singwi et al.[6] proposed by Hedin and Lundqvist[7].
For a single impurity in jellium, the spherical symmetry implies
for the radial wave function $u_{kl}(r) = rR_{kl}(r)$:

$$[- \frac{1}{2} \frac{d^2}{dr^2} + V_{eff}(r) + \frac{l(l+1)}{2r^2}] u_{kl}(r) = E_k u_{kl}(r). \qquad (2)$$

These equations have been solved using the Numerov method[8] for partial waves up to $l = 8$, from the origin to a radius of 20 or 30 a.u. (depending on r_s) where the solutions have been matched to those of the free-particle case. The density $n(r)$ has been calculated using about 50 k-points. After each iteration the new effective potential was determined using the feedback and additional screening technique proposed by Manninen[9,10]. In addition, the convergence was checked for many Friedel oscillations by a

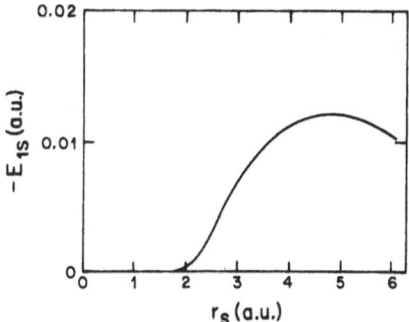

Fig. 1. Bound state energy versus the jellium parameter r_s.

very large number of iterations, typically one hundred for large r_s values. Special care was also given to the determination of the bound state wave-function, since for $r_s \sim 2$, very shallow bound states occur, as can be seen in Fig. 1.

The calculations have been performed for jellium densities corresponding to $r_s = 2.07, 2.2, 2.5, 3, 3.5, 4, 4.5, 5, 5.5$, and 6. In the next Sec. the resulting charge distribution $\Delta n(r) = n(r) - n_o$ is discussed.

RESULTS AND PARAMETRIZATION

Contact density. The behavior of the density at the origin, $\Delta n(0) = n(0) - n_o$, is shown in Fig. 2. Since one expects that for large r_s values the contact density approaches that of hydrogen ($\Delta n(0) \cong 1/\pi$), it is natural to parametrize $\Delta n(0)$ by a function that tends asymptotically to $1/\pi$. An inspection of a log-log plot of $\Delta n(0) - 1/\pi$ versus r_s indicates that a very good parametrization for $1.8 \leq r_s \leq 6$ is given by

$$\Delta n(0) = \frac{1}{\pi} + \exp[- 0.720 - 1.28 \ln r_s - 0.385(\ln r_s)^2] \qquad (3)$$

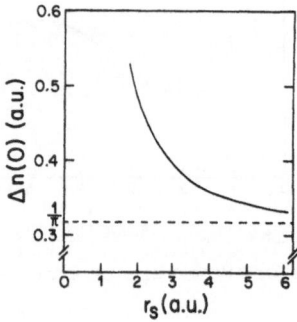

Fig. 2. Behavior of the contact density as a function of r_s.

Behavior near the origin. An inspection of Eq. (2) shows that for $V_{eff}(r) = - Z/r$ as $r \to 0$, the correct boundary condition for $u_1(r)$ is

$$u_1(r) = Ar^{1+1} - AZr^{1+2}/(1+1).$$ (4)

Whereas the first term is well-known, the second term is usually not considered. It implies however, that the probability density at zero must fulfill the condition

$$\lim_{r \to 0} \frac{d}{dr} |\psi(r)|^2 = - 2Z |\psi(0)|^2.$$ (5)

For our case with $Z = 1$ we therefore have close to the origin

$$\Delta n(r) = \Delta n(0) e^{-2r}.$$ (6)

This equation implies an exponential decrease of $\Delta n(r)$ close to $r = 0$.

Parametrization of $\Delta n(r)$. A convenient parametrization of $\Delta n(r)$ can be obtained using the following function

$$\Delta n(r) = \frac{1}{\pi} e^{-2r} + (\Delta n(0) - \frac{1}{\pi}) e^{-2r(1+r)} + f(2k_F r)$$ (7)

which consists of the hydrogen 1s-state density, a contribution giving the correct contact density and slope at the origin, and a function f which accounts for the oscillations. We exploited the fact that the zeros of $f(2k_F r)$ are strictly linear functions of r_s to parametrize f using different expressions for $r \gtrless z_2$, where z_2, the second zero of f, is given by

$$z_2 = 1.52 \, r_s + 0.464.$$ (8)

With $x = 2k_F r$, $f(x)$ can be written as

$$f(x) = \frac{A_o}{x^4 + 1} \hat{j}_o(x)(1-e^{-x}) + \frac{1}{x^3 + 1} \sum_{1=1}^{3} A_1 \hat{j}_1(x) \quad \text{for } r < z_2$$ (9a)

$$f(x) = \frac{1}{x^3 + 1} \sum_{1=2}^{5} B_1 \hat{j}_1(x) \quad \text{for } r > z_2.$$ (9b)

Here $\hat{j}_1(x)$ denote the Riccati-Bessel functions which behave asymptotically as $\sin(x+l\pi/2)$. Thus the asymptotic behavior of $\Delta n(r)$ is correctly given by Eq. (9b).

The amplitudes A_1 and B_1 are smooth functions of r_s as shown in Figs. 3 and 4. Their values, in the range $2 \leq r_s \leq 6$, have been fitted to a power series in $1/r_s$ which is given in Table I.

Table I

a) $r<Z_2$: $A_1 = a_1/r_s^4 + b_1/r_s^3 + c_1/r_s^2 + d_1/r_s + e_1$

1	a_1	b_1	c_1	d_1	e_1
0	− 9.879	10.795	− 4.422	0.696	− 0.018
1	0.347	2.257	− 1.711	0.927	− 0.103
2	14.900	−20.780	10.200	− 2.769	0.233
3	−15.040	17.681	− 8.380	1.946	− 0.156

b) $r>Z_2$: $B_1 = a_1/r_s^4 + b_1/r_s^3 + c_1/r_s^2 + d_1/r_s + e_1$

1	a_1	b_1	c_1	d_1	e_1
2	− 6.197	5.882	− 1.256	− 0.379	0.047
3	− 4.056	6.326	− 6.186	1.631	− 0.120
4	2.388	− 6.313	6.083	− 1.688	0.122
5	−16.430	19.463	− 9.391	1.820	− 0.114

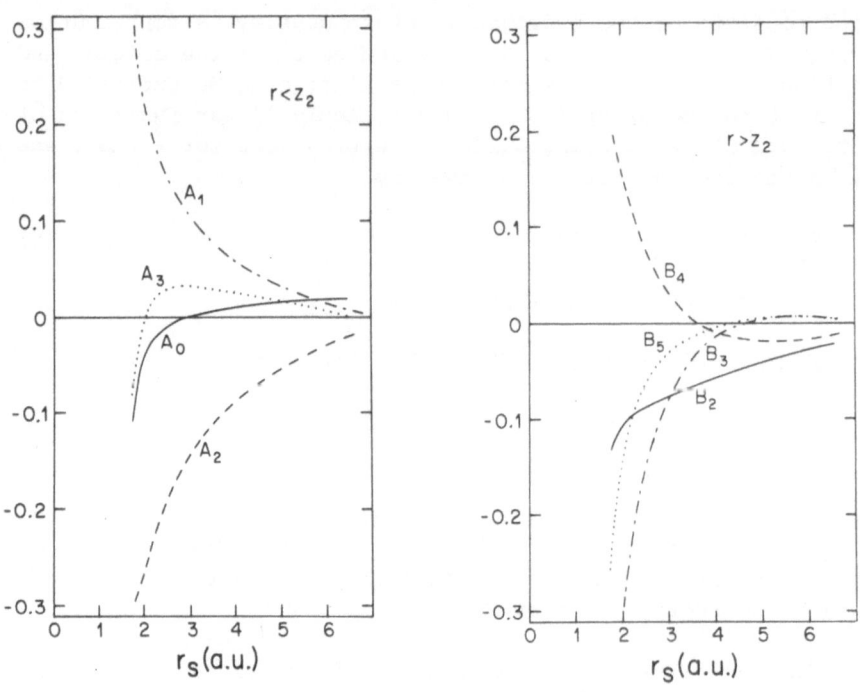

Figs. 3 and 4. Parameters A_1 and B_1 versus r_s.

DISCUSSION

The electron densities obtained from the parametrized form have been compared to the results of other authors[9,11,12] and a good agreement has always been found. Excellent agreement exists also between the Fourier transforms of the calculated Δn and the parametrized expressions. Furthermore, the displaced charge calculated with the values from Table I was found to be very close to one for all values of r_s from 1.5 to 6.5. More details and a discussion of the influence of a different choice of the exchange and correlation potential will be published elsewhere.

ACKNOWLEDGMENTS

We would like to thank E. Holzschuh and M. Manninen for instructive discussions and K. Petzinger for suggesting the problem.

REFERENCES

1. H. Hohenberg, W. Kohn, Phys. Rev. B 136: 864 (1963).
2. W. Kohn, L.J. Sham, Phys. Rev. A 140: 1133 (1965).
3. L.J. Sham, W. Kohn, Phys. Rev. 145: 561 (1966).
4. Z.D. Popović, M.J. Stott, Phys. Rev. Lett. 33: 1164 (1974).
5. O. Gunnarsson, B.I. Lundqvist, Phys. Rev. B 13: 4274 (1976) and references cited therein.
6. K.S. Singwi, A. Sjölander, M.P. Tosi, R.H. Land, Phys. Rev. B 1: 1044 (1970).
7. L. Hedin, B.I. Lundqvist, J. Phys. C 4: 2064 (1971).
8. G.W. Pratt, Phys. Rev. 1217 (1952).
9. M. Manninen, private communication.
10. G.P. Kerker, Phys. Rev. B 23: 3082 (1981).
11. K. Petzinger, private communication.
12. P. Jena, Hyperfine Interactions 6: 5 (1979).

ELECTRONIC STRUCTURE OF VANADIUM-HYDROGEN CLUSTERS

G. Bambakidis[*], M. P. Guse and G. C. Abell[**]

*Wright State University and
**Mound Facility

The electronic structure and total energy of clusters of vanadium atoms of various sizes, both with and without a hydrogen atom, have been calculated using the self-consistent multiple scattering $X\alpha$ method. The smallest cluster considered, V_4, represents the four vanadium atoms nearest the tetrahedral interstitial site in the body-centered-cubic lattice. Recent theoretical work by Abell has ascribed to this and analogous isolated four-atom clusters of the other Group V metals, an inherent instability with respect to a Jahn-Teller distortion. This instability arises from the existence of a two-fold degenerate electronic state in the isolated cluster. In this theory, the presence of hydrogen stabilizes the Jahn-Teller distortion. We have investigated this by performing our calculations for hydrided and unhydrided clusters, for both the undistorted and distorted structures, with the aim of identifying states near the Fermi level whose energies may be especially sensitive to the distortion.

*Supported by USDOE through the Mound Research Participation Program.
**Operated by Monsanto Research Corporation for USDOE under Contract No. DE-AC04-76-DP00053.

SUPERCONDUCTIVITY IN METAL-HYDROGEN SYSTEMS

B. Stritzker

Institut für Festkörperforschung
Kernforschungsanlage Jülich
D-5170 Jülich, W. Germany

ABSTRACT

The experimental situation for superconductivity in metals charged with hydrogen or deuterium is reviewed. The original idea for a definite modification of the electronic properties of the metal by the addition of H(D) interstitials has to be modified, since the H(D) atoms also dramatically change the phonon system as well as the electron-phonon interaction itself. The different influences of the H(D) addition are demonstrated for the well studied case of H(D) charged Pd alloys. The difficulties in charging these alloys have been solved with several experimental techniques, i.e. with high pressure or electrolytic charging at lower temperatures as well as by H(D) implantation into metals at liquid helium temperatures. These methods have been applied not only to other transition metals but also to a series of non-transition metals. The results of these measurements, mostly performed with the low temperature implantation technique, are discussed. It will become obvious that the model developed for the PdH system cannot unambiguously explain all these data.

INTRODUCTION

The interest in metal-hydrogen systems, with respect to superconductivity, was initiated by the discoveries of relatively high superconducting transition temperature,

T_c, in Th_4H_{15} by Satterthwaite and Toepke[1] and in the Pd-H system by Skoskiewicz[2] for ratios H/Pd $\gtrsim$ 0.8. Further interest in the latter system was aroused by the remarkable inverse isotope effect,[3] i.e. a higher $T_c \simeq$ 11K in Pd compared to $T_c \simeq$ 9K in PdH, and the astonishing increase of T_c to 17K in hydrogenated PdCu alloys[4]. Comparably high T_c values have not been found in many other materials[5]. All these observations caused a lot of experimental and theoretical work which has been reviewed several times in the past[6].

In the following the influence of H on the superconducting properties will be briefly summarized for the well studied PdH system. The other examples, both transition and non-transition metal hydrogen systems where hydrogen exerts a positive influence will be described. It will become evident that both electronic and phonon effects, especially coupling to optic H phonons, determine the superconductivity in metal hydrogen systems. Due to the limited space this paper will describe some common features for a very limited, personal choice of experimental results. References will be only given for recent results; older results are cited in preceeding reviews[6].

Since the theory of superconductivity for metal-hydrogen systems will be covered in the following paper by M. Gupta[7] only a very short introduction into the theoretical considerations will be given. Below T_c the attractive electron-phonon interaction λ overcomes the repulsive Coulomb interaction μ^* between electrons and the pair breaking effect of spin fluctuations μ_{sp}.

In order to obtain reasonably high T_c values, first of all μ_{sp}, i.e. the spin fluctuation effect, has to be reduced. In addition, high values of λ have to be obtained. This may be accomplished by

1) a high electron-phonon coupling λ itself
2) a high electronic density of states at the Fermi surface, $N(0)$
3) a "weak" phonon density of states, $F(\omega)$, or additional phonon modes, i.e. optic phonons in metal hydrogen systems.

In the presence of optic H(D)-phonon modes the electron-phonon coupling constant can be split into an acoustic and an optic part: $\lambda = \lambda_{ac} + \lambda_{op}$. In general the occurrence of λ_{op} should lead to an enhancement of T_c as long as other properties remain unchanged. However, the influence of optic phonon modes due to H(D) intestitials decreases with increasing frequency ω_{op} of the optic modes.

PREPARATION METHODS

For the preparation of superconducting metal-H systems mainly two different classes of preparation methods have been applied:

a) Equilibrium methods:

The conventional H(D) charging methods, like electrolytic or pressure methods depend on the H solubility of the metal. In general the solubility increases whereas the diffusion decreases with decreasing temperature. Thus the maximum obtainable H(D) concentration is limited, but accurately known.

b) Non-equilibrium methods:

The lack of H(D) diffusion at 4K is used to accumulate high H(D) concentrations in metals which do not dissolve much H(D) under equilibrium conditions. The H(D) atoms are either introduced by implantation into metal targets at 4K or incorporated during co-condensation with the metal atoms onto substrates at 4K. In both cases the H(D) concentration is totally independent of the equilibrium solubilty. However, the concentrations cannot be determined very accurately and the H(D) charging is accompanied by the introduction of severe lattice disorder.

Pd-H SYSTEM

A large variety of both experimental and theoretical work yielded a good understanding of the microscopic mechanisms causing superconductivity in PdH(D). These mechanisms will be briefly summarized in the following.

Pure crystalline Pd is not a superconductor (at least above the lowest applied temperature of 1.7 mK) although its high electronic density of states at the Fermi surface, $N(0)$, and its suitable phonon spectrum should result in a reasonable T_c. The reason is that Pd is a strongly exchange enhanced paramagnet. The resulting pair breaking effect of the spin fluctuations μ_{sp} is sufficient to explain the absence of superconductivity. The addition of H(D) to the Pd decreases $N(0)$ because, besides the creation of new s-states far below the Fermi energy, the s and d bands of the Pd at the Fermi energy are filled. Thus μ_{sp} decreases with increasing H content and the system becomes diamagnetic for concentrations H/Pd $\gtrsim$ 0.65. However, this decrease of μ_{sp} is only a precondition for the occurrence of superconductivity, and it is not sufficient to explain the high observed T_c

values. These results are caused by an especially high
electron-phonon coupling due to the H(D) interstitials.
These light interstitials give rise to optic phonon modes
at comparable low energies, i.e. $\omega_{opH} \simeq 45$ meV for H and
$\omega_{opD} \simeq 31$ meV for D. By means of superconducting tunneling
experiments and measurements of the temperature dependence
of the electrical resistance it was demonstrated that
the coupling to the optic modes dominates the attractive
electron-phonon interaction. This strong electron-phonon
coupling to the optic modes is caused by the large density
of s-electrons, $N(O)_H$, at the site of the H or D interstitials. Whereas the overall value for $N(O)$ decreases,
$N(O)_H$ increases with increasing H(D) concentration. Thus
the coupling of s electrons to the optic phonons results
in a large value of $\lambda_{op} = (1-3) \cdot \lambda_{ac}$. The herewith calculated dependence of T_c^{op} on the H(D) concentration agrees
qualitatively with the measured values shown in Fig. 1.
However, there is some disagreement with regard to the
observed inverse isotope effect.

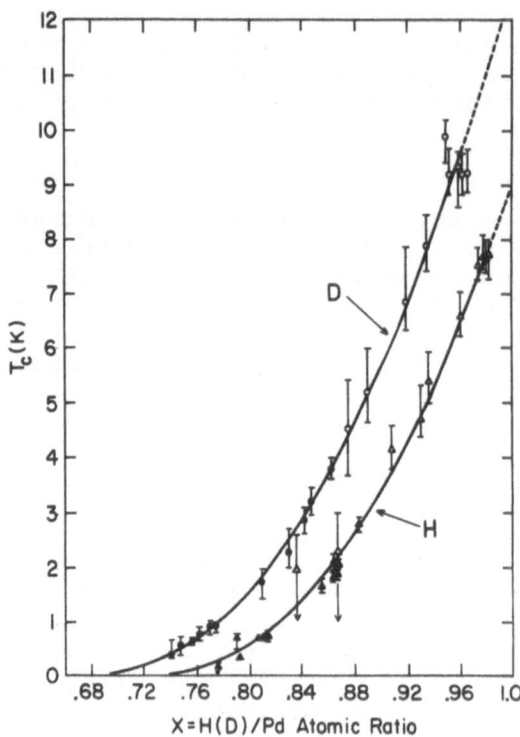

Fig. 1. T_c as a function of H(D) concentration for hydrogenated
         and deuterated Pd (after ref. 8).

There are two main models to explain the higher T_c of PdD compared to PdH:

a) Anharmonic effects lead to a ration $\omega_{opH}/\omega_{opD} > \sqrt{2}$. Thus the optic phonons due to D interstitials couple more effectively to the electrons and $\lambda_{opD} > \lambda_{opH}$. Ratios $\omega_{opH}/\omega_{opD} > \sqrt{2}$ have been detected by superconducting tunneling, neutron scattering and resistance measurements.

b) The electronic properties of PdD and PdH are different because more bonding states are created in the H case compared to the D case. Consequently the electronic density of states at the interstitial site are different for the two isotopes: $N(0)_H < N(0)_D$. This argumentation is in agreement with NMR and de Haas van Alphen measurements as well as with the observed scaling of T_c (compare Fig. 1)

In my opinion there is not too much of a controversy since the different force constants, as required in model a, are a result of different bondings, i.e. different electronic properties as in model b. Besides the disagreement between model a and b the reasons for superconductivity are quite obvious:

1) $\mu_{sp} = 0$

2) substantial coupling to the optic H(D) phonons

3) high s-electron density of states at the interstitial site.

The situation is not as clear in the case of hydrogen-generated Pd-noble metal alloys where T_c increases up to 17K for PdCu alloys. Up to now, these rather high superconducting transition temperatures could be only achieved with ion implantation methods. A systematic study of the variation of T_c with Cu concentration was performed by low temperature implantation of H and D. It was found that the inverse isotope effect not only vanished but also changes sign with increasing Cu content. In a recent experiment about 25 at% Cu has been implanted into the surface layer of rather thick Pd foils. Then the Pd foils were charged electrolytically with H at 77K. Broad superconducting transitions starting at about 16K have been observed[9]. The increase of T_c with increasing Cu content is not yet fully understood. There are some hints that the acoustic phonon spectrum is softened and thus λ_{ac} increases with the addition of noble metals. A more important effect might be that $N(0)_H$ increases with increasing noble metal content. However, the reasons for a maximum of T_c with respect to noble metal concentration

and the change of sign, of the isotope effect are not
fully understood.

Recent assumptions that the disorder produced by
the implantation is responsible for the high T_c seem to
be rather unlikely for the following reasons. In pure
PdH(D) the measured T_c values are independent of the H(D)
charging method. In addition it has been shown that ordering effects influence T_c only very slightly[10]. On the
other hand, high pressure H charging (up to 20 kbar) of
PdCu alloys yielded no T_c values above 9K, the value for
PdH[11].

THORIUM$_4$ HYDROGEN$_{15}$

Although the T_c values are very similar for Th_4H_{15}
and PdH other superconducting properties and presumably
the reasons for superconductivity are quite different.
For instance in Th_4H_{15} there is no isotope effect at all
and T_c does not depend significantly on the H concentration as long as the crystal structure remains unchanged.
The latter property leads to the assumption that H or D
only stabilize a crystal structure, which is more favorable for superconductivity[12]. Thus $T_c \simeq 1.4K$ for pure Th
increases to $T_c \simeq 9K$ for Th_4H_{15} or Th_4D_{15}. The lower hydride, ThH_2, which crystallizes in a different structure,
is not a superconductor above 1.2K. The two different
hydride phases have been compared by incoherent neutron
scattering[13]. The result showed that only the superconducting Th_4H_{15} has optic phonon modes at rather low energy (~80 meV) which could contribute noticeably to the
electron-phonon coupling. Hence it seems reasonable that
H in Th_4H_{15} enhances the electron-phonon coupling in
analogy to PdH. However, λ_{op} will be smaller in Th_4H_{15}
compared to PdH, since the energy of the optic modes is
lower in PdH. Nevertheless the H and D atoms not only
stabilize the Th_4H_{15} structure but also cause an enhanced electron-phonon coupling.

OTHER TRANSITION METAL-HYDROGEN SYSTEMS

The only new transition metals where an enhancement
of T_c by addition of H or D has been found in recent
years are the elements of groups IV B, i.e. Ti, Zr and
Hf[14]. No improvement of the superconducting properties
have been found for Pt[15], V[16], Cr[17] and Ni[17]. The results for Ti, Zr and Hf are shown in Table I. The H(D)
charging was performed by low temperature implantation.
The effect of lattice disorder on T_c was simulated by
implantation of inert He atoms. It can be seen that T_c

Table 1

Maximum T_c values for group IVB elements implanted with H(D) and corresponding experimental paramaters

Element (T_c/K)	Implanted Species	Maximum T_c/K	Corresponding H(D)/metal	Isotope effect
Ti	H_2^+	4.95	0.14	Normal
(0.4)	D_2^+	4.89	0.13	sign
	He	1.00	–	
Zr	H_2^+	3.14	0.13	Inverse
(0.6)	D_2^+	4.65	0.13	
	He	1.49	–	
Hf	H_2^+	1.75	0.18	Inverse
(0.13)	D_2^+	2.23	0.16	
	He	1.00	–	

increases substantially by the addition of H or D whereas disordering affects T_c much less. Although the implanted H(D)/metal concentrations amount to ~ 0.15, preliminary x-ray measurements show only the presence of α-phase. For both Zr and Hf a remarkable inverse isotope effect is observed, whereas a slight normal isotope effect occurs in the Ti case. Similar results have been obtained by co-deposition of H(D) and Zr onto targets at 4K[18]. Recent neutron scattering experiments on α-phase hydrogenated and deuterated Zr showed no sign of any anharmonic effects in this system[19]. Thus the model developed for the explanation of superconductivity in PdH does not seem to be appropriate. It is assumed that mainly electronic effects dominate the superconducting properties of H(D) implanted Ti, Zr and Hf.

NON TRANSITION METAL-HYDROGEN SYSTEMS

A very interesting system, which had been studied in the past, is the Al-H(D) system[20]. H or D implantation at low temperatures yielded a remarkable increase of T_c from 1.17K to 6.75K or 6.05K. Simulation of the influence of lattice defects by He implantation resulted only in a $T_c \simeq 3.7K$. Recently we started a systematic study of the H(D) influence on the superconducting non-transition metals in order to get a better understanding of the ef-

fect of H(D) interstitials on superconductivity[21]. Since
most of the non-transition metals have no solubility for
hydrogen we used the low temperature implantation method.
Besides H and D, He also was implanted to simulate the
plain radiation effect on T_c. In all cases we found a T_c
enhancement due to H or D implantation far exceeding the
increase caused by lattice defects.

In this connection, Pb is an especially good example
since it is the only non transition metal superconductor
whose T_c value is not enhanced but slightly reduced by
lattice defects. Fig. 2 shows the result of the low temperature implantation experiment. T_c of a Pb-foil is
plotted as a function of the implanted dose of H, D or
He[22]. It is obvious that the hydrogen isotopes considerably increase T_c in contrast to He. A pronounced inverse
isotope effect, i.e. a higher T_c for the deuterated sample
compared to the hydrogenated sample, can be detected. However, it is not yet clear if this isotope effect is real
or if D can be accumulated to higher concentrations in the
implanted region due to a lower D mobility at 4K compared
to H.

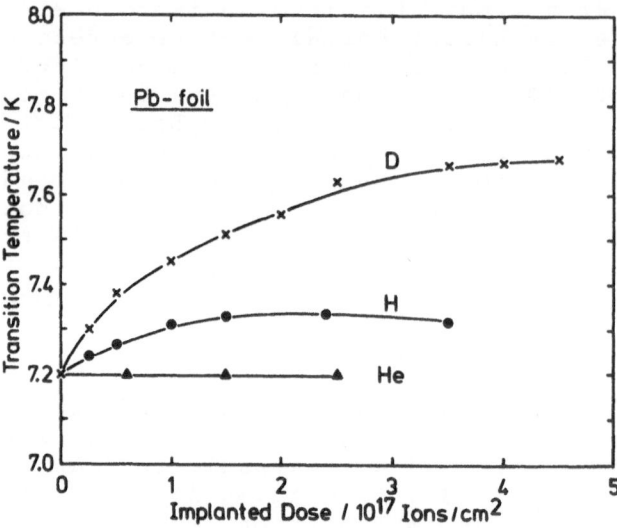

Fig. 2. T_c of a Pb-foil as a function of the implanted
          dose of H, D, and He.

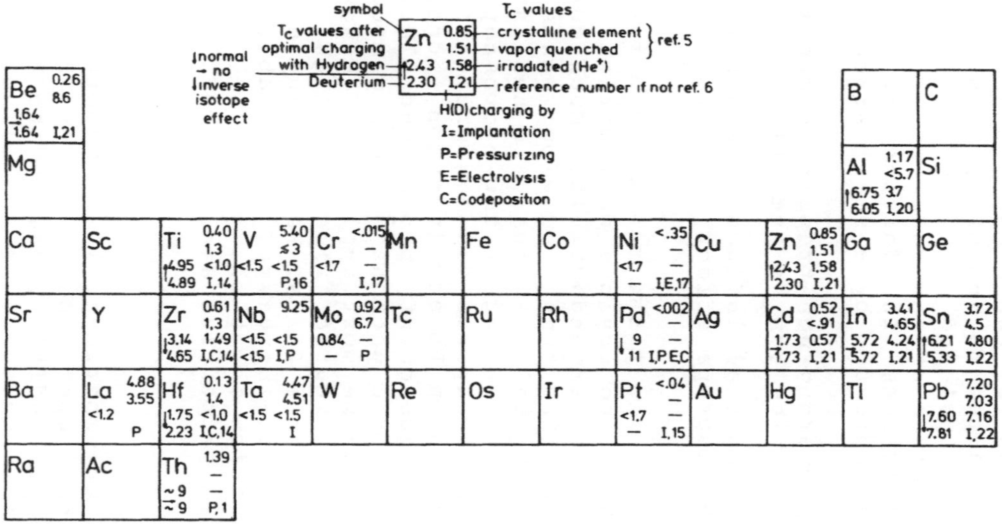

Legend:

symbol

T_c values after optimal charging with Hydrogen

Deuterium

↑normal → no ↓inverse isotope effect

Zn 0.85 / 1.51 / 1.58 / 2.43 2.30 1.21

T_c values:
— crystalline element
— vapor quenched
— irradiated (He⁺) } ref. 5
— reference number if not ref. 6

H(D)charging by
I=Implantation
P=Pressurizing
E=Electrolysis
C=Codeposition

Element entries:

Be 0.26 / 8.6 / 1.64 1.64 1.21
Mg

Al 1.17 / <5.7 37 / 16.75 6.05 1.20

Ca — Sc — Ti 0.40 / 1.3 / <1.0 <1.5 / 4.95 4.89 1.14 — V 5.40 / ≤3 / <1.5 <1.5 / P.16 — Cr <.015 / — / <1.7 / 1.17 — Mn — Fe — Co — Ni <.35 / — / <1.7 / — — Cu — / 1.E.17 — Zn 0.85 / 1.51 / 1.58 / 12.43 2.30 1.21 — Ga — Ge

Sr — Y — Zr 0.61 / 1.3 / 1.49 <1.5 / 3.14 4.65 1.C.14 — Nb 9.25 / 1.3 / <1.5 <1.5 / 1.C.14 1.P — Mo 0.92 / 6.7 / 0.84 — / — P — Tc — Ru — Rh — Pd <.002 / — / 9 11 I.P.EC — Ag <.002 / — / — — Cd 0.52 / <.91 / 1.73 0.57 / 1.73 1.21 — In 3.41 / 4.65 / 5.72 5.72 / 1.21 — Sn 3.72 / 4.5 / 4.21 4.80 / 5.33 1.22

Ba — La 4.88 / 3.55 / <1.2 / P — Hf 0.13 / 1.4 / 1.75 <1.0 / 2.23 1.C.14 — Ta 4.47 / 4.51 / <1.5 <1.5 / I — W — Re — Os — Ir — Pt <.04 / — / <1.7 — / 1.15 — Au — Hg — Tl — Pb 7.20 / 7.03 / 7.16 / 7.60 7.81 1.22

Ra — Ac — Th 1.39 / — / ~9 ~9 P.1

Fig. 3. Part of the periodic chart including data for those elements which have been investigated for superconductivity after H(D) charging

Besides Pb various other non transition metals like Be, Zn, Cd, In and Sn, were previously investigated[21]. The results of these implantation experiments are included in Fig. 3, showing part of the periodic table. The results for all elements which have been investigated for the influence of H or D according to my present knowledge are included. For these elements the known T_c values for the crystalline, the vapor quenched and the He irradiated metal as well as for the H and D charged metal are plotted. In addition the method of H, D charging and the reference number is included. For the older data former review papers are given as reference in order to shorten the listing of references. In addition the sign of the isotope effect is indicated by a small arrow.

For the H(D) charged non-transition metals the following common properties can be summarized:

a) H and D exert a special influence, enhancing T_c.
b) The sign of the isotope effect seems to be quite accidental.

In the present situation these results are rather puzzling because there is no obvious explanation for all the observed effects on superconductivity. Much more detailed information, like tunneling experiments and detailed band structure calculations, are necessary to obtain a detailed understanding of the microscopic mechanisms determining the superconducting properties. Recent tunneling experiments on Pb and In films, doped with H by the method of vapor quenchint, proved the electron-phonon coupling to be the optic phonons[23]. Thus a model as developed for PdH seems to be most applicable. However, the T_c values of these investigated films are slightly lower compared with the H free films. Perhaps the H concentration was too low to increase T_c noticeably. In any case, more information about changes in the phonon specra and the electronic properties are required.

CONCLUSION

Based on large experimental and theoretical efforts, the reasons for the superconductivity of PdH(D) are well known. After the precondition, namely the depression of spin fluctuations in pure Pd, is fulfilled, superconductivity is caused by the strong electron-phonon coupling via the optic H(D) phonon modes. The situation is less clear in the case of hydrogenated, α-phase Ti, Zr and Hf, where mainly changes of the electronic properties seem to be responsible for the increase of T_c. For Th_4H_{15} it seems also important that the H(D) atoms stabilize a crystal structure favoring high T_c values. Due to the lack of detailed information, the positive influence of H(D) on the superconductivity in non-transition metals is not yet explained. Probably reasons similar to the PdH(D) system are important, i.e. coupling to optic phonons, as well as a high electronic density of states at the sites of the H(D) interstitials.

REFERENCES

1. C.B. Satterthwaite and I.L. Toepke, Phys.Rev.Lett. 25:741 (1970)
2. T. Skoskiewicz, Phys.Status Solididi (a) 11:K 123 (1972)
3. B. Stritzker and W. Buckel, Z.Physik 257:1 (1972)
4. B. Stritzker, Z.Physik 268:261 (1974)
5. B.W. Roberts, J. Physical and Chemical Reference Data 5:581 (1976) and NBS Technical Note 983
6. Review papers with additional references:
    B. Stritzker and H. Wühl in: Topics in Applied Physics, Vol. 29, Hydrogen in Metals II, G. Alefeld and J. Völkl, eds., Springer-Verlag, Berlin (1978) p. 243
    D.G. Westlake, C.B. Satterthwaite and J.H. Weaver, Physics Today 31:32 (1978)
    W. Buckel, Z.Phys.Chem. 116:135 (1979)
7. M. Gupta, this conference proceedings
8. R.W. Standley, M. Steinback and C.B. Satterthwaite, Solid State Comm. 31:801 (1979)
9. A. Leiberich, W. Scholz, W.J. Standish and C.G. Homan, Phys.Lett. 87A:57 (1981) and this conference proceedings
10. R.W. Standley and C.B. Satterthwaite, this conference proceedings and references therin
11. V.E. Antonov, I.T. Belash, E.G. Ponyatovskii and V.I. Rashupkin, JETP Lett. 31:422 (1980)
12. R. Caton and C.B. Satterthwaite, J.Less-Common Met. 52:307 (1977)
13. M. Dietrich, W. Reichardt and H. Rietschel, Solid State Comm. 16:1085 (1975)

14. J.D. Meyer and B. Stritzker, Nucl.Instr. and Methods
    182/183:933 (1981)
15. A. Traverse, H. Bernas and J. Chaumont, Solid State
    Comm. 40:725 (1981)
16. W. Däumer, K. Lüders, Z. Szücs and H. Weber, J.Less-
    Common Metals 78:91 (1981)
17. H. Bernas and P. Nedellec, Nucl.Instr. and Methods
    182/183: 845 (1981) and references therein
18. P. Plein and S. Ewert, to be published
19. R. Khoda-Bakhsh and D.K. Ross, J.Phys.F. 12:15 (1982)
    R. Hempelmann, D. Richter and B. Stritzker, J.Phys.F.
    12:79 (1982)
20. A.M. Lamoise, J. Chaumont, F. Meunier and H. Bernas,
    J.Phys.Lett. (Paris) 36:L 271 and L 305 (1975)
21. F. Ochmann and B. Stritzker, to be published
22. F. Ochmann, J.D. Meyer and B. Stritzker, Physica
    107B:655 (1981)
23. B.W. Nedrud and D.M. Ginsberg, Physica 108B:1175
    (1981)

SUPERCONDUCTIVITY IN METAL HYDRIDES

Michèle Gupta

Centre de Mécanique Ondulatoire Appliquée du C.N.R.S.
23 rue du Maroc, 75019 Paris, France and
Université Paris-Sud, Bâtiment 506, 91405 Orsay, FRANCE

ABSTRACT

Using the results of a systematic study of the electronic structure of stoichiometric metal hydrides the electron-phonon coupling parameter has been evaluated, within the McMillan approximation, for a series of mono and dihydrides. The electronic term η is calculated using the rigid-ion approximation while experimental data are used to estimate the phonon contribution. Systematic trends are observed in the variation of η due to the metal site M and hydrogen site H. Sizeable values of η_H are obtained for the metal hydrides with filled d bands such as PdH ; η_H is also large when a metal-hydrogen antibonding band crosses the Fermi level, a case which happens in AlH and may happen for unstable dihydrides. The electronic contribution η_M is found to be small for all stable mono and dihydrides such as PdH, NiH, ZrH_2, NbH_2, LaH_2 and for FeTiH and $FeTiH_2$, although nothing prevents in principle η_M from being large in some metal hydrides, as the Fermi level sweeps through the metal d band. A good agreement is obtained with available experimental data for the occurrence of superconductivity in tbe hydrides under study.

INTRODUCTION

After the discovery of superconductivity in thorium[1] and palladium hydrides[2] with fairly high values of the superconducting transition temperature T_c, numerous metal-hydrogen (and deuterium) systems have been investigated[3] in the hope of finding high T_c materials. Two basic ideas have guided the search for superconductivity: (1) one expects, as it has been shown to be the case for PdH, to obtain large values of the electron-optical phonon coupling parameter; and (2) by alloying effects on the metal matrix or by varying

the hydrogen concentration, the average electron-per atom ratio can
be continuously varied and one can thus expect to modulate the
strength of the electron-acoustic phonon coupling.
In this paper we present an evaluation of the electron-phonon coupling parameter λ for several stoichiometric metal hydrides within
the formalism of McMillan[4]. For this evaluation we used our electronic band structure results in conjunction with experimental data on
the lattice vibrational properties. After briefly summarizing the
method, we shall address ourselves several questions : (1) can the
electron-optical phonon coupling λ_H be large for other hydrides, as
it is for PdH ? Why NiH, which is isoelectronic to PdH, is not
superconducting down to 1K ? To answer these questions we studied
the variation of λ_H for transition metal (TM) and rare earth hydrides
such as TiH_2, ZrH_2, NbH_2, LaH_2 ; for simple metal hydrides AlH, AlH_2
which can be prepared by ion implantation technique[5] and for hydrides
of intermetallic compounds FeTiH, $FeTiH_2$. (2) why does the presence
of H in the lattice kill superconductivity in the high T_C group V TM
such as Nb(T_c=9.2K) or in La (T_c = 6 K for the fcc phase) ? (3) can
we possibly make predictions of high T_C superconductors by considering the effects of alloying on the metal matrix ? A summary and
concluding remarks will be given in the last section.

METHOD

    We have evaluated the electron-phonon coupling constant λ using
McMillan's approximation[4] according to which λ can be written as the
ratio of an electronic contribution η to a phonon contribution. We
will further assume that for compounds with large mass difference
between the constituent atoms such as in the TM dihydrides, λ can be
written in the following form :

$$\lambda \simeq \frac{\eta_{Metal}}{M_{Metal}<\omega^2>_{acoustic}} + \frac{\eta_H}{M_H<\omega^2>_{optic}} = \lambda_{Metal} + \lambda_H \qquad (1)$$

where M is the atomic mass and the second moment of the renormalized
phonon frequencies $<\omega^2>$ has been defined by McMillan[4]. This approximation which has been always used for metal-hydrogen systems ignores
the so-called interference terms. To calculate η we used the formulation proposed by Gaspari and Gyorffy[6] which is based upon the rigid
ion approximation ; these authors have shown that for each atomic
site K, η_K can be conveniently expressed in terms of quantities
obtained from band structure calculations :

$$\eta_K \sim \frac{E_F}{N_\uparrow(E_F)\pi^2} \sum_1 2(1+1)\sin^2(\delta_{1+1}-\delta_1) \frac{n_1^K \quad n_{1+1}^K}{n_1^{K(1)} \quad n_{1+1}^{K(1)}} \qquad (2)$$

where δ_1^K is the phase shift of angular character 1 of the potential
at site K. n_1^K and $n_1^{K(1)}$ are respecitively the partial density of states
(DOS) and the free scatterer DOS at site K, $N_\uparrow$ is the total DOS of
the one spin ; all the quantities are calculated at the Fermi

energy E_F. In this work we used the augmented plane wave (APW) method to calculate the electronic contribution η.

ELECTRON-OPTICAL PHONON COUPLING λ_H

For the Pd-H system, neutron scattering[7], superconducting tunneling[8] and the study of the temperature dependence of the resistivity[9] as well as theoretical results[10] show that the electron-optical phonon coupling is responsible in large part for the high value of T_c. We will first discuss the variation of λ_H. From Eq.(1), we can expect substantial values of λ_H if a large electronic term η_H is associated with a low value of the phonon contribution.

1. Variation of η_H

Since the hydrogen potential of the hydrogen atom is characterized by large s phase shifts, we expect η_H to be dominated by a s-p scattering mechanism. From the results listed in Table II, it appears that for all the hydrides considered, δ_0^H is always close to a resonance, $\delta_0^H \sim \Pi/2$, and does not vary much from one hydride to the next. Thus, the trends in the variation of η_H will be determined by the magnitude at the H site and at the Fermi level of the partial DOS of s and p type, relative to the total DOS. The position of the H-s states relative to the Fermi energy E_F will thus play an important role. From the results of the studies of the electronic structure of metal hydrides[11,12] it appears that for all the series of stable TM, rare earth and intermetallic hydrides, the overwhelming contribution of the H-s states are essentially found at low energies, below the metal d band. The H-s contribution inside the metal d bands remains always very weak. Nevertheless, for monohydrides of TM with filled d bands such as PdH, and to a lesser extent NiH[13], a non negligible fraction of H-s states are found around the Fermi level. It is to be notices also that, in this energy range, n_s^H increases with energy. This explains several results:(i) since the Fermi energy of NiH is closer to the top of the d bands than in PdH, n_s^H will be smaller as shown in Table I and thus, η_H^{NiH} is smaller than η_H^{PdH} (see Table II) (ii) Although the rigid band model for Pd-Ag alloys is not quantitatively correct, the Fermi level of the Pd-Ag-H alloys falls in an energy range where n_s^H is larger than in PdH and this feature[14] can explain the observed increase of T_c. We wish to point out that the H-s states found at E_F for PdH do not form part of the metal-hydrogen antibonding band, which lies at still higher energies and contains a larger proportion of H-s states.

Since for the dihydrides of the TM as well as for LaH_2, FeTiH and $FeTiH_2$. Figs 2 to 5 show that the Fermi energy falls in the metal d bands, we expect small values of n_s^H (see Table I) ; this leads as shown in Table II to small values of η_H.

At the H sites, n_p^H arises from the tails of the metal states which have a p symmetry at the H interstitial site. Thus, the s $\rightarrow$ p scattering mechanism should not be viewed as an intra atomic scattering process ; it is rather indicative of the metal-hydrogen

Table I. The partial-wave analysis n_l of the DOS inside the muffin-tin metal and hydrogen spheres at the Fermi energy. $N_\uparrow(E_F)$ is the total DOS at E_F.

		n_s	n_p	n_d	n_f	$N_\uparrow(E_F)$
AlH	Al	0.621	0.757	0.217	0.037	3.203
	H	0.325	0.168	0.011	0.0005	
AlH$_2$	Al	0.684	0.520	0.124	0.016	3.155
	1xH	0.073	0.107	0.005	0.0003	
LaH$_2$	La	0.017	0.168	3.250	0.014	7.550
	1xH	0.017	0.154	0.0168	0.0007	
TiH$_2$	Ti	0.0015	0.0590	18.7405	0.035	23.519
	1xH	0.015	0.8755	0.0535	0.0015	
ZrH$_2$	Zr	0.004	0.073	10.805	0.035	16.460
	1xH	0.008	0.432	0.021	0.0008	
NbH$_2$	Nb	0.004	0.051	4.369	0.034	6.440
	1xH	0.016	0.119	0.010	0.0012	
NbH$_0$	Nb	0.337	1.464	11.405	0.1652	15.41
PdH	Pd	0.058	0.163	2.618	0.012	3.405
	1xH	0.255	0.029	0.0016	0.001	
NiH	Ni	0.048	0.133	4.741	0.009	5.390
	H	0.161	0.035	0.0025	0.001	
FeTiH	Fe	0.241	0.4545	13.25	0.007	
	Ti	0.1195	0.2870	4.619	0.0195	11.965
	H	0.043	0.2388	0.0045	0.0002	
FeTiH$_2$	Fe	0.2995	0.3697	12.7325	0.0135	
	Ti	0.0028	0.0793	0.0973	3.525	11.41
	1xH	0.019	0.088	0.0032	0.0003	

Table II. Values of the various parameters entering the calculation of λ. Symbols are defined in Eqs. (1) and (2). The angular momentum dependent phase shift δ_l are given in radians.

		δ_0	δ_1	δ_2	δ_3	η (eV/Å^2)	$M\langle\omega^2\rangle$ (eV/Å^2)	λ
AlH	Al	0.3365	0.4014	0.0528	0.0015	0.294		
	H	1.1856	0.0404	0.0012	0.0	2.292		
AlH$_2$	Al	0.2536	0.3714	0.0556	0.0018	0.224		
	1xH	1.1738	0.0379	0.0010	0.0	0.744		
LaH$_2$	La	-1.1006	-0.4556	0.5326	0.0006	0.753	7.35	0.103
	1xH	1.5290	0.0400	0.009	0.0	0.043	3.35	0.013
TiH$_2$	Ti	-0.7913	-0.2262	0.3156	0.0016	3.898		
	1xH	1.2502	0.0374	0.0009	0.0	0.067		
ZrH$_2$	Zr	-0.9504	-0.3662	0.9042	0.0049	2.352 $(3.87)^2$	6.24	0.377
	1xH	1.2462	0.0373	0.0009	0.0	0.088	4.92	0.018
NbH$_2$	Nb	-1.0219	-0.4075	1.3749	0.0067	2.975 $(7.39)^2$		
	1xH	1.1593	0.0408	0.0011	0.0	0.102		
NbH$_0$	Nb	-0.5105	-0.1402	1.3396	0.0032	3.6848		
PdH	Pd	-0.5115	-0.1094	2.8066	0.0030	0.886	5.971	0.15
	H	1.1931	0.0280	0.0006	0.0	0.641	1.062	0.60
NiH	Ni	-0.3857	-0.0235	2.7916	0.0025	0.810	7.06	0.08
	H	1.0604	0.0318	0.0008	0.0	0.275	3.44	0.08
FeTiH	Fe	-0.6230	-0.1276	2.6874	0.0018	1.1473	12.50	0.092
	Ti	-0.9733	-0.3293	0.7323	0.0036	0.3926	10.72	0.037
	H	1.0608	0.0264	0.0005	0.0	0.234	3.78	0.062
FeTiH$_2$	Fe	-0.6193	-0.1228	2.6434	0.0023	1.2178	12.50	0.097
	Ti	-1.0495	-0.3693	0.6276	0.0028	0.4575	10.72	0.043
	1xH	1.0027	0.0279	0.0006	0.0	0.056	2.24	0.025

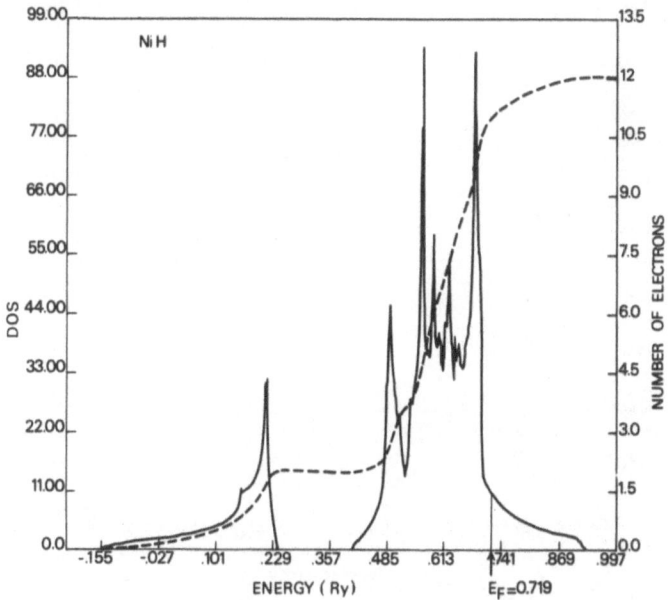

Fig. 1. The total DOS (states of both spin per Rydberg unit cell)
of NiH : full line curve and left hand side scale. The
total number of electrons : dashed curve and right hand
side scale.

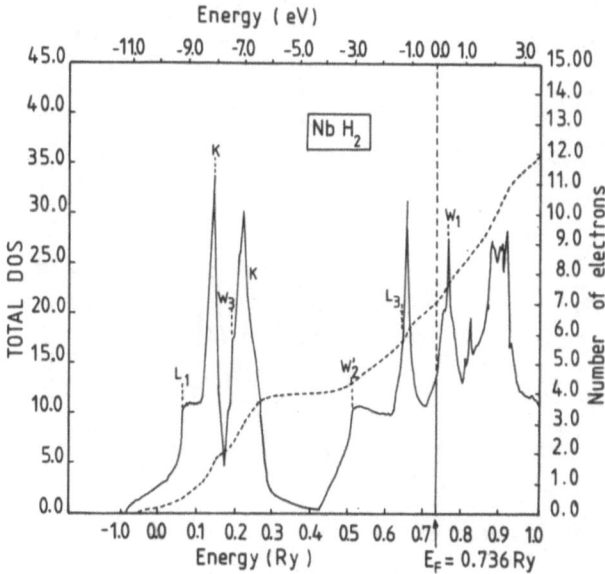

Fig. 2. The total DOS (states of both spin per Rydberg unit cell) of
NbH$_2$: full line curve and left hand side scale. The total
number of electrons : dashed curve and right hand side scale.

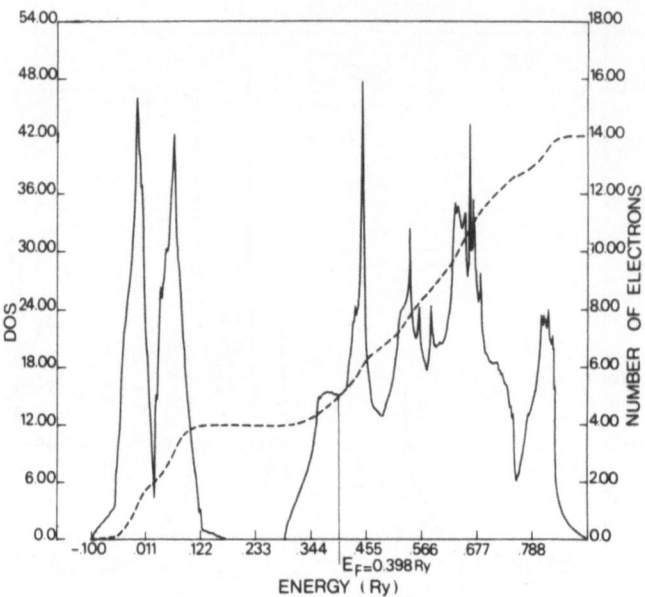

Fig. 3. The total DOS (states of both spin per Rydberg unit cell) of LaH₂ full line curve and left hand side scale. The total number of electrons : dashed curve and right hand side scale.

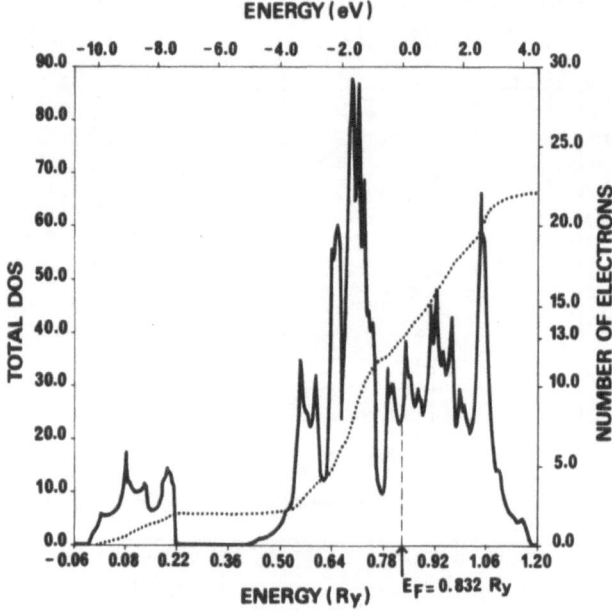

Fig. 4. The total DOS (states of both spin per Rydberg FeTiH) of β-FeTiH full line curve and left hand side scale. The total number of electrons : dotted curve and right hand side scale.

interaction. From the results listed in Table I we notice that n_p^H is larger than n_s^H in the energy range spanned by the metal d states ; this is not surprising since the metal d states having a p symmetry at the H site are not affected by the H potential and thus remain practically unaltered from the pure metal to the hydride.

For hydrides of simple metals, n_s^H becomes large ; this is the case particularly of AlH where a metal-hydrogen antibonding band shown in Fig. 6, crosses the Fermi level and gives,as it can be seen in Table I, large values of n_s^H at E_F.

2. Variation of the optical phonon contribution

No theoretical work has yet been performed on the vibrational properties of metal hydrides. From a review of the experimental results[15] it appears that the value of the optic mode frequencies is governed by several factors : the metal-hydrogen distances, the spatial extension of the metal-d orbitals as well as their orienta- tion which depends upon the symmetry of the interstice occupied by the H atom, certainly plays a role in determining the value of the force constants. Surprisingly, all the dihydrides of the TM series have optic phonons of comparable frequencies, an observation which suggests that the optic modes are not much influenced by the value of the DOS at E_F. For the dihydrides of the beginning of the TM series, the optic modes are found to be rather hard ($\hbar\omega_{opt} \simeq 140$ meV for NbH_2 ; $\hbar\omega_{opt} = 103$ meV for LaH_2 ; $\theta_E^{ZrH_2} \sim 1550K$). For the intermetallic hydri- des $FeTiH_x$, neutron scattering data[16] as well as the analysis of high temperature specific heat measurements[17] lead to $\hbar\omega_{opt} \simeq 100$ meV for $\beta-FeTiH$ and $\hbar\omega_{opt} \simeq 77$ meV for $\gamma-FeTiH_2$.

In the case of PdH, where H occupies octahedral interstices, the optic mode frequencies are very low[7], $\hbar\omega_{opt} \sim 56$ meV. This factor is particularly favorable for superconductivity. The optic modes of H in NiH appear to be much harder than in PdH ; from the study of the temperature dependence of the resistivity which shows the onset of the optical phonon scattering, McLaclan et al. [18] showed that the Einstein temperature is a factor of 1.8 larger in NiH than in PdH.

3. Variation of λ_H

As we can see in Table II, the values of λ_H obtained for the TM dihydrides for LaH_2 and $FeTiH_x$ are found to be small for two reasons : η_H is small due to the position occupied by E_F in these hydrides and large values of the optic phonon frequencies further contribute to the reduction of λ_H.

For NiH we obtain a smaller value of the electron-optic phonon coupling than in PdH due to the difference in the Fermi level posi- tion and to a drastic hardening of the optic phonon.

For AlH, we do not have experimental data concerning $\hbar\omega_{opt}$. However, the large values obtained for the electronic contribution make these compounds good candidates for superconductivity. This mechanism could explain why T_c of H implanted Al samples ($T_c \sim 6.6K$) is larger[5] than for pure Al (Tc $\sim 2.2K$).

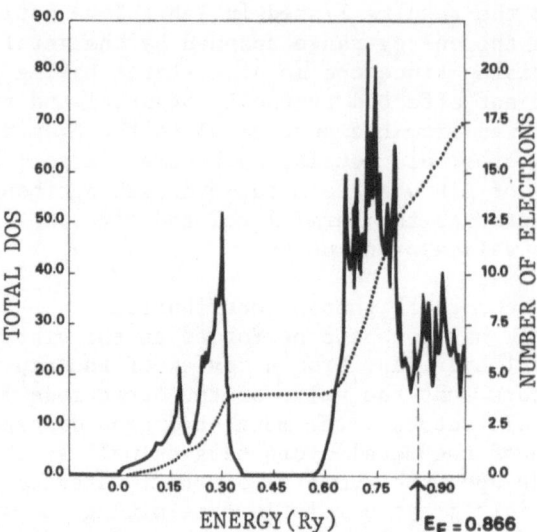

Fig. 5. The total DOS (states of both spin per Rydberg FeTiH$_2$) of
FeTiH$_2$ full line curve and left and side scale. The total
number of electrons : dotted curve and right hand side scale.

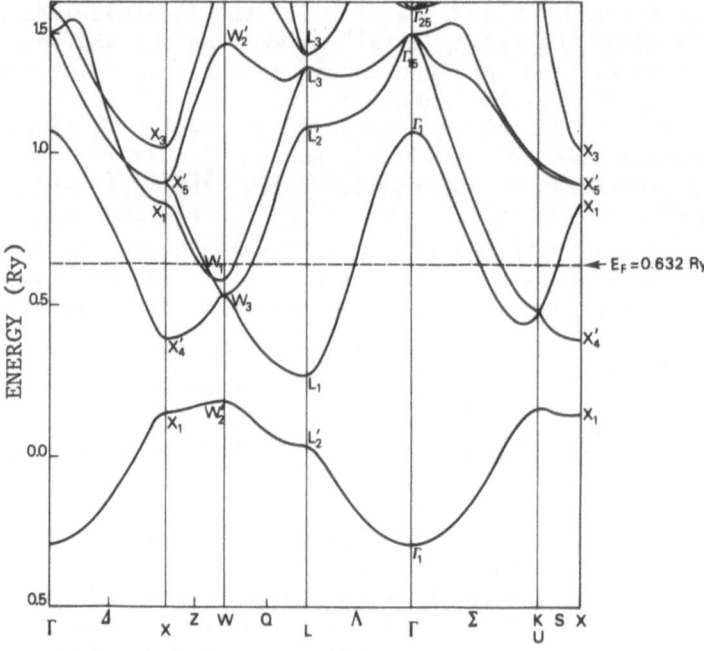

Fig. 6. The energy bands of AℓH (NaCℓ structure) in several high
symmetry directions of the fcc Brillouin zone. The second
which crosses the Fermi level contains numerous M-H
antibonding states.

ELECTRON-ACOUSTIC PHONON COUPLING λ_M

   For the TM hydrides, Table I shows that the d phase shifts
increase with the filling of the metal d bands of the hydrides. Thus,
δ_2^{Metal} is small for hydrides whose Fermi level falls at the bottom
of the d bands (like LaH$_2$), it increases and reaches a resonance
$\delta_2^{Metal} \sim \Pi/2$ in the middle of the d bands ; it is large and close to
the value of Π for the hydrides with filled d bands. Since δ_2^{Metal}
varies considerably with the filling of the d bands, we expect the
relative importance of the p-d and d-f scattering mechanisms to vary
considerably in a series; the latter mechanism becomes overwhelming
for hydrides of the right of the Periodic Table, while the p-d
mechanism provides also a non negligible contribution to η_{Metal} for
the early TM hydrides. In contrast to the TM hydrides, only the s and
p phase shifts at the Al site are large at E_F. Nevertheless, the s-p
scattering term will be small at the Al site, due to a near cancellation of the phase shift dependent term in Eq.(2). The d-phase shift
is, as expected, small for simple metal-hydrides.
   We would like to emphasize here that for the TM hydrides, the f
contribution at one site arises from the tails of the d states at
neighboring sites ; thus the d-f mechanism corresponds, as discussed
in the framework of the tight binding representation[19] to the d-d
interaction.
   We can see, from the results listed in Table II, that the values
of η_{Metal} obtained for the hydrides are always smaller than the
corresponding values for pure metals. We shall discuss here the cases
of NbH$_2$ and LaH$_2$ since the corresponding pure metals are good
superconductors while the dihydrides are not superconducting down
to 1K.

1. NbH$_2$
   In order to understand the role of the structural change in the
metal lattice (from bcc in the pure metal to fcc in the dihydride)
which is also accompanied by a substantial lattice expansion, we calculated η_{Nb} for the fcc NbH$_o$ phase which corresponds to the NbH$_2$
lattice where the H atoms have been removed. Although the DOS at E_F
of NbH$_o$ is larger than in the bcc phase, the value of η_{Nb} decreases
by ~ 50 %. This drastic reduction can be ascribed in part to the
lattice expansion effect which leads to an increase of the Nb-Nb
nearest neighbor distance from 5.369 a.u. in the bcc phase to 6.088
a.u. in the fcc lattice and thus reduces the strength of the d - d
interaction. This effect is also accompanied by the structural change
which induces a different orientational dependence of the interacting
lobes of the neighboring d orbitals.
   A further but small decrease is observed from $\eta_{Nb} = 3.7$ eV/$\overset{\circ}{A}^2$
in the fcc lattice to $\eta_{Nb} = 2.97$ eV/$\overset{\circ}{A}^2$ in NbH$_2$, due to the further
change in the electronic structure induced by the dihydride formation
namely, the deformation of the metal - d bands and the change in the
Fermi level position due to the additional electrons brought by the
H atoms.

Neutron scattering data for stoichiometric NbH_2 are not available at the present time, however it has been shown[20] that the acoustic modes of Nb harden with hydrogen uptake : the low frequency anomalous phonons of the pure bcc metal which certainly contribute to the large value of λ disappear ; the anomalously low slope of the acoustic branch in the [100] direction becomes larger. Thus, the large decrease of n_{Nb} from the pure metal to the dihydride does not appear to be compensated by an increase in the phonon contribution. We do not give an estimate of $M<\omega^2>$ in Table II since neither neutron scattering nor values θ_{Debye} are available for stoichiometric NbH_2. Nevertheless, as mentioned previously, the experimental data point to an increase of the phonon contribution ; this fact in conjunction with the very low value of n_{Metal} should lead to very small value of λ_{Metal} and thus to the disappearence of superconductivity due to a weakening of the electron-acoustic phonon coupling in NbH_2.

2. LaH_2

LaH_2 is not superconducting[21] above 1K. As shown in Table II, the value of n_{La} is 0.753 eV/$\mathring{A}^2$; it is much smaller than that obtained for the pure fcc metal n_{La} = 2.62 eV/$\mathring{A}^2$ using similar approximation[22]. This decrease of n_{La} is due to (i) the reduction of the DOS at E_F from 13.74 states/per rydberg spin unit cell to 7.55 states/per rydberg spin unit cell. This is due to the deformation in the dihydride of the lower portion of the metal d band and the apparent depopulation of the d states in favor of low-lying metal-hydrogen states in the dihydride (ii) the 6.7 % increase in the lattice constant which leads to a change in the metal d - d interaction. A similar effect is observed in the pure metal since n_{La} increases under pressure (iii) the low value of δ_2, the Fermi level being to the left of the d resonance.
The drastic decrease in n_{La} from the pure metal to the dihydride is further accompanied by a hardening of the acoustic phonon frequencies. The low temperature specific heat data show a large increase of θ_D from 140 K in pure La[23] to 243K in $LaH_{2.03}$[24]. To give an estimate of the phonon contribution we used $M_{La}<\omega^2>_{La} \simeq \frac{1}{2} M_{La} \theta_D^2$. With this estimate, we obtain λ_{La} = 0.103 in the LaH_2 compared to λ_{La} = 1.42 in the pure metal. Thus, our results show that the change in the electronic structure upon formation of a dihydride leads to a small electron-acoustic phonon coupling which is not favorable to superconductivity in LaH_2.

POSSIBILITY OF HIGH T_c SUPERCONDUCTORS

1. Can the electron acoustic phonon coupling be large ? The electronic structure studies have shown that for a fixed hydrogen concentration, the metal d states of two hydrides of the same TM series crystalizing within a given structure have similar metal d bands; thus, one can obtain reasonable information by shifting the Fermi level according to the number of electrons to be accomodated. We have thus studied the variation of n_{Metal} as a function of energy

for dihydrides with the CaF_2 structure and monohydrides with a NaCl
structure. The corresponding results are plotted in Figs 7 and 8.
These results give an indication of the possible effect of alloying
on the metal matrix. From Figs 7 and 8 we can see that fairly large
values of η could be obtained as the Fermi level sweeps through the
metal d bands. This could be the case for example, for alloys of NbH_2
(for example with Mo), corresponding to 7.6 valence electrons. Our
investigation of the NaCl phase was motivated by the experimental
results of Robbins et al.[25] on hydrides of Nb-Pd, Nb-Pd-Mo and Nb-Ru
which indicate the existence of an fcc phase for e/at.>5.8; and sizeable values of T_c~5.5K were reached for alloys such as $Nb_{0.8} Ru_{0.2}$
$H_{0.5}$. Fig. 7 indicates that a large value of η_{Metal} could be obtained
for a mono hydride with e/at.~7.26. Of course, one could hope to
modify also the electron-acoustic phonon coupling by varying the H
concentration. This is beyond the scope of the present investigation
since, in that case, the rigid band model is known to be erroneous
except maybe for slight variations of the H concentration around
the stoichiometric composition.

2. Can the electron-optical phonon coupling be large ?
Besides the cases previously discussed of the TM with filled d bands
such as PdH and its alloys with noble metals and the case of the
simple metal hydrides prepared by ion implantation, we can surmise
that high T_c superconductors could possible be obtained for dihydrides of the middle of the TM series if these hydrides, such as
MoH_2..., could be stabilized. Indeed, electronic structure
studies[11,12] have shown that if one considers a series of TM dihydrides, as one goes from left to right of the periodic Table, the
antibonding H-H states, which for stable dihydrides such as LaH_2,
TiH_2... are found at low energy, below the metal d bands, are
destabilized. This is due to conjugated effects of the decrease in
the H-H distance and of differences in the hybridization with metal
states as Z_{metal} increases. For Z_{Metal}^{val}>6, the antibonding H-H states
cross the Fermi level, a situation unfavorable for the stability of
the dihydride but which could lead to sizeable values of n_g^H and thus,
possibly to a large value of the electron-optical phonon coupling if
these compounds could be prepared, by implantation techniques for
example.

CONCLUSIONS

Our theoretical results of the electron phonon coupling in
stoichiometric metal hydrides are in satisfactory agreement with
available experimental data. We have shown that the difference in
the electron-optical phonon coupling between PdH and NiH explain the
experimental data of McLachlan et al.[18] who did not observe superconductivity down to 1K in NiH. We have also shown that hydrogen
destroys the superconductivity of Nb and La in agreement with the
experimental data on the dihydrides[26,21]. This is due to a large
decrease of the electronic contribution at the metal site and a

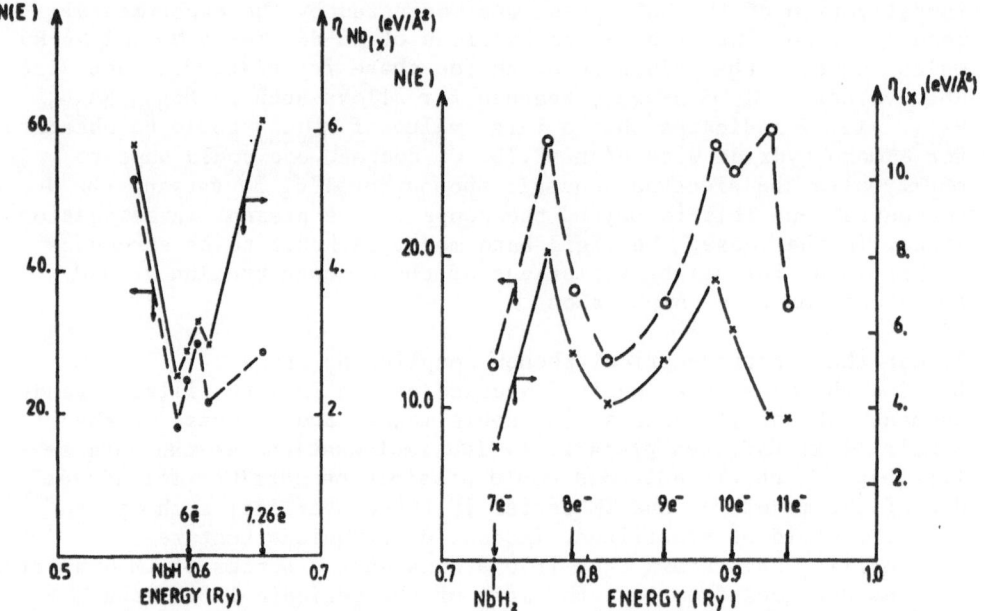

Fig. 7. The electronic contribution η at the metal site in MH [(x) and full line, right hand side scale]. The total DOS [(o) and dashed line, left hand side scale] units are states of both spin per rydberg unit cell. The number of electrons is indicated by the arrows on the horizontal scale.

Fig. 8. The electronic contribution η at the metal site in MH_2 [(x) and full line, right hand side scale]. The total DOS [(o) and dashed line, left hand side scale] units are states of both spin per rydberg unit cell. The number of electrons is indicated on the horizontal scale.

hardening of the acoustic modes ; this decrease of λ_{Metal} is not compensated by the contribution of the electron-optical phonon coupling which is very small for these hydrides. Our results on the simple metal hydrides such as the hydrogen implanted Al samples can provide an explanation for the increase of T_c in these compounds. We indicate also the possible existence of high T_c hydrides by alloying effects on the metallic matrix.

REFERENCES

1. C.B. Satterthwaite and I.L. Toepke, Phys. Rev. Lett. 25, 741 (1970)
2. T. Sckoskiewicz, Phys Status Solidi A11, K123 (1972) ;
   B. Stritzker and W. Bückel, J. Phys. 257, 1 (1972).
3. B. Stritzker and H. Wühl in Hydrogen in Metals I, Vol. 23, edited by G. Alfeld and J. Völkl, Springer, Berlin (1978) p 243, and references therein.
4. W.L. McMillan, Phys. Rev. 167, 331 (1968).
5. A.M. Lamoise, J. Chaumont, F. Meunier and H. Bernas, J. Phys. Lett. 36, L271 (1975).
6. G.D. Gaspari and B.L. Gyorffy, Phys. Rev. Lett. 29, 801 (1972).
7. J.M. Rowe, J.J. Rush, H.G. Smith, M. Mostoller and H. E. Flotow, Phys. Rev. Lett. 33, 1297 (1974).
8. A. Eichler ; H. Wühl and B. Stritzker, Solid State Commun. 17, 213 (1975).
9. D.S. Mc Lachlan, R. Mailfert, J.P. Burger and B. Souffaché, Solid State Commun. 17, 281 (1975).
10. D.A. Papaconstantopoulos and B.M. Klein, Phys. Rev. Lett. 35, 110 (1975).
11. A.C. Switendick in Hydrogen in Metals I, Vol. 23, edited by G. Alefeld and J. Völkl, Springer, Berlin (1973) chap 5 and references there in.
12. M. Gupta in Proceedings of the NATO Summer Institute on Metal Hydrides 1980, edited by G. Bambakidis, Plenum, in print and references there in.
13. M. Gupta and J.P. Burger, J. Phys. F : Metal Phys.10, 2649 (1980)
14. D.A. Papaconstantopoulos, E.N. Economou, B.M. Klein and L.L. Boyer, Phys. Rev. B 20, 177 (1979).
15. T. Springer in Hydrogen in Metals I, Vol. 23, edited by G. Alefeld and J. Völkl, Springer, Berlin (1973) and references there in.
16. J. Eckert, J.A. Goldstone, and D. Richter, J. Phys. F : Metal Phys. 11, L101 (1981).
17. H. Wenzl and S. Pietz Solid State Comm. 33, 1163 (1980).
18. D.S. McLachlan, I. Papadopoulos and T.B. Doyle, J. Physique Coll 6, 430 (1978).
19. S. Barisic, J. Labbé and J. Friedel, Phys. Rev. Lett. 25, 919 (1970).
20. J.M. Rowe, N. Vagelatos, J.J. Rush and H.E. Flotow, Phys. Rew B 12, 2959 (1975).

21. M.F. Merriam and D.S. Schreiber, J. Phys. Chem. Solids 24, 1375
    (1963).
22. W.E. Picket, A.J. Freeman and D.D. Koelling, Phys. Rev B 22,
    2695 (1980).
23. D.L. Johnson and D.K.Finnemore, Phys. Rev. 158, 376 (1967).
24. Z. Bieganski, D. Gonzalez-Alvarez and F.W. Klaaysen, Physica
    37, 153 (1967).
25. C.G. Robbins and J. Muller, J. Less. Common. Met. 42, 19 (1975) ;
    C.G.Robbins ,M. Ishikawa, A. Treyvand and J. Muller, Solid State
    Comm. 17, 903 (1975).
26. C.B. Satterthwaite and D.T. Peterson, J. Less. Common Met. 26,
    361 (1972).
27. W.H. Butler, Phys. B 15, 5267 (1977).
28. D.A. Papaconstantopoulos, L. L. Boyer, B.M. Klein, A.R. Williams,
    V.L. Morruzzi and J.F. Janak, Phys. Rev. B 15, 4221 (1977).

EFFECT OF STRUCTURAL PHASE TRANSITIONS ON THE SUPERCONDUCTIVITY OF

$PdH_x(D_x)$

R. W. Standley* and C. B. Satterthwaite[†]

Physics Department and Materials Research Laboratory
University of Illinois at Urbana-Champaign
Urbana, IL 61801

We have examined how the order-disorder structural transitions associated with the "50 K anomaly" in palladium hydride (deuteride) affect the superconducting transition temperature. In $PdD_{0.817}$ and $PdH_{0.837}$ samples, the long-range ordering of vacancies in the H(D) sublattice leads to a 7-9% reduction in T_c. While no long range ordering occurs in a $PdD_{0.742}$ sample, the enhancement of short-range order results in a 7.5% increase in T_c. In all cases the changes in T_c are reversible, the transition temperatures reverting to their original values after the samples are again disordered.

INTRODUCTION

Palladium hydride (deuteride), $PdH_x(D_x)$, provides an interesting system in which to investigate the interplay between superconductivity and lattice instability. The dominant contribution to the electron-phonon coupling in $PdH_x(D_x)$ is from the H(D) atom vibrations, and the H(D) atoms are also involved in structural phase transitions at low temperatures. The question of how these phase transitions affect the superconductivity thus naturally arises.

Recent neutron diffraction work[1-3] has demonstrated that the "50 K anomaly" in $PdH_x(D_x)$ is due to ordering of vacancies in the partially occupied H(D) sublattice. When PdD_x samples, $0.7 \leq x \leq 0.8$ are cooled below 110K diffuse scattering intensity appears around $(1, \frac{1}{2}, 0)$[2], which has been attributed to the formation of short-

_____

*present address: Amoco Research Center, P. O. Box 400,
                   Naperville, Illinois 60566
[†]present address: Physics Dept., Virginia Commonwealth University
                   Richmond, VA. 23284

range order (SRO) consisting of a mosaic of microdomains of Ni_nMo
-type structures (n=2, 3, 4). When samples in the range $0.7 \lesssim x \lesssim 0.75$
are cooled from 110 K to 50K this diffuse intensity increases by a
factor of five. In this composition range PdD_x is superconducting,
while the superconducting T_c of PdH_x becomes vanishingly small[5].
For $0.75 \lesssim x \lesssim 0.8$, cooling below 80K results in a decrease in the diffuse intensity around $(1,\frac{1}{2},0)$ and the growth of a sharp superlattice
peak at $(4/5,2/5,0)$ as the SRO is replaced by a long-range ordered
(LRO) Ni_4Mo-type structure[1,2]. At these compositions, both PdH_x and
PdD_x are superconducting. For compositions $x \lesssim 0.7$, other ordered
structures appear[3] but superconductivity no longer occurs.[5]

EXPERIMENTAL

Samples for this work were prepared from Pd foils, approximately
25.4mm x 5.5mm x 0.05mm, which were loaded with H(D) electrolytically,
as described elsewhere.[5] H(D) concentrations were determined by
thermal decomposition of the samples and volumetric measurement of
the $H_2(D_2)$ evolved. The uncertainty in the resulting H(D)/Pd ratio,
x, is 0.005. Sample resistances were determined by 4-terminal d.c.
measurements, and superconducting transitions were measured by an
a.c. mutual inductance technique.

Each sample was mounted in a He^3 cryostat and quickly cooled
to 100K to prevent H(D) loss. After 48 hours at 100K, the samples
were quenched to 4.2K, and the superconducting T_c's of the disordered
states were determined. They were reheated to 100K for another 48
hours and then cooled to 72K to initiate the ordering. The temperature was changed periodically to maximize the decrease in sample
resistances; the actual thermal histories of each sample are given
at the tops of figures 1, 2, 4, 6 and 7. In the $PdD_{0.817}$ and
$PdH_{0.837}$ cases, the samples were quenched to 4.2K and T_c's measured
at intermediate points in the ordering process (denoted by "T_c" on
figures 1 and 4) and were then returned to their original temperatures to continue the ordering. After about 650 hours of annealing,
all samples were quenched to 4.2K and the ordered state T_c's were
measured. The samples were then heated to 100K for 48 hours and
quenched to 4.2K to remeasure the disordered state T_c's. The samples
were then removed from the cryostat and analyzed for H(D) content.

RESULTS

The $PdD_{0.817}$ sample had a disordered state T_c (at 50% of maximum
diamagnetism) of 2.211K. During the subsequent low temperature annealing, it displayed resistance changes typical of samples forming
the Ni_4Mo-type LRO (Fig 1). After the resistance had decreased by
$\sim$20%, the sample was quenched and T_c was found to have been depressed
by about 7%, to 2.056K (Fig. 3). The annealing was then continued
and at a slightly higher temperature the resistance fell by an additional $\sim$20% (Fig. 2). It was found that this additional ordering had

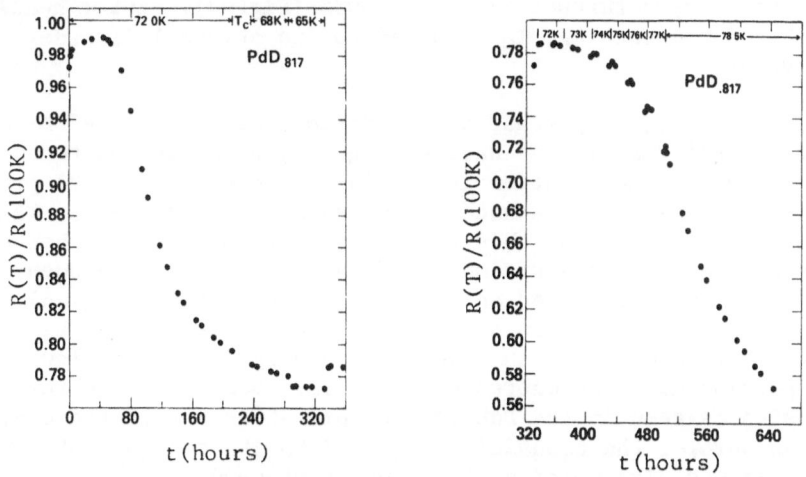

Figures 1 and 2.   Resistance ratio of PdD$_{0.817}$ during ordering.
Thermal history shown at top of graph.

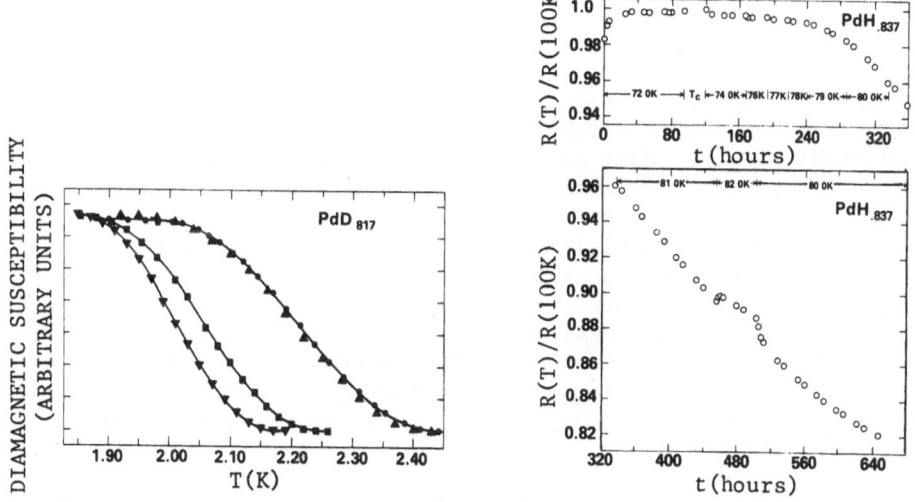

Fig. 3. Superconducting transitions
        of PdH$_{0.817}$

    (●) Initial disordered T$_c$
    (■) Ordered T$_c$, R/R(100K)=0.80
    (▼) Ordered T$_c$, R/R(100K)=0.57
    (▲) Final disordered T$_c$.

Fig. 4. Resistance ratio of
        PdH$_{0.873}$ during
        ordering.

depressed T_c by a further $\sim$2%, to 2.008K (Fig. 3). After heating
to 100K and requenching, T_c returned to its original disordered
state value.

The $PdH_{0.837}$ sample had a disordered state T_c of 1.436K and a
slight "double transition" nature (Fig. 5), probably due to an
inhomogeneous H distribution in the foil. Annealing at 72K produced
a small increase in resistance but no subsequent decrease. When the
sample was quenched from this temperature, no measurable change in
the superconducting transition was found (Fig. 5). Subsequent
annealing at $\sim$80K produced an 18% drop in resistance (Fig. 4), after
which the sample was again quenched and T_c measured. The high temperature portion of the transition was unchanged, but the low temperature portion had been depressed by $\sim$7.5%, about the same relative
depression as seen in the $PdD_{0.817}$ sample after a resistance drop of
20%. As before, the transition reverted to its original shape after
the sample was disordered at 100K and requenched.

The $PdD_{0.742}$ sample had a disordered state T_c=0.401K. During
annealing, the resistance behaved in a complex way which we have
found to be typical of samples in this concentration range as they
form SRO (Fig. 6,7). After $\sim$680 hours of annealing, the sample
resistance had decreased by 3%, and upon quenching, T_c was found to
have increased by 7.5%, to 0.431K)Fig. 8). T_c reverted to 0.401K
after the sample had been heated to 100K and requenched.

DISCUSSION

The results of our experiments, taken together with the neutron
diffraction results, are consistent with the local structural excitation (LSE) model of Ngai and Reinecke[6] This model proposes that
the ground state of the lattice is separated by a small energy
difference from other local configurations, and that electrons and
phonons can excite transitions (LSE's) between these states. When
the local states become nearly degenerate, the lattice undergoes a
structural distortion which splits the configurational states,
suppressing the LSE's. Ngai and Reinecke have shown that the LSE-
phonon interaction can account for the unusual lattice dynamics of
$PdH_x(D_x)$ and that LSE's can also increase T_c via the LSE-induced
softening of the phonon modes, and an electron pairing mediated
directly by the LSE's.

We propose that the "local configurations" of the LSE model are
identical with or intimately related to the Ni_nMo-type structures
proposed by Blaschko, et. al.[4] The decrease in T_c seen in the
$PdD_{0.817}$ and $PdH_{0.837}$ upon formation of LRO may then be attributed
to the suppression of LSE's between these states as they are split
and the Ni_4Mo-type state becomes energetically favored. This
suppression of LSE's is also consistent with the observed increase
in optic mode frequencies in the ordered state of $PdD_{0.78}$[7]. The

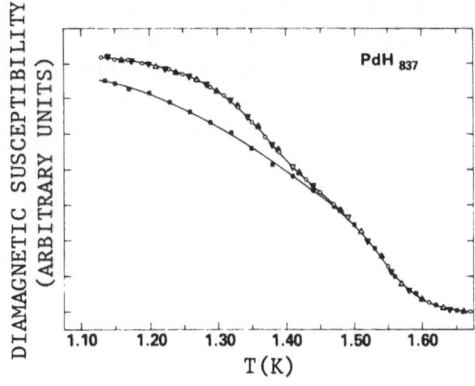

Fig. 5. Superconducting transi-
        tions of PdH$_{0.837}$
   (O) Initial disordered T$_c$
   (▲) Ordered T$_c$, R/R(100K)=1.0
   (■) Ordered T$_c$, R/R(100K)=0.82
   (▼) Final disordered T$_c$.

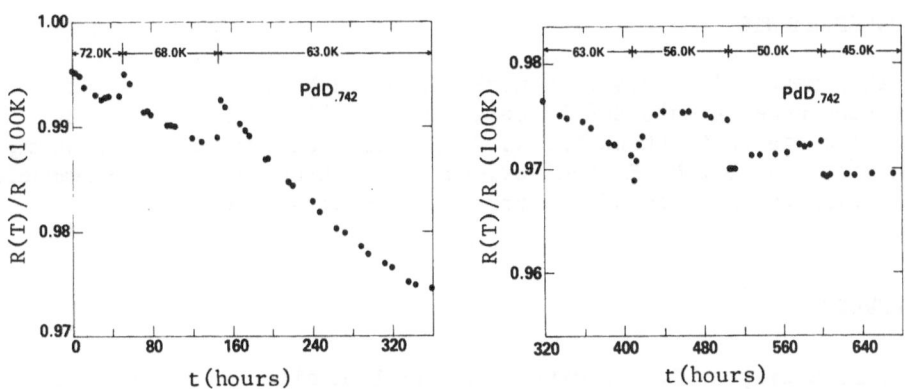

Figures 6 and 7. Resistance ratio of PdH$_{0.742}$ during ordering.

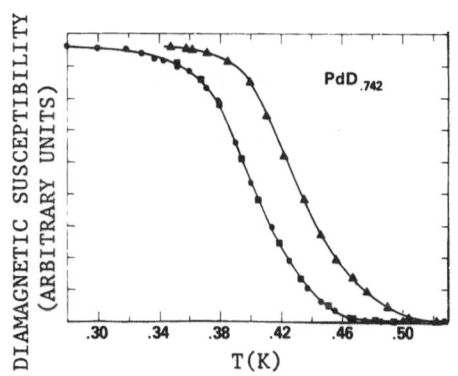

Fig. 8. Superconducting transitions
        of PdH$_{0.742}$
   (●) Initial disordered T$_c$
   (▲) Ordered T$_c$, R/R(100K)=0.97
   (■) Final disordered T$_c$.

increase in SRO observed when samples of lower H(D) content are annealed may promote LSE's between the Ni_nMo-type states, which would increase T_c, as we observe in our $PdD_{0.742}$ sample. On the basis of the LSE model, we would also expect a softening of the optic mode frequencies with increasing SRO, but this has not been experimentally verified.

If the ordered phases of PdH_x and PdD_x are similar, then we may draw an additional conclusion from our results. The splitting of the $PdH_{0.837}$ transition after annealing implies that the low T_c (relatively H-poor) part of the sample developed LRO, while the high T_c (H-rich) part did not. Evidently, this composition, x=0.837, falls in a disordered β + Ni_4Mo-type two phase region of the low temperature phase diagram. Since the $PdD_{0.817}$ transition was not split, the upper boundary of the Ni_4Mo-type phase must lie between x=0.817 and x=0.837, and is separated from the disordered β-phase near x=1.0 by a two-phase region, as required by the Gibbs phase rule.

ACKNOWLEDGEMENTS

This material is based on work supported by the National Science Foundation under grants DMR 77-08463 and DMR 77-23999. One of us (RWS) also wishes to thank Dr. A. C. Anderson and Dr. D. M. Ginsberg for advice, and Dr. M. H. Mueller, Dr. T. O. Brun and Dr. R. Klemencic for sharing the results of neutron diffraction work with us.

REFERENCES

1   T. E. Ellis, C. B. Satterthwaite, M. H. Mueller and T. O. Brun, Phys. Rev. Lett. 42, 456 (1978)

2   O. Blaschko, R. Klemencic, P. Weinzierl, O. J. Eder and P. von Blanckenhagen, Acta Crystallog. A 36, 605 (1980) and references therein

3   I. S. Anderson, C. J. Carlile and D. K. Ross, J. Phys. C 11, L381 (1978)

4   O. Blaschko, P. Fratzl and R. Klemencic, Phys. Rev. B 24, 277 (1981) and Phys. Rev. B 24, 6486 (1981)

5   R. W. Standley, M. Steinback and C. B. Satterthwaite, Solid State Commun. 31, 801 (1979)

6   K. L. Ngai and T. L. Reinecke, J. Phys. Chem. Solids 39, 793, (1978) also Phys. Rev. B 16, 1077 (1977)

7   O. Blaschko, R. KLemencic, P. Weinzierl and L. Pintschovius, Phys. Rev. B 24, 1552 (1981)

SUPERCONDUCTIVITY AND HYDROGEN DEPTH PROFILES

IN ELECTROLYTICALLY CHARGED Cu-IMPLANTED Pd

W.J. Standish, A. Leiberich, W. Scholz and C.G. Homan[*]

Department of Physics, SUNY/Albany
Albany, NY 12222
*ARRADCOM, Benet Weapons Laboratory
Watervliet, NY 12189

Superconductivity has been observed in Pd foils implanted with Cu at room temperature and subsequently charged electrolytically with H at dry ice temperature. Superconducting transition temperatures, T_c, with onsets as high as 16.6 K have been observed. Hydrogen concentration profiles have been measured to a depth of $\sim$0.2 μm for similar samples at 77 K using the 6.4 MeV resonance of the nuclear reaction $^{15}N(^{1}H,\alpha\gamma)^{12}C$. A step-wise warmup procedure between 77 K and 273 K produces considerable variations both in the superconducting transition curves and in the H depth profiles. The observed concentration changes evolve from an initial ratio H/Pd$\sim$3 and show a strong positive correlation with T_c.

INTRODUCTION

The Pd-H system has been studied extensively.[1] Subsequent to the discovery[2] of superconductivity in this system, superconductivity has also been reported[3] for Pd-M-H(D), where M is Cu, Ag or Au. These alloys show T_c's which are substantially higher than the 9 K seen in PdH, a maximum value $T_{c,max}$= 16.6 K having been reported for an alloy with the composition H/Pd$_{55}$Cu$_{45}$$\sim$0.7.

Several methods have been used to achieve the high H concentrations required for superconductivity in Pd and its alloys,[4] but until now H concentrations have generally been determined[1] from bulk measurements such as weight changes, H desorption, and resistivity, or, in the case of H implantation, inferred from the implant dose and theoretical range-energy relations. Recently, however, electro-

lytic charging at -78° C has been used successfully for an alloy
made by implantation of Cu into a Pd substrate.[5] In this paper we
present additional results and report for the first time on measurements of the H concentration and depth distribution in H-charged
Cu-implanted Pd samples by nuclear reaction analysis.

EXPERIMENTAL

Sample Preparation and Characterization

    Singly charged Cu ions were implanted to a dose of.
8×10^{16} ions/cm^2 at 100 keV into 25-μm thick Pd foils. The foils
had a nominal purity of 99.99%; analysis by flame spectroscopy
showed no metallic impurities with concentrations above 10 ppm.

    The range and range straggling of 100 keV Cu ions in Pd are
expected to be $<x>$ = 0.023 μm and σ = 0.013 μm, respectively.[6]
Because of sputtering, the implant profile will be given approximately by a complementary error function erfc $[(x-<x>)/\sqrt{2}\sigma]$, where
x is the depth into the sample.[5] The maximum Cu concentration
(atom fraction) of 0.23±0.04 at the surface is determined by the
inverse of the Cu self-sputtering yield for which we have adopted[5]
the value S_{Cu}=4.3±0.8. The calculated total remaining Cu implant
dose is $(3.7±0.7) \times 10^{16}$ atoms/cm^2. Independent analysis of the
implant dose using Rutherford backscattering (RBS) yielded
$(2.5±0.5) \times 10^{16}$ atoms/cm^2, a reasonable agreement in view of the
uncertainties of the sputtering yields and range-straggling values.

    In order to make resistance measurements, four leads were spot-
welded onto the Cu-implanted foils. Samples to be examined for
superconductivity were epoxied onto a hollow Plexiglas tube which
contained the electrical leads and a thermocouple. Samples to be
profiled were epoxied onto a hollow aluminum cylinder attached to
the bottom of a LN$_2$ dewar. A thermocouple, and resistors for heating the sample block, were also attached to the cylinder. In both
cases charging with H was accomplished at a temperature of 200 K
using a solution of 38% HCl in methanol. Charged samples were
immediately placed in LN$_2$ after electrolysis, and did not exceed
77 K until deliberately warmed up during the measurement process.

    In order to be assured of a high H concentration after electro-
lysis, the resistance ratio R/R_0 of H-charged-sample resistance R
to uncharged-sample resistance R_0 was monitored continuously during
the electrolysis. Pure Pd samples prepared in this fashion typically show a $T_c \approx 9$ K corresponding to an average H/Pd$\approx$1. Thus one
can be confident that the samples used for the T_c determinations as
well as those used in the profiling measurements were prepared in an
identical state and with an initial average H/Pd$\approx$1.

Measurement of T_c

Measurement of the superconducting transition curves has previously been described.[5] To reiterate, a four-probe dc resistance
technique was used, employing currents from 40 mA to 2 A. A calibrated thermocouple was used to measure stable temperatures. T_c
was defined as the average of the temperatures at 10% and 90% of
the normal state resistance. Following initial measurement of the
superconducting transition curve, samples were warmed up to increasingly higher temperatures and the transition curves remeasured.
The warming procedure involved withdrawing the sample from the LHe
bath until it reached a given temperature, then quickly quenching it.
Samples annealed at 77 K were simply transferred to a dewar of LN_2.

Hydrogen Profiling

Hydrogen profiling by nuclear reaction analysis makes use of
isolated resonances in the reaction cross section. In our experi-
ments, the reaction $^{15}N(^{1}H,\alpha\gamma)^{12}C$ has been used.[7] This reaction has
a narrow (FWHM $\approx$ 6 keV) resonance at 6.4 MeV in the laboratory
frame. The yield of characteristic 4.43 MeV γ-rays is measured with
a NaI detector. For concentration distributions that do not appreciably change over the half-width of the Lorentzian-shaped resonance,
this yield is given by

$$Y(E_B) = kQn(x)/(dE/dx). \qquad (1)$$

In this equation Q is the number of incident ^{15}N ions, n(x) is the
concentration at depth x, (dE/dx) is the stopping power of the sample
material, and k is a constant characteristic of the experimental
setup (including resonance parameters, detection efficiency and solid
angle of the detector, and absorption of γ-rays between sample and
detector). The calibration constant k has been determined experimentally for our arrangement using several targets of known H concentration; e.g., ice. The experimental calibration constant is in
good agreement with a value calculated from resonance parameters
and the experimental geometry.

In Eq. (1) the ^{15}N beam energy E_B is related to the depth x by
$E_R = E_B - x (dE/dx)$, where E_R is the resonance energy. Thus, by
varying the bombarding energy, the H concentration with depth can be
determined, provided that (dE/dx) is known[8] for the sample material
and the calibration substance. The depth resolution depends on the
resonance width and ^{15}N beam energy spread through (dE/dx) and is
estimated to be $\sim$3 nm.

Hydrogen profiles were measured on different samples (prepared in a manner identical to that used in the T_c measurements)
following the same warmup procedure, using the sample-block heaters.

Comparability of the samples was insured by monitoring R/R_0 throughout all experiments. Profiles were taken at LN$_2$ temperatures to minimize the effects of beam heating on the samples. Typical beam currents used were ∿5 nA.

RESULTS

Superconducting Transitions

        Results for superconducting transition curves in Cu-implanted H-charged Pd have already been presented elsewhere;[5] a summary is included here for completeness. Measurements were done on a single sample, immediately after transfer to the LHe bath and also after warming the sample to increasingly higher temperatures. An increase in T_c from 11.1 K to 12.4 K is observed as the sample is warmed to 113 K. Subsequent annealing at increasingly higher temperatures reduces T_c until there is another increase, from 10.9 K to 12.4 K, upon warming to 193 K. Continuing the annealing at increasingly higher temperatures finally produces a monotonic decrease in T_c. The normal state resistance of the sample above the critical temperature remains fairly constant for warmup steps up to 213 K, indicating a small loss of H overall from the bulk up to this point.[2,5] Above 213 K, H is released from bulk as indicated by the increase in normal state resistance. After warmup to 273 K the onset of the superconducting transition is just barely visible at 4.2 K.

        In more recent experiments, the same general behavior of T_c with warming has been seen by us, with T_c reaching 13.6 K and with onsets as high as 16.6 K. In addition, annealing samples overnight in LN$_2$ has produced increases in T_c of samples which had previously been warmed to 113 K. These increases can be fairly large; e.g., from 10.4 K to 13.2 K in one particular sample.

Hydrogen Profiling Measurements

        The results of a series of profile measurements for one sample are shown in Fig. 1. Background count rates are small (<5%) and have already been subtracted. Immediately after electrolysis and even after a warmup to 113 K (Fig. 1a), the profile shows a small increase of counts for the implant region as compared with the bulk Pd foil (beyond 0.05 μm). This is due in part to an increased H concentration and in part to stopping power effects. In the bulk region, the atom ratio H/Pd∿3; a more accurate determination must await a more precise determination of the stopping powers of the sample and the calibration substance. Warming the sample to 143 K (Fig. 1b) produces a large reduction in H concentration in the implant region and a slight increase in the region just beyond the implant. The observed H profile closely traces the calculated implant profile. Further warming to 178 K (Fig. 1c) replenishes H

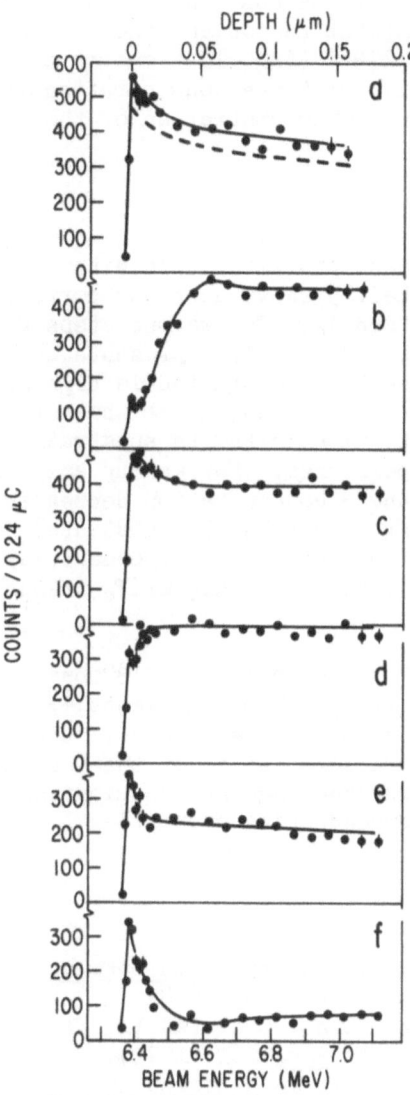

Fig. 1. H profiles measured at 77 K on a Cu-implanted Pd sample.
(a) Immediately after electrolysis (dashed line) and after
warmup to 113 K (solid line). Profiles labeled (b)-(f) were
measured after warmup to 143 K, 178 K, 203 K, 233 K, and
273 K, respectively. Depth scale assumes $(dE/dx) = 4$ MeV/μm.

in the implant and slightly reduces the concentration beyond the implant. A warmup to 203 K (Fig. 1d) affects only the surface and near-surface H, reducing it by about 1/3. Warming to 233 K (Fig. 1e) returns some H to the surface and considerably reduces the H concentration in both the implant region and the Pd substrate. Finally, warming the sample to 273 K (Fig. 1f) removes a large fraction of H from the bulk. At this point the concentration in the bulk appears to be close to the minimum concentration of the β-phase,[1] $H/Pd \sim 0.6$.

CONCLUSIONS

Comparison of these profiling data with the superconducting transition curves of Ref. 5 shows a strong correlation between H in the implant region and T_c. For warmup steps with temperatures up to 203 K, where $T_c > 9$ K and clearly represents superconductivity in the implant region, the H concentration is high at those warmup temperatures where T_c is high and low at those temperatures where T_c is low. The H concentration in the Pd substrate remains nearly constant at these temperatures. For warmup steps with increasingly higher temperatures, the H concentration decreases monotonically in both the Cu-implant region and the Pd substrate. The fact that both the H and Cu distributions are depth-dependent may account for the broadness observed in some of the higher T_c transition curves of Ref. 5.

The high concentration ratios $H/Pd \sim 3$ observed in this experiment are at variance with bulk measurements,[1,4] implying that they may not extend beyond a thin surface layer. Conceivably both the octahedral and the tetrahedral sites of the fcc lattice could be occupied by H in the surface region, contributing in different ways to H mobility and superconducting properties.

REFERENCES

1. For a recent review, see Hydrogen in Metals, edited by G. Alefeld and J. Volkl (Springer, Berlin, 1978).
2. T. Skoskiewicz, Phys. Stat. Sol. (a) 11, K123 (1972).
3. B. Stritzker, Z. Physik 268, 261 (1974).
4. B. Stritzker and H. Wuhl, in Hydrogen in Metals, edited by G. Alefeld and J. Volkl (Springer, Berlin, 1978).
5. A. Leiberich, W. Scholz, W.J. Standish, and C.G. Homan, Phys. Lett. 87A, 57 (1981).
6. G. Dearnaley, J.H. Freeman, R.S. Nelson, and J. Stephen, Ion Implantation (North-Holland, Amsterdam, 1973) Vol. 8, p. 766.
7. W.A. Lanford, H.P. Trautvetter, J.F. Ziegler, and J. Keller, Appl. Phys. Lett. 28, 566 (1976).
8. The Stopping Power and Ranges of Ions in Matter, edited by J.F. Ziegler (Pergamon, New York, 1980), Vol. 5.

SUPERCONDUCTING TUNNELING INTO Pd-Ni-D ALLOYS

P. Nedellec, L. Dumoulin and J.P. Burger

Laboratoire de Physique des Solides
Université de Paris-Sud - Bât. 510
91405 ORSAY (France)

ABSTRACT

We present experimental data concerning the tunneling density of states of two superconducting $Pd_{1-y}Ni_y D_x$ films with different Ni concentrations and different T_c. The data concern the acoustical and optical contribution to the electron-phonon coupling parameter λ, the electron-electron parameter μ^*, and the energy position of the optical mode. The results are discussed in light of the differences existing between the electronic and atomic structure of PdH and NiH.

INTRODUCTION

The discovery of superconductivity[1] in Palladium hydrides speeded up the interest of experimentalists and theorists in the electronic and atomic structure of such compounds and for their physical properties. Tunneling [2,3] and normal state electrical resistivity [4] measurements showed that there was in PdH_x a strong contribution of the hydrogen or optic vibrations to the superconducting state. Investigations on several other transition metal hydrides demonstrated that this is not always the case. The trends concerning the electron-optical phonon coupling for different hydrides have since been analyzed [5] and a good correlation with the experimental data could be obtained. Particularly interesting is the case of NiH which is isoelectronic to PdH, but which is not a superconductor. Progressively substituting Ni for Pd atoms leads to a strong decrease of T_c, which extrapolates to zero for a Ni concentration of about 15 %[1,6]. One

may ask if this is an electronic effect, due for instance, to changes
in the Fermi level density of states, or an atomic effect due to
changes in the vibration spectrum. The fact that pure Ni is a ferromagnet while pure Pd is a superparamagnet raises also the question if
some kind of residual magnetic fluctuation may not be responsible in
part for the absence of superconductivity in NiH and in concentrated
$Pd_{1-y}Ni_yH_x$ alloys despite all these compounds are considered as non
magnetic.

Here we will present experimental data concerning the tunnel
density of states of two superconducting $Pd_{1-y}Ni_yD_x$ with $y = 2,5$ %
(T_c . $7,7^{\circ}K$) and $y = 8$ %
($T_c = 7,7^{\circ}K$) and $y = 8$ % ($T_c = 3,3^{\circ}K$). Using an inversion program of
the Eliashberg equation we calculate $\alpha^2(\omega) F(\omega)$, $\lambda = 2\int \frac{\alpha^2(\omega) F(\omega)}{\omega}d\omega$
and μ^*. λ is the electron-phonon coupling parameter, $F(\omega)$ is the pho-
non density of states, $\alpha^2(\omega)$ is the electron-phonon coupling matrix
element and μ^* the repulsive electron-electron interaction. We are
also able to separate the acoustical and optical contribution to λ.

EXPERIMENTAL TECHNIQUE

The measurements of tunneling density of states always meets
two difficulties : the preparation of clean and reliable tunnel
junctions and the weakness of the phonon induced anomalies in the tunneling conductance (of order 10^{-3} -10^{-4}). These difficulties are
increased in our case because the superconducting electrode is a
Pd-Ni alloy which has to be hydrogenated (or deuterated) in order to
create the superconducting state : this adds several possible
sources of inhomogeneities. We will now describe briefly some of
the precautions we took for the preparation of the $Al/Al_2O_3/Pd$-Ni-D
junctions. The first electrode is evaporated on a glass substrate and
the tunneling barrier is formed by a glow-discharge oxydation of a few
minutes in dry atmosphere of pure oxygen at a pressure around 5×10^{-2}
torr. The useful area of the junction is limited by a window obtained
by evaporating four strips of SiO. The second Pd-Ni electrode is then
evaporated on top of the junction, the surface of this second
electrode being larger than the surface of the barrier in order to
avoid edge effects. Two ingots of Pd and Ni are evaporated
with two electron guns and the evaporation rates are regulated in
order to give the desired Ni concentration y. The mean concentration of Ni is measured on a test film using the fact that
these films are ferromagnetic : the corresponding ordering temperature, which is obtained through a measurement of the extraordinary
Hall Coefficient gives us a first value of y. Rutherford backscattering of α particles permits us to estimate both y and the homogeneity of the concentration profile : good agreement with the ferromagnetic behavior is obtained and the concentration y is constant
within 5 % of y.

The charging of the Pd-Ni electrode with deuterium is then done directly inside the cryostat in order to reduce the number of manipulations on the sample : the deuterium is introduced electrolytically using a $C_2H_5OD + D_2SO_4$ mixture at a temperature of - 80° C and the sample is cooled to helium temperatures immediately after the charging. We control the charging procedure by monitoring the resistivity of the films. The samples considered here have been charged to the maximum value of deuterium concentration x which is very near to the stoichiometric value of one.

RESULTS AND DISCUSSION

The analysis of the tunneling conductance for low applied voltages (ev $\sim$ Δ) shows deviations compared to the classical B.C.S. behavior and we think this is related to small inhomogeneities in the deuterium concentration x near the tunneling junction. We can explain the deviations if we admit that x is slightly smaller near the barrier leading there to a somewhat weaker superconductivity.

This kind of proximity effect, which has also been observed in other circumstances[7] does not permit an unambiguous determination or the true gap. That is why we use two gaps for each alloy : a best fit gap to the measured conductance at low voltages which is characteristic of the situation near the barrier and a higher B.C.S. gap obtained by using the $2\Delta = 3.52 \, kT_c$ relation which is more characteristic of the bulk properties. Some of our results, such as the absolute values of λ and μ^*, depend on the exact choice of Δ, but the overall shape of $\alpha^2 F(\omega)$ as well as the trends of variation with y do not. In Fig. 1, we plot the shape of $\alpha^2 F(\omega)$ for the two alloys with the acoustical phonon contribution extending up to about

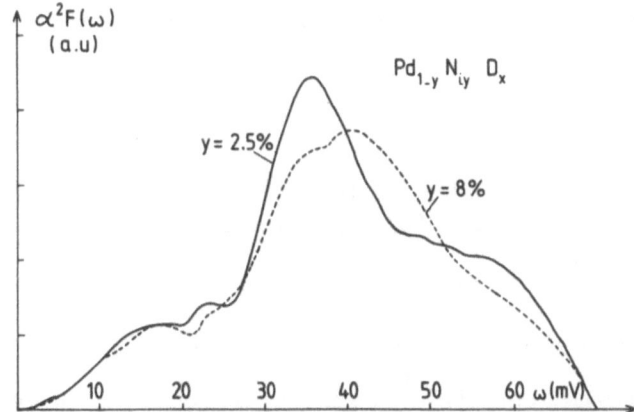

Figure 1.   Effective phonon spectrum of two $Pd_{1-y}Ni_yD_x$ alloys obtained from tunneling data.

26 meV and the optical phonons beyond. We will summarize now our main results. The acoustic contribution to λ is slightly smaller in the y = 8 % alloy but the difference is within the experimental errors. We obtain λ_{ac} = 0.19 ± 0.05 and λ_{ac} = 0.165 ± 0.04 respectively for y = 2.5 % and 8 %. Two peaks corresponding probably to the transverse and longitudinal phonons are visible for the lowest Ni concentration. The second peak is somewhat smeared for the higher Ni concentration. As concerns the optical phonons, we observe a decrease of λ_{op} with y and we obtain λ_{op} = 0.32 ± 0.07 and λ_{op} = 0.285 ± 0.08 for y = 2.5 % and 8 % (the large error bars are related to the different values of Δ used).

One should notice also that the width of the optical phonon peak and its mean position in energy are somewhat larger in the 8 % alloy. The absolute values of $\mu^\ast$ depend largely on the value used for the gap : it is in the range 0.1 - 0.15 for the lowest gap and in the range 0 - 0.05 for the second gap choice but in all cases we have an increase of $\mu^\ast$ with y. Most of the observed trends agree reasonably well with the theoretical predictions as they have been developped recently [8] by comparing the superconducting parameters of PdH and NiH. For instance one expects λ_{op} to decrease with increasing Ni concentration for two reasons : first, the s type density of states is smaller in NiH because the Fermi level is nearer to the top of the d band and second the optical phonons are expected to be higher in energy in NiH compared to PdH, in agreement with experimental information gained from resistivity data[9]. It remains to explain why $\mu^\ast$ increases with the Ni concentration. Recent calculations [10] concerning the possible importance of spin fluctuations (paramagnons) may be the clue for such an effect. In a preceding work [6] it was also noticed that the decrease of T_c with Ni concentration could not be entirely explained by a decrease of λ alone but a non phonon mechanism, probably magnetic in origin, is present.

REFERENCES

1. B. Stritzker, H. Wühl, "Hydrogen in Metals", Topics in Applied Physics (Springer Verlag), Vol 29 II, 243, (1978).
2. A. Eichler, H. Wühl, B. Stritzker, Sol. St. Com 17, 213 (1975)
3. L. Dumoulin, P. Nédellec, C. Arzoumanian, J.P. Burger, Phys. St. Sol. 90, 207 (1978).
4. C. Arzoumanian, J.P. Burger, L. Dumoulin, P. Nédellec, Zeits f. Phys. Chem. 116, 117 (1979).
5. M. Gupta, J.P. Burger, Phys. Rev. B 24 7099 (1981)
    L. Dumoulin, E. Guyon, P. Nédellec, Phys. Rev. B 16 1086

6. L. Sniadower, L. Dumoulin, P. Nédellec, J.P. Burger, J.
    Physique (Paris 42 L 13 (1981).
7. D.F. Moore, M.R. Beasley, J.M. Rowell, Jour. Phys. (Paris)
    C6 39 1390 (1978).
8. M. Gupta, J.P. Burger, J. Phys. F 10 2649 (1980)
9. D.S. Mac Lachlan, I. Papadopoulos, T.B. Doyle, J. Phys.
    Coll. (Paris) 6 430 (1978).
10. J.M. Daams, B. Mitrovic, J.P. Carbotte, Phys. Rev. Lett.
    46, 65 (1981).

RESISTIVITY AND ELECTRON-PHONON COUPLING IN PALLADIUM

ALLOY HYDRIDES*

A. F. Rex, J. Ruvalds, and B. S. Deaver, Jr.

Department of Physics
University of Virginia
Charlottesville, VA 22901

ABSTRACT

Palladium copper hydrides have been fabricated by coevaporating palladium with copper and subsequently implanting hydrogen ions. The change in high temperature resistivity as a function of temperature upon implantation is measured and correlated to the electron-phonon coupling parameter λ. This correlation is in good agreement with the observation of high superconducting temperatures in $Pd_{1-x}Cu_xH_y$.

INTRODUCTION

Since the discovery of superconductivity in palladium hydrides and deuterides[1,2], a great amount of theoretical and experimental work has been done in an attempt to understand the superconducting behavior of these systems. Much work has been devoted to understanding the sharp increase of T_c in PdH_x for $x \gtrsim 0.75$, particularly since both theoretical[3] and experimental[4] results indicate the Pd alone would have a T_c of no more than about 3K in the absence of spin fluctuations. The elimination of the strong paramagnetic behavior of Pd is therefore clearly a necessary but not sufficient cause for the high T_c's of 9K for PdH_x and 11K for PdD_x. For this reason it has become clear that understanding the nature of the enhancement of the electron-phonon interaction produced by the presence of hydrogen or deuterium is of paramount importance.

---

*Research supported by NSF Grant DMR78-25791.

A traditional approach has been to consider the relationship between the electron-phonon coupling parameter λ and the phonon resistivity. By solving Boltzmann's equation for electrons in metals, Allen[5] derived the relationship for the phonon relaxation time

$$\tau^{-1}_{PHONON} \propto \lambda_{TR} \qquad\qquad\qquad (1)$$

valid at high temperatures, where λ_{TR} differs from the superconducting parameter λ only by velocity factors. This result was used by Pinski, Allen and Butler[3] to analyze superconductivity in pure Pd. Ruvalds, Kahn and Tutto used a similar analysis along with optical measurements of the dielectric function to calculate λ for other transition metals and several A-15 compounds.[6,7] More recently Burger[8] has used an analysis of the high temperature resistivity of PdH_x and PdD_x to conclude that the coupling of the electrons to optical phonons is mainly responsible for the high T_c's observed in these materials.

We have applied these principles to one of the more interesting characteristics of palladium alloy hydrides, namely the exceptionally high transition temperatures found in palladium-noble metal-hydrides. To study this property we shall use resistivity data from the palladium-copper-hydride system.

EXPERIMENTAL

Our samples were fabricated by implanting hydrogen ions into thin films of $Pd_{1-x}Cu_x$. The binary alloys were produced by electron-beam co-evaporation from two independently controlled sources. Of the starting materials used, only the Pd contained more than 1 ppm magnetic impurities, containing 6 ppm nickel and 12 ppm iron. The films produced were all 0.3μ thick and were deposited onto sapphine substrates at room temperature. The substrates were then placed on a cryotip apparatus which could be maintained at liquid helium temperature during implantation. The samples were all implanted with a 60 KV beam of H_3^+ with beam current densities of 2-10 $\mu A/cm^2$. Thus the heat load on the sample was minimal, ensuring a lack of mobility of both implanted atoms and induced defects. Most of the sample was shielded from the beam by a thick copper mask which contained a single rectangular opening 1/4" by 1/32". Therefore only an area that size on the sample was irradiated. The sample resistance was measured (using a four-probe technique) as a function of temperature both before and after implantation for temperatures ranging from 3K to 300K. Because we know both the area and depth of the implanted region[8,9], we were able to determine the resistivity of the implanted region alone by measuring the resistance of the entire

sample. This also allowed us to determine resistively the
superconducting transition temperature of each sample as a
function of ion dose.

RESULTS AND DISCUSSION

     In order to use the resistivity data obtained, we begin by
considering the McMillan strong coupling formula[10]

$$T_c \simeq 0.7 \; \theta_D \; \exp \; [- \frac{1+\lambda}{\lambda - \mu^*}] \tag{2}$$

If we use the normal free electron relationship between the
resistivity and lattice relaxation time

$$\rho = \frac{4\pi}{\omega_{p\ell}^2 \tau} \tag{3}$$

and consider only the phonon contribution to the resistivity,
for which[7]

$$\frac{1}{\tau_{PH}} = 2\pi\lambda T \Big|_{T \simeq 300K} , \tag{4}$$

we obtain immediately the relationship between the electron-phonon
coupling parameter and the high temperature resistivity

$$\lambda = \frac{\omega_{p\ell}^2}{8\pi^2} \; \frac{\partial\rho}{\partial T} \Big|_{T \simeq 300K} . \tag{5}$$

     Using the known value $\omega_{p\ell} \simeq 7$ eV for pure Pd along with $\partial\rho/\partial T$
which we measured for pure Pd ($\partial\rho/\partial T \simeq 0.012$ $\mu\Omega$-cm-K^{-1}), we obtain
$\lambda \simeq 0.2$, which is in reasonable agreement with the observed $T_c =$
0. Since $\omega_{p\ell}$ is not well defined for palladium alloy hydrides
which we have studied, we cannot calculate T_c directly for the
hydrides.

     The other factors in eq. (2) should be constant. There is
no reason to expect μ^* to be appreciably different from 0.1.
Further, one may assume that $\theta_D \simeq 270K$ (the debye temperature of
pure Pd[11]) will be close to the debye temperature of all alloys
before implantation and $\theta_D \simeq 220K$ (the debye temperature of
PdH[12]) will be close to the debye temperature after implantation,
since the addition of up to 50 at % noble metal to palladium
does not change the debye temperature significantly.[13] Therefore
the important parameter is the high temperature $\partial\rho/\partial T$.

     Figure 1 shows $\rho(T)$ for several representative $Pd_{1-x}Cu_xH_y$

alloys as well as Pd and PdH, and the sharp increase in $\partial\rho/\partial T$ can
be seen clearly for those alloys with high T_c. The increase
in $\partial\rho/\partial T$ upon addition of H to Pd is consistent qualitatively
with the increase in T_c, but one would expect a much greater

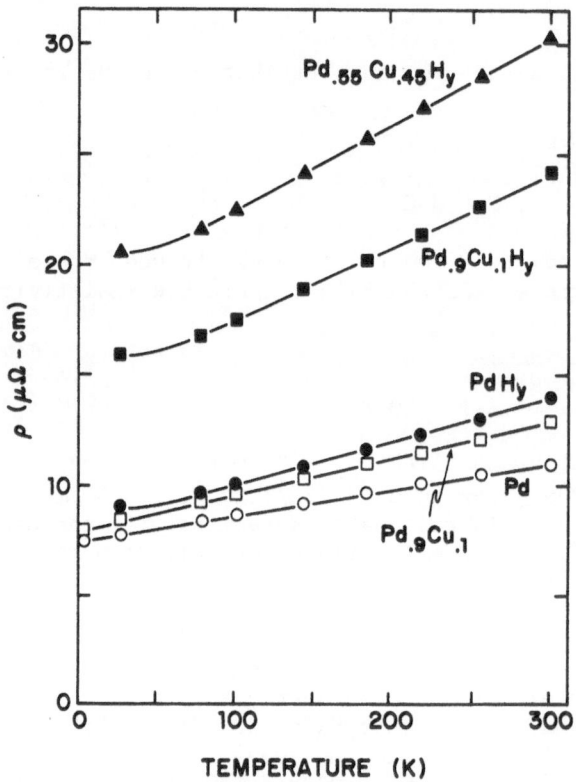

Figure 1.   Temperature dependence of resistivity ρ for
            selected $Pd_{1-x}Cu_xH_y$ alloys for temperatures up
            to 300K. The steepest slopes are clearly for
            $Pd_{.9}Cu_{.1}H_y$ and $Pd_{.55}Cu_{.45}H_y$, which also have the
            highest transition temperatures.

increase in $\partial\rho/\partial T$ to correspond with the sharp rise in T_c for
PdH. It is also interesting to note that the resistivity curve
for PdH$_y$ is not much different from the one for unimplanted
Pd$_{.9}$Cu$_{.1}$, particularly since hydrogen and copper both contribute
one conduction electron. A further indication of a discrepancy
for PdH is shown in Figure 2, where the measured T_c is shown as
a function of the corresponding measured $\partial\rho/\partial T$. There is an
almost linear dependence of T_c on $\partial\rho/\partial T$ for the Pd$_{1-x}$Cu$_x$H$_y$ alloys
shown, which are representative of all of our data. The
measured $\partial\rho/\partial T$ for PdH$_y$, however, is almost the same as for the
unimplanted Pd$_{1-x}$Cu$_x$ alloys. The data for PdH$_y$ would actually
seem to contradict the idea that the phonon contribution to the
resistivity is chiefly responsible for the sharp increase in
λ while the data for the palladium copper hydrides would seem to
confirm it.

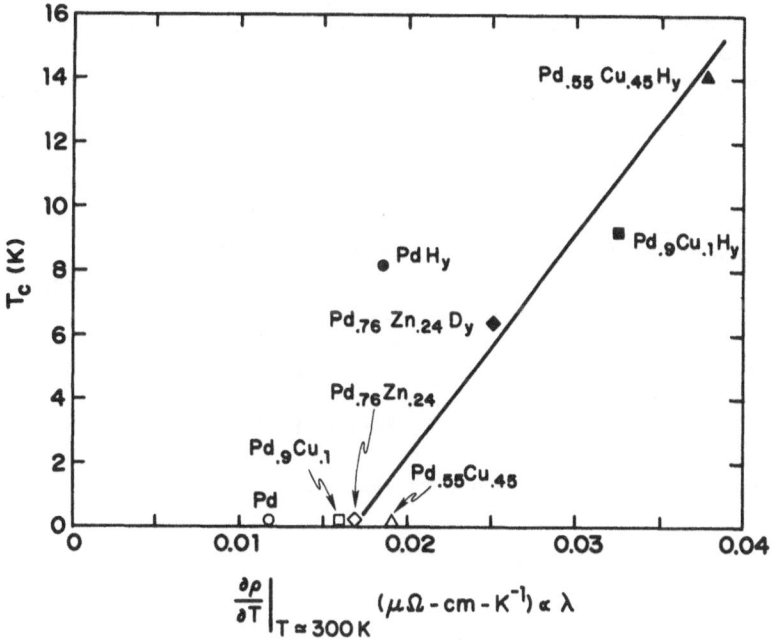

Figure 2.   Variation of the transition temperature as a
            function of high temperature $\partial\rho/\partial T$. The depen-
            dence is nearly linear for the palladium alloy
            hydrides shown.

A possible resolution of the anomalous variation in T_c relative to $\partial\rho/\partial T$ may be an unusual change in the plasma frequency $\omega_{p\ell}$ which enters in eq. (5) in the determination of λ. However, to reconcile the T_c discrepancies would require extraordinary changes in the plasma frequency, namely from $\omega_{p\ell}(Pd) \simeq 7$ eV and $\omega_{p\ell}(Pd_{.55}Cu_{.45}) \simeq 7$ eV to $\omega_{p\ell}(PdH) \simeq 12$ eV and $\omega_{p\ell}(Pd_{.55}Cu_{.45}H) \simeq 9.5$ eV. Such variation would certainly be difficult to understand in terms of a rigid band model, but surely cannot be ruled out. In the case of Pd <u>vis a vis</u> PdH, another factor may be involved, namely that the measured $\partial\rho/\partial T$ is slightly less for our thin films than for bulk samples. The width of the superconducting region after implantation is only about 1000 Å, which may also have a slight effect upon the magnitude of $\partial\rho/\partial T$. This immediately suggests that it would be useful to make similar measurements using high pressure or electrolytic hydration techniques so that one could be more certain of measuring bulk effects.

REFERENCES

1.   T. Skoskiewicz, Phys. Stat. Sol. <u>11K</u>, 123 (1972).
2.   B. Stritzker and W. Buckel, Z. Phys. <u>257</u>, 1 (1972).
3.   F. J. Pinski, P. B. Allen, and W. H. Butler, Phys. Rev. Lett. <u>41</u>, 431 (1979).
4.   B. Stritzker, Phys. Rev. Lett. <u>42</u>, 1769 (1979).
5.   P. B. Allen, Phys. Rev. <u>17B</u>, 3725 (1978).
6.   J. Ruvalds and L. M. Kahn, Phys. Lett. <u>70A</u>, 477 (1979).
7.   I. Tutto, L. M. Kahn, and J. Ruvalds, Phys. Rev. <u>B20</u>, 952 (1979).
8.   H. H. Anderson and J. F. Ziegler, Hydrogen Stopping Powers and Range in All Elements, Pergamon (1977).
9.   W. S. Johnson and J. F. Gibbons, Projected Range Statistics in Semiconductors, Stanford University Press (1970).
10. W. C. McMillan, Phys. Rev. <u>167</u>, 331 (1968).
11. U. Mizutani, T. B. Massalski, J. Berk, J. Phys. F. Metals <u>6</u>, 1 (1975) and references cited therein.
12. M. Zimmerman, G. Wolf, K. Bohnhammel, Phys. Stat. Sol. (a) <u>31</u>, 511 (1975).
13. M. R. Chowdhury, J. Phys. F: Metals <u>4</u>, 1657 (1974).

SUPERCONDUCTIVITY OF Pd-H$_x$ AND ITS

CONNECTION WITH THE PHASE DIAGRAM

I.S. Balbaa and F.D. Manchester

Department of Physics, University of Toronto
Toronto, Ontario, Canada M5S 1A7

ABSTRACT

Measurements have been made of the superconducting transition temperature and of the bulk magnetization for superconducting Pd-H$_x$ alloys in the concentration range $0.81 \leq X = H/Pd \leq 0.92$. The observed magnetization behaviour is consistent with the presence of two superconducting forms of Pd-H$_x$ in the concentration range $0.84 \lesssim X = H/Pd \lesssim 0.88$ which can be regarded as a mixed phase region giving behaviour similar to that observed for mixed phase regions of other superconducting alloy systems. In the Pd-H$_x$ case, for the above concentration range, disordered β-phase and ordered β-phase are the most likely constituents of this mixed phase region. Considerations of candidate structures for the mixed phase region and of the properties of the two superconducting forms of Pd-H$_x$ will be given. A full account of these experiments is being published elsewhere.

MAGNETISM OF Mn, Fe, Co IMPURITIES IN PdH$_x$ AS A FUNCTION OF HYDROGEN

CONCENTRATION;   RELATION WITH THE ELECTRONIC STRUCTURE OF THE HOST

B. Souffaché and J.P. Burger

Laboratoire de Physique des Solides
Université de Paris-Sud, Bât 510
91405 ORSAY (France)

ABSTRACT

We present experimental data concerning the magnetic ordering
temperature T_C of PdH$_x$ compounds containing small amounts of magnetic impurities M (= Mn, Fe, Co). We consider the case x = 0 and the
entire homogeneous β phase with $0,6 \leqslant x \leqslant 1$. We show that the interactions are strongly reduced if one goes from the x = 0 case to the
β phase. Most of the trends concerning the changes of T_C with x are
explained by variation of the Fermi level conduction electrons and of
the Stoner exchange enhancement factor of the host conduction
electrons.

INTRODUCTION

Homogeneous monophased PdH$_x$ hydrides spanning the concentration
range $0,6 \leqslant x < 1$ can now be prepared and it is well known that the
metal-hydrogen interaction changes drastically the electronic structure of the starting pure metal [1]. For instance the Fermi level is
shifted upwards from a region of high 4d density of states when
x = 0 to a region of low density of states when $0,6 \leqslant x < 1$. s and
p type electrons become important in this region just like in noble
metals. Within this low density of states range one observes [2] in
agreement with theoretical speculations that the density of states
is a smooth decreasing function of the hydrogen concentration x. All
these variations in electronic structure have deep consequences on
the physical properties of Pd and PdH$_x$; the hydrides for instance
are superconducting but this is not the case for pure Pd which is
a strongly exchange enhanced, nearly ferromagnetic metal. It becomes
thus interesting to follow the evolution with hydrogen concentration

361

x of the magnetic ordering temperature of $Pd_{1-y}M_yH_x$ alloys where M is a 3d magnetic impurity (M = Mn, Fe, Co). These magnetic ordering temperature T_c have been extensively studied in pure Pd [3] but only a few data exist up to now for the hydrides [4,5]. For the range of impurity concentration y considered (y $\simeq$ few percent) one expects the magnetic interaction to be dominated by the oscillating long range RKKY type of interactions. The magnetic interaction energy for two impurity spins separated by a distance r_{ij} is then given by :

$$\varepsilon_{ij} \simeq J^2 \frac{N(\varepsilon_F)}{1-\alpha} \cos \frac{2k_F r_{ij}}{r_{ij}^3} \vec{S}_i \vec{S}_j$$

(J is an exchange integral between a conduction electron spin and an impurity spin, $N(\varepsilon_F)$ is the Fermi level density of states of the host, $(1-\alpha)^{-1}$ is the Stoner exchange enhancement factor resulting from the interactions between conduction electrons).

Large values of T_c are thus expected and observed [3] in pure Pd because $N(\varepsilon_F)$ and $(1-\alpha)^{-1}$ are large. For the hydrides we will split the results and discussion in two parts : we will just make a gross feature comparison between the T_c observed in the hydrides and in pure Pd and in a second part look at the more detailed variation of T_c within the low density of states range.

EXPERIMENTAL

The $Pd_{1-y}M_y$ alloys are melted in an induction furnace and rolled down to a thickness of about 50 μ before being charged with hydrogen at a temperature of - 80° C. The details of the charging procedure are given elsewhere [6]. Due to the intrinsic instability of the compounds with x $\gtrsim$ 0,8 they can only be handled at temperatures below the preparation temperature. The concentration x is determined by measuring the volume of hydrogen released by the samples at high temperature. The in phase susceptibility χ' is obtained with the help of an alternating susceptibility bridge.

RESULTS

1) Gross feature comparison between Pd and PdH_x

In fig. 1 we show the magnetic behaviour of an y = 0,5 % alloy for x = 0 and x $\simeq$ 0,95. A large susceptibility peak, essentially ferromagnetic in nature, is observed for pure Pd or x = 0 case while no transition shows up above 1° K for the hydride : the reduction in T_c is thus by a factor more than ten, a result also observed for the case of Fe impuriries. Less dramatic but nevertheless strong decrease of T_c are observed for Mn : for a 2 % Mn alloy the decrease is by a factor between three and four. One can also notice that there is a change in the nature of the magnetic transition which is

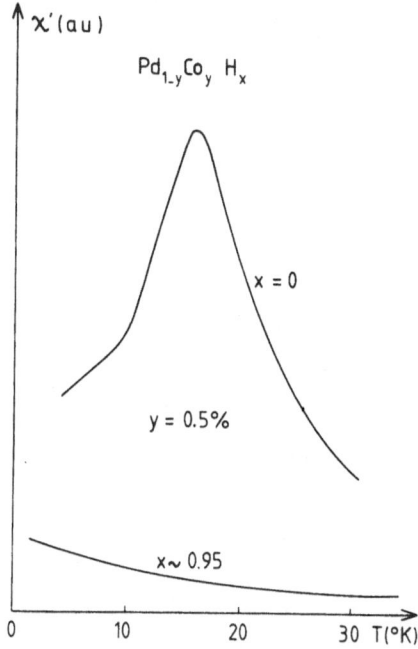

Fig. 1. Susceptibility of an y = 0.5 % Co
for x = 0 (pure Pd) and x = 0,95

essentially ferromagnetic for the x = 0 case but becomes spin glass
type in the hydrides.

2) Evolution of T$_C$ within the β phase 0,6 ≤ x ≤ 1

In Fig. 2 we show the evolution of T$_C$ for Mn and Co with y and x
A simple behaviour is observed for all Mn concentrations : at fixed y
we have $\frac{dT_C}{dx} > 0$ and roughly independent of y. For the case of Co, $\frac{dT_C}{dx}$
is also positive for low enough values of y but it changes to more
and more negative values as y increases : the same behaviour is
observed for Fe. The nature of the transition is mainly of the
spin glass type but a weak ferromagnetic component develops when
$\frac{dT_C}{dx}$ gets more and more negative.

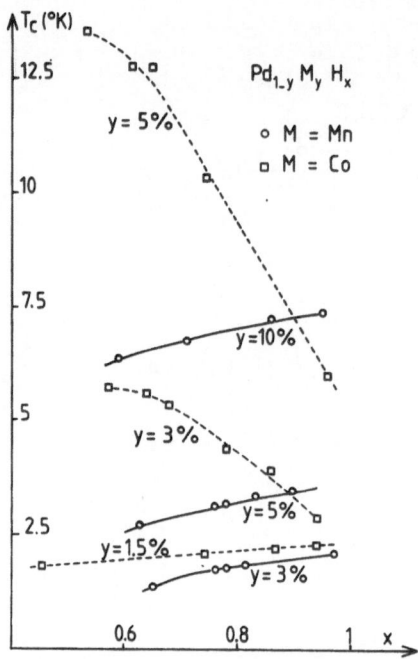

Fig. 2.   Variation of T_c of Mn and Co
impurities as a function of x for different values of y. Only the β phase
is considered.

DISCUSSION

The dramatic and systematic decrease of T_C as one goes from the
Pd to the PdH_x host matrix must be attributed to the variations with
x of the three parameters $N(\varepsilon_F)$, $(1 - \alpha)^{-1}$ and J. As concerns the
density of states $N(\varepsilon_F)$ one knows from theory [1] and experiment [2]
that it decreases roughly by a factor four. For the Stoner factor
$(1 - \alpha)^{-1}$ one can assume it to be of order unity in PdH_x, just like
in noble metals compared to a value of seven in pure Pd. One sees
thus that combining the first two factors can explain by itself the
whole variation of T_C. Less quantitative information are available
at the moment concerning the variation with x of J. Electron spin
resonance measurements [7] done on Mn show that J varies indeed with
x, changing even sign as one goes from Pd where it is positive to
PdH_x. It may thus be premature to conclude that the whole variation
of T_C is related to the variation of $N(\varepsilon_F)$ and $(1 - \alpha)^{-1}$.

If we consider now the concentration range $0,6 < x < 1$ our
main result is that $\frac{dT_C}{dx}$ is positive for all impurities considered

provided the impurity concentration y is low enough. This general
result suggests that the long range interaction should depend only
on parameters related to the host matrix but not on the impurity
itself. This is indeed the case if one admits that one can apply
to these low density of states materials the familiar picture of
impurity virtual found states developped by Friedel for the noble
metal hosts. In this model it is possible to show [8] that the negative exchange integral J varies essentially like $1/N(\varepsilon_F)$ so that
ε_{ij} and T_c should vary also like $\frac{1}{N(\varepsilon_F)}$. As $N(\varepsilon_F)$ decreases with x,
we expect $\frac{dT_c}{dx}$ to be positive in agreement with observation. It
remains nevertheless to explain why $\frac{dT_c}{dx}$ changes sign for Co and
Fe but not for Mn when the impurity concentration y increases.
Two explanations can be put forward. First direct short range
interactions between impurities may become increasingly important
as y increases. Such an effect may be particularly important if
the impurity potential spreads over the neighbouring host atoms.
Such interactions, due to direct covalent admixture between two
impurities have been considered [9] and they appear to be strong
and ferromagnetic for Co and Fe, but weak and antiferromagnetic
for Mn. One expects such interactions to be also very sensitive to
the impurity-impurity distance. The continuous dilatation of the
lattice with increasing x may be responsible for the sign of
$\frac{dT_c}{dx}$. All these features are in agreement with the experimental
findings. On the other hand one should also consider the possibility
that the hydrogen distribution is inhomogeneous around a given
impurity, because of a repulsive impurity-hydrogen interaction.
One expects the degree of inhomogeneity to decrease as one forces
the hydrogen concentration to reach its maximum value of one. It
is difficult to assert what could be the consequences of such a
possibility. We feel that the first picture has more to do
with the facts because such direct interaction seems also to
be at work[10]in hydrogen free Pd-Mn alloys as soon as the manganese
concentration growths beyond about 2 %.

REFERENCES

1. M. Gupta, A.J. Freeman, Phys. Rev. B 17, 3029 (1978)
2. M. Zimmerman, G. Wolf, K. Bohmhammel, Phys. St. Sol. 31 A
   511, (1975)
3. G.J. Nieuwenhuys, Adv. Phys. 24, 515 (1975)
4. S. Senoussi, B. Souffaché, J.P. Burger, Jour. Less Com. Mat.
   49, 213, (1976)
5. G. Alquié, R. Kreister, B. Souffaché, J.P. Burger, Jour Magn.
   and Magn. Materials 15, 89 (1980)
6. B. Souffaché, D.S. Mac Lachlan, J.P. Burger, Revue Phys.
   Appl. 14, 749 (1979)

7. G. Alquié, Thesis, Paris (1977)
8. B. Coqblin, M.B. Maple, G. Toulouse, Int. J. Magnetism $\underline{1}$, 333 (1971)
9. T. Moriya, Prog. Theor, Phys. $\underline{33}$, 157 (1965)
10. J. Rault, J.P. Burger, C.R.A.S. $\underline{269}$, 1085 (1969)

THE MAGNETIC PROPERTIES OF $Gd_xY_{6-x}Fe_{23}$ (x = 0,1,2,3) COMPOUNDS
AND THEIR HYDRIDES[*]

A. T. Pedziwiatr[†], E. B. Boltich, W. E. Wallace and
R. S. Craig

Department of Chemistry, University of Pittsburgh
Pittsburgh, PA 15260

The magnetic properties of the compounds $Gd_xY_{6-x}Fe_{23}$ (x = 0,1,
2,3) and their hydrides were studied. In the parent intermetallics,
progressive substitution of Gd in place of Y in Y_6Fe_{23} led to a reduction in the magnetic moment ($\sim$ 7 μ_B per Gd substituted) and an
increase in the Curie temperature. For the hydrided materials a
similar effect was observed (Gd substitution into $Y_6Fe_{23}H_{16}$ resulted
in a decrease in magnetic moment and an increase in Curie temperature). In all cases the hydride exhibited a larger moment and a
higher Curie temperature than the corresponding intermetallic. This
behavior indicates that in both the parent compounds and the hydrides
the Fe sublattice is dominant and antiparallel to the Gd sublattice.
Furthermore, the Fe moment is increased by hydrogenation, which seems
to indicate that hydrogen creates vacancies in the minority spin
d-band (i.e., hydrogen behaves as an electron acceptor). All hy--
drides retained the crystal structure of the parent materials
(Th_6Mn_{23} type), with a volume expansion of about 11%.

INTRODUCTION

It has been observed in the past that adsorbed hydrogen can have
dramatic effects on 6:23 systems, a family of compounds consisting
of R_6Mn_{23} and R_6Fe_{23} where R is a rare earth. The R_6Mn_{23} systems
have been extensively studied,[1,2] and it has been found that the

---

*This paper is dedicated to Prof. S. Methfessel, Ruhr University,
 Bochum, on the occasion of his 60th birthday. The work was
assisted by a contract with the U.S. Army Research Office.
†On leave from Institute of Physics, Jagellonian University,
 Cracow, Poland.

367

effect of hydrogenation is particularly strong in these materials. For instance, Y_6Mn_{23} is ferrimagnetic, but upon hydrogenation becomes antiferromagnetic.[1,2] Meanwhile, Th_6Mn_{23}, which is a Pauli paramagnet, becomes ferrimagnetic upon hydrogenation.[1]   It has also been observed that hydrogenation of certain solid solutions of Th_6Mn_{23} in Y_6Mn_{23} produces what appears to be a spin-glass behavior.[1] The effect of hydrogen on R_6Mn_{23} when R is a magnetic rare earth is equally dramatic. For instance, in those cases where R is a heavy rare earth the parent compounds are strongly ferrimagnetic, while the hydrides are either very weakly ferrimagnetic or antiferromagnetic.[1]

, Information regarding the rare earth 6:23 Fe compounds is neither as consistent nor as complete as that for the Mn compounds. For instance, the Curie temperature of Gd_6Fe_{23} has been reported as both 468 K[3] and 659 K.[4]   This inconsistency is most likely a result of the extreme difficulty in preparing pure compounds. It was therefore one of the purposes of this investigation to obtain information regarding Gd_6Fe_{23} through indirect means, i.e., by replacing Y with Gd in Y_6Fe_{23}. For example, the present investigation lends additional credibility to the reported Curie temperature at 659 K as opposed to the value of 468 K.

The second purpose of this investigation was to acquire structural and magnetic information regarding the hydrides of the $Gd_xY_{6-x}Fe_{23}$, i.e., are the crystal structure and coupling scheme retained in the hydrides? Interest in the structural features of the hydrides has been aroused by the recent discovery that Er_6Fe_{23} becomes tetragonally distorted upon hydrogenation.[5]

Both the R_6Mn_{23} and R_6Fe_{23} compounds crystallize in the Th_6Mn_{23} structure (space group Fm3m), with 116 atoms per unit cell.[1] This structure possesses 4 crystallographically inequivalent transition metal sites and a unique rare earth site. Although they are crystallographically similar, R_6Mn_{23} and R_6Fe_{23} are quite different magnetically. Neutron diffraction studies of Y_6Mn_{23} indicate that the Mn moments are coupled ferrimagnetically.[2] Similar studies of Y_6Fe_{23} indicate that the Fe moments are coupled ferromagnetically.[2] Single crystal studies of several R_6Mn_{23} (R = heavy rare earth) compounds indicate that the resultant rare earth moment is coupled parallel with the net Mn moment.[1]   On the other hand, neutron diffraction studies of Ho_6Fe_{23} indicate that all Fe moments are coupled parallel with each other and antiparallel with the rare earth.[5]

EXPERIMENTAL

The parent materials were obtained by r.f. induction melting under a flowing stream of purified argon gas in a water-cooled copper boat. Stoichiometric amounts of high purity metals were melted

several times in order to insure the homogeneity. No weight loss
during melting was observed. Long-term annealing at 1100°C was required in order to obtain single phase materials (varying from 3
hours for Y$_6$Fe$_{23}$ to 25 days for Y$_3$Gd$_3$Fe$_{23}$). The quality of the
samples was verified not only by x-ray method but also by high temperature magnetic analysis.

The hydrides studied were those which are stable at 1 atm with
respect to desorption. All samples were hydrogenated according to
the following procedure. Approximately 1 to 2 g of each sample was
polished with a rotary grinder to remove surface oxide, and weighed.
The sample was then placed in a system at known volume and heated to
$\sim$ 200°C under constant pumping for a period of $\sim$ 12 hrs. The sample
was then cooled to room temperature and exposed to a measured amount
of hydrogen at a pressure of $\sim$ 35 atm. Once absorption occurred and
equilibrium was attained, the excess hydrogen was removed by bubbling
through water into an inverted buret, until the pressure within the
system was 1 atm. SO$_2$ was then introduced into the system to poison
the surface of the hydride according to the method of Gualtieri et
al.[1]   The hydride was then transferred, in a glove bag containing
1 atm. hydrogen pressure, to the sample-holder used for the low temperature magnetic measurements. The hydrogen content of the sample
was established from the known amount of hydrogen in the preparation
chamber and the amount of hydrogen escaping into the buret as the
pressure was reduced to 1 atm. Due allowance was made for the
hydrogen retained in the dead space.

Sample quality and lattice constants for the hydrides were established from x-ray (CuK$_\alpha$) powder diffraction patterns obtained
using a Diano XRD-700 powder diffractometer equipped with a graphite
diffracted beam monochromator. All intermetallics and their hydrides
(with the exception of GdY$_5$Fe$_{23}$H$_{16}$ and Gd$_2$Y$_4$Fe$_{23}$H$_{17}$, which showed
about 5% unhydrided material) were found to be single phase within
the limits of x-ray detection.

Magnetic measurements were made on loose powders, using the
Faraday technique. The temperature dependence of the magnetization
was measured in an applied field of 4 kOe, and the field dependence
of the magnetization was measured at a temperature of 4.2 K. Two
Faraday units were used, one covering the range 4.2 to 300 K and the
other capable of covering the range 300 to 1100 K. The latter was
used to determine the Curie temperatures, while the former was used
to determine the saturation magnetization at 4.2 K and the temperature
dependence of the magnetization at low temperatures. Saturation
moments were obtained from Honda plots (M vs 1/H).

RESULTS AND DISCUSSION

Results and magnetic measurements are given in Fig. 1 and sum-

marized in Table 1. The lattice parameters of the various compounds
are shown in Table 2. In regard to the magnetic coupling in the
parent materials, it is quite obvious from Fig. 1 that Gd couples
antiparallel to Fe (at least within the composition range studies),
as evidence by the ~ 7 μ_B drop in saturation magnetization per Gd
atom substituted into Y_6Fe_{23}. These results corroborate Kirchmayr's
original hypothesis regarding the Gd-Fe coupling in Gd_6Fe_{23}.[3] The
fact that this investigation and that originally performed by
Kirchmayr, while being different in approach, both lead to the same
conclusion is fairly conclusive evidence that Gd is coupled antiparallel to Fe in Gd_6Fe_{23}.

From Fig. 1 it can be seen that for every system studied, absorption of hydrogen increases the net magnetic moment. Since the
direction of magnetization is in the direction of the Fe moments,
this increase in magnetization represents an increase in Fe moment
upon hydrogenation. This behavior (increase in Fe moment upon hydrogenation) is quite typical and has been observed by bulk magnetization measurements for numerous systems - $Y_6Fe_{23}H_x$, $Er_6Fe_{23}H_x$ and
$Ho_6Fe_{23}H_x$.[5] This increase has also been observed in neutron diffraction studies of the cubic Laves phase $ErFe_2$ and $HoFe_2$ compounds.[5]
Since atomic Fe contains more than five d electrons, the local moment
model and any model which assumes an intra-atomic Hund's rule behavior, i.e., that of Stearns[6], would require hydrogen to behave as
an electron acceptor in order to account for this increase in magnetic moment.

To very briefly summarize the model advocated by Stearns, she
asserts that the magnetic moment arising from the 3-d electrons is
localized in nature, originating from spin-polarized tight-binding

Table 1.   Magnetic Properties of $Gd_xY_{6-x}Fe_{23}$ Compounds
           and Their Hydrides

Compound	$M_{sat}(\mu_B/f.u.)$ 4.2 K	T_C (K)
Y_6Fe_{23}	45.4	480
$Y_6Fe_{23}H_{16}$	59.8	-
GdY_5Fe_{23}	39.4	519
$GdY_5Fe_{23}H_{16}$	50.7	-
$Gd_2Y_4Fe_{23}$	32.9	544
$Gd_2Y_4Fe_{23}H_{17}$	41.5	-
$Gd_3Y_3Fe_{23}$	25.4	572
$Gd_3Y_3Fe_{23}H_{17}$	35.2	-

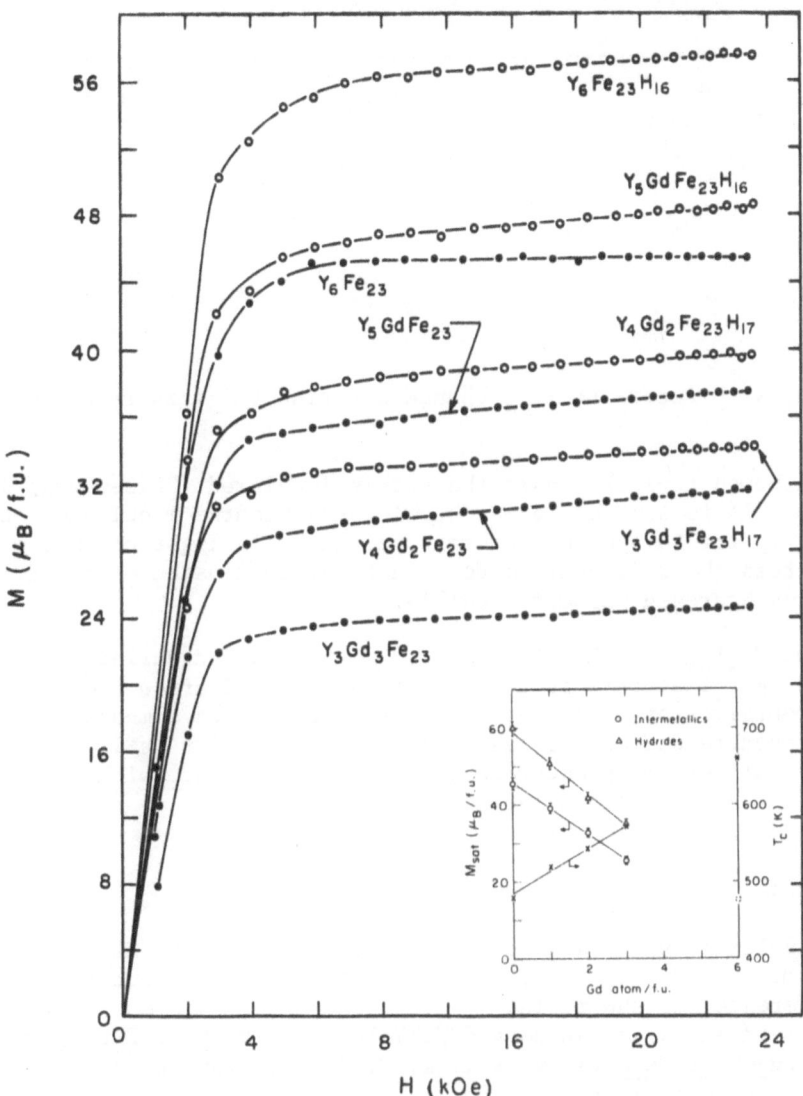

Fig. 1. Field dependence of $Y_{6-x}Gd_xFe_{23}$ alloys and their
         hydrides. Inset – Saturation magnetization at
         4.2°K versus composition.

Table 2. Lattice Parameters of $Gd_xY_{6-x}Fe_{23}$ Compounds
          and Their Hydrides

Compound	Lattice Parameter (Å)	$\Delta V/V$
$Y_6Fe_{23}H_{16}$	12.508(12.072)[a]	11%
$GdY_5Fe_{23}H_{16}$	12.520(12.097)[a]	11%
$Gd_2Y_4Fe_{23}H_{17}$	12.535(12.114)[a]	11%
$Gd_3Y_3Fe_{23}H_{17}$	12.559(12.127)[a]	11%

a. Value for the host metal.

d-bands, with inter-atomic exchange occuring via itinerant d electrons.

Although application of the purely localized (Heisenberg) model to 3d metals is somewhat questionable, the Stearns model was designed explicitly for 3d systems. Therefore, it is in light of this model, rather than the Heisenberg model, that the conclusion of electron-accepting hydrogen becomes credible.

The $Gd_xY_{6-x}Fe_{23}$ hydrides show a magnetization variation of 8.5 μ_B/Gd atom. That this is not 7 μ_B/Gd atom could arise from: (1) experimental error, (2) non-collinearity of the Gd moments or (3) band structure differences as observed recently for $LaNi_5$ and $GdNi_5$.[7] All hybrides occur in the Th_6Mn_{23} structure with the lattice parameters given in Table 2.

REFERENCES

1. E. Boltich, W. E. Wallace, F. Pourarian and S. K. Malik, J. Phys. Chem., to appear, Feb., 1982. See this for numerous earlier references in the field.
2. K. Hardman, W. J. James and W. Yelon, in The Rare Earths in Science and Technology, Vol. 1, G. J. McCarthy and J. J. Rhyne, eds. Plenum (1978), p. 403.
3. H. R. Kirchmayr and W. Steiner, J. de Physique 32, 45 (1971).
4. L. R. Salmans, Wright-Patterson Air Force Base Rept. AFML-TR-68-159 (1968).
5. E. B. Boltich, F. Pourarian, W. E. Wallace, H. K. Smith and S. K. Malik, Sol. State Commun. 40, 117 (1981).
6. M. B. Stearns, Phys. Rev. B9, 2311 (1974) and B13, 1183 (1976).
7. S. K. Malik, F. Arlinghaus and W. E. Wallace, Phys. Rev., submitted.

MEASUREMENTS OF THE MAGNETIZATION OF ACTIVATED FeTi[†]

D. Khatamian, F.D. Manchester, G.C. Weatherly and
C.B. Alcock
Departments of Physics and Metallurgy and Materials
Science, University of Toronto, Toronto, Ontario
Canada M5S 1A7

ABSTRACT

We have measured the magnetization of activated FeTi samples
from room temperature to 800°C. The magnetization measurements,
together with structural information from diffraction experiments,
indicate that metallic iron is not formed in the activation process
and, apparently, is not responsible for observed increases in the
hydrogen absorption rate of FeTi after activation treatment.

INTRODUCTION

We have been investigating the effect of activation for hydrogen absorption on the magnetic properties of the FeTi surface.
Recently there have been reports[1-3] that after each activation
process the magnetization of this Pauli paramagnet[4] irreversibly
increases and this increase in magnetization has been attributed
to the formation of metallic or superparamagnetic iron particles
on the surface. Schlapbach et al[2] have suggested that the dissociation of the H_2-molecule is catalyzed by such segregated iron clusters
or by the metallic FeTi interface before its absorption into the
bulk. This view point has also been proposed by Shaltiel et al[3],
who have investigated the magnetic properties of this material by
Ferromagnetic Resonance (FMR) techniques.

The present work is an attempt to investigate the macroscopic
magnetic properties of activated and unactivated samples of FeTi

---

[†]Work supported in part by a Strategic Grant from the Natural
Sciences and Engineering Research Council of Canada.

from room temperature to 800°C. It will be shown that metallic
iron does not form in the activation process and is not responsible
for the observed increase in the hydrogen absorption rate of this
intermetallic compound after activation treatment.

EXPERIMENTAL

All the alloys were prepared by arc melting iron and titanium
under a helium atmosphere. The ingots were then crushed to an
average size of ∿ 100 microns, immediately before use in a set of
measurements. For this experiment, after any preliminary process
(e.g., activation), the samples were sealed in quartz tubes under
vacuum of ∿ 10^{-5} mm Hg. We have also obtained data under dynamic
vacuum conditions which will be published elsewhere[5].

A Foner type[6] vibrating-sample magnetometer with a sensitivity
of about 10^{-8} Amp. m^{-1} was used to measure the magnetization.

RESULTS AND DISCUSSION

In Fig. 1 magnetization versus temperature of a sample, which
had been given 10 cycles of activation is presented. Here, one
cycle of activation means to heat the sample under ∿ 10^{-6} mm Hg
vacuum to 500°C, hold this temperature for ∿ 1 hour under ∿ 10 atm.
of H_2 pressure then cool to room temperature while the hydrogen is
pumped out and subsequently exposed to 60 atm. of H_2 pressure for
one hour. Fig. 1 is a typical result and represents all of the other
temperature dependent magnetization data obtained in our laboratory.
Generally these results show that as the temperature is increased
above room temperature the magnetization of the sample decreases
very slowly until ∿ 300°C. At this temperature we see a very sharp
increase in the magnetization which continues up to ∿ 575°C. Our
early measurements showed that if the temperature went higher than
575°C, the magnetization dropped very fast and the process slowed
down as the temperature approached ∿ 800°C. Therefore in our later
measurements (given in Fig. 1), we stopped heating the sample at
575°C and started to cool it. As it is clear from the figure, the
magnetization increases almost linearly as the temperature is
decreased from 575°C to room temperature. This type of gradual
change in magnetic moment may be caused by a mixture of many
titaniferrous ferrites (magnetites and/or maghemites) in which the
chemical composition varies continuously as cited by T. Nagata and
S. Akimoto[7]. Consequently it is almost impossible to determine a
unique value of the Curie temperature and other physical constants.
This linear behaviour is exactly reversed if the sample is reheated
up to 575°C. Beyond this temperature, as we mentioned before, the
magnetization drops sharply and irreversibly. This irreversible
transition suggests that material with high magnetic moment has
meta-stable structure and at this temperature it transforms to a
more stable structure with much lower magnetic moment. This

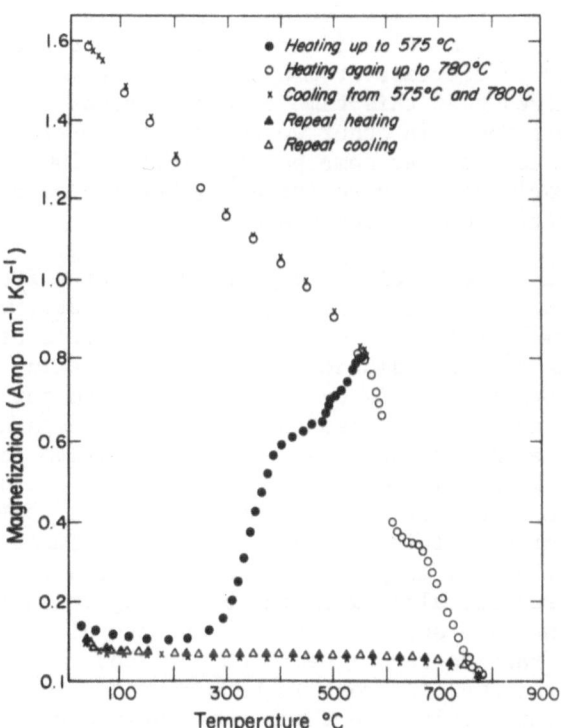

Fig. 1.   Magnetization of FeTi versus temperature after 10
         activation cycles. The applied field was kept constant
         at H_a = 37.7 Amp m^{-1}.

resembles very much the transition of titanomaghemites to titano-
hematites with transition temperature between 400 to 800°C[7]. If
the sample is exposed to air, by just breaking and resealing the
quartz container, the upper branch of the magnetization curve will
be reproduced upon a repeat heating otherwise the lower part of
the curve will be obtained. The kinks at ∿ 400, 500 and 650°C are
caused by changes in the rate of heating the sample. It should be
mentioned that the points on the figure are not equilibrium points
and are obtained while the sample is heated continuously. There
is another feature of Fig. 1 which should be made more clear. This
is that, at ∿ 600°C the temperature of the sample was kept constant
for ∿ 5 minutes, but as is indicated by missing points the magnetization continued to decrease from 0.6 to 0.4 Amp m^{-1} Kg^{-1}. Other
measurements showed that this time dependent process, also, exists
at lower temperatures. In other words if the temperature of the
sample was kept constant at some point between ∿ 300 – 575°C, the
magnetization would continue to increase, but the rate of increase
will depend directly on the temperature.

The field dependence of the magnetization of a fresh
(unactivated) sample is given in Fig. 2. These data, which were
obtained at room temperature, show that the magnetization of the
sample increases rather dramatically by merely heating it to 500°C.
This is not surprising and is quite in accord with our temperature
dependent data and with the results obtained by Busch et al[8]. The
saturation of the magnetization at rather low fields indicates,
again, that the material under observation is a ferro or ferrimagnetic
type. The reminant magnetization, shown in Fig. 2, following a
simple cycling of the field H_a, relaxes back to almost zero over
night and thus there is no permanent magnetisation in the FeTi
powder. These results along with our other magnetization measurements[5] and the surface and near surface structural data obtained
by electron diffraction techniques[9], do not show any evidence for
the formation of elemental iron clusters near the surface of FeTi
samples. From our results, we think that the presence of pure iron
near the surface of these samples is not a necessary condition for
their ability to absorb hydrogen, which our FeTi samples do readily
after just a few activation cycles[9]. At the present time our
observations indicate that, as a result of exposure to air, some
oxides of Ti form on the surface of FeTi, and these oxides prevent
the hydrogen from penetrating into the bulk of the alloy.

CONCLUSION

In none of the magnetic measurements we have made, is there
any indication of the presence of metallic iron showing ferromagnetic
or super-paramagnetic behaviour as we can interpret it. For the
$M(T)$ curves (T being temperature) and M^{-1} (T) (see reference [5]
find the functional form is more appropriate for mixtures of ferri-
magnetics. Also, the temperature cycling pattern displayed in

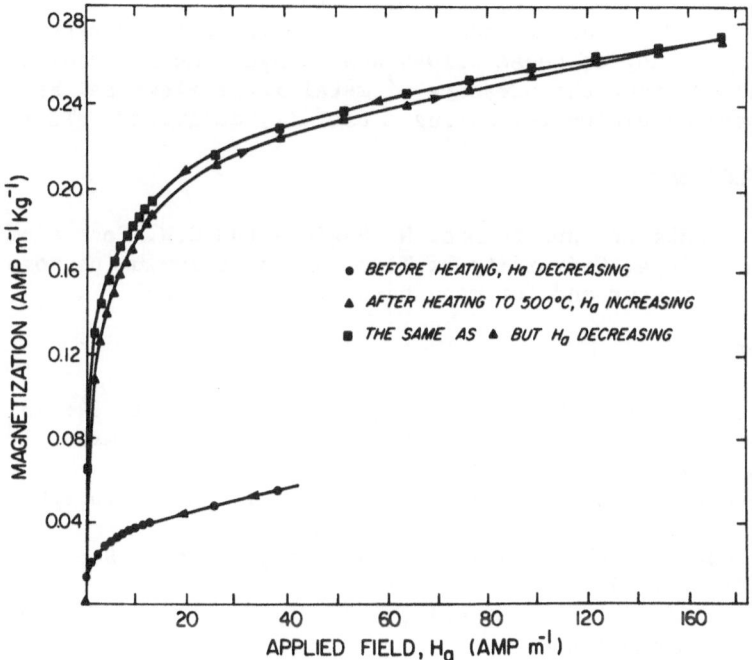

Fig. 2. Magnetization of unactivated FeTi versus applied field
which is taken at room temperature before and after
heating to 500°C.

Fig. 1 does not represent the M(T) behaviour for a ferromagnet or a super-paramagnetic system as we understand it – included in our view of this comparison are our time–temperature behaviour and irreversibility of the upper branch beyond 575°C, as described earlier in this paper. We should emphasize that we took trouble to ensure that the thermal, hydrogen and magnetic cycles to which the FeTi powder in our experiments was subjected, did not cause the bulk FeTi material inside a powder grain to change. Thus we have confidence that the magnetic observations we have made come from regions close to the grain surfaces. These observations taken together with other work in our laboratory in structural features of activated FeTi surfaces lead us to the conclusion that metallic or elemental Fe is not present in any significant quantity in our FeTi alloys, and, as these alloys absorb hydrogen successfully, we also conclude that the presence of metallic or elemental Fe is not a necessary condition for having a usefully activated FeTi alloy.

ACKNOWLEDGEMENTS

    Our thanks are due to Drs. N. Sugiura and G.W. Pearce of Erindale College, University of Toronto, for allowing us to use their magnetometer and for many helpful discussions.

REFERENCES

1. F. Stucki and L. Schlapbach, J. Less. Com. Metals 74:143 (1980).
2. L. Schlapbach, A. Seiler, F. Stucki and H.C. Siegmann, J. Less. Com. Metals 73:145 (1980).
3. D. Shaltiel, Th. von Walkrich, F. Stucki and L. Schlapbach, J. Phys. F: Metal Phys. 11:471 (1981).
4. K. Ikeda, T. Nakamichi and M. Yamomoto, J. Phys. Soc. Japan 37:652 (1974).
5. D. Khatamian, G.C. Weatherly, F.D. Manchester and C.B. Alcock, (to be published).
6. S. Foner, The Rev. Sc. Ins. 30:548 (1959).
7. T. Nagata and S. Akimoto, Magnetic Properties of Rock-Forming Ferromagnetic Minerals, in: "Rock Magnetism", T. Nagata, ed., Maruzen Company Ltd., Tokyo (1961).
8. G. Busch, L. Schlapbach, P. Fischer and A.F. Andresen, Int. J. Hydrogen Energy 4:29 (1979)
9. D. Khatamian, N.S. Kazama, G.C. Weatherly, F.D. Manchester and C.B. Alcock (to be published in the proceedings of the International Symposium on the Properties and Applications of Metal Hydrides, May 30 – June 4, 1982, Toba, Japan).

JAHN-TELLER RESONANCE STATES IN THE

Vb METAL HYDRIDES

G. C. Abell

MRC-Mound*
Miamisburg, Ohio 45342

*A hamiltonian with a localized Jahn-Teller electron-phonon in-
teraction plus a band-coupling term belonging to a metal cluster
state adjacent to interstitial hydrogen in the Vb metals is presented. It is shown that in a low-frequency approximation, this
hamiltonian is analogous to the Anderson Hamiltonian for magnetic
impurities in metals. The role of hydrogen in triggering the Jahn-
Teller instability is explained in the context of the Hartree-Fock
solutions of the low-frequency hamiltonian, using a 'phase diagram'
in the Hartree-Fock parameter space. An elastic susceptibility
expression is derived and compared with an observed elastic relaxation. The empirical connection of the Jahn-Teller distortion to
the observed elastic dipole tensor is reconsidered.*

The concept of Jahn-Teller (JT) resonance states was recently introduced to explain the unusual behavior of hydrogen in the Vb metals
(V, Nb and Ta).[1] A model calculation[2] showed this idea to be plausible. One interesting result of that study was the similarity of
the full hamiltonian (JT state interacting with d-band) to the
Anderson Hamiltonian for magnetic impurities in metals.[3] In this
paper we exploit that similarity and discuss conditions under which
JT resonance stabilization can occur. We argue that interstitial
hydrogen acts indirectly to trigger an instability in the adjacent
JT orbital. A relaxation mechanism is described and an expression is
given for the corresponding susceptibility. This mechanism quantitatively explains a dispersion step observed in an acoustic phonon
branch of both NbH_x and TaH_x.[4] Before proceeding with the above
agenda, we present a brief discussion of the empirical ground from
which is the basic concept arises, partly in order to correct an error
in I (reference 1).

*MRC-Mound is operated by Monsanto Research Corporation for the U.S.
Department of Energy under Contract No. DE-AC04-76-DP00053.

The hydrogen elastic dipole tensor $\underline{P}$ is unusual because it is apparently cubic,[5] in spite of the tetragonal symmetry of the occupied site. Various ideas have been advanced to explain this puzzle, and the general consensus now has a relatively long-range impurity potential extending to the second shell of metal atoms about the impurity.[6] This idea rests on the reasonable assumption of central impurity forces. But to explain the facts with a long-range potential requires a strong (~ 1 eV/Å) direct H-M interaction at the second shell separation $R_{HM} = 0.90$ a_o. This is inconsistent with x-ray and neutron diffraction experiments showing that the direct H-H interaction at $R_{HH} = 0.87$ a_o (the nearest neighbor hydrogen separation in ordered hydride phases) is very weak.[7] What sort of impurity potential would allow this dichotomy?

If the idea of a long-range potential must be abandoned, one is forced to conclude that the cubic elastic dipole tensor is somehow manifesting a noncentral potential. The JT model[1] assumes that the hydrogen screening perturbation triggers a local instability in a degenerate first-shell molecular orbital. This idea can explain the existence of noncentral forces, but what specific JT (symmetry-lowering) distortion is consistent with the facts? The particular distortion in which two nearest-neighbor first-shell metal atoms are displaced along the <111>-bond connecting them does not give a cubic elastic dipole tensor as claimed in I. In spite of this error, the simple <111>-distortion may still be consistent with the facts. To see this, note that there are four equivalent <111>-bonds for a given four-atom t-site cluster. Dynamical averaging of the distortion over these four bonds (dynamic JT effect) in fact gives a cubic elastic dipole tensor. But apparently even this is not required because, according to Yamada et al[8] in an X-ray study of magnetite based on an analogous JT-model, such dynamical behavior would have little effect on the intensity distribution from Huang scattering. In other words, for a given kind of t-site (x-, y- or z-), spatial averaging of the symmetry-lowering distortion would give essentially the same experimental result as dynamical averaging.

Before writing down the hamiltonian, it will be useful to reiterate the basic hypothesis of I and II (reference 2). In the Vb metals there is a degenerate molecular orbital, ϕ_{JT}, made up of atomic t_{2g} d-orbitals belonging to the four-atom metal cluster surrounding a given H-site that has localized JT interaction comparable to its band-coupling interaction. Band coupling quenches the JT degeneracy in the pure metal, whereas in the hydride, occupancy of the t-site by hydrogen tips the scales in favor of a localized symmetry-lowering distortion of the occupied cluster. Because hydrogen couples only to the fully symmetric subspace, its role is indirect.[2] Thus the following treatment does not explicitly include the hydrogen potential. The hamiltonian for a single JT defect in second-quantized notation is

$$H = \sum_k \varepsilon_k n_k + \varepsilon_o (n_+ + n_-) + V \sum_{\sigma = \pm} \psi_\sigma^\dagger c_\sigma + \lambda (n_+ - n_-)(b^\dagger + b) + \omega_o b^\dagger b. \quad (1)$$

The first term describes band states; the second term represents
the JT defect; the third term is the band-coupling interaction; the
fourth term is the JT electron-phonon interaction; and the last term
is the kinetic energy of the local oscillator, with frequency ω_o.

To gain insight into the behavior of the full hamiltonian (1),
we invoke the Born-Oppenheimer (BO) approximation that the electrons
adjust adiabatically to the instantaneous defect nuclear configuration.
The following is an HF type of theory.[9] The operator $(n_+ - n_-)$ of Hamiltonian (1) is replaced by its expectation value $S \equiv \langle n_+ - n_- \rangle$ to calculate the boson states. Using a simple displaced oscillator transformation, the bose operators for the new ground state are found to
be $a^\dagger = b^\dagger + \lambda S/\omega_o$. An effective one-electron Hamiltonian, valid
for frequency $\omega < \omega_o$, is now obtained by replacing $(b^\dagger + b)$ by its expectation value $\langle b^\dagger + b \rangle = -(2\lambda/\omega_o)S$, giving

$$\tilde{H} = \sum_k \varepsilon_k n_k + V \sum_{\sigma = \pm} \psi_\sigma^\dagger c_\sigma + \varepsilon_+ n_+ + \varepsilon_- n_- + \tfrac{1}{2}\varepsilon_{JT}S^2 + \omega_o a^\dagger a. \qquad (2)$$

In this expression $\varepsilon_\pm = \varepsilon_o \pm \varepsilon_{JT}S$, where $\varepsilon_{JT} \equiv 2\lambda^2/\omega_o$.

Apart from the oscillator energy, Hamiltonian (2) is essentially
the same as the HF approximation to the Anderson Hamiltonian for magnetic
impurities in metals.[3] The HF method of solution is to calculate
self-consistent values of $S \equiv \langle n_+ - n_- \rangle$, using a Green's function formalism with some particular modeling of the free-electron (band)
states and of the band-coupling term proportional to $\underline{V}$. There are two
distinct solutions: a nondegenerate one corresponding to S=0 and a
doubly degenerate one wiht $S=\pm p$, where p is a positive number satis-
fying $p_c \leq p \leq 2$. The nondegenerate solution always gives self-consistency, but the degenerate solution is the stable one if it exists.
In the present case, $S \neq 0$ corresponds to orbital- rather than to
spin-polarization.

The results are most succinctly expressed in terms of an HF 'phase
diagram' in the parameter space. The important parameters for the
low-frequency Hamiltonian (2) are, in reduced form: $\varepsilon_r = (\varepsilon_F - \varepsilon_o)/\varepsilon_{JT}$
and $V_r = V/\varepsilon_{JT}$, where ε_F is the Fermi level. Figure 1 is a schematic
HF phase diagram in (ε_r, V_r)-space showing regions of broken symmetry
HF solutions $(S=\pm p)$ and of quenched HF solutions (S=0). Figure 1
also illustrates the role of hydrogen: in the pure Vb metals (represented by Nb), the JT degeneracy is quenched. When a single hydrogen
impurity occupies a t-site in Nb, the self-energy ε_o of the JT orbital
belonging to the occupied cluster is renormalized,[2] forcing the system
across the boundary. Quite apart from the details, this role of hydrogen -- indirect though it may be -- is essential to the integrity
of the concept of JT resonance states as applied to the Vb hydrides.[1,2]
This suggests an important test of the model: if the system is so
close to the boundary that a relatively small change in ε_r can force
it across, then it should be possible to restore the system to its
initial quenched state by effecting a change in the global parameter

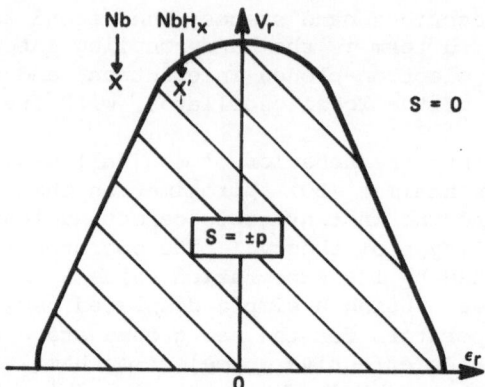

Fig. 1.   Hartree Fock 'phase diagram' in reduced parameter space
         (schematic).

ε_F so as to compensate the hydrogen-induced change in the local parameter ε_0. Specifically, the addition of a group IV metal (e.g., Ti)
would change ε_F in the right direction, with the possibility of an
abrupt change in hydrogen related properties -- especially those having to do with local symmetry (e.g., lattice distortion).

    In the following, we consider the response of an ensemble of hy-
drogen-induced JT defects to an externally imposed elastic tetragonal
deformation, achieved by exciting the T_{1A} phonon branch.   The existence of a dispersion step in the T_{1A} branch between ultrasonic
(10^7 Hz) and neutron (10^{12} Hz) frequencies in both NbH_x and TaH_x im-
plicates a relaxation mechanism in this frequency range.[4] The following expressions give the static and high frequency limits of the
elastic susceptibility due to what we have termed isoconformational
relaxation:[10]

$$\chi_0 = 2\rho_{JT}\beta^2/(1-2\rho_{JT}\varepsilon_{JT}), \quad \omega << \omega_0 \tag{3a}$$

$$\chi_\infty = 2\rho_{JT}\beta^2, \omega >> \omega_0, \tag{3b}$$

where $\rho_{JT} \equiv \rho_+(\varepsilon_F) + \rho_-(\varepsilon_F)$ is the projection of total density of
states onto the Jahn-Teller states and β is the strain coupling co-
efficient.

    To show that the isoconformational relaxation can explain the
observed dispersion step, we first relate the susceptibility (3) to
a difference in the T_{1A} elastic constant $C' \equiv \frac{1}{2}(C_{11}-C_{12})$ for a
given hydrogen composition, between its high frequency (neutron diffraction) and low frequency (ultrasonic) values. Assuming that the
characteristic frequency of isoconformational relaxation lies well
with this range,

$$\Delta C'_{relax} \equiv [C'(10^{12}Hz) - C'(10^7Hz)]_{MH_x} = -\frac{1}{3}N_H(\chi_\infty-\chi_0) \tag{4}$$

where N_H is the hydrogen concentration (which is also the concentration of JT defects). In equation (4) we have used the relations[10] $E = E_0 - e^2 \chi N_H$ and $C' \equiv (E - E_0)/e^2$ (e is the strain). The observed dispersion step[4] is given in terms of $\Delta\nu/\nu_M$ vs. $k(110)$, where $\Delta\nu$ is the frequency of the hydride relative to ν_M, the frequency of the pure metal. Using the relation $C \propto \nu^2/k^2$ (valid for small k) and the fact that $\Delta\nu << \nu_M$ gives $(C'_{MH_x} - C'_M)/C'_M \simeq 2\Delta\nu/\nu_M$. Since there is no dispersion step for the pure metal, the observed relaxation is

$$\Delta C'_{relax} = 2C'_M (\lim_{k \to 0}[(\Delta\nu/\nu_M)_{neu.diff.} - (\Delta\nu/\nu_M)_{ultras.}]_{MH_x}) \qquad (5)$$

Precise determination of the susceptibility (3) requires knowledge of β, ρ_{JT}, and ϵ_{JT}. Given that the system is reasonably close to the boundary (see the discussion of Figure 1) such that $2\rho_{JT}\epsilon_{JT} \overset{<}{\sim} 1$, a reasonable estimate may be obtained by taking $\rho_{JT} = (4\epsilon_{JT})^{-1}$. This latter value is close to values obtained in model calculations[2] and gives $\chi_\infty - \chi_0 = -\beta^2/2\epsilon_{JT}$. Since $\beta \simeq \lambda \cdot R$ and $\epsilon_{JT} = \lambda^2/2\mu\omega_0^2$, where $R \simeq 3A$ is the nearest-neighbor distance and μ is the mass of a cluster metal atom, we obtain $\chi_\infty - \chi_0 \simeq -\mu\omega_0^2 R^2$. Taking[2] $\omega_0 \simeq 10^{-2} eV$ gives $\chi_\infty - \chi_0 \simeq -22 eV$ and by equation (4) the calculated isoconformational relaxation for NbH_x is $\Delta C'_{relax}$(per 1% H) = $\underline{0.6 \times 10^{10} \text{ erg cm}^{-3}}$. The observed relaxation, given by equation (5) (adjusted to $N_H = 5 \times 10^{20} \text{ cm}^{-3} = 1\% H$, and using[11] $C'_M = 55 \times 10^{10} \text{ erg cm}^{-3}$) is $\Delta C'_{relax}$ (per 1% H) = $\underline{0.4 \times 10^{10} \text{ erg cm}^{-3}}$ for NbH_x and only slightly larger for TaH_x. In addition to this numerical agreement, the theory also predicts that $\Delta C'_{relax}$ should be temperature-independent, as observed.[11]

REFERENCES

1. G. C. Abell, Phys. Rev. B <u>20</u>, 4773 (1979) (hereafter referred to as I).

2. G. C. Abell, Phys. Rev. B <u>22</u>, (1980) (hereafter referred to as II).

3. P. W. Anderson, Phys. Rev. <u>124</u>, 41 (1961).

4. A. Magerl, W. D. Teuchert and R. Scherm, J. Phys. C <u>11</u>, 2175 (1978).

5. H. Peisl, <u>Hydrogen in Metals I</u> (Berlin: Springer-Verlag, 1978) pp. 53-74.

6. G. Bauer, E. Seitz, H. Horner and W. Schmatz, Solid State Commun. <u>17</u>, 161 (1975).

7. T. Springer, <u>Hydrogen in Metals I</u> (Berlin: Springer-Verlag, 1978) pp. 75-100; M. A. Pick and R. Bausch, J. Phys. F <u>6</u>, 1751 (1976).

8. Y. Yamada, N. Wakabayashi and R. M. Nicklow, Phys. Rev. B <u>21</u>, 4642 (1980).

9. A. C. Hewson and D. M. Newns, J. Phys. C <u>12</u>, 1665 (1979).

10. G. C. Abell, J. Phys. F, 1982 (in press).

11. A. Magerl, B. Berry and G. Alefeld, Phys. Stat. Sol. (a) <u>36</u>, 161 (1976).

HYDROGENATION ENTROPIES OF THE $ZrMn_{2+y}$ SYSTEM[*]

F. Pourarian, V. K. Sinha, W. E. Wallace, A. T.
Pedziwiatr[**] and R. S. Craig

Department of Chemistry, University of Pittsburgh
Pittsburgh, PA 15260

ABSTRACT

Hydrogenation of Mn-containing intermetallics leads to striking changes in magnetic properties. For example, Th_6Mn_{23} is a Pauli paramagnetic but becomes ferrimagnetic upon hydrogenation. Vapor pressure studies of $ZrMn_{2+y}$ hydrides have suggested unusual ΔH and ΔS values for hydrogen release. The values are about 40 and 50%, respectively, lower than the corresponding values for the paradigm hydride material $LaNi_5H_6$. The unusual thermodynamics could originate with either a large magnetic contribution or irreversibility for the system. Recent calorimetric measurements indicate that irreversibility is the major factor. Entropies of absorbed hydrogen derived from the calorimetric results are about 17 J/K g.atom of H, which is in good agreement with the computed values, 18.3 J/K g. atom H.

INTRODUCTION

Manganese forms many intermetallic compounds with strongly electropositive metals, e.g., Y_6Mn_{23} and Th_6Mn_{23} (discussed elsewhere in this conference), RMn_2 (where R is a rare earth), $ZrMn_2$, etc. These materials readily hydrogenate to form very hydrogen-rich systems. For example, $GdMn_2$ forms[1] the hydride $GdMn_2H_3$, and $ZrMn_2$ may

---

[*]This work was assisted by a contract with the Koppers Co., Inc. Pittsburgh, and by Grant No. CHE-7908914 from the National Science Foundation.
[**]On leave from the Institute of Physics, Jagellonian University, Cracow, Poland.

be hydrogenated[2] to $ZrMn_2H_{3.6}$.

At saturation the several Mn intermetallics cited contain hydrogen to a proton density exceeding that of liquid hydrogen.[3] Because of the high concentration of hydrogen, substantial changes occur in the systems during hydrogenation. For example, Th_6Mn_{23}, which is a Pauli paramagnet, becomes[4] ferrimagnetic during hydrogenation, whereas Y_6Mn_{23} is transformed[5,6] from a ferrimagnet to an antiferromagnet during hydrogenation.

$ZrMn_2$ was examined by Shaltiel et al.[2] as a hydrogen storage material. It has excellent hydrogen capacity but is found to be too stable for hydrogen storage. Following the discovery by Buschow and Van Mal[7] that the stability of $LaNi_5$ hydride could be significantly affected by making the intermetallic non-stoichiometric, Van Essen and Buschow[8] and Pourarian et al.[9] examined non-stoichiometric $ZrMn_2$ and $ZrMn_2$-based systems in which the Mn/Zr ratio exceeded 2, viz., $ZrMn_{2+y}$. Hydrogenation characteristics of systems were studied with y ranging from 0 to 1.8. It was shown by Sinha and Wallace[10] that the deviation from stoichiometry was achieved through the partial replacement of Zr by Mn so that $ZrMn_{2+y}$ is better represented as $Zr_{1-x}Mn_xMn_2$. (There is, of course, a simple relationship between x and y.) Pourarian et al.[9] found that as x increased to about 0.4 the material acquired characteristics making it significant for hydrogen storage. The non-stoichiometric material exhibits not only good hydrogen capacity but excellent kinetics of hydrogenation and dehydrogenation, the reasons for which are discussed elsewhere.

In regard to hydrogen storage materials, it is often asserted[11,12] that one needs a material whose hydride decomposes endothermically with $\Delta H \approx 30$ kJ/mole H_2. The rationale behind this is the assumption that $\Delta S \approx 130$ J/°K mole H_2, i.e., the entropy gain in dehydrogenation equals the entropy of gaseous hydrogen. This is equivalent to attributing zero entropy to the dissolved hydrogen. Actually, in the case of $LaNi_5H_6$ (see Table 1), hydrogen carries a configurational entropy of about 4.5 $J°K^{-1}$ per gram atom of hydrogen because of statistical site occupancy (see ref. 12) and there are, in addition, some vibrational changes in the lattice[13], so that $\Delta S = 106.4$ instead of 130 $J°K^{-1}$ per mole of H_2.

As indicated in the preceding paragraph, it has been generally accepted that a good hydrogen storage material will be characterized by a ΔH of about 30 kJ/mole H_2. In our earlier studies of $Zr_{1-x}Mn_xMn_2$ alloys it was found[9] that ΔH is in the range 18 to 20 kJ/mole H_2. Were ΔS comparable with that in $LaNi_5H_6$, the pressure of $Zr_{1-x}Mn_xMn_2$ hydride at 300°K would be about 4600 atm. Experimentally, it is found[9] that the pressure is about 10^4 smaller. This indicates one of two things - either ΔH as obtained earlier is incorrect or ΔS is very different for this system than for $LaNi_5H_6$. These two possibilities are explored in the remainder of this paper.

Table 1.   Thermodynamics of Hydrogenation
(obtained using equation 2)

	ΔH(kJ/mole H$_2$)	ΔS(J/K mole H$_2$)
ZrMn$_{2.8}$H	17.8	50.6
ZrMn$_{2.8}$H$_2$	18.4	52.3
ZrMn$_{3.8}$H$_{1.5}$	17.0	51.0
ZrMn$_{3.8}$H$_2$	19.7	61.5
LaNi$_5$	30.0	106.4

ENTHALPIES OF HYDRIDE DECOMPOSITION

A hydride develops a vapor pressure P at temperature T.   The familiar Clausius-Clapeyron equation

$$\frac{d\ell nP}{dT} = \frac{\Delta H}{RT^2} \tag{1}$$

can be applied in this instance.   Neglecting the temperature variation of ΔH, this integrates to

$$\ell nP = -\frac{\Delta H}{R}\frac{1}{T} + B \tag{2}$$

where B, the constant of integration, is $\Delta S/R$.   Hence a plot of ℓnP versus T^{-1} should be linear, with slope and intercept equalling $-\Delta H/R$ and $\frac{\Delta S}{R}$, respectively.   The ΔH and ΔS refer, in the case that the plateau pressure is used in equation (2), to the conversion of a hydrogen-rich hydride phase into a hydrogen-poor solid solution phase.   If the latter is rather dilute, the thermodynamic quantities are virtually equivalent to the negative of the values obtained in inserting hydrogen into the pure storage material.

The values obtained from equation (2) for ZrMn$_2$-based systems are shown in Table 1, along with, for purposes of comparison, those obtained for LaNi$_5$.   As noted in the Introduction, ΔH for hydrides of ZrMn$_{2+y}$ alloys is abnormally low.   These values are derived through the use of equation (2), which originates from equation (1). The Clausius-Clapeyron equation applies only to systems at equilibrium, i.e., those exhibiting reversible behavior.   Essentially no systems studied to date meet the reversibility criterion.[14]   Almost all systems exhibit hysteresis.   Therefore, almost all ΔH and ΔS values published to date, including our own, are suspect.   They may be correct, but this cannot be assumed á priori.

The validity of the ΔH values listed in Table 1 can be tested by calorimetrically measuring the enthalpy of hydrogenation.   Results obtained by this technique are listed in Table 2.   Data were

Table 2.   Calorimetrically Determined ΔH Values

$Zr_{0.79}Mn_{0.2.}Mn_2H_x$		$Zr_{0.62}Mn_{0.38}Mn_2H_x$	
x	ΔH(kJ/mole H_2)[a]	x	ΔH(kJ/mole H_2)[a]
1.27	33.5		
1.60	34.7		
1.78	33.4		
1.84	31.2	1.28	31.9
2.6	33.2	1.97	31.2
2.7	31.9	2.7	28.1
2.89	32.4	2.8	28.6,29.1
3.39	32.3		
Average 32.9		Average 29.8	

a. These are heats of absorption.

also acquired for $LaNi_5H_x$ as a test of the calorimetric procedure. For this material ΔH (absorption) was measured as -32.2 kJ/mole H_2 (average of 11 measurement) for x ranging from 1.74 to 5.54 and ΔH (desorption) = 29.8 kJ/mole H_2 (average of 3 measurements) for x ranging from 2.15 to 2.32.   These compare favorably with recent results of Murray et al.,[15] viz. -32.3 and 31.8 kJ/mole H_2, respectively.

ENTROPIES OF $ZrMn_{2+x}$ HYDRIDES

The fact that ΔH determined from the temperature coefficient of vapor pressure is in disagreement with the calorimetrically determined enthalpy change indicates that the utilization of equation (2) is unsatisfactory.   The ΔS values in Table 1 indicate that the effective entropy of one gram atom of hydrogen in $ZrMn_{2+y}$ hydride is about 40 J/K, whereas that computed from the measured vapor pressure (0.55 atm at 303 K) and the value of ΔH for $Zr_{0.62}Mn_{0.38}Mn_2$ hydride is 17.2 J/K g.atom of H.   A similar calculation for $Zr_{0.79}Mn_{.21}Mn_2$ hydride gives 16.4 J/K g.atom of H.

It is of interest to ascertain whether or not the computed configurational entropy (S_c) is in this range. S_c for $ZrMn_{2+y}$ hydride is computed using the formalism employed earlier for β $LaNi_5$ hydride.[12]

$$S_c = -R \Sigma g_i[\theta_i \ln\theta_i + (1-\theta_i)\ln(1-\theta_i)] \qquad (4)$$

In equation (4) g_i is the multiplicity of the ith site and θ_i is its fractional occupancy.   We use the structural information for $ZrMn_2D_{2.75}$ provided by Didisheim et al.[16] to calculate S_c for

$Zr_{0.79}Mn_{0.21}Mn_2H_{2.8}$. The neutron diffraction data of Didisheim et al. show that the unit cell contains $Zr_4Mn_8D_{10.99}$ with D located as follows: 4.30 in 24 ℓ, 4.51 in 12 ℓ, 1.87 in 6 h_1 and 0.31 in 6h_2. Using these, we calculate S_c = 201 J/K mole of $Zr_4Mn_8D_{11.0}$ or 18.3 J/K per gram atom of D. It is to be noted that this is in close agreement with experiment for $Zr_{1-x}Mn_xMn_2$ hydrides.

The close agreement between the effective entropy of hydrogen obtained using the calorimetric values and S_c calculated using equation (4) may be in part fortuitous. Use of equation (4) together with the data of Didisheim et al. probably leads to an overestimate of the configurational entropy of the $Zr_{1-x}Mn_xMn_2$ hydrides. Switendick maintains[17] that configurations which involve H-H distances less than 2.14 Å are too energetically unstable to occur because under these circumstances the antibonding orbitals formed by LCAO of the hydrogen 1s states are too energetic to be populated. Hence, some of the configurations involved in the use of equation (4) do not occur. Additionally, the work of Didisheim et al. shows that only interstices involving 2 Zr's as near neighbors are populated. Those with less - i.e., 1 or 0 Zr's - are energetically unfavorable and are not populated. In the non-stoichiometric $ZrMn_2$ certain of the sites which are energetically favored in $ZrMn_2$ become less favorable in $Zr_{1-x}Mn_xMn_2$ because there are no longer 2 Zr's coordinate to the site. This, too, has been ignored in the computation, and this leads to an overestimate of S_c.

There is, of course, a vibrational contribution to ΔS, but this is probably small. Ohlendorf and Flotow[13] estimated this to be about 1 J/K g. atom of H in hydrided $LaNi_5$, and it is expected that the vibrational contribution would be of this order in the present system.

CONCLUDING REMARKS

From the arguments in the preceding section it appears that the entropy values derived using the calorimetrically obtained ΔH values are in reasonable accord with that expected from the structural evidence at hand. It should be appreciated that the error in ΔS due to irreversibility is probably very slight. For example, even a 100% error in vapor pressure produces only a 5% error in the calculated ΔS. This is because the $\Delta H/T$ term is the major contribution to ΔS.

It is interesting to note that the ΔH values obtained calorimetrically for the $Zr_{1-x}Mn_xMn_2$ hydrides are lower than that obtained[2] for the hydrogenation of stoichiometric $ZrMn_2$, 53 kJ/mole H_2. This implies that the higher energy sites in the non-stoichiometric material involving less than 2 Zr near neighbors are not avoided in the hydrogenation process, which is also indicated by the configurational entropy of the system.

REFERENCES

1. I. Jacob and D. Shaltiel, J. Less-Common Metals <u>65</u>, 117 (1979).
2. D. Shaltiel, I. Jacob and D. Davidov, <u>ibid</u>., <u>53</u>, 117 (1977).
3. W. E. Wallace, R. S. Craig and V. U. S. Rao, Advances in
    Chemistry Series, No. 186, "Solid State Chemistry: A Contem-
    porary Overview," Smith L. Holt, Joseph B. Milstein and Murray
    Robbins, eds. American Chemical Society, 1980, p. 207.
4. S. K. Malik, T. Takeshita and W. E. Wallace, Solid State Commun.
    <u>23</u>, 599 (1977).
5. K. Hardman, J. J. Rhyne, H. K. Smith and W. E. Wallace, in <u>The</u>
    <u>Rare</u> <u>Earths</u> <u>in</u> <u>Modern</u> <u>Science</u> <u>and</u> <u>Technology</u>, Vol. 3, G. J.
    McCarthy, J. J. Rhyne and H. Silber, eds. Plenum Press (1982),
    to appear.
6. C. Crowder, B. Kebe, W. J. James and W. Yelon, <u>ibid</u>., to appear.
7. K. H. J. Buschow and H. H. Van Mal, J. Less-Common Metals <u>29</u>,
    203 (1972).
8. R. M. Van Essen and K. H. J. Buschow, Mater. Res. Bull. <u>15</u>,
    1149 (1980).
9. F. Pourarian, H. Fujii, W. E. Wallace, V. K. Sinha and H.
    Kevin Smith, J. Phys. Chem. <u>85</u>, 3105 (1981).
10. V. K. Sinha and W. E. Wallace, unpublished.
11. H. H. Van Mal, K. H. J. Buschow and A. R. Miedema, J. Less-
    Common Metals <u>35</u>, 65 (1974). See also R. L. Cohen and J. H.
    Wernick, Science <u>214</u>, 1081 (1981).
12. W. E. Wallace, Howard E. Flotow and D. Ohlendorf, J. Less-
    Common Metals <u>79</u>, 157 (1981).
13. D. Ohlendorf and H. E. Flotow, J. Chem. Phys. <u>73</u>, 2987 (1980).
14. See, for example, Biays S. Bowerman, C. A. Wulff and Ted B.
    Flanagan, Zeit. f. Phys. Chem. N.F. <u>116</u>, 197 (1979).
15. J. J. Murray, M. L. Post and J. B. Taylor, J. Less-Common
    Metals <u>80</u>, 201 (1981).
16. J.-J. Didisheim, K. Yvon, D. Shaltiel and P. Fischer, Sol.
    State Commun. <u>31</u>, 47 (1979).
17. A. C. Switendick, Adv. in Chem., No. 167, p. 281 (1978). See
    this for references to other APW calculations by Switendick.

ON THE ELECTRONIC STRUCTURE OF HYDROGEN IN METALS

IN THE LIGHT OF MUON KNIGHT SHIFT MEASUREMENTS

Alexander Schenck

Laboratorium für Hochenergiephysik
der ETH Zürich, c/o SIN
5234 Villigen, Switzerland

The Knight shift of positive muons has been measured up to now in 18 elemental nontransition (simple) metals and in the transition metals V, Nb, Ta, Pd and Ni. In addition data are available from the hydrogen storage system $LaNi_5H_x$. Except for the latter substance all data correspond to the limit of zero hydrogen concentration. Since the bare muon like the proton carries a single elementary positive charge and introduces a strongly disturbing Coulomb potential into the host lattice it is expected that both particles induce the same local electronic structure, subject perhaps to isotope effects. The total Knight shift, consisting of paramagnetic and diamagnetic contributions, allows thereby to test certain features of the local electronic structure. The data will be compared with existing predictions obtained from jellium, cluster and band structure calculations. Some interesting systematics of the data and possible isotope effects will be pointed out.

INTRODUCTION

Much could be learned about the local electronic structure of hydrogen in metals from measurements of the proton Knight shift (KS) and the related electron induced longitudinal relaxation time T_{1e}. In practice such measurements appear difficult for various reasons

391

and in fact only a few reliable KS data are available, restricted to systems that can absorb sufficient quantities of hydrogen (see e.g. Cotts[1]). The most reliable proton KS data have been obtained by Kazama and Fukai[2] in the α-phase of VH_x, NbH_x and TaH_x. No data at all are known for protons in simple metals, such as Mg, Cu or Al or in transition metals such as Ni, Fe etc. Even though the hydrogen solubility is generally low in these elements information on the electronic structure should be very important in order to gain some general insight into the metal-hydrogen problem. In contrast positive muons (μ^+) can easily be implanted in any kind of metal or alloy, independent of whether the hydrogen solubility is high or low or even zero and the μ^+ Knight shift can be measured straightforwardly by muon spin rotation (μSR) spectroscopy[3]. In this context the μ^+ can be considered a light isotope of the proton, carrying like it a single elementary positive charge and introducing the same strongly disturbing Coulomb potential into the metal matrix. To the extent that isotope effects can be neglected ($m_\mu \sim 1/9\ m_p$) the local electronic structure around the two particles should be the same. Taking possible isotope effects into account a comparison of proton and μ^+ KS data in the same metal host should prove of extreme interest in checking on theoretical models, if available.

Note, that in a μSR experiment there is usually only one μ^+ present at a given instant of time, so that the situation corresponds to the limit of zero hydrogen concentration. Effects of hydrogen-hydrogen interactions are therefore absent. One is confronted with the model case of just one hydrogen impurity in an infinite metal lattice.

The KS constant K is defined by the following expression for the total magnetic field B_μ at a μ^+ inside the metal when an external field B_{ext} is applied:

$$B_\mu = B_{ext} + (-N \cdot \chi_b + \frac{4\pi}{3}\chi_b) \cdot B_{ext} + K \cdot B_{ext} \qquad (1)$$

Here $-N\chi_b B_{ext}$ is the demagnetization field depending on the sample shape, $4\pi/3\ \chi_b\ B_{ext}$ is the so called Lorentz field, χ_b is the bulk magnetic susceptibility and $K \cdot B_{ext}$ the total KS. It follows that

$$K = \frac{B_\mu}{B_{ext}} - 1 - (\frac{4\pi}{3}-N)\ \chi_b \qquad (2)$$

For spherical samples, which are usually used in μSR experiments, the demagnetization field is canceled by the Lorentz field and the KS constant is directly determined from the internal field B_μ and the applied field B_{ext}. The bulk susceptibility χ_b need not to be known.

The total KS is composed of several contributions, each of which is dependent on the established local electronic structure:

$$K = K_s + K_{cp} + \sigma_d \tag{3}$$

The first term originates from the Fermi contact interaction with Fermi surface conduction electrons which are spin polarized by the applied field:

$$K_s = \frac{8\pi}{3} <|\psi(r_\mu)|^2>_F \Omega_e \chi_s = \frac{1}{\mu_B} <B_{hf}(r_\mu)>_F \Omega_{at} \chi_s \tag{4}$$

χ_s is the conduction s-electron spin susceptibility (the Pauli susceptibility), $<|\psi(r_\mu)|^2>_F$ the average density of Fermi surface conduction electrons at the μ^+ and $\Omega_e^{-1} = n$ the conduction electron density. Alternatively one can define an induced hyperfine field per unpaired Fermi surface electron per atom $<B_{hf}(r_\mu)>_F$, Ω_{at} is now the atomic volume. This KS term is usually called direct KS and is always positive and temperature independent. The second term in eq. (3) is a "core polarization" term with either positive or negative sign. It is particularly important in transition metals. It involves electronic states below the Fermi energy which have different spin up and spin down densities at the μ^+ due to exchange interactions with the spin polarized Fermi-surface electrons. In analogy to eq. (4) K_{cp} may be written as

$$K_{cp} = \frac{1}{\mu_B} B_{hf}^{cp}(r_\mu) \Omega_{at} \chi_{cp} \tag{5}$$

$B_{hf}^{cp}(r_\mu)$ is again an induced hyperfine field per unpaired electron per atom and χ_{cp} the spin susceptibility of those Fermi surface electron that participate in the exchange mechanism. In transition metals χ_{cp} can be identified with the d-electron spin susceptibility χ_d. Since in metals like paramagnetic Ni, Pd etc. χ_d is strongly temperature dependent it becomes possible to distinguish between K_s and K_{cp}. In contrast to the host nuclear KS the d-electrons in transition metals may also produce directly a contact KS at the interstitial

μ^+ due to the tails of the d wave functions.

The third term in eq. (3) concerns the diamagnetic screening of the μ^+. It is analogous to the chemical shift in molecules and is given by the Lamb expression

$$\sigma_d = - \frac{e}{3mc} <o|\frac{1}{r}|o> \tag{6}$$

In contrast to K_s and K_{cp} it is determined by the entire charge distribution around the μ^+. Its contribution to the nuclear KS is generally negligibly small in relation to K_s and K_{cp}. Not so in the case of an interstitial proton or μ^+. Choosing a jellium model approach Zaremba and Zobin[4] have calculated σ_d for μ^+ for a wide range of metallic electron densities. Only a weak dependence on electron density is found with $\sigma_d \simeq - 20$ ppm. This is close to the diamagnetic screening of protons in atomic hydrogen (- 17.8 ppm). These values are a sizable fraction of at least K_s.

In summary the various terms contributing to the total KS reflect the local electronic structure in various parts: K_s and K_{cp} depend on the charge or rather the spin density at the μ^+ site and test therefore only a local property, while K_{dia} is effected by the total charge distribution around the μ^+.

It is to be noted, however, that K_s and K_{cp} appear always as a product of a bulk property, namely χ_s or χ_{cp}, respectively, and a local property, namely $<|\psi(r_\mu)|^2>_F$, $<B_{hf}(r_\mu)>_F$, or $B_{hf}^{cp}(r_\mu)$, respectively. To gain access to the local property, the spin susceptibility has to be known with some confidence. This is the case for the alkali metals, Be, Al and Cu[3]. For all other metals χ_s is determined indirectly from other quantities[3]. In transition metals χ_d is identified by its temperature dependence. Since the study of the μ^+ KS concerns the infinite dilute concentration limit one can at least be sure, that the bulk properties of the host metal are unchanged.

Theoretical predictions for the KS are available from different model calculations of the electronic structure of hydrogen in metals, ranging from non linear screening jellium calculation[5,6], to cluster calculations[7], to band structure calculations[8]. Results will be listed and compared with experimental data in the following sections.

MUON KS IN NONTRANSITION (SIMPLE) METALS

A collection of available results[3,9] is presented in Fig. 1. The data are plotted versus the main quantum number of the valence electrons of the host metal atoms. This allows to display the general trend of the KS as one moves down a column in the periodic table. All the data were obtained at SIN by the stroboscopic μSR technique, which allows a precision determination of the μ^+ Larmor frequency in the ppm region[3]. Not shown in Fig. 1 are results obtained in Zn and Cd single crystals, which are unusual in may respects[10]. The Cd data are presented separately by Studer at this conference[11].

In a few cases the KS has also been measured in the liquid phase of the metal (Rb, Cs, Ga, Hg). Generally the KS in the liquid phase is larger than in the solid phase. Most remarkable are the change from a negative KS in solid Ga to a positive KS in liquid Ga, and the huge jump of the KS in Hg upon melting. In Cs and less so in Rb, the KS increases only little upon melting. Fig. 1 shows further that a negative KS is observed in Be, Sr as well as in Ga. This fact demonstrates that it is not sufficient to consider the direct KS (eqs. 3, 4) alone, but that indeed diamagnetic screening and core polarization effects have to be taken into consideration.

In the case of Cu the KS has been measured between 30 K and 800 K. No significant dependence on temperature was noticed[3]. The bulk of the data were obtained at room temperature.

Let us now compare the experimental results with theoretical predictions. For nontransition metals almost all theoretical calculations in the past were based on the jellium model, treating the screening of the μ^+ in a nonlinear selfconsistent fashion by either adopting the charge density functional- or HKS-formalism or the spin density functional (SDF) formalism[5]. Crystalline effects were not considered at all or were taken into account only in a rough way by the so-called spherical solid model[12].

Using the SDF-formalism the direct KS can be calculated as follows:

$$K_s = \frac{8\pi}{3} \rho_s(r_\mu) \cdot \chi_s \tag{7}$$

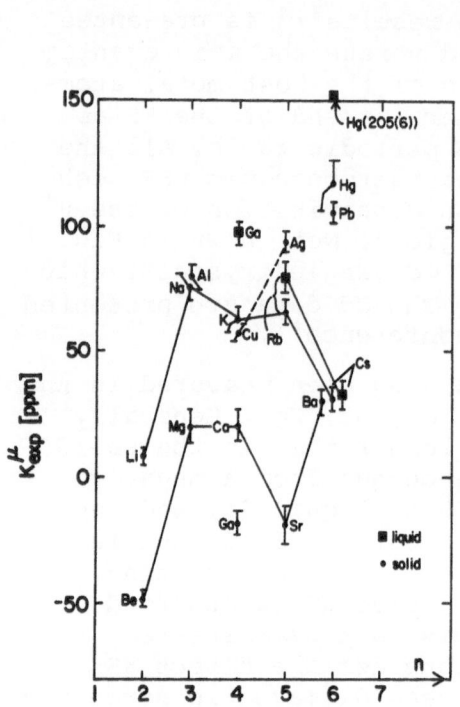

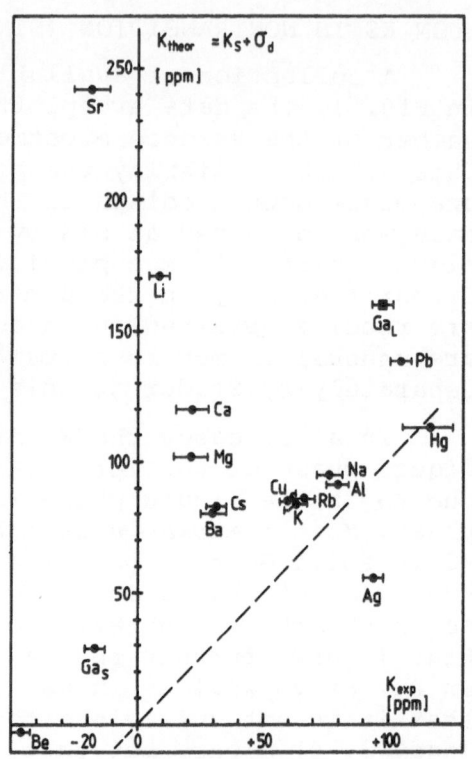

Fig. 1. μ^+-Knight shift in simple metals versus the main quantum number of the valence electrons.

Fig. 2. Comparison of experimental K_μ and theoretical K_μ from jellium calculations.

where $\rho_S(r_\mu)$ is the spin density enhancement factor, which will be provided by theory.

$\rho_S(r_\mu)$ does not only include the contribution from the Fermi surface electrons ($<|\psi(r_\mu)|^2>_F$) but also contributions from electron states below ε_F, thus including a "core polarization" effect. For χ_S we use consistently only values of experimental origin, collected in ref. 3. The total KS is calculated by adding σ_d, taken from ref. 4. Theoretical predictions and experimental values are compared in Fig. 2.

It is seen that the predictions reproduce very poorly or not at all the experimental numbers. For most of the data not even the trend is correctly predicted.

The discrepancies would be even larger if the KS would have been calculated on the basis of the HKS-formalism. Interestingly, all data with the exception of Ag and perhaps Hg are smaller than the predictions.

Next we compare the results with more realistic approaches that incorporate also structural effects.

Available results from spherical solid model (SSM), cluster and band structure calculations are collected in Tab. 1 together with the experimental numbers. The SSM predictions, including a diamagnetic contribution, are in fact quite close to the experimental numbers in Na and in Al, when placing the muon in the latter case in the octahedral interstitial site. This, however, seems to be inconsistent with proton channeling[13] and muon diffusion and trapping studies[14] in Al which both suggest a tetrahedral site occupation of protons and muons in Al. For Cu it would seem that a tetrahedral site assignment is also the more favorable one, which is in contradiction with the octahedral site occupation observed in a more direct way[15].

Table 1.   Theoretical KS predictions from SSM, cluster and band structure calculations. Values in parenthesises include diamagnetic shielding. * octahedral,   ** tetrahedral site.

		Ref.	Na	Cu	Be	Al
K_{exp} [ppm]		3	76.5(5.0)	60.0(2.3)	-46.7(3.0)	70.6(4.0)
K_{theor} [ppm]	Spherical solid model (SSM)	12 *	91.3 (71.3)	118.5 (102.5)		93.4 (78.4)
	calculated with use of $\rho_s(r_\mu)$ and experimental χ_s (Diamagnetic corrections from ref. 4)	**	91.3 (71.3)	98.0 (82.0)		50.4 (35.4)
	Cellular cluster	16,17 *		+71 (+61)	+9 (-18)	
	KKR-CPA band structure σ_d=-10 ppm σ_d=-16 ppm	8 * (63) (16) (4)		55 (39) (45)		

Cellular cluster calculations are available for μ^+ in Cu and Be. The predicted KS in Cu[16] is in quite good agreement with experiment. The calculations also yield $\sigma_d = - 10$ ppm, which is quite small compared with the atomic hydrogen value (- 17.8 ppm). The direct KS in Be is found to be very small (+ 9 ppm) but σ_d is now the dominating contributions ($\sigma_d = - 27$ ppm) and the total KS becomes negative[17]. There is still quite a gap to the experimental value, which could point to a "core polarization" contribution[2]. Analysis of the results in liquid and solid Ga points also to the possibility of abnormally high diamagnetic screening[9]. Finally a KKR band structure calculation for μ^+ in Cu yields a direct KS, which appears to be a little bit on the low side when diamagnetic screening is included. The cluster and band structure calculation produce doubly occupied bonding states between the μ^+ and its neighbors, as has been noted previously for transition metals. It appears that also in simple metals a covalent-hydrogen model is applicable.

Finally we mention a curious empirical correlation that is observed for cubic metals, when the induced hyperfine field per unpaired electron per atom, calculated from K by means of eq. 4 after correcting for diamagnetic screening, is plotted versus the molar electronic specific heat γ_{cp} (Fig. 3). With the exception of Li and Pb the data are correlated exponentially with γ_{cp} or, in terms of a free electron gas description with the atomic density of states at the Fermi energy. This empirical relationship is used to calculate an empirical spin density factor ρ_s using appropriate free electron gas formulas. Fig. 4 displays ρ_s versus the electron density parameter r_s. Different curves are obtained for different valencies of the host metal. The heavy solid line is the SDF prediction. The latter one seems to provide a boundary on the low r_s-side. Hence it seems, that the spin density at the μ^+ in high electron density cubic metals is well accounted for in a jellium model, while interesting deviations occur at lower densities.

MUON KS IN TRANSITION METALS

Measurements are available for the group VB metals V, Nb, and Ta[3] and for Pd[18] and paramagnetic Ni[19]. Results for the VB metals are listed in Tab. 2 together with results for the proton KS extrapolated to zero

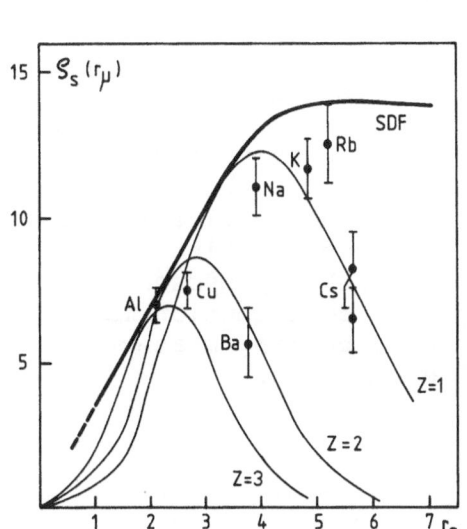

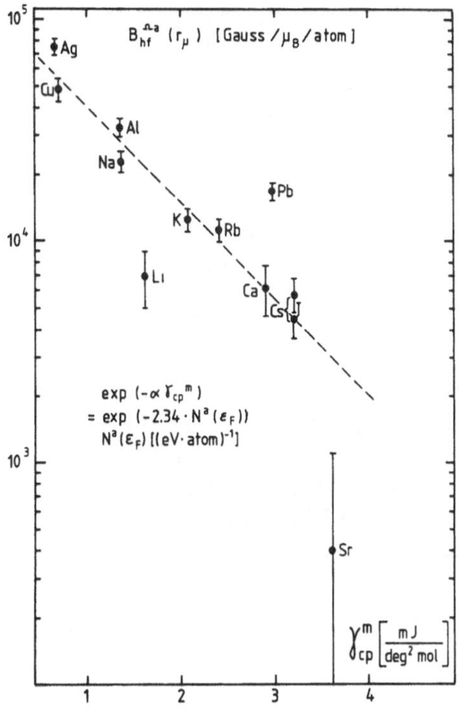

Fig. 3. Display of the correlation of the induced $B_{hf}(r_\mu)/\mu_B/$atom with the molar electronic heat γ_{cp}^m in cubic metals.

Fig. 4. Plot of the empirical spin density enhancement factor versus the electron density parameter r_s.

Table 2. Muon and Proton Knight shift in group VB metals[3,2].

Element	μ^+SR		p NMR	
	Temp [K]	K [ppm]	Temp [K]	K_p [ppm]
V	~290	$-(88\pm8)$	370-470	-66
Nb	100 ~290 423 average	$-(150\pm6)$ $-(16.4\pm4.5)$ $-(15.0\pm4.5)$ $-(15.6\pm3.0)$	370-470	-25
Ta	~290	$+(5.5\pm10)$	370-470	0

hydrogen concentration[2]. In Nb the μ^+ KS has been measured at three different temperatures between 100 K and
423 K yielding the same value. The μ^+ KS follows closely
the proton KS, the absolute magnitude being about
(25 ± 9) % larger in V, (38 ± 12) % smaller in Nb and
essentially in agreement with the proton KS in Ta. The
deviations point to the presence of isotope effects,
which can be of either sign. With the exception of Ta
the KS is negative and, considering the value of the KS
in V, must be partially induced by the d-electrons (via
core polarization). This is also suggested by the observation that the proton and μ^+ KS in the three metals
shows roughly a linear dependence on the total spin susceptibility, calculated from γ_{cp}, which contains a large
d-electron contribution. Kazama and Fukai proposed[2],
that the d-electron induced negative contribution may
be partially due to hydrogen induced bonding states
composed of H-1s and host metal d-orbitals.

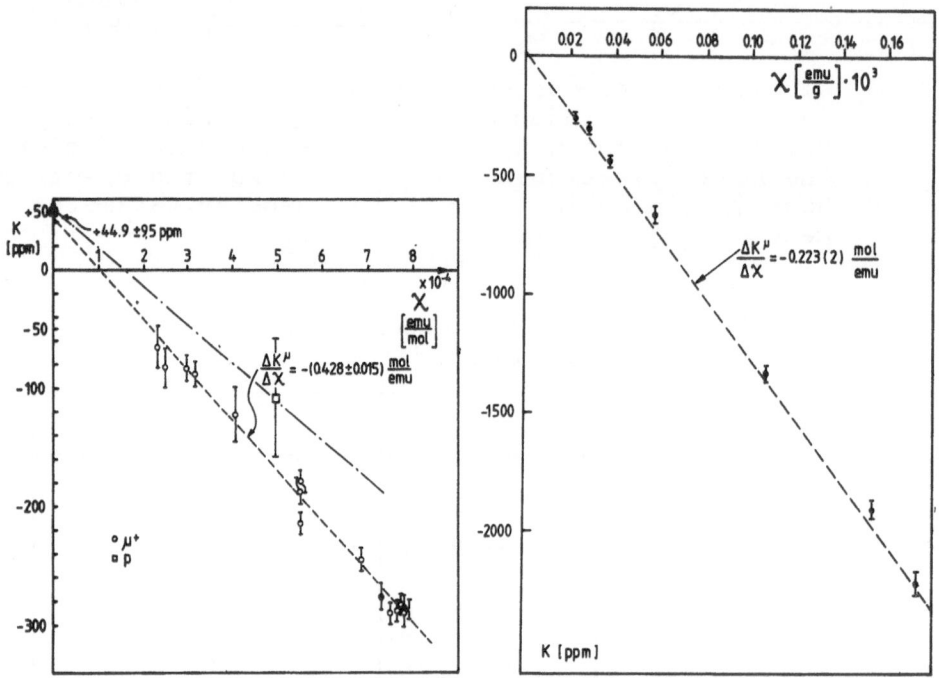

Fig. 5. μ^+ and proton KS in   Fig. 6. μ^+ KS in paramagnetic
        Pd versus the bulk   Ni versus the bulk
        susceptibility χ.   susceptibility χ.

The μ^+ KS in Pd and paramagnetic Ni has been measured as a function of temperature ranging from 20 K
to 880 K in Pd, and from 637 K to 906 K in Ni. The results are plotted versus the bulk susceptibility, with
temperature as an implicit parameter in Figs. 5 and 6.
The data are well reproduced by an expression of the
form of eq. 3.

$$K = \frac{\Omega_{at}}{\mu_B} \{ <B_{hf}(r_\mu)>_F \chi_s + B_{hf}^{cp}(r_\mu) \chi_d(T) \} + \sigma_d \qquad (8)$$

and the usual assumption that the temperature dependence
of the total susceptibility originates solely from the
d-electron contribution $\chi_d(T)$. The slopes of the curves
in Figs. 5 and 6 are then given by $1/\mu_B B_{hf}^{cp}(r_\mu)$, i.e.
for Pd: $B_{hf}^{cp}(r_\mu) = - 2.39(11)$ kG/μ_B/atom and for Ni:
$B_{hf}^{cp}(r_\mu) = - 1.246(6)$ kG/μ_B/atom. For T → ∞,
$\chi_{tot} = \chi_s + \chi_{orb} + \chi_{dia}$ and $K = K_s + \sigma_d$. Using estimates
for χ_s, χ_{orb} and χ_{dia} from refs. 20, 21 one obtains
$K_s + \sigma_d = + 42(13)$ ppm in Pd and $K_s + \sigma_d = - 11(23)$ ppm
in Ni. The results show that the d-electrons contribute
significantly to the μ^+ KS. It is also realized that the
induced hyperfine field per unpaired d-electron per atom
is relatively small compared to $<B_{hf}(r_\mu)>_F \simeq 50$ kG/μ_B/
atom[3] observed, e.g. in Cu. This suggests an indirect
mechanism, i.e. core polarization by an antiferromagnetic exchange interaction. The induced hyperfine field
$B_{hf}^{cp}(r_\mu)$ in paramagnetic Ni can be compared with the
corresponding spontaneous hyperfine field in ferromagnetic Ni[22,3]: $- 1.14(2)$ kG/μ_B/atom. The two values
agree with each other within 10%, indicating that the
induced hyperfine field at the μ^+ is largely independent
of how a d-electron spin polarization is created. The
effective hyperfine field at μ^+ in ferromagnetic Ni has
been calculated by various authors using different
methods. A compilation of results is shown in Tab. 3.
The data are most closely reproduced by a KKR band
structure calculation[23]. This calculation produces
again a doubly occupied bonding state well below the
bottom of the conduction band which is responsible for
the negative sign of the total hyperfine field. A covalent hydrogen-metal picture is again emerging.

Fig. 5 shows also the result of a proton KS measurement in Pd at 75 °C[24]. The large error bars are due
to systematic uncertainties concerning the demagnetization effect. Fig. 5 contains also an estimate for the
proton KS (~ 50 ppm) for T → ∞ (χ_d → 0) by applying the

Table 3. Compilation of theoretical predictions
          for the hyperfine field at μ^+ in ferro-
          magnetic Ni.

Source	Ref.	B_{hf} Gauss
Daniel-Friedel model	28	− 600
Cellular cluster model	29	− 680
KKR-CPA-band structure	23	− 720
Supercell band structure calc.	30	− 463
Experiment ferromagnetic Ni paramagnetic Ni	22 29	− 710(10) (− 770(3))*

* calculated from $B_{hf}^{cp}(r_\mu)$

Korringa relation to a T_{1e} measurement in PdH[25] assuming
that the s-electron density of states at ε_F changes only
little upon hydrogenation, and that $\sigma_d \simeq$ − 25 ppm. Since
it was also found that the spin density at a deuteron in
PdD is slightly larger than at a proton in PdH[25] it
can be argued that the s-electron related spin density
increases with the isotopic mass. The opposite is then
true for the d-electron induced spin density, which becomes larger with decreasing isotopic mass ($B_{hf}^{cp}(r_\mu) >$
$B_{hf}^{cp}(r_p)$, see Fig. 5). Clearly, a consistent understanding
of these trends would imply a fairly complete understanding of the local electronic structure problem.

MUON KS IN LaNi$_5$H$_x$

LaNi$_5$ is one of the best metallic hydrogen storage
systems known. It absorbs with ease up to 6-hydrogen
atoms per formular unit. The proton KS has been measured
at 300 K in LaNi$_5$H$_6$ with the result K $\simeq$ + (25 ± 10) ppm[26].
Results from µSR measurements are displayed in Figs. 7
and 8[27]. In the first figure the KS is plotted versus
the hydrogen-concentration x. We see that the number of
hydrogenation cycles has a significant impact on the

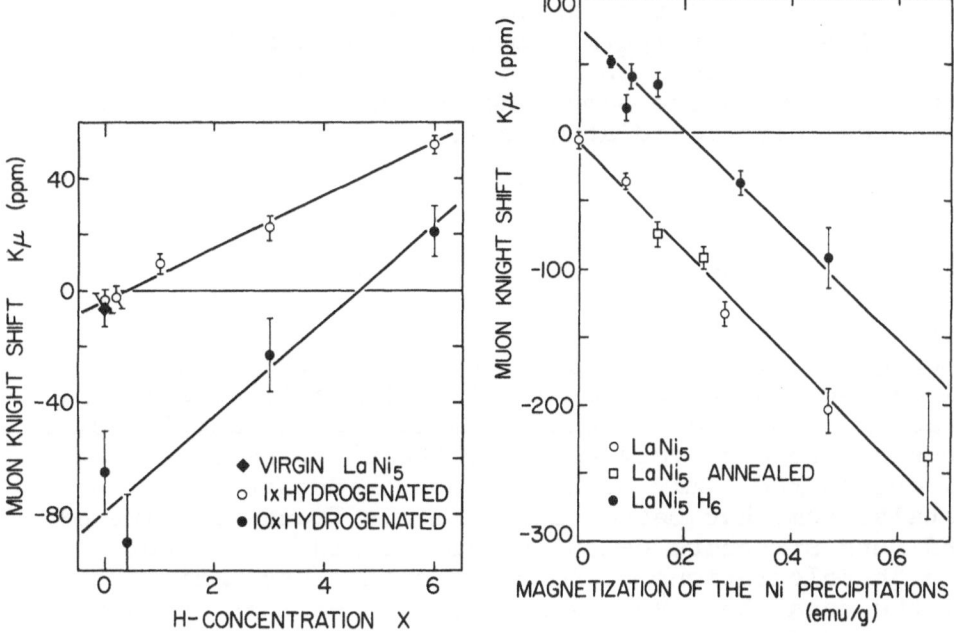

Fig. 7. μ^+ KS in LaNi$_5$H$_x$ as a function of hydrogen-concentration x.

Fig. 8. μ^+ KS in LaNi$_5$H$_x$ plotted versus the magnetization of the Ni surface precipitation. The number of hydrogenated cycles is an implicit parameter.

data. The reason becomes apparent in Fig. 8 where the KS is plotted versus the magnetization of the Ni-surface precipitations with the number of hydrogenation cycles as an implicit parameter[27]. Neglecting the effect of Ni surface precipitations can easily shift the observed KS away from the true value. This puts also the proton KS value[26] into doubt, which was obtained from samples that had gone through many hydrogenation cycles.

Extrapolating the results for the LaNi$_5$H$_x$ sample to zero magnetization a μ^+ KS of ~ + 80 ppm is observed (Fig. 8). This value is temperature independent down to 20 K. The μ^+ KS in pure virgin LaNi$_5$ amounts to - (6 ± 6) ppm. The susceptibility changes from 4.5 · 10^{-6} emu/g in LaNi$_5$ to 2.0 · 10^{-6} emu/g in LaNi$_5$H$_6$. Ascribing the change in the KS and the susceptibility to d-electrons, one calculates an induced

hyperfine field per unpaired d-electron per Ni atom of
$B_{hf}^{cp}(r_\mu)$ = - 2.2 kG/μ_B/Ni-atom. This number is quite
close to the values obtained in Pd and Ni and one is
tempted to draw the same conclusions with respect to
the local electronic structure.

CONCLUSIONS

A tremendeous amount of μ^+ KS data have become
available over the past four years, involving simple
metals, as well as transition metals and the hydrogen
storage compound LaNi$_5$H$_x$. It was found that the non-
linear screening jellium (protonic) model, usually
thought to provide a decent description in simple metals
does not account very well for the μ^+ KS results,
except in the limit of high electron density cubic
metals. Somewhat better agreement is found for cluster
and band structure calculations, but there are not
enough calculations to compare to the systematic trend
of the data. In transition metals as well as in LaNi$_5$H$_x$
the μ^+ KS displays a strong d-electron dependence,
pointing to a core polarization mechanism in conjunction
with the possible formation of local "covalent" bonding
states favoring a "covalent" hydrogen picture. Subtle
isotope effects on the KS have been observed in Nb, V
and Pd by comparison with proton KS data. Their explanation has yet to await the development of refined and
more complete theoretical models.

REFERENCES

1.   R.M. Cotts, in Hydrogen in Metals I, eds. G. Ale-
        feld and J. Völkl, Springer Verlag (1978).
2.   S. Kazama, Y. Fukai, J. Less-Common Met. 53:25
        (1977).
3.   A. Schenck, Helv. Phys. Acta (in press).
4.   E. Zaremba, D. Zobin, Phys. Rev. B22:5490 (1980).
5.   P. Jena, Hyperfine Interactions 6:5 (1979).
6.   R.M. Nieminen, Hyperfine Interactions 8:437 (1981).
7.   J. Keller, Hyperfine Interactions 6:15 (1979).
8.   J. Kanamori, H.K. Yoshida, K. Terakura, Hyperfine
        Interactions 8:573 (1981).
9.   M. Camani et al, Phys. Rev. Lett. 42:679 (1979).
10. F.N. Gygax et al, Hyperfine Interactions 8:479
        (1981).

11. W. Studer et al, this conference.
12. M. Manninen, R.M. Nieminen, J. Phys. F9:1333 (1978).
13. J.P. Bugeat et al, Phys. Lett. 71A:93 (1979).
14. O. Hartmann et al, Phys. Rev. Lett. 41:1055 (1978).
15. M. Camani et al, Phys. Rev. Lett. 39:836 (1977).
16. J. Keller et al, preprint (U.N.A.M. Mexico, 1981).
17. J. Keller, A. Schenck, Hyperfine Interactions 6:39 (1979).
18. F.N. Gygax et al, Solid State Communic. 38:1245 (1981).
19. F.N. Gygax et al, J.M.M.M. 15-18:1191 (1980).
20. J.A. Seitchik et al, Phys. Rev. 136:A1119 (1964).
21. P.J. Segransan et al, J. Phys. F6:L153 (1976).
22. A.B. Dennison, Helv. Phys. Act. 52:460 (1979).
23. H. Katayama et al, Solid State Communic. 29:431 (1979).
24. P. Brill, J. Voitländer, Ber. Bunsenges. Phys. Chem. 77:1097 (1973), and private communication.
25. C.L. Wiley, F.Y. Fradin, Phys. Rev. B17:3462 (1978).
26. R.G. Barnes et al, J. Less-Common Met. 49:483 (1976).
27. F.N. Gygax et al, in Recent developments in condensed matter physics, Vol. 2, Eds. J.T. Devreese et al, (Plenum Press 1981).
28. P. Jena, Solid State Communic. 15:1509 (1974).
29. J. Keller, M. Castro, preprint (U.N.A.M. Mexico, 1979).
30. O. Jepsen et al, Solid State Communic. 34:575 (1980).

ANOMALOUS MUON KNIGHT SHIFT BEHAVIOR

IN A Cd SINGLE CRYSTAL

F.N. Gygax, A. Hintermann, W. Rüegg, A. Schenck
W. Studer, A.J. van der Wal and H. Wehr*

Laboratorium für Hochenergiephysik der ETHZ
c/o SIN, 5234 Villigen, Switzerland

*Institut für Physik, Universität Mainz
D-6500 Mainz, Germany

For the positive muon implanted in a metal the
precession frequency shift due to hyperfine fields can
be measured with high precision. This provides means to
obtain information about the local electronic structure
of a hydrogen like impurity in any metal in the inde-
finitely dilute impurity concentration. Ref. 1 gives a
summary of the muon Knight shift (KS) investigations in
18 nontransition (simple) metals and some transition
metals and discusses the results in the context of the
electronic structure of hydrogen in metals.

Quite exceptional and puzzling compared to the
cases of simple metals cited above is the strongly
temperature dependent KS in Cd. Cd, a divalent metal
with the atomic electron configuration $4d^{10}5s^2$ has a
hcp structure with a large c/a ratio of 1.87. It is a
small anisotropic diamagnet [Ref. 2]. Its hydrogen
solubility is too small for NMR or photoemission experiments. Band structure calculations have been
published among others by [Ref. 3].

The present measurements were done with two samples,
both of them single crystals of high purity (6N resp.
5N), but different origin. Sample 1 is a sphere with a
diameter of 20 mm, whereas sample 2 is a cylinder, 26 mm
in diameter and 34 mm in length. Special care was taken

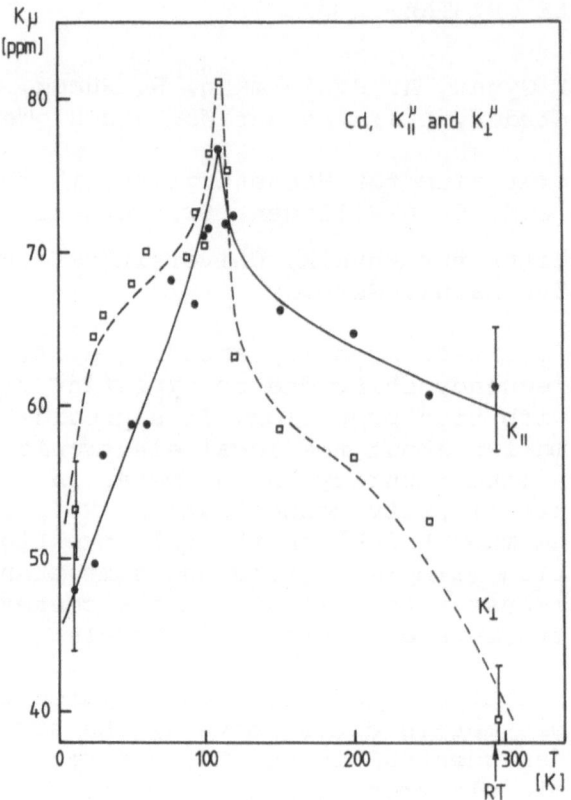

Fig. 1. (sample 1) shows the temperature dependence of the Muon Knoght shift for two
orientations: c-axis parallel ($K_{\parallel}$) resp.
perpendicular ($K_{\perp}$) to the applied field.

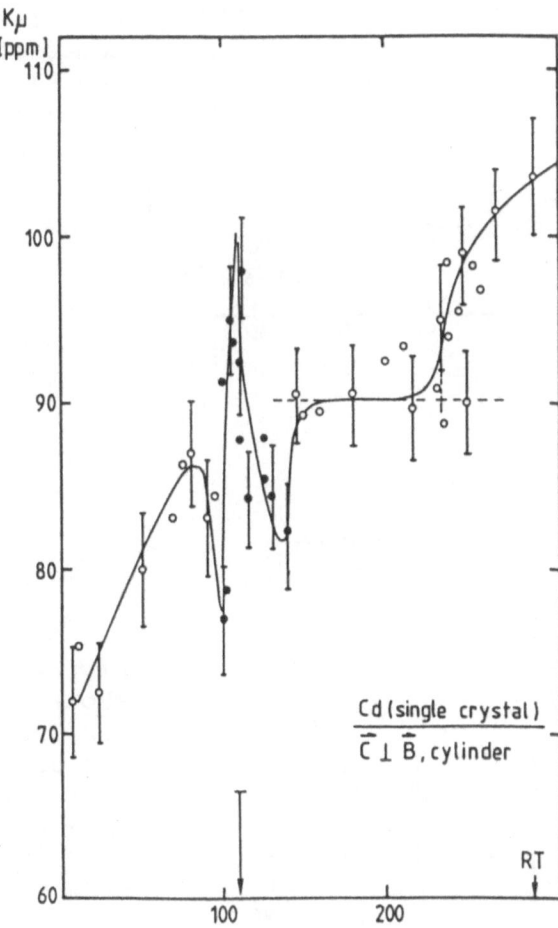

Fig. 2. (sample 2) shows the temperature dependence of the Muon Knight shift for the
perpendicular orientation. The temperature was partly scanned in finer steps
compared to Fig. 1.

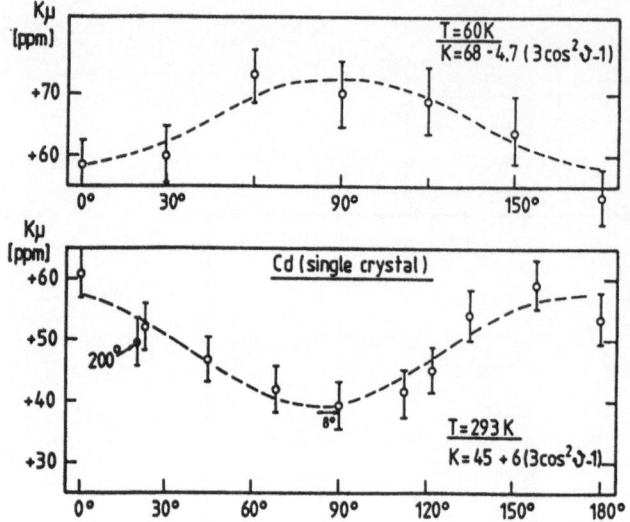

Fig. 3 (sample 1) shows the angular dependence of the Knight shift for a temperature below (60 K) and above (room temperature) the cusp singularity.

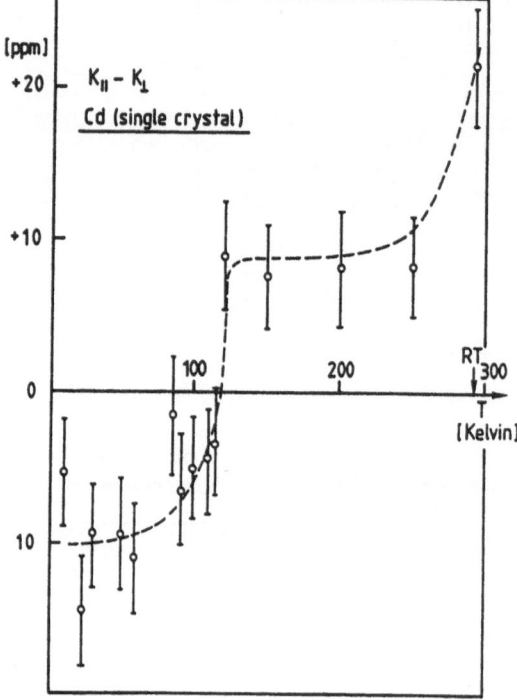

Fig. 4 (sample 1) shows the temperature dependence of the anisotropic part defined as $K_\parallel - K_\perp$ deduced from the data of Fig. 1.

during data taking to change the temperature of the samples very slowly to avoid thermal stresses.

All curves in the drawings except those of Fig. 3 are to be understood as guides to the eyes only.

The striking features of the data are the strong temperature dependence, the presence of a cusp like singularity centered at about 110 K and the change in sign of the anisotropy which accompanies the cusp. Since part of the presented measurements were completed a few days before this conference a comprehensive discussion appears premature at this time.

The strong temperature dependence is unusual for a diamagnetic nontransition metal. The increase of the KS in sample 2 with increasing temperature follows the trend observed in a polycrystalline sample (not shown here). Partly it might be correlated to a change of the interstitial or atomic volume leading to a larger number of conduction electrons screening the charge of the impurity.

The mechanism which leads to the cusp like enhanced anomaly of the KS at about 110 K is so far not known. It resembles closely a cusp singularity of a phase transition, but also some Fermi surface effect at the interstitial site, some resonance phenomenon or some local muon host complex interaction might provide a possible explanation. No other measurements (NMR [Ref. 4], susceptibility [Ref. 2], specific heat) indicate a phase transition.

In a simple model the change in sign of the KS which accompanies the singularity could mean that the charge or spin density around the muon passes from a cigar shape through a sperical to an oblate disk-like distribution. A similar change, although at lower temperature,was observed with NMR at the Cd nucleus. The change in sign cannot be explained directly by an anomalous temperature dependence of the c/a ratio, since this ratio increases only smoothly with temperature.

REFERENCES

1. A. Schenck, this conference;
    A. Schenck, Helv. Phys. Act., in press.
2. J.A. Marcus, Phys. Rev. 76:621 (1949).
3. R.W. Stark, L.M. Falicov, Phys. Rev. Lett. 19:795
        (1967).
4. R.V. Kasowski, L.M. Falicov, Phys. Rev. Lett. 22:
        1001 (1969).

THE HYPERFINE FIELD OF THE POSITIVE MUONS IN THE FERROMAGNETIC TRANSITION METALS

Kiyoyuki Terakura, Hisazumi Akai[*], Masako Akai[#] and
Junjiro Kanamori[#]

The Institute for Solid State Physics, Roppongi
Minatoku, Tokyo, 106 Japan
[*] Nara Medical University, Kashiwara, 634 Japan
[#] Osaka University, Toyonaka, Osaka, 560 Japan

ABSTRACT

Some basic aspects of the electronic structure around the
positive muons in the ferromagnetic transition metals are discussed
with the aim of elucidating the mechanism of the muon hyperfine
field. The importance of the s-d hybridization is emphasized. An
ab initio calculation on the energetics concerning the location of
the positive muons in the ferromagnetic iron is also presented.

INTRODUCTION

The positive muons μ^+ have spin 1/2, charge +e and mass 0.11 m_p
(m_p: proton mass), and therefore they can be regarded as light
protons. The polarized muons produced by the pion decay are
implanted into a material, where they arrive at thermal equillibrium
and their spins undergo the Larmor precession under the given local
magnetic field. In the muon decay process with the lifetime τ =2.2
μs, the positrons are emitted preferentially in the muon spin
direction. Thus the observation of the time variation of the
emitted positrons provides us with the information about the local
field and the relaxation time. This technique named as μ^+SR is
characterized by its high sensitivity and as μ^+ is chemically
equivalent to a proton, it gives us an insight to the electronic
structure of the dilute hydrogens in materials. μ^+SR is also useful
in the study of lattice defects and the diffusion process in solids.
Several reviews on μ^+SR are available[1-4] and the present state of
the art can be seen in Ref.5.

This article deals with mainly two subjects in μ^+SR. The first one is on the hyperfine fields seen by μ^+ in the ferromagnetic transition metals, especially Fe and Ni. Although there is no definite experimental evidence about the location of muons in Fe and Ni, it is commonly believed that the muons reside mainly at the tetrahedral interstitial sites (T-sites) in iron (bcc) and the octahedral interstitial sites (O-sites) in nickel (fcc). The observed hyperfine fields H_{hf} are -11.1 kG in Fe[6] and -0.71 kG in Ni[7] and the recent ab initio calculations reproduced these values quite satisfactorily. The negative H_{hf} are rather common for muons and protons in the transition metals. Related examples are the muon Knight shifts in V, Nb and Ta (5th group transition metals)[8] and also in Pd.[9,10] The Knight shifts are mostly negative. The microscopic origin of the negative hyperfine fields will be discussed with an emphasis on the role of the Fano effect.[11-13] However, some important factors, e.g., the zero point motion of muons and the lattice relaxation around the muons, have not been fully taken account of in the ab initio calculations of H_{hf}. A qualitative discussion will be given on these effects.

The second subject is on the energetics with regard to the positions of muons. Recent experiments[14] on the electron irradiated iron showed the presence of the resonance with H_{hf} of -9.5 kG, which was assigned to muons trapped at vacancies. However, this H_{hf} is not so different from the H_{hf} for the unirradiated iron and the theoretical prediction of H_{hf} of a muon at the center of a vacancy is much larger than the experimental value.[15] These facts suggest that the center of a vacancy is not a favorite position for muons and the muons reside among T-sites around the vancancy. Our preliminary calculation[15] of the adiabatic potential for a muon does indicate that the center of the vacancy corresponds to the highest position of the adiabatic potential and that T-site has a lower energy than O-site in a perfect crystal of iron. (We have not succeeded in determining the most stable position in the presence of a vacancy.) The calculation is based upon the local spin density functional formalism[16,17] with the KKR method for the impurity problem.[18] The results will be compared with the model potential introduced by Sugimoto and Fukai.[19]

CALCULATION OF MUON HYPERFINE FIELDS

Several attempts have been made for calculating H_{hf} for muons in ferromagnetic transition metals. The calculation with the use of a finite cluster[20] gave H_{hf} as -10.1 kG for T-site of Fe, -5.7 kG for O-site of Co and -0.59 kG for O-site of Ni. The corresponding experimental values are -11.1 kG,[6] -6.1 kG[21,22] and -0.71 kG,[7] respectively. The thin-film supercell method for Ni_5H predicted -0.463 kG for protons (or muons) at O-site.[23] These calculations seem to be fairly successful. An approach using the impurity formalism based upon the Green function method with the muffin-tin

potential model[18] is also very powerful. Yoshida (his former name
was Katayama) et al applied this method to the calculation of H_{hf} of
muons at O-site of Ni and obtained the value of -0.72 kG.[24] As the
result depends on the choice of the potential for the muon to some
extent, the very beautiful agreement of this result with the
experimental value may be rather fortuitous. However, they carried
out calculations of H_{hf} for various kinds of impurities in Fe[25] and
Ni[26-28] and reproduced the systematic variation of H_{hf} with respect
to the impurity valency successfully. They gave a general
discussion on the mechanism underlying the systematic variation of
H_{hf}, and the dependence on the impurity sites of H_{hf}. The general
agreement between theory and experiment is not affected by the small
change in the impurity potential. Recently Akai et al[15] extended
Yoshida et al's calculation by attaining selfconsistency in the
framework of the local spin density functional formalism.[16,17]
Their result of H_{hf} for a muon at T-site in Fe is -10.3 kG. Before
presenting these works, we make some general remarks about the
electronic structure of transition metals to make our standpoint
clear.

General Remarks

The transition metals are characterized by the coexistence of
the narrow d band and the free electron like sp band. In the
following, we make a comment on the hybridization between the two
bands, which plays an important role in the hyperfine field. s
(l=0), p (l=1) and d (l=2) symmetries of wave functions are uniquely
defined with regard to a given point. However, a wave function of d
symmetry with regard to a given site can have components of s, p, d,
f and g symmetries with regard to the neighboring sites.
Accordingly, once we construct eigenstates of the crystal, even the
states of s symmetry within a given inscribed sphere have an
appreciable contribution from the tails of the d wave functions at
the neighboring sites. We also note that the d band in a transition
metal has a width of some electron volts and the d wave function has
an appreciable energy dependence within the energy range of the d
band. The tail contribution and the energy dependence of the d wave
function significantly modify the local densty of states (LDOS) of s
symmetry at the given site from that of the nearly free electron
system and will turn out to be substantial in producing the spin
polarization at muons and/or impurities in general. Therefore a
theory with the aim of discussing the impurity hyperfine field in
transition metals should include the following factors: 1) the free
electron like sp band; 2) the narrow d band; 3) the s-d
hybridization; 4) the energy dependence of the d wave function.

The Electronic Structure around Muons

The local density of states (LDOS) at a given position $\underset{\sim}{r}$, which
is an important quantity in the following argument, is defined by

$$\rho^{\sigma}(\underset{\sim}{r},E) = \sum_{i} |\psi_i^{\sigma}(\underset{\sim}{r})|^2 \delta(E - E_i^{\sigma}), \qquad (1)$$

where the superscript σ denotes the spin state of an electron, and $\psi_i^{\sigma}(\underset{\sim}{r})$ and E_i^{σ} are the eigen function and the eigen value of the system. In the following, we set the origin of $\underset{\sim}{r}$ at the impurity nucleus.

Figure 1a shows $\rho^{\sigma}(0,E)$ associated with the muon at T-site in the ferromagnetic iron (thin solid lines).[15] Because of the attractive potential of the muon, $\rho^{\sigma}(0,E)$ has an appreciable peak near the bottom of the band, which corresponds to the bonding state between the 1s state associated with the muon and the d states of the surrounding host atoms. Another characteristic feature of $\rho^{\sigma}(0,E)$ is the presence of the depressed region around the Fermi energy E_F. The appearance of a peak and a valley in LDOS is formally described by the Fano effect,[11] which is a common effect observed in a system with the coexistence of a broad band and a narrow band. This problem in transition metals with impurities of a non-transition element was treated previously[12,13] and the relation between the Fano effect and the impurity hyperfine field was also discussed.[29] Here we will first proceed along the similar line to that in Refs. 30 and 31.

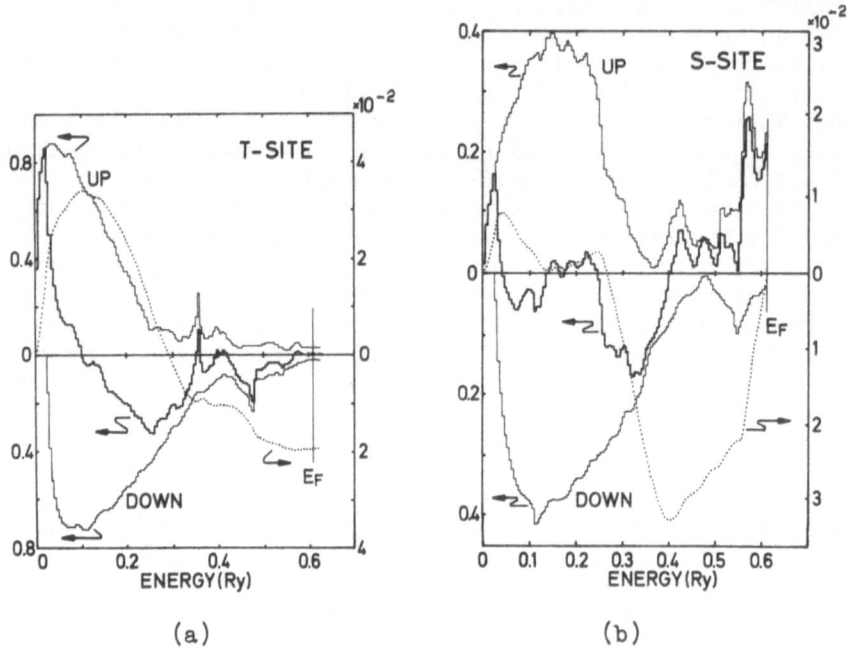

(a)   (b)

Fig. 1. $\rho^{\sigma}(0,E)$ (thin solid lines) and $\delta\rho(0,E)$ (bold solid lines) for a muon in the ferromagnetic iron. The dotted lines are the integral of $\delta\rho(0,E)$ over E. All quantities are in Rydberg atomic units.

Figure 2 is a schematic illustration of the tail contribution and the energy dependence of the d wave function. r_{ws} is the radius of the Wigner-Seitz sphere of the host metal and E_b^σ and E_t^σ correspond roughly to the bottom and the top of the d band. For $E = E_a^\sigma$, the tail of the d wave function vanishes at the impurity nucleus, making no contribution to $\rho^\sigma(0,E)$. As the distance between T-site and its nearest neighbor host site is larger than r_{ws} only by a factor of 1.135, E_a^σ is only a little bit smaller than E_t^σ. Thus $\rho^\sigma(0,E)$ is depressed near the top of the d band. $E_t\uparrow$ and $E_t\downarrow$ are about 0.62 Ryd and 0.75 Ryd, respectively. The energy E_a^σ is called the antiresonance energy[12,13] and as is guessed from Fig.2, the bonding nature between the d orbitals at the host sites and the s state at the impurity site changes from bonding to antibonding at E_a^σ.[13] $\rho^\sigma(0,E)$ for the substitutional case (the muon is at the center of a vacancy) are shown in Fig.1b by thin solid lines. The antiresonance energy E_a^σ are about 0.37 Ryd for $\sigma = \uparrow$ and 0.48 Ryd for $\sigma = \downarrow$ and are much smaller than the corresponding ones for T-site. This is because the distance to the nearest neighbor site for S-site (substitutional case) is much larger than the one for T-site. The enhanced density of states near E_F in the up spin band for S-site comes from the contribution from the antibonding states. In contrast to this situation for S-site, the stronger admixture between the impurity orbital and the surrounding d orbitals for T-site pushes the antibonding states to a much higher energy region.

The Muon Hyperfine Field

The hyperfine field of the muons comes predominantly from the Fermi contact interaction and is calculated by

$$H_{hf} = 524.2 \int^{E_F} \delta\rho(0,E) \, dE, \quad \text{(in kG)} \qquad (2)$$

where

$$\delta\rho(0,E) = \rho^\uparrow(0,E) - \rho^\downarrow(0,E), \qquad (3)$$

and $\rho^\sigma(0,E)$ and $\delta\rho(0,E)$ are in Rydberg atomic units. $\delta\rho(0,E)$ for

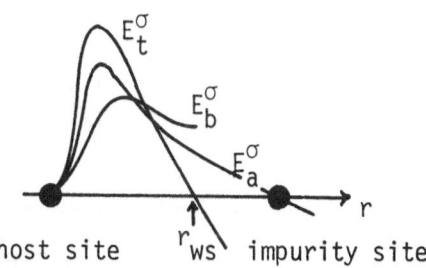

Fig. 2. A sketch of the energy dependence of d wave functions.

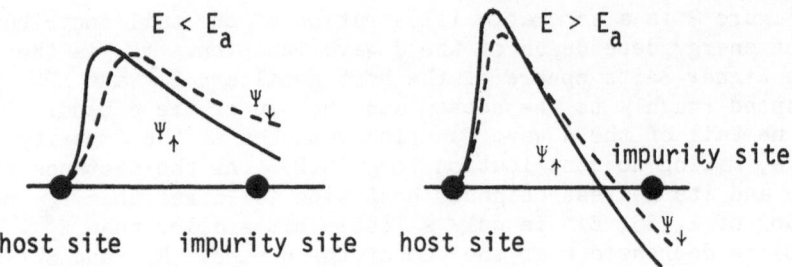

Fig. 3. The mechanisms producing a negative (the left figure) and
a positive (the right figure) polarizations.

T-site and S-site are shown in Figs.1a and 1b by bold solid lines.
By using these $\delta\rho$ in Eq.(2), we obtain -10.3 kG and -0.24 kG for
T-site and S-site, respectively. ρ^0 for O-site is not so different
from that of T-site and the corresponding H_{hf} is -7.98 kG. As was
discussed in earlier works,[24,27,29] $\delta\rho(0,E)$ is mostly negative for E
< E_a = $(E_a^\uparrow + E_a^\downarrow)/2$ and positive for E > E_a. This is clearly seen
in $\delta\rho$ for S-site. The mechanism responsible to this fact is
schematically shown in Fig.3. The different behavior of the d wave
functions between the up and down spin states comes from the fact
that the potential seen by d electrons is deeper for the up spin
state than for the down spin state. The up spin d wave function is
spatially more contracted than the down spin one. This gives rise
to different consequences with regard to the tail contribution to
$\delta\rho(0,E)$ depending on the situation whether E < E_a or E > E_a. The
above argument is related to Daniel and Friedel's theory[32a] to some
extent but ours is more transparent for the real transition metals.
The importance of the antiresonance was first pointed out in our
previous papers.[26,29] The more negative H_{hf} for T-site compared
with H_{hf} for S-site is partly due to the fact that E_F is closer to

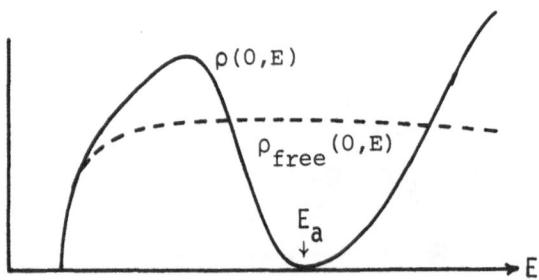

Fig. 4. LDOS without the s-d hybridization (broken line) and
with the s-d hybridization (solid line).

E_a for T-site. The antibonding states near E_F make a significant positive contribution to H_{hf} for S-site. More detailed discussions on the difference between S-site and T-site were given in Ref.28 for the nickel host case.

Following the argument in Ref.29, we present a different approach which is useful for understanding the overall aspect. We devide $\rho^\sigma(0,E)$ into two contributions

$$\rho^\sigma(0,E) = \rho^\sigma_{free}(0,E) + \rho^\sigma_{sd}(0,E) , \tag{4}$$

where ρ^σ_{free} is the contribution from the unhybridized sp band and ρ^σ_{sd} comes from the s-d hybridization.[13] For a very qualitative understanding, we show in Fig.4 a rough sketch of $\rho(0,E)$, i.e., $\rho^\sigma(0,E)$ for the non-magnetic case, and $\rho_{free}(0,E)$. $\rho_{free}(0,E)$ has a smooth E-dependence and the pronounced peaks and dips come from $\rho^\sigma_{sd}(0,E)$. We assume that ρ_{free} is independent of the spin state and that the spin dependence of $\rho^\sigma_{sd}(0,E)$ is simply expressed as

$$\rho^\sigma_{sd}(0,E) = \rho_{sd}(0, E+\sigma\Delta) , \tag{5}$$

where σ in the r.h.s. of Eq.(5) takes + for the up spin state and – for the down spin state, and 2Δ is the exchange splitting of the host d band. Expanding $\rho^\sigma(0,E)$ in power series of Δ and retaining only up to the first order in Δ, we obtain

$$\delta\rho(0,E) = 2\Delta (\partial/\partial E)\rho_{sd}(0,E) , \tag{6}$$

Therefore,

$$H_{hf} = 1048.4 \, \Delta \, \rho_{sd}(0,E_F)$$
$$= 1048.4 \, \Delta \, \{\rho(0,E_F) - \rho_{free}(0,E_F)\} \quad \text{in kG} \tag{7}$$

Although Eq.(7) does not bear a quantitative analysis because of the approximations introduced in the above, it leads us to the following

Table I. Knight shift of a muon and a proton

host metal	K_μ (in ppm)	K_p (in ppm)
V	-88 ± 8[a]	-66[c]
Nb	-13 ± 3[a]	-25[c]
Ta	2 ± 4[a]	0[c]
Pd	-155[b]	-122[d]

[a]Ref. 8. [b]Ref. 10. [c]Refs. 34 and 35. [d]Ref. 36.

qualitatively important conclusions: H_{hf} is positive if E_F is
situated in the region where $\rho(0,E)$ is enhanced by the s-d
hybridization and negative if E_F is situated in the region where
$\rho(0,E)$ is depressed by the s-d hybridization. The conclusion is
consistent wth the argument presented in the previous paragraph,
explains the systematic variation of H_{hf} with regard to the impurity
valency,[29] and furthermore has some generality applicable to other
problems.[33]

Before ending this subsection, we give a brief discussion on
the muon Knight shift K_μ in V, Nb, Ta and Pd. The observed K_μ are
shown in Table I, where the proton Knight shifts K_p are also
shown.[34,35] Except for the possibly small positive K_μ and K_p for
Ta, most of them are negative. The important factor that we have to
take account of is the exchange enhancement in the host d band.

The impurity Knight shift in the exchange enhanced host metal
is physically very similar to the impurity hyperfine field in the
ferromagnetic host metal. An approximate expression for the Knight
shift is given by[29]

$$K_\mu = 446\ \rho_{ex}(0,E_F) \qquad \text{in ppm,} \qquad (8)$$

where

$$\rho_{ex}(0,E_F) = \rho(0,E_F) + (\alpha_d - 1)S , \qquad (9)$$

α_d is the exchange enhancement factor for the d band and S is
defined as the spin density at the impurity nucleus induced by a
unit exchange splitting of the host d band. With the same
approximation as that in deriving Eq.(7), S is given by

$$S = \rho(0,E_F) - \rho_{free}(0,E_F) . \qquad (10)$$

α_d is estimated by the band structure calculation based on the local
spin density functional formalism,[37,38] and the obtained values are
2.34 (V),[37] 1.72 (Nb),[37] and 4.45[37] or 13.6[38] (Pd). (As Janak
pointed out,[37] the calculation is somewhat delicate in the case of
Pd. No calculation of U is available for Ta.) Protons and muons
reside at T-site for V, Nb and Ta and O-site for Pd. (See, however,
Ref.19 for the muon location in V, Nb and Ta.) The calculated LDOS
$\rho^\sigma(0,E)$ for the muon at T-site of Fe (Fig.1) and O-site of Ni[24]
suggest that E_F is situated in the depressed region of $\rho(0,E)$ also
for V, Nb, Ta and Pd. Therefore $\rho(0,E_F)$ is small and S is negative
in Eq.(10). The trend in Table I that K_μ and K_p are more negative
for the larger α_d is understandable.

Effect of Lattice Relaxation and Muon Zero Point Motion

Although it is generally very difficult to determine the
location of the muon in a metal, the experiment for Cu by Camani et
al[39] is a beautiful example which determined the location of the

muons and the lattice dilation around the muon at the same time. They studied the field dependence of the transverse relaxation rate and showed that the theoretical result taking accout of 5% lattice expansion agrees well with the experimental one. About 10% expansion of the first neighbor site of the muon at T-site was predicted for Nb theoretically by Sugimoto and Fukai.[19] Several percents of local lattice expansion may be common to other cases. The effect of the lattice dilation on the impurity hyperfine field was discussed by Katayama et al[26] and Manninen and Nieminen[40] and both of them suggested a possibly large effect. According to the argument presented so far, the lattice expansion will pull down E_a^σ to a lower energy to enhance the positive contribution to H_{hf}. Also the bonding between the 1s orbital around the muon and the d orbital at the neighboring host sites becomes weaker, so that the antibonding states will have lower energies. This effect will enhance $\delta\rho(0,E)$ above E_a. The third effect concerns the potential associated with the muon. As is consistent with the renormalized atom picture,[41] the lattice expansion will reduce the electron density around the muon, making the potential associated with the muon deeper. This effect will also enhance $\delta\rho(0,E)$ above E_a. All the three effects mentioned above have a tendency of enhancing H_{hf} with the local lattice expansion. Although there remain some ambiguities concerning the change in the electronic structure at the surrounding sites, we believe that the above conclusion may be correct. Note, however, that Manninen and Nieminen[40] obtained an opposite conclusion. They aimed to discuss the situation of Gd and Dy and the theory is not necessarily relevant to transition metals.

Because of its light mass, the zero point motion of the muon is also an important factor to be taken account of. Some theoretical studies have been done on the effect of the zero point motion to the muon hyperfine field.[40,42,43] A simple extrapolation of our arguments to this problem is as follows. The zero point motion or the finite extension of the muon wave function will give the muon a chance of coming closer to the host atom. This may have a qualitatively similar effect to the local lattice contraction. Therefore, as for the muon hyperfine field, the effect of the local lattice expansion and that of the muon zero point motion will cancel to some extent.

CALCULATION OF ADIABATIC POTENTIAL

The determination of the stable position of the muons and their diffusion process in solids are important subjects of μ^+SR. The ab initio calculation of the adiabatic potential for a muon will provide us with useful information for these studies. We present some results of our preliminary calculation of the adiabatic potential in the case of iron host.[15] The whole process of the calculation is devided into three steps. In the first step, the potential associated with a particular site seen by an electron is set to be zero and the electron Green function is calculated within

this empty sphere by the standard method for an impurity problem.[18]
A partial wave expansion of the Green function with respect to an
off center O', which is displaced from the center of the site by δ,
can be carried out within the empty sphere. The second step is to
solve an impurity problem, in which a muon is put at the off center
O'. In the third process, we calculate the change in the total
energy $E(\delta)$ within the framework of the local spin density
functional formalism[16,17] by using LDOS and the generalized phase
shift[18] obtained in the second step. We obtain the δ dependence of
H_{hf} at the same time. In the actual calculation, we took an average
with respect to the angle of δ in the first step to make the
calculation simpler. Figure 5 shows $E(\delta)$ (solid line) and $H_{hf}(\delta)$
(broken line). We can see that the center of S-site is the unstable
position for muons. As for T- and O-sites in a perfect crystal, the
energy change was calculated only for $\delta = 0$. We found that T-site
has a lower energy than O-site and that the energy difference
between T- and O-sites is about 0.1 Ryd. Therefore our calculation
supports the common assumption that the muons reside at T-site in a
perfect crystal of iron. Calculations have not been performed for
T- and O-site near a vacancy. A theoretical determination of the
most stable position in a crystal with vacancies should be done to
analyze the experimental data of Ref.14. Here we should mention the
work by Sugimoto and Fukai.[19] They introduced an adiabatic
potential with a double Born-Mayer type and determined the
parameters in it by using experimental information. In the case of
unrelaxed Nb, they found the energy difference of 0.12 Ryd between
T- and O-sites, which is nearly equal to our result for the iron
case. However, after they took account of the lattice dilation
around the muon in a selfconsistent way, they found that the energy
difference between T- and O-sites almost vanishes and that the

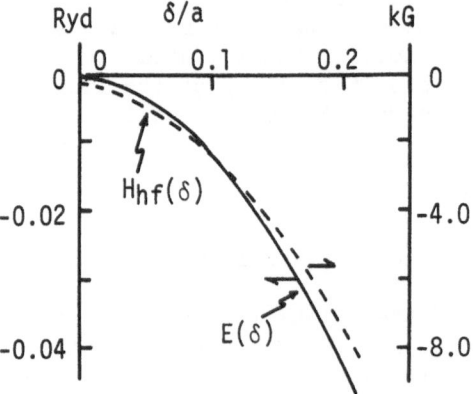

Fig. 5. The adiabatic potential for a muon $E(\delta)$ and the muon
hyperfine field $H_{hf}(\delta)$ as functions of the displacement
from the center of S-site. a is the cubic lattice constant.

density distribution of the muon has a broad maximum at O-site rather than at T-site. These are very interesting results and in near future a similar analysis based upon ab initio electronic structure calculations will be performed. Our calculation presented here may be the first step toward this direction.

Figure 5 shows that H_{hf} decreases rapidly as δ increases. This trend will not be confined only to S-site and may support our argument about the effect of the zero point motion of muons to H_{hf}.

REFERENCES

1. A. Seeger, Chapter 13, in:"Hydrogen in Metals," G. Alefeld and J. Völkl, ed., Springer, Berlin-Heidelberg-New York (1978) Vol. I.
2. A. B. Denison, H. Graf, W. Kündig and P. F. Meier, Helv. Phys. Acta 52:460 (1979).
3. E. Karlsson, in:"Recent Developments in Condensed Matter Physics," J. T. Devreese, L. F. Lemmens, V. E. Vandoren and J. Vandoren, ed., Plenum, New York-London (1981) Vol. I, p.69.
4. D. Herlach, in:"Recent Developments in Condensed Matter Physics," J. T. Devreese, L. F. Lemmens, V. E. Vandoren and J. Vandoren, ed., Plenum, New York-London (1981) Vol. I, p.93.
5. Proc. Second Int. Top. Meeting on Muon Spin Rotation, Vancouver, B.C., Canada, August 11-15, 1980, Hyperfine Interactions Vol. 8, Nos. 4-6.
6. A. Schenck, Hyperfine Interactions 4:282 (1978).
7. H. Graf, E. Holzschuh, E. Recknagel, A. Weidinger, T. Wichert, Hyperfine Interactions 6:245 (1979).
8. F. N. Gygax, unpublished data (from the talk at Int. Symp. Basic Research in Science using Mesons, Tokyo, Japan, November 16-19, 1981).
9. J. Imazato, Y. J. Uemura, N. Nishida, R. S. Hayano, K. Nagamine and T. Yamazaki, J. Phys. Soc. Japan 48:1153 (1980).
10. F. N. Gygax, A. Hintermann, W. Rüegg, A. Schenck and W. Studer, Hyperfine Interactions 8:487 (1978).
11. U. Fano, Phys. Rev. 134:1866 (1961).
12. K. Terakura and J. Kanamori, Prog. Theor. Phys. 46:1007 (1971).
13. K. Terakura, J. Phys. F 7:1773 (1977).
14. A. Möslang, H. Graf, E. Recknagel, A. Weidinger, Th. Wichert and R. I. Grynszpan, to be published in Proc. Yamada Conf. on Point Defects and Defects Interactions (Kyoto, Japan, November 16-20, 1981).
15. H. Akai, M. Akai and J. Kanamori, to be published.
16. U. von Barth and L. Hedin, J. Phys. C 5:1629 (1972).
17. O. Gunnarsson and B. I. Lundqvist, Phys. Rev. B 15:6006 (1977).
18. M. Hamazaki, S. Asano and J. Yamashita, J. Phys. Soc. Japan 41:378 (1976).

19. H. Sugimoto and Y. Fukai, Phys. Rev. B 22:670 (1980).
20. J. Keller, Hyperfine Interactions 6:15 (1979).
21. H. Graf, W. Künding, B. D. Patterson, W. Reichart,
    P. Roggwiller, M. Camani, F. N. Gygax, W. Rüegg,
    A. Schenck, H. Schilling and P. F. Meier, Phys. Rev.
    Lett. 37:1644 (1976).
22. N. Nishida, K. Nagamine, R. S. Hayano, T. Yamazaki,
    D. G. Fleming, R. A. Duncanz, J. H. Brewer, A. Ahktar
    and H. Yasuoka, Hyperfine Interactions 4:318 (1978).
23. O. Jepsen, R. M. Nieminen and J. Madsen, Solid State Commun.
    34:575 (1980).
24. H. Katayama, K. Terakura and J. Kanamori, Solid State Commun.
    29:431 (1979).
25. H. K. Yoshida, K. Terakura and J. Kanamori, J. Phys. Soc.
    Japan 50:1942 (1981).
26. H. Katayama, K. Terakura and J. Kanamori, J. Phys. Soc.
    Japan 46:822 (1979).
27. H. K. Yoshida, K. Terakura and J. Kanamori, J. Phys. soc.
    Japan 48:1504 (1980).
28. H. K. Yoshida, K. Terakura and J. Kanamori, J. Phys. Soc.
    Japan 49:972 (1980).
29. K. Terakura, J. Phys. F 9:2469 (1979).
30. J. Kanamori, H. K. Yoshida and K. Terakura, Hyperfine
    Interactions 8:573 (1981).
31. J. Kanamori, H. K. Yoshida and K. Terakura, Hyperfine
    Interactions 9:363 (1981).
32. E. Daniel and J. Friedel, J. Phys. Chem. Solids 24:1601 (1963).
33. K. Terakura, N. Hamada, T. Oguchi and T. Asada, to be
    published in J. Phys. F.
34. S. Kazama and Y. Fukai, J. Phys. Soc. Japan 42:119 (1977).
35. S. Kazama and Y. Fukai, J. Less-Common Metals 53:25 (1977).
36. P. Brill and J. Voitländer, Ber. Bunsenges. Physik. Chem.
    77:1097 (1973).
37. J. F. Janak, Phys. Rev. B 16:255 (1977).
38. A. H. MacDonald, J. M. Daams, S. H. Vosko and D. D. Koelling,
    Phys. Rev. B 23:6377 (1981).
39. M. Camani, F. N. Gygax, W. Rüegg, A. Schenck and
    H. Schilling, Phys. Rev. Lett. 39:836 (1977).
40. M. Manninen and R. M. Nieminen, J. Phys. F 11:1213 (1981).
41. L. Hodges, R. E. Watson and H. Ehrenreich, Phys. Rev.
    B 5:3953 (1972).
42. M. Manninen and P. Jena, Phys. Rev. B 22:2411 (1980).
43. S. Estreicher, A. B. Denison and P. F. Meier, Hyperfine
    Interactions 8:601 (1981).

HYPERFINE FIELD OF POSITIVE MUONS

TRAPPED AT MONOVACANCIES IN IRON

H. Graf, A. Möslang, E. Recknagel and A. Weidinger

Universität Konstanz, Fakultät für Physik
Postfach 5560
7750 Konstanz, West-Germany

ASTRACT

A muon spin rotation (μSR) experiment was performed on electron irradiated iron. A new spin precession frequency (ν_2 = 30 MHz) was found which could be attributed to muons trapped at monovacancies. The hyperfine field of the trapped muons is B_{hf} = - 0.956 T at 140 K. Between 90 K and 140 K an unusual increase of $|B_{hf}|$ is observed.

INTRODUCTION

The hyperfine field B_{hf} of positive muons is known for many magnetic materials at *interstitial sites*[1] but so far no *substitutional* hyperfine field was reported. B_{hf} at the substitutional site can be measured if muons are trapped in a single vacancy present in the material e.g. from a preceding electron irradiation. In this contribution, we report on the first experiment of this kind on electron irradiated iron.

EXPERIMENTAL

The μSR experiment was performed at the low momentum (29.8 MeV/c) muon beam at the Swiss Institute for Nuclear Research (SIN) in zero external magnetic field. Single crystal Fe targets with an area of about 2 cm^2 and 1.8 mm thickness were irradiated at 10 K with 3 MeV electrons at the Kernforschungsanlage Jülich with different doses: 3 x 10^{18}, 7 x 10^{18} and

425

1.1×10^{19} e$^-$/cm^2. The targets were transfered and loaded into the
μSR–cryostat at 77 K. The resistivity recovery was measured parallel to the μSR experiments. Due to the small range of the low
momentum muon beam it is assured that the muons are stopped in a
homogeneously damaged layer. An isochronal annealing program was
performed in situ for 10 minutes at temperatures T_A. μSR experiments were carried out at different temperatures T_M well below T_A
so that no significant recovery took place during the experiment.

RESULTS AND DISCUSSION

In Fig.1 μSR time spectra taken at two characteristic temperatures T_M = 79 K (top) and 169 K (bottom) are shown. For both
spectra, the annealing temperature T_A was 196 K. In the upper
spectrum, only one precession signal with a frequency ν_1 of about
50 MHz is observed. This signal is well known from experiments on
unirradiated samples and is attributed to muons diffusing over interstitial sites in an undisturbed Fe lattice. In the lower spectrum this frequency is observed only at the very beginning, but after
a short time (about 50 ns), the μ$^+$ spin precession changes and a
new frequency ν_2 of about 30 MHz shows up. This change in frequency is a direct indication that the muon is trapped. The solid
lines are fits with one and two frequencies in the upper and lower
spectrum, respectively.

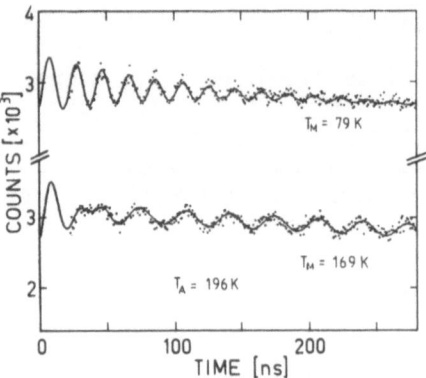

Fig.1.   Time spectra of μSR experiments on electron irradiated
         iron. The sample was annealed at 196 K. At 79 K (upper
         spectrum) only the 50 MHz frequency due to muons at
         interstitial sites is seen. At 169 K (lower spectrum)
         a new precession frequency (30 MHz) shows up. The
         transition from 50 MHz to 30 MHz occurs around 50 ns and
         is due to trapping of muons at monovacancies.

In Fig. 2 the corresponding Fourier transforms are shown.
The upper part is dominated by the defect line with ν_2 = 30 MHz.
The small line width reflects the small relaxation rate
($\lambda_2 \approx 0.3 \; \mu s^{-1}$) seen in Fig.1 for the second frequency. In the
lower part of Fig.2 the normal 50 MHz line of iron is seen.

For the interpretation of the data the identification of the
defect at which the muon is trapped is of importance. Fortunately
here the experiment itself gives a clear answer. In all spectra
only *one* defect line was found (Fig.1 and 2). Lines at $\nu \approx$ 0 MHz
or at high frequencies, which could have been overlooked, can be excluded since in a heavily irradiated sample (fast trapping) almost
all muons were processing with 30 MHz indicating that no big
fraction is missing. In ferromagnetic iron, the observation of a
single line indicates, that the muon is at a site of cubic symmetry.
This requirement is fullfilled for the substitutional site but not
for other trapping positions. However, it would be sufficient for
the present experiment if cubic symmetry is maintained in the
average. Thus, the substitutional site and an off-center position
with fast motion around the substitutional site are undistinguish-
able. In the latter case a jump frequency beyond 10^9 s^{-1} is required. Sine these two possibilities cannot be distinguished by the
experiment we will talk only of muons trapped at monovacancies in

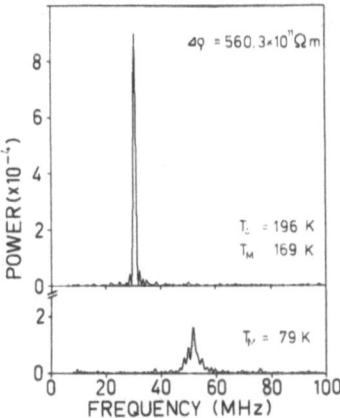

Fig.2.   Fourier transforms of the μSR spectra shown in Fig.1. The
upper spectrum shows the defect line at 30 MHz. In the
lower part the rather broad (due to relaxation) 50 MHz
line from muons at interstitial sites is seen.

this paper.

In zero external field the local magnetic field B_μ and the hyperfine field B_{hf} at the muon site can be calculated from the measured frequency ν_2 by the following formulas

$$B_\mu = 2\,\pi\nu_2/\gamma$$

$$B_{hf} = B_\mu - 4\,\pi/3\,M_S$$

Thereby the dipolar field contribution was neglected because of the cubic symmetry discussed above. γ is the gyromagnetic ratio and M_S the saturation magnetization. M_S was calculated as $M_S = \rho(T) \cdot \sigma_{ST}$, where σ_{ST} is the saturation magnetization per unit mass at temperature T (taken from Ref.2), and $\rho(T)$ is the sample density. The density as a function of temperature was calculated assuming $\rho(293\ K) = 7.874\ g/cm^3$ and using the temperature dependent lattice parameters of Ref.3.

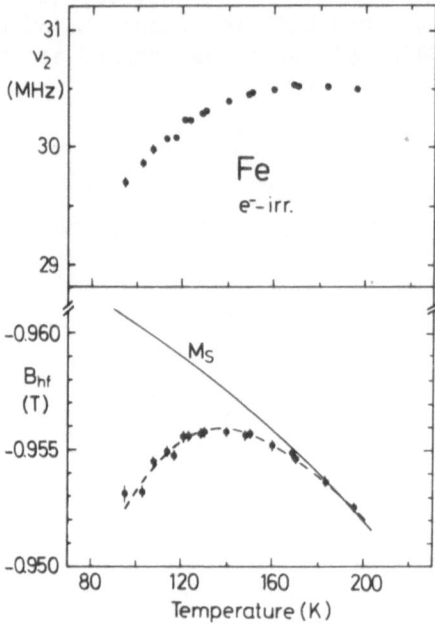

Fig.3.   Upper part: Precession frequency of muons trapped at mo-
         novacancies. Lower part: Hyperfine field B_{hf} calculated
         from the measured frequencies. The solid line shows
         the relative temperature dependence of the saturation mag-
         netization M_S, normalized at 183 K. On the x-axis the tem-
         perature during the μSR experiment is given.

Fig.3 shows the measured frequencies ν_2 in the upper part and the derived hyperfine field B_{hf} in the lower part. The solid line represents the saturation magnetization normalized to B_{hf} at T = 183 K. The experimental data are limited at the low temperature side by the fact, that the muons are too slow and do not reach the defect in a sufficient short time and on the high temperature side by the annealing of monovacancies in stage III around 220 K.

The hyperfine field (Fig.3) shows an unusual temperature dependence with an increase of B_{hf} between 90 K and 140 K. This increase is not understood at present. At 140 K the following values were obtained:

$$\nu_2 = 30.40(1) \text{ MHz}$$
$$B_{\mu} = -0.2243(1) \text{ T}$$
$$B_{hf} = -0.956(1) \text{ T}$$

The sign of B_{μ} was measured by observing the frequency shift in an external magnetic field above saturation.

The muon hyperfine field at the substitutional site was calculated by Kanamori et al. [4]. These authors obtain a positive value for B_{hf} which is in complete disagreement with the present negative value. Whether the discrepancy is due to omissions in the calculation (no lattice relaxation and no zero point motion was taken into account) or rather indicates that the muon sits at an off-center position remains an open question. Recently, S. Estreicher and P.F. Meier [5] reported a value for B_{hf} between - 0.6 T and - 0.9 T, in good agreement with the present experiment. A negative value of the right magnitude was also obtained by B.G. Lindgren and D.E. Ellis [6] in a cluster calculation.

REFERENCES

1. A.B. Denison, H. Graf, W. Kündig and P.F. Meier,
    Helv. Phys. Acta 52 (1979) 460

2. J. Crangle and G.M. Goodman,
    Proc. Roy. Soc. London A321 (1971) 477

3. Thermophysical Properties of Mater., Vol.12, ed.
    Y.S. Touloukian (IFI/Plenum, New York, 1975) p. 157

4. J. Kanamori, H.K. Yoshida and K. Terakura,
    Hyperfine Interactions 8 (1981) 573

5. S. Estereicher and P.F. Meier,
    Phys. Rev. B25 (1982) 297

6. B.G. Lindgren and D.E. Ellis, Preprint

PULSED μSR STUDIES ON μ[+] IN METALS

K. Nagamine, K. Nishiyama, T. Natsui[†], Y. Kuno,
J. Imazato, H. Nakayama and T. Yamazaki

Meson Science Laboratory and Department of Physics
University of Tokyo, Bunkyo-ku, Tokyo, Japan
† Department of Metallurgy and Materials Sciences
University of Tokyo

INTRODUCTION

Intense and sharply pulsed muons with 50 ns pulse width and 50 ms pulse distance are now available in our laboratory (UT-MSL) at KEK Booster Utilization Facility[1,2]. Strong backward muons with a focussing property are generated by the superconducting muon channel with 0° take-off angle from the pion production target, while surface muons can be obtained simultaneously by the parasite pion channel with 102.5° take-off angle from the same target. With the aid of multi-telescope and multi-stop TDC system, we are able to measure decay e[+] time spectrum from instantaneously intense stopped μ[+] with a rate of 10[4] μ[+]/pulse. Using beam pulse as a start pulse and multiple e[+] as a stopping signal, one can measure μSR time spectrum. Thus, μSR time spectrum can be observed with 50 ns time resolution, so that muon spin rotation can be straightly measured for the frequency up to 10 MHz.

There are two excellent features in this pulsed μSR method: 1) time range for μSR measurement can be extended up to 20 μs or longer without any background contributions from accidental coincidences between μ and e pulses. This feature is useful for μ[+] slow relaxation and fast or ultra-slow diffusion phenomena; 2) muon spin resonance can be realized with a strong pulsed r.f. field[3], useful for direct observation of the time dependent state change of μ[+] in metals such as transition from interstitial to defect states.

Concerning the subjects related to hydrogen in metals, several important results have been obtained in μ[+]SR studies. During commissioning period of muon channel (October '80 to October '81) and the

431

first term of regular experimental beam time (November '81 to March '82).

μ^+ PROBING MAGNETISM OF Ni

μ^+ in ferromagnetic Ni has been one of the major interests in the history of μSR researches. However, due to the limited time range of d.c. μSR (practically below 5 μs), the experiments so far realized have had serious restrictions so that , as a typical example, no systematic studies have been done for the μ^+ behaviour around T_C in Ni.

Critical Phenomena in μ^+ Fields and Knight Shifts[4]

Pulsed μSR is particularly effective for the measurement of a very slow spin rotation, which enables us to measure critical phenomena under very weak external transverse field, without destroying critical behaviour of magnetic ordering.

A single crystal Ni which has a spherical shape of 4.0 cm diameter was placed in an oven. An external field down to 30 G was applied in transverse μSR set-up[1]. Typical pulsed μSR spectra are seen in Fig. 1, which shows clearly the onset of damping in μSR spectrum after 10 μs, which is usually very difficult to be observed by the conventional d.c. μSR method.

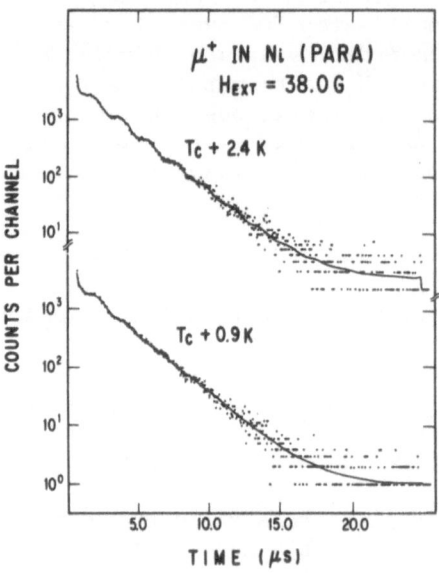

Fig. 1. Typical μSR time spectrum for
μ^+ in paramagnetic Ni close to T_C.

Precise measurements were made on the temperature dependences around T_c for paramagnetic Knight shifts under a weak external field as well as local fields in ferromagnetic phase under zero external field. The results are shown in Fig. 2. The critical indexes are obtained to be 0.336 $\sim$ 0.345 for the μ^+ field in ferromagnetic phase (β) and 1.25 $\sim$ 1.30 for paramagnetic Knight shift (γ). Both results are more or less consistent with macroscopic magnetization and susceptibility measurements.

After correcting Lorentz field contributions, both ferromagnetic local field and paramagnetic Knight shift can be converted into hyperfine field H_{int} and Knight shift K_{int} due to contact interaction. If we define reduced contact hyperfine field and Knight shift as $H_{int}/\frac{8\pi}{3}M_s$ and $K_{int}/\frac{8\pi}{3}\chi$, both results should be identical, if the origin of the field is same in both para- and ferro-magnetic regions. These values are -0.15 at 0 K, while they are -0.175 above T_c. A slight difference, if it is real, should be explained by the basic mechanism contributing to the Knight shift in transition metals[5].

Critical Index in Transverse Relaxation

As it is partly seen in Fig. 1, temperature dependent transverse relaxation was observed in both above and below T_c. The result is summarized in Fig. 3. The result on paramagnetic relaxation rate is systematically smaller than the old data[6]. It is probably due to the fact that the old data at stronger field had some contributions from Knight shift inhomogeneity due to temperature non-uniformity inside the sample. As for the present data in ferromagnetic region, particularly close to T_c, the observed relaxation might be partly due to such a temperature non-uniformity effect.

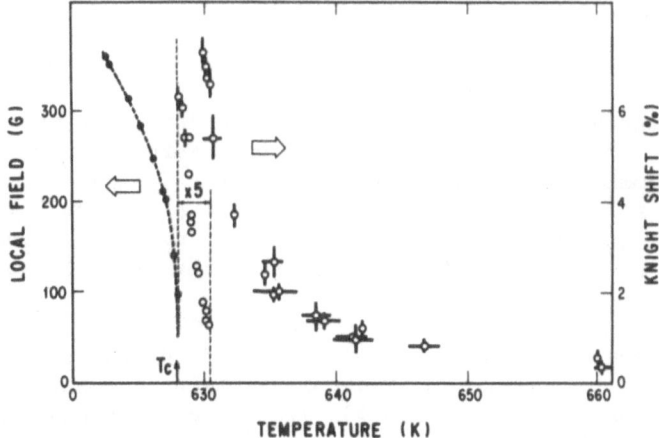

Fig. 2. Temperature dependence of paramagnetic Knight shift and ferromagnetic local field of μ^+ in Ni around T_c.

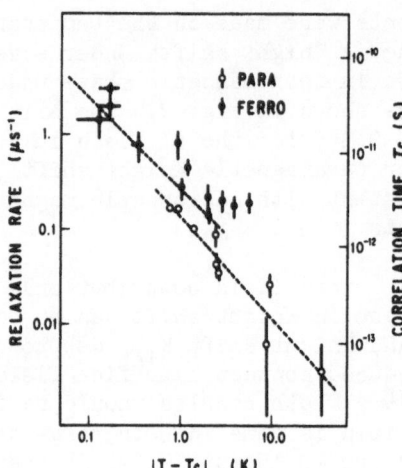

Fig. 3. Transverse relaxation rate of μ^+ in Ni for both para- and
ferro-magnetic regions close to T_c.

By using the formula $T_2^{-1} = N\omega_f^2\tau_c$ and setting ω_f to be random
dipolar fields from surrounding atoms, temperature change in τ_c is
obtained. Thus, critical index due to critical slowing-down was
obtained to be 1.1 ± 0.1 in both phases, which should be compared
with the corresponding numbers obtained by neutron scattering (1.4)
and by PAC method (0.67 ∿ 0.70)[7].

μ^+ DIFFUSION AND TRAPPING IN TYPICAL METALS

By using the capability of long time-range measurement, pulsed
μSR was applied to observe a tiny change in μ^+ transverse relaxation
due to a change in μ^+ diffusion.

Transverse Relaxation

Pulsed surface muons were stopped in various thin metallic
samples. So far, almost all the experiments were done to search μ^+
trapping phenomena at elevated temperatures[8]. Typical examples are
shown for μ^+ relaxation in Indium in Fig. 4. A tiny but significant
change can be seen by approaching to the melting point. The result
indicates increase of μ^+ diffusion at the temperature close to the
melting point.

Careful measurements have been done on deformed Cu sample. It
is aimed to see the change in μ^+ depolarization rate which depends
upon deformation dgree (50 and 80%), measuring temperature and
annealing temperature. After various tedious runs, we have concluded
that no evidence has been seen so far for the μ^+ depolarization rate
to increase above 0.01 μs^{-1} by raising temperature. Contrary to the

existing data [9], the result indicates that, as far as the observation is made at room temperature and slightly above (up to 200 C), the μ^+ depolarization is quite insensitive to the induced defects and dislocations due to deformations.

Almost the same conclusion was drawn for the deformed Al, while a clear increase of the μ^+ depolarization rate (0.02 μs^{-1} level) was observed at room temperature for irradiated Al containing $10^{16}/$ cc voids[10]. The result suggests that diffusing μ^+ is trapped at voids in Al even at room temperature.

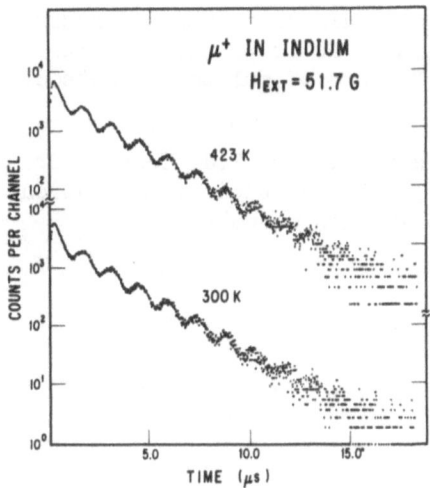

Fig. 4.   Typical example of transverse relaxation (a tiny change in depolarization rate is visible).

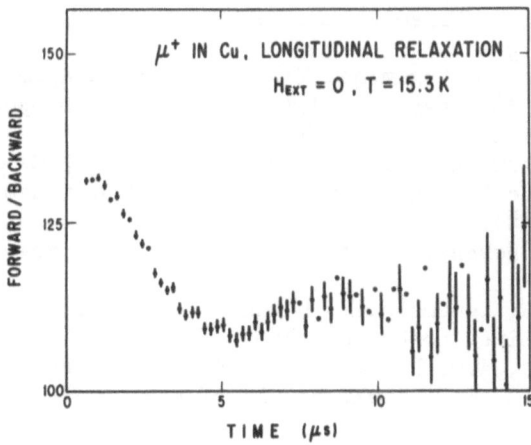

Fig. 5. Typical example of zero-field relaxation

Longitudinal Relaxation

Longitudinal relaxations at zero field have been measured on various materials to see slow modulation due to the onset of the μ^+ slow diffusion from the Kubo-Toyabe type relaxation. Typical result is shown in Fig. 5.

So far, μ^+ localization at room temperature is confirmed in isotopically separated light elements such as ^{9}Be, ^{10}B and ^{11}B. The onset of slow diffusion was observed for μ^+ in Cu at 15.3 K with a correlation time of around 20 μs^{11}. The result indicates existence of the μ^+ slow diffusion which corresponds to the transverse relaxation measured as lower temperature. The result also demonstrates a powerfulness of pulsed μSR method for ultra-slow diffusion phenomena.

The authors acknowledge Dr. Y.J. Uemura, Dr. E. Yagi, Dr. S. Tanigawa and Mr. K. Ishida for their contributions to a part of the experiments reported in this manuscript.

REFERENCES

1. UT-MSL Newsletter No. 1 (May, 1981) ed. K. Nagamine and T. Yamazaki.
2. K. Nagamine, Hyperfine Interactions $\underline{8}$, 787 (1981).
3. Y. Kitaoka, M. Takigawa, H. Yasuoka, M. Itoh, S. Takagi, Y. Kuno, K. Nishiyama, R.S. Hayano, Y.J. Uemura, J. Imazato, H. Nakayama, K. Nagamine and T. Yamazaki, Hyperfine Interactions, in press.
4. K. Nishiyama, K. Nagamine, T. Natsui, S. Nakajima, Y. Kuno, J. Imazato, H. Nakayama and T. Yamazaki, to be published.
5. K. Terakura, private communication.
6. B.D. Patterson, K. Nagamine, C.A. Bucci and A.M. Portis, Magnetism and Magnetic Materials - AIP Conference Proceedings, $\underline{24}$, 281 (1974).
7. L. Chow, C. Hohenemser and R.M. Suter, Phys. Rev. Lett. $\underline{45}$, 908 (1980).
8. T. Natsui, R. Yamamoto, M. Doyama, H. Sasaki, K. Nishiyama and K. Nagamine, to be published.
9. W.B. Gauster, A.T. Fiory, K.G. Lynn, W,J, Kossler, D.M. Parkin, C.E. Stronach, W.F. Lankford, J. of Nucl. Materials $\underline{69/70}$, 197 (1978).
10. S. Tanigawa, private communication.
11. Y.J. Uemura, private communication.

MAGNETIC SUSCEPTIBILITY AND PROTON NMR STUDY OF TiCr$_{1.8}$H$_x$

James F. Lynch,[†] John R. Johnson & Robert C. Bowman, Jr.[*]

Brookhaven National Laboratory
Upton, New York 11973
* California Institute of Technology
  Pasadena, California 91125

The electronic properties of the hexagonal (C-14) and cubic (C-15) allotropes of the TiCr$_{1.8\pm0.1}$/H$_2$ system have been examined via magnetic susceptibility and proton NMR studies. The results indicate an increase in N(E$_F$), the electron density of states at the Fermi level, with increasing H/M for both allotropes; the trend extends into the α'-hydride phase. The results suggest a basis for thermodynamic anomalies reported for the solution of hydrogen in TiCr$_{1.8}$.

INTRODUCTION

Titanium and chromium form two intermetallic Laves phase compounds at the composition TiCr$_{1.8\pm0.1}$, a high temperature hexagonal (C-14) phase and a low temperature cubic (C-15) phase. Both allotropes[1,2] exothermically absorb hydrogen to form very unstable hydrides[1,2]. The thermodynamic behavior of the solid solution of hydrogen in TiCr$_{1.8}$ (i.e., the α-phase) exhibits several unusual aspects[3] when compared to most other metal-hydrogen systems[3] (eg., Pd/H$_2$[4]). The most striking of these is a pronounced positive deviation from Sieverts' Law of ideal solubility, even as H/M → 0. This is a consequence of a decline in exothermicity of absorption with increasing hydrogen content; as the H-content increases, the partial molar enthalpy of solution tends towards more positive values. This trend is opposite that exhibited by most exothermic hydride forming systems. Recalling that the hydrogen absorption process requires insertion of an H-atom into an appropriate lattice interstitial site with simultaneous accomodation of the

---

* Permanent Address: Mound Facility-MRC, Miamisburg, Ohio 45342
† Permanent Address: Allied Corporation, Morristown, N.J. 07960

H electron by the metal conduction band, the present study was
undertaken to investigate the extent to which electronic factors
influence the thermodynamic behavior of the $TiCr_{1.8}/H_2$ system.

EXPERIMENTAL

Most of the experimental procedures employed in the present
work have been described in detail eslewhere. Thus, details concerning preparation of the starting intermetallics[1,2], composition[5],
activation, hydrogen charging, poisoning techniques[6], etc., will
not be reiterated here, save to note that the poisoning technique
utilized CO at -196°C. The Faraday-type susceptibility apparatus
at Brookhaven has been described[7]; measurements were made at fields
from 2500 Oe to 8650 Oe and at temperatures from 78-298K. Several
CO poisoned samples were sealed into evacuated tubes for NMR investigation. Proton Knight shifts were measured via the multiple-
pulse zero-crossing technique[8] at a resonance frequency of 56.4
MHz at the Jet Propulsion Laboratory[9]; σ_K values are relative to
an external reference of TMS, uncorrected for any demagnetization
effects[10,11]. T_1 values were obtained via the standard inversion
recovery method at 56.4 MHz and 34.5 MHz. Vacuum outgassing at
the conclusion of the NMR investigations confirmed hydrogen
content.

RESULTS AND DISCUSSION

Typical magnetic susceptibility data exhibited by several
hydrogen-free specimens are summarized in Table 1. Both allotropes
are Pauli paramagnets exhibiting nearly identical susceptibilities,
regardless of crystal structure or of small variations in Cr content. We detect only a slight increase in susceptibility with increasing temperature, as evidenced by the single 80K result included in Table 1; a similar effect was found for the hydrogen-
containing specimens.

The variation of room temperature magnetic susceptibility
with hydrogen content is displayed in Figure 1 for both allotropes
of $TiCr_{1.8}$. All of the data refer to the region of single phase
solid solution (the α-phase) except for the result at H/Ti = 1.59,

TABLE 1

Susceptibility and X-ray Diffraction Results Exhibited by $TiCr_{1.8}$

Sample	Lattice Parameter, Å	298K χ_o X 10^6 emu/g	80K χ_o X 10^6 emu/g
(C-15) $TiCr_{1.8}$	6.940±0.004	2.568±0.001	------
(C-15) $TiCr_{1.87}$	6.934±0.005	2.582±0.015	2.540±0.005
(C-15) $TiCr_{1.9}$	6.932±0.004	2.565±0.020	------
(C-14) $TiCr_{1.9}$	a = 4.932 b = 7.983	2.636±0.010	------

which is a two-phase (α' + α) specimen. (The line drawn through the data points is only a guide to the eye and has no analytical significance; thus, the break suggested at H/Ti≅0.8 is speculative.) The data clearly indicate that χ_0 increases with increasing hydrogen content in the region of solid solution. Extrapolation from H/Ti≅0.9 to higher hydrogen contents suggests a value χ_0 = 5.6 X 10^{-6} emu/g for the α'-phase hydride, $TiCr_{1.8}H_{2.3}$, an increase of more than two-fold over the value exhibited by the hydrogen-free metal.

It is of interest that most transition metal/H_2 systems exhibit a trend opposite that observed here; i.e., χ_0 decreases as H/M increases and χ_0 (hydride)< χ_0 (H-free)[12]. Indeed, the only exceptions among the transition metals, particularly noteable in the present context, are Ti and Cr, both of which exhibit χ_0 (hydride)> χ_0 (H-free). The decline in χ_0 with increasing H/M is often interpretted as reflective of a decrease in the density of electron states at the Fermi level, even though χ_0 is the sum of several contributions, of which only one, the Pauli term, depends directly on $N(E_F)$[13]. With this reservation in mind, the variation of χ_0 with H/M exhibited by $TiCr_{1.8}$/H_2 suggests an increase in $N(E_F)$ with increasing hydrogen content. The results of the proton NMR study support this view.

Due to the high hydrogen mobility in both allotropes of $TiCr_{1.8}$[14], only σ_K and T_1 data obtained below 120K can be related to the electronic structure parameters without complication[8,11] from diffusion contributions to the proton relaxation times. For T < 120K we assume T_1 = T_{1e}, the conduction electron component[11]; the variation of σ_K and of $(T_{1e}T)^{-\frac{1}{2}}$ with hydrogen content for the single phase (α- and α'-) samples examined here are summarized in Figure 2. Both parameters increase linearly with hydrogen content and there is little difference between the two allotropes.

Making conventional assumptions[9,11] to neglect electron-electron coupling, interband mixing and other higher order terms, both the σ_K and T_{1e} parameters can be separated into three

$$\sigma_K = 2\mu_B[H_{hf}(s)N_s(E_F) + H_{hf}(d)N_d(E_F)] + (N_A\mu_B)^{-1}H_{hf}(0)\chi_0$$

$$(T_{1e}T)^{-1} = 4\pi\gamma_n^2 k_B \{[H_{hf}(s)N_s(E_F)]^2 + [H_{hf}(d)N_d(E_F)]^2q + [H_{hf}(0)N_d)E_F)]^2p\}$$

contributions, where $H_{hf}(s)$, $H_{hf}(d)$ and $H_{hf}(0)$ are the hyperfine fields for the Fermi contact interaction with unpaired s-electrons, the "core" polarization form the s-d exchange mechanism and the orbital motion of d-electrons, respectively. $N_s(E_F)$ and $N_d(E_F)$ are the densities of s and d electrons at E_F, χ_0 is the Van Vleck orbital paramagnetic susceptibility, p and q are reduction factors that depend upon the d-electron orbital degeneracy at E_F

and the rest of the terms have their usual meaning. Because χ_O represents the mixing of the unoccupied states above E_F into the occupied states below E_F it is not related to $N_s(E_F)$ or $N_d(E_F)$.

Recent NMR investigations of TiH_x[15], $Ti/V/H_2$[9] and $Ti/Nb/H_2$[16] have indicated that the orbital terms represent a negligible contribution to σ_K and T_{1e}; we assume similar behavior in the present case. Note that $H_{hf}(s)$ is always positive but $H_{hf}(d)$ is usually negative[10,11]. Thus, although the contact and core polarization terms represent an additive contribution to $(T_1 T)^{-1}$, their contribution to σ_K may effectively cancel. Negative σ_K values, as are observed in $TiCr_{1.8}/H_2$, indicate that the core polarization contribution is dominant and suggest that $N_d(E_F) \gg N_s(E_F)$, a similar result to that observed in previous NMR studies of Ti-based hydrides[9,15,16]; small contributions from the contact and orbital terms cannot, of course, be totally excluded.

The trends in σ_K and $(T_{1e}T)^{-\frac{1}{2}}$ observed in Figure 2 indicate an increase in $N_d(E_F)$ with increasing H/M (note that an increase in $N_s(E_F)$ would drive σ_K towards smaller, i.e., more positive, values). Neglecting completely the s-electron contact term, FIgure 2 suggests that $N_d(E_F)$ is increased by about threefold between the α-phase $TiCr_{1.8}H_{0.2}$ and the α'-hybride $TiCr_{1.8}H_{2.8}$. This is in reasonable agreement with the susceptibility result χ_O ($TiCr_{1.8}$) 2.6 X 10^{-6} emu/g and χ_O ($TiCr_{1.8}H_{2.3}$) = 5.6 X 10^{-6} emu/g, and supports our view that the increase in χ_O with H/M reflects an increase in $N(E_F)$.

We have already noted that for most metal/hydrogen systems, both experimental[12] and theoretical[13] considerations indicate that hydrogen solution yields a decrease in the electron density of states at the Fermi level. The experimental evidence offered here suggests the opposite for $TiCr_{1.8}/H_2$; hydrogen solution yields an <u>increase</u> in $N(E_F)$, and specifically in $N_d(E_F)$. This trend might well explain the thermodynamic anomalies reported earlier[3] for $TiCr_{1.8}/H_2$. Such a view should be accepted with some reservation, however. The magnetic susceptibility and NMR results indicate similar electronic structures near E_F for the α-phase solid solution and α'-phase hydrides of both allotropes, but the thermal stabilities of the two hydride phases are significantly different[1,2]. Thus, considerations in addition to electronic factors play a role in determining hydride stability in $TiCr_{1.8}/H_2$, and the same is likely true in the region of solid solution.

ACKNOWLEDGEMENTS

This work was supported by the U.S. Dept. of Energy, Division of Chemical Sciences and Office of Basic Energy Sciences, and by the Caltech's President's Fund. Mound is operated by Monsanto Research Corp. for the U.S. Dept. of Energy under Contract No. DE-ACO4-76-DP00053. Brookhaven National Lab. is operated for the U.S. Dept.

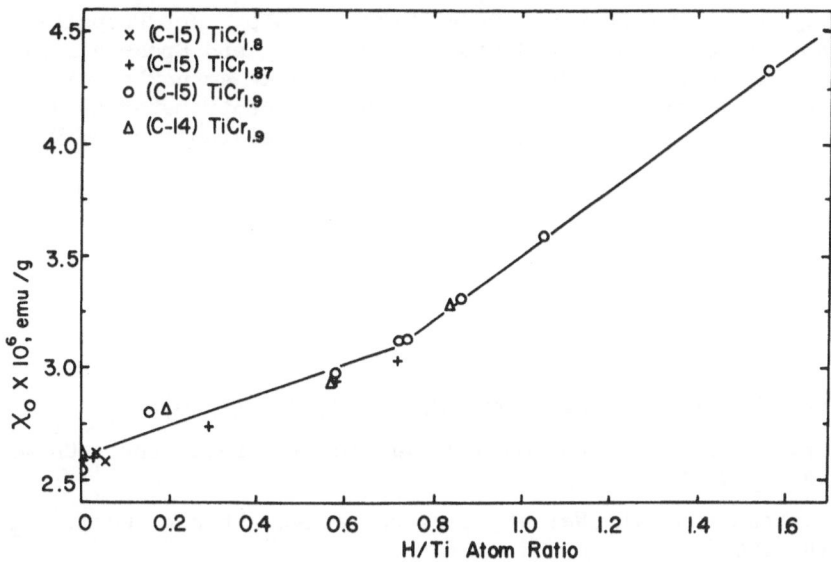

Fig. 1.   Variation of room temperature magnetic susceptibility
with hydrogen content exhibited by $TiCr_{1.8}/H_2$.

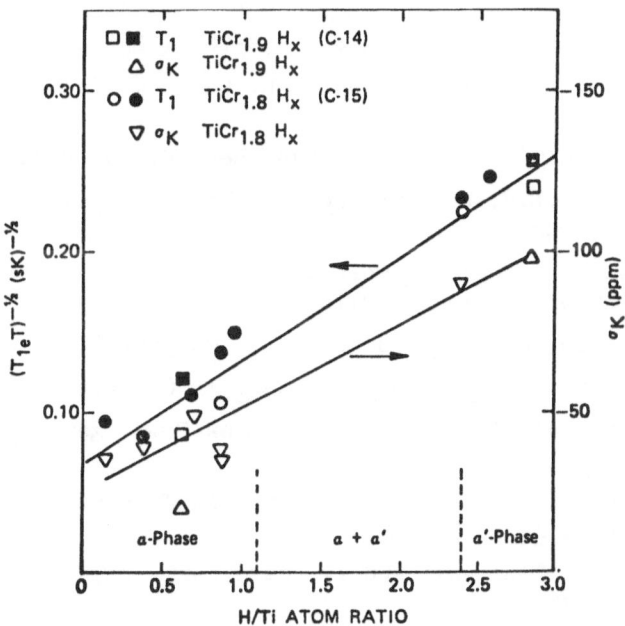

Fig. 2.   Variation of σ_K and $(T_{1e}T)^{-\frac{1}{2}}$ with hydrogen content
exhibited by $TiCr_{1.8}/H_2$.
Open symbols: 56.4 MHz   Filled Symbols: 34.5 MHz

of Energy under Contract No. DE-AC02-76-CH00016. Jet Propulsion
Laboratory is operated for National Aeronautic and Space Adminis-
tration under Grant No. NAS7-100. R. C. Bowman acknowledges
Dr. W.-K. Rhim for making available his NMR spectrometer at Jet
Propulsion Laboratory and the experimental assistance of B. D.
Craft.

REFERENCES

1. J.R. Johnson and J.J. Reilly, Inorganic Chem., 17, 3103 (1979).

2. J.R. Johnson, J. Less-Comm. Met., 73, 345 (1980).

3. J.F. Lynch, J.R. Johnson and J.J. Reilly, Zeit. Phys. Chem.
    N.F., 117, 229 (1979).

4. E. Wicke and G.H. Nernst, Ber. Bunsenges, Physik. Chem., 68,
    224 (1964).

5. J.R. Johnson, J.J. Reilly, J.F. Lynch, F. Reidinger, to be
    published.

6. J.F. Lynch, R. Lindsay and R.O. Moyer, Solid State Comm.,
    41, 9 (1982).

7. R.O. Moyer and R. Lindsay, J. Less-Comm. Met., 70, P57 (1980).

8. D.P. Burum, D.D. Elleman and W.-K. Rhim, J. Chem. Phys., 68,
    1164 (1978).

9. R.C. Bowman and W.-K. Rhim, Phys. Rev. B, 24, 2232 (1981).

10. G.C. Carter, L.H. Bennett and D.J. Kahan, Metallic Shifts in
    NMR, 1977, Pergamon Press, Oxford.

11. R.M. Cotts, Hydrogen in Metals I, G. Alefeld and J. Volkl,
    eds, Springer,-Verlag, Berlin 1978 Chapter 9

12. W.E. Wallace, Hydrogen in Metals I, Chapter 7.

13. A.C. Switendick, Hydrogen in Metals I, Chapter 5.

14. R.C. Bowman and J.R. Johnson, J. Less-Comm. Met., 73, 254
    (1980).

15. R. Göring, R. Lukas and K. Bohmhammel, J. Phys. C: Solid
    State Phys., 14, 5675 (1981).

16. B. Nowak, N. Pislewski and W. Leszcynski, Phys. Stat. Sol.(a)
    37, 669 (1976).
    B. Nowak, O.J. Zogal and M. Minier, J. Phys. C: Solid State
    Phys., 12, 4591 (1979).

PROTON NMR STUDIES OF THE ELECTRONIC

STRUCTURE OF ZrH_x

A. Attalla*, R. C. Bowman, Jr.*, B. D. Craft*,
E. L. Venturini**, and W.-K. Rhim***

*Monsanto Research Corporation-Mound
Miamisburg, Ohio 45342
**Sandia National Laboratories
Albuquerque, New Mexico 87185
***Jet Propulsion Laboratory
California Institute of Technology
Pasadena, California 91103

ABSTRACT

The proton spin-lattice relaxation times and Knight shifts have been measured in f.c.c. (δ-phase) and f.c.t. (ε-phase) ZrH_x for $1.5 \leq x \leq 2.0$. Both parameters indicate that $N(E_F)$ is very dependent upon hydrogen content with a maximum occurring at $ZrH_{1.83}$. This behavior is ascribed to modifications in $N(E_F)$ through a fcc-fct distortion in ZrH_x associated with a Jahn-Teller effect.

The electronic properties of the non-stoichiometric dihydrides of the IVB metals Ti, Zr, and Hf have been the subjects of numerous theoretical[1-3] and experimental[4-10] studies. Much of this interest has focused on the temperature and composition dependent fcc to fct phase transition that has been associated[1,2,4] with a Jahn-Teller type mechanism, as well as the more general problem of the character of the metal-hydrogen bonds.[1,5-8] Although there have been several recent nuclear magnetic resonance (NMR) studies related to the electronic structure of TiH_x,[6-8] only limited NMR results[9,10] address the electronic properties of δ-phase (fcc) and ε-phase (fct) ZrH_x. In the present work, the temperature and composition dependences of the proton spin-lattice relaxation times (T_1) and Knight shifts (σ_K) have been measured in high-purity polycrystalline ZrH_x for $1.5 \leq x \leq 2.0$. These parameters, which are related[11] to the densities of electron states $N(E_F)$ at the Fermi energy E_F, show the dominance of the core-polarization hyperfine interaction with the Zr d-electrons and yield a maximum in $N(E_F)$ near $x = 1.83$. These observations support the Jahn-Teller mechanism[1,4] for the tetragonal distortion and are consistent with recent APW band-theory calculations[3] of fcc ZrH_2 and photoemission spectra[5] for ZrH_x.

The ZrH_x samples were prepared by direct reactions between purified H_2 gas and zone-refined Zr foils (Materials Research Corporation - MARZ grade). The ZrH_x foils were ground under an argon atmosphere to produce powders that were sealed in evacuated glass tubes. The T_1 values were obtained by the standard inversion-recovery method at the proton frequency of 34.5 MHz. The previously described[8] zero-crossing method produced the σ_K values, which are relative to an external reference of tetramethylsilane, for a resonance frequency of 56.4 MHz. The σ_K values have a precision of ±2 ppm while the exponential T_1 recoveries yielded T_1 with a precision of ±3%. The T_1 values were measured over the temperature range 100 K to 300 K and the σ_K values were obtained between 170 K and 310 K, as shown in Figs. 1 and 2, respectively. There was no unusual temperature behavior for either parameter. Below 310 K, the proton T_1 values are assumed to be dominated by the conduction electron component T_{1e} as found previously.[9,10]

The composition and temperature behavior of the proton σ_K and $(T_{1e}T)^{-\frac{1}{2}}$ parameters for ZrH_x are summarized in Fig. 3 where T is the absolute temperature. The major feature for $(T_{1e}T)^{-\frac{1}{2}}$ is an increase above $x = 1.65$ to reach a maximum near $x = 1.83$ before decreasing smoothly up to $x = 2.00$. $(T_{1e}T)^{-\frac{1}{2}}$ exhibits the largest temperature dependence for $1.80 \leq x \leq 1.85$ (i.e., at the peak). Similar behavior is noted for the proton σ_K parameters; however, neither the composition nor temperature dependence of σ_K is as large as seen for $(T_{1e}T)^{-\frac{1}{2}}$.

The proton T_{1e} and σ_K parameters are related to the electronic structure of a metal hydride through the hyperfine fields[11] produced at the proton sites. Since the d-electron orbital terms[11] appear to make insignificant contributions to the proton parameters in several metal hydrides,[7,8] the σ_K and T_{1e} parameters can be separated as

$$\sigma_K = 2\mu_B[H_{hf}(s)N_s(E_F) + H_{hf}(d)N_d(E_F)] \tag{1}$$

$$\frac{1}{T_{1e}T} = 4\pi \gamma_H^2 k_B\{[H_{hf}(s)N_s(E_F)]^2 + [H_{hf}(d)N_d(E_F)]^2 q\}. \tag{2}$$

Here, H_{hf} (s) and H_{hf} (d) are the hyperfine fields for the Fermi contact interaction with unpaired s-electrons and transferred "core" polarization from the s-d exchange with metal d-states, respectively; μ_B is the Bohr magnetron, γ_H is the proton gyromagnetic moment; k_B is the Boltzmann's constant; and the reduction factor $q = 1/3[f(t_{2g})]^2 + 1/2[1-f(t_{2g})]^2$ in a cubic structure where $f(t_{2g})$ is the fractional

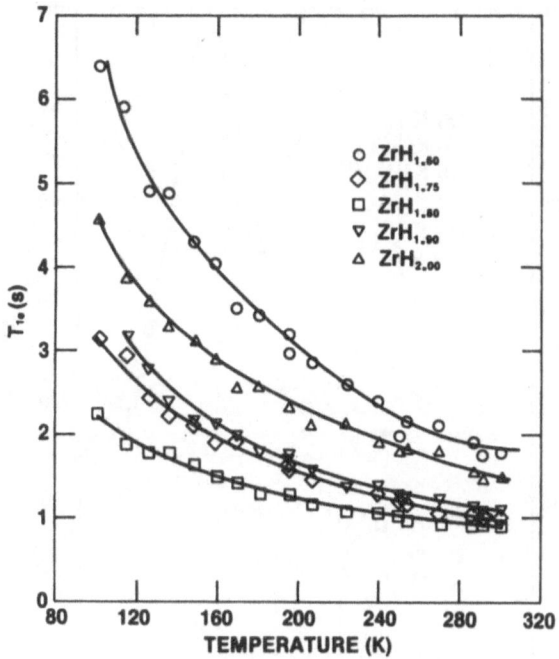

Fig. 1.   Proton T_1 for ZrH$_x$ at ν_H = 34.5 MHz.

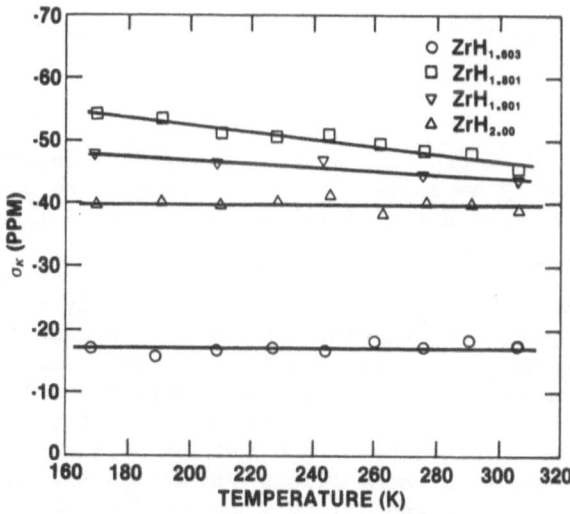

Fig. 2.   Proton Knight shifts for ZrH$_x$ measured by multiple-pulse
zero-cross technique at 56.4 MHz.

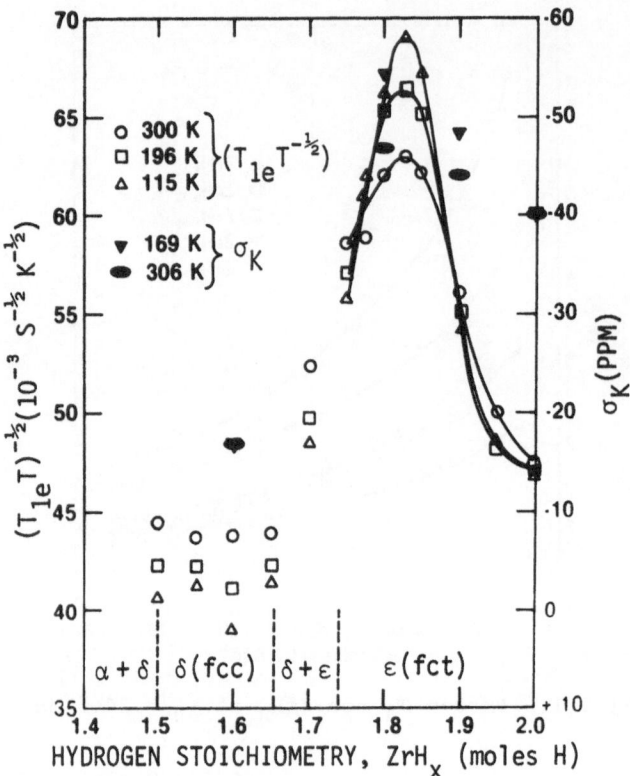

Fig. 3.   Composition dependence of proton $(T_{1e}T)^{-\frac{1}{2}}$ and σ_K for ZrH_x.

character of the t_{2g} d-orbitals at E_F. The $H_{hf}(s)$ is always positive, but $H_{hf}(d)$ is usually negative.[11] Thus, although the contact and core polarization terms are additive for $(T_{1e}T)^{-\frac{1}{2}}$, a large cancellation can occur in σ_K.

The negative proton σ_K values in Fig. 2 clearly indicate that the core-polarization term exceeds the contact term in δ-ZrH_x and ε-ZrH_x, which implies $N_d(E_f) \gg N_s(E_F)$ in these hydrides. Similar conclusions have been made for several other hydrides including γ-TiH_x,[7] $Ti_{1-y}V_yH_x$,[8] $TiCuH_x$,[12] and $TiCr_2H_x$;[13] hence, a relatively large $N_d(E_F)$ seems to be a general property of the hydride phases formed by group IVB metals and of at least some alloys containing Ti. A more detailed analysis of the proton σ_K and $(T_{1e}T)^{-\frac{1}{2}}$ values for ZrH_x suggests substantial s-electron contact (or, perhaps, some orbital) contribution in δ-phase $ZrH_{1.60}$ while the core-polarization seems to be the exclusive hyperfine interaction in ε-phase ZrH_x.

The composition dependence of $(T_{1e}T)^{-\frac{1}{2}}$ and σ_K in Fig. 3 is consistent with the available band theory calculations for TiH$_2$,[1,2] and ZrH$_2$,[3] as well as the Jahn-Teller mechanism[2,4] for the fcc-fct transition. Briefly, $N_d(E_F)$ is apparently constant in fcc δ-ZrH$_x$ ($x \leq 1.65$) as the E_F level moves upward with x, but adding more hydrogen above 1.65 causes E_F to enter the band region where $N_d(E_F)$ increases rapidly until $x \stackrel{\sim}{=} 1.83$. However, because of the electronic instability of a large $N_d(E_F)$ the tetragonal distortion continually increases, which will tend to reduce the $N_d(E_F)$ value as x becomes larger. The competition between $N_d(E_F)$ increasing with more hydrogen and decreasing with the Jahn-Teller tetragonal distortion produces the $(T_{1e}T)^{-\frac{1}{2}}$ maximum near $x = 1.83$. This general behavior corresponds to changes seen in the photoemission spectra[5] of ZrH$_x$ as hydrogen content varies. A more detailed discussion of the electronic structure and proton parameters will be published elsewhere.

ACKNOWLEDGEMENTS

This work was partially supported by the Division of Chemical Sciences, U.S. DOE, and the Caltech's President's Fund. Mound is operated by Monsanto Research Corporation for the U.S. DOE under Contract No. DE-AC04-76-DP00053. Sandia National Laboratories are supported by the U.S. DOE under Contract No. DE-AC04-76-DP00789. JPL is operated for the National Aeronautics and Space Administration under Grant No. NAS 7-100.

REFERENCES

1. A. C. Switendick, Zeit. Physk. Chem. N.F. 117: 89 (1979).
2. M. Gupta, Solid State Commun. 29: 47 (1979).
3. M. Gupta and J. P. Burger, Phys. Rev. B24: 7099 (1981).
4. F. Ducastelle, R. Caudron and P. Costa, J. Physique 31: 57 (1970).
5. J. H. Weaver, D. J. Peterman, D. T. Peterson and A. Franciosi, Phys. Rev. B23: 1692 (1981).
6. C. Korn, Phys. Rev. B17: 1707 (1978).
7. R. Göring, R. Lukas and K. Bohmhammel, J. Phys. C: Solid State 14: 5675 (1981).
8. R. C. Bowman, Jr. and W.-K. Rhim, Phys. Rev. B24: 2232 (1981).
9. C. Korn, in "Hydrides for Energy Storage," A. F. Andresen and A. J. Maeland, Eds., Pergamon, Oxford, 1978, p. 119.
10. K. R. Doolan, P. P. Narang and J. M. Pope, J. Phys. F: Metal Phys. 10: 2073 (1980).
11. A. Narath in "Hyperfine Interactions," A. J. Freeman and R. B. Frankel, Eds., Academic, New York, 1967, p. 287.
12. R. C. Bowman, Jr., A. J. Maeland, W.-K. Rhim and J. Lynch, paper at this meeting.
13. J. F. Lynch, J. R. Johnson and R. C. Bowman, Jr., paper at this meeting.

NUCLEAR MAGNETIC RESONANCE OF HYDROGEN IN CeH_x FOR x>2

*D. Zamir, **N. Salibi and R. M. Cotts, +Tan-Tee Phua
and R. G. Barnes

*Soreq Nuclear Research Center, **Cornell University
+Ames Laboratory, USDOE, Iowa State University

ABSTRACT

Experimental evidence reviewed by Libowitz (1) indicates that
at room temperature, the CeH_x system undergoes a metal-to-nonmetal
(MNM) transition with increasing hydrogen content at about x=2.75.
For x>2 CeH_x is cubic, and hydrogen fills the octahedral (or nearby
off-center sites) of the CeH_2 fluorite structure. The existence
of an MNM transition is consistent with band structure calculations
in the literature (2).

To further the study of the MNM transition, we have measured
^1H (proton) and ^2H (deuteron) NMR spin-lattice relaxation times T_1
for 2 < x < 2.9 in CeH_x. NMR parameters are sensitive to changes
in electronic structure through hyperfine interactions. At room
temperature the proton T_1 is weakly dependent upon concentration
up to x≃2.5, but for x>2.5, T_1 increases rapidly with x. This increase in T_1 is consistent with the compositional MNM transition.
From temperatures of about 100K to 300K, T_1 increases linearly
with temperature for samples with x>2.5 as reported (3) for x=2.0
and x=2.53. T_1 data for deuterons show that the relaxation mechanism is principally magnetic so that quadrupolar contributions to
deuteron relaxation can be neglected.

The strong magnetic relaxation mechanism is attributed mostly
to indirect (RKKY) and direct (dipolar) coupling between hydrogen
nuclear spins and the local cerium 4f electron magnetic dipole
moment. It is expected that the MNM transition affects the RKKY
coupling through the change in electronic structure.

*Sponsored by the U.S.-Israel Binational Science Foundation, the National Science Foundation, and the U.S. Department of Energy.

REFERENCES

1.   G.G. Libowitz, Ber. Bunsenges, Phys. Chem., 76, 837-45 (1972).
2.   A. Fujimori and N. Tsuda, J. Phys. C: Solid State Phys. 14, 1427-34 (1981).
3.   L. Shen, J.P. Kopp, and D.S. Schreiber, Phys. Letters, 29A, 438-39 (1969).

USE OF NUCLEAR MAGNETIC RESONANCE TO OBSERVE

DIFFUSION OF HYDROGEN IN METALS*

Robert M. Cotts

Cornell University
Ithaca, NY 14853

ABSTRACT

Utilization of measured NMR relaxation rates and diffusion coefficients are described. With the recent availability in the literature of lattice-specific calculations relating relaxation rates to mean residence times, details of the diffusion process of H in metals can be deduced. None of the lattice-specific theories predicts the asymmetry in the temperature dependences of relaxation rates observed in some hydrides. Observed and theoretical frequency dependences of relaxation rates on the high and low temperature sides of T_1 minmima are compared.

INTRODUCTION

In the utilization of nuclear magnetic resonance (NMR) to study atomic motion in solids, two different types of NMR experiments are usually done. In the first, measurements of one or more of the several spin system relaxation times are analyzed from theory to obtain the value of the mean residence time, τ_D for hydrogen atoms on their interstitial sites. Values of τ_D ranging from about 10^{-3} to 10^{-10} seconds can be established by this technique. In the second experiment, one measures the attenuation of spin echoes caused by pulsed magnetic field gradients applied in time intervals between the radio frequency (RF) pulses of the spin echo experiment. The echo attenuation can be related directly to the translational diffusion coefficient, D, of H atoms by the solution to the diffusion equation and the results are independent of any other model or theory concerning diffusion processes in the lattice. With accurate values of D and τ_D, it is possible to study some of the details of atomic transport in hydrogen-metal systems. The

451

relationship between relaxation times and atomic motion was first
described by Bloembergen, Purcell, and Pound (BPP) (1). For recent
reviews on the application of NMR to hydrogen in metals see
references (2), (3), (4), and (5).

In II, the methods for direct measurement of D are reviewed.
In III the theory and some experiments for determining τ_D are
discussed. Finally, in IV, some of the agreeable and disagreeable
aspects of experiment and theory are aired.

MEASUREMENT OF HYDROGEN DIFFUSION COEFFICIENTS

The basic two-pulse magnetic field gradient NMR experiment (6)
is subject to an error caused by a distribution of "background"
gradients, G_0, associated with sample heterogeneity (3). Since
G_0 scales with the applied magnetic field, the experiment is done
in as low a value of magnetic field as is compatible with
signal-to-noise requirements. This technique, and the closely
related stimulated echo (7) technique have been used in a number of
metal-hydrogen systems to measure D for ^1H. These systems include
PdH_x(8), $Pd_yAg_{1-y}H_x$(9), $NbH_{0.6}$(10), $Nb(^1H_x{}^2H_y)$(11), and δ-TiH$_x$(12).

Two modifications of the basic experiment have been used to
eliminate systematic errors due to G_0. In the first, (13) the
amplitude of one echo in a Carr Purcell echo train is measured as
the spacing, Δ, is varied between two fixed amplitude gradient
pulses inserted in the train. The echo amplitude remains independent of G_0 during an experiment. In this technique one must
sacrifice some signal to gain independence from G_0. The technique
has seen application in measuring D of ^1H in group V transition
metals NbH_x(13,14) and TaH_x(14). The second technique is a
little more difficult to instrument, but is much more effective in
that the $G \cdot G_0$ term is eliminated as an attenuation factor. This
technique of Karlicek and Lowe (15) uses an alternating pulse field
gradient in the Carr-Purcell pulse train consisting of five π RF
pulses. The technique can be extended to longer CP pulse trains
(15), and it is more effective than the first technique (13) in that
it produces a greater attenuation of the echo due to diffusion.

In a convincing demonstration of the efficacy of this technique, Karlicek and Lowe (16) measured the temperature dependence of
D of ^1H in LaNi$_5$ hydride where D is in the range of 10^{-7} cm^2/sec
and G_0 was found to be approximately 2.9×10^3 Gauss/cm! Values of
G_0 are typically under 100 Gauss/cm in most transition metals.
Karlicek and Lowe (15) describe circuits and coils for producing
gradients in excess of 10^3 Gauss/cm. They use a single power source
and directed the current pulses to one of two identical coils wound
in opposite senses around the sample chamber. Mauger (17) used two
power sources of opposite polarity driving a single coil for D
measurements on ^1H in Nb-V alloys.

These NMR techniques can be used to measure values of $D > 10^{-8}$ cm^2/sec, and depending upon signal strength, even lower values of D might be realized.

MEASUREMENT OF SPIN SYSTEM RELAXATION TIMES

Four different relaxation times have been measured and related to ^{1}H diffusive motion. These are the longitudinal or spin-lattice relaxation time T_1, the transverse relaxation time T_2, the rotating frame spin-lattice relaxation time T_{1r}, and the dipolar relaxation-time T_{1D}. Most frequently, the first three are measured, and of these T_1 and T_{1r} are usually selected as being parameters most familiar and best understood. The apparent value of T_2 can be, for example, affected by the background gradient G_0 discussed in Sec. II, while T_1 and T_{1r} are insensitive to G_0. T_1 measurements are most effective at high temperatures where the mean residence time, τ_D, of an ^{1}H atom is less than 10^{-6} sec, and T_{1r} measurements are made at low temperatures where $\tau_D < 10^{-3}$ sec. Observations of T_{1D} are effective for $\tau_D > 10^{-4}$ sec.

Theoretical expressions for the first three relaxation rates are given in terms of the power spectra of local magnetic dipole fields $J^{(p)}(\omega)$ caused by relative motion of nuclei of the resonant spin system with respect to other nuclear dipole moments, including members of the resonant spin system. The equations, for H-H interactions only, are

$$(T_1)^{-1} = (3/2)\gamma^4 h^2 I(I+1)[J^{(1)}(\omega) + J^{(2)}(2\omega)] \tag{1}$$

$$(T_{1r})^{-1} = (3/8)\gamma^4 h^2 I(I+1)[J^{(0)}(2\omega_1) + 10J^{(1)}(\omega) + J^{(2)}(2\omega)] \tag{2}$$

$$(T_2)^{-1} = 3/8\gamma^4 h^2 I(I+1)[J^{(0)}(0) + 10J^{(1)}(\omega) + J^{(2)}(2\omega)] \tag{3}$$

where γ is the gyromagnetic ratio; $I = 1/2$, ω is the Larmor frequency in the applied field, $\omega = \gamma B_0$, and $\omega_1 = \gamma B_1$ when B_1 is the rotating RF field, and ω_d is defined in the section on BPP model. The $J^{(p)}(\omega)$ are Fourier transforms of correlation functions of the dipole-dipole interaction.

$$J^{(p)}(\omega) = \int_{-\infty}^{\infty} G^{(p)}(t)e^{-\omega t} dt, \tag{4}$$

$$G^{(p)}(t) = \sum_j \langle F_{ij}^{(p)}(t') F_{ij}^{(p)*}(t'+t) \rangle. \tag{5}$$

The question that remains is, what is the form of $G^{(p)}(t)$ for any given metal hydride?

There are two responses to the question. The first is the BBP (1) model in which $G^{(p)}(t) = G^{(p)}(0)\exp(-t/\tau_c)$, for any solid or liquid. Then,

$$J^{(p)}(\omega) = G^{(p)}(0) \, 2\tau_c/(1+\omega^2\tau_c^2), \tag{6}$$

where $\tau_c = \tau_D/2$ for H-H interactions only.

In averaging $G^{(p)}(0)$ over all crystal orientations, it is found that $G^{(0)}(0): G^{(1)}(0): G^{(2)}(0) = 6:1:4$ so that, for example, in a polycrystalline sample,

$$(T_1)^{-1} = \frac{2}{3} \frac{\omega_d^2}{\omega} [\frac{\omega\tau_c}{1+\omega_c^2\tau^2} + \frac{4\omega\tau_c}{1+4\omega_c^2\tau^2}] \tag{7}$$

where $\omega_d^2 = 3/5 \, h^2\gamma^4 I(I+1) \sum_j (r_j^{-6})$ and the sum extends over all occupied H sites from one typical H site. The inclusion of M-H interactions is straightforward (3). The correlation time τ_c is, for thermally activated jumps, $(1/2)\tau_{Do} \exp(E/kT)$, where E is the activation energy and τ_{Do} is the reciprocal of the pre-exponential attempt frequency factor.

The above model has been widely applied to survey the temperature dependence of τ_D. The τ_D dependence (assume ω is held constant) of the relaxation rates for T_1, T_{1r}, and T_2 are shown in Fig. 1. The $(T_1)^{-1}$ curve peaks near $\omega\tau_D \approx 1$, where $\omega = \gamma B_o$. $(T_{1r})^{-1}$ peaks near $\omega_1\tau_D \approx 1$ and since B_o is usually of the order of one tesla and B_1 of the order of 10^{-3} tesla, a large range of τ_D is spanned by the combination of the two measurements. Since $\log \tau_D \propto (E/kT)$, the graph of $\log (T_1)^{-1}$ vs $(1/T)$ should have the same form as the graph of $\log (T_1)^{-1}$ vs $\log (\omega\tau_D)$ in Fig. 1, for constant ω. The asymptotic dependence of the rates are

$$(T_1)^{-1}, \; (T_{1r})^{-1} \propto \omega_d^2\tau_D, \text{ for } \omega\tau_D \ll 1,$$

$$(T_1)^{-1} \propto \omega_d^2/(\omega^2\tau_D), \text{ for } \omega\tau_D \gg 1, \text{ and}$$

$$(T_{1r})^{-1} \propto \omega_d^2/(\omega_1^2\tau_D), \text{ for } \omega_1\tau_D \gg 1.$$

Therefore, a graph of $\log T_1$ or $\log T_{1r}$ vs $(1/T)$ should be straight lines with equal and opposite slopes in the asymptotic regions. The slope is then a measure of E. From a fit of relaxation rates to the theory one can also obtain the temperature dependence of τ_D, and presumably a better determination of E over a

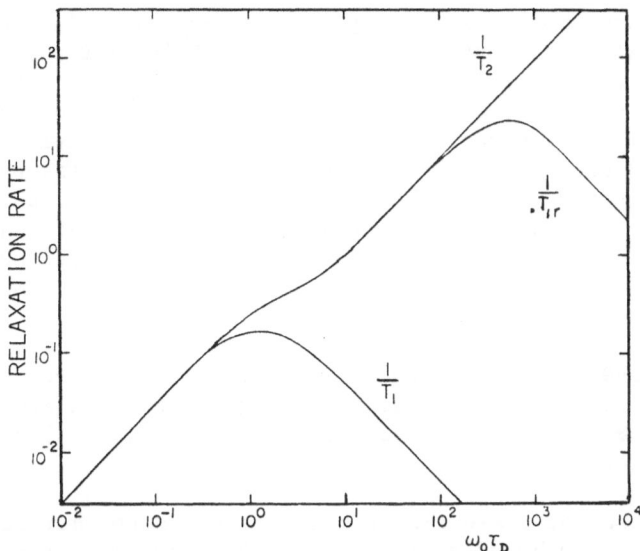

Fig. 1. Relaxation rates $(T_1)^{-1}$, $(T_{1r})^{-1}$, and $(T_2)^{-1}$ calculated for
BPP theory with $\omega_1 = \omega_0/500$, where ω_0 is the Larmor
frequency, γB_0.

larger range in $(1/T)$ than in one of the asymptotic regions. Without a lattice specific calculation for $G^{(p)}(t)$, as is the case
with the BPP model, the absolute value of τ_D so obtained will not
have an accuracy better than about 50%. However, values of E obtained from the BPP model agree with other methods to better than
10%.

While there is confidence in the use of BPP theory to analyze
T_1 and T_{1r} data for a survey of the mean residence times of
hydrogen in metals, it is recognized that improved theory must be
used for study of diffusion in detail.

Lattice-Specific Theory for $G^{(p)}(t)$ and T_1 or T_{1r}

Torrey (18) improved the BPP model with his theory of random
flights. In the 1970's there came a resurgence of interest in this
problem in the lattice-specific calculations of Sholl (19) on random
walk, Wolf (20) in the monovacancy limit including spatial correla-
lation effects, and Fedders and Sankey (21) who identified significant problems in random walk theories. In a departure from prior
techniques Fedders and Sankey used a reciprocal space formalism in a
mean field theory which included correlation effects in an average
way. Their more accurate multiple scattering approximaton includes
effects of correlations of particles in D as well as in calculating
T_1 and T_{1r}.

Barton and Sholl (22) incorporated the mean field theory and the reciprocal space approach to calculate $J^{(P)}(\omega)$ for BCC, FCC, and SC lattices. Their tables of $J^{(P)}(\omega)$ for a full range of values of $(\omega\tau_D)$ can be applied to any (hydrogen) concentration in a simple way. They show that spatial correlations have been included exactly in the low temperature $(\omega\tau_D \gg 1)$ regime. For the multiple scattering theory the c dependence has been tabulated for the high and low temperature limits (23, 24). Most of the work assumes atomic jumps to the nearest neighbor (NN) site, but Barton and Sholl (25) compared effects of first and third NN jumps. Bustard (26) simulated an SC lattice (20x20x20) in a computer and in a Monte Carlo calculation, he calculated $G^{(P)}(t)$ for c=0.1, 0.85, 0.775, and 0.999625 (monovacancy). His results compared favorably with the multiple scattering approximation calculated explicitly (21) for c=0.85 for first NN jumping. Bustard also calculated $G^{(P)}(t)$, and thus $(T_1)^{-1}$, for third NN jumps.

The results of these recent calculations show that the curve of $(T_1)^{-1}$ vs $(\omega\tau_D)$ is broader with a less sharp maximum near $(\omega\tau_D)$ than in the BPP theory. Nevertheless the numerical value of the maximum relaxation rate is relatively insensitive to the theory selected. The asymptotic limits are reached more gradually in the lattice-specific theories. This point has been addressed specifically by Sholl (27). In all lattice-specific theories, the ω and τ_D dependences well into the asymptotic regions are the same as in the BPP theory.

The diffusion coefficient is given by

$$D = fL^2/(6\tau_D) \tag{8}$$

where L is the jump distance and f is the tracer correlation factor. With independent experimental knowledge of τ_D and of D, the value of L can be determined since f is known for most jump processes or can be calculated (21, 28). Bustard, Seymour and Cotts (12) measured T_1 and D in γ-TiH$_{1.55}$ and γ-TiH$_{1.71}$. Two diffusion models were tested, and it was found using Bustard's lattice-specific calculations that H atoms jump to the nearest neighbor vacancy on the SC lattice in γ-TiH$_x$.

A second method for combining D and T_1 data is to form the ratio (D/T_1) for one temperature in the high temperature region where $(\omega\tau_D) \gg 1$. (D/T_1) depends only upon the distribution of hydrogen atoms in the metal and the jump process. (D/T_1) is independent of τ_D and, of course, of the NMR frequency in this region. Recently Mauger (14) completed, using NMR, a survey of temperature and concentration dependence of D and T_1 in α-TaH$_x$. Using activation energies obtained from D measurements he separated the diffusion-dependent $(T_1)^{-1}$ from the electronic contribution.

The ratio (D/T_1) was then expressed as the ratio of pre-exponential factors from the two analyses, (D_0/T_{1o}), and these values are shown for seven hydrogen concentrations in Fig. 2. The theoretical curve is from Sankey and Fedders (24), who included H-H and Ta-H nuclear spin interactions. Their calculation assumed that H atoms occupy tetrahedral sites randomly and jump to the first nearest neighbor site. They assume no interaction between hydrogen atoms. There is acceptable agreement between theory and experiment at low values of x, but the experimental values of (D/T_1) are much less than theoretical values as x increases. The deviation from theory is just what would be expected from the short range repulsive inter-action between H atoms. In the BPP theory it is seen that $(D/T_1) \propto \sum r^{-6}$ and if simultaneous occupancy of H atoms on NN sites is excluded, the largest terms in the sum would have to be omitted. While this interpretation is valid qualitatively, Fedders (29) warns that a new (D/T_1) curve must be calculated for each model of site occupancy to be tested.

In a similar test of site occupancy, Sankey and Fedders (23) compared theory and experimental values of (D/T_1) for $PdH_{0.7}$ (FCC) with good agreement for octahedral site occupancy and no strong evidence for a repulsive interaction.

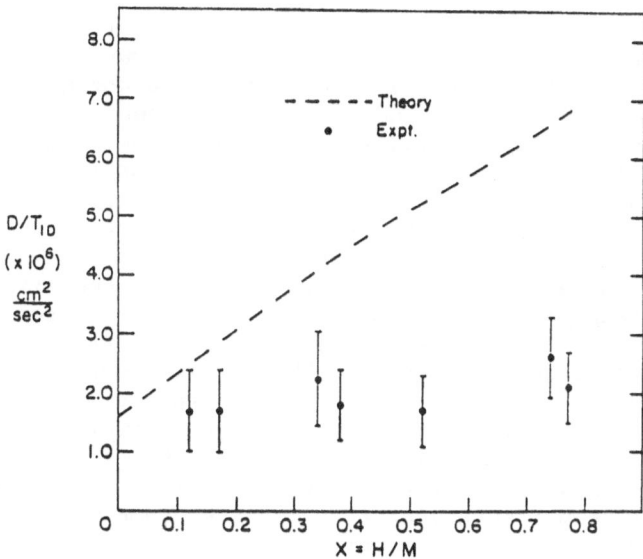

Fig. 2. D/T_1 vs hydrogen concentration in $\alpha'TaH_x$ (14). The uncertainties in experimental points are associated with the values of the pre-exponential factors Do and T_{1o} used in forming D/T_1.

EXPERIMENTS ON $J^{(p)}(\omega)$

The preceeding discussion might paint too rosy a picture of the current status of T_1 measurements and interpretation for ^{1}H in metal hydrides. In both the lattice-specific and the BPP theories, the experimental plots of $\log (T_1)^{-1}$ or $\log (T_{1r})^{-1}$ vs $(1/T)$ should have equal and opposite slopes in the asymptotic regions. In 1980 Jones and Halstead (30) compiled a list of some 12 various metal-hydrogen systems in which the apparent E from the high temperature side exceeds the E from the low temperature side of the curve. These include PdH_x where H-H dipolar interactions dominate and NbH_x where H-M interactions are large. The shape of $J^{(p)}(\omega)$ is thus questioned.

In a study of ^{1}H NMR in the intermetallic hydrides $Hf_2RhH_{2.2}$ and $Hf_2CoH_{3.8}$, Jones and Halstead demonstrated an iterative technique for combining frequency and temperature dependence of measured T_1, T_{1r}, and T_2 to devise a single $J(\omega\tau_c)$ vs $(\omega\tau_c)$ curve. Their empirical technique assumes no prior knowledge of the detailed shape of $J(\omega\tau_c)$. These hydrides have complex structures, but in both systems the dipolar coupling is dominated by H-H interactions. No lattice-specific calculation of $J(\omega)$ is available. Jones and Halstead give $J(\omega\tau_c)$ in the form, $J(\omega\tau_c) = A(\tau_c)$ $B(\Omega) F(\omega\tau_c)$ where $A = \tau_c^{1-m}$, $F(\omega\tau_c \ll 1) = 1$, $F(\omega\tau_c \gg 1) = (\omega\tau_c)^{n-2}$, and $B(\Omega)$ is a material dependent factor independent of ω and τ_c.

Their graphs of $\log F(\omega\tau_c)$ vs $(\omega\tau_c)$ are shown in Figs. 3 and 4. Their fitting procedure yields m = 0.36 and 0.14 for the Rh and Co samples and n = 0.87 and 0.53 for the same samples, respectively. Theoretically the values of m and n should be zero. The expectation that $J(\omega\tau_D \ll 1) \propto \tau_c$ follows from an assumption that the diffusion process is independent of temperature. That is to say that there is a single $G^{(p)}(t)$ that would apply at all temperatures with time scaled by τ_D. In these systems $J(\omega\tau_D \ll 1) \propto \tau^{0.64}$ and $\tau^{0.86}$ for Rh and Co systems, suggesting that the diffusion process is not temperature-independent. At low temperatures $J(\omega\tau_D \gg 1)$ $\propto \omega^{-1.13}$ and $\omega^{-1.47}$ for Rh and Co systems respectively, when ω^{-2} dependence would be expected. The weaker experimental frequency dependences that are expressed here are essentially equivalent to the lower magnitudes of slope of $(T_1)^{-1}$ on the low temperature side of the maximum in $(T_1)^{-1}$ vs $(\omega\tau_D)$. It is noted that Jones and Halstead did not directly measure the ω dependence of T_1 on the low temperature side at a constant temperature, and that the ω dependence follows from their data analysis.

Sholl (27) has examined the low and high temperature limits of spin relaxation in liquids and solids. He finds that the low frequency form of $J(\omega\tau_D)$ is $J(\omega\tau_D < 1) \approx J(0) - A\tau_D(\omega\tau_D)^{1/2}$ for liquids and crystals. The frequency dependence differs from a BPP

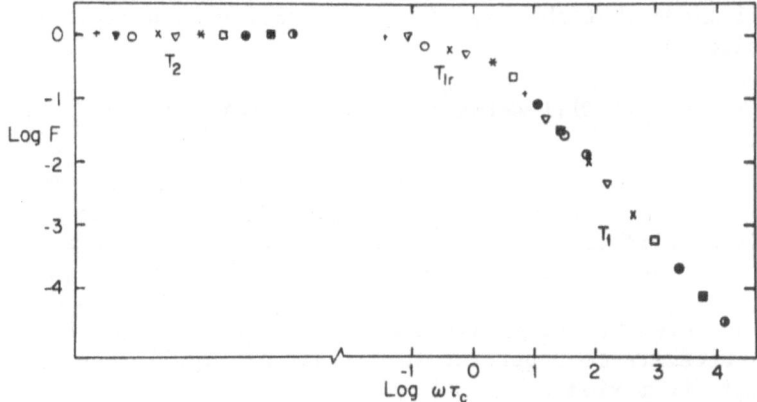

Fig. 3. $F(\omega\tau_c)$ <u>vs</u> $(\omega\tau_c)$ for $Hf_2RhH_{2.2}$. The frequency
dependence of $J(\omega)$ is contained in the factor $F(\omega\tau_c)$ of
Reference (30). Data from T_1, T_{1r}, T_2 used to construct
the curves in the regions shown, from (30).

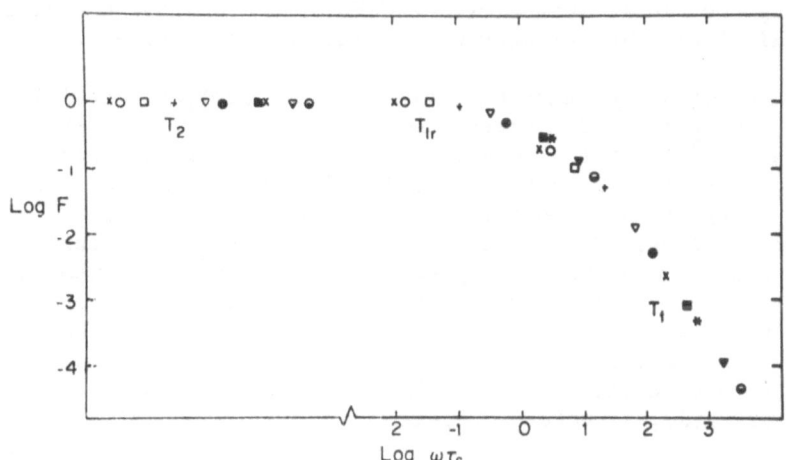

Fig. 4. $F(\omega\tau_c)$ <u>vs</u> $(\omega\tau_c)$ for $Hf_2CoH_{3.8}$, from (30).

model in which $J(\omega\tau_D<1) \approx J(0) - B\omega^2\tau_D^3$. The weaker ω dependence in
Sholl's expression is responsible for the broader, flatter shape of
the maximum in $(T_1)^{-1}$ vs $(\omega\tau_D)$ on the high temperature side.
The $(\omega\tau)^{1/2}$ dependence in liquids was shown by Abragam (31, 27) to
be a direct consequence of the application of the diffusion equation
to calculation of $J(\omega\tau_D)$ for low values of ω and, correspondingly,
long times in atomic motion. Harmon and Muller (32) noted the
$(\omega\tau)^{1/2}$ dependence in $(T_1)^{-1}$ in liquid ethane.

Sholl shows that the same form must obtain in a crystal lattice. For a solid,

$$(T_1)^{-1} = A_1 \, (f/D) \, [1-A_2 (\omega L^2/D)^{1/2}]; \qquad (\omega \tau_D) < 1 \qquad (9)$$

where A_1 and A_2 are known constants for a given sample. From mean-field theory results, one set of constants A_1 and A_2 can be used for any cubic crystal (27). It is therefore suggested that a measurement of the frequency dependence of T_1 and T_{1r} on the high temperature side could be used to obtain values of D without the need to do pulsed field gradient experiments. The equation is given as being accurate to 3 percent for values of $(T_1)^{-1}$ less than about half the maximum.

Abragam (31) shows in Chapt. VIII.E. that when the diffusion equation is applicable, $G^{(p)}(t)$ should have a $t^{-3/2}$ dependence at long times, $t \gg \tau_D$. In his Monte Carlo calculations, Bustard found that the $t^{-3/2}$ dependence gave a good fit to $G^{(p)}(t)$ for $t \gg \tau_D$ (33). However, this particular form was not used (26) in calculating $J(\omega \tau_D)$ at low frequencies because it was not practical in taking the Fourier transform of $G^{(p)}(t)$ and because the Monte Carlo calculation had greater uncertainty at long times.

On the high frequency side (low temperature) of the $(T_1)^{-1}$ maximum, $J(\omega \tau_D) \propto \omega^{-2}$ for three-dimensional solids (21,27). Sankey and Fedders (21) noted a considerably more gradual approach to this asymptotic limit than in BPP. Sholl (27) states that the limiting form of $J(\omega \tau_D)$ is reached for $(\omega \tau_D)^2 \gg 4(1+c)/(1-c)^3$ where c is the concentration. This is a severe requirement rarely achieved on the low temperature side for systems with $c > 0.95$. For example with 1% vacancy concentration the asymptotic limit would not have been reached by $(\omega \tau_D) = 2500$. While this behavior affects the shape of the $(T_1)^{-1}$ maximum at the peak it is probably not sufficient to cause the reduced slopes (and values of E observed on the low temeprature side, as noted earlier.

The frequency dependence of $(T_1)^{-1}$ on the high temperature (low frequency) side of the $(T_1)^{-1}$ maximum has been observed by Salibi and Cotts (34) in γ-TiH$_{1.63}$. T_1 and T_{1r} were measured at a fixed temperature of 712 K. Frequency was varied between 6.57 MHz and 47.6 MHz. Assuming the $(\omega)^{1/2}$ dependence in $(T_1)^{-1}$ it is readily shown that the T_{1r} measurement at $\omega_1 \ll \omega_0$ is equivalent to a T_1 measurement at $0.346 \, \omega$. Therefore $(T_{1r})^{-1}$ and $(T_1)^{-1}$ measurements can be plotted on a common graph with $\omega_{eff} = 0.346 \, \omega$ for $(T_{1r})^{-1}$ data. Salibi's data in Figure 5 show that $(T_1)^{-1}$ falls off from its zero frequency value as $\omega^{1/2}$. Only the diffusional component of the relaxation rate is plotted. The conduction electron relaxation rate, which is frequency independent, has been subtracted off in the usual way. For TiH$_{1.63}$, (T_{1e}^{-1}) = (T/C_e) where C_e = 100 sec°K.

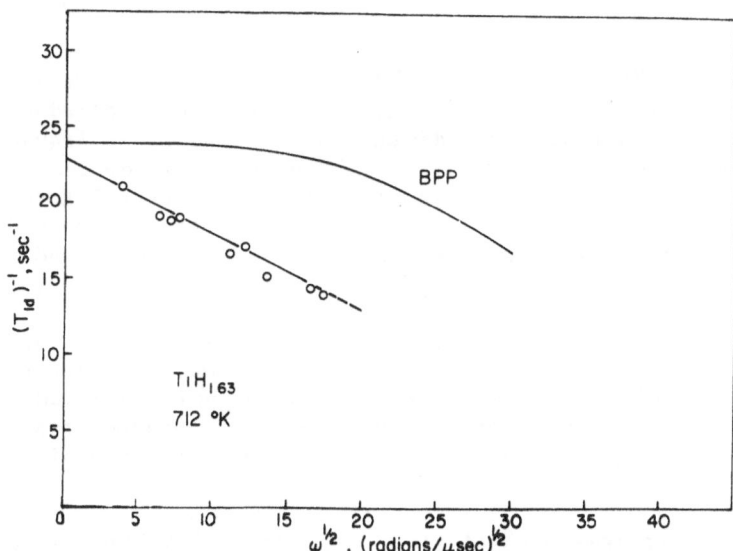

Fig. 5. The diffusional relaxation rate $(T_{1d})^{-1}$ vs $\omega^{1/2}$ for γ-TiH$_{1.63}$. Open circles are experimental points and the straight line is the prediction of mean field theory.

The mean temperature varied from run to run but by no more than 8.5 K from low to high. All measured $(T_1)^{-1}$ were adjusted to the temperature 712 K by using Equation (9), below, and an activation energy of 0.54 ev. (12). These adjustments were typically about 2% of $(T_1)^{-1}$. To fit Equation (9) we used a value of D at 712 K obtained by interpolating between values of D for TiH$_{1.55}$ and TiH$_{1.71}$ (12). Using this value of D = 1.49 x 10^{-7} cm^2/sec, f = 0.73 (12), and mean field theory (22,27), the theoretical dipolar relaxation rate is

$$(T_1)^{-1} = \left[23.1 - 0.52 \times 10^{-3}(\omega)^{1/2} \right] sec^{-1}$$

A straight line fit to the data gives

$$(T_1^{-1}) = \left[22.75 - 0.53 \times 10^{-3}(\omega)^{1/2} \right] sec^{-1}$$

in good agreement with Equation (9). At temperature 712 K, the value of the dipolar relaxation rate at the highest frequency used is less than 40% of the maximum rate at that frequency, so that no measured rate was within 50% of the maximum rate.

SUMMARY

The long-time behavior of $G^{(p)}(t)$ and the corresponding low frequency dependence of $J^{(p)}(\omega)$ are seen to be controlled by continuum diffusion in three-dimensional solids as in liquids.   In all lattice-specific theory for simple hopping of non-interacting atoms in a solid, the high frequency (low temperature) form of $J^{(p)}(\omega)$ is proportional to $\omega^{-2}\tau^{-1}$.   The asymmetries observed in the $(T_1)^{-1}$ maximum vs $\log \tau_D$ and the analysis of Jones and Halstead are consistent with a weaker frequency dependence.   It is sometimes suggested that an $\omega^{-3/2}\tau^{-1/2}$ dependence might be obtained for continuum diffusion controlled $G^{(p)}(t)$ at short times under circumstances in which the step length might be much less than the distance of closest approach.   However both Sholl (27) and Hwang and Freed (35) have independently shown the ω^{-2} dependence must apply to this case as well when there exists a minimum distance of closest approach.

In a paper presented at this conference, T-T Phua, et al (36) show that small amounts of paramagnetic impurities (a few ppm) can contribute a very effective relaxation mechanism that could cause the weakened temperature dependences (30) on the low temperature side of the $(T_1)^{-1}$ and $(T_{1r})^{-1}$ maxima.   This mechanism is analogous to effects recently observed in fast ionic conductors such as PbF_2:Mn (37), (38).   The maxima in $(T_1)^{-1}$ and $(T_{1r})^{-1}$ would not be appreciably shifted at low impurity levels, but it is clear that precautions must be taken if the shape of the relaxation rate curves are to be analyzed.

If samples are made non-uniform due to severe internal strain, a distribution of hopping rates would cause a distortion of these curves.   But for pure homogeneous samples, the theories based upon no interactions between H atoms predict symmetric relaxation rate curves.   Fedders found (29) that for low H concentration on the SC lattice, the inclusion of interactions between H atoms did not change the symmetry in a noticeable way even though $(T_1)^{-1}$ was changed by a large factor.   It is not known how a jump process other than first NN would affect the symmetry.   Since the symmetry can be affected by a few ppm magnetic impurity it is possible that many of the asymmetries already observed could be due to this mechanism. Since diffusion is being studied by NMR in many hydride systems, we can expect that more attention will be given to application of lattice-specific theories.   If the asymmetries in rate curves are observed and cannot be attributed to impurities, we may find in their study the opportunity to identify detail in the diffusion process which should be most evident in observations on the low-temperature, high frequency side of the rate curve maxima.

ACKNOWLEDGEMENT

Supported by the National Science Foundation Grant DMR-79-25383.

REFERENCES

1. N. Bloembergen, E.M. Purcell, and R.V. Pound, Phys. Rev. 73, 679 (1948).
2. R.G. Barnes, in "Nuclear and Electron Spectroscopies Applied to Materials Science," E.N. Kaufmann and G. Shenoy, Ed., Elsevier, North Holland (1981) pages 19-30.
3. R.M. Cotts, in "Hydrogen in Metals I," G. Alefeld and J. Volkl, Ed., Springer-Verlag, Berlin (1978) pgs. 227-265.
4. B. Pedersen, "Hydrides for Energy Storage, Proceedings of an International Symposium at Geilo" (1977) pgs. 83-95.
5. R.C. Bowman, Jr., in "Metal Hydrides," Vol. 76 NATO Advanced Study Institute Series B: Physics, Plenum, New York (1981) pages 109-44.
6. E.O. Stjeskal and J.E. Tanner, J. Chem. Phys. 42, 288 (1965).
7. J.E. Tanner, J. Chem. Phys. 52, 2523 (1970).
8. E.F.W. Seymour, R.M. Cotts, and W.D. Williams, Phys. Rev. Letters 35, 165 (1975).
9. P.P. Davis, E.F.W. Seymour, D. Zamir, W.D. Williams, and R.M. Cotts, J. the Less-Common Met. 49, 159 (1976).
10. O.J. Zogal and R.M. Cotts, Phys. Rev. B11, 2443 (1975).
11. Y. Fukai, K.T. Kubo, and S. Kazama, Z. Phys. Chem. Wiesbaden 115, 181 (1979).
12. D.L. Bustard, R.M. Cotts and E.F.W. Seymour, Phys. Rev. B22, 15 (1980).
13. W.D. Williams, E.F.W. Seymour, and R.M. Cotts, J. Mag. Resonance 31, 271 (1978).
14. P.E. Mauger, W.D. Williams, and R.M. Cotts, J. Phys. Chem. Solids 42, 821 (1981).
15. R.F. Karlicek, Jr. and I.J. Lowe, J. Mag. Resonance 37, 75 (1980).
16. R.F. Karlicek, Jr. and I.J. Lowe, J. Less-Common Met. 73, 219 (1980).
17. P.E. Mauger, Ph.D. Thesis, Cornell University, 1981 (unpublished).
18. H.C. Torrey, Phys. Rev. 96, 690 (1954), H.A. Resing and H.C. Torrey, Phys. Rev. 131, 1102 (1963).
19. C.A. Sholl, J. Phys. C: Solid St. Phys. 7, 3378 (1974) and C.A. Sholl, J. Phys. C: Solid State Phys. 8, 1737 (1975).
20. D. Wolf, J. Phys. C: Solid St. Phys. 10, 3545 (1977).
21. P.A. Fedders and O.F. Sankey, Phys. Rev. B18, 5938 (1978).
22. W.A. Barton and C.A. Sholl, J. Phys. C: Solid St. Phys. 13, 2579 (1980).
23. O.F. Sankey and P.A. Fedders, Phys. Rev. B20, 39 (1979).
24. O.F. Sankey and P.A. Fedders, Phys. Rev. B22, 5135 (1980).
25. W.A. Barton and C.A. Sholl, J. Phys. C: Solid St. Phys. 9, 4315 (1976).
26. L.D. Bustard, Phys. Rev. B22, 1 (1980).
27. C.A. Sholl, J. Phys. C: Sol. St. Phys. 14, 447 (1981).
28. H.J. de Bruin and G.E. March, Phil. Mag. 27, 1475 (1973).

29. P.A. Fedders, Phys. Rev. B25, 78 (1982).

30. T.C. Jones and T.K. Halstead, J. Less-Common Met. 73, 209
     (1980).

31. A. Abragam, 1961 "The Principles of Nuclear Magnetism", Chapter
     VIII, Oxford, Clarendon.

32. J.F. Harmon and B.H. Muller, Phys. Rev. 182, 400 (1969).

33. L.D. Bustard, Private Communication.

34. N. Salibi and R.M. Cotts (unpublished).

35. L.P. Hwang and J.H. Freed, J. Chem. Phys. 63, 4017 (1975).

36. T-T. Phua, R.G. Barnes, D.R. Torgeson, D.T. Peterson, M.
     Belhoul, and G.A. Styles, This Conference Proceedings.

37. P.M. Richards, Phys. Rev. B18, 635B (1978).

38. S.P. Vernon and V. Jaccarino, Phys. Rev. B24 3756 (1981).

DIFFUSION OF HYDROGEN IN VANADIUM-BASED BCC ALLOYS[*]

D. J. Pine and R. M. Cotts

Cornell University

ABSTRACT

A resistiometric technique for measurement of large diffusion coefficients of hydrogen in metals is described. Diffusion is along the length of a foil (7.0 cm x 0.15 cm x 100 μm) which has been loaded electrolytically to a hydrogen concentration of approximately 0.5 atomic per cent. Initially the hydrogen is held at a uniform concentration in essentially half the length of the foil by the effect of a DC electrotransport current of 2500 A/cm^2 in the other half. After this current is removed, the diffusion of H into the empty half is monitored by its effect on the resistivity of the foil. The sensitivity to hydrogen concentration is better than ± 10ppm. Recovery of equilibrium resistivity at essentially one point in the foil measures recovery of equilibrium hydrogen concentration because of the linear dependence of resistivity upon concentration. Good fits to the diffusion equation are obtained though 99 percent of full recovery. Values of diffusion coefficients of ^{1}H and ^{2}H have been measured in the vanadium-based BCC alloy system, V-Ti and V-Cr up to about ten percent Ti and Cr. Before adding hydrogen, the V-Ti foils are purified and annealed at 1800°K in a pure argon atmosphere using electrotransport to remove O, N, and C impurities. V-Cr foils are annealed in a similar way but cannot be similarly purified because of high Cr vapor pressure at the temperature needed for electrotransport. The addition of Ti or Cr to V causes an increase in the activation energy of diffusion for both hydrogen isotopes.

*Supported by the National Science Foundation through the Cornell Materials Science Center Grants DMR 76-81083 and DMR 79-24008.

THE EFFECT OF MAGNETIC IMPURITIES ON THE APPARENT DIFFUSION CO-
EFFICIENT OF HYDROGEN IN METAL HYDRIDES DEDUCED FROM NMR

T-T. Phua, R.G. Barnes, D.R. Torgeson, D.T. Peterson
M. Belhoul* and G.A. Styles*

Ames Laboratory, USDOE and Departments of Physics and
Materials Sciences, Iowa State University, Ames, Iowa
*Physics Department, University of Warwick
Coventry CV4 7AL U.K.

ABSTRACT

Observations of the well-known minimum in proton spin-lattice relaxation time, T_1, as a function of temperature, T, caused by diffusion-induced fluctuations of nuclear dipole-dipole inter-actions, have often been used to determine hydrogen diffusion coefficients, D, in metal-hydrogen systems. In several cases the minimum has appeared asymmetric with a shallower low-temp-erature slope interpreted as a change to a diffusion mechanism with a lower activation energy. In a few cases a subsidiary low-temperature minimum has been observed or inferred and interpreted in terms of two coexisting mechanisms. We believe that these complications are likely to be due to interaction with paramagnetic impurities, analogous to the case of doped fast-ion conductors. We have measured $T_1(T)$ for ^1H in $YH_{1.98}$ containing controlled amounts of gadolinium and observed growth of the above features with increasing impurity content, together with a low-temperature region (T_1 independent of T) controlled by spin-diffusion, and also the beginning of a third, high-temperature minimum not previously found in metal-hydrogen systems. It is concluded that some D(T) values deduced from existing NMR data must be treated with caution until measurements on specimens with con-trolled impurity content are complete.

INTRODUCTION

Nuclear magnetic resonance has been used for many years to investigate hydrogen diffusion in transition metals. In most cases

the proton spin-lattice relaxation time $T_1(T)$ passes through a
minimum when $\omega_0 \tau_c \sim 1$ and has been used to deduce the temperature
dependence of the correlation time τ_c (half the mean jump time τ_D,
for proton-proton interactions), which is then fitted to an Arrhenius
relation to deduce the activation energy E_a. For many systems a
discontinuous change in E_a has been found, usually on the low-
temperature side of the T_1 minimum, and sometimes an additional
minimum, perhaps only partially resolved, has been observed in the
same region. These effects are illustrated in fig. 1.

     Similar behaviour has been seen in yttrium dihydride. In this
system the metal atoms form an f.c.c. lattice with $a_0 = 5.205$Å and
the hydrogen atoms predominantly occupy tetrahedral interstitial
sites [5]. However a significant octahedral site occupation at
temperatures as low as 160-200K has been deduced from NMR second
moment measurements [6] in agreement with neutron experiments[6][7]. A
non-Arrhenius behaviour of τ_c giving a discontinuity in E_a [6] and a
partially resolved double minimum in $T_1(T)$[6] have been interpreted as
due to different types of motion on the tetrahedral and octahedral
sublattices. In this paper we present some new observations of T_1

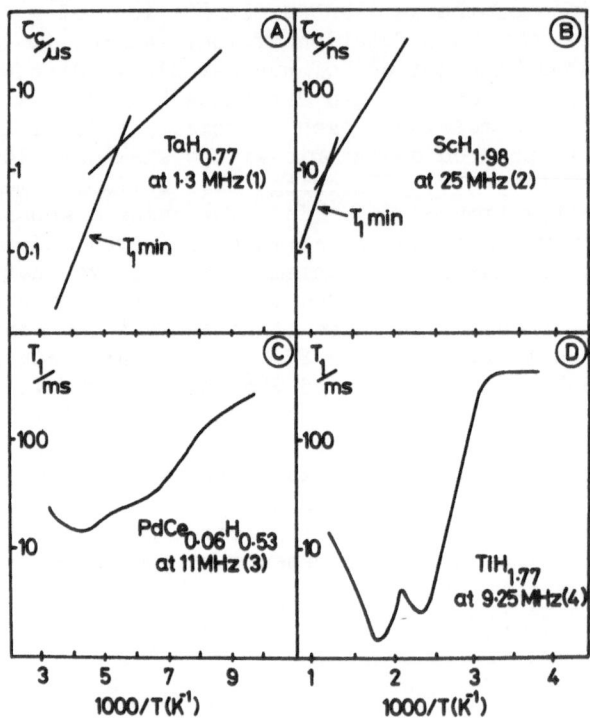

Fig. 1.   Variations of τ_c (A and B) and T_1 (C and D) with temper-
           ature in some metal hydrides.

in yttrium dihydride containing controlled amounts (3 ppm and 20 ppm) of Gd^{3+} impurities which show that the latter may have a strong effect on T_1. On the basis of these observations we suggest that many of the above effects must be reinterpreted in terms of the previously unsuspected influence of paramagnetic impurities.

THEORY

The relaxation rate of protons in a metal hydride containing paramagnetic impurities is given by

$$\frac{1}{T_1} = \frac{1}{T_{1D}} + \frac{1}{T_{1E}} + \frac{1}{T_{1P}} \tag{1}$$

where $(T_{1D})^{-1}$ and $(T_{1E})^{-1}$ are the only contributions normally considered in a pure system. $(T_{1D})^{-1}$ arises from proton-proton (and, where appropriate, metal nucleus-proton) dipolar interactions and assuming an exponential correlation function is given by the BPP formula

$$\frac{1}{T_{1D}} = \frac{3}{10} \gamma^4 \hbar^2 \sum_i r_i^{-6} \left[\frac{\tau_c}{1+\omega_o^2 \tau_c^2} + \frac{4\tau_c}{1+4\omega_o^2 \tau_c^2} \right] \tag{2}$$

where γ is the proton paramagnetic ratio, $\hbar$ is Planck's constant, ω_o is the resonance frequency and the sum is usually taken over all nearest neighbour proton sites.

$(T_{1E})^{-1}$ is due to the interaction with the conduction electrons and is generally found to follow the Korringa relation

$$T_{1E} T = K \tag{3}$$

where K is a constant which depends on the density of states at the Fermi level. Normally $(T_{1E})^{-1}$ is the only significant contribution at very high or very low temperatures.

$(T_{1P})^{-1}$ is an additional contribution due to the hyperfine interaction with paramagnetic impurities analogous to that recently observed in the superionic conductor PbF_2: Mn [9,10]. In a metal the protons will be relaxed by the impurities through the indirect exchange interaction via the conduction electrons and through dipolar interactions. While this interaction may be of fairly long range, we assume for the sake of simplicity that protons only experience significant relaxation when they diffuse to sites nearest neighbour (nn) to a paramagnetic ion.

In the nn position a proton experiences a hyperfine field

$$B_{HF} = \frac{|A|S}{\gamma} \tag{4}$$

where $\underline{I} \cdot \underline{A} \cdot \underline{S}$ is the hyperfine interaction. In the slow hopping regime $(A \tau_D \gtrsim \pi)$ the proton is relaxed during a single encounter with a paramagnetic ion and one has

$$\frac{1}{T_{1P}} = \frac{Zc}{\tau_D} \tag{5}$$

where Z is the coordination number and c the concentration of paramagnetic impurities. At the other extreme in the fast hopping regime $(A \tau_D \lesssim \pi)$ several encounters are required to relax the proton and one then has

$$\frac{1}{T_{1P}} = Zc |A|^2 \tau_D \tag{6}$$

It can be seen from equations (5) and (6) that as τ_D varies with temperature T_{1P} will pass through a minimum when $A\tau_D \sim \pi$.

At high temperature the fluctuation of the hyperfine field produced by the transverse component of the electron spin will have

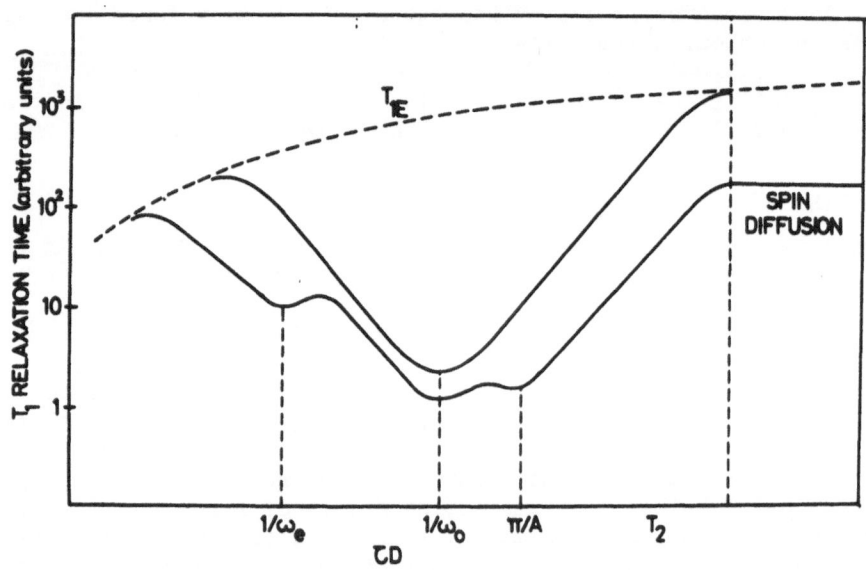

Fig. 2.   Illustrative T_1 behaviour in a metal hydride. The upper curve is for a pure material while the lower curve includes paramagnetic impurity effects as described in the text.

an effect on $T_{1\rho}$ producing a further minimum when $\omega_e \tau_D \sim 1$ corresponding to a maximum in the component of the spectral density at the electron frequency ω_e. At low temperature spin diffusion takes over from atomic diffusion as the agent of relaxation when $T_2 \stackrel{<}{\sim} \tau_D$ where T_2 is the rigid lattice spin-spin relaxation time due to the proton-proton dipolar interaction. Thus $(T_{1\rho})^{-1} = Zc/T_2$ and is independent of temperature.

All the above features are illustrated in fig. 2.

YTTRIUM DIHYDRIDE

Our observations of the proton T_1 in $YH_{1.98}$ containing Gd^{3+} are shown in fig. 3. The principal features of these results are (a) a main minimum, attributed to proton-proton dipolar interactions occurs at T = 633K for ν_0 = 40 MHz and at T = 536K for ν_0 = 7 MHz, (b) a subsidiary minimum at T = 450K, occurring when $A\tau_D \sim \pi$ from which we deduce that $A \sim 3$ MHz (compared with 7.8 MHz for ^{19}F in PbF_2:Mn), (c) a decrease in T_1 at the highest temperature for the 3 ppm Gd^{3+} sample at 7 MHz which we interpret as due to a minimum for $\omega_e \tau_D \sim 1$ at approximately 1480K, using an activation energy E_a = 0.47eV and (d) a decrease of T_1 with increasing Gd^{3+} concentration at very low

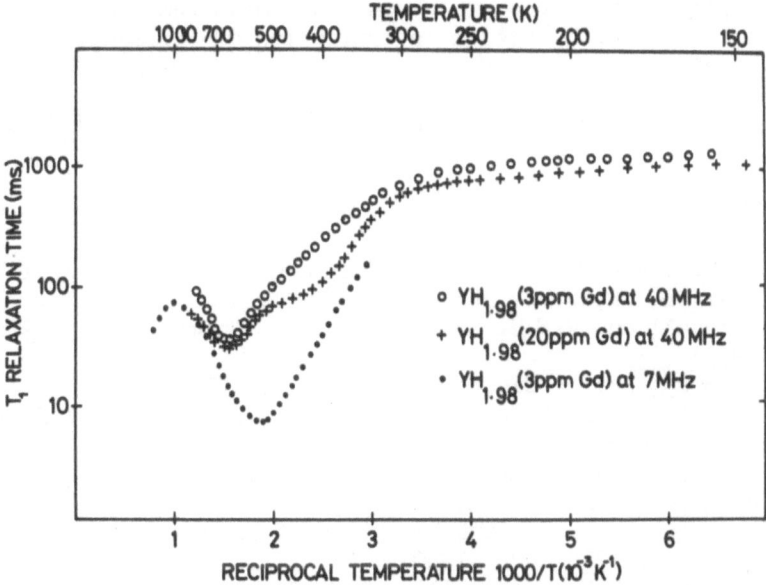

Fig. 3. Variation of T_1 with temperature in $YH_{1.98}$ containing Gd^{3+}.

temperatures where T_1 is becoming independent of temperature.

All these features are consistent with the predictions of the preceding section.

REINTERPRETATION OF EARLIER DATA

On the basis of our results in yttrium dihydride we suggest that conclusions drawn from existing data which ignore $T_{1\rho}^{-1}$ should be treated with caution, particularly in cases where a departure from expected behaviour has been observed. For example, the apparent change in activation energy in figs. 1(A) and 1(B) may be due to unsuspected paramagnetic impurities giving rise to an unresolved minimum at $A\tau_D \sim \pi$ which distorts the main dipolar minimum. In figs. 1(C) and 1(D) this minimum is resolved. In addition in fig. 1(D) the effects of impurities at low temperature are evident and a value of T_{1E} obtained by fitting a Korringa relation will be subject to error. The high temperature T_1 minimum has not been observed before in metal hydrides because of the lack of data at higher temperatures and because of the common preference for the use of large magnetic fields which shift the minimum to yet higher temperatures. A fortunate consequence is that activation energies derived from data in the region $\omega_0 \tau_D \lesssim 1$ should be reliable because here the $T_{1\rho}$ contribution is relatively small. It is clear, however, that in a number of cases reported in the literature, the lower temperature data are open to a reinterpretation which takes possible low-level paramagnetic impurities into account. Future NMR measurements of this nature will call for a careful determination of the sample impurity content.

This work was supported by U.S., DOE, U.K., SERC and NATO.

REFERENCES

1. E.P. Apgar, PhD Thesis (Rutgers University, 1957) unpublished.
2. H.T. Weaver, Phys. Rev. B5, 1663 (1972).
3. D.A. Cornell and E.F.W. Seymour, J. Less-Common. Met., 39, 43 (1975).
4. H.C. Torrey, Nuovo. Cimento. Suppl. 9, 95 (1958).
5. W.M. Mueller, J.P. Blackledge and G.G. Libowitz, Metal Hydrides, Academic Press, N.Y. (1968).
6. D.L. Anderson, R.G. Barnes, T.V. Hwang, D.T. Peterson and D.R. Torgeson, J. Less-Common. Met. 73, 243, (1980).
7. D. Khatamian, W.A. Kamitakahara, R.G. Barnes and D.T. Peterson, Phys. Rev. B21, 2622 (1980).
8. R.S. Kashaev, E.F. Guduidullin, A.N. Gil'manov and M.E. Kost, Sov. Phys. Solid State, 22, 530 (1980).
9. M. Richards, Phys. Rev. B18, 6358 (1978).
10. S.P. Vernon and V. Jaccarino, Phys. Rev. B24, 3756 (1981).

NMR STUDIES OF HYDROGEN DIFFUSION

IN PALLADIUM AND β-LaNi$_5$ HYDRIDES

K. L. Ngai, A. K. Rajagopal[*] and R. W. Rendell[**]

Naval Research Laboratory
Washington, D.C. 20375
[*]Department of Physics
Louisiana State University
Baton Rouge, Louisiana
[**]Naval Research Laboratory
Sachs/Freeman Associate

INTRODUCTION

Nuclear magnetic resonance (NMR) techniques have been employed extensively to study the dynamics of hydrogen diffusion in metal hydrides. The pulsed field gradient (PFG) spin-echo technique has been used[1] to deduce the hydrogen diffusion coefficient, D, in transition metals. To overcome difficulties encountered with conventional PFG spin-echo technique in β-LaNi$_5$ hydride, an alternating pulsed field gradient technique[2] has also been introduced. Proton spin-lattice relaxation time T_1 and the rotating frame, $T_{1\rho}$ measurements[2-5] have also been used to obtain the correlation time which corresponds to the mean proton diffusion jump time, τ_D. However, the T_1 and $T_{1\rho}$ data for many hydrides including LaNi$_5$H$_7$, Hf$_2$RhH$_{2.2}$, Hf$_2$CoH$_{3.8}$, PdH$_x$, NbH$_x$, Th$_4$H$_{15}$ etc. are anomalous and difficult to interpret. The slope on the high temperature side of the T_1 or $T_{1\rho}$ minimum is always steeper than the slope on the low temperature side. The frequency dependence on the low temperature side is not quadratic and the departure can be large.[2-5] The reported attempt frequencies as deduced from T_1 and $T_{1\rho}$ data for different hydrides differ from each other by orders of magnitude. Several suggestions have been made to explain these anomalous results. These include: (a) two separate activation energies[2] governing two types of hydrogen motion which is discarded subsequently[4] because it leads to a ω^2 dependence; (b) translational diffusion relaxation[4] of nuclear spin model of Torrey[6] but unfortunately no comparison of Torrey's model predictions with data has been made; and (c) the reference made by Jones et al.[3] and by Chang et al.[4] to a recently proposed unified low frequency response model[7] for possible explanation; but again no comparisons have been made.

In this work we carry out the comparisons of the experimental data[4] with the predictions of both Torrey's model and the unified model to examine the degree of applicability of each. Before presenting these results, we examine first what modification the unified model has on the classical expression of the spin-echo height in pulsed-gradient method.

SPIN ECHO

Stejskal and Tanner (ST)[8] consider the case of a time dependent gradient $\vec{G}(t)$ described by a gradient of magnitude g and time width δ applied between 90° and 180° pulses and between the 180° pulse and the echo. The first gradient pulse occurs at a time t_1 and the second at a time $t_1+\Delta$ where $t_1 + \delta < \tau < t_1 + \Delta$. For this choice of $\vec{G}(t)$, the echo occurs at $t=2\tau$. Background field gradients are considered equal to zero here.

ST determine the time dependence of the function $\psi(\vec{r},,t)$, which governs the behavior of the magnetization in the rotating frame, from the standard equation of motion.

$$\partial\psi/\partial t = -i\ \gamma(\vec{r}.\vec{G})\psi + D\nabla^2\psi \tag{1}$$

where γ is the gyromagnetic ratio. In the absence of diffusion, ψ takes the form: $\psi = A\ \exp\{-i\gamma\vec{r}.[\vec{F}(t) - (z-1)\vec{F}(\tau)]\}$ where $\vec{F}(t) = \int_0^t dt'\ \vec{G}(t')$ and $z=+1$ for $0<t<\tau$ and $z=-1$ for $t>\tau$. The effect of diffusion can now be found by integrating (1). ST find that $\ln[A(2\tau)/A(0)] = -g^2\gamma^2 D\delta^2(\Delta-\delta/3)$, which is the spin echo amplitude.

It is the diffusion of the nuclei positions between the two gradient pulses which is responsible for the resulting amplitude decay. The unified model predicts[7,10,11] that the diffusion will be accompanied by a characteristic transient response which causes D in (1) to be replaced by $Da_n t^{-n}$, where $0<n<1$. Here $a_n = (\exp(0.577)\omega_c)^{-n}/(1-n)$ and ω_c, a cut-off frequency, is typically $\sim 10^9$Hz for glasses and polymers. From this we can derive a new result for the amplitude at the spin echo:

$$\ln[A(2\tau)/A(0)] = -(2g^2\gamma^2Da_n/n_1n_2n_3)\{-\delta n_3[(t_1+\delta)^{n_2}+(t_1+\Delta)^{n_2}]$$

$$+ (t_1+\delta)^{n_3}-t_1^{n_3} + (t_1+\Delta+\delta)^{n_3}-(t_1+\Delta)^{n_3}\} \qquad (2)$$

where $n_1 = 1-n$, $n_2 = 2-n$ and $n_3 = 3-n$. This reduces to the ST result when $n\rightarrow0$. Comparison with ST is made clear by considering the case where $\delta<<\Delta$, t_1. Then $\ln[A(2\tau)/A(0)] = -g^2\gamma^2D^*\delta^2\Delta$ where $D^* = Da_n[(t_1+\Delta)^{n_1}-t_1^{n_1}]/n_1\Delta$. The unified model result takes the same form as ST in this limit except with D replaced by D^*. Note that if D is thermally activated, then plots of log D and log D^* vs $1/T$ will yield the same value for the activation energy E_A. The two plots will however be shifted vertically, depending on the value of n.

The limiting case of a constant field gradient can be obtained from (2) by setting $t_1=0$ and $\Delta=\delta=\tau$. This yields $\ln[A(2\tau)/A(0)] = -(2g^2\gamma^2Da_n/n_1n_2n_3) \tau^{n_3}[2^{n_3}-2n_3]$, which is the unified model modification of Hahn's famous result.[9] Results for other field gradient histories can be evaluated in a similar manner.

TORREY'S MODEL AND THE UNIFIED MODEL

Torrey[6] has considered a random walk model of the motion of nuclei in an effort to examine more microscopic details than is allowed by classical diffusion. In the limit where the mean squared step length $<r^2>$ is small and the mean time τ between steps is small, Torrey's random walk model reduces to classical diffusion with $D = <r^2>/6\tau$. The time correlations of the nuclei can be formulated within this model to provide an expression for the relaxation time: $T_1^{-1} = (3/2)\gamma^4h^2I(I+1) [2S_1(2\omega)+S_1(\omega)]$.

The time correlation spectral density $S_1(\omega)$ is given by Torrey's Eq. (27), for the case of isotropic diffusion, in terms of the distribution function $A(\rho)$. Torrey is able to obtain a solution for T_1^{-1} in closed form. This is the case where $A(\rho) = 1/[1+D\tau\rho^2]$, which corresponds to a spin diffusing classically between a set of potential wells at which it can become trapped. Torrey labels his solutions to this case using the parameter α, where $\alpha\equiv<r^2>/12a^2$ and a is interpreted as approximately the closest possible distance of approach of two nuclei before relaxation. Thus $\alpha\rightarrow0$ corresponds to motion which is classically diffusive and larger α corresponds to a hopping motion. The solutions for T_1 show a minimum near $\omega\tau=1$. For fixed $\alpha\neq0$, T_1^{-1} is independent of ω as $\omega\tau\rightarrow0$ and $T_1 \sim\omega^2$ as $\omega\tau\rightarrow\infty$. There exists a region on the low temperature side of the minimum where $T_1 \sim\omega^{3/2}$ and the size of this region increases as α decreases. For $\alpha\equiv0$, the entire region varies as $\omega^{3/2}$. Plots of this solution have been given by Krüger[12] in which these limiting behaviors can be seen.

The modification of Torrey's solution due to the unified model can be obtained simply by replacing t by $a_n t^{1-n}$ in Torrey's Eq. (25). The form of $A(\rho)$ corresponding to a particle diffusing classically between trapping sites is not altered by the unified model. However, the solutions for T_1^{-1} can now no longer be obtained in closed form. Numerical solutions can be easily worked out by formulating the problem as follows.

$$S_1(\omega) = (8\pi N/15a^3)\omega^{-1} \int_0^\infty \rho^{-1} d\rho J_{3/2}^2 (a\rho)\chi''(\omega Q(\rho)) \qquad (3)$$

where

$$\chi''(\omega) = 2\omega \int_0^\infty dt \cos\omega t \exp-(t/\tau_p)^{1-n} \qquad (4)$$

$\tau_p = \tau_\infty^* \exp(E_A^*/RT)$ and $Q(\rho) = 2[1-A(\rho)]^{-1/(1-n)}$. The susceptibilities χ'' are calculated numerically for a wide range of ω and then tabulated. These values are then interpolated at the values of the shifted frequencies ωQ during the ρ integration. This formulation for the NMR relaxation time can also be used in other applications such as diffusion in glasses and polymers where n is expected to be significantly different from zero and the modification of Eq. (3) important.

RESULTS AND DISCUSSIONS

We have evaluated T_1 and $T_{1\rho}$ and compared with experimental data of $LaNi_5H_7$ of Chang et al. [1] Sample comparisons are given in Figs. (1a)-(1e) for various combinations of α and n values. Other combinations (not shown) with n larger than 0.2 and α values either larger than 0.05 or smaller than 0.01 can be ruled out for poor agreement with data. The optimum or near optimum combinations shown in Figs. (1b) and (1e) are not in full accord with the data either. The large discrepancies at the low temperature are probably due to temperature dependence of the hopping activation energy consistent with the small polaron model. In Fig. 1, data of Chang et al. are replotted as log $T_{1\rho}$ versus E_A^*/RT with $E_A^* = 36KJ/$ gm-Atom H. The four solid curves in each plot are calculated from Eqs. (3)-(4) with this same E_A^* value and for the four rotating magnetic field values of 10.6G, 15.9G, 21.0G and 26.7G. Figs. (1a), (1b), and (1c) all with n=0 are basically Torrey's model prediction with α=0.01, 0.05 and 0.001 respectively. For α=0.01, near optimum fit to data is obtained. Fig. (1e) with the combination (α=0.05, n=0.2) gives as good a fit as this. The $T_{1\rho}$ data thus is not sufficient to determine uniquely the size (measured by n) of the contribution from the mechanism of the unified model. In any case n cannot have values larger than 0.2. Thus the small value of n inferred from $T_{1\rho}$ data is consistent with intuition that transition metal hydrides are rather well ordered solids in contrast to

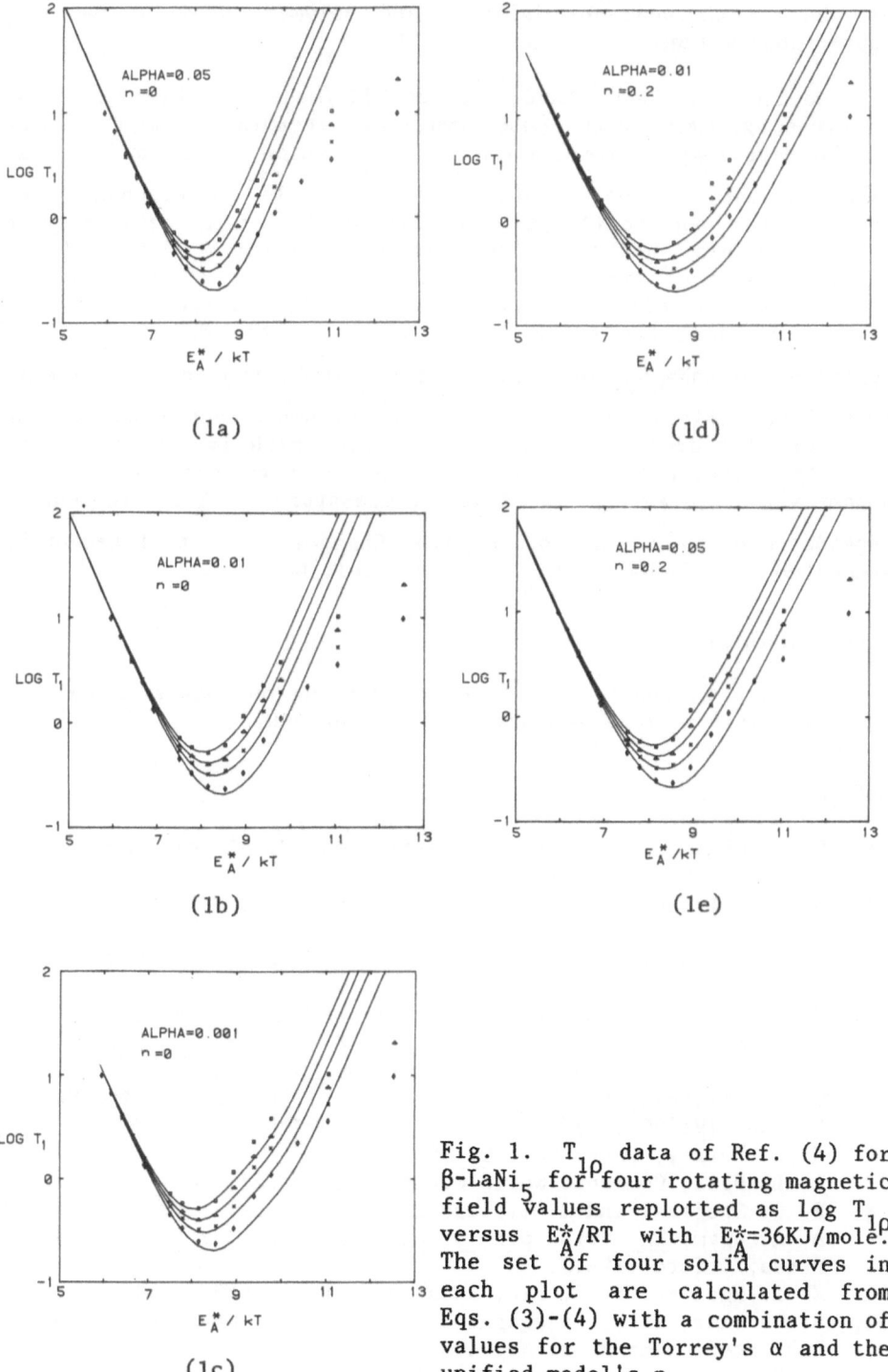

(1a)

(1d)

(1b)

(1e)

(1c)

Fig. 1. $T_{1\rho}$ data of Ref. (4) for β-LaNi₅ for four rotating magnetic field values replotted as log $T_{1\rho}$ versus E_A^*/RT with E_A^*=36KJ/mole. The set of four solid curves in each plot are calculated from Eqs. (3)-(4) with a combination of values for the Torrey's α and the unified model's n.

amorphous semiconductors, glasses and polymers where n can have appreciable values.

Karlicek and Lowe[2] have measured 39KJ/mode by spin-echo method in close agreement with $E_A^*=36kJ/mole$ as determined from the slope of the high temperature arms of the $T_{1\rho}$ minima. From our discussions in the section on spin-echo, 39KJ/mole corresponds to E_A which is the product $(1-n)E_A^*$. It follows that n is zero or close to zero. This should lay to rest the speculation[3,4] that the mechanism of the unified model is responsible for NMR anomalies in transition metal hydrides. We can conclude that Torrey's diffusion relaxation model can explain to some extent the NMR T_1 and $T_{1\rho}$ anomalies in $LaNi_5H_7$ for $T \gtrsim 180K$. Below 180K, too sharp a rise is predicted. This disagreement could be evidence for small polaron behavior of hydrogen in transition metals which implies a transition from thermally activated hopping at high temrperautres to nonactivated behaviors at lower temperatures. The frequency dependence of $\omega^{-1.35}$ as deduced from Chang et al. still cannot be accounted for adequately by Torrey's model, however.

ACKNOWLEDGEMENT

We wish to thank I. J. Lowe for drawing our atention to this problem. This work is supported in part by ONR.

REFERENCES

1. R. M. Cotts, in Hydrogen in Metals, ed. G. Alefeld and J. Völkl, Springer-Verlag (1978).
2. R. F. Kalicek and I. J. Lowe, J. Less Common Metals 73:219 (1980).
3. T. C. Jones, T. K. Halstead and K. H. J. Buschow, ibid., 209 (1980).
4. H. Chang, I. J. Lowe and R. J. Kalicek, in Nuclear and Electron Resonance Spectroscopies Applied to Materials Science, ed. Kaufmann and Shenoy, Elsevier, p. 331 (1981).
5. T. K. Halstead, N. A. Aboud and K. H. J. Buschow, Solid State Commun. 19:425 (1976).
6. H. C. Torrey, Phys. Rev. 92:962 (1953).
7. K. L. Ngai, Comments Solid State Phys. 9:127, 141 (1980).
8. E. O. Stejskal and J. E. Tanner, J. Chem. Phys. 42, 288 (1965).
9. A. Abragam, The Principles of Nuclear Magnetism, Clarendon Press, Oxford (1961).
10. K. L. Ngai and F. S. Liu, Phys. Rev. B24, 1049 (1981).
11. S. Teitler, A. K. Rajagopal and K. L. Ngai, Naval Research Laboratory Memo Rept. No. 4757 (1982).
12. G. Krüger, Z. Naturf. 24a:560 (1969).

HYDROGEN DIFFUSION AND ELECTRONIC STRUCTURE

IN CRYSTALLINE AND AMORPHOUS Ti_yCuH_x

R.C. Bowman, Jr.[*], A.J. Maeland[†], W.-K. Rhim[‡]
and J.F. Lynch[§**]

Division of Chemistry and Chemical Engineering
California Institute of Technology, Pasedena CA 91125
[†]Materials Research Center, Allied Chemical Corporation
Morristown, NJ 07960, [‡]Jet Propulsion Laboratory
California Institute of Technology
[§]Brookhaven National Laboratory, Upton, NY 11973

ABSTRACT

Hydrogen diffusion behavior and electronic properties of crystalline $TiCuH_{0.94}$, $Ti_2CuH_{1.90}$, and $Ti_2CuH_{2.63}$ and amorphous $a\text{-}TiCuH_{1.4}$ are studied using proton relaxation times, proton Knight shifts, and magnetic susceptibilities. Crystal structure and hydrogen site occupancy have major roles in hydrogen mobility. The density of electron states at E_F is reduced in amorphous $a\text{-}TiCuH_{1.4}$ compared to the crystalline hydrides.

The crystalline intermetallics TiCu and Ti_2Cu and the amorphous $Ti_{1-y}Cu_y$ ($0.3 \leq y \leq 0.7$) alloys directly react with gaseous hydrogen to form crystalline and amorphous ternary hydrides,[1-4] respectively, providing the temperature is maintained below 200°C. A recent nuclear magnetic resonance (NMR) study of the proton relaxation times[5] indicated a much higher hydrogen mobility in amorphous $a\text{-}TiCuH_{1.3}$ compared

---

*On leave from current address: Monsanto Research Corporation, Mound, Miamisburg, Ohio 45342.
**Current address is Materials Research Center, Allied Chemical Corporation, Morristown, New Jersey 07960.

to polycrystalline $TiCuH_{0.94}$. The increased disorder[3] of interstitial hydrogen occupancy in a-$TiCuH_x$ has been suggested[5] for the enhanced mobility in the amorphous phase. The present paper describes further NMR studies of diffusion in crystalline and amorphous $TiCuH_x$ as well as crystalline Ti_2CuH_x. Furthermore, the electronic structure has been investigated using magnetic susceptibility, proton spin-lattice relaxation time (T_1), and proton Knight shift (σ_K) measurements.

Table 1 summarizes the structural properties of the $TiCuH_x$ and Ti_2CuH_x samples as deduced from x-ray diffraction, neutron scattering, and proton lineshape parameters. The preparation procedures have been previously described.[1-5]

Table 1. Descriptions of $TiCuH_x$ and Ti_2CuH_x and Hydrogen Diffusion Activation Energies (E_a)

Sample	Metal Sublattice Structure	Probable Hydrogen Site Occupancies	E_a (eV)	Temperature Range (K)
$TiCuH_{0.94}$	Tetragonal	94% Ti_4 only	0.84±0.02	465 − 560
a-$TiCuH_{1.4}$	Amorphous	Mixed (mostly Ti_4 with some $Ti_{4-y}Cu_y$ and octahedral)	0.45±0.02	357 − 410
			0.185±0.01	208 − 357
			0.09±0.01	150 − 207
$Ti_2CuH_{1.9}$	Orthorhombic (?)	~95% Ti_4 (some Ti_2Cu_4 likely)	0.35±0.02	290 − 519
$Ti_2CuH_{2.63}$	Orthorhombic (?)	100% Ti_4 and 63% Ti_2Cu_4	0.29±0.02	290 − 395

Hydrogen diffusion behavior has been evaluated using the temperature dependence of the proton rotating-frame spin-lattice relaxation times[6] ($T_{1\rho}$) where the spin-locking field was about 7.3G and the proton resonance frequency was 34.5 MHz. The temperature dependences of the $T_{1\rho}$ data for Ti_yCuH_x are shown in Fig. 1. Table 1 summarizes the diffusion activation energies (E_a) that have been deduced from the $T_{1\rho}$ data. Although a single E_a corresponding to Arrhenius behavior represents proton mobility in the crystalline $TiCuH_{0.94}$ and Ti_2CuH_x, three E_a values are required for amorphous a-$TiCuH_{1.4}$, which confirms the behavior previously seen[5] in a-$TiCuH_{1.3}$. Furthermore, E_a is greatly reduced when protons occupy sites in addition to the tetrahedral Ti_4 interstitials. This effect is seen in both crystalline Ti_2CuH_x and amorphous a-$TiCuH_{1.4}$. From a consideration of the TiCu and Ti_2Cu crystal structures,[1,4] hydrogen diffusion in crystalline TiCuH can only occur by nearest neighbor jumps between the Ti_4 sites,

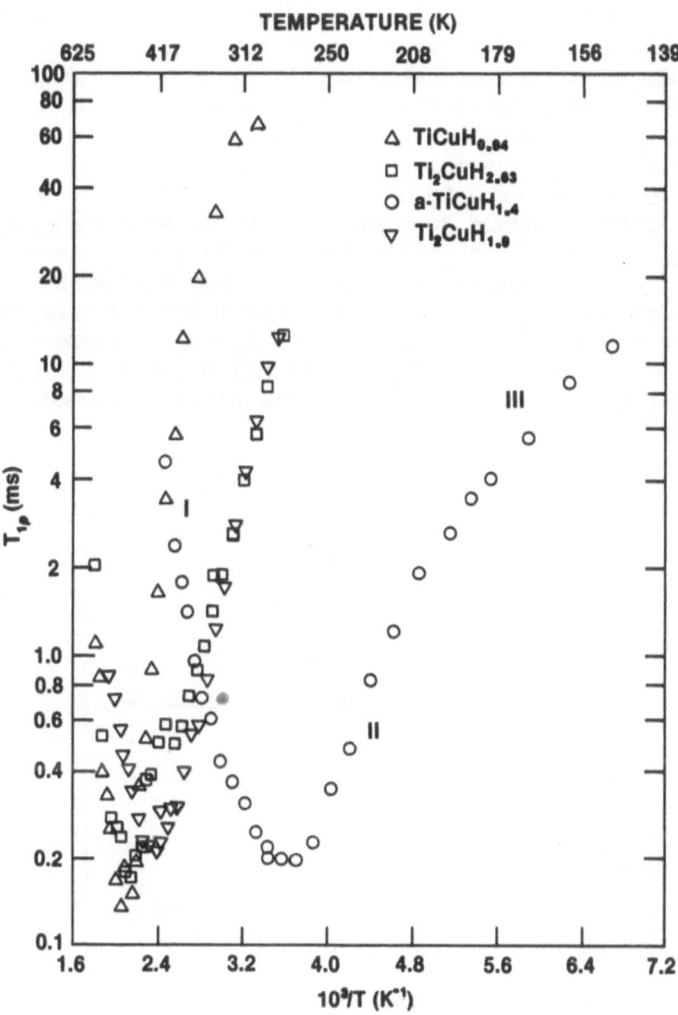

Fig. 1. Proton $T_{1\rho}$ relaxation times with H_1 = 7.3 G at ν_H = 34.5 MHz.

while jumps through the intermediate Ti_2Cu_4 sites become possible in
Ti_2CuH_x. This probably accounts for the lower E_a values for Ti_2CuH_x
and similar (or even easier) jump-paths are available in the more
disordered amorphous phase.

The magnetic susceptibilities (χ_m) for Ti_yCuH_x were measured between 80 K and 300 K and are summarized in Fig. 2. Although the χ_m
values in Fig. 2 have been extrapolated to infinite magnetic field,
the field-dependent ferromagnetic contribution was negligible except
for a-$TiCuH_{1.4}$, which appears to have some magnetic impurities as
well as an opposite temperature dependence for χ_m. There are several
contributions[7] to χ_m, but only the paramagnetic term χ_p is directly
related to $N(E_F)$, the density of electron states at the Fermi level
E_F. Hence, caution should be exercised in correlating χ_m differences
only to $N(E_F)$ changes. In particular, the larger χ_m for a-$TiCuH_{1.4}$
compared to $TiCuH_{0.94}$ probably reflects either ferromagnetic or orbital contributions[7] and not a greater $N(E_F)$ for the amorphous phase.
However, the unusual[7,8] χ_m increase with hydrogen content for Ti_2CuH_x
is believed to actually correspond to $N(E_F)$ becoming larger since the
proton T_1 and σ_K parameters also indicate $N(E_F)$ increasing from
$Ti_2CuH_{1.9}$ to $Ti_2CuH_{2.63}$.

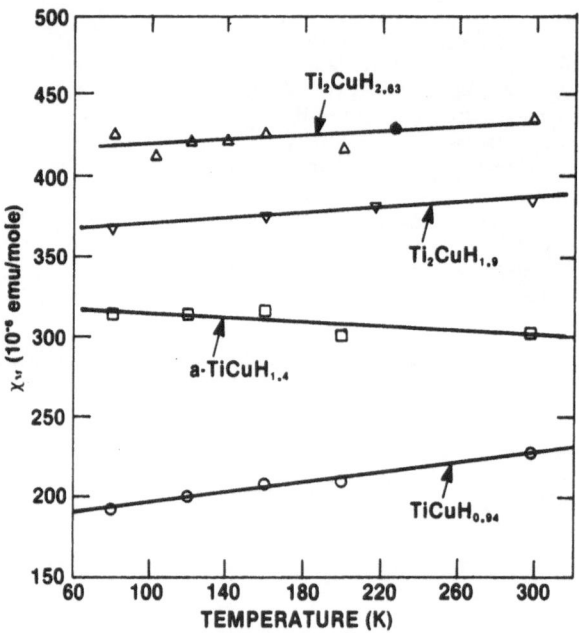

Fig. 2. Magnetic susceptibility values for $TiCuH_x$ and Ti_2CuH_x.

The proton T_1 and σ_K were measured at 56.4 MHz using methods previously described.[9] The σ_K values are referenced to tetramethyl-silane. Table 2 summarizes the σ_K and $(T_{1e}T)^{-\frac{1}{2}}$ parameters, which are directly proportional[6,8,9] to $N(E_F)$, at the upper and lower temperature limits of the present NMR measurements. The negative σ_K values in Table 2 indicate that core-polarization[6] with d-electrons dominates proton hyperfine interactions in Ti_yCuH_x where the population of d-states is much larger than s-states as has been previously found in other Ti-based hydrides.[8,9] Futhermore, the proton parameters suggest $N(E_F)$ is significantly reduced in $a-TiCuH_{1.4}$ compared to crystalline $TiCuH_{0.94}$, while $N(E_F)$ increases with content in crystalline Ti_2CuH_x. However, a more detailed analysis based upon generalized Korringa relations[6] shows increased s-electron contact hyperfine interactions in the Ti_2CuH_x samples.

More extensive discussions of hydrogen diffusion and the electronic structures of Ti_yCuH_x will be published elsewhere.

Table 2. Proton Parameters $(T_{1e}T)^{-\frac{1}{2}}$ and Knight Shifts σ_K

Sample	T (K)	$(T_{1e}T)^{-\frac{1}{2}}$ $(sK)^{-\frac{1}{2}}$	σ_K (ppm)
$TiCuH_{0.94}$	300	0.163	-120
	80	0.150	-107
$a-TiCuH_{1.4}$	210	0.113	- 77
	80	0.108	- 87
$Ti_2CuH_{1.9}$	300	0.118	- 67
	115	0.115	- 69
$Ti_2CuH_{2.63}$	300	0.140	- 85
	80	0.145	- 91

ACKNOWLEDGEMENTS

This work was partially supported by the Division of Chemical Sciences, Office of Basic Energy Sciences, U. S. Department of Energy, and the Caltech's President's Fund. Mound is operated by Monsanto Research Corporation for the U. S. Department of Energy under Contract No. DE-AC04-76-DP00053. Brookhaven National Laboratory is operated for the U. S. Department of Energy under Contract No. DE-AC-02-76-CH00016. Jet Propulsion Laboratory is operated for the National Aeronautics and Space Administration under Grant No. NAS7-100.

REFERENCES

1. A. Santoro, A. Maeland, and J. J. Rush, Acta Cryst. B34: 3059 (1978).
2. A. J. Maeland, L. E. Tanner and G. G. Libowitz, J. Less-Common Met. 74: 279 (1980).
3. J. J. Rush, J. M. Rowe and A. J. Maeland, J. Phys. F: Metal Phys. 10: L283 (1980).
4. A. J. Maeland and G. G. Libowitz, J. Less-Common Met. 74: 295 (1980).
5. R. C. Bowman, Jr. and A. J. Maeland, Phys. Rev. B24: 2328 (1981).
6. R. M. Cotts, in "Hydrogen in Metals I: Basic Properties," G. Alefeld and J. Völkl, Ed., Springer-Verlag, Berlin, 1978, p. 227.
7. J. F. Lynch, R. Lindsay and R. O. Moyer, Jr., Solid State Commun. 41: 9 (1982).
8. J. F. Lynch, J. R. Johnson and R. C. Bowman, Jr., Paper at this meeting.
9. R. C. Bowman, Jr. and W.-K. Rhim, Phys. Rev. B24: 2232 (1981).

LOCAL DIFFUSION AND TUNNELING OF TRAPPED HYDROGEN

IN METALS

H. Wipf and K. Nuemaier*

Physik-Department der Technischen Universität
München, D-8046 Garching
*Zentralinstitut für Tieftemperaturforschung
der Bayerischen Akademie der Wissenschaften
D-8046 Garching

ABSTRACT

Hydrogen interstitials bound by impurity atoms in metals can locally diffuse or tunnel between equivalent trap sites around its impurity-atom trap center. For hydrogen trapped by N or O interstitials in Nb, these local dynamics were studied in specific heat and neutron scattering measurements demonstrating the existence of low-temperature tunneling eigenstates for hydrogen in metals. Hydrogen impurity-atom complexes represent a unique system for studies of the dynamical behavior especially in the low-temperature range where, for impurity-free metals, hydrogen diffusion or tunneling can not be experimentally investigated because of precipitation. They offer also the possibility to study for the first time the transition from coherent to incoherent quantum transport of hydrogen in metals by powerful experimental techniques such as neutron spectroscopy.

INTRODUCTION

The decisive influence of quantum effects on transport processes of interstitial H in metals is well established, although microscopic models /1-5/ developed so far (and basing essentially on the polaron concept) are still far from reproducing satisfactorily quantitative experimental results. In order to test theoretical

predictions, low-temperature experimental studies are of
special importance since, with decreasing temperature,
quantum effects predominate more and more those caused
by thermal fluctuations. Any demand for low-temperature
experiments requires, however, consideration of precipitation which reduces the amount of unprecipitated H in
the α-phase exponentially with decreasing temperature.
Therefore, except for quenching experiments /6-8/ (probing essentially precipitation or trapping kinetics) and
for muon spin rotation /9/ (the muon can be considered
to be a "fourth" H isotope of very small mass), no
studies on long-range H diffusion were successfully undertaken below temperatures of about 120 K. That precipitation does not necessarily represent an insurmount-
able obstacle for low-temperature studies on H mobility
was shown by Baker and Birnbaum /10/. In internal friction measurements on H-doped Nb containing also small
amounts of interstitial O or N, they observed relaxation peaks which they correctly associated with diffusional jumps of the H around O or N atoms acting as trap
centers. These measurements demonstrated the possibility
to study low-temperature transport processes of H in
metals by its "local" diffusion or tunneling around an
impurity atom trap center, providing that the binding
enthalpy between impurity and H is large enough to prevent the H from forming precipitates.

That O and N impurities represent sufficiently effective trap centers for H in Nb was later shown by a
large number of experimental studies /11-19/. The binding enthalpies, which can be deduced from these studies,
are between 100 and 120 meV. The influence of trapping
is especially important at low temperatures where,
barring effects of quenching, H either forms precipitates
or becomes trapped by impurities. The reciprocal action
of precipitation and trapping is particularly well
studied for N trap centers in Nb /14,16/. For this system, the H atoms within a sample are trapped at low
temperatures under formation of N-H pairs as long as un-
depleted N trap centers remain available. The H atoms
which are in excess of the N will form precipitates. It
can be assumed that O trap centers in Nb behave in the
same way since, in all the studies cited above, no principal differences were observed in the trapping behavior
of N and O impurities.

The local dynamics (local H diffusion and tunneling)
of O-H and N-H pairs in Nb were investigated in internal friction /10-13,15,18-21/, heat capacity /22-24/,

thermal conductivity /25,26/, and neutron spectroscopic
/27,28/ studies. The discussion in this paper will mainly be devoted to the heat capacity and neutron spectroscopic results. These results (and those of thermal conductivity as long as resonant phonon scattering is involved) are directly determined by the energy differences
between tunneling eigenstates (i.e. by the tunneling matrix elements). They provide therefore an immediate access to the most significant and crucial quantity characterizing the properties of a tunneling H atom. On the
other hand, relaxation times obtained from the internal
friction measurements above describe the transition rate
between different tunneling eigenstates (of a delocalized
hydrogen) or, at elevated temperatures, between states
corresponding to a localization of the H on different
interstitial sites (in other words, the relaxation time
describes the lifetime of tunneling eigenstates, or the
mean residence time of the H on interstitial sites). To
determine the value of tunneling matrix elements from
relaxation times implies, therefore, to presuppose a
specific model for the transition rates, which means at
the same time that any result derived for matrix elements depends crucially on the model chosen. For this
reason, relaxation times obtained from internal friction
need be considered to provide only a very indirect access
to the determination of tunneling matrix elements.

SPECIFIC HEAT AND NEUTRON SPECTROSCOPIC
MEASUREMENTS

    Fig. 1 shows the low-temperature specific heat data
of Ref. 23 in a conventional log-log plot versus temperature. The measurements were made on five samples part of

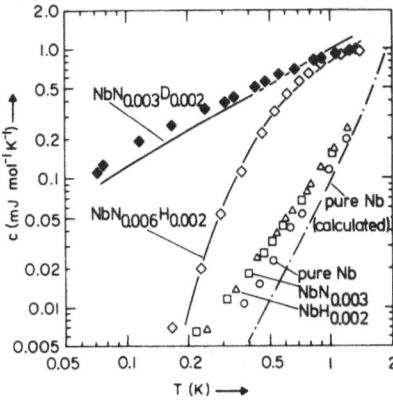

Fig.1. Specific heat c of five
(superconducting) Nb samples
partially doped with N, H and
D in a log-log plot versus temperature T. The dash-dotted curve
represents calculated specific
heat values for pure Nb, and the
full lines fit results explained
in Specific Heat section. The
data are taken from Ref. 23.

which were doped with N, H and D. The data demonstrate
that the samples simultaneously doped with both N and H
(or N and D) show a large and isotope-dependent specific-
heat anomaly, which is not observed for pure Nb and for
Nb doped with either N or H. According to these results,
it was concluded in Ref. 23 that N-H and N-D pairs in Nb
provide tunneling systems for the H or D, and that tunneling of (precipitated) H(D) in pure Nb is not observable. From the size of the measured specific heat anomaly, and from the known N-H (N-D) pair concentration, it
was additionally concluded that the individual tunneling
systems are strongly influenced by stress-induced interaction effects between different N-H (N-D) pairs.

     The neutron scattering results of Ref. 27 and 28
are shown in Fig. 2. The measurements were made at two
different temperatures (0.09 and 5 K), and on a polycrystalline Nb sample containing 1.3 at% of O-H pairs,
and with a neutron momentum transfer $q = 2.5$ $Å^{-1}$. At the
lower temperature (0.09 K), the intensity shoulder at
$\sim$0.2 meV on the energy loss side demonstrates the presence of inelastic scattering processes (energy gain
processes do not occur at this low temperature). At 5 K,
the intensity on the energy gain side increases, whereas it decreases on the energy loss side. The observed
temperature dependence is typical for neutron scattering

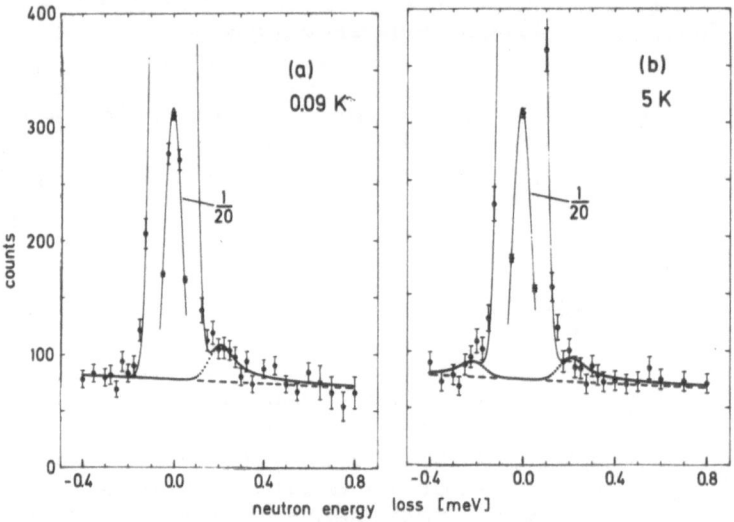

Fig. 2.   Inelastic neutron spectra of $NbO_{0.013}H_{0.016}$ at
0.09 K (a) and 5 K (b). The full, dotted and broken lines
indicate fit curves explained in Neutron Scattering
section. The data are taken from Ref. 27.

on tunneling systems, proving at the same time that the
inelastic intensity at ∿0.2 meV on the energy loss side
is not due to phonons since, in such a case, it would
have to increase between 0.09 and 5 K at least by a factor of three.

TUNNELING IN THE PRESENCE OF STRESS-INDUCED
INTERACTION

The specific-heat and neutron-scattering results
can quantitatively be explained within a tunneling model
developped in Ref. 27 and 28. In this model, the H(D)
trapped by N or O is tunneling between two equivalent
interstitial sites whose respective energies are randomly shifted relative to each other because of stress-
induced interaction between different N-H(D) or O-H(D)
pairs. For such a tunneling system, a one-dimensional
potential contour for the H(D) is shown in Fig. 3, where

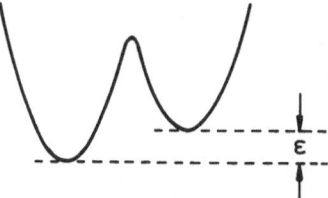

Fig. 3. One-dimensional potential contour for two-site tunneling (ε specifies an energy
shift between the two sites).

ε specifies the energy shift between the two sites.
Tunneling between only two sites represents the most
simple tunneling system possible, the reason for which
it is also discussed for glasses /29,30/. From the discussion of possible geometrical configurations for a
N-H or O-H pair in Nb (Section 4), it will also follow
that tunneling systems consisting of more than two sites
need be considered unlikely. An additional justification
for the present model is that, even in the case of more
complex tunneling systems, the qualitative aspects of
the subsequent discussion will not change.

For a two-site tunneling system as shown in Fig. 3,
the energy difference ΔE between the two vibrational
ground states of the H(D) is given by /29-31/

$$\Delta E = \sqrt{J^2 + \varepsilon^2} \tag{1}$$

where J is the tunneling matrix element, and ε the energy shift between the two sites due to random elastic stresses caused by neighboring N-H or O-H pairs. For a statistical pair arrangement, the distribution function $Z(\varepsilon)$ of the random elastic stresses ε is a Lorentzian line /32/

$$Z(\varepsilon) = \frac{1}{\pi} \frac{\varepsilon_0}{\varepsilon_0^2 + \varepsilon^2} \qquad (-\infty < \varepsilon < +\infty) \qquad (2)$$

The width ε_0 of this line characterizes also a typical energy shift between the two sites. According to Eq. 1, the distribution $Z(\varepsilon)$ in the energy shifts ε implies a corresponding distribution function $Z(\Delta E)$ for the energy differences ΔE,

$$Z(\Delta E) = \frac{2}{\pi} \frac{\Delta E}{\sqrt{\Delta E^2 - J^2}} \frac{\varepsilon_0}{\varepsilon_0^2 + \Delta E^2 - J^2} \qquad (\Delta E \geqslant J) \quad (3)$$

which shows an integrable singularity at $\Delta E = J$.

Although the present model is closely analogous to the two-site tunneling systems discussed for glasses /29,30/, there is also a distinct difference. Since the spatial configuration of all the tunneling systems is identical (except for symmetry operations), essentially the same matrix element J is expected for all systems since this quantity is usually not significantly modified by stress-induced energy shifts /33,34/. Contrary to glasses, the energy difference ΔE in Eq. 1 has therefore a lower bound value J for those tunneling systems for which the energy shift ε if fortuitously zero.

Specific Heat

For a given tunneling system with an energy difference ΔE, the specific heat $C(\Delta E)$ is given by

$$C(\Delta E) = k_B \cdot \left(\frac{\Delta E}{k_B T}\right)^2 \cdot \frac{e^{-\Delta E/k_B T}}{\left(1 + e^{-\Delta E/k_B T}\right)^2} \qquad (4)$$

Starting from zero temperature, $C(\Delta E)$ increases at first exponentially, passes then through a maximum at $T_{max} \sim$ $\sim 0.42 \cdot \Delta E/k_B$, and decreases finally proportional to T^{-2}. The entropy increase over the entire temperature range is $k_B \ln 2$.

The measured specific heat C_{meas} is the sum of the specific heat contributions of all the N-H(D) pairs (tunneling systems). Under consideration of the distribution function $Z(\Delta E)$ for ΔE, it can be written as

$$C_{meas} = N \cdot \int d(\Delta E) \cdot Z(\Delta E) \cdot C(\Delta E) \tag{5}$$

where N is the total number of N-H(D) pairs.

The specific-heat results in Fig. 1 show fit curves (full lines) according to Eq. 5, valid for the data obtained from the samples containing N-H and N-D pairs. The values derived for the two fit parameters J and ε_0 were J = (0.14 ± 0.03) meV and $\varepsilon_0 = (2.2 \pm 1.0)$ meV for the N-H sample, and J $\overset{\sim}{\sim}$ 0.013 meV and $\varepsilon_0 = (1 \pm 0.5)$ meV for the N-D dample. It is emphasized that the matrix element J obtained for the N-D sample represents only a very crude estimate since the fit curves turned out to be extremely insensitive on this quantity.

The above widths ε_0 of the energy shift distributions (they are also typical energy shifts) are an order of magnitude larger than the tunneling matrix elements J. The principal consequence is that, according to Eq. 1, the energy difference ΔE of most of the tunneling systems exceeds the thermal energy $k_B T$ in the temperature range of the measurements (1 meV corresponds to $\sim$10 K). This means also that, in the investigated temperature range, most of the tunneling systems practically do not contribute to the measured specific heat, which therefore is considerably smaller than expected for tunneling systems with identical energy differences ΔE (equivalently, the entropy increase observed over the temperature range investigated is smaller than expected for identical energy differences ΔE). The fact that the measured specific heat, or the entropy increase, was an order of magnitude smaller than under conditions of identical energy differences was the reason to postulate in Ref. 23 the presence of strong stress-induced interaction effects between neighboring N-H(D) pairs.

Neutron Scattering

For the considered tunneling model, and for a polycrystalline sample, the inelastic and elastic (incoherent) scattering cross sections are /27,28/ (the scattering intensity observed results mainly from scattering processes on the H because of its extremely large incoherent cross section)

$$\left(\frac{d^2\sigma}{d\Omega d(\hbar\omega)}\right)_{inel} = \frac{\sigma_{inc}}{4\pi} \cdot \frac{k_f}{k_i} \cdot e^{-2W} \cdot \left(\frac{1}{2} - \frac{\sin(qd)}{2qd}\right)$$

$$\cdot \int_{-\infty}^{+\infty} d\varepsilon Z(\varepsilon) \cdot \frac{J^2}{J^2+\varepsilon^2} \frac{\delta(\hbar\omega+\Delta E) + \delta(\hbar\omega-\Delta E)}{1+e^{-\hbar\omega/k_B T}} \qquad (6a)$$

and

$$\left(\frac{d\sigma}{d\Omega}\right)_{el} = \frac{\sigma_{inc}}{4\pi} \cdot e^{-2W} \cdot \frac{1}{\varepsilon_0+J} \left[\varepsilon_0 + J\left(\frac{1}{2} + \frac{\sin(qd)}{2qd}\right)\right] \quad (6b)$$

In this equation, σ_{inc} is the incoherent cross section of H, d is the distance between the two tunneling sites, e^{-2W} is the Debye-Waller factor, k_i and k_f are the wave vector of the incident and scattered neutrons, $\hbar\omega$ is the energy and $\hbar q = \hbar(k_i-k_f)$ is the momentum transferred during scattering. From these equations, the ratio between the integral inelastic and total scattering intensities I_{inel} and I_{tot} found in constant q-scans is

$$\frac{I_{inel}}{I_{tot}} = \frac{J}{\varepsilon_0+J} \left(\frac{1}{2} - \frac{\sin(qd)}{2qd}\right) \qquad (7)$$

where the term k_f/k_i is neglected since it is close to one for the relevant $\hbar\omega$-range. According to Eq. 7, the inelastic scattering approaches a maximum for $q \approx 4.5/d$ and $\varepsilon_0 \ll J$.

The neutron scattering data in Fig. 2 show fit results obtained according to Eq. 6. The fits are indicated as dotted, full and broken lines representing the calculated inelastic, total and background scattering, respectively. The parameters obtained from the fits were the tunneling matrix element J for O-H pairs (J = (0.19±0.04)meV) and the ratio between the integral inelastic and total scattering intensities (I_{inel}/I_{tot} = 0.011±0.003). The characteristic energy shift ε_0 was not an explicit fit parameter since the fits were entirely insensitive on this quantity as long as $\varepsilon_0 \gg J$. However, in Ref. 27 or 28, a rough estimate could be derived for ε_0 in an indirect way yielding the value $\varepsilon_0 \approx 3.7$ meV. This shows also for the O-H sample of the neutron-scattering measurements a typical energy shift ε_0 about an order of magnitude larger than the tunneling matrix element J.

The most important consequence of a large ε_0

value is seen from Eq. 7. The equation shows that the
inelastic scattering cross section decreases drastically
as soon as ε_0 becomes larger than J. This expresses the
fact that inelastic scattering is essentially only possible on tunneling systems whose matrix elements J ex-
ceed the energy shifts ε whereas, for J<ε, inelastic
scattering becomes negligible (the reason for this is
that, for J<ε, the eigen-wave-functions of the H tend
to become increasingly localized on one of the two
tunneling sites /31/). The discussed inelastic scattering behavior can immediately be recognized from Fig. 2
where inelastic intensity is only observed closely
around a neutron energy loss (or gain) of $\sim$0.2 meV ($\approx$J),
in spite of the fact that, according to Eq. 1, most of
the tunneling system have much larger energy differences
ΔE between their eigenstates (the neutron energy loss or
gain corresponds to ΔE). That inelastic intensity actually is not observed on tunneling systems whose ΔE
values exceed noticeably 0.2 meV reflects therefore the
fact that the inelastic cross section becomes negligible
as soon as ε>J (which implies $\Delta E > \sqrt{2} \cdot J \approx 0.2$ meV). Note
also finally that this insensitivity of inelastic scatter-
ing for ε>J is in contrast to specific heat measurements
which do probe thermal excitations of a tunneling system
irrespectively of whether ε is smaller or larger than J
as long as the thermally-induced transition rate between
the two eigenstates energetically separated by ΔE exceeds
the speed of the measurement.

GEOMETRY OF THE N-H(O-H) PAIR

It is known that O and most probably also N impurity
atoms in Nb occupy octahedral interstitial sites /35,36/,
and that H(D) in pure Nb is located on tetrahedral sites
(see, e.g., /37,38/ and references therein). The interstitial sites occupied by trapped H around an O or N
trap center are, however, not yet established, although
some information does exist that allows a significant
reduction of the great number of potential geometries.

First, neutron scattering measurements /39/ show
that, at low temperatures, the vibrational frequencies of
H trapped by N or O in Nb are identical to those of free
(untrapped) H on tetrahedral sites. This result indicates
very convincingly that the trapped H also occupies tetrahedral sites since these frequencies depend very sensitively on interatomic potentials and, therefore, on the
type of interstitial site occupied. It excludes also the
possibility that, besides tetrahedral sites, additional

sites, such as e.g. triangular ones /18,21,40/, are occupied at low temperatures by any measurable fraction.

The second information available results from a
comparison of the tunneling matrix elements J derived
from the low temperature (T<1 K) specific heat and neutron spectroscopic studies with the relaxation times observed in internal friction at higher temperatures
(40 K < T < 200 K) for reorientational jumps of the trapped
H /10-13, 15,18,19/. At the lowest temperature measured
($\sim$40 K), the relaxation time τ_{rel} is about 100 s /15/
corresponding to an energy $\hbar/\tau_{rel} \stackrel{\sim}{\sim} 10^{-14}$ meV, which is
13 orders of magnitude smaller than the tunneling matrix
element J (see Section 3.1). This fact makes it unlikely that the reorientational jumps occur between sites
located within one tunneling system. It was therefore
proposed in Ref. 23 that the tunneling of the trapped H
takes place within tunneling systems consisting of two
(or possibly even more) sites, and that the reorientational relaxations correspond to diffusional jumps of
the hydrogen out of one tunneling system into another
one located around the same O or N trap center. As a consequence of these considerations, it need be assumed
that there exist several independent tunneling systems
around each O or N impurity trap center.

A third fact to be taken account of is that, because of elastic and electronic interaction, the presence of an O or N interstitial defect changes severely
the energetic levels of the surrounding potential tunneling sites (obviously, trapping itself is one result of
these energetic changes). The strength of this interaction, and therefore the energy change, depends sensitively on both distance and crystal direction. This
means specifically that energy differences will exist
between all surrounding sites that cannot be transformed
into each other by (point-group) symmetry operations
around the O or N. The size of these energetic differences is expected to be typically in the 100 meV range -
an energy range that also characterizes the binding enthalpy. On the other hand, the neutron spectroscopic
results in Fig. 2 clearly show the existence of tunneling systems comprising sites whose energetic differences
necessarily are smaller than the matrix element $J \stackrel{\sim}{\sim} 0.2$
meV since, otherwise, no inelastic scattering would be
observed (see the discussion in Section 3.2). The conclusion following from these considerations is that the
sites of these tunneling systems need to be transformable
into each other by point-group symmetry operations around

the trap center. This condition reduces very effectively
the number of possible tunneling systems, and it excludes
in particular the existence of "tunneling rings" as suggested in Ref. 18 and 21.

Finally, in order to yield matrix elements J in the
energy range of the measured ones, the relevant distances
between the sites of a tunneling system need to be about
1 Å /5,27,28/. This results in the plausible conclusion
that tunneling occurs between nearest-neighbor tetrahedral sites.

The foregoing discussion provides an a posteriori
justification for the two-site tunneling model since it
seems indeed difficult to conceive potential systems
comprising more than two sites, and being in compliance
with the criteria stated above. This can also be realized
from Fig. 4, which shows a bcc Nb unit cell with an O or

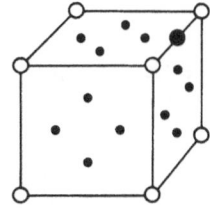

O Nb host lattice atoms
● O, N trap center
• tetrahedral sites

Fig. 4.   bcc-Nb unit cell with
an O(N) trap center at (1,1/2,1).
The figure shows also 12 tetrahedral sites on the cell surface.

N trap center in (1,1/2,1)-position, and 12 potential
tetrahedral tunneling sites on its surface planes. Two-
site tunneling systems, as discussed in this paper, can
be imagined to be composed either of tetrahedral sites
at (1/4,0,1/2) and (1/2,0,1/4), or of sites at
(1/2,0,3/4) and (3/4,0,1/2). For an undisturbed Nb
lattice, the distance between the two respective sites
would be 1.17 Å, but it may be that this distance actually is changed (reduced) somewhat because of the lattice
disturbance caused by the closely located O or N trap
center (a reduction by 10 or 20% would, e.g., result in
a better agreement with the estimate given in Ref. 27
and 28).

OUTLOOK

    In the authors' opinion, one of the most challeng-
ing aspects of the experiments discussed is that N-H or
O-H pairs in Nb provide the possibility for neutron spectroscopic studies on H tunneling in metals, and for
studies on how the tunneling eigenstates are destroyed
by coupling to phonons at elevated temperatures. At very
low temperatures $(T \ll J/k_B)$, the tunneling eigenstates
are expected to be well-defined with a lifetime significantly larger than the reciprocal matrix element $\hbar/J$.
With rising temperatures, this lifetime becomes increasingly shorter because of a coupling to phonons, until
finally at high temperatures, it ends up being smaller
than h/J. In hydrogen diffusion theory /1-5/, this change
of the relationship between lifetime and matrix element
corresponds to a transition from "coherent" to "incoherent" diffusion. N-H or O-H pairs in Nb open therefore
for the first time the way to investigate this transition
by powerful experimental methods such as neutron spectroscopy.

ACKNOWLEDGEMENT

    It is a pleasure for the authors to acknlowledge
discussions with G. Alefeld, A. Magerl, J.M. Rowe,
J.J. Rush and S.M. Shapiro.

REFERENCES

1/   C.P. Flynn and A.M. Stoneham, Phys. Rev. <u>B1</u>, 3966
      (1970)
2/   Yu. Kagan and M.J. Klinger, J. Phys. C: Solid State
      Phys. <u>7</u>, 2791 (1974)
3/   D. Emin, M.I. Baskes and W.D. Wilson, Phys. Rev.
      Lett. <u>42</u>, 791 (1979)
4/   E.G. Maksimov and O.A. Pankratov, Sov. Phys. Usp.
      <u>18</u>, 481 (1976)
5/   K.W. Kehr, in <u>Topics in Applied Physics</u>, vol. 28,
      <u>Hydrogen in Metals I</u>, eds. G. Alefeld and J. Völkl,
      Springer-Verlag, Berlin - Heidelberg - New York,
      1978, p. 197
6/   K. Faber and H. Schultz, Scripta Met. <u>6</u>, 1065
      (1972)
7/   R. Hanada, Scripta Met. <u>7</u>, 681 (1972)
8/   J. Engelhard, J. Phys. F: Metal Phys. <u>9</u>, 2217
      (1979)
9/   A. Seeger, in <u>Topics in Applied Physics,</u> vol. 28,
      <u>Hydrogen in Metals I</u>, eds. G. Alefeld and J. Völkl,

Springer-Verlag, Berlin - Heidelberg - New York, 1978, p. 549

10/ C.C. Baker and H.K. Birnbaum, Acta Met. 21, 865 (1973)

11/ R.F. Mattas and H.K. Birnbaum, Acta Met. 23, 973 (1975)

12/ P. Schiller and A. Schneiders, phys. stat. sol.(a) 29, 375 (1975)

13/ P. Schiller and H. Nijman, phys. stat. sol.(a) 31, K77 (1975)

14/ G. Pfeiffer and H. Wipf, J. Phys. F: Metal Phys. 6, 167 (1976)

15/ C.G. Chen and H.K. Birnbaum, phys. stat. sol.(a) 36, 687 (1976)

16/ R. Hanada, Proc. 2nd Int. Congr. on Hydrogen in Metals, Paris 1977 (Pergamon, Oxford) vol. 3, 1B6

17/ D. Richter and T. Springer, Phys. Rev. B18, 126 (1978)

18/ P.E. Zapp and H.K. Birnbaum, Acta Met. 28, 1275 (1980)

19/ P.E. Zapp and H.K. Birnbaum, Acta Met. 28, 1523 (1980)

20/ E.L. Andronikashvili, Melik-Shaknazarov and I.A. Naskidashvili, J. Low Temp. Phys. 23, 1 (1976)

21/ D.B. Poker, G.G. Setser, A.V. Granato and H.K. Birnbaum, Z. Phys. Chem. N.F. 116, 39 (1979)

22/ G.J. Sellers, A.C. Anderson and H.K. Birnbaum, Phys. Rev. B10, 2771 (1974); in this paper the reported specific heat anomalies were (wrongly) attributed to tunneling eigenstates of free (untrapped) hydrogen (see Ref. 23).

23/ C. Morkel, H. Wipf and K. Neumaier, Phys. Rev. Lett. 40, 947 (1978)

24/ H. Wipf and K. Neumaier, to be published

25/ S.G. O'Hara, G.J. Sellers and A.C. Anderson, Phys. Rev. B10, 2777 (1974); as in Ref. 22, the thermal conductivity anomalies in this paper were attributed to free hydrogen.

26/ M. Locatelli, K. Neumaier and H. Wipf, J. Physique 39, C6-995 (1978)

27/ H. Wipf, A. Magerl, S.M. Shapiro, S.K. Satija and W. Thomlinson, Phys. Rev. Lett. 46, 947 (1981)

28/ H. Wipf, A. Magerl, S.M. Shapiro, S.K. Satija and W. Thomlinson, Proc. Miami Int. Symp. on Metal-Hydrogen Systems, Miami 1981

29/ P.W. Anderson, B.I. Halperin and C.M. Varma, Phil. Mag. 25, 1 (1972)

30/ W.A. Phillips, J. Low Temp. Phys. 7, 351 (1972)

31/ Y. Imry, in Tunneling Phenomena in Solids, eds.

E. Burnstein and S. Lundquist, Plenum Press, New York 1969, p. 563. Note that the ratio given for α/β in Eq. 3 of this reference is misprinted (see, e.g., R. Blinc and D. Hadzi, Mol. Phys. <u>1</u>, 391 (1958)).

32/   A.M. Stoneham, Rev. Mod. Phys. <u>41</u>, 82 (1969)

33/   V. Narayanamurti and R.O. Pohl, Rev. Mod. Phys. <u>42</u>, 201 (1970)

34/   F. Bridges, Crit. Rev. Solid State Science <u>5</u>, 1 (1975)

35/   P.P. Matyash, N.A. Skakun and N.P. Dikii, JETP Letters <u>19</u>, 18 (1974)

36/   H.D. Carstanjen, phys. stat. sol.(a) <u>59</u>, 11 (1980)

37/   T. Schober and H. Wenzl, in <u>Topics in Applied Phys.</u> vol. 29, <u>Hydrogen in Metals II</u>, eds. G. Alefeld and J. Völkl, Springer-Verlag, Berlin - Heidelberg-New York, 1978, p.11

38/   V. Lottner, U. Buchenau and W.J. Fitzgerald, Z. Physik <u>B35</u>, 35 (1979)

39/   J.J. Rush, J.M. Rowe, A. Magerl, D. Richter and H. Wipf, to be published

40/   H.K. Birnbaum and C.P. Flynn, Phys. Rev. Lett. <u>37</u>, 25 (1976)

EXTRACTING MAXIMAL INFORMATION ON

POSITIVE MUON DIFFUSION IN METALS[*]

M. Leon

Los Alamos National Laboratory
Los Alamos, NM 87545

ABSTRACT

We discuss how combining zero- and longitudinal-field μSR can give more information on the motion of the μ^+ in the target than zero-field studies alone.

INTRODUCTION

Transverse-field μSR has been used for several years to provide information of the mobility of the μ^+ in various materials. In nonmagnetic metals, depolarization of the μ^+ is caused by the nuclear magnetic moments surrounding the muon interstitial site; when the muon is hopping from site to site, the depolarization rate is reduced by motional narrowing. Furthermore, in many materials, the depolarization rate is found to be a very complicated function of temperature which has been interpreted in terms of capture and release of the μ^+ by traps.[1] However, since the finding of traps and the release from them can produce the same polarization fuction for transverse-field μSR, significant ambiguity of interpretation remains. Recently, Petzinger pointed out that zero-field μSR can readily resolve these ambiguities, since a muon that is trapped and therefore stationary at long times will have its polarization return to one-third of its initial value, while one that is escaping from traps will have its polarization approach zero at long times.[1]

The purpose of the present paper is to point out how combining zero- and longitudinal-field data can yield more information about the muon's motion at some temperatures than can be extracted from

---

[*]Work supported by the U.S. Department of Energy.

H = 0 data only. One instance is where the muons are both finding
and escaping from a given type of trap, and the escape rate ε is
<u>not</u> smaller than Δ, the Gaussian width of precession frequencies at
the trap site. In that case, a family of (ε,ν) values--ν being the
trapping rate--all give the same zero-field polarization function.
A second instance is where the muons are escaping from one type of
trap and finding a second. Then there are families of (ε_o, ν_1)
values for which the zero-field polarization functions differ only
at long times, where the experimental information is poor because
of the muon decay. In each case, supplementing the H = 0
measurement by an $H_{\parallel}$ one can resolve the ambiguity and determine
the trapping and escape rates separately. We discuss these
situations in detail in the following.

TRAPPING AND ESCAPE FROM A SINGLE TYPE OF TRAP

     The integral equation for the polarization function P(t) (for
H = 0, $H_{\perp}$ or $H_{\parallel}$) and its solution have been discussed in a recent
paper.[2] We assume that depolarization occurs only when the muon is
trapped. Typical functions with ε large enough to produce motional

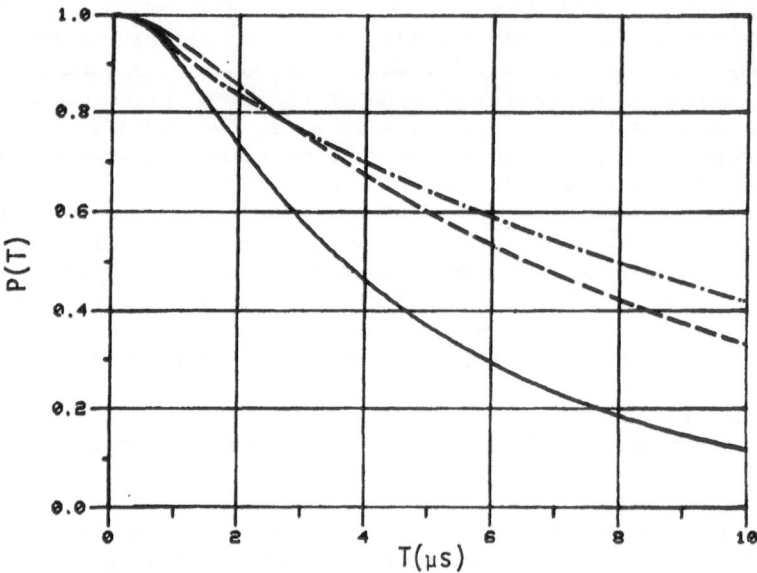

Fig. 1. P(t) for H = 0 (solid curve), $H_{\perp}$ (dashed curve), and $H_{\parallel}$
        corresponding to muon precession angular frequency ω = 2
        (dot-dashed curve). These curves were calculated for
        Δ = 0.5 μs^{-1}, ν = 5 μs^{-1}, ε = 1.45 μs^{-1}.

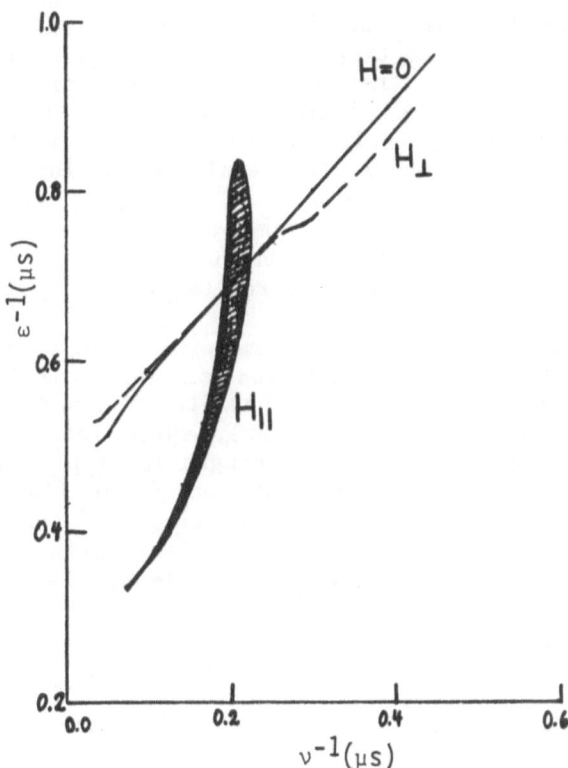

Fig. 2. Points in the $(\nu^{-1}, \varepsilon^{-1})$-plane which give rise to $P(t)$
curves essentially identical to those of Fig. 1 (Δ, ω fixed
at the values of Fig. 1). The $H = 0$ and $H_\perp$ curves nearly
coincide, while the airfoil-shaped area for $H_\parallel$ intersects
them at a significant angle.

narrowing are shown in Fig. 1. Essentially identical $P(t)$'s (over
the entire 10 μs range shown!) are generated by families of (ε, ν)
values in each case. The crucial point is that while the curves of
"degeneracy" (i.e., identical $P(t)$) practically coincide for the
$H = 0$ and $H_\perp$ cases, the area for $H_\parallel$ intersects these at a
significant angle; this is shown in Fig. 2 (where for convenience
the reciprocals of ε and ν are plotted). Thus combining an $H = 0$
measurement with a properly chosen $H_\parallel$ one will determine both ε and
ν, while an $H_\perp$ experiment does not add information to an $H = 0$ one.

ESCAPING FROM ONE TYPE OF TRAP AND FINDING A SECOND

Here we suppose that the escape rate ε_o from trap 1 is <u>not</u> extremely large so that the depolarization in the "free state" (i.e., going from trap to trap of type 1) is <u>not</u> negligible. When the muons are finding the second type of trap, $P(t)$ can be calculated by solving a "hierarchy of integral equations" as discussed in ref. 2. A resulting $P(t)$ for $H = 0$ is shown in Fig. 3 (solid curve). Also shown is the result neglecting depolarization in the free state, i.e., taking $\varepsilon_o = \infty$, with ν_1 adjusted to make the curves nearly coincide for time $t < 5$ µs. These two cases would be difficult to distinguish because they differ mainly for $t > 5$ µs where the data have large error bars.

An analogous pair of curves for $H_\parallel$ is shown in Fig. 4. Again the value of ν_1 has been adjusted to make the curves nearly coincide for short times, but now they differ significantly for $t > 2$ µs. The asymptotic value of $P(t)$, which reflects the static environment of the muon once it is caught in trap 2, is reduced by the depolarization in the free state before it finds trap 2. The presence of $H_\parallel$ makes this change of asymptotic value easier to observe by moving it to smaller t. Thus here again combining $H = 0$ and $H_\parallel$ measurements will allow the experimenter to determine <u>two</u> diffusion parameters, now ε_o and ν_1.

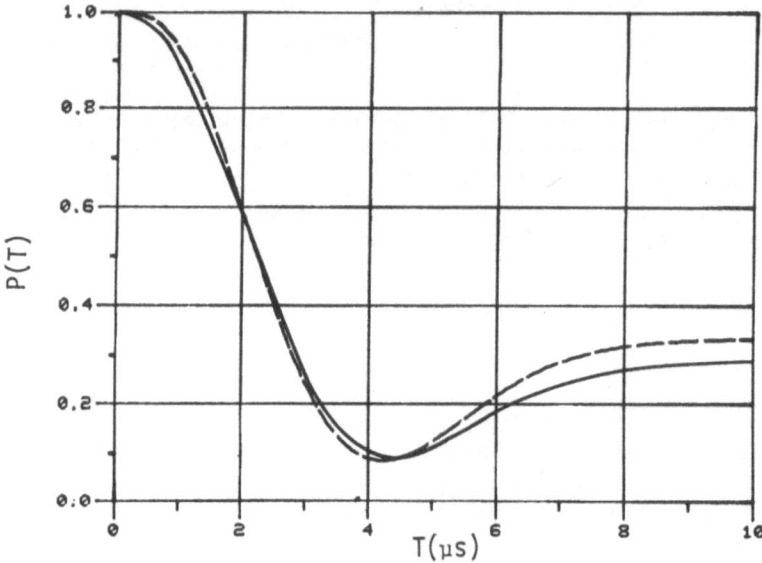

Fig. 3. $P(t)$ for $H = 0$ and $\Delta_o = \Delta_1 = 0.5$ µs^{-1}. Solid curve: $\varepsilon_o = 2$, $\nu_1 = 1$; dashed curve: $\varepsilon_o = \infty$, $\nu_1 = 1.5$.

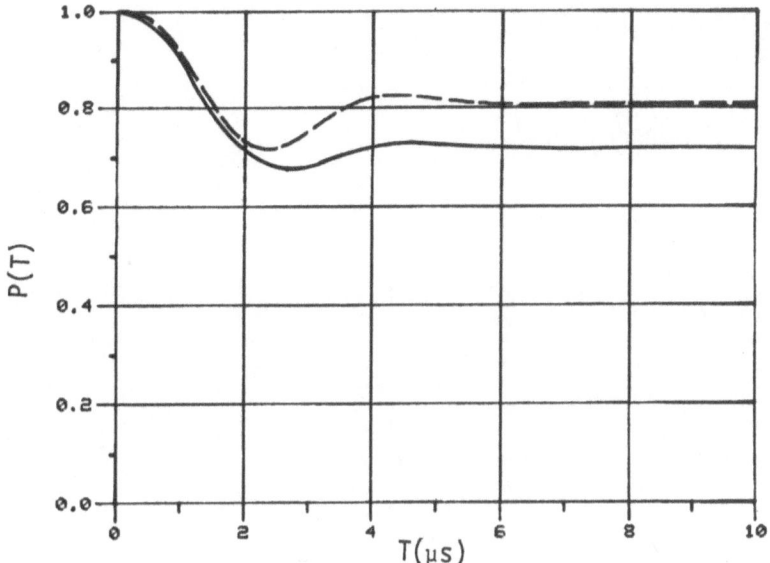

Fig. 4. P(t) for $H_{\parallel}$ such that $\omega = 1.5$, $\Delta_0 = \Delta_1 = 0.5$ μs^{-1}. Solid curve: $\varepsilon_0 = 2$, $\nu_1 = 1$; dashed curve: $\varepsilon_0 = \infty$, $\nu_1 = 2$.

SUMMARY

We have shown how combining longitudinal-field with zero-field μSR experiments can significantly increase the amount of information that can be extracted on the motion of the μ^+ in the target.

REFERENCES

1. K. G. Petzinger, Hyperf. Int. 8 (1981) 639; Phys. Lett. 75A (1980) 225.

2. C. Boekema, R. H. Heffner, R. L. Hutson, M. Leon, M. E. Schillaci, W. J. Kossler, M. Numan, S. A. Dodds; to be published.

EFFECTS OF IMPURITIES ON THE DIFFUSION OF POSITIVE MUONS IN NIOBIUM

O. Hartmann, E. Karlsson, R. Wappling[*], E. Holzschuh,
W. Kündig, B. D. Patterson[**], K. Schulze[+], D. Richter,
R. Hempelmann[++], and S. Cox[+++]

[*]Univ. of Uppsala (S), [**]Univ. of Zürich. (CH),
[+]MPI Stuttgart, (FRG), [++]KFA Julich, (FRG), [+++]Rutherford
Laboratory, (U.K.)

ABSTRACT

The diffusion and trapping of positive muons in different
niobium samples have earlier been studied by the μSR method and
it is known that muons are sensitive even to very small concentrations of impurities (of the order of a few ppm). The present study
has aimed at an understanding of the role of some typical impurities
and starts out from the purest material obtainable which is then
doped with selected impurities at levels of 10 - 50 ppm.

In the purest material (< 1 ppm substitutional impurities,
$\lesssim$ 2 ppm interstitial N,C,O) the majority of the muons are found to
be mobile over all temperatures down to about 1 K, but a fraction
of about 1/5 still seem to be trapped by the remaining impurities
and released from these traps only for T > 80 K. Material doped
with 50 ppm Ta shows strong trapping in the range T = 10 - 20 K,
but release of muons for T > 20 K with some indication of retrapping
around T = 50 K, probably by residual N-atoms. Doping with 15 ppm
N, which is an interstitial impurity, shows on the other hand both
an increased trapping in the 10 - 30 K region and a well developed
maximum (with possible fine structure) for T = 30 - 80 K.

In trying to explain all these effects we have developed a
model which allows a certain fraction of the muons to be self-
trapped thermally by the Nb-lattice, while the remaining ones are
self-trapped only if there are impurities present. They can then
move further as small polarons to high -T traps. In sufficiently

pure materials there would then still exist a fraction of muons
in non-self-trapped, propagation states which do not contribute to
the μSR-damping.

MUON DIFFUSION IN NOBLE METALS[*]

M.E. Schillaci, C. Bokema, R.H. Heffner, R.L. Hutson,
M. Leon, C.E. Olsen, S.A. Dodds[†], D.E. MacLaughlin[‡],
and P.M. Richards[§]

Los Alamos National Laboratory, Los Alamos, NM 87545
[†]Rice University, Houston, TX 77001
[‡]University of California, Riverside, CA 92521
[§]Sandia National Laboratories, Albuquerque, NM 87185

ABSTRACT

We have measured diffusion-induced muon depolarization in dilute AgGd and AgEr in the temperature range 200–700 K and have thereby determined the muon diffusion parameters in Ag. The diffusion parameters for μ^+ in Cu, Ag, and Au are compared with those of hydrogen. For Ag and Au, the μ^+ parameters are similar to those of hydrogen, whereas for Cu, the μ^+ parameters are much smaller. Lattice-activated tunneling and over-barrier hopping are investigated with computational models.

INTRODUCTION

The similarity of the positive muon (μ^+) to a light proton (p) in a material makes the study of muon diffusion and trapping an ideal complement to the investigation of hydrogen in metals. Because the muon is so much lighter than the proton ($m_\mu \approx m_p/9$), and, because the muon diffusion can be studied at lower temperatures than hydrogen diffusion, the possibility of studying different diffusion mechanisms is enhanced.

----
[*]Work supported by the U. S. Department of Energy and National Science Foundation.

Muon diffusion parameters can be derived from the temperature dependence of the muon depolarization rate, Λ_2, in a transverse magnetic field. The muon spin rotation (μSR) technique is described in several review articles.[1] In nonmagnetic metals, depolarization of the diffusing muon is caused by the inhomogeneous magnetic fields of the host nuclear dipole moments. As the muon hops more rapidly with increasing temperature, motional narrowing of the line width results--i.e., Λ_2 is reduced. Obviously, this method does not apply to metals having negligible nuclear moments, such as Ag and Au.

Another method for studying muon diffusion, reported earlier,[2] overcomes this limitation by doping the metal with small amounts of paramagnetic impurities and measuring Λ_2 resulting from a muon's interaction with the electronic moments of the impurity ions. This mechanism is effective only if a muon moves to the vicinity of an impurity within its lifetime, and thus provides a measure of the μ^+ hopping time τ_h. These measurements are also sensitive to the impurity ion spin dynamics and to the muon-ion interaction; the diffusion parameters thus determined, however, are not sensitive to the nature of the interaction, provided that it is reasonably short range.[2]

We report on measurements of muon diffusion in Ag, a continuation of our previous studies in Au.[2] Diffusion parameters for muons and hydrogen in Cu, Ag, and Au are then compared. A discussion of possible diffusion mechanisms and how they relate to the observed parameters follows.

RESULTS

Measurements were made on polycrystalline samples of Ag doped with either the S-state ion Gd^{3+} or the crystal-field-split ion Er^{3+} and were carried out at the Clinton P. Anderson Meson Physics Facility (LAMPF) at Los Alamos. The reader is referred to our earlier work[2] for the details of the experiment. Figure 1 shows the data for Λ_2 as a function of temperature, T, for AgGd and AgEr, taken in a transverse applied magnetic field of 80 Oe. In each case, the higher concentration data is scaled by the concentration ratio, showing that the concentration dependence of Λ_2 is linear. The observed peaks are clearly identified with muon-impurity ion interactions by comparison with data taken in the pure host, where $\Lambda_2(T)$ is negligible. The data begins to rise out of the background ($T \sim 200$ K) at the temperature at which the muons are hopping fast enough to reach a magnetic impurity within a muon lifetime. Since the rising portions of the data are almost identical, muon diffusion in the host is not affected by the nature of the impurity.

The curves shown in Fig. 1 are model fits to the data. A description of the model used has been given elsewhere[2,3]; we give only a brief outline here. The muon is assumed to hop between adjacent octahedral interstitial sites which are arranged in shells surrounding an impurity out to a radius corresponding to the mean volume per impurity. The muon-ion interaction is assumed to have two components--dipolar and contact, which applies only to sites adjoining the impurity. The spin-lattice relaxation rate is assumed to have the Korringa form. The muon hopping rate is assumed to have an Arrhenius form, $\tau_h^{-1} = \nu_o \exp(-\varepsilon/T)$. Thus there are four adjustable parameters; (1) muon-ion contact interaction strength; (2) impurity ion spin-lattice relaxation coefficient; (3) the activation energy ε; and (4) the pre-exponential factor ν_o.

While the muon-ion interaction and the impurity spin dynamics are certainly interesting topics, we shall discuss only the diffusion parameters here. Any ambiguity in the determination of ν_o and ε is substantially reduced by the fact that in the rising portion of the data, ε and ν_o are not very sensitive to the values of the other two parameters.[2] The muon diffusion parameters extracted from the model fits to the Ag data are listed in Table I along with those for Au and Cu. Also included for comparison in Table I are the hydrogen diffusion parameters for all three metals. We also list the Debye temperature, Θ_D, and the temperature range studied, ΔT, for each case.

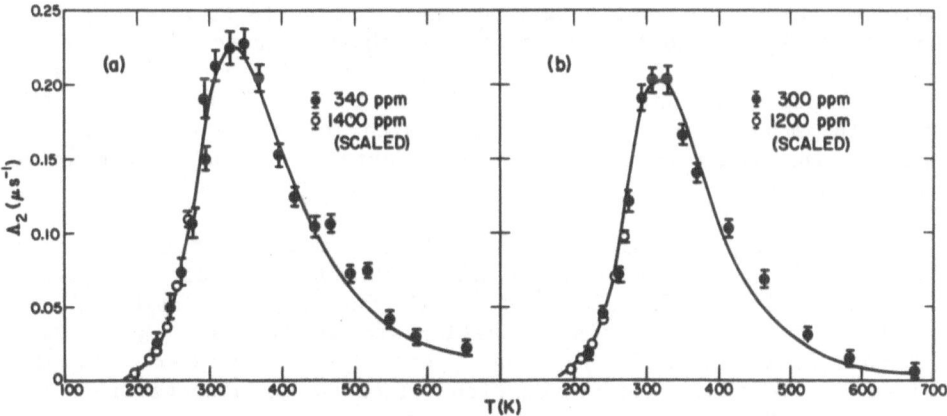

Fig. 1. Temperature dependence of muon depolarization rate in (a) AgGd and (b) AgEr at 80 Oe transverse applied field. The higher-concentration data has been scaled by the concentration ratios. Curves are model fits as discussed in text.

Table I. Diffusion parameters for muons and proton in noble
         metals

μ^+ in:	$\nu_o(s^{-1})$	$\varepsilon(K)$	$\Delta T(K)$	$\Theta_D(K)$
Cu[a]	$10^{7.46 \pm .04}$	551 ± 15	80–250	343
Ag[b]	$10^{13.5 \pm .5}$	3200 ± 200	200–700	225
Au[c]	$10^{13.5 \pm .5}$	1350 ± 100	85–230	165
p in:				
Cu[d]	10^{14}	4640 ± 30	720–1200	343
Ag[e]	$10^{13.8}$	3620	950–1120	225
Au[f]	$10^{12.6}$	2840	770–1200	165

[a]Ref. 4; [b]this work; [c]Ref. 2; [d]Ref. 5; [e]Ref. 6; [f]Ref. 7

DISCUSSION

The obvious observation to be made from Table I is that the diffusion parameters for muons in Ag and Au are similar to the hydrogen parameters, while both ν_o and ε for μ^+ in Cu are very much smaller. This suggests significant incoherent tunneling for μ^+ in Cu,[4] but all of the other parameters are consistent with over-barrier hopping. Indeed, Teichler[8] has shown that the tunneling model gives good agreement with the μ^+ data for Cu. We examine the remaining cases in the context of the over-barrier hopping model.

We have calculated the potential of a point unit charge as a function of position within the unit cell of each metal following the calculation of Teichler,[9] to which the reader is referred for details. Electron screening densities around the point charge were obtained from the self-consistently calculated curves of Jena and Singwi.[10] The host ion-conduction electron pseudopotentials of Gubanov and Nikulin[11] were used, but their energy parameter, α, was varied as described below. Two general features of all of the potentials studied are: (1) only the octahedral (O) well is deep enough for the μ^+ ground state--the tetrahedral (T) well is too shallow; and (2) the O–O barrier is higher than the O–T barrier. Experimental evidence indicates that μ^+ is localized at the O site in Cu[12] and Au,[13] so we assume it is in Ag also.

The pseudopotential energy parameter, α, was varied in order to adjust the difference, Δ_μ (or Δ_p), between the 0-T barrier height and the μ^+ (p) ground state. The over-barrier activation energy is then given by $\varepsilon_\mu = \Delta_\mu + B_\mu$, where B_μ (B_p) is the μ^+ (p) self-trapping energy. In the harmonic approximation, the ratio would be given by $B_\mu/B_p = \sqrt{m_p/m_\mu} \approx 3$; however, we shall use the ratio calculated by Teichler[9]--viz., $B_\mu/B_p = 90/40 = 2.25$. The two unknown quantities, α and B_p, are determined from the two measured activation energies, ε_p and ε_μ. In Table II, we list the values of α, B_p, and B_μ together with the value, α_o, which yields the correct host atom nearest-neighbor separation in lowest order perturbation theory within the pseudopotential scheme.[14] We note that the α's are within $\sim 5\%$ of the α_o's. The self-trapping energies required to fit the data in Au and Ag are 2 to 4 times greater than those calculated[9] for Cu. To actually calculate B_μ (B_p) is difficult because the result depends very sensitively on both the host-host potential and the host-μ^+ (p) potential. Better potentials are clearly needed for this task.

If tunneling were the dominant mechanism rather than over-barrier hopping, a much smaller pre-exponential factor, ν_o, would result unless the transfer integral, J, is large.[15] For μ^+ 0-0 tunneling, we calculate J values for Ag and Au that are similar to that for Cu--on the order of 10 μeV. Since only μ^+ in Cu exhibits a small ν_o, we conclude that incoherent tunneling is not important for μ^+ in Ag and Au.

Table II. Self-trapping energies and pseudopotential parameters determined from the diffusion data.

	Cu	Ag	Au
α_o-eV	14.22	12.19	12.23
α-eV	13.25	11.73 $\pm$ .03	11.72 $\pm$ .02
B_μ-meV	90[a]	344 $\pm$ 26	177 $\pm$ 13
B_p-meV	40[a]	153 $\pm$ 12	79 $\pm$ 6

[a]These are computed values from ref. 9.

SUMMARY

   We have presented muon diffusion parameters in Ag determined
from measurements of the muon depolarization rate versus
temperature in samples doped with dilute magnetic impurities.
Comparison of the diffusion parameters for muons and hydrogen in
the noble metals leads to the conclusion that only for μ^+ in Cu is
incoherent tunneling a dominant mechanism. Model studies of
over-barrier hopping indicate that this mechanism can explain the
remaining data if the self-trapping energies for Au and Ag are 2 to
4 times greater than those calculated[9] for Cu.

REFERENCES

1.   See, for example, A. Seeger in Hydrogen in Metals I, edited by
     G. Alefeld and J. Völkl (Springer-Verlag, Berlin, 1978).
2.   J. A. Brown, R. H. Heffner, R. L. Hutson, S. Kohn, M. Leon, C.
     E. Olsen, M. E. Schillaci, S. A. Dodds, T. L. Estle, D. A.
     Vanderwater, P. M. Richards, and O. D. McMasters, Phys. Rev.
     Lett. 47, 261 (1981).
3.   M. E. Schillaci, R. L. Hutson, R. H. Heffner, M. Leon, S. A.
     Dodds, and T. L. Estle, Hypf. Int. 8, 663 (1981).
4.   V. G. Grebinnik, I. I. Gurevich, V. A. Zhukov, A. P. Manych,
     E. A. Meleshko, I. A. Muratova, B. A. Nikoskii,V. I.
     Selivanov, and V. A. Suetin, Zh. Eksp. Teor. Fiz. 68, 1548
     (1975) [Sov. Phys. JETP 41, 777 (1969)].
5.   L. Katz, M. Guinan, and R. J. Borg, Phys. Rev. B4, 330 (1971).
6.   H. Katsuta and R. B. McLellan, Scripta Metallurgica 13, 65
     (1979).
7.   W. Eichenauer and D. Liebscher, Z. Naturforsch. 17A, 355
     (1962).
8.   H. Teichler, Phys. Lett. 64A, 78 (1977).
9.   H. Teichler, Phys. Lett. 67A, 313 (1978).
10. P. Jena and K. S. Singwi, Phys. Rev. B17, 3518 (1978).
11. A. I. Gubanov and V. K. Nikulin, Fiz. Tver. Tel. 7, 2701
     (1965), [Sov. Phys. Solid State 7, 2184 (1966)].
12. M. Camani, F. N. Gygax, W. Rüegg, A. Schenck and H. Schilling,
     Phys. Rev. Lett. 39, 836 (1977).
13. K. Maier, G. Flik, D. Herlach, G. Jünemann, A. Seeger, and H.
     -D. Carstanjen, Phys. Lett. A (in press).
14. N. W. Ashcroft and D. C. Langreth, Phys. Rev. 159, 500 (1967).
15. D. Emin, M. I. Baskes, and W. D. Wilson, Phys. Rev. Lett. 42,
     791 (1979).

DIFFUSION OF POSITIVE MUONS IN ALUMINUM

AND ALUMINUM-BASED ALLOYS

K.W. Kehr, D. Richter, J.M. Welter, O. Hartmann[†]
E. Karlsson[†], L.O. Norlin[†], T.O. Niinikoski[*] and
A. Yaouanc[**]

Institut für Festkörperforschung, KFA Jülich, W-Germany
[†]Institute of Physics, Uppsala, Sweden
[*]CERN, Geneva, Switzerland
[**]CERG, Grenoble, France

ABSTRACT

Results from muon spin rotation (µSR) measurements on doped aluminum samples are reported. The doping elements were Mn at various concentrations, and Li, Mg, Ag at concentrations of about loo ppm. The µSR linewidth exhibits typical trapping maxima above 1 K, whereas below 1 K the linewidth increases monotonically with decreasing temperature. The analysis of the trapping peak above 1 K reveals a capture rate proportional to temperature. The increase of the linewidth below 1 K is attributed to capture controlled by coherent diffusion. The coefficient of coherent diffusion is found proportional to $T^{-0.6}$. The temperature dependence of the diffusion coefficient in both regions is compared with current theories of diffusion of light interstitials in metals.

INTRODUCTION

Preliminary results for diffusion and localization of μ^+ in some pure and doped Al samples were presented in ref. [1]. This investigation has now been complemented with further measurements and a more detailed interpretation of the data.

The primary aim has been to find out which are the main diffusion mechanisms working at low and intermediate temperatures, respectively. Since the muons are known to be mobile in the purest Al samples down to temperatures as low as 0.03 K [2] this study is based on the observation of diffusion in aluminum samples which have been doped with selected impurities. The muons are supposed to diffuse towards the impurities, which form trapping centers, so that the probability for trapping reflects the rate of intrinsic diffusion in the pure material. In this way, it is possible to observe indirectly muon diffusion at rates which would have led to complete motional narrowing in a pure metal sample. The primary experimental data are, as before, the muon depolarization rates, which are equivalent to linewidths in NMR.

EXPERIMENTS

Polycrystalline Al samples were prepared from a base materials with less than one ppm metallic impurities and doped with 5, 10, 42 or 70 ppm Mn, or 75 ppm Li, 42 ppm Mg, 117 ppm Ag. In addition, single crystals were prepared with 57 or 1300 ppm Mn impurities. The doping concentration was checked with atomic-absorption spectrometry, and the residual impurity level by spark-source mass spectrometry.

The measurements were performed at the 600 MeV synchrocyclotron at CERN using a conventional muon spin rotation set-up with transverse field geometry. The experiments above 2 K were carried out with a conventional He^4 cryostat, whereas below 2 K a He^3–He^4 dilution refrigerator was used. The temperatures were measured by a calibrated carbon resistor (below 30 K) and a Pt resistor (above 30 K). The temperature stability was about $0.2 - 1.6\%$.

The linewidth parameters displayed in the figures in this paper were obtained by least-square fits to the μSR spectra, taking the polarization decay P(t) as a Gaussian function, $P(t) = P_0 \exp(-\sigma^2 t^2)$. All spectra were corrected for instrumental background which was determined by running antiferromagnetic stainless steel dummy samples (giving no μSR signal) of mass and size equal to the Al samples. Fig. 1 presents an overview of the polycrystalline $\underline{Al}Mn_x$ data and Fig. 2 shows in detail the data for T<3 K for Mn-, Li- and Ag-doping. It is clear that the muon depolarization in the $\underline{Al}Mn_x$ samples exhibits two different regions:

(i) Above 1 K a peak in the depolarization rate evolves around 17 K. While its position does not depend on the Mn concentration, its height and width are correlated to it. Similarly for $\underline{Al}Ag$, $\underline{Al}Li$ and $\underline{Al}Mg$ peaks develop around 20 K, 40 K and 50 K respectively. (ii) Below 1 K the muon depolarization increases with decreasing temperature. A rise of the Mn concentration enlarges the damping at a given temperature.

The peak in the depolarization rate around 17 K in the $\underline{Al}Mn_x$ data is assigned to trapping of muons by impurities. In this regime we analyze the linewidth data applying the two-state model for μ^+ depolarization in the presence of traps [3]. Apart from τ_c^{-1}, the parameters of this model are the capture rate $1/\tau_1$ and

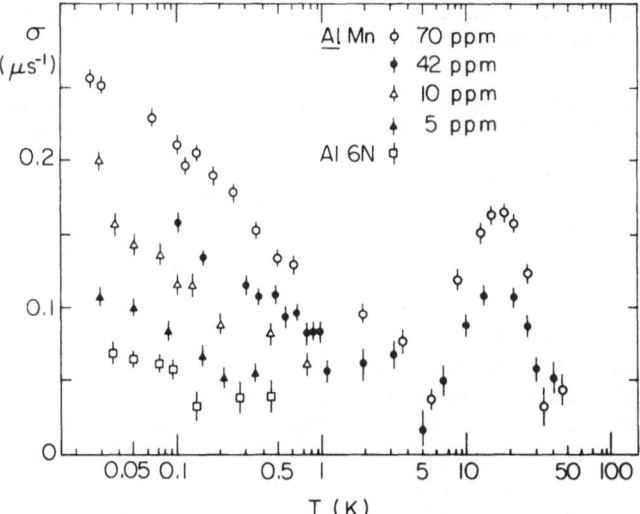

Fig. 1 Linewidth as function of temperature for Mn-doped
        Al samples compared to pure Al (6N).

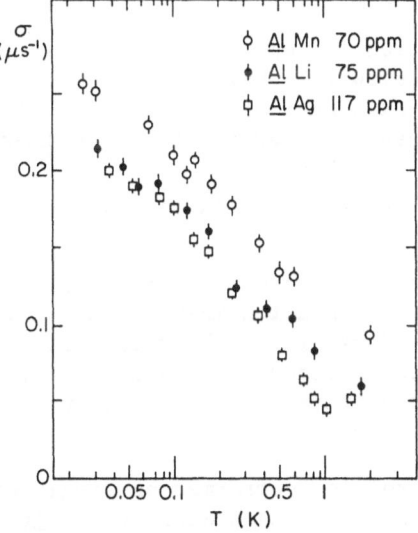

Fig. 2 Linewidth vs. temperature for Al with different
        doping elements.

the escape rate $1/\tau_0$ of the traps. The results of the fitting procedure can be summarized as follows:

(i) the capture rate $1/\tau_1$ depends only weakly on temperature, $1/\tau_1 \sim \Gamma_1(c)(T/T_0)^\beta$, $\beta = 0.89(3)$. (ii) $1/\tau_1$ is proportional to the Mn concentration c. Values for $\Gamma_1(c)$ are (T=1 K): 42 ppm: $1.62(12) \times 10^4$ s^{-1}, 57 ppm: $2.02(14) \times 10^4$ s^{-1}, 70 ppm: $2.43(14) \times 10^4$ s^{-1}. (iii) The escape rate $1/\tau_0 \sim \exp(-E_a/kT)$ shows an activation energy of 120(6) K.

The data below 1 K in A̲l̲Mn were first fitted with the usual Abragam formula for motional narrowing. The static linewidth σ_0 was set to the saturation value of 0.265 μs^{-1} found for high Mn concentrations [2]. The measurements on the 42 ppm Mn sample were carried out at a higher transverse field (500 G vs. 160 G). Since σ_0 is field dependent the correspondingly smaller value of $\sigma_0 = 0.22$ μs^{-1} was used in this case. The τ_c data taken on AlMn$_x$ below 1 K can be described with good accuracy by the equation

$$\tau_c^{-1} = 6.9(1.2)10^5 \times T^{0.60(4)} \times c^{-0.76(4)} \quad s^{-1} \tag{1}$$

where T is given in mK and c in ppm. This approach corresponds to a picture where muons become more mobile as the temperature increases from 0.03 to about 1 K. The c-dependence cannot be reasonably explained by such an assumption since the only diffusion process that could be expected to increase with T in this range is the one-phonon assisted process [4] which would require independence (or increase) with c.

An interpretation in terms of trapping is therefore favored also for the data below 1 K. Using again the two-state model of Ref. [4], with motional narrowing in the diffusion toward the traps, and a negligible chance of escape, the experimental data now lead to a trapping rate

$$\tau_1^{-1} = 3.0(6) \times 10^5 \, T^{-0.61(4)} \, c^{0.76(4)} \quad s^{-1} \tag{2}$$

using the same units as before. Proportionality with respect to c is expected when c is a small number so that different trapping centers do not interfere with each other. This inverse T-dependence of the trapping rate corresponds to a bulk diffusion which increases on lowering the temperature and therefore suggests the onset of a coherent motion below 1 K.

Experiments were also performed on single crystals oriented with their (100)- or (111)- axes along the magnetic field (Fig. 3). These are interpreted in terms of the theory by Hartmann [5]. The behaviour of $\sigma(B)$ at T=0.04 K indicates clearly that the muons are situated in octahedral sites for most of their lifetime, whereas

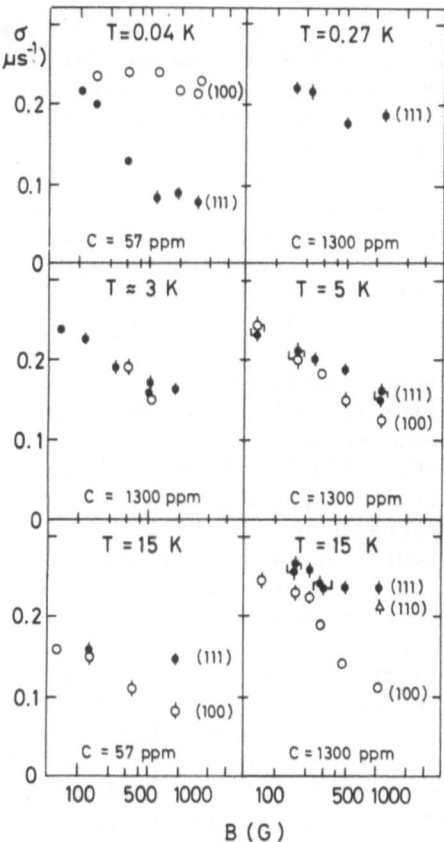

Fig. 3 Linewidth vs. applied magnetic field in the 100- and
       111-directions for single crystal A̲lMn samples at
       different temperatures.

at 15 K the sites are essentially tetrahedral. At intermediate
temperatures there is no clear preference for any of these two
types of sites.

DISCUSSION

(i) Above 1 K the diffusion process leading to trapping is
thermally assisted incoherent hopping. The diffusion coefficient
deduced from the capture rate is nearly proportional to T. The conventional small-polaron theory predicts a T^7-behaviour [6] as a
result of a two-phonon process, in evident disagreement with the

observations. Linear behaviour with T is typical for a one-phonon process, which is only possible between energetically inequivalent sites due to momentum and energy conservation. One possible cause for energetically inequivalent interstitial sites are strains induced by the impurities in the crystals. This would lead to a concentration dependent diffusion coefficient [7], whereas the diffusion coefficient deduced from the capture rate is practically independent of concentration. Another possibility is given by alternating transitions between tetrahedral and octahedral sites, if they are energetically almost degenerate in aluminum. This conjecture is not contradicted by the experiments on site determination in Al, which allowed no clear distinction between both types of sites at intermediate temperatures. Since this possible explanation applies specifically to Al, experiments on other metals could lead to different results.

(ii) Below 1 K the diffusion coefficient derived from diffusion-controlled trapping is found proportional to $T^{-0,6}$. The current theory of coherent diffusion of a light interstitial in metals predicts a T^{-9}-behaviour [8], as a result of phonon scattering on the interstitials. This is in clear contradiction to the experimental findings. Another process which gives approximately the correct behaviour is scattering of electrons on muons. It has been pointed out by Andreev and Lifshitz [9] that this process should delimit coherent diffusion since it is mainly determined by the number of thermally excited electrons, which is proportional to T . Simple estimates show that this process has the correct order of magnitude, hence it represents a likely explanation of the low-temperature behaviour of the diffusion coefficient.

A more detailed account of the investigations and discussions will be given in a forthcoming publication [10].

REFERENCES

[1] K.W. Kehr, D. Richter, J.M. Welter, O. Hartmann, L.O. Norlin, E. Karlsson, T.O. Niinikoski, J. Chappert, and A. Yaouanc, Hyperfine Interactions $\underline{8}$, 681 (1981).

[2] O. Hartmann, E. Karlsson, L.O. Norlin, T.O. Niinikoski, K.W. Kehr, D. Richter, J.M. Welter, A. Yaouanc, and J. LeHéricy, Phys.Rev.Lett. $\underline{44}$, 337 (1980).

[3] K.W. Kehr, G. Honig, and D. Richter, Z.Physik B$\underline{32}$, 49 (1978).

[4] J.A. Sussmann, Phys.Kondens.Materie $\underline{2}$, 146 (1964).

[5] O. Hartmann, Phys.Rev.Lett. $\underline{39}$, 832 (1977).

[6] C.P. Flynn and A.M. Stoneham, Phys.Rev.B $\underline{1}$, 3966 (1970).

[7] H. Teichler and A. Seeger, Phys.Lett. $\underline{82A}$, 91 (1981).

[8] Yu. Kagan and M.J. Klinger, J.Phys..C $\underline{7}$, 2791 (1974).

[9] A.F. Andreev and I.M. Lifshitz, Sov.Phys.-JETP $\underline{29}$, 1107 (1969).

[10] K.W. Kehr, D. Richter, J.M. Welter, O. Hartmann, E. Karlsson, L.O. Norlin, T.O. Niinikoski, A. Yaouanc, subm. to Phys.Rev.B.

ANOMALOUS BEHAVIOR OF DIFFUSION COEFFICIENTS OF HYDROGEN AND DEUTERIUM IN HIGH CONTENT PALLADIUM HYDRIDE AND DEUTERIDE

S. Majorowski, B. Baranowski

Institute of Physical Chemistry, Polish Academy
of Sciences
01-224 Warsaw, Poland

Fick's diffusion coefficients of hydrogen and deuterium in palladium hydride and deuteride were determined under high equilibrium pressures of gaseous hydrogen and deuterium using an electrical resistance relaxation method.The measurements were made for pressures ranging from 1 MPa to 1 GPa (10 to 10000 atm) and temperatures ranging from 208 to 338 K.This covered the H,D/Pd concentration range from about 0.8 to nearly 1.Einstein's diffusion coefficients were calculated and the activation volumes and activation enthalpies were evaluated.An attempt is made to interpret the anomalous behavior of the diffusion coefficients in the low temperature range.The results are presented in a compact formula.

Isothermal diffusion of hydrogen and deuterium in metallic palladium were investigated within the range of low interstitial concentrations (α-phase) and in the high concentration hydride and deuteride β-phase.Reviews of previous results can be found elsewhere[1,2] Measurements in the range of higher concentrations require very high hydrogen and deuterium pressures.The region of hydrogen and deuterium concentration (n=H,D/Pd) greater than 0.75 has been largely ignored until now[3].In our work diffusion of hydrogen and deuterium in the concentration range n from about 0.8 to nearly stoichiometry was studied.A special high pressure hydrogen apparatus was used for this purpose[4].Electrical resistance of palladium wires kept in equilibrium with gaseous hydrogen and deuterium was measured as a function of time. This enabled us to determine diffusion coefficients of hydrogen and deuterium in palladium hydride and deuteride phases.A careful purification of gases and activation of the surface of samples with Pd-black had to be carried out in order to measure the bulk diffusion as the rate determining step.Two wires of different diameter were used simul-

519

taneuosly for checking.As mentioned above,the diffusion experiments
started from palladium hydride samples of minimum hydrogen concentration n about 0.8.These hydrides were prepared by two procedures:
-A treatment in gaseous hydrogen(deuterium) of about 20 MPa and temperature around 700 K was carried out.Both variables are supercritical
with respect to the hydride formation[5].This means that the transition
from the α to β-phase can be realized without any discontinuous change
of the lattice parameter.Later the temperature was decreased,keeping
the pressure unchanged.Results obtained for these samples are marked
by the symbol o in fig.1.
-The transition to the hydride was carried out at room temperature.
A phase discontinuity took place here,leading probably to a much more
defected matrix of the metallic palladium.Repeated cycling from the
α to the β-phase and back did not lead to any remarkable changes.
Results for these samples are denoted by symbol ∇ in fig.1.

Fick's diffusion coefficients of hydrogen and deuterium in
β -palladium hydride and deuteride were determined for the pressures
up to 1 GPa (10000 atm) and temperatures between 208 and 338 K.Results
obtained for the two kinds of samples do not differ at temperatures
above 273 K.Differences appear at lower temperatures as can be seen
in fig.1.Within the whole temperature and pressure range the inverse
isotope effect was observed,i.e. the diffusion coefficient of deuterium was greater in the same p-T conditions than for hydrogen.At lower
temperatures maxima of the both diffusion coefficients as a function
of pressure appear.

Examining the influence of hydrogen equilibrium pressure changes
on the diffusion coefficient of hydrogen in the hydride phase one can
distinguish two main factors: the first one is the change of hydrogen
concentration in the solid phase,which changes the values of the thermodynamic and blocking factors and second is the influence of pure
hydrostatic pressure on diffusion.In the present work the two factors
were separated using pressure-concentration isotherms for the $Pd-H_2$
and $Pd-D_2$ systems[6,9]. In this way thermodynamics factors were determined,what enabled us to calculate Einstein's diffusion coefficients
and evaluate the influence of pure hydrostatic pressure on diffusion.
The parameter describing influence of hydrostatic pressure on diffusion: activation volume is presented in fig.2.Numerical values of the
activation volumes are not high,which is in agreement with the fact
that the partial molar volume of hydrogen in β-palladium hydride is
rather small: about 0.5 ccm/mol[8].Extrapolations of the data as pressure and concentration n approach zero allow us to compare the present
results with literature data. The agreement is good. Fig. 3 presents
Arrhenius' plots of diffusion coefficients extrapolated in this way.
The results shown in fig.3 satisfy the following equations:

(1.) $D_H(n,p=0) = 0.0113 \exp(-27.1(kJ/mol)/RT)$ for hydrogen

(2.) $D_D(n,p=0) = 0.0105 \exp(-25.9(kJ/mol)/RT)$ for deuterium

values of D are expressed in cm^2/s.

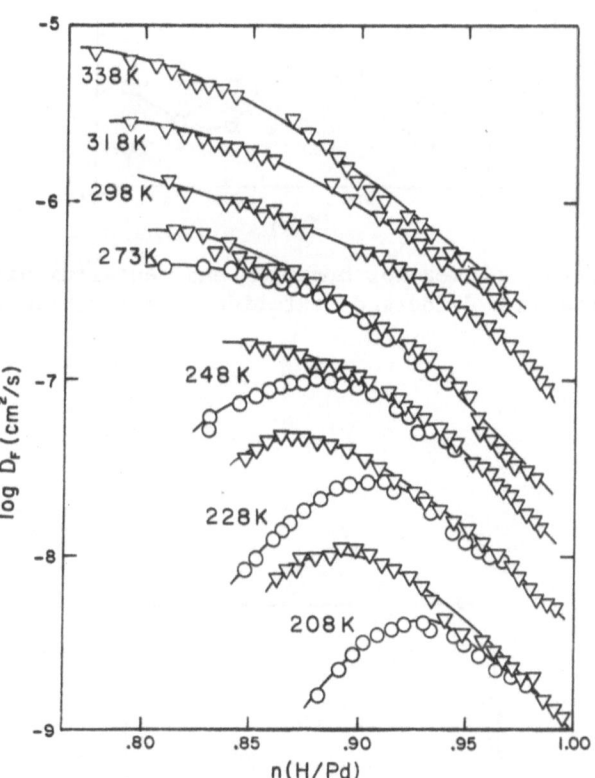

Fig.1. Fick's diffusion coefficient of hydrogen in β-palladium
hydride versus equilibrium pressure of gaseous hydrogen.
Symbols used are explained in the text.

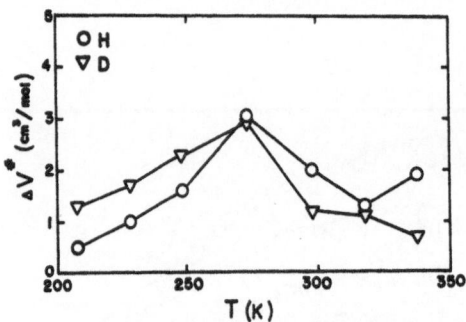

Fig.2. Activation volumes for hydrogen and deuterium diffusion in
       β-palladium hydride and deuteride as a function of temperature.

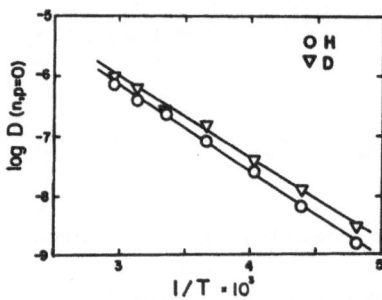

Fig.3. Arrhenius' plot of the diffusion coefficients of hydrogen and
       deuterium in β-palladium hydride and deuteride for n,p=0.

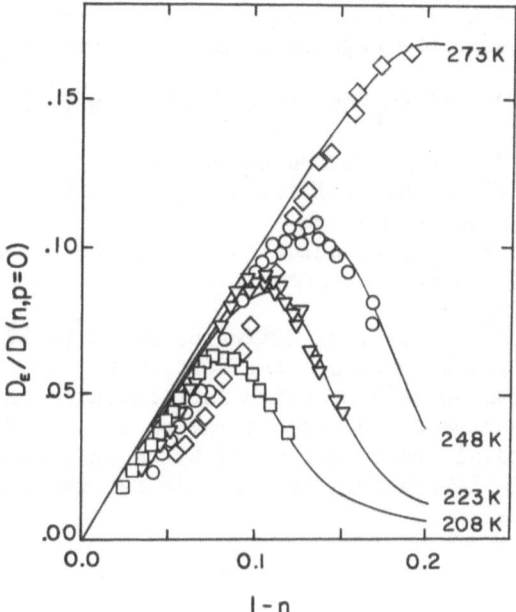

Fig.4. Normalized Einstein's diffusion coefficient $D_E/D(n,p=0)$ as a function of deviation from stoichiometry (1-n). Continuous lines are the result of the clustering model fitted to the experimental data. Values of parameter x are: 0.91 for 208 K, 0.89 for 228 K, 0.83 for 248 K, 0.76 for 273 K, ΔG for reaction (3.) is -11.8 kJ/mol H.

The reasons responsible for the appearance of the maxima are not clear. One of many explanations is based on the fact of hydrogen clustering in palladium hydride lattice. In this way only a part of hydrogen can be assumed as mobile. For a quantitative description knowledge of the numerical value of the derivative dn_H/dn is necessery; n_H denotes concentration of "free", mobile hydrogen. The following formal quasi-chemical equation of hydrogen clustering was used for this purpose:

$$(3.) \quad Pd + xH_{(Pd)} = PdH_x$$

with two fitting parameters: ΔG- free enthalpy change per mol H and the parameter x. Result of this attempt are shown on fig.4.

Finally we present the compact formula describing our data:

$$(4.) \quad D_F = D_o \, \Upsilon (1-n) \; dn_H/dn \; \exp \left(-(\Delta H^{\#} + p\Delta V^{\#})/RT \right)$$

D_F - measured Fick's diffusion coefficient
D_o - constant preexponential factor, equal to 0.0113 for hydrogen
      and 0.0105 for deuterium
Υ - thermodynamic factor, calculated from the p-n-T relationships
      for $Pd-H_2$ and $Pd-D_2$ systems
$(1-n)$ - blocking factor
$\Delta H^{\#}$ - activation enthalpy of diffusion, equal to 27.1 kJ/mol for
      hydrogen and 25.9 kJ/mol for deuterium
$\Delta V^{\#}$ - activation volumes of diffusion(presented in fig.2)
dn_H/dn - derivative of the concentration of the "free" hydrogen
      n_H over total concentration of hydrogen n (discussed in
      text), this is the term responsible for the appearance of
      the maxima

REFERENCES

1. J.Völkl, C.Alefeld,"Hydrogen Diffusion in Metals," in:"Diffusion in Metals:Recent Developments",A.S.Nowick,J.J.Burton,eds, Academic Press,New York (1975)
2. J.Völkl, G.Alefeld,"Diffusion of Hydrogen in Metals,"in:"Hydrogen in Metals I",G.Alefeld,J.Völkl,eds,Springer Verlag,Berlin-Heidelberg-New York (1978)
3. M.Küballa, B.Baranowski, Ber.Bunsenges,78:335(1974)
4. S.Majorowski, Thesis,Warsaw (1980)
5. F.A.Lewis,"The Palladium-Hydrogen System",Academic Press,London (1967)
6. E.Wicke, G.Nernst,Ber.Bunsenges.68:224(1964)
7. M.Tkacz, B.Baranowski,Roczn.Chem.50:2159(1976)
8. B.Baranowski, T.Skośkiewica, A.W.Szafrański,Fiz.Nizk.Temp.1:616 (1975)

PROTON JUMP RATES IN PALLADIUM-HYDRIDE 50-100 K

S. R. Kreitzman, R. L. Armstrong

Department of Physics
University of Toronto
Toronto, Ontario, Canada M5S 1A7

ABSTRACT

Proton jump rates in β-Palladium-Hydride (H/Pd=.82, .65) have been measured using the NMR technique of dipolar relaxation in the strong collision, low temperature regime. The results show distinct negative curvatures in the logarithm of the jump rate versus temperature. This is in complete contrast to the positive curvature encountered in a classical sum of activated processes. Specifically we find the behaviour of the relaxation rate to be of the form $1/T_{1D} = \text{const.} T^{\nu} \cdot \exp(-T_a/T)$, with

$$\nu(.82) = -8.4 \pm 1, \quad T_a(.82) = 1410 \pm 100 \text{ K}$$
$$\nu(.65) = -5.7 \pm .7, \quad T_a(.65) = 1284 \pm 60 \text{ K}.$$

These results are discussed in the light of the small polaron theory of quantum diffusion.

INTRODUCTION

In this work we present the first experimental results which measure directly the low temperature proton jump rates for a metal-hydride (MH) system. The method used relies on the fact that below a certain characteristic temperature (120K for β-PdH) the proton jump times become slow with respect to the average precession period (T_2) of a proton located in the local magnetic field of its neighbours. Thus the jumping of a proton to a neighbouring vacancy is seen by the surrounding nuclei as a sudden, but infrequent change in the spin Hamiltonian which governs the time evolution of the system.

Insofar as the nuclear spin system is very weakly coupled to the
lattice, the internal spin degrees of freedom establish a 'quasi-
equilibrium' (i.e. a state of maximum spin entropy consistent with
a given spin energy) and can be described by a spin temperature,
$1/\beta$. Modulations of the spin Hamiltonian, on a time scale slow
with respect to the time required to establish the internal equilibrium ($\sim 5T_2$), result in the quasi-equilibrium evolution of the
nuclear spin temperature toward the temperature of the lattice.
The theoretical analysis has been applied first to the self-diffu-
sion of Li^2 [3], and subsequently to a host of slow motional effects
in solids[4].

The dipolar relaxation rate ($1/T_{1D}$) follows directly from the
fact that the dipolar energy reservoir rethermalizes after every
jump. This leads[2] [3] [4] to the equation

$$1/T_{1D} = \text{const} \cdot (1/\tau) \qquad \qquad (i)$$

with τ the mean jump time for a single proton. Although the exact
value of the constant depends very much on the jumping process, one
can immediately see that it is of order unity. In the time τ, all
the protons will probably have jumped once, and therefore the dipolar order, which is basically a nearest neighbour spin correlation, will have been randomized for almost all of the spins. We
take the result (i) as fundamental. Below 100 K in β-PdH the dipo-
lar rate is less than 10^3 s^{-1} compared to $1/T_2 \sim 5 \times 10^4$ s^{-1}, thus fully
justifying the assumption leading to equation (i).

EXPERIMENTAL

Measurement of the dipolar signal has been carried out using
a Jeener-Broekaert[5] three pulse sequence. Data simulation, digiti-
zation and signal averaging were carried out under the control of
an IEEE 488-1975 based system controlled by a Tektronix 4051. The
magnetic field was stabilized using an external NMR lock. Sample
preparation and the technique of using the pulse sequence have been
described previously[6] [7] in works dealing with other aspects of the
proton NMR in this system. All the data have been analysed using
a non-linear three parameter fit to a single exponential.

RESULTS

Generally an NMR relaxation rate contains contributions from
a number of sources. For hydrogen in metals the only contributors
are motion (of the hydrogen) and the electron proton electromagne-
tic coupling. Fourtunately the two rates have a completely different temperature behaviour, thereby allowing their separation in
most circumstances. The data, Figure 1, is presented with the
electronic contribution to the rate removed[7] so that attention can
be focussed on the motion. Multiple runs are plotted on top of
themselves. Errors in the points are smaller than the symbol size.

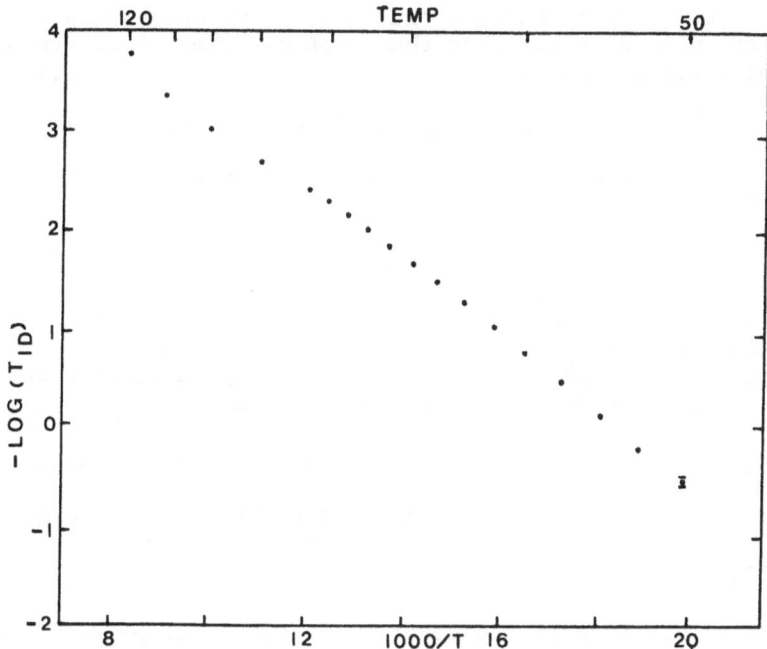

Figure 1.   The dipolar relaxation rate for PdH(.82) between 45 and
120K.   The electronic terms, dominant below 55K, have been removed.

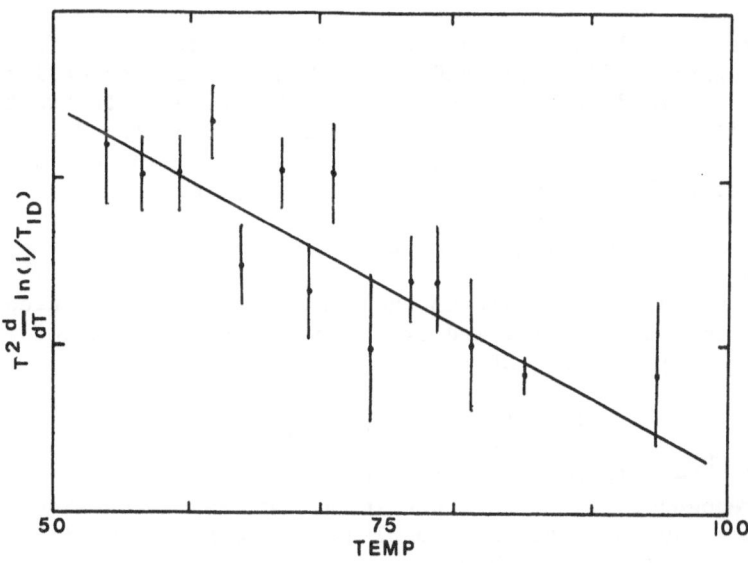

Figure 2.   Dipolar data recast to display a temperature dependence
of the form $T^{\nu} \cdot \exp(-T_a/T)$.

In attempting to extract the temperature dependence of the data we are guided by the fact that most theories of diffusion, both semi-classical and quantum[8], display a transition rate, W, of the form

$$W = const \cdot T^{\nu} \cdot exp(-T_a/T)$$

Assuming this is the form of $1/T_{1D}$ below 100 K we can extract the coefficients ν and T_a by plotting

$$T^2 \frac{d}{dT}[\ln(1/T_{1D})] \text{ vs } T.$$

The resulting plot is shown in Figure 2 for the H/Pd(.82) data below 100 K. The derivative of the logarithm is taken as $[\ln(1/T_{1D}^i) - \ln(1/T_{1D}^j)]/[T^i - T^j]$, for i and j successive data points. From the slope and intercept we find

$$\nu(.82) = -8.4 \pm 1, \; T_a(.82) = 1410 \pm 100 \text{ K},$$

and

$$\nu(.65) = -5.7 \pm .7, \; T_a(.65) = 1284 \pm 60 \text{ K}.$$

DISCUSSION

Interpretations of the negative curvature displayed in the data, must at present be speculative. To our knowledge no such curvature has not been previously reported. Furthermore, similar data on other metal-hydride systems is also completely lacking for comparison. However, two theoretical treatments of small polaronic motion may shed light on the processes.

Kagan and Klinger[9] have presented a density matrix formalism. For $T << (\hbar\omega_D/k)$, $\equiv 325$ K for PdH, they show that the so-called coherent diffusion dominates the movement of protons. In treating coherent diffusion, Kagan and Klinger have distinguished the band motion, at very low temperatures, with motion arising out of localized states. The diffusion coefficient is given as

$$D \propto (\frac{\hbar\omega_D}{T})^9 exp[-2\Phi(T)]. \tag{ii}$$

The quantity in the exponent is the probability for a coherent translation of the interstitial and its accompanying lattice deformation to a neighbouring site. It is given approximately by

$$\Phi(T) = (.5E_D/\hbar\omega_D)[2n(\omega_D) + 1]$$

$n(\omega) = 1/(e^{\beta\hbar\omega}-1)$ is the lattice occupation number of a typical phonon frequency. The function Φ is an increasing function of temperature, and we therefore conclude that expression (ii) cannot describe the diffusion in PdH.

A recent calculation by Teichler[10], however, may have more bearing. This author has included the effect of the lattice defor-

mation and the phonon fluctuations on the overlap integral. His
basic result may be written as

$$W = \frac{1}{\tau} = \frac{1}{h} \frac{\pi}{4E_a kT} <|J^2|>e^{-E/kT}$$

with E containing a dependence on E_a, the energy required to move
the distortion field. The form of $<|J^2|>$ is calculated by the
author for temperatures around and below the Debye temperature.
His results are reproduced in Figure 3. The parameter q measures
the correlation between fluctuations of the barrier height and the
interstitial energy levels; q=0 is the completely correlated case.

    Teichler has modeled the lattice with an Einstein frequency
spectrum. Taking ω_E to be the Debye frequency, we find that in
the 50-100 K region the value of J^2 is a decreasing function of
temperature for small q. As the correlation decreases, the
transition rates increase and show less negative curvature, with
the bend to positive curvature moving toward lower T. The
behaviour is similar to that observed in PdH with decreasing concentration.

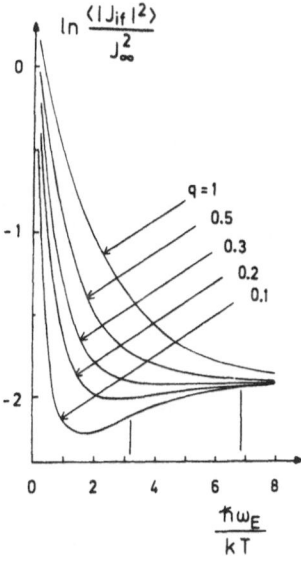

Figure 3[10]. Logarithmic plot of $<|J^2|>$ as a function of $\hbar\omega_E/kT$ for
              different values of q. The vertical bars indicate the
              50 to 100 K interval for PdH.

CONCLUSION

    We have presented the first direct measurements of proton jump
rates in a metal-hydride system. The negative curvature of the
logarithm of the jump rate is at first a surprising result. As

shown by Teichler, it is consistent with a correlation between the phonon fluctuations and the interstitial energy levels. It is hoped these results will stimulate a further interest in obtaining such experimental data from similar systems.

REFERENCES

1. M. Goldman. Spin Temperature and Nuclear Magnetic Resonance in Solids. Oxford University Press, London (1970).

2. C. P. Slichter, D. Ailion. Phys. Rev. 135, A1099 (1964).

3. D. C. Ailion, C. P. Slichter. Phys. Rev. 137, A235 (1965).

4. D. C. Ailion. Advances in Mag. Res. 5, 177 (1971).

5. J. Jeener, P. Broekaert. Phys. Rev. 151, 232 (1967).

6. S. R. Kreitzman, R. L. Armstrong. Phys. Rev. B. Rapid Comm. To appear February, 1982.

7. S. R. Kreitzman, R. L. Armstrong. Phys. Rev. B. Rapid Comm. To appear February, 1982.

8. J. A. Sussman. Ann. Phys. (France) 6, 135 (1971).

9. Yu Kagan, M. F. Klinger. J. Phys. C. 1, 2791 (1974).

10. H. Teichler. Z. Phys. Chem. Neue Folge 114, 201 (1979).

THEORY OF THE DIFFUSION OF HYDROGEN IN METALS

K.W. Kehr

Institut für Festkörperforschung
Kernforschungsanlage Jülich
517o Jülich, Federal Republic of Germany

INTRODUCTION

Hydrogen in metals has large diffusion coefficients compared to other interstitials.[1] They are particularly large in the bcc metals V, Nb, and Ta; see Fig.1. In Nb and Ta the isotope H shows a lower activation energy for diffusion below 25o K, resulting in a large isotope effect at lower temperatures. In the last few years the study of positive muons has added new, often unexpected, information on the motion of a light isotope of hydrogen ($m_\mu/m_H \approx 1/9$). This gives strong incentives for considering the possible diffusion mechanisms of light particles in a metal and to compare theoretical models with experiments on actual systems. Many idealizations are customary in the theoretical description, for instance the restric-

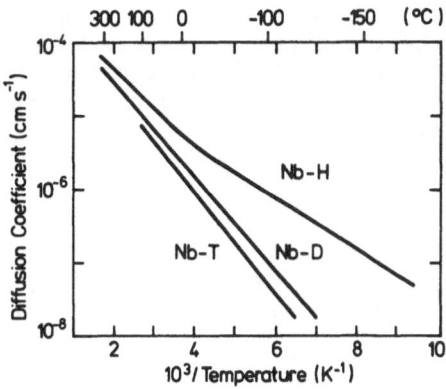

Fig.1. Diffusion coefficients of H, D, and T in niobium according to the most recent measurements of Qi et al.[2]

tion to the motion of one particle in an ideal crystal. This
idealization will also be made here. This restriction also implies
that only long-range diffusion is discussed, and not motion around
defects, although both problems have many common features.

OVER-BARRIER PROCESSES

Separation between Transition Rate and Lattice Diffusion

It is instructive to consider the motion of a light interstitial in a metal first from the point of view of classical stochastic
transport theory. Classically one has to study the motion of a particle in a periodic potential, such as sketched in Fig.2. This picture is only applicable for an infinitely heavy particle in an
ideal, classical crystal at zero temperature. At finite temperature
the lattice atoms perform thermal vibrations; this provides additional stochastic forces, which enable the motion of the particle
over the potential barriers. The appropriate theoretical description
is the Fokker-Planck equation, which has been intensively studied
for the periodic potential.[3]

A common simplification in studying diffusion in metal physics
is the restriction to the jumps of a particle between neighboring
potential minima. This is justified when the potential barriers are
large compared to thermal energies. The particle will then spend
only a small fraction of time in continuous motion near the tops
of the potential barriers. It is further assumed that a particle
while performing a jump has no memory to its previous jump. This
requires essentially that lattice relaxation processes are rapid
compared with the jump rates. If these assumptions hold the interstitial diffusion problem can be separated into two parts: (i) the
problem of determining the rate of jumps between two potential
minima, and (ii) the problem of diffusion in the lattice of interstitial sites when the transition rates are known. The latter
problem is almost trivial in ideal Bravais lattices; a master equation is set up and easily solved.[4] Also the extension to non-Bra-
vais lattices is well-known.[5]

Fig.2. Potential energy of an interstitial particle along one
      representative coordinate.

The result of the application of master equations to the diffusion of hydrogen are incoherent dynamical structure functions $S_{inc}(\vec{q},\omega)$ where $\hbar\vec{q}$ is the momentum transfer and $\hbar\omega$ the energy transfer. A comparison of $S_{inc}(\vec{q},\omega)$ with the experimental quasielastic scattering cross sections yielded information on the lattice of interstitial sites on which the hydrogen atoms move. It turned out that hydrogen jumps between octahedral sites in the fcc lattice of Pd[6] with jump distance d of about 2.7 Å while the equilibrium positions in the bcc lattices are the tetrahedral sites[7] with d of about 1.1 Å.

Modified Classical Rate Theory

The classical rate theory for transitions of a system between two potential minima was originated by Eyring[8] and applied to diffusion in solids by Wert and Zener[9] and Vineyard.[10] It is assumed that all energy values of the system occur with probabilities deduced from Boltzmann statistics. The transition rate Γ is obtained from the probability of finding the system near a saddle-point configuration, multiplied by the current over the saddle-point where return processes of the particle are neglected. The result has the form of an Arrhenius law, $\Gamma = \Gamma_0 \exp(-E_a/k_B T)$, where Γ_0 is a renormalized attempt frequency and E_a is given by the difference between potential energy at the top of the barrier (saddle-point) and the minimum, when the difference between activation enthalpy and energy is disregarded.

As a light particle hydrogen can perform localized vibrations with widely separated energy levels $\hbar\omega_0$, well above the vibrational energies of the lattice. Hence the classical rate theory must be modified for hydrogen in order to take the discreteness of the energy levels into account. This modification has been done by Le Claire[11] and Ebisuzaki et al.[12] by replacing classical with quantum-mechanical partition functions. The main features of this theory are: (i) The activation energy is the difference of potential energy between saddle-point and minimum configuration, plus corrections due to zero-point energies in the saddle-point and minimum configurations. (ii) The prefactor Γ_0 shows a classical isotope effect for $\hbar\omega_0 \ll k_B T$ provided that the ΔK-effect is small,

$$\Gamma_0^H/\Gamma_0^D \approx (m_H/m_D)^{1/2}. \tag{1}$$

ΔK is the fraction of kinetic energy contained in the particle's motion over the saddle-point, relative to the kinetic energy in all degrees of freedom.[13] Arguments have been given by Katz et al.[14] that ΔK is nearly 1 for hydrogen diffusion, leading to the classical isotope effect Eq.(1). Indeed the isotope effect expressed by this equation has been confirmed by experiments on fcc metals,[14] for a review see Ref.1. The observations have shown a reversed isotope effect for the activation energy, i.e. the lowest activation energy for T, followed by D and H. An exception is Pd-T.[15] The reversed isotope effect can be fitted with the modified classical theory by an appropriate choice of the vibrational frequencies[16] in the minimum and saddle-point configurations, however the physical origin is not well understood.

In the limit of very widely spaced localized vibrations when $\hbar\omega_0 \gg k_B T$, the modified classical theory leads to a universal prefactor[12,17] $\Gamma_0 = k_B T/2\pi\hbar$. The frequencies of the localized vibrations in the bcc metals are of the order of 100 meV, hence the limiting conditions is fulfilled at room temperature. Indeed all group Vb-metals have approximately the same prefactor above room temperature within the uncertainties of the experiments, independent of the isotope.[2] Also $Pd_{0.47}Cu_{0.53}$ which forms a bcc phase shows[18] a comparable prefactor. Further the observed isotope effects of the activation energies can be qualitatively understood from the differences of the zero-point levels in the minimum and saddle-point configurations.[17] However, there are serious objections to such a simple interpretation of the bcc results.

Failure of Assumption of Uncorrelated Jumps

The jump rates of hydrogen in the bcc metals can be extremely large, e.g., a proton in niobium performs about 10^{12} jumps per second at 125 °C. Although this rate is still smaller than typical lattice frequencies, it is not well separated from them, hence one expects memory effects in the jump process. Also the mean transit time of a hydrogen atom between two sites is not orders of magnitude smaller than the mean residence time at a site.[19] A clear experimental demonstration of the breakdown of the simple jump model has been given by Lottner et al.[20] They have attempted to fit the quasielastic neutron scattering data on Nb and other bcc metals with $S_{inc}(\vec{q},\omega)$ resulting from the model of nearest-neighbor jumps between tetrahedral sites. The result of this attempt is a jump rate Γ which depends apparently on the momentum transfer $\vec{q}$ whereas all $\vec{q}$-dependences should be included in other parts of $S_{inc}(\vec{q},\omega)$, and not appear in Γ. Satisfactory fits of the data are possible with a model which includes jumps to all topologically second neighbors, and other models with correlated jumps.[20] The contributions of second-neighbor jumps become large at elevated temperatures. This feature indicates that there is a tendency to persistence of motion of the hydrogen atoms: once they have left an equilibrium site they continue to diffuse before they come to rest again. It is clear that a description of hydrogen diffusion in those metals in terms of transition rates between neighboring sites is no longer appropriate at elevated temperatures. A theoretical description of the correlated jumps of a hydrogen atom at higher temperatures beyond the phenomenological level seems to be an extremely difficult task.

UNDER-BARRIER PROCESSES

Standard Small-Polaron Theory

The motion of a light particle in a crystal is fundamentally a quantum-mechanical problem. If one estimates the transfer matrix element J for tunneling of a proton between interstitial sites one obtains values well below 1 meV in the bcc metals and 10^{-6} meV in the fcc metals. The values also depend strongly on isotope mass. On the other hand, when a hydrogen atom is localized at a particular site it exerts forces on the neighboring lattice atoms which lead to a relaxation of the lattice around it, and a corresponding energy gain of the order of 100 meV. Direct tunneling of the hydrogen atom together with the relaxed lattice configuration is still possible, however, the effective tunneling transfer element J_{eff} is greatly reduced compared to J. In this situation the "small-polaron process" can induce transitions of a hydrogen atom between adjacent interstitial sites at moderate temperatures. This process is depicted schematically in Fig.3. The small-polaron process has been comprehensively studied for electrons by Holstein[21] and adapted to hydrogen diffusion by Flynn and Stoneham.[22] The predictions of the theory are: (i) At high temperatures $k_BT \gtrsim \hbar\omega_D$ where ω_D is the Debye frequency of the host lattice

$$\Gamma = \left(\frac{\pi}{4E_a k_B T} \right)^{1/2} \frac{J^2}{\hbar} \exp \left(- \frac{E_a}{k_B T} \right) \tag{2}$$

Here E_a is the activation energy necessary to create the configuration shown in Fig.3(b). It can be estimated from the Kanzaki forces of a hydrogen atom and the host phonon frequencies.[17] Two typical estimates are 27 meV for Nb and 40 meV for Al. Several phonons are necessary to create the configuration of Fig.3(b) and Eq.(2) represents a multiphonon process. (ii) At low temperatures $k_BT \ll \hbar\omega_D$ for an isotropic solid in the Debye approximation

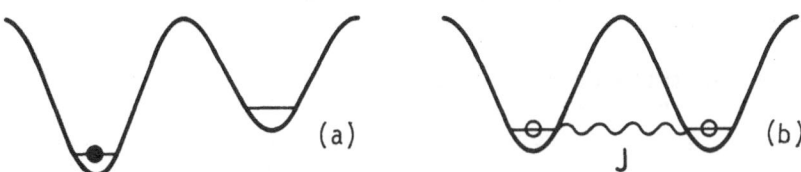

Fig.3. Small-polaron process. (a) The ground-state level of a hydrogen atom localized at a site is lowered compared to an adjacent site. (b) Thermal fluctuations have produced a coincidence of both sites and the atom can tunnel.

$$\Gamma = 5.76 \times 10^5 \frac{E_a^2 J_{eff}^2}{\hbar^4 \omega_D^3} \left(\frac{k_B T}{\hbar \omega_D} \right)^7 \tag{3}$$

where E_a is the activation energy appearing in Eq.(2). Eq.(3) represents a two-phonon process. One phonon is absorbed and creates a virtual coincidence configuration; it is emitted at the end of the transition process. It is important to point out that the assumption of uncorrelated jumps between nearest-neighbor sites is implicitly made in small-polaron theory. This assumption becomes violated when the tunneling matrix elements become sufficiently large. It might also be questionable in the case that lattice-activated processes – which will be discussed in the following section – become important.

Standard small-polaron theory cannot describe the experimental results for hydrogen diffusion in the fcc metals, and in the bcc metals above room temperature. The prefactor Γ_0 of the experimental jump rates shows the classical isotope effect in fcc metals or nearly no isotope effect in bcc metals, whereas Eq.(2) would predict a strong isotope effect in the prefactor. The observed prefactor in the fcc metals is many orders of magnitude larger than estimated from Eq.(2). Also the detailed comparison of experimental with theoretical activation energies gives arguments against the application of Eq.(2) in both cases.[17]

Standard small-polaron theory seems to be applicable to the diffusion of positive muons in copper, aluminum, and iron above about 50 K. Teichler[23] could describe the results on copper with an interpolation formula between Eqs.(2) and (3), with E_a = 75 meV and J = 0.018 meV. This should be compared with the activation energy of 403 meV of hydrogen diffusion in copper.[14] Herlach[24] could make an indirect inference on the diffusion coefficient of muons in Al above 50 K from an analysis of trapping on irradiation-produced vacancies. He also found Arrhenius behavior with an activation energy of 32 meV; application of Eq.(2) suggests $J \approx 1$ meV. Graf et al.[25] and Möslang[26] found activated hopping of muons in iron between 35 and 150 K; the activation energy is 46 meV above 95 K.

While thermally activated behavior according to Eq.(2) has been observed with muons in several instances, the T^7-behavior predicted by Eq.(3) has not yet been found. Our group[27] concluded from diffusion-controlled trapping on substitutional impurities in Al that the diffusion coefficient of muons in Al is proportional to T between 1 and 15 K, in sharp contrast to the predictions of Eq.(3). It is unclear at present whether there is an intrinsic deficiency in the small-polaron theory at low temperatures, or whether some special circumstances such as alternating transitions between tetrahedral and octahedral sites allow one-phonon processes in Al, leading to linear behavior of the diffusion coefficient with temperature.

Extensions of Small-Polaron Theory

Already Flynn and Stoneham[22] pointed out that the dependence of the tunneling matrix element on the lattice coordinates Q should be included in the applications of small-polaron theory to hydrogen diffusion. In pictorial terms, the motion of lattice atoms near the path of the moving atom might "open the door" for a large transfer matrix element. According to these authors, the "lattice-activated processes" are important for fcc metals, not for transitions between octahedral sites in bcc metals. It seems unlikely that the hydrogen diffusion data - except for muons - in fcc metals can be explained by this extension, without invoking higher excited states. However, lattice-activated processes should be especially important for the jumps of hydrogen between the tetrahedral sites in bcc metals, where the estimated tunneling transfers are much larger than in fcc metals already for fixed lattice coordinates. Several authors[28,29] have considered the contributions arising from the dependence of J on Q. The most advanced calculation with respect to an explanation of hydrogen diffusion data in the bcc metals is the one by Emin et al.[28] The authors use the occurrence probability approach[21] where the (classical) probability of having a coincidence configuration (see Fig.3(b)) is multiplied by the transition probability in case of coincidence. The contributions of excited levels are also included in their approach. They use a phenomenological interaction between the hydrogen and the host atoms and derive the Q-dependence of J by solving a (one-dimensional) Schrödinger equation for various lattice configurations. The result is a qualitative agreement with the experiments in bcc metals, especially niobium, where the decrease of the activation energy of H below room temperature is obtained.

The approach of Emin et al. is not only valid in the non-adiabatic limit where J is the smallest parameter, it also applies to the transition region to the adiabatic case, when J becomes large. Their high-temperature results are dominated by transitions in the excited levels, in the adiabatic limit, where their theory yields $\Gamma_0 \propto \omega_D$. The diffusion data in bcc metals above room temperature are described by the authors in this way, however, there are doubts that this part of the theory is complete. The possibility of correlated motion has been disregarded as well as the transitions between energetically different levels. The experimental results do not indicate the trend with ω_D suggested by this theory. Moreover the relation to the modified classical rate theory which should emerge as a limiting case from the quantum-mechanical description, is not yet understood.

Coherent Diffusion

As mentioned above, there is the possibility of tunneling of a hydrogen atom, together with the relaxed lattice configuration, without thermal activation and in ideal crystals. At zero temperature the particle would spread out over the whole crystal. At finite temperature a finite diffusion coefficient exists, determined by scattering processes with thermal excitations. This diffusion coefficient decreases with increasing temperature. Kagan and Klinger[30] derived a T^{-9}-behavior for the diffusion coefficient from scattering processes with phonons while the author[31] derived an explicit estimate for the coefficient of the T^{-9}-law using a slightly different phonon interaction process. The estimated scattering rates turn out to be extremely small at low temperatures.

Coherent diffusion – here always understood as a long-range diffusion process – has not yet been observed for hydrogen in metals and it will perhaps never be observed in quasi-stationary experiments. One practical reason is that one cannot investigate small concentrations of hydrogen because of precipitation. Also interaction effects between the hydrogen atoms, and with imperfections, would negatively influence the coherent transfer. In the dilute limit, atoms would always be caught by inevitable imperfections. Positrons do exhibit coherent propagation, but they have too short lifetimes and too small mass for a meaningful comparison with hydrogen.

Positive muons have offered the opportunity to study diffusion in metals in the dilute limit and at very low temperatures. In addition, extremely pure and well characterized metal samples have been prepared. The technique of random implantation provides a non-equilibrium situation in which transport properties are investigated. It is discussed whether the muon finds itself in a metastable state after the implantation process and whether the transition from the metastable to the stable state affects the apparent transport processes.[32,33] These possibilities will be disregarded here. The purest samples of aluminum and niobium show absence of a damping signal, which would indicate immobile muons, down to 30 mK (Al)[34] and 1 K (Nb).[35] This indicates that motional processes occur at temperatures where thermal activation is extremely unlikely, and gives indirect evidence for coherent diffusion. Graf et al.[25] have also obtained evidence for coherent diffusion of muons in iron below 35 K.

Additional information on coherent diffusion is obtained by studying controlled amounts of impurities in Al.[27] The observed increase of the damping parameter in the samples with impurities below 1 K can be attributed to trapping, limited by coherent diffusion. The analysis yields a coefficient of coherent diffusion which shows proportionality with $T^{-0.6}$, cf. also Fig.4. This behavior is in evident disagreement with the predicted T^{-9}-law.[30,31] It can be

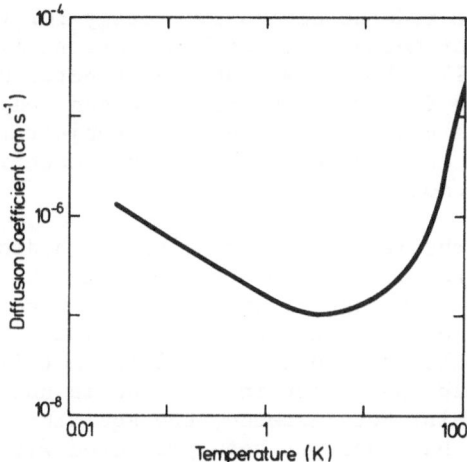

Fig.4. Diffusion coefficient of muons in aluminum as deduced indirectly from diffusion-limited trapping.

qualitatively understood from the scattering of electrons on the
muons. Long ago Andreev and Lifshitz [36] pointed out a T^{-1}-behavior
of the coefficient of coherent diffusion in a metal, as a result of
this scattering process. More quantitative calculations are necessary before the applicability of this process to the observations
can be ascertained.

CONCLUSIONS

The main qualitative features of the diffusion of hydrogen (H,
D, T) in the dilute limit are understood. Quite different processes
can dominate, depending on isotope mass, potential barriers, and
temperature. A summary of the main processes is given in the Table.
Some distinctions are not incorporated in this Table, especially
the diffusion process of H in bcc metals below room temperature
where probably lower states contribute more strongly than for D and

Table 1. Diffusion processes oh hydrogen metals.

Crystal structure Distance of sites	H, D, T	Muons
fcc $d \approx 2.5$ Å	Over-barrier	Small polaron Coherent diffusion
bcc $d \approx 1.1$ Å	Higher excited states Lattice activated	Small polaron Coherent diffusion

T; resulting in the reduced activation energy for this isotope. The diffusion process of muons at lower temperatures is only partially clarified. Especially the transition region between coherent and thermally activated diffusion and the observed temperature dependences are not well understood. One might conjecture that coherent diffusion always occurs in sufficiently pure materials, but this has not yet been demonstrated.

Quantitative theoretical descriptions of hydrogen or muon diffusion in metals are almost completely missing. Most urgently needed is a better knowledge of the interaction potentials between hydrogen atoms or muons and the host metal atoms. While this is certainly a difficult task in the transition metals of interest for hydrogen, much of the pertinent information is not even available in the simpler metals. For instance, the equilibrium positions of a muon in aluminum have not yet been predicted with reliability. If the interaction potentials were more precisely known, the parameters of the diffusion theories could be calculated for specific metals from first principles, for example activation energies for over-barrier or small-polaron diffusion, or tunneling transfer elements. Especially tunneling matrix elements are very sensitive to the exact form of the interaction potentials. Hopefully such calculations will be performed in some exemplary cases in the near future.

ACKNOWLEDGMENTS

I have greatly benefitted from the collaboration with O. Hartmann, E. Karlsson, T.O. Niinikoski, L.O. Norlin, D. Richter, J.M. Welter, and A. Yaouanc. I thank Zh. Qi, J. Völkl, R. Lässer, and H. Wenzl for the permission to use unpublished data.

REFERENCES

1. For a review see J. Völkl and G. Alefeld, in "Hydrogen in Metals I," Topics in Applied Physics, Vol.28, G. Alefeld and J. Völkl, eds., Springer, Berlin (1978).
2. Zh. Qi, J. Völkl, R. Lässer, and H. Wenzl, to be published.
3. For a review see T. Geisel, in "Superionic Conductors," Topics in Current Physics, Vol.15, M.B. Salamon, ed., Springer, Berlin (1979).
4. C.T. Chudley and R.J. Elliott, Proc.Phys.Soc.(London) 77:353 (1971).
5. J.M. Rowe, K. Sköld, and H.E. Flotow, J.Phys.Chem.Solids 32:41 (1971).
6. J.M. Rowe, J.J. Rush, L.A. de Graaf, and G.A. Ferguson, Phys. Rev.Lett. 29:125o (1972).
7. V. Lottner, U. Buchenau, and W.J. Fitzgerald, Z.Physik B 35:35 (1979).

8. H. Eyring, J.Chem.Phys. 3:1o7 (1932).
9. C. Wert and C. Zener, Phys.Rev. 76:1169 (1949).
1o. G.H. Vineyard, J.Phys.Chem.Solids 3:121 (1957).
11. A.D. Le Claire, Phil.Mag. 14:1271(1966).
12. Y. Ebisuzaki, W.J. Kass, and M. O'Keeffe, J.Chem.Phys. 46:1373 (1967); Phil.Mag. 15:1o71 (1967).
13. J.G. Mullen, Phys.Rev. 121:1649 (1961).
14. L. Katz, M. Guinan, and R.J. Borg, Phys.Rev.B 4:33o (1971).
15. G. Sicking and H. Buchold, Z.Naturforsch. 26a:1973 (1971).
16. W. Jost and A. Widmann, Z.Phys.Chem. B29:247 (1935); B45:285 (194o).
17. K.W. Kehr, in "Hydrogen in Metals I," Topics in Applied Physics, Vol.28, G. Alefeld and J. Völkl, eds., Springer, Berlin (1978).
18. G. Lang, Diploma thesis, unpublished, Techn.Univ.München (1976).
19. L.A. de Graaf, J.J. Rush, H.E. Flotow, and J.M. Rowe, J.Chem. Phys. 56:4574 (1972).
2o. V. Lottner, J.W. Haus, A. Heim, and K.W. Kehr, J.Phys.Chem. Solids 4o:557 (1979).
21. T. Holstein, Ann.Phys.(NY) 8:325;343 (1959).
22. C.P. Flynn and A.M. Stoneham, Phys.Rev.B 1:3966 (197o).
23. H. Teichler, Phys.Lett. 64A:78 (1977).
24. D. Herlach, in "Recent Developments in Condensed Matter Physics," Vol.1, J.T. Devreese, ed., Plenum, New York (1981).
25. H. Graf, G. Balzer, E. Recknagel, A. Weidinger, and R.I. Grynszpan, Phys.Rev.Lett. 44:1333 (198o).
26. A. Möslang, unpublished (198o).
27. K.W. Kehr, D. Richter, J.M. Welter, O. Hartmann, L.O. Norlin, E. Karlsson, T.O. Niinikoski, J. Chappert, and A. Yaouanc, in "Proceedings of the Second International Topical Meeting on Muon Spin Rotation," J.H. Brewer and P.W. Percival, eds., North Holland, Amsterdam (1981).
28. D. Emin, M.I. Baskes, and W.D. Wilson, Phys.Rev.Lett. 42:791 (1979).
29. H. Teichler, phys.stat.sol.(b) 1o4:239 (1981).
3o. Yu. Kagan and M.I. Klinger, J.Phys.C 7:2791 (1974).
31. K.W. Kehr, Suppl.Trans.Japan Inst. Met. 21:181 (198o).
32. D. Emin, in "Proceedings of the Second International Topical Meeting on Muon Spin Rotation," J.H. Brewer and P.W. Percival, eds., North Holland, Amsterdam (1981).
33. A.M. Browne and A.M. Stoneham, to be published.
34. O. Hartmann, E. Karlsson, L.O. Norlin, T.O. Niinikoski, K.W. Kehr, D. Richter, J.M. Welter, A. Yaouanc, and J. LeHéricy, Phys.Rev.Lett. 44:337 (198o).
35. T.O. Niinikoski, O. Hartmann, E. Karlsson, L.O. Norlin, K. Pernestål, K.W. Kehr, D. Richter, E. Walker, and K. Schulze, in "Proceedings of the First International Topical Meeting on Muon Spin Rotation," F.N. Gygax, W. Kündig, and P.F. Meier, eds., North Holland, Amsterdam (1979).
36. A.F Andreev and I.M. Lifshitz, Sov.Phys.-JETP 29:11o7 (1969).

SIMPLE SMALL POLARON MODEL FOR THE HYDROGEN CONCENTRATION

DEPENDENCE OF HYDROGEN DIFFUSION IN Nb

D. L. Tonks and R. N. Silver

Theoretical Division and Physics Division
Los Alamos National Laboratory*
Los Alamos, NM 87545

ABSTRACT

We have generalized simplified small polaron models for the
tunneling of interstitials in solids to include interactions
between the diffusing particles. This is applied to the calculation of the average hopping rate of hydrogen in Niobium as a
function of hydrogen concentration for small concentrations, i.e.
for an H/Nb ratio (c) of .06 or less. The hopping of a single H
under the influence of nearby, stationary H's was treated. The
interactions between interstitials include a hard-core repulsion
and a lattice-mediated strain interaction. The tunneling transfer
integral was taken to depend on the displacements of nearby Nb
atoms. Only tunneling via the ground vibrational level of the
interstitials was treated. For $c \leq .06$, the calculated increase
in the hopping activation energy as a function of c was linear in
c and comparable in magnitude with the experimental increase.
Our simplified-model results show that the strain interaction
between H's in Nb is important for their diffusion and that this
interaction needs to be included in whatever more elaborate
diffusion models are developed.

INTRODUCTION

For several years,[1] it has been known that pronounced hydrogen
(H) concentration effects exist in the room temperature diffusion
of H interstititals in Nb. No detailed theoretical analysis of
such effects has yet been made, to our knowledge. However, for
the dilute-H case, a recently proposed modified small polaron
tunneling model,[2] appears to be a promising candidate for the

543

diffusion mechanism, and so it would be of interest to generalize this model to include diffusion concentration effects. The purpose of this paper is to so generalize a simplified version of the proposed model for low concentrations, i.e., c(H/Nb ratio) < .06. Our simplified model includes defect-lattice coupling linear in Nb displacements, and tunneling between the ground vibrational levels of the H. The effects on the motion of a given H due to the strain fields of nearby H's will be included. Both changes in the transfer integral and the difference in initial and final defect energies caused by these strains will be included. Our calculations do not include all the modifications of Ref. 2, in particular, no tunneling between the excited vibrational states of the H is included. However, our simplified model can account for a major part of the observed increase in the activation energy with higher H concentrations. Hence, the strain field effects are sufficiently important to require inclusion in whatever more complete models are proposed. This is the main conclusion of this paper.

The increase in activation energy arises in part because the strain interaction between H's is attractive and stronger at close range (causing H-clustering) and in part because the hard cores of nearby H's have a blocking effect on the motion of a given H. In the former effect, due to the clustering and blocking, an H must, on the average, move away from a nearby H to a site of higher interaction energy. The blocking raises the effective or average activation energy because the blocking becomes less pronounced at higher temperatures where the clustering is reduced and this behavior is consistent with an average activation energy increase.

An additional effect cancelled part of the above-mentioned increase in activation energy, but did not dominate it. We assumed that the transfer integral between two H-sites is increased by an expansion of the lattice about these sites, i.e., a "non-Condon effect," and that a sufficiently large expansion is necessary before hopping occurs. This has the result that the strain fields of the other H's near the two sites reduce the activation energy. This is so because the nearby H's will, roughly speaking, already dilate the lattice near the two sites in question and bring the system closer to the hopping condition to begin with.

Nb H_c SYSTEM MODEL AND HOPPING RATE FORMULA

Central to our model for the Nb H_c system is the assumption that, since the H's are lighter and faster than the Nb atoms, they establish quantum vibrational states that adiabatically follow the Nb atomic motions. We further assume, in light of supporting experimental evidence, that the H's vibrate in local modes. The H's tunnel sufficiently seldom that this is a good

approximation. Thus, for the situation in between hops, one can
treat the Nb lattice quasistatically and define an adiabatic
potential energy (APE) for the Nb H$_c$ system in which the Nb
atoms' kinetic energy is neglected but in which the H's are taken
to be in their quantum mechanical ground vibrational states. We
approximate the adiabatic potential by expanding it in a Tyalor
series in the Nb displacments and keeping terms linear and quadratic in the displacements. The linear term has the significance
of constant, H-induced forces on the Nb atoms while the quadratic
term is the lattice harmonic potential energy. We take the
latter to be that of the infinite, pure Nb host lattice. We
assume the same model for the linear term as Horner and Wagner,[3]
i.e., each H induces constant, equal longitudinal forces on its
Nb second nearest neighbors equal in magnitude to .23 of that of
equal, longitudinal forces on the nearest Nb neighbors. The
magnitudes are fixed using the observed, macroscopic H-induced
volume expansion of the lattice, which is caused by the longitudinal forces. A direct hard-core interaction U_{ij} between H's at
sites i and j is assumed that exludes the occupation of up to the
third nearest-neighbors of the interstitial lattice. This and
the longitudinal force model were used by Horner and Wagner[3] to
model the α-α' phase transition in Nb H. The H-induced forces
cause interactions among different H's to arise as the strain
field of one H does work against the longitudinal forces induced
by another. The energy of interaction, ε_{ij}, between two H's at
sites i and j, plays a central role in our formalism later on.
For values, we will use those calculated by Horner and Wagner,[3]
using a Born von Karman model for the Nb lattice, but we will
average those involving the same i-j separation to obtain ε_{ij}
dependent only on the separation. For two H's lying zero through
twelve n.n. shells apart, we use the following averaged values
(the zeroth shell corresponds to the self energy) -4174, -2907,
-1967, -1218, -589, -426, -212, 86, 97, 500, 314, -110, and -16.

To describe the tunneling process, the occurrence probability
approximation of Holstein[4] is used in which the lattice is treated
classically and quasistatically. Within this approximation, an H
vibrates rapidly in a well of the quasistatically moving lattice
until a thermal fluctuation induces a quasistatic lattice distortion such that the system's total adiabatic energy would be
unchanged if the H were moved to the vacant final site. For each
such "coincidence event," the H has a certain probability of
tunneling to the potential final site. The adiabatic potential
energy increase involved in this lattice distortion constitutes
the activation energy (AE) appropriate to the given H-configuration
and the given lattice distortion. If a hard core is blocking the
hop, the AE is infinite and the hop doesn't occur. In the simplest
polaron models, the average AE, i.e., the result after the rate
is averaged over all possible coincidence events, is equal to the
lowest possible coincidence-event AE. In our model, we take not

the coincidence event, of lowest energy, but we assume that a dilation in addition to that of the lowest-lying coincidence event is necessary to sufficiently increase the transfer integral to make a hop possible. This additional dilation, lattice activation, or higher-lying coincidence event requires an additional adiabatic potential energy increase beyond that of the lowest possible, which means a larger average activation energy.

We will average over all dilations beyond a certain higher-lying threshold dilation, which we now describe. It will involve equal dilations about the initial and final H-sites, the dilation about each site having equal n.n. Nb and equal second n.n. Nb longitudinal displacements in the same proportion as the forces of the ε_{ij}. This plausible assumption allows the ε_{ij} to be used in the formalism to also describe the dilation. The overall magnitude of the dilation necessary is determined from the dilute-H case by taking the additional activation energy of lattice activation to be the experimental dilute-H AE (106 meV for T > 290K) minus the calculated minimum AE, e_o, which equals $-(\varepsilon_{JJ}-\varepsilon_{JK})/4$ for a hop between sites J and K. When other nearby, static H's are present, they will cause the extra dilation needed for the hop of a given H to be partially present initially before any thermal fluctuations.   We assume that the transfer integral is sufficiently large during a coincidence event that the probability of a hop during a coincidence event is unity. This is the so-called adiabatic hopping regime.[4] We also assume that all memory of previous hops is lost before another occurs.

We now briefly describe how our hopping rate formula is derived and then write it. We first found the hopping rate for a given initial H-configuration and given higher-lying coincidence event and then integrated the result over all higher-lying events followed by an average over initial H-configurations. The rate for a given event has a prefactor, ω, independent of c, but with a $T^{-\frac{1}{2}}$ dependence. This can be established by working through the coincidence event formalism in Appendix II of Ref. 4, with W_c set to 1 (adiabatic hopping) and with the additional dilation condition added. The AE of the given higher-lying event is obtained by setting the prehop and posthop APE's equal to obtain one condition and then by minimizing the prehop APE subject to this and the condition that the extra dilation be a certain amount. After making use of the approximation $\mathrm{erfc}(x) \approx (x\pi^{\frac{1}{2}})^{-1} \exp(-x^2)$ in the integration (which we numerically showed to be valid in our case) we obtain for $W^{JK}(\{n_\ell^a\})$, the rate for an interstitial n.n. hop from site J to site K with initial H-configuration $\{n_\ell^a\}$:

$$W^{JK}(\{n_\ell^a\}) = \omega' \exp\{-[(E^b-E^a)/4+e_o]^2/(e_o kT) - B^2/kT\}/B$$

where E^b and E^a are the pre- and posthop APE's respectively, and

B is $\Delta e_h + \frac{1}{2}\Sigma_{i\neq J}(\varepsilon_{iJ}+\varepsilon_{iK})n_i^a/|\varepsilon_{JJ}+\varepsilon_{JK}|^{\frac{1}{2}}$. Δe_h is the additional AE in the dilute-H case necessary for the additional transfer-integral-increasing dilation and the rest of B involves the static strains that initially provide part of the additional dilation. E^b-E^a can be written $-\Sigma_{i\neq J}(\varepsilon_{iJ}-\varepsilon_{iK}+U_{iJ}-U_{iK})n_i^a$. The form of the exponent above clearly shows that the initial strains modify the lowest-lying-event activation energy through E^b-E^a.

The rate W^{JK} should be averaged over the initial H-configurations. We approximate the probability of occurrence of a given H-configuration as $\Pi_\ell P_J^\ell(n_\ell^a)$, i.e., $P_J^\ell(1)$ and $P_J^\ell(0)$, the prehop probabilities of a static H being or not being at site ℓ depend only on the H (about to hop) at site J. We set $P_J^\ell(1)$ to $(c/6)\cdot \exp(-Y_{J\ell}/kT)/D$ and $P_J^\ell(0)$ to $(1-c/6)/D$, where $Y_{J\ell} = \varepsilon_{J\ell}+U_{J\ell}$ and $P_J^\ell(1) + P_J^\ell(0)$ equals 1. We made the additional approximation in the exponent of W^{JK} of keeping only terms linear in the n_ℓ^a. Similarly, we expanded the denominator in a Taylor series and kept only linear-in-n_ℓ^a terms. In our numerical calculations, we estimated the above approximations to be good for $c \leq .06$, when 1% or less of the available interstitial sites are occupied. We also estimated the error in W^{JK} due to using ε_{ij} values for only twelve n.n. shells (setting the rest to zero) to be 10%. To do this the twelve ε_{ij} values were extrapolated to larger distances using a modified Debye model for the phonons.

We consider only the c-dependent part of the hopping rate, Z_p, defined by $\ln Z_p = |n<W^{JK}>-\ln(\omega') + (e_o+\Delta e_h)/kT$. We fitted our calculated $\ln Z_p$ as a function of T by least squares to the formula $\ln f - \Delta E/kT$, where ΔE is a c-dependent change in the effective or average AE and f is a multiplicative factor describing how the effective prefactor changes with c. The calculated $\ln Z_p$ fitted this form fairly well over the limited T range involved in both the calculation and in the data of Fig. 1, i.e. 400-600K. Had we not subtracted the $T^{+\frac{1}{2}}$-dependent ω' out of $\ln Z_p$, it would have dropped out anyway. ΔE and f are plotted in Fig. 1.

DISCUSSIONS AND CONCLUSIONS

Fig. 1 shows that our hopping model predicts an increase in the activation energy with c, as observed experimentally. For $c \leq .06$, where our approximations are valid, the increases in activation energy ΔE, predicted by our model is roughly comparable with the experimental increase. The linear extrapolation of the calculated ΔE to higher c, yields ΔE values also roughly comparable to the experimental ones. We conclude that the effect of the H-H strain interaction on the diffusion rate is sufficiently strong that it needs to be included in any more realistic model for the H-concentration effect on H diffusion in Nb.

For $c < .06$, our calculated prefactor multipliers f were not

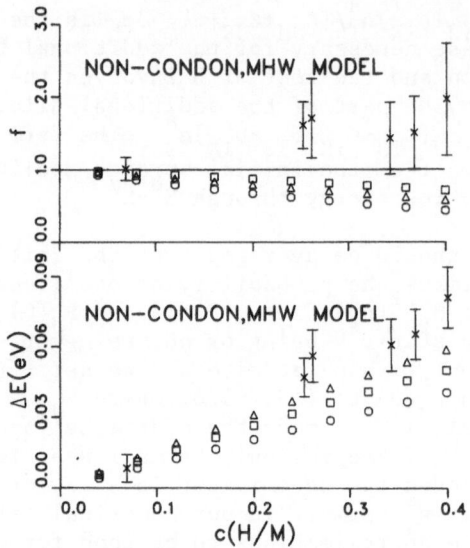

Fig. 1.   The X's and error bars give experimental values for
          ΔE, and for f. The squares, circles, and triangles
          give our calculated values for an interstitial n.n.
          hop, a 2nd n.n. hop, and a 3rd n.n. hop, respectively.
          The latter two are not discussed in the text. MHW
          indicates that our averaged ε_{ij} values were used.

incompatible with the experimentally determined f. However, the
linear extrapolation to higher c of the calculated f did not fit
the data at all. These discrepancies could be due to the crude
assumptions made concerning the transfer integral dependence on
lattice displacements, i.e., the sudden turn-on for the threshold
coincidence event, and the assumption that adiabatic hopping
obtained during a coincidence event.

REFERENCES

1.   H. C. Bauer, J. Völkl, J. Tretkowski, and G. Alefeld, Z.
     Physik B29, 17 (1978).
2.   D. Emin, M. I. Baskes, and W. D. Wilson, Phys. Rev. Lett.
     42, 791 (1979).
3.   H. Horner and H. Wagner, J. Phys. C 7, 3305 (1974).
4.   T. Holstein, Ann. Phys. (NY) 8, 343 (1959).

TRANSPORT BEHAVIOR OF DENSE PROTONS IN A SLAB

M. H. Lee and J. Hong

Department of Physics
University of Georgia
Athens, Georgia   30602

ABSTRACT

Absorbed hydrogen in a transition metal may be to a first
order approximation regarded as a system of protons interacting
pairwise via long-range forces. By limiting momentum transfers in
proton-proton scattering to small wavevectors, the relaxation function for a many-proton system is obtained analytically by a method
of recurrence relations recently developed by Lee. Other dynamical
quantities including diffusivity are also obtained analytically.

INTRODUCTION

Absorbed hydrogen in open transition metals e.g. Pd may be to
first order regarded as a system of dense protons interacting pair-
wise via long-range forces.[1] In this first approximation, the
effects of the host metal and the electrons released by hydrogen
are incorporated into the form of the proton-proton interaction.[2]
We consider a gas of these "dressed" protons confined to a two-
dimensional space such as a very thin slab. If these protons are
quantized in the direction perpendicular to the surface of a slab,
the system is quasi-two-dimensional. We include such a quasi-two-
dimensional system in our discussion.

We shall denote by x the number of protons per host atom. If
x is sufficiently large, the Fermi energy ϵ_F becomes large compared
with thermal energy kT. Then, dynamical properties of a two-dimensional or quasi-two-dimensional many-fermion system is governed to
a large degree by quantum many-body consideration taken at T = 0.[3]

One can write down an appropriate interaction Hamiltonian H for a gas of dressed protons under an external field in the following form:

$$H = H_o + H_{ext}(\omega) \; , \tag{1}$$

where

$$H_o = \Sigma(k^2/2m)C_k^+C_k + \Sigma v_k \rho_k \rho_{-k} \; , \tag{2}$$

$$H_{ext}(\omega) = \Sigma \rho_k(t) u_k e^{i\omega t} \; , \tag{3}$$

$$\rho_k(t) = e^{iH_o t} \rho_k e^{-iH_o t} \; , \quad \rho_k = \Sigma C_{p+k}^+ C_p \; , \tag{4}$$

where C_k^+ and C_k are, respectively, Fermion creation and annihilation operators at wavevector k. Also, m is the mass of the proton, v_k is the lattice-mediated proton-proton interaction, u_k is the Fourier component of the external field modulated at frequency ω such as to permit the use of linear response theory,[4] $\hbar = 1$. We neglect inessential constant terms in H_o. The sum on k will be limited to $|k| << k_F$, where k_F is the 2d Fermi wavevector. This amounts to limiting the interaction terms to "particle-hole" excitations near the Fermi surface only.

Here ρ_k denotes the density fluctuations at wavevector k. Since $[H_o, \rho_k] \neq 0$, it will have time evolution $\rho_k(t)$. Transport properties of our system will be given by an explicit knowledge of the time evolution ρ_k.[5] One can obtain the time evolution of ρ_k by solving the Liouville eq. at $t = 0$,

$$\dot{\rho}_k = iL\rho_k \; , \tag{5}$$

where $L\rho_k = [H_o, \rho_k]$. It is generally difficult to solve the Liouville eq. analytically. But for a 2d gas of protons at small wavevectors it is possible to give analytic solutions.[6]

For the method of solution, we shall follow a recent work by Lee based on Hilbert space theory.[7] The method of recurrence relations is equivalent to the well known Mori formalism.[8] The new method, however, permits analytic solutions directly from forms of recurrence relations (RR's) which the Liouville eq. satisfies. Hence this method is more useful and in some cases more powerful than the Mori formalism. The Mori formalism has been applied by a number of people for a variety of physical problems.[9] To apply this formalism one usually resorts to an ab initio approximation, i.e., truncating certain continued fractions which naturally arise in the analysis of the Liouville eq. This truncation is argued on physical grounds.[10] Lee's method does not require this kind approximation at all.

Following Lee, we expand $\rho_k(t)$ in terms of an infinite sequence of orthogonal basis vectors $\{f_n\}$

$$\rho_k(t) = \sum_{n=0}^{\infty} a_n(t) f_n ,\qquad (6)$$

where $a_n(t)$ is a coefficient, a time-dependent function given by $a_n(t) = (\rho_k(t), f_n)/g_n$ and the orthogonality is given by $(f_n, f_m) = g_m \delta_{mn}$, where $|f_n| = g_n^{\frac{1}{2}}$ is the length of f_n. Here the inner product of two arbitrary vectors F and G are defined by (F, G), i.e., for the inverse temperature β,

$$(F, G) = \int_0^{\beta} d\lambda <F(\lambda) G^+ >\qquad (7)$$

where $F(\lambda) = e^{\lambda H_O} F e^{-\lambda H_O}$ (8)
and the angular bracket denotes an ensemble average, + the hermitian conjugation. In the language of Hilbert space theory,[11] $\rho_k(t)$ is a vector in an infinite-dimensional vector space in which $a_n(t)$ represent projections. The time evolution is determined by relative magnitudes of these projection coefficients.

Following the linear response theory of Kubo[4] we choose $f_O = \rho_k$. Then it follows directly that $a_O(0) = 1$ and $a_n(0) = 0$ for $n \geq 1$. Thus, $a_O(t)$ is the relaxation function, $\equiv (t)$ in the notation of Mori[8], and the other a_n's describe nonlinear effects.

Lee obtained RR's for $\{f_n\}$ and $\{a_n(t)\}$ in the following form:

$$f_{n+1} = \dot{f}_n + \Delta_n f_{n-1}\qquad (9)$$

and

$$\Delta_{n+1} a_{n+1}(t) = -\dot{a}_n(t) + a_{n-1}(t)\qquad (10)$$

where $\Delta_n = g_n/g_{n-1}$, $f_{-1} = 0$, $a_{-1} = 0$, $\Delta_O = 1$, and $n \geq 0$.
From (10) one can obtain a continued fraction representation for a $a_O(z) = T[a_O(t)]$, where T is the Laplace transform operator. Thus, one can obtain the relaxation function for our system by evaluating Δ_n explicitly with our interaction Hamiltonian (1).

RELAXATION FUNCTION

For small k, our evaluation[12] of Δ_n with H given by (1) shows:

$$\Delta_1 = \omega_o^2 + 2k^2\varepsilon_F^2 \tag{11a}$$

$$\Delta_2 = k^2\varepsilon_F^2 + O(k^4) \tag{11b}$$

$$\Delta_3 = k^2\varepsilon_F^2 + O(k^4) \tag{11c}$$

where k is expressed in units of k_F and ω_o is a kind of classical plasma frequency. We immediately note that to lowest order k, $\Delta_n = \Delta$ for n = 2, 3, 4, ... where we define $\Delta = k^2\varepsilon_F^2$.

If $\Delta_1 = 2\Delta$, i.e., $\omega_o = 0$, then eq. (10) represents the RR for the Bessel function. That is,

$$a_o(t) = (\rho_k(t),\rho_k)/(\rho_k,\rho_k) = J_o(\mu t) \tag{12}$$

where J_o is the Bessel function of order zero and $\mu = 2\Delta^{\frac{1}{2}}$.

From this ideal solution, one can obtain the relaxation function for the interacting system (i.e., $\omega_o \neq 0$),

$$a_o(t) = S\sum_{n=0}^{\infty}(-\alpha)^n(\partial/\partial\mu t)^{2n}J_1(\mu t)/\mu t + P\cos\omega_p t \tag{13}$$

where $S = 1 - (1-\alpha)^{\frac{1}{2}}$, $P = \{(1-\alpha)^{\frac{1}{2}} - (1-\alpha)\}/\frac{1}{2}\alpha$, $\alpha = (x^2 + \frac{1}{4})/(x^2 + \frac{1}{2})^2$, $x = \omega_o/\mu$, and $\omega_p = \alpha^{-\frac{1}{2}}\mu$.

The diffusivity can be calculated by the Kubo formula.[14] From the relaxation function, we obtain the frequency-dependent diffusivity $D_k(\omega)$

$$Re\tilde{D}_k(\omega) = (\mu^2 - \omega^2)^{\frac{1}{2}}/\mu, \quad 0<\omega<\mu . \tag{14}$$

It vanishes for $\omega>\mu$. Here $\tilde{D}_k(\omega) = D_k(\omega)/D_k$, $D_k = 2\rho/m^2\mu$, ρ the proton density.

DISCUSSION

We draw an interesting conclusion from eq. (14). The mass diffusivity is finite up to μ. If the frequency of a modulating field is greater than μ, the protons are unable to follow through.

That the mass-current behaves Brownian-like is expected. The Liouville eq. is equivalent to the Langevin eq.[8] Hence, eq. (13) is also a solution for the Langevin eq. Because the protons are quantum particles, there are fluctuations intrinsic to quantum mechanics. This implies that a similar system of deuterons would behave differently since they must obey Bose statistics. It is suggestive that the isotope effects[13] may be bound in statistics.[15] We are pursuing ramifications of this idea.

This work was supported in part by the U. S. Department of Energy under Contract Number DE-AS09-77-ER01023. Our numerical work was supported by Research Corporation.

REFERENCES

1. G. Alefeld and J. Volkl, "Hydrogen in Metals," Springer, NY (1978).
2. H. Wagner and H. Horner, Adv. Phys. 23: 587 (1974).
3. A. Fetter and J. Walecka, "Many-Particle Systems," McGraw-Hill, NY (1971).
4. W. Marshall and R. Lowde, Repts. Prog. Phys. 31: 705 (1968).
5. J. Duderstadt and W. Martin, "Transport Theory," Wiley, NY (1979).
6. M. H. Lee and J. Hong, Phys. Rev. Letts. 48: 634 (1982).
7. M. H. Lee, to be published.
8. H. Mori, Prog. Theo. Phys. 34: 399 (1965).
9. S. Yip, Ann. Rev. Phys. Chem. 30: 546 (1979).
10. M. H. Lee and R. Dekeyser, Physica 86 B: 1273 (1977).
11. P. Halmos, "Hilbert Space," Chelsea, NY (1951).
12. J. Hong, Ph.D. Thesis, University of Georgia, Athens (1982).
13. Y. DeRibaupierre and F. Manchester, J. Phys. C 6: L930 (1973); J. Volkl, et al, Naturf. A 26a: 922 (1971); G. Sicking, Ber. Buns. Phys. Chem. 76: 750 (1972).
14. R. Kubo, Repts. Prog. Phys. 29: 255 (1966) and in "Transport Phenomena," G. Kirczenow and J. Marrow, eds., Springer-Verlag, NY (1974).
15. M. H. Lee in: "Hydrogen in Metals," T. N. Veziroglu, Ed., Pergamon, Lond. (1982).

ANALYSIS OF HYDROGEN PERMEATION EXPERIMENTS IN 403 STAINLESS STEEL

Barbara Okray Hall, Richard J. Jacko, and James A. Begley

Westinghouse Research and Development Center
1310 Beulah Road
Pittsburgh, PA   15235

ABSTRACT

A series of room temperature hydrogen permeation experiments was performed with AISI Type 403 martensitic stainless steel. Permeation transients were measured for specimens electrochemically charged in an 8.0 M NaOH solution under constant current (galvanostatic) conditions. Results were compared with a model that describes time-dependent diffusion of hydrogen in a material containing saturable and nonsaturable hydrogen trapping sites. At low charging currents, the microstructural traps are reversible and saturable and values for the trapping parameters, specifically trap density and binding energy, were obtained by fitting to model calculations. At high charging currents, nonsaturable traps appear to be present, indicative of either surface damage induced by the charging current or changing surface conditions.

INTRODUCTION

The diffusion of hydrogen in iron and steels is not adequately described by Fick's laws for temperatures below approximately 400°C. Darken and Smith[1] first attempted an explanation of anomalous effects by proposing a trapping mechanism; they assumed that hydrogen atoms diffuse randomly through the crystal lattice, but are delayed periodically at certain fixed sites or traps, which are uniformly distributed throughout the metal. A firm mathematical

555

formulation of these ideas was developed by McNabb and Foster[2], who derived a set of two coupled nonlinear differential equations that describe the time evolution of the mobile and trapped hydrogen concentrations in a material with a uniform density of traps, which are saturable, that is, capable of trapping only a single hydrogen atom. Subsequent workers have obtained numerical solutions for various values of the trapping parameters and for several geometries.[3-6] The formalism has been extended to include non-saturable trapping[7,8] and stress effects.[9]

In this paper we use the modified McNabb-Foster model to analyze hydrogen permeation in Type 403 stainless steel in order to determine appropriate trapping parameters for the material.

MODEL

The diffusion equation for interstitial hydrogen in the absence of stress can be written generally[9] as

$$\frac{\partial C_L}{\partial t} = D_L \nabla^2 C_L - \frac{\partial C_T}{\partial t}, \tag{1}$$

where C_L is the interstitial hydrogen concentration, C_T is the concentration of hydrogen in traps, and D_L is the diffusion coefficient of hydrogen in a trap-free material.

The rate of change of the trapped hydrogen concentration is given by

$$\frac{\partial C_T}{\partial t} = \sum_{i=1}^{n} [k_i C_L^{m_i} (N_i - C_i \delta_{m_i,1}) - p_i C_i], \tag{2}$$

where C_i is the concentration of hydrogen in traps of type i, k is the capture rate constant, N is the trap density, and p is the release rate constant. $m_i = 1$ if the traps are saturable; for non-saturable traps, the trapping rate is generally assumed to be proportional to some power (m > 1) of the interstitial hydrogen concentration. When p is nonzero, the traps are reversible, and when p is zero, they are irreversible.

Since the differential equations for the interstitial and trapped hydrogen concentrations contain nonlinear terms, they must in general be solved numerically[5,9]. Oriani[4] and Kumnick and Johnson[7] have, however, analyzed Eqs. (1) and (2) under the assumption that local equilibrium obtains between the mobile and trapped populations. An effective hydrogen diffusion coefficient can then be defined:

$$D_{eff} = D_L \left\{ 1 + \sum_i \left[\frac{m_i k_i C_L^{m_i-1} N_i}{p_i \left(1 + \frac{k_i}{p_i} C_L \delta_{m_i,1}\right)} + \frac{k_i^2 C_L N_i \delta_{m_i,1}}{p_i^2 \left(1 + \frac{k_i}{p_i} C_L\right)^2} \right]^{-1} \right\} \tag{3}$$

If saturable traps dominate, D_{eff} increases as C_L increases and approaches D_L at high concentrations. If nonsaturable traps dominate, D_{eff} decreases from D_L as C_L increases.

In permeation experiments time lag measurements are routinely used to determine effective diffusion coefficients. In the presence of saturable traps, the permeation time lag τ is given by[2]

$$\tau = t_L \left[1 + \frac{3N_T}{C_o} + \frac{6N_T}{C_o^2} \left(\frac{k}{p}\right)^{-1} - \frac{6N_T}{C_o^3} \left(\frac{k}{p}\right)^{-2} \left(1 + C_o \frac{k}{p}\right) \ln\left(1 + C_o \frac{k}{p}\right) \right], \tag{4}$$

where C_o is the hydrogen concentration at the input surface of the permeation membrane, and t_L is the time lag in the absence of trapping, given by $a^2/6D_L$, where a is the thickness of the membrane. The equivalent expression for nonsaturable traps[8] is

$$\tau = t_L \left[1 + \frac{6N_T k C_o^{m-1}}{p(m + 1)(m + 2)} \right]^{-1} \tag{5}$$

Equations (1) through (5) can be used in various ways to interpret permeation data. The first two can be solved for a number of trapping parameters and the results used in a direct least squares fitting of the permeation transient. Knowledge of the effective diffusion coefficient as a function of lattice hydrogen concentration can be used with Eq. (3) to determine trap type. If this is known, Eqs. (4) and (5) can be fit to time lag data to obtain N_T and k/p.

MATERIAL AND EXPERIMENTAL PROCEDURE

The material studied in this investigation was AISI Type 403 martensitic stainless steel that had been solution heat-treated, quenched, and tempered at 230°C. The chemical composition of the steel is given in Table 1. The typical microstructure is a mixture of martensite lathes, precipitated carbides, inclusions and delta ferrite. In this metallurgical condition, the alloy is susceptible to environmentally-assisted cracking under the hydrogen conditions used in this study.

Permeation transients were obtained using the electrochemical method of Devanathan and Stachurski.[10] Hydrogen was introduced into the membrane by electrolytic decomposition of water in an electrolyte of 8M NaOH at current densities of 1, 10, and 100 A/m^2. Hydrogen was extracted on the opposite side of the membrane by potentiostatically controlling the surface at a potential of 0.050 V SHE in 0.1M NaOH and the resultant current was recorded. In order to prevent corrosion reactions from masking the hydrogen oxidation current, the membrane was sputtered clean and a Pd coating was vapor deposited onto the exit surface. Both cells were deoxygenated by continuous purging with ultra-high purity nitrogen. The potential at the membrane surfaces was measured against Hg/HgO/NaOH reference electrodes. Successive transients were recorded using increasing and decreasing amounts of cathodic polarization on the input side of the membrane.

RESULTS AND DISCUSSION

A typical permeation curve is shown in Fig. 1 for an increasing transient and the integrated charge is plotted as a function of time in Fig. 2. On this latter curve, the graphical method of determining the time lag is indicated: the slope of the curve at long times, when steady-state has been attained, is projected back to the time axis, with the intercept giving τ. Steady-state permeation fluxes and time lag values for these and other transients are given in Table 2, which also contains calculated values for the effective diffusion coefficient ($D_{eff} = a^2/6\tau$). These are in good agreement with others in the literature. Bockris, et al.[11] have reported 1.5 x 10^{-12} m^2/s for a Fe-12% Cr alloy, and Sakamoto and Hanada[12] measured 3.0 x 10^{-12} m^2/s for 403 stainless steel given a similar heat treatment. Careful examination of the data reveals several important trends. Calculated values of D_{eff} change as a function of normalized flux. In most experiments, D_{eff} decreased during the course of the transient, and decreased continuously in successive transients at the same charging current.

At the lower charging currents, D_{eff} increased with charging current, while at the highest i_{ch}, D_{eff} decreased. Since C_0 increases as i_{ch} increases, reference to Eq. (3) indicates that trapping occurs at saturable sites in the former case. The appearance of nonsaturable trapping at the highest charging current could

Table 1. Chemical Composition of Type 403 SS Test Material

			Nominal Composition, Wt. Percent						
C	Mn	P	S	Si	Cr	Ni	Mo	Cu	Fe
0.10	0.44	0.016	0.028	0.26	12.49	0.25	0.45	0.08	Bal.

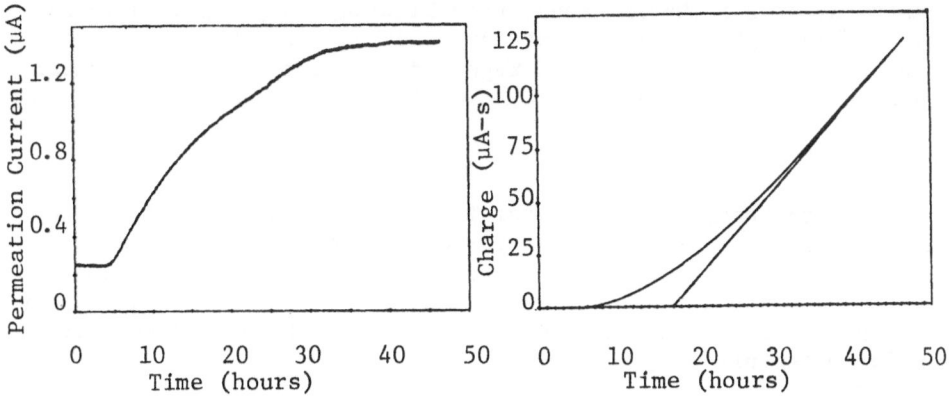

Fig. 1 Rise transient for
i_{ch} = 10 A/m^2.

Fig. 2. Integrated charge vs.
time for rise transient.

arise: 1) if both saturable and nonsaturable traps are present in
the virgin material, and at high i_{ch} the effect of the latter on
D_{eff} overtakes that of the former; 2) if surface damage is induced
at high charging currents; or 3) if the surface boundary condition
changes. At present, these possibilities are under investigation.

A least-squares analysis yields

$$P_{ss} = 0.911 \, i_{ch}^{0.23}, \qquad (6)$$

where P_{ss}, the steady-state permeation current, is in mA/m^2 and i_{ch}
is in A/m^2. Since $P_{ss} = D_L C_o/a$, Eq. (6) was used with Eq. (4) to
determine sets of values of D_L, N_T, and the ratio k/p that are consistent with the time lag data. These are listed in Table 3. Eqs.
(1) and (2) were solved using these parameter sets. The best fit
to the permeation transient shown in Fig. 1 was obtained for a trap

Table 2. Experimental Data for Permeation Transients (a=8.36x10^{-4}m).

Run Number	Charging Current Density, i_{ch} A/m^2	Steady-State Permeation Current Density, P_{ss} mA/m^2	Time Lag, τ 10^4s	Effective Diffusivity, D_{eff} 10^{-12} m^2/s
2	10	1.84	5.84	1.99
3	100	2.72		
4	10	1.68	5.76	2.02
5	100	2.48	4.80	2.43
6	10	1.31	5.20	2.24
7	100	2.34	5.08	2.29
8	1000	4.79	10.70	1.09
9	100	2.36		
12	1000	4.57	6.88	1.69
13	100	2.80	4.32	2.70
14	10	1.55	6.30	1.85

density N_T of $3 \times 10^{24}/m^3$ and a trap binding energy E_B of 31.8 kJ/ mole. This high density is not inconsistent with the microstructure and E_B is in the range of reported values for microstructural traps in iron and steels.

CONCLUSIONS

From the permeation data for Type 403 stainless steel, we conclude that: 1) at low charging currents, trapping is saturable and reversible; 2) the McNabb-Foster model parameters which best fit the data are a trap density of $3 \times 10^{24}/m^3$ and a binding energy of 31.8 kJ/mole, values not inconsistent with those reported for similar material.

Table 3.   Trap Parameters Consistent With Time Lag Data.

Trap Density, N_T $10^{23}/m^3$	Ratio k/p $10^{-23}m^3$	Time Lag, t_L 10^4s	Diffusion Coefficient, D_L 10^{-12} m^2/s	Binding Energy, E_B kJ/mole
4.87	≳286	3.49	3.34	≳50.0
7	1.02	3.23	3.61	36.2
8	0.62	3.12	3.74	35.0
9	0.51	3.02	3.86	34.5
10	0.39	2.92	3.99	33.9
20	0.18	2.15	5.42	31.9
30	0.17	1.57	7.42	31.8
40	0.17	1.03	11.27	31.9
50	0.26	0.59	19.90	32.8
60	0.69	0.18	63.00	35.3
65	∞	0	∞	∞

REFERENCES

1. L. S. Darken and R. P. Smith, Corrosion 5, 1 (1949).
2. A. McNabb and P. K. Foster, Met. Trans. 227, 618 (1963).
3. P. K. Foster, A. McNabb, and C. M. Payne, Met. Trans. 233, 1022 (1965).
4. R. A. Oriani, Acta Met. 18, 147 (1970).
5. G. R. Caskey, Jr. and W. L. Pellinger, Met. Trans. 6A, 467 (1975).
6. W. M. Robertson, Scripta Met. 15, 137 (1981).
7. A. J. Kumnich and H. H. Johnson, Met. Trans. 6A, 1087 (1975).
8. H. H. Johnson and R. W. Lin, in Hydrogen Effects in Metals, (I. M. Bernstein and A. W. Thompson, eds.)(Metallurgical Society of AIME, Warrendale, PA 1981) 3.
9. B. O. Hall, to be submitted to Scripta Met. for publication.
10. M. A. V. Devanathan and Z. Stachurski, Proc. Royal Soc., A270, 90 (1962).
11. J. O'M. Bockris, M. A. Genshaw, and M. Fullenwider, Electrochimica Acta, 15, 47 (1970).
12. Y. Sakamoto and U. Hanada, 2nd Inter. Congress on Hydrogen in Metals, Paris, 1977, paper 1A7.
13. J. P. Hirth, Met. Trans. 11A, 861 (1980).

THERMOPOWER OF β-PHASE PdH_y *

C. L. Foiles

Physics Department
Michigan State University
East Lansing, MI   48824

ABSTRACT

Thermopower data can provide confirming tests for the separation of electrical resistivity into acoustic and optical phonon contributions. Two tests are applied and are found to be inconsistent with the present separation. Possible sources of the inconsistency are noted.

Motivated by a goal of obtaining information on the electron-phonon coupling in PdH_y, several studies of electrical resistivity $\rho(T,y)$ have attempted to identify the acoustic and optical phonon contributions[1-4]. After confronting serious experimental and analysis problems, a prevailing concensus has emerged: (i) the temperature dependent portion of electrical resistivity can be separated into a Debye spectrum contribution, acoustic phonons, and an Einstein spectrum contribution, optical phonons and (ii) the electron-phonon coupling for the optical phonon contribution decreases as $1-y$ increases.

The preceding features of electrical resistivity behavior have definite implications for thermoelectric behavior and thus thermopower data should provide a sensitive test for the correctness of the resistivity interpretation. An earlier attempt to use this test[5] supported the resistivity interpretation but the creditibility of that conclusion is seriously limited by attendant assumptions. In this brief report we use thermopower data to provide two separate tests of the resistivity interpretation.

---

*Research supported in part by NSF grant DMR80-05865.

The relationship between electrical resistivity (ρ) and thermopower (S) is expressed through the well-known Gorter-Nordheim relation

$$S = \sum_i \rho_i S_i / \sum_i \rho_i \qquad (1)$$

where i denotes an identifiable scattering process. Both ρ_i and S_i can be functions of temperature and ρ_i can be a function of y. However, S_i, most simply associated with the energy dependence of scattering process i, would be expected to be independent of y.

The first test using thermopower data is for varying y values at a single temperature. The most convincing application of this test, one which minimizes the desorption of hydrogen problem and maximizes the number of data points, uses the equilibrium data of Baranowski for thermopower[6]. Combining his figures 4.5 and 4.21 gives S(300K,y) for $0.9 \leq y \leq 1.0$. Resistivity data comes from the previously noted concensus, in particular references[1-3], and is summarized as follows:
   (a)   Acoustic phonon scattering, denoted as i = 1, is 3$\mu\Omega$-cm and independent of y.
   (b)   Optical phonon scattering, denoted as i = 2, is 7$\mu\Omega$-cm for y = 1 and linearly decreases to zero at y = 0.7.
   (c)   Residual scattering, denoted as i = 3, is 85y(1-y)$\mu\Omega$-cm.
When these variations of ρ_i with y and the assumption that S_i is independent of y are used in equation (1) the predicted behavior for Z = (1-y) becomes

$$\rho S = A + BZ - CZ^2 \qquad (2)$$

where A, B and C are straightforward combinations of ρ_i and S_i parameters. A comes directly from experimental data for y = 1 (Z = 0) and thus a plot of (ρS - A)/Z completes this first test.

Figure 1 provides the requisite plot and the data are consistent with equation (2). The slope and intercept from figure 1 are readily deciphered to give the T = 300K values of S_1, S_2 and S_3 as -64.3, 33.4 and 16.5 μV/K respectively. With the possible exception of S_3 these values are larger than those found for transition metals and thus they seem unreasonably large.

The second test comes from a comparison of predicted and observed temperature dependence for thermopower. The predicted behavior is a straightforward application of equation (1), the temperature dependencies which earlier studies associated with each ρ_i and the usually valid assumption that $S_i(T) = a_i T$. The deabsorption of hydrogen above T > 200K for larger y values[1,2] presents a serious impediment to obtaining reliable experimental data. To mitigate this problem we suggest that combining S(T) for $T \leq 170K$ from Kopp, et. al.[5] with the equilibrium values of

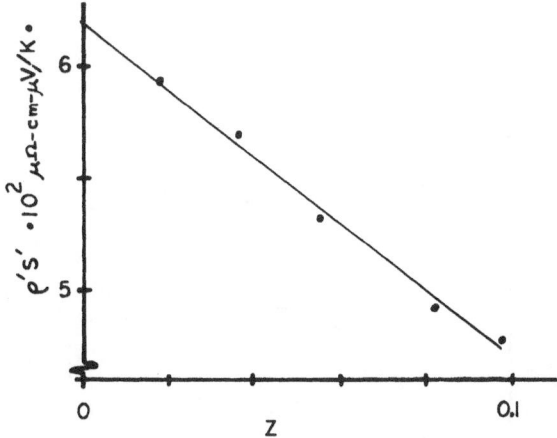

Fig. 1. $\rho'S'$ versus Z. Terms from equation (2) are recast into the form $(\rho S - A)/Z \equiv \rho'S'$ and plotted as a function of Z. A was determined by $\rho(300K, y = 1.0)$ and $S(300K, y = 1.0)$. The 5 data points represent nearly equal spaced data from Baranowski's smooth curve fit for $S(300K, y)$; approximately 28 data points were used in obtaining that smooth curve fit.[6]

Baranowski[6] for T = 300K provides a reliable estimate of experimental behavior. This suggestion is reinforced by actual data over the entire temperature range for y $\cong$ 0.63 where desorption is not a problem.[7] Figure 2 shows a summary of experimental data.

We begin the comparison by commenting on the trend in experimental data suggested by figure 2. For each alloy S(T) is positive and increases at a decreasing rate as T increases. For the y $\cong$ 0.63 alloy a plateau occurs above 200K. The predicted behavior using S_i parameters deduced from figure 1 are shown as dashed lines. The predicted behavior is totally incorrect and thus the parameters deduced from figure 1 are not valid. The analysis of Kopp, et. al.[5] was also at a constant temperature, T = 170K, and their parameters produce similar conflicts with the experimental data in figure 2. First, they used only ρ_1, ρ_2 and S_1 in their analysis and found S_1 (170K) = 13.2 μV/K. Thus, the loss of ρ_2 in the y $\cong$ 0.63 sample would predict a value twice as large as the observed value. Second, their qualitative prediction of a decreasing rate of increase for S(T) as T increases is associated with the influence of a ρ_2

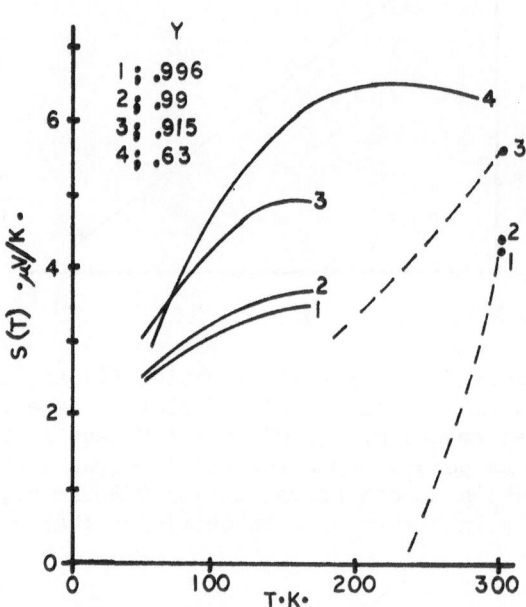

Fig. 2.   S versus T.   The solid curves are data from references 5
         and 7 and the points are equilibrium data from reference 6.
         The dashed lines are behaviors predicted using the
         parameters from figure 1.

contribution.   Since ρ_2 is required to be missing in the y $\cong$ 0.63
sample their model can give no consistent explanation for the
plateau in S(T) for this sample.

    The present analysis leads us to offer one warning and to draw
one general conclusion.   The warning: attempts to use thermopower

data to confirm the identification of acoustic and optical phonons contributions to electronic transport must be applied carefully and comprehensively. Use of limited data can give apparent but false confirmations. The general conclusion: existent thermopower data are inconsistent with the present division of electrical resistivity into acoustic phonon, optical phonon and residual scattering contributions[1,2,3]. The cause of this inconsistency is unclear. One origin may lie in the methods used to separate acoustic and optical phonon contributions. The simple ansatz that ρ_1 and ρ_2 are independent of y and reasonable thermopower parameters[2] do produce the trends of experimental data in figure 2. This ansatz seems unjustified but, for completeness, it must be noted that in their initial study Chiu and Devine[4] concluded that the optical phonon term did not change significantly for y > 0.8. Another origin of the inconsistency may lie in the assumptions made for $S_2(T)$. Optical phonon scattering in semiconductors is not as simply characterized as acoustic phonon scattering[8]. This may also be true in metals and may cause an unusual temperature and/or y dependence in the associated thermopower. Further theoretical developments related to the recent studies of Engquist and Grimvall[9] are needed to explore this possibility.

REFERENCES

*Research supported in part by NSF grant DMR80-05865.

1. C. Arzoumanian, J.P. Burger, L. Dumoulin and P. Nedellec, _Zeit. Phys. Chem._ 116:S117 (1979).
2. J.P. Burger, D.S. MacLachlan, R. Mailfert and B. Souffache, _Solid State Comm._ 17:281 (1975).
3. J.C.H. Chiu and R.A.B. Devine, _Solid State Comm._ 22:631 (1977).
4. J.C.H. Chiu and R.A.B. Devine, _J. Phys. F_ 6:L33 (1976).
5. J. Kopp, D.S. MacLachlan and G. Vekinis, _J. de Phys._ 39:C6-439 (1978).
6. B. Baranowski, Hydrogen in Metals II ("Topics in Applied Physics," Vol. 29), Chapter 4.
7. C.L. Foiles, _Solid State Comm._ 33:125 (1980).
8. Rowland W. Ure, Jr., "Semiconductors and Semimetals," Vol. 8, Chapter 2 (1972).
9. H-L. Engquist and G. Grimvall, _Phys. Rev. B_ 21:2072 (1980).

DIRECT EVIDENCE FOR INTERSTITIAL AND VACANCY-ASSOCIATED HYDROGEN

TRAPPING NEAR IMPURITY ATOMS IN F.C.C. METALS

P. Boolchand, C.T. Ma, M. Marcuso and P. Jena[*]

Physics Department
University of Cincinnati
Cincinnati, Ohio 45221
[*]Department of Physics
Virginia Commonwealth University
Richmond, VA 23284

A new method is developed to probe the interaction of hydrogen (H) or deuterium (D) with lattice defects in metals. The method makes use of Mössbauer spectroscopy to detect microscopic trapping of interstitial H (D) or vacancy associated H (D) in the immediate vicinity of selected impurity atoms.

To illustrate the type of new microscopic information that can be obtained from this method, we have performed experiments on Cu. In-situ Mössbauer spectra of a Cu (^{57}Co) target foil proton (p) or deuteron (d) irradiated have been studied as a function of irradiation dose and post-irradiation isochronal annealing. The data reveal population of three new Co sites in addition to the original substitutional site. From their thermal annealing character, the sites are identified respectively as interstitial H(D) trapped near Co, a mono or di-vacancy associated H (D) that is trapped near Co, and a multi-vacancy associated H (D) that is trapped near Co.

The method can be applied generally to the study of hydrogen-metal-defect interactions and further can be readily extended to include studies of He-metal defect interactions as well.

INTRODUCTION

The study of hydrogen in metals has attracted widespread attention[1] for reasons which are motivated both from a basic as well as applied point of view. Many metals and intermetallics reversibly absorb large quantities of hydrogen offering prospects as hydrogen storage media. The unusually high diffusivity of the light interstitial atoms (μ, p, d) in metals makes possible trapping with lattice defects at low temperatures. Indeed, the interaction of hydrogen with vacancies, voids, grain boundaries, dislocations and impurity atoms has profound consequences on the mechanical behaviour of materials. Embrittlement of steels, radiation hardening, first wall problems associated with nuclear fusion reactors, degradation of the fuel cladding in fission reactors, are but some of the many technological problems that are really manifestations of the underlying fundamental problem viz. the interaction of H(D) and He with metal lattice-defects.

The interaction of hydrogen with metal atoms has been studied theoretically by several groups.[2-4] These calculations predict a net transfer of charge from metal atoms to hydrogen. Recently it has been suggested[4] that hydrogen may be trapped by metal vacancies. The interaction of hydrogen and its isotopes with lattice defects has been studied by numerous experimental methods which include low temperature resistivity[5], electron microscopy, quasi-elastic neutron scattering internal friction, Gorsky effect, positron annihilation spectrosocpy[6] (PAS), spin precession methods[7] and resonance methods[8]. It will not be possible to treat all these methods in this brief review. The interested reader is referred to a number of excellent review articles available in the literature[1].

In this paper we describe a new experimental method to investigate the interaction of dilute hydrogen and deuterium with metal defects. The method utilizes Mössbauer spectroscopy to detect microscopic trapping of interstitial H(D) or lattice defect associated H (D). Our approach has similarities to the one used by Mansel and Vogl[9], who nearly a decade ago first demonstrated impurity trapping of metal defects.

We chose to focus on the fcc host Cu, primarily because the solubility[5] and diffusion[10] of H, and the kinematics of lattice defect annealing[11] in this material are well documented. Equally significant is the fact that Mossbauer experiments on Cu ([57]Co) performed following high-energy neutron or electron irradiations, indicate[12] that neither Cu self-interstitials nor Cu vacancies are trapped at Co impurities. We discuss results of our experiments on Cu in conjunction with other measurements on this material which include ion-beam channeling, low temperature resistivity, PAS and μSR.

THE EXPERIMENTAL METHOD

We have developed an experimental facility[13] to obtain in-situ Mössbauer spectra of cooled target foils irradiated with p or d using a 2-MeV Van de Graaff accelerator. Fig. 1 schematically shows the basic components of this facility which includes a cold finger assembly forming part of the tail of a liquid He dewar, on which the target of interest is mounted at 45° to both the Van de Graaff beam line and the motion of a Mössbauer velocity transducer.

In the present experiments a high purity target foil of 54 μm thickness, doped with 1/3 ppm ^{57}Co impurity atoms, was subjected to p or d irradiations at a beam current of ≤ 10 μA and beam

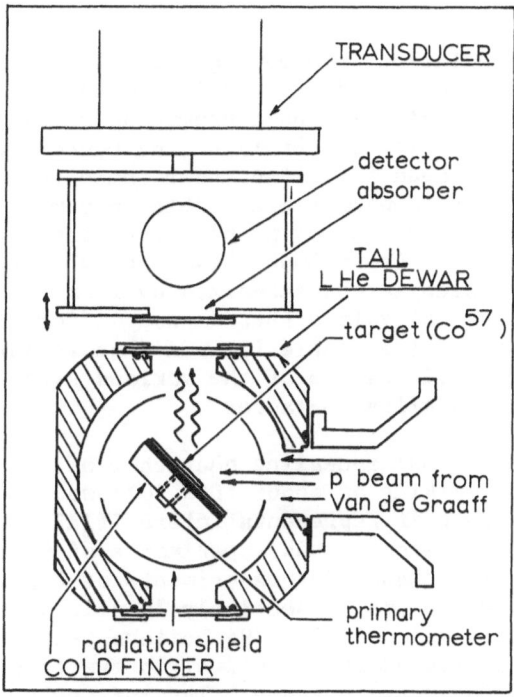

Fig. 1 Principal features of facility used for in-situ Mössbauer studies on p or d irradiated targets.

energies ranging from 2 MeV to 0.5 MeV. Since the range of a
2 MeV p in Cu is 19 μm, we believe that nearly all H implanted comes
to rest in the target. The chosen target area of 1.5 cm x
1.5 cm was somewhat larger than the beam cross section of 1.2 cm
diameter. Particular care in design of the target holder was
necessary to measure as closely as possible the temperature rise
of the target during the p or d irradiations. For the cold finger
cooled to 80K, we estimate that a temperature rise of about 60K
occurs at the exposed surface of the target with ~3 Watts of
(corrected) incident irradiated power. In a typical experiment, a
total of 5 hours of irradiations were carried out during which the
beam energy was successively reduced starting from 2 MeV. This
resulted in a total integrated dose of approximately 4×10^{17} particles on the target corrected for secondary electron emission.[14]
Spectra of the target were taken as a function of irradiated dose
and post-irradiation isochronal annealing (30 min. duration) in
the range 100K < T < 800K. Fig. 2 reproduces selected spectra of
the Cu target taken at various stages of annealing.

RESULTS AND DISCUSSION

Copper

On p or d irradiation, two new features in the spectrum of
the Cu target appeared, a weakly populated quadrupole doublet,
henceforth denoted as site 1, and a prominent satellite feature on
the negative velocity side of the main line henceforth denoted as
site 2. We have followed the areas under these features as a
function of annealing temperature and this information is plotted
in Figs. 3 and 4 for the case of p and d irradiations performed
on the same Cu target. Above 300K these data show the growth of a
new site (site 3) which is characterized by isomer-shift (δ) and
quadrupole splitting (Δ) values that do not differ significantly
from those of site 1. (See table 1). In what follows we will
identify the microscopic nature of the various sites based primarily on the thermal annealing character.

Site 1. The thermal annealing character of this site, particularly its narrow peaking at about 160K followed by a slow decay,
strongly suggests that it represents interstitially trapped H (D)
in the vicinity of Co atoms. μSR experiments of Camani et al[15]
and independently ion-channeling experiments of Bugeat and Ligeon[16]
have shown that μ and p are respectively localized at the octahedral interstitial location in Cu. We believe that site 1 represents H(D) that is trapped at octahedral interstices in the immediate vicinity of Co impurity atoms. Interestingly, low temperature resistivity measurements on H doped Cu, exhibit a local minimum near 170K, a temperature indeed quite close to where site 1

shows a maximum in population (Figs. 3 and 4). This resistivity minimum was taken as evidence for H trapping at impurities by Wampler, Schober and Lengeler.[5] These authors developed a model to quantitatively describe the observed behaviour of resistivity. The principle components of this model are: (i) that the H induced resistivity in Cu is smaller when H is trapped as opposed to when it is free (ii) at low temperatures 150K < T < 170K, the spontaneous decrease in resistivity is due to a diffusion limited impurity trapping of H (iii) while for T > 170K, thermally activated detrapping dominates and causes the resistivity to go back to its original value as H becomes free again. In this model, the temperature location of the resistivity minimum directly yields the binding enthalpy (H^b) of H to the impurities. We have adapted this model to quantitatively describe the temperature dependence of site 1 population $P_1(T)$ and find that

$$P_1(T) \quad \alpha \quad (1 - B^{-1})\ (1 - e^{-\lambda t}) \qquad (1)$$

where $B = 1 + 6C_t \exp H^b/kT$; $\lambda = (8BD_o/a^2)\exp-(H^b+H^m)/kT$; C_t is the concentration of Co solutes; $(D_o/a^2)\exp-H^m/kT$ is the diffusion constant of H or D in Cu; H^b is the binding enthalpy of Co to H or D; and t is the annealing time per step. The smooth continuous lines (Figs. 3 and 4) drawn through the data of site 1 represent fits to equation (1) and yield

$$H^b\ (Co-H) = 0.122(4)eV; \quad H^b\ (Co-D) = 0.127(4)eV$$

Table 1. Isomer shift (δ) and quadrupole splitting (Δ) of various $^{57}Co(^{57}Fe)$ sites observed in p or d irradiated Cu. δ is quoted relative to substitutional Co in Cu. A + sign of δ indicates a larger $|\psi(o)|^2$.

Co site in Cu	δ (mm/s)	Δ (mm/s)
Site 1	+0.04 (3)	0.50 (3)
Site 2	−0.30 (3)	0.16 (3)
Site 3	+0.08 (3)	0.50 (3)

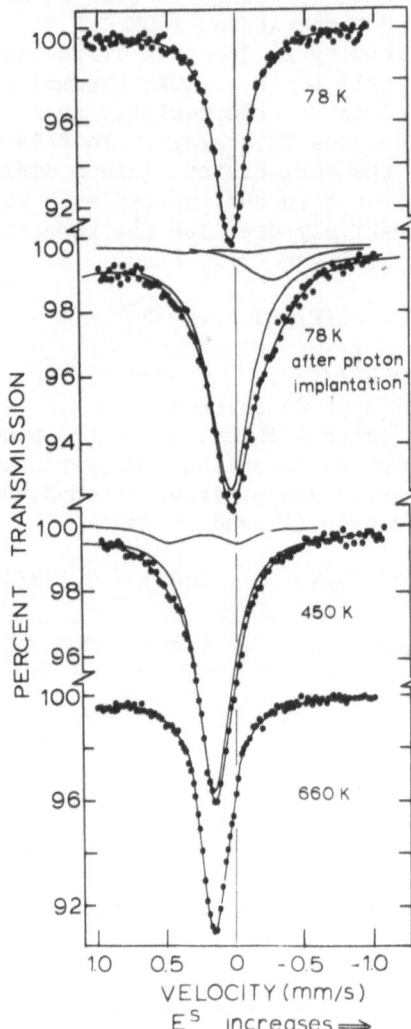

Fig. 2 ^{57}Fe spectra of p-irradiated Cu taken before and after different stages of isochronal annealing.

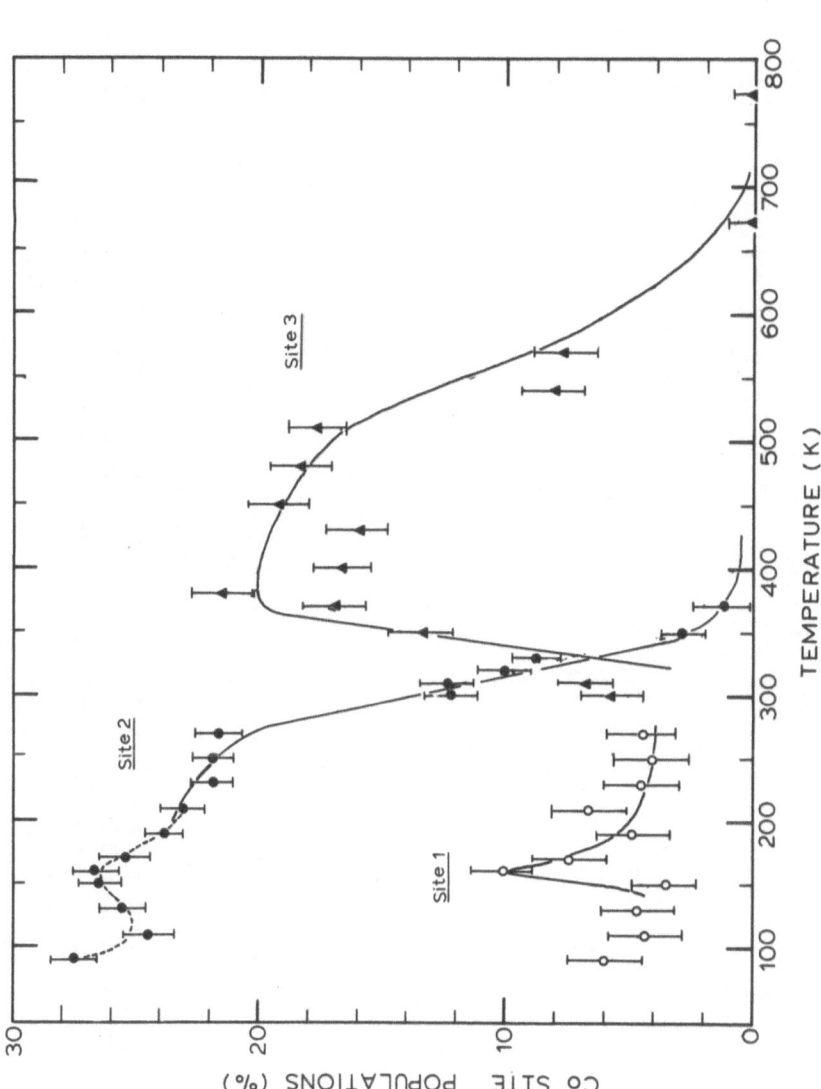

Fig. 3   Co-site population as a function of annealing temperature for d irradiated Cu. See text for site identification. The smooth continuous lines represent a model dependent theoretical fit to the data and yield migration (H^m) and binding (H^b) enthalpies.

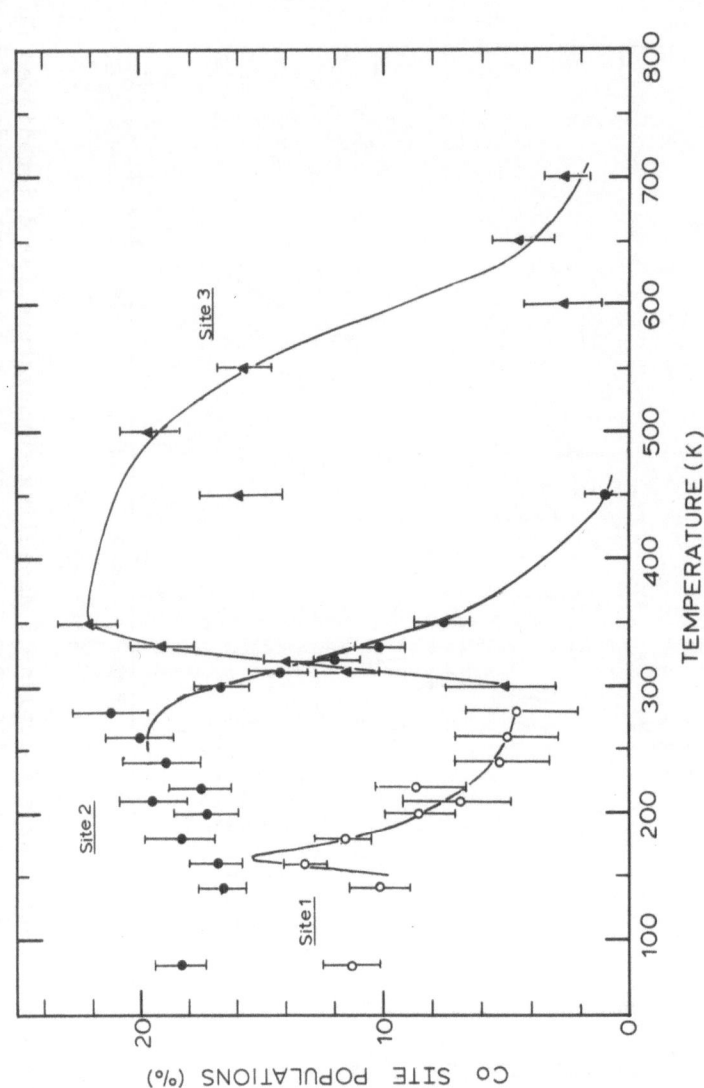

Fig. 4 Co-site population as a function of annealing temperature for p irradiated Cu. See text for site identification. The smooth continuous lines represent a model dependent theoretical fit to the data and yield migration (H^m) and binding (H^b) enthalpies.

In our fits we have fixed the diffusion constants of H and D in Cu
to their measured value.[10] It is not surprising that the H^b values
found here are nearly the same as those reported by Wampler et al[5].
These values are strongly correlated to the temperature location
of the peak population or resistivity minimum. The present data
thus provides direct microscopic evidence for the trapping of in-
terstial H and D at Co impurities in Cu.

We can also derive information on the nature of bonding between H and Co from the isomer-shift in Table 1. The positive δ
for site 1 implies that the contact electron density, $|\psi(0)|^2$ at
the Co nucleus is increased when H is present in its vicinity.
This can be qualitatively understood within the charge transfer
scheme mentioned earlier[2]. The transfer of d-like charge from Co
to hydrogen would give rise to a shift of the inner s-core states
towards higher binding, thus causing the charge density to increase
at the Co-nucleus. This analysis is consistent with the core level
shifts obtained from XPS experiments.

Site 2. As mentioned earlier Mössbauer experiments[12] on Cu
performed following fast neutron and electron irradiation do not
reveal evidence for either self-interstitial or vacancy trapping
as was observed by Mansel and Vogl in their pioneering experiments[9]
on Al. We consider this strong evidence to suggest that all sites
observed in the present experiments on Cu, are intrinsic to H or D,
and further, that these sites do not represent trapping of a metal
defect per se. In particular, site 2 (Fig. 3 and 4) does not
represent Cu self-interstitials trapped near Co as was suggested
by us earlier[13]. We propose instead that site 2 actually represents a H or D atom associated with a mono-vacancy or possibly
even a di-vacancy that is trapped in the vicinity of the Co impurity atoms. The present measurements indicate that this site is
strongly populated after the p or d irradiations and further that
it appears to precipiteously anneal near 300K, a temperature where
mono-vacancies in pure Cu are known[11] to become mobile.

Although we have not measured the residual resistivity of the
p and d irradiated Cu sample, this quantity has been measured by
Echt et al.[17] for conditions that are similar to ours. Based upon
their data, we estimate that for every p or d implanted, there are
at least 10 Frenkel pairs left in our Cu sample at 100K. Because
of the high concentration of defects in the implanted region, and
further because the target temperature increases to near 150K
during the implantations, we believe that the majority of H (D)
entering the target becomes vacancy associated (both mono and di-
vacancies are possible) and further that the vacancy-H (D) complex
in question migrates at those temperatures and is trapped at impurity atoms. The high population of site 2 in relation to site 1
can thus be understood. The growth kinetics of site 2 as a function
of annealing temperature are needed to reliably establish the mi-

gration enthalpy of the complex.

The strong affinity of H for mono-vacancies in Cu has been observed elsewhere. The PAS experiments of Lengeler, Mantl and Triftshauser[6] on vacancy and H loaded Cu have shown a striking reduction in the lineshape parameter (ΔIv) as a function of annealing temperature above 150K, H becomes mobile and is vacancy associated. This precludes positrons from seeking vacancies causing ΔIv to reduce sharply. Ion-beam channeling and nuclear reaction profile experiments[16] also indicate that H implanted in Cu at low temperature (<250K) is vacancy associated.

The negative δ for site 2 implies that $|\psi(0)|^2$ at the Co nuclear site associated with a nearest neighbor hydrogen-vacancy complex is less than that at a substitutional Co site in Cu. Unlike the situation in site 1, there are two competing mechanisms which contribute to $|\psi(0)|^2$ at site 2. The charge transfer mechanism discussed in site 1 also applies to site 2 and would give rise to an increase in $|\psi(0)|^2$. However, the magnitude of this increase may be somewhat smaller in site 2 than in site 1 in view of the increased distance between Co and hydrogen-vacancy complex in site 2. The other contribution, which we believe to be the dominant one, corresponds to the removal of electron density at the Co-site due to the absence of a nearest neighbor Cu-atom.

Site 3. The data of Figs. 3 and 4 clearly show that site 3 grows rapidly at the expense of site 2 in the temperature range 300K < T < 400K. We further note that the peak population of site 3 is commensurate with that of site 2 at 285K, the temperature where vacancies in Cu are known[11] to migrate. It is on this basis, we propose that site 3 represents a multi-vacancy associated H (D) that is formed when a vacancy gets attached to site 2. The growth kinetics of site 3 would then be described by a first order rate process determined by the migration enthalpy H^m of mono-vacancies in Cu. The decay characteristics of this site would be determined by the binding enthalpy H^b of the said complex to the Co trap. The correctness of these ideas is supported by the continuous lines (Figs. 3 and 4) drawn through the site 3 data which represent fits to equation 1 and yield

H^m = 0.75(3)eV; H^b = 0.63(3)eV; for the p irradiations

H^m = 0.69(3)eV; H^b = 0.61(3)eV; for the d irradiations.

It is notable indeed that the values of H^m obtained above coincide with the migration enthalpy of monovacancies[17] in Cu (0.70(3)eV) deduced from resistance recovery (stage III) in quenching as well as irradiation experiments. These data provide strong support for the view that site 3 derives from site 2 when a mobile mono-vacancy gets trapped at the latter.

In the above picture, the annealing of site 2 for T > 300K is not visualised to be a thermally activated process. Consequently it is not feasable to obtain H^b of site 2 except to point out that this quantity must be larger than 0.33 eV. This lower estimate for H^b is obtained by fitting the site 2 data in the temperature range 220K < T < 400K to equation 1, for the case where $e^{-\lambda t} \to 0$. The binding enthalpy characteristic of site 3 is on the other hand well deduced and its value of 0.62(3)eV is rather large. This large value lends support to the notion of a progressively stronger $Co^+ -H^-$ bond as H becomes more anionic H^- (D^-) in character[4] when it is localized in a large vacancy cluster.

Aluminum

Although the nature of lattice defects responsible for the different recovery stages of residual resistivity in quenched and irradiated Al has been a subject of extensive investigations[11,18] we are aware of only a few studies that deal directly with the interaction of light interstitial atoms (μ^+, p, d) with defects in this material. Evidence for μ^+ localization in Mn doped Al has been reported[19]. Nuclear reaction profiles as well as ion-channeling experiments[20] on p and d irradiated Al indicate that the implaneted elements occupy tetrahedral positions and are associated with vacancies for temperatures T < 175K.

Mössbauer experiments following fast neutron and high energy electron irradiations of Al (^{57}Co), and independently ^{57}Co ions implanted[21] in Al, have clearly shown that Co impurity atoms act as deep traps for vacancies, mono and multi-interstitials. We have also performed Mössbauer experiments on d irradiated Al (^{57}Co) and find evidence for multi-interstitial trapping. On annealing the target T > 200K, new ^{57}Co sites are populated which appear to anneal away near 340K. Because of the weak trapping observed at the studied dose, it is at this stage not possible to unambiguously decide if these sites represent the trapping of a vacancy and/or D-vacancy complex. Further experiments on Al and other metals are currently in progress.

CONCLUSIONS

We have described a new method to study the interaction of H and D with metal defects. The defects and H (D) are incorporated in thin metal targets by p or d irradiations using a Van de Graaff accelerator. Information concerning the interaction of H(D) with vacancies is obtained by microscopic trapping of such complexes at impurity atoms through the use of Mössbauer spectroscopy.

In the case of Cu, evidence for the trapping of interstitial H (D) as well as vacancy associated H (D) complexes at ^{57}Co impurity

atoms is presented. The identification of the trapped species is
inferred from isochronal annealing measurements which provide the
thermal annealing behaviour. This behaviour can be quantitatively
described using a first order rate process from which migration
enthalpies (H^m) and binding enthalpies (H^b) are deduced. These
enthalpies are used as an aid in site identification.

The present method can be extended to the study of He-metal
defect interactions by the use of α particles. Because of its
microscopic nature the present method, in conjunction with resistivity measurements is likely to provide basic information regarding the interaction of metal-defects with H, D and He. The use
of hyperfine interaction methods in addressing these problems will
undoubtedly provide a contribution towards understanding impurity-
defect interactions in materials.

ACKNOWLEDGEMENTS

It is a pleasure to acknowledge valuable correspondence with
Professor J. Moteff, Professor G. Vogl, Dr. W.R. Wampler and Dr.
Sisson during the course of this work. We are particularly grate-
ful to Professor J. Anno for making available the use of the Van
de Graaff accelerator and to Howard Boeing for technical assistance.
This work was made possible by a grant from the University Research
Council.

REFERENCES

1. G. Alefeld and J. Volkl, "Hydrogen in Metals I and II,"
    Springer-Verlag, New York, (1978).
2. P. Jena, F.Y. Fradin, and D.E. Ellis, Phys. Rev. B20: 3543
    (1979); Z.D. Popovic, M.J. Stott, J.P. Carbotte, and G.R.
    Piercy, Phys. Rev. B13: 590 (1976), P. Jena and K.S. Singwi,
    Phys. Rev. B18: 2712 (1978).
3. C.D. Gelatt, Jr., H. Ehrenreich, and J.A. Weiss, Phys. Rev.
    B 17; 1940 (1978).
4. J. Friedel, Bev. Bunsenges 76; 828 (1972); A. Seyer, Phys. Lett.
    53A; 324 (1975); P. Jena, Solid St. Commun. 27; 1249 (1978)
    M.J. Stott and E. Zaremba Phys. Rev. B 22: 1564 (1980). W.R.
    Wampler, T. Schober, and B. Lengeler, Phil. Mag. 34: 129 (1976).
6. B. Lengeler, S. Mantl, and W. Triftshauser, J. Phys. F: Metal
    Phys. 8: 1691 (1978).
7. J.A. Brown, R.H. Heffner, M. Leon, M.E. Schillaci, D.W. Cooke,
    and W.B. Gauster, Phys. Rev. Lett. 43: 1513 (1979); also see A.
    Seeger, in ref. 1; F. Pleiter and C. Hohenemser, Phys. Rev. B
    (in press).
8. R.M. Cotts in ref. 1.
9. W. Mansel and G. Vogl, J. of Phys. F, 7: 253 (1977), see also

W. Mansel, G. Vogl and W. Koch, Phys. Rev. Lett., 31: 359 (1973).

10. L. Katz, M. Guinan, and R.J. Borg, Phys. Rev. B4: 330 (1971).

11. W. Schilling and K. Sonnenberg, J. Phys. F: Metal Phys. 3:322 (1973), and also see F.W. Young, Jr., J. Nucl. Mat. 69-70: 310 (1978).

12. G. Vogl, W. Mansel, W. Petry, and V. Groger, Hyper, Int. 4:681 (1978), and private communciation from G. Vogl.

13. C.T. Ma, B. Makowski, M. Marcuso, and P. Boolchand, in: "Nuclear and Electron Resonance Spectroscopies Applied to Material Science," E. Kaufmann and G.K. Shenoy, ed., Elsevier North Holland, Inc., New York, (1981), also see T.L. Marcuso, Ph.D thesis, University of Cincinnati (unpublished, 1979).

14. T.A. Thornton and J.N. Anno, J. of Appl. Phys., 48: 1718 (1977).

15. M. Camani, F.N. Gygax, W. Ruegg, A. Schenck, and H. Schilling, Phys. Rev. Lett., 39: 836 (1977).

16. J.P. Bugeat and E. Ligeon, Phys. Lett. 71A 93 (1979).

17. O. Echt, H. Graf, Holzschuh, E. Recknagel, A. Weidmger, and Th. Wickert, Phys. Lett. 67A; 427 (1978).

18. R.W. Balluffi, J. Nucl. Mat. 69-70: 240 (1978).

19. O. Hartmann, E. Karlsson, and L.O. Norlin, D. Richter, and T.O. Niinikoski, Phys. Rev. Lett., 41: 1055 (1978).

20. J.P. Bugeat, A.C. Chami, and E. Ligeon, Phys. Lett., 58A: 127 (1976).

21. E. Verbiest, H. Pattyn, and J. Odeurs, J. de Phys. 41: C1-431 (1980), see also K. Sassa, W. Petry, and G. Vogl, Proceedings of the Yamada Conference V, Kyoto (1981).

INTERACTION OF HYDROGEN WITH SUBSTITUTIONAL SOLUTE METALS IN THE ß-PHASE OF THE PALLADIUM-HYDROGEN SYSTEM

F.E. Wagner, M. Karger, F. Pröbst and B. Schüttler

Physics Department, Technical University of Munich
D-8046 Garching, Germany

ABSTRACT

Mössbauer data obtained with a number of resonances provide information on the local hydrogen distribution around more than a dozen substitutional solute elements in ß-PdH$_x$. From these data and a model for the description of the line positions and intensities in the Mössbauer spectra, one can obtain the interaction energies of Co, Ru, Rh, Sn, Os, Ir, Pt and Au solutes with different nearest neighbor hydrogen configurations. Except for Rh, all these elements clearly repel hydrogen. The energies necessary to bring the first hydrogen atom into the nearest neighborhood of a solute were found to lie between about 20 meV for Ru and at least 150 meV for Au and Sn. For higher numbers of nearest neighbors the repulsive interaction per hydrogen usually becomes weaker, and in some cases the configurations with 5 or 6 nearest neighbors even seem to be energetically favored over those with only 3 or 4.

INTRODUCTION

The ß-phase of the palladium-hydrogen system has been studied by Mössbauer spectroscopy with a number of resonances /1-5/, notably with those of 57-Fe, 99-Ru, 119-Sn, 151-Eu, 191,193-Ir, 195-Pt and 197-Au. In all these cases the Mössbauer atoms were introduced in concentrations of less than about one percent as substitutional solutes into the fcc Pd lattice. With the resonant nuclei probing mainly their nearest environment, such experiments yield microscopic information on the hydrogen distribution near the solute atoms. Actually one can investigate not only the environment of the Mössbauer elements but also that of their radioactive parents, i.e. the ele-

581

ments right or left of the Mössbauer atom in the periodic table,
depending on the type of nuclear decay that feeds the Mössbauer
transition. This is so because at low temperatures the diffusion of
hydrogen is so slow that the hydrogen distribution typical for the
parent element is expected to be retained for times much longer
than the time that elapses between the nuclear transformation and
the emission of the Mössbauer gamma radiation. This is usually determined by the lifetime of the Mössbauer level and lies between
10^{-7} and 10^{-9} s. Low temperature source experiments, in which the
radioactive isotope is introduced into the hydride matrix, will thus
represent the hydrogen environment typical for the parent element.
However the radioactive decay may itself stimulate hydrogen
jumps in its vicinity, which may arise due to the energy
deposition resulting from the Auger cascade following an electron
capture decay. There is, however, no experimental evidence for such
aftereffects of the nuclear decay and they appear to be improbable
due to the rapid heat dissipation in metallic systems. In the following, it will therefore be taken for granted that emission Mössbauer
experiments represent the unaltered distribution of nearest neighbor
hydrogen configurations typical for the source elements.

   As has been pointed out previously /5/, the isomer shift /6/ is
the most important Mössbauer parameter in the spectra of the non-
magnetic hydrides of Pd. The experimental data show /5/ that for
all studied solute elements but Sn the electron density $\rho(0)$ at the
probe nucleus decreases on going from the α to the β phase of PdH_x,
and further decreases with increasing x within the β phase. The
change of $\rho(0)$ on hydrogenation can be visualized as the sum of a
long-range and a nearest neighbor contribution. The long-range
term represents hydrogen-induced changes of gross volume and
electronic structure, while the hydrogen atoms in the nearest
neighborhood of the probe produce an additional change of $\rho(0)$ that
depends on their specific configuration. In PdH there are six nearest-neighbor octahedral sites that can be occupied by hydrogen. In
the nonstoichiometric, disordered hydrides one expects to have a
distribution of nearest-neighbor configurations. Hence the Mössbauer spectrum will be a superposition of components with different
isomer shifts, each of which represents a certain number i and spatial arrangement of nearest neighbors. The resulting structure in
the Mössbauer spectra turns out to be at best partially resolved.
As an example, 99-Ru spectra obtained /4/ with absorbers of Ru in
PdH_x are shown in Fig. 1. Spectra of comparable quality can be obtained with the 57-Fe and 193-Ir resonances /2,5/. The dependence
of the shapes of the Mössbauer patterns on hydrogen concentration
shows that the electron density $\rho(0)$ decreases monotonically with
the number i of nearest neighbors. When (i =)2, 3 or 4 out of the six
available nearest-neighbor octahedral sites are occupied, two different geometrical arrangements may occur for each i. There is no experimental evidence that this cis-trans isomerism causes noticable
differences in the electron density $\rho(0)$ at the metal nucleus.

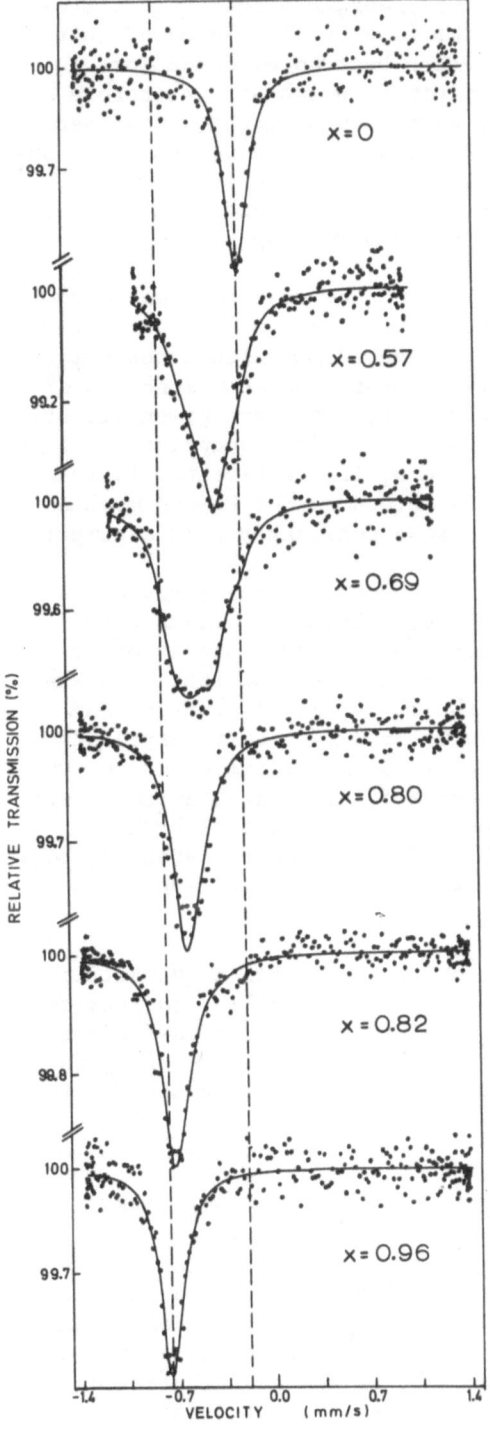

Fig. 1.
Mössbauer spectra obtained
with absorbers of 99-Ru in
PdH_x and single line sources
of 99-Rh in Ru metal at 4.2 K
/4/. The Ru content of the absorbers was typically 1 at. %.
The curves drawn to the data
points were obtained by least-
squares fitting the model
parameters described in the
text simultaneously to all
spectra.

Evidence from coordination complexes /7/ suggests that such differences should indeed be negligibly small, as will be assumed in the
following evaluations. Also, the spectra bear no evidence for
substantial electric quadrupole splittings, which will therefore
also be neglected in the following, even though small, unresolved
splittings cannot be completely ruled out. It is, however, well con-
ceivable that the electric field gradients produced by the non-
cubic arrangement of the protons in the otherwise cubic lattice are
rather small because of strong electronic shielding.

By applying the phenomenological concepts described above to
the interpretation of the hydride Mössbauer spectra, one can derive
the probabilities $P_i(x)$ with which configurations with i nearest
neighbors occur for different solute elements and as a function of
the hydrogen concentration x. These probabilities are given directly by the relative intensities with which the corresponding components contribute to the Mössbauer pattern, if the f-factor is the
same for all configurations. Since f-factors do not change much on
hydrogenation /3,8/, this appears to be a reasonably good assumption.

The most direct evidence for differences in the hydrogen distribution around different solute elements in ß-PdH$_x$ stems from
comparisons of the isomer shifts observed in source and absorber
Mössbauer experiments with the same Mössbauer transition. Such experiments have been performed with the 57-Fe, 99-Ru, 193-Ir, and
197-Au resonances /2-5,8/. They have sofar only been interpreted
qualitatively in terms of the attractive or repulsive nature of the
solute-hydrogen interaction. The outcome of this has been the most
of the studied impurity elements repel the interstitial hydrogen.
The only exception found sofar is Rh, which has more nearest hydrogen
neighbors than the statistical average throughout the ß-phase and
hence may be considered as a hydrogen trap.

In the present work a simple model approach for a quantitative
evaluation of such hydride Mössbauer spectra is applied to previous
/2-5,8/ as well as some new experimental data. In this way values
for the energies needed to form certain solute-hydrogen configurations in the palladium hydride matrix could be obtained for Co, Ru,
Rh, Sn, Os, Ir, Pt and Au impurities.

MODEL DESCRIPTION OF MÖSSBAUER SPECTRA OF DISORDERED HYDRIDES

As has been pointed out, the Mössbauer patterns obtained for solutes in Pd hydride are at best partially resolved. For a quantitative analysis of such spectra one has to make certain assumptions
concerning (i) the dependence of the isomer shift on the number of
nearest neighbors and (ii) the dependence of the distribution of
nearest-neighbor configurations on the hydrogen concentration x.
Spectra that reveal the presence of nearest hydrogen neighbors have

been observed with the 57-Fe, 99-Ru, 193-Ir and 197-Au resonances. These results are all compatible with the notion that the electron density $\rho(0)$ at the probe nuclei decreases monotonically with the number of nearest hydrogen neighbors, i. In the following we will assume a linear dependence,

$$S(i,x) = S(0,x) + i \cdot \Delta S, \qquad (1)$$

where i takes values from 0 to 6 and $S(0,x)$ represents the long-range influence of volume and electronic changes. The results for probes that have no nearest hydrogen neighbors up to very high hydrogen contents, like Sn and Au, show that $S(0,x)$ can be approximated by a linear function of x, whence

$$S(i,x) = S_O + x \cdot (\partial S/\partial x)_{i=0} + i \cdot \Delta S. \qquad (2)$$

Using this relation, one can fit the Mössbauer patterns obtained for probes like Co, Ru, Rh, Os or Ir at different values of x with super-positions of seven Lorentzian lines, whose relative intensities represent the probabilities $P_i(x)$ with which configurations with i nearest neighbors occur.

Comparing these values for $P_i(x)$ with the probabilities calculated for a binomial distribution, one finds that, with the exception of Rh, all impurity elements studied sofar have fewer hydrogen neighbors than are expected for a purely statistical interstitial site occupancy. To describe the probability to find i nearest neighbors to an impurity, we use the relation

$$P_i(x) = g \cdot \binom{6}{i} \cdot x^i \cdot (1-x)^{-i} \cdot \exp(-U_i/kT), \qquad (3)$$

where g is a normalization factor. This is in fact a binomial distribution multiplied by a Boltzmann factor which takes into account the fact that formation of a state with i nearest neighbors requires an excess energy U_i. A similar model has been used successfully to describe short-range order phenomena in random alloys. Eq. (3) implies the assumption that the cis and trans configurations occurring for i=2, 3, and 4 have the same excess energies. There is little justification for this assumption, but since cis and trans configurations cannot be distinguished in the spectra there is little point in introducing them separately into the expression for the P_i. As a consequence, the corresponding U_i should be regarded as averages over both configurations. Eq. (3) also does not take into account that the excess energies will depend on the occupation of sites beyond the first neighbor shell because of hydrogen-hydrogen interactions. Again, one has to consider the U_i as appropriate averages over the actual distributions. The fact that Eq. (3) holds only for low probe concentrations and hydrogen contents not too close to x = 1 also appears to be of minor importance in view of the other approximations that have to be made in order to arrive at a model description of the spectra.

RESULTS AND DISCUSSION

For the case of Ru solutes, Fig. 1 shows the fitted curves obtained by a simultaneous adjustment of the model parameters to the Mössbauer data obtained at different hydrogen concentrations x. The spectra for Co, Rh, Os, Ir, and Pt solutes can be fitted in a similar manner. For Sn and Au solutes the only information from the Mössbauer spectra is, that no hydrogen is found in their nearest neighbor shell up to the highest hydrogen concentrations attainable. Consequently, in these cases one can only determine lower limits for U_i/kT from Eq. (3).

The Boltzmann factor in Eq. (3) implies that the probabilities P_i depend on temperature. A temperature dependence roughly as predicted has, indeed, been observed with sources of 57-Co in PdH . For the present purposes, it is important that the 57-Co spectra change only at temperatures above about 100 K, but not below. Apparently, the thermal equilibrium in the population of sites in the vicinity of the impurities freezes out near 100 K for a wide range of cooling rates. It is, indeed, to be expected that hydrogen diffusion is by far too slow for equilibrium to be reached at 4.2 K, where most of the Mössbauer spectra were taken. On the other hand, a freezing temperature of 100 K appears rather high since the hydrogen jump rates at this

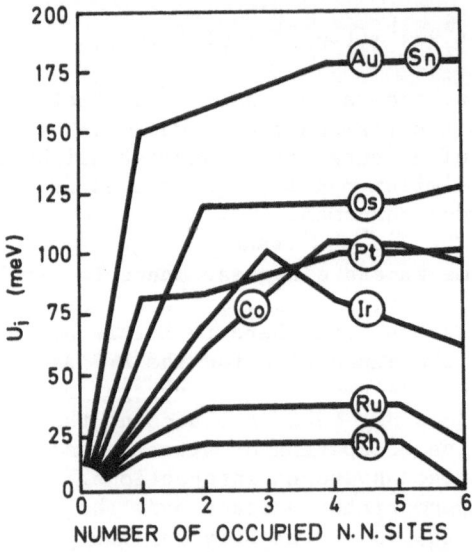

Fig. 2. Excess energies U_i for impurity-hydrogen configurations with i nearest hydrogen neighbors in ß palladium hydride.

temperature are still of the order of 10^3 s^{-1} /11,12/, but possibly
the jump rates near impurities are reduced by higher-than-normal potential barriers or blocking effects. In the absence of evidence to
the contrary, we have assumed that freezing of the equilibrium occurs
at 100 ± 10 K for all impurity elements and used this temperature in
order to obtain the configuration energies U_i from least-squares fits
of the spectra. The results are summarized in Fig. 2. The errors of the
configuration energies thus obtained are estimated to be about 20 %
in most cases, but may be substantially higher for configurations
that are not or weakly represented in the spectra. For Au and Sn, for
example, only lower limits necessary to keep the hydrogen out of the
nearest neighbor shell even at the highest hydrogen concentrations
$(x \approx 0.96)$ can be given.

As expected, the U_i values of Fig. 2 confirm the qualitative
classification of the impurity elements given previously /5/ in terms
of the strength of the repulsive interaction with hydrogen. A new
feature emerging from Fig. 2 is that the U_i increase steeply with i
for the first few neighbors, but level out for higher i values. This
means that once one or a few hydrogens occupy nearest neighbor sites
to the impurity, additional ones are accommodated more easily. In
some cases configurations with 5 or 6 neighbors even appear to be
energetically favored over those with 3 or 4. Generally the repulsive
interaction per hydrogen neighbor, U_i/i, seems to decrease with i.
In the case of Rh, it is not clear whether it is weakly attractive
for high or merely weakly repulsive for medium i as shown in Fig. 2,
since the latter, together with statistics, also ensures a higher
than average population of the nearest neighbor shell.

There does not seem to be a straightforward explanation for the
differences in the behavior of the various probe elements. Solutes
that are definitely too large to fit into the Pd lattice, like Au and
Sn, appear to be highly repulsive with $U_i > 150$ mV. The behavior of
elements with atomic radii close to or smaller than that of Pd, however, cannot be explained merely by size arguments. In these cases
electronic effects that cannot easily be understood seem to dominate.

CONCLUSIONS

We have shown that one can obtain the energies of formation
of certain impurity-hydrogen configurations in metal hydrides from
Mössbauer data. The limited resolution of the spectra as well as the
limited amount and statistical accuracy of the Mössbauer data are
still a major problem. At least the latter point, however, can be improved upon in the future. In order to arrive at a quantitative
interpretation of the Mössbauer spectra, several assumptions had to
be made. To some extent these have found an a posteriori justification by the resulting coherent interpretation of all existing data
with a reasonable and relatively small set of parameters. Neverthe-

less further experiments, for instance to clarify beyond doubt the role of electric quadrupole interactions or f-factors, appear desirable.

REFERENCES

1. F. E. Wagner and G. Wortmann, in: "Hydrogen in Metals I", Top. Appl. Phys. 29, G. Alefeld and J. Völkl, eds., Springer, Berlin (1978).
2. F. Pröbst, F. E. Wagner, M. Karger and G. Wortmann, Z. Phys. Chem. 114:195 (1979); - J. Physique Colloque 40-C2:635 (1979).
3. F. Pröbst, F. E. Wagner and M. Karger, J. Phys. F: Metal Physics 10:2081 (1980).
4. M. Karger and F. E. Wagner, Hyperfine Interactions 9:553 (1981).
5. M. Karger, F. Pröbst, B. Schüttler and F. E. Wagner, Proceedings of the Miami International Symposium on Metal-Hydrogen Systems, Pergamon Press, Oxford (in print).
6. "Mössbauer Isomer Shifts", G. K. Shenoy and F. E. Wagner, eds., North Holland, Amsterdam (1978).
7. N. N. Greenwood and T. C. Gibb, "Mössbauer Spectroscopy", Chapman and Hall, London (1971), p. 185.
8. M. Karger, F. E. Wagner, J. Moser, G. Wortmann and L. Iannarella, Hyperfine Interactions 4:849 (1978).
9. T. E. Cranshaw, J. Phys. F: Metal Physics 10:1323 (1980).
10. R. R. Arons, Y. Tamminga and G. de Vries, phys. stat. solidi 40:107 (1970).
11. S. R. Kreitzmann and R. L. Armstrong, contribution to this conference.

This work has been supported by the Federal Ministry of Research and Technology of the Federal Republic of Germany.

DEUTERIUM DESORPTION AND HOST INTERSTITIAL

CLUSTERING IN d-IRRADIATED Ni*

Carl Allard, Gary S. Collins and Christoph Hohenemser

Department of Physics
Clark University
Worcester, MA 01610

ABSTRACT

Perturbed angular correlations measurements were made on ^{111}In doped Ni after deuteron irradiation or deformation at 77 K. The irradiated samples provide evidence of deuteron desorption at 220 K as well as host interstitial clustering on ^{111}In during irradiation.

INTRODUCTION

In recent years perturbed angular correlations (PAC) of gamma rays have been widely applied to study the migration, trapping and clustering of lattice defects in damaged metals. Essential features of these experiments are distinctive precessional signals arising in the nuclear hyperfine interaction. These signals serve as convenient labels for specific defects located in the neighborhood of the gamma emitting probe, and in certain cases permit interpretations based on defect-probe structure.

In a recent review of ^{111}In PAC experiments in fcc metals Pleiter and Hohenemser[1] presented strong arguments that most defects observed so far in Ag, Al, Au, Cu, Ni, Pd and Pt have a vacancy character. There are defects, however, for which the vacancy character is not clearly established; for instance recent low temperature PAC work on Al[2] and Cu[3] are almost certainly of an interstitial character. Interstitial trapping is also quite generally observed when small probe atoms such as [57]Co and [100]Rh are used.[4] Finally, there has been recent work in which the effect of He loading of vacancies[5] and the presence of interstitial H[6] have been observed

---

*   Supported by National Science Foundation Grant DMR 81-08307.

via ^{111}In PAC in fcc metals. The light atom effects are now under intensive study.

In the present paper we present recent PAC results on ^{111}In doped Ni which has been (a) deuteron irradiated at 77 K, and (b) cold-worked at 77 K. Subsequent isochronal annealing of the irradiated samples gives evidence for deuterium migration and escape at ~220 K, and relates to previous work on hydrogen loaded Ni.[6] The irradiated samples also indicate that interstitial clusters are trapped on the ^{111}In during irradiation, followed by subsequent evaporation and anti-defect annihilation at higher temperature. In contrast, cold-worked samples show only vacancy trapping near stage III, as observed in all previous ^{111}In PAC experiments in Ni.[1,7]

EXPERIMENTAL METHODS

A detailed discussion of the perturbed angular correlation technique, as applied to defects, has been given in the review by Pleiter and Hohenemser[1] and the references therein. In brief, from the correlation function $G_2(t)$ between emission of two gamma rays in a cascade, one can detect fractional populations of probe nuclei precessing in internal perturbing fields caused by the set of local atomic environments. For ferromagnetic Ni, the simplest case encountered is a constant hyperfine field H_{hf} which results in a sinusoidal precession. Another case found in the present work is a combined magnetic/electric interaction. This has a more complicated character, and has been described in detail by Pleiter.[8] Examples of observed correlation functions are given in Fig. 1.

In these experiments sources were prepared from 99.998% pure, polycrystalline Ni foil. Carrier free ^{111}In was electro-deposited onto these foils and diffused to a region approximately 10μm thick. Two sources were irradiated at 77 K with a 0.6 MeV d$^+$ beam, to doses of 43 x 10^{17} and 1 x 10^{17} d$^+$/cm^2. Range and straggling of the deuterons were approximately 2.0 μm and 0.2 μm respectively. For cold-working experiment sources, deformation at 77 K was achieved with a jeweler's roller completely submerged in liquid N$_2$.

Initial PAC experiments were done at 77 K under liquid N$_2$ before and after introduction of defects, without intervening warming. Subsequent PAC measurements were done at 77 K after isochronal annealing for 15 minutes at temperatures ranging up to 800 K. These indicate that before irradiation, nearly the full signal is described by a nuclear larmor frequency ω_L = 104 Mrad/s (Fig. 1a). According to previous work[9] this arises from defect-free ^{111}In in substitutional lattice locations. In contrast, after irradiation, approximately 25% of the substitutional signal is replaced by an inhomogeneous, combined interaction signal having a quadrupole frequency of (80-20) Mrad/s (Fig. 1b,c). The latter must be attributed to a defect or defects trapped on the ^{111}In during irradiation, but

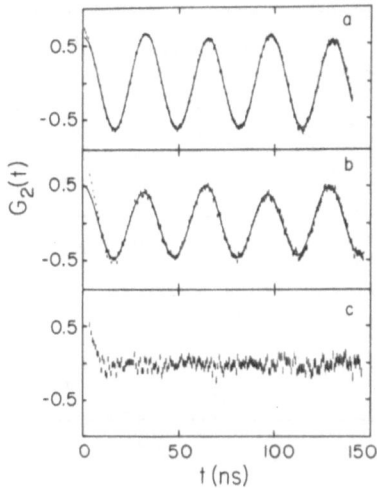

Fig. 1.   Precessional correlation functions observed (a) before the
high dose d$^+$ irradiation, and (b) after the high dose irradiation.   The sinusoidal frequency results from ^{111}In atoms
on defect-free (substitutional) lattice sites.   The amplitude of this signal, proportional to the fraction of defect-
free In atoms, is reduced by about 25% during irradiation
by introduction of lattice defects.   Curve (c) illustrates
the form of the mixed interaction signal observed in (b),
and was obtained by subtraction of the periodic signal from
(b).

since the frequency and amplitude of this signal is not well defined,
our discussion below is focussed on changes in the defect-free
signal.

ISOCHRONAL ANNEALING RESULTS

     The results of isochronal annealing sequences for d$^+$ irradiated
sources are shown in Fig. 2a,b.   Results for cold-worked sources are
described in Fig. 2c.   For comparison, previous results for [111]In
implanted Ni[9] are shown in Fig. 2d.

     Irradiated samples.   For the d$^+$ irradiated samples three fea-
tures summarize the defect-free fraction.   First there is a sharp
drop immediately after irradiation at 77 K, followed by slow recovery that does not reach completion even at 800 K.   Secondly, for the
high dose case, there is a well-defined recovery stage at 220 K.
(For the low dose case a similar stage may exist, but this is
unclear because of an insufficient number of points.)   Lastly, also
for the high dose, there is evidence of a trapping stage near 300 K,
very similar to that seen in implanted sources (Fig. 2d).

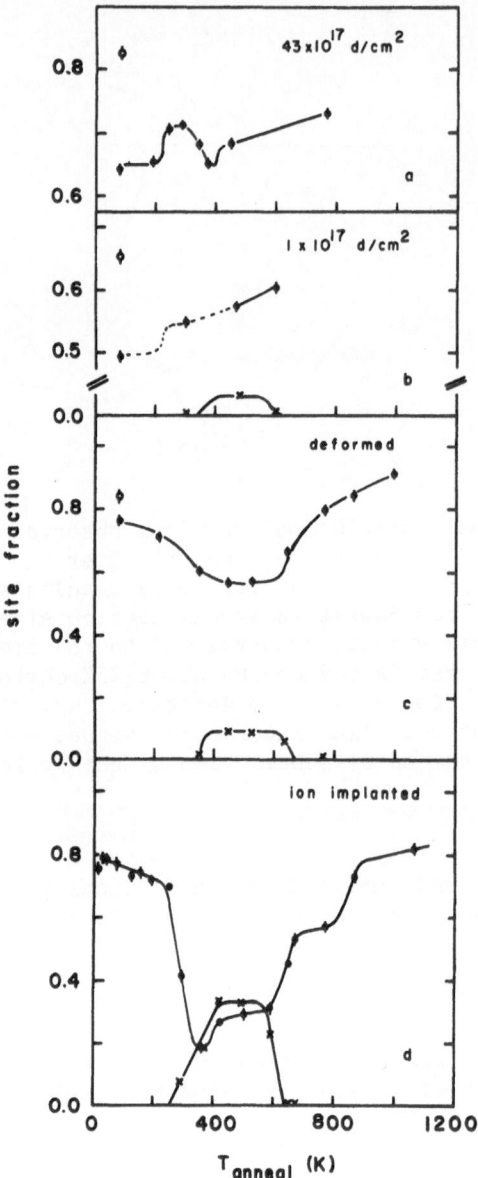

Fig. 2.   Site fractions of ^{111}In in Ni observed after deuteron irra-
          diation at high dose (a) and low dose (b), and after
          deformation (c) and ion implantation (d). Open and closed
          circles identify defect-free fractions before damaging and
          after isochronal anneals at the indicated temperatures,
          respectively. Crosses indicate the fraction of a 40 Mrad/s
          trapped vacancy defect; this defect was not observed after
          the high dose irradiation (a).

Beyond the defect-free fraction, we find that following irradiation and during subsequent annealing, the lost defect-free fraction appears as an inhomogeneous combined interaction with a quadrupole frequency of 80 ± 20 Mrad/s. Also seen in the low dose case is the well known ω_L = 40 Mrad/s state, previously identified as a vacancy defect.[1] It appears just as it does in implanted sources (Fig. 2d).

Cold-worked samples. Here, in contrast to the irradiated case we find only a small drop following defect introduction at 77 K and no recovery stage at 220 K (Fig. 2c). Near 300 K, the defect-free fraction declines, and the distinctive ω_L = 40 Mrad/s stage appears, exactly as in implanted sources.

DISCUSSION

Of the features observed here, the behavior near room temperature is consistent with vacancy defect trapping, as described in earlier work on Ni.[9] What is new in the present work is the recovery stage in irradiated sources at 220 K; and the creation of an apparent defect trap during irradiation, followed by slow recovery through 800 K. We interpret the 220 K recovery as resulting from deuterium desorption, and we ascribe the 77 K trap to host interstitial clusters.

Deuterium desorption. The range of the implanted deuterons was overlapped by the ^{111}In distribution completely. For the high dose we estimate that the deuterium concentration in the region of stopping is near one per Ni atom. Accordingly we expect that for the stopping region the PAC signal will be determined by an inhomogeneous combined interaction which replaces a corresponding fraction of the defect-free signal as observed. The fraction of ^{111}In so affected is not expected to be more than 5% because of the narrowness of the deuteron stopping region (<1 μm). Primary confirmation of this picture arises out of the observed stage at 220 K, where significant migration of d is expected according to diffusion measurements,[10] and where 5% of the substitutional fraction is observed to recover. Additional confirmation is obtained from the work of Thomé and Bernas,[6] who observed a deuterium desorption stage at 220 K in resistivity measurements, and who also found recovery of the PAC-determined substitutional fraction at an undetermined temperature somewhere in the range 77<T<300 K.

Interstitial trapping. According to above model of the recovery at 220 K, most of the 25% decrease in the defect-free fraction during irradiation must be explained in other ways. Assuming that vacancies are immobile, and hence incapable of being trapped during irradiation (according to Pleiter and Hohenemser[1] vacancies have never been observed to trap athermally during irradiation), we attribute the loss of the defect-free fraction to trapping of freely

migrating interstitials. Previous PAC studies of Al^2 and Cu^3 have
shown sharply defined detrapping of interstitials from ^{111}In, in
contrast to the present case. Since the former experiments were
done at relatively low dose, it is natural to presume that the
present work involves accretion of extended interstitial defects.
Confirmation of this picture is obtained by noting that recovery of
the defect-free fraction occurs over a wide range of temperature in
both sources studied, but that the low dose case shows more rapid
recovery (i.e. smaller interstitial clusters). Added confirmation
of this model comes from the fact that the well known 40 Mrad/s
vacancy trap above 350 K is observed only in the low dose case,
where, presumably, not all available ^{111}In atoms are already trapped
in interstitial clusters.

SUMMARY AND CONCLUSION

The present study on Ni has given evidence for a well-defined
deuterium desorption stage at 220 K as well as accretion of interstitial clusters around ^{111}In during deuteron irradiation at 77 K.
Further confirmation of these observations will come in future work,
in which we plan to study lower dose irradiations for which the
deuteron stopping range lies beyond the ^{111}In doped region.

REFERENCES

1. F. Pleiter and C. Hohenemser, Phys. Rev. B25, 106 (1982).
2. R. Sielemann, H. G. Müller, W. Semmler, R. Butt, H. Metzner,
    Yamada Conference V on Point Defects and Defect Interactions
    in Metals, Kyoto 1981, in press.
3. M. Deicher, O. Echt, E. Recknagel and Th. Wichert, Geometrical
    Structure of Lattice Defect-Impurity Configurations Deter-
    mined by TDPAC, in: "Nuclear and Electron Resonance Spectros-
    copies Applied to Materials Science," E. N. Kaufmann and G.
    K. Shenoy, ed., North-Holland, New York (1981).
4. W. Mansel, G. Vogl and W. Koch, Phys. Rev. Lett. 31, 359 (1973).
5. H. de Waard, Hyperfine Interaction Investigations of Helium
    Trapping in Metals, in: Ref. 3, loc. cit.
6. A. Traverse, T. Kachnowski, L. Thomé and H. Bernas, Phys. Rev.
    B22, 4355 (1980); L. Thomé, A. Traverse, L. Brassard and
    H. Bernas, Hyp. Int. 9, 559 (1981).
7. G. S. Collins, G. P. Stern and C. Hohenemser, Phys. Lett. 84A,
    289 (1981).
8. F. Pleiter, A. R. Arends and H. G. Devare, Hyp. Int. 3, 87
    (1977).
9. C. Hohenemser, A. R. Arends, H. de Waard, H. G. Devare, F.
    Pleiter and S. A. Drentje, Hyp. Int. 3, 297 (1977).
10. J. Völkl and G. Alefeld, Hydrogen Diffusion in Metals, in:
    "Diffusion in Solids," A. S. Nowick and J. J. Burton, ed.,
    Academic Press, New York (1975).

TRAPPING BY DISLOCATIONS IN α'-PdD$_x$ OBSERVED BY NMR

A.J. Holley, W.A. Barton and E.F.W. Seymour

Physics Department
University of Warwick
Coventry CV4 7AL, UK

ABSTRACT

Spin-spin relaxation of ^{2}D in α'-PdD$_{0.64}$ measured by a CPMG sequence exhibits two coexisting relaxation times, $T_{2b} < T_{2a}$, both much less than the spin-lattice relaxation time T_1 at 20°C, when the sample is loaded from the gas phase so that it passes through the α-α' two-phase region. T_1 is explained by a fluctuating quadrupolar interaction with electric field gradients due to the motion of neighbouring deuterons. T_{2a} must be predominantly due to some other mechanism, identified as interaction with field gradients due to long-range strain fields of dislocations, with a much longer correlation time than that characterizing the T_1 mechanism. T_{2b} is tentatively ascribed to deuterons trapped in dislocation cores, since this decay is absent in a sample loaded under high pressure so as to avoid the mixed-phase region and thus greatly reduce the dislocation density. The signal amplitude associated with T_{2b} is temperature dependent with an activation energy ~ 40 meV which would represent the deuterium-dislocation interaction energy. A difficulty with this interpretation is the large apparent fraction of trapped deuterium.

INTRODUCTION

The interaction of hydrogen with dislocations has been demonstrated by a variety of means, including a cold-work internal friction peak[1], and the effect of mechanical deformation on the solubility and diffusion of hydrogen[2,3,4]. Results have been interpreted in terms of trapping at dislocation cores, and a weaker interaction of hydrogen with the long-range elastic strain fields

of dislocations. A solubility enhancement observed on cycling a
sample through the $\alpha-\alpha'$ mixed phase region[5] results from a large
dislocation density produced by the mismatch in lattice constants
of the two phases.

By loading with hydrogen under pressure, α'-phase samples can
be produced without passing through the two-phase region; in this
case the dislocation density is much reduced. In this paper the
effect of dislocations is studied through ^2D spin relaxation in
samples loaded through and over the miscibility gap. Deuterium was
chosen rather than hydrogen since ^2D has an electric quadrupole
moment and is therefore sensitive to lattice distortions through
electric field gradients produced.

RESULTS

Foil samples were loaded with deuterium from the gas phase.
Sample A was cycled three times through the two-phase region;
electron microscopy showed a dislocation density $\sim 10^{12}$ cm^{-2}.
Sample B was loaded at 35 atm. and 280°C and was essentially dislocation free.

T_1 passes through the familiar minimum when $\omega_o \tau_c \sim 1$, where
$\omega_o/2\pi$ is the resonance frequency (7 MHz) and τ_c the correlation
time of the fluctuating interaction responsible (Fig. 1). For
sample A two co-existing spin-spin relaxation times, T_{2a} and T_{2b},
were observed using a Carr-Purcell-Meiboom-Gill pulse sequence
(Fig. 2), both much shorter than T_1, even though one expects
$T_1 = T_2$ for $\omega_o \tau_c < 1$. The initial signal heights, x_1 and $x_2 - x_1$,
for the separate relaxation processes are expected to be proportional
to numbers of deuterons contributing. The fraction x_1/x_2 increases
with decrease of temperature yielding an apparent activation energy
~ 40 meV. For sample B T_{2a} is approximately doubled and the T_{2b}
decay is essentially removed.

DISCUSSION

The spin-lattice relaxation rate, T_1^{-1}, is attributed to a
combination of a Korringa conduction electron term, a nuclear
dipole-dipole term (both calculable from the known ^1H values in
$\alpha' - PdH_{0.64}$) and a quadrupolar term arising from fluctuating
short-range electric field gradients due to neighbouring deuterium
ions. For this last contribution $\tau_c = \frac{1}{2}\tau_D$, where τ_D is the mean
diffusional jump time, and $\omega_o \tau_c$ is such that the associated T_2 is
approximately the same as T_1.

The above contribution to T_{2a}^{-1} is a small fraction of that
observed. The dominant contribution is attributed to a quadrupolar

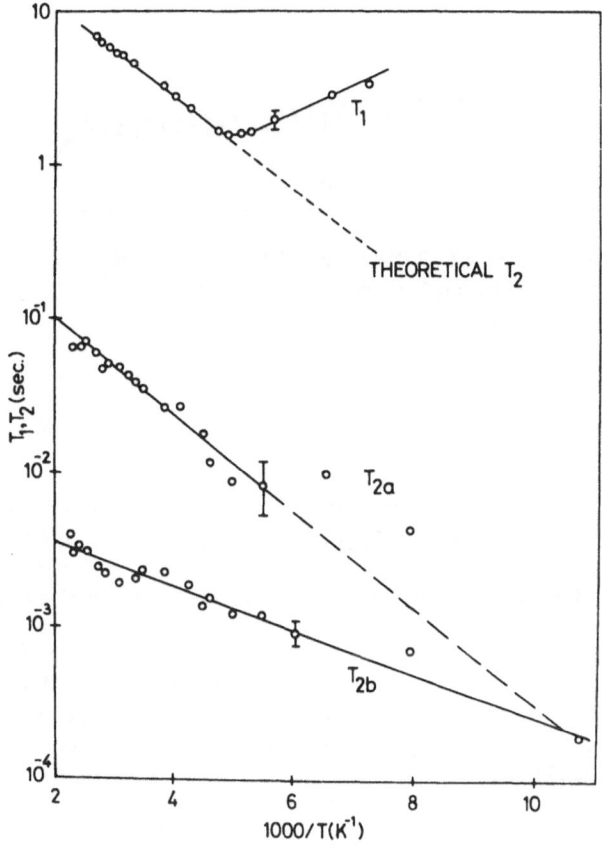

Fig. 1.   Temperature dependence of T_1 and T_2 for PdD$_{0.63}$ at 7 MHz (Sample A).

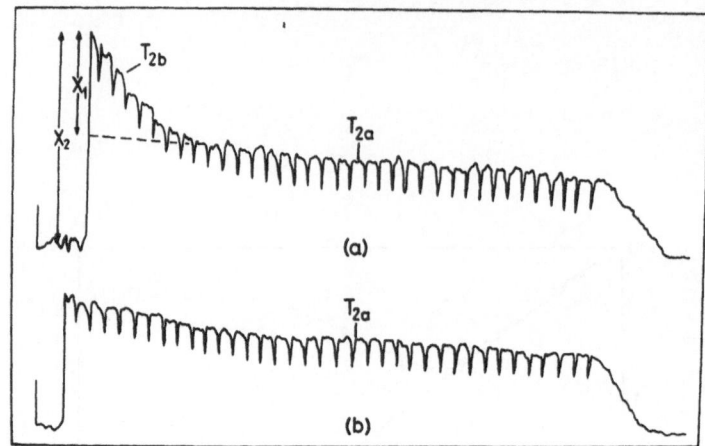

Fig. 2.   CPMG traces at 7 MHz and 295 K for PdD$_{0.63}$ (a) Sample
          A cycled through mixed phase region   (b) Sample B
          avoiding mixed phase region.

interaction with the long-range strain fields of dislocations proportional to the reciprocal of the distance from a dislocation core.
For a dislocation density $\sim 10^{12}$ cm^{-2}, the appropriate $\tau_c \sim 600$ μs
at 295K, estimated from $n^2\tau_D$ where n is the mean number of step
lengths between dislocations. Using this value, the mean quadrupolar
coupling deduced from T_{2a} is ~ 1 kHz, in satisfactory agreement with
the value estimated from the known elastic strain. Since $\omega_0\tau_c \gg 1$,
$T_1 \gg T_{2a}$ and the associated T_1 contribution is negligible (25 s
at 295 K). T_{2a} increases by about a factor of two for sample B, as
expected qualitatively for a reduced dislocation density.

     The T_{2b} process is essentially removed by loading without
passing through the miscibility gap. It is tempting therefore to
attribute T_{2b} to quadrupolar interaction for nuclei trapped within
dislocation cores. Without information about either the magnitude
of the interaction or τ_c one cannot find these two quantities
separately. Kirchheim[4] finds that at 295 K the dislocation pipe
diffusivity is about ten times the bulk value in α-PdH$_x$. If this
were so here, we would need a somewhat unlikely quadrupole coupling
of 300 kHz and moreover T_1 for these nuclei would be equal to T_{2b},
which we have not so far observed. We prefer to interpret T_{2b} as
due to a smaller coupling ~ 30 kHz (similar to that observed in
distorted tetrahedral sites in bcc metals) and a longer τ_c perhaps
arising from blocking of filled core interstitial sites.

Although the evidence indicates that the T_{2b} process is specifically connected with dislocations, a serious difficulty arises if one interprets the observed ratio x_1/x_2 of initial signal amplitudes for sample A with the fraction of deuterium trapped within dislocation cores. Although this fraction increases with decreasing temperature with an apparent activation energy ~ 40 meV, which is of the same order as the hydrogen-dislocation interaction energy found by Hasegawa and Nakajima[6], its value is 40% at 295 K. If one takes a core radius of ten Burgers vectors[3] and assumes only octahedral site occupation within cores, the $x_1/x_2 < 37\%$. One would need to suppose some additional tetrahedral site occupation within cores or a larger core radius.

This work was supported by UK SERC, UKAEA Harwell and NATO.

REFERENCES

1. C. Baker and H.K. Birnbaum, Scripta Met. 6:851 (1972).
2. T.B. Flanagan and J.F. Lynch, J. Less-Common Met. 49:25 (1976).
3. W.R. Tyson, J. Less-Common Met. 70:209 (1980).
4. R. Kirchheim, Acta Met. 29: 835 (1981).
5. T.B. Flanagan, B.S. Bowerman and G.E. Biehl, Scripta Met. 14:443 (1980).
6. H. Hasegawa and A.W. Nakajima, J. Phys. F, 9:1035 (1979).

POSITIVE PI-MESONS AND POSITIVE MUONS AS LIGHT ISOTOPES OF HYDROGEN

K. Maier and A. Seeger

Max-Planck-Institut für Metallforschung, Institut für
Physik, and Universität Stuttgart, Institut für theore-
tische und angewandte Physik, Postfach 800665
D-7000 Stuttgart 80, Germany

INTRODUCTION

It must be considered as a fortunate coincidence that in a
period of great scientific and technological interest in the behaviour
of hydrogen in metals and alloys it has become possible to perform
experiments on short-lived elementary particles of the same "nuclear
charge" as hydrogen. Of these particles positrons (e^+) have been
employed longest in solid-state physics and chemistry. However, since
the positron mass equals the electron mass m_e, the dynamic behaviour
of a positron in a crystal may be quite different from that of a
proton or of a heavier hydrogen nuclide. Therefore, it is often difficult to relate the behaviour of positrons in metals quantitatively
to that of protons, particularly when the gap of a factor of 1836
between the proton mass m_p and m_e exists.

In recent years this gap has been partially bridged by investi-
gations using positive muons (μ^+, of mass $m_\mu = 207 m_e \approx m_p/9$, lifetime
$\tau_\mu = 2.2 \cdot 10^{-6}$s) and positive pions ($\pi^+$, of mass $m_{\pi^\pm} = 273 m_e \approx m_p/7$,
lifetime $\tau_{\pi^\pm} = 2.6 \cdot 10^{-8}$s). On the one hand, these two "radio-isotopes"
of hydrogen are much lighter than the hydrogen nuclides and should
therefore exhibit quantum effects more clearly than the latter. On
the other hand, they are still heavy compared with the electrons, so
that the theoretical treatment of their behaviour in condensed matter
may take advantage of the same simplification as that of hydrogen.

Some of the areas in which experimental information useful for
the understanding of hydrogen in condensed matter may be obtained
with the help of μ^+ and π^+ are the following:
(i) Quantum theory of diffusion, extension of diffusion measurements

601

to very low temperatures and isotope effects.
(ii) Lattice sites and vibration amplitudes.
(iii) Interaction with foreign atoms or intrinsic defects.
(iv) Materials with low hydrogen solubility.

By now the use of positive muons as "light isotopes of hydrogen" by means of a variety of so-called μ^+SR (muon spin rotation) techniques may be considered as well established [1,2]. In the present review we confine ourselves to what is presumably the most powerful μSR technique for studying muon diffusion in crystals, namely the measurement of the logitudinal spin relaxation of muons in ferromagnetic single crystals to which a large magnetic field has been applied. In Sect. 3 we shall report on such experiments in progress on α-Fe and on their interpretation.

Compared with muons, the application of positive pions in solid state physics is very young [3,4]. Nevertheless, a number of very promising and encouraging results have been obtained, and a short review of these will be given in Sect. 2. Since pions possess zero spin and are thus without magnetic moments, techniques analogous to μ^+SR are not available. It has recently been realized [3], however, that the behaviour of π^+ in crystals can be studied very efficiently by means of the lattice steering (channelling) of the positive muons resulting from the decay (ν_μ = muon neutrino)

$$\pi^+ \rightarrow \mu^+ + \nu_\mu \quad . \tag{1}$$

In a sense, positive pions allow us to perform virtually "ideal" channelling experiments, since the decay of muons is monoenergetic with a kinetic energy $E_{kin} = 4.12\text{MeV}$ that is convenient for lattice-steering, since the "source" π^+ decays without leaving any impurities in the crystal, and since the number of implanted π^+ needed in a given experiment is usually small enough for the accompanying radiation damage to be negligible. Similar experiments can be done on the μ^+/e^+ pair, making use of the muon decay (ν_e = electron neutrino)

$$\mu^+ \rightarrow e^+ + \overline{\nu}_\mu + \nu_e \tag{2}$$

and the lattice steering of the relativistic positrons (with energies up to 53 MeV). The feasibility of such experiments has been demonstrated [5]. However, as discussed elsewhere [5], the μ^+/e^+ experiments are much more difficult to perform than the π^+/μ^+ lattice-steering experiments, and we shall not consider them further in the present paper.

For a long-range perspective of the application of π^+ and μ^+ to solid-state problems the reader is referred to a recent conference report [6], for further background information on lattice steering to the review literature [7,8].

CRYSTAL PHYSICS USING π^+/μ^+ LATTICE STEERING

Introductory Remarks

Fast π^+ and μ^+ injected into crystals are slowed down to thermal energy ("thermalized") within about 10^{-12}s, i.e., in times that are very short compared with their lifetime. Subsequently they either occupy interstitial sites in the crystal or, in the presence of impurities or other crystal defects, special trapping sites. Information on the implanted particles is obtained through the charged particles (μ^+ or e^+) in the decays (1) or (2). The emission probability of μ^+ in the decay (1) is isotropic, but the probability for the muons to leave the crystal is not (Fig. 1). This is due to the fact that the trajectories of μ^+ emitted in directions close to one of the major crystallographic directions are subject to lattice-steering effects. Lindhard's critical angle Ψ_{cr} [9] for the existence of lattice-steering phenomena depends on the atomic number Z and the lattice spacing d in the crystallographic direction. For muons with $E^\mu_{kin} = 4.12$ MeV we obtain

$$\Psi_{cr} = 0.15(Z/d)^{1/2} \quad (\Psi_{cr} \text{ in degrees, d in Å}) \quad (3)$$

The Ψ_{cr} are of the order of magnitude of a few tenth of a degree (Ta <111> : 0.8°; Ge <110> : 0.4°). The muon flux distributions around the lattice-steering directions contain detailed information on the behaviour of pions in the crystal.

On the material side the only requirement is the need for fairly large and perfect single crystals with a mosaic spread that is small compared with the critical nagles. The growth of such crystals does not present serious problems for valence crystals and for the majority of the metals. With the help of carefully aligning small crystals to a reasonable large 'quasi-single crystal', a technique which has been developed for neutron scattering experiments, π^+/μ^+ measurements on

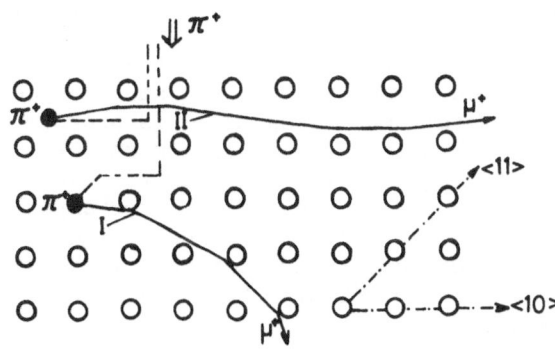

Fig. 1.

Lattice steering of decay muons (I) from a pion located on a lattice site and (II) from a pion located in an interstice.

nearly all crystalline solids are possible.

Experimental [10]

The experimental set-up has to satisfy the following requirements:
(i) In order to resolve the muon intensity profiles the angular resolution of the detecting apparatus has to be better than about 0.2°.
For a crystal size of about $2\times2\text{cm}^2$ this implies a minimum distance
crystal-detector of D = 6m. A crystal size of not less than about
$2\times2\text{cm}^2$ is highly desirable for intensity reasons, since at the Swiss
Institute of Nuclear Research (SIN), at which all experiments discussed
in this review have been carried out, the incident μ^+ flux of about
$10^8 \pi^+ \text{s}^{-1}$ is distributed over a cross-section of about that magnitude.
(ii) In order to reduce the counting time and to eliminate the effects
of fluctuations in the primary μ^+ intensity, the entire angular
distribution of the emitted muons has to be measured at the same time
and cannot be determined by angular scans as is commonly done in
channelling experiments. Thus a two-dimensional position-sensitive
detector capable of accepting an angular range of about $4\Psi_{cr}$ is
required.
(iii) The channelled decay muons undergo dechannelling when penetrating
the crystal. According to an estimate by Van Vliet [11] the mean
multiple-scattering angle $\Delta\Psi$ after a penetration length of $10\mu\text{m}$ at
room temperature amounts to 0.18° (Ta) or 0.12° (Au) for best-
channelled muons in a <100> channel, and to 0.58° (Ta) or 0.53° (Au)
for muons with high transverse energy (directional straggling). A
comparison with the expected critical angles shows that only muons
coming from depths of not more than $20\mu\text{m}$ should be counted. In order
to deposit as many π^+ as possible within a layer of this thickness
from the entrance surface, the pion beam (typically with a kinetic
energy of 65 MeV, corresponding to a momentum p = 150 MeV/c) has to
be moderated (Fig. 2).

The maximum stopping density in the surface layer is obtained
if the thickness of the moderator approximately corresponds to the
pion range. However, on account of the fluctuations of the energy
there is a fluctuation in the range of the individual pions ("range
straggling"). The width of this range-straggling distribution is too
large for the directional straggling of the majority of the muons
reaching the surface to be tolerated. Since for a given small-angle
scattering mechanism the square of the scattering angle is proportional to the energy loss, we may eliminate undesirable muons by
monitoring their energy loss. This requires the detector to resolve
the muon energy with a resolution   better than 400 keV (corresponding to a depth of about $15\mu\text{m}$ in Ta). By a proper choice of the
detector thickness, the large flux of background positrons (mainly
from the μ^+-decays in the crystal) can be discriminated by using
the detector in a dE/dx arrangement for positrons.

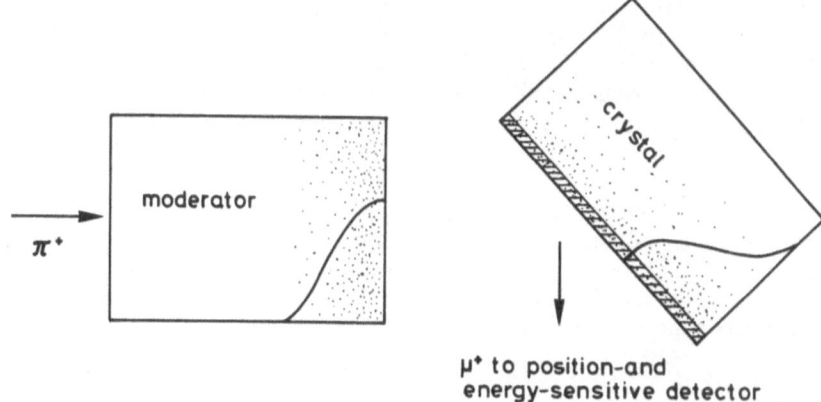

Fig. 2. Range straggling of 150 MeV/c π^+ in the moderator and the crystal. The full lines and the dots represent the density of deposited pions. Only about 1% of the π^+ are stopped within the volume on which the measurements are performed (hatched area); the decay muons coming from this volume are identified by measuring their kinetic energy.

The crystal is mounted on a goniometer allowing angular tilts in two directions by about 3° and rotations by about 45°. The samples are preoriented by standard X-ray techniques and mounted in such a way that the crystallographic direction to be investigated points (within 0.1°) in the direction of the detector. Because of the relatively low kinetic energy of the muons of about 4 MeV the entire muon path (from crystal to detector) of about 8m length has be be in vacuum, since even a helium-gas atmosphere would lead to considerable small-angle scattering of the muons. The scattering chamber is shielded in such a way that only muons from the crystal can reach the detector. Rutherford scattering of the pions gives only a small uniform background. The majority of the scattered pions decay in flight within the first two meters; therefore their "blocking structure" is completely averaged out.

The muon intensity profiles are measured with a windowless position-sensitive scintillation detector (1.5mm thick) with a counting area of 300 mm in diameter divided into four sectors. About 70% of the scintillation light is collected in four photomultiplier tubes. The coordinates X, Y of an event are determined by

$$X = x_1/(x_1+x_2) \quad , \qquad Y = y_1/(y_1+y_2) \quad , \qquad (4)$$

where x_1, x_2, y_1, and y_2 denote the pulse heights of the four tubes. The division according to (4) is done on-line with analogue dividers. The spatial resolution of the detector is better than 2cm, its energy resolution for 4 MeV muons corresponds to FWHM = 300 keV. In front of and behind the scintillators are anti-counter tubes in order to

reduce the background due to high-energy positrons crossing
the detector under small angles (thus having a prolonged path length
and giving rise to scintillation light similar to that of 4 MeV
muons). For a pion momentum of 150 MeV/c and a sample area of $4cm^2$
a signal-to-background ratio of 15 : 1 has been obtained.

Results

There are many potential applications of the technique. We give
here a number of examples; further results are expected in near
future. The results to be described below were obtained with a crystal detector distance of 8.15m.

a) Pion sites

The pion lifetime is short enough for pions in "pure" materials
not to be affected by the impurities present, so that, disregarding
the isotope effect for the time being, we can gain information on the
hydrogen sites, irrespective of the hydrogen solubility. Fig. 3 shows
the μ^+ intensity profiles along <100> and <111> directions of a tantalum crystal. Heights and widths of the two profiles are compatible
with tetrahedral interstices, in agreement with the determination of
deuteron sites [12]. The muon flux distribution is sensitive to the
vibrations (thermal or quantum mechanical) of the pions. This allows
us to study the potential in the neighbourhood of the pion sites and
thus to gain further information relevant for the understanding of
hydrogen in condensed matter.

From the measured profiles in Fig. 3 we may conclude that in
Ta the vibration amplitudes of the pion are rather small (<0.2 Å
at 200 K).

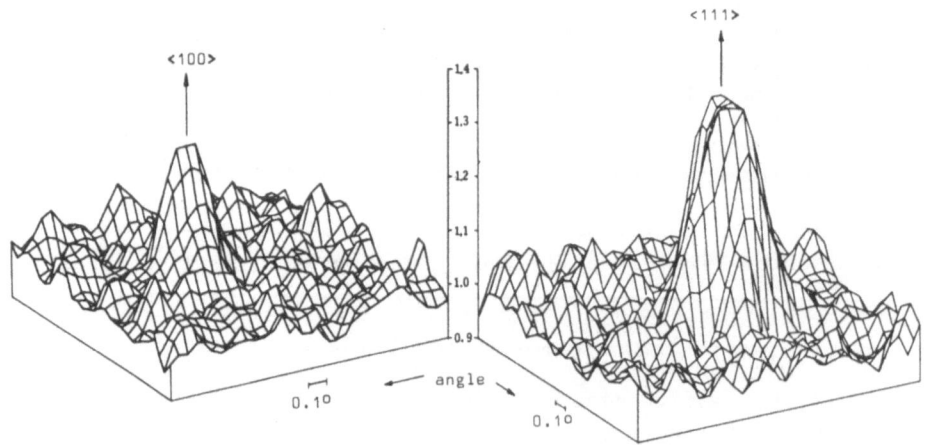

Fig. 3. Normalized intensity of decay muons from positive pions
        stopped in a Ta single crystal at 200 K along <100> and
        <111> directions.

Similar μ^+ profiles were found in Au <110>, Ge <110>, and Mo <100>. Here measurements in other directions have yet to be performed in order to determine the occupied sites with certainty.

b) Temperature dependence of energy loss
The dependence of the flux distribution on the energy of the emerging muons (i.e., on the energy loss during the channelling) gives us information on the dechannelling by phonon scattering.

Fig. 4 shows the temperature dependence of the number of muons within the half-width of the <100> channelling peak of Ta (normalized with respect to the same solid angle in a random direction) for the first or second energy window of the detector. In the first energy window there is very little interference of phonon scattering up to 350 K. This suggests that if the measurements are limited to muons with very small energy losses we can carry out the π^+/μ^+ lattice steering experiments even at very high temperatures.

c) Trapping of π^+ at oxygen dissolved in tantalum
From μ^+SR experiments we know that in many crystals positive muons are highly mobile even down to very low temperatures. We may expect a qualitatively similar behaviour for positive pions. This means that in spite of their short lifetime π^+ may be used to "decorate" lattice defects present in sufficiently high concentrations. In most cases the trapping affects the channelling pattern rather strongly, so that pions constitute probes for studying defects in crystals in much the same way as has been demonstrated for positive muons [13,14,15].

Fig. 5 shows that at room temperature the muon profile of a tantalum crystal containing 0.17 at % oxygen is very similar to that of a pure crystal, but at lower temperatures the width of the profile increases and the height decreases. These experiments indicate that

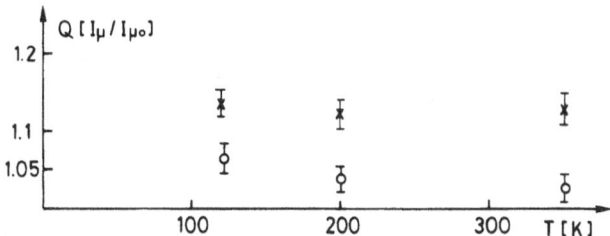

Fig. 4. Normalized muon intensity (within FWHM) in Ta <100> as a
        function of temperature. X: Muons with 0-0.5 MeV energy loss.
        o: Muons with 0.5-1 MeV energy loss. With larger energy losses
        channelling profiles could not be observed.

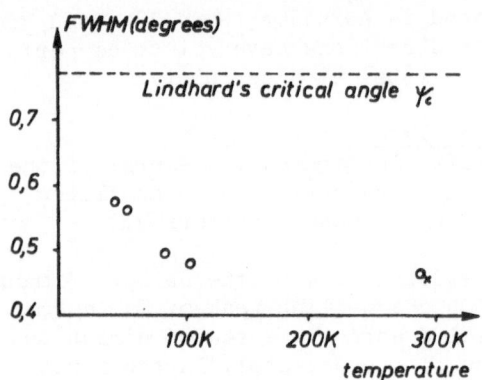

Fig. 5.

Trapping of π^+ at oxygen impurities in Ta. X: Pure crystal.

the pions may get trapped at the oxygen atoms at sites that are displaced from the pion sites in a perfect environment. The free enthalpy of binding between pions and oxygen atoms is such that at low temperatures the pions stay most of their lifetime in the traps, whereas at room temperature they are virtually free and thus spend most of the time at their "normal" sites.

d) Germanium

From the μ^+SR studies on Si and Ge it is known that in these materials at low temperatures muonium atoms (Mr) are formed [16]. In spite of intensive research the mechanism of muonium formation and the lattice sites involved have remained unclear [17]. In ultra-high-purity Ge cyrstals a maximum of the muon intensity has been observed in the <110> directions. Below 80 K this channelling peak broadens markedly over a quite narrow temperature interval (Fig. 6). We suspect that this is a consequence of pionium Pi = (π^+e^-) formation and the establishment of a Pi-Ge bond which draws the pion closer to one of the lattice rows forming the channel. If this interpretation is confirmed these experiments constitute the first observation of the new "element" pionium Pi.

Time-differential lattice steering

Making use of the time structure of the SIN pion beam it is possible to measure the lattice-steering pattern as a function of the "pion age" (i.e., the time a pion has spent in the crystal before decaying), provided the cyclotron is operated in the 17 MHz mode or, even better, in a 10 MHz mode.

The pulse widths of the beams SIN/πE1 and SIN/πE3 are about 1.10^{-9}s. The time of flight of the decay muons between the crystal and the muon detector is known from the muon energy. The time resolution of the scintillation detector employed is better than 2.10^{-9}s; its energy resolution better than 8%. This means that we are able to

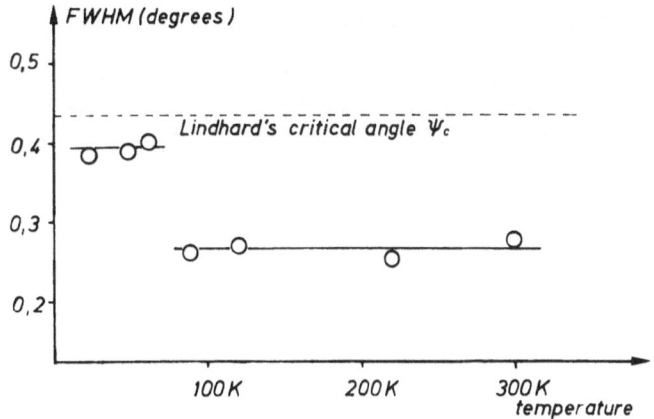

Fig. 6.

Full width at half maximum of lattice steering peak along <110> in ultra-high purity germanium as a function of temperature.

measure the time the pion stays in the crystal with a precision of at least $5 \cdot 10^{-9}$s.

The application of the technique to time-differential lattice steering is planned for the near future. It should be possible to detect the kinetics of trapping of π^+ at crystal defects or of pionium formation (e.g., in valence crystals) on the time scale $5 \cdot 10^{-9}$s to $6 \cdot 10^{-8}$s (for the 17 MHz mode of SIN).

LONGITUDINAL MUON-SPIN RELAXATION IN α-IRON IN HIGH MAGNETIC FIELDS

Introductory Remarks

The ideal method to study diffusion and localization of positive muons in metals is the measurement of the decay of muon-spin polarization in logitudinal magnetic fields (i.e. fields parallel or anti-parallel to the direction of initial polarization of the muon beam). The longitudinal relaxation rates Γ_1 are far less sensitive to impurities or lattice imperfections than the spin-phase relaxation rates Γ_2 obtained from transverse-field experiments. This is due to the fact that as long as no exterior alternating fields act on the muon magnetic moments, only mobile muons give rise to a logitudinal relaxation rate. By contrast, immobilized muons (e.g., muons trapped by impurities) do contribute to the spin relaxation rate Γ_2 in transverse fields.

Due to the strong variation of local magnetic fields between

interstitial sites in ferromagnetic α-iron the logitudinal relaxation rates may be measured over a very wide temperature range
extending from about room temperature down to well below liquid-
helium temperature. This offers the possibility to study quantum-
mechanical effects on the diffusion of light positively charged
particles in metals.

Theory

The magnetic field acting on the muon magnetic moments is given by

$$\underline{B}_\mu = \overline{\underline{B}}_\mu + \underline{B}_{dip} \quad , \tag{5}$$

where

$$\overline{\underline{B}}_\mu = \underline{B}_{appl} + \underline{B}_{Lorentz} + \underline{B}_{Fermi} \tag{5a}$$

comprises the applied field $\underline{B}_{appl}$, the demagnetizing field $\underline{B}_{demag}$,
the Lorentz field $\underline{B}_{Lorentz}$, and the Fermi field $\underline{B}_{Fermi}$, which
is due to the contact interaction of the muon spins with the spin
polarization of the conduction electrons at the muon sites. Via
the last term $\overline{\underline{B}}_\mu$ depends on the type of muon site (octahedral)
or tetrahedral); it is, however, independent of the tetragonality
axes of these sites. The last-mentioned property of $\overline{\underline{B}}_\mu$ is not shared
by the dipolar field $\underline{B}_{dip}$, which depends on both the interstitial
site and on the direction of its tetragonality axis with respect to
the magnetization. We characterize $\underline{B}_{dip}$, by the field $\underline{B}_{dip}^{\parallel}$ at interstitial sites with tetragonal axes parallel to the magnetization direc-
tion <100>. By symmetry $\underline{B}_{dip}^{\parallel}$ must be parallel or antiparallel to the
magnetization direction. The field at the interstices with tetragonal-
ity axes perpendicular to the <100> magnetization direction are
given by $-\frac{1}{2} \underline{B}_{dip}^{\parallel}$. The logitudinal relaxation rate Γ_1 depends strongly on the crystallographic directions of $\overline{\underline{B}}_\mu$ [1,18]. (Since we shall
consider only $\underline{B}_{appl}$ large enough to saturate the sample magnetically we need not distinguish between the direction of $\overline{\underline{B}}_\mu$ and of the
magnetization). If we confine ourselves to $\underline{B}_\mu$ lying in (011) planes
and characterize the direction of $\overline{\underline{B}}_\mu$ by the angle θ with respect to
<100>, we have for muon sites with tetragonal symmetry

$$\Gamma_1^{(\theta)} = \frac{3}{4} \sin^2\theta \ (1+3 \cos^2\theta)\Gamma_1^{<111>}. \tag{6}$$

For $\overline{\underline{B}}_\mu$ parallel to <100> it follows from elementary considerations
that $\Gamma_1^{<100>} \equiv 0$.

Γ_1 assumes its maximum value for $\overline{\underline{B}}_\mu$ parallel to <111> since in
this case the dipolar fields at the interstitial site lie in <2$\bar{1}\bar{1}$>
directions perpendicular to the magnetization direction and thus
exert the maximum torque on the muon moment in a longitudinal-
polarization arrangement. Under the assumption that the muon motion
may be described by a correlation time τ_c we find [1]

$$\Gamma_1^{<111>} = \frac{1}{2}(\gamma_\mu B_{dip}^\parallel)^2 \frac{\tau_c}{1+\omega_\mu^2 \tau_c^2} \quad , \tag{7}$$

where $\gamma_\mu = 8.52 \cdot 10^8 T^{-1} s^{-1}$ denotes the gyromagnetic ratio of the muons and $\omega_\mu = \gamma_\mu \bar{B}_\mu$ the precession frequency of the muon spins. According to (7), Γ_1 exhibits a maximum at $\omega_\mu \tau_c = 1$. From the dependence of this maximum on B_{appl} we may determine $B_{dip}^\parallel$ and hence the <u>site</u> of the μ^+. The temperature dependence of Γ_1 at fixed B_{appl} provides us with the temperature dependence of the muon <u>mobility</u>. Most transverse relaxation measurements in iron have been carried out in zero applied fields. In this case $\bar{B}_\mu = B_{Lorentz} + B_{Fermi}$ is parallel to <100> and the transverse relaxation rate is given by

$$\Gamma_2^{<100>} = \frac{1}{2}(\gamma_\mu B_{dip}^\parallel)^2 \tau_c \quad , \tag{8}$$

which coincides with the logitudinal relaxation rate $\Gamma_1^{<111>}$ for $\omega_\mu \tau_c \ll 1$.) In the present model the muon diffusion coefficient is given by

$$D = \alpha a_o^2/\tau_c \quad , \tag{9}$$

where a_o = edge length of the elementary cube and $\alpha = 1/36$ for octahedral interstices and $\alpha = 1/72$ for tetrahedral interstices.

<u>Experiment</u> [19]

A schematic view of the apparatus designed for the use of surface muons is shown in Fig. 7. The superconducting magnet for fields up

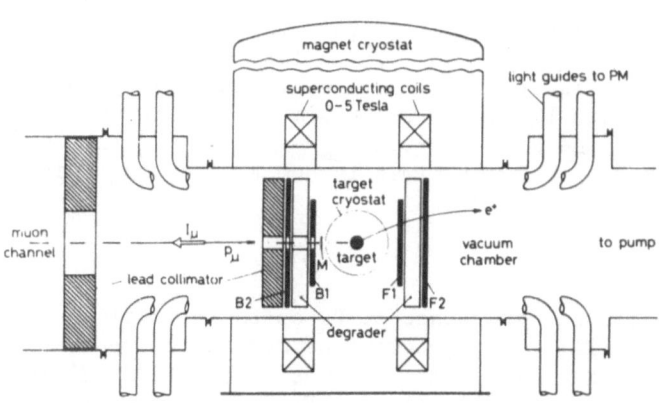

Fig. 7.

Schematic view of the apparatus for the measurement of longitudinal muon-spin relaxation in high magnetic fields.

scintillators M muon counter (0 2 mm thick for surface muons)
B1,B2, F1,F2 positron telescopes

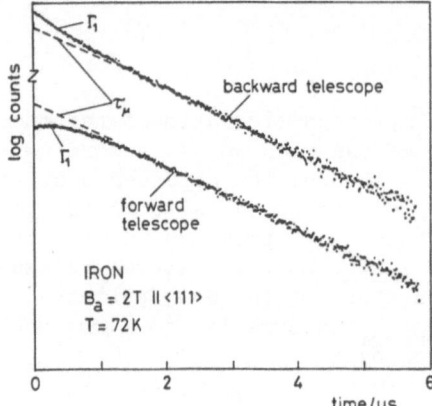

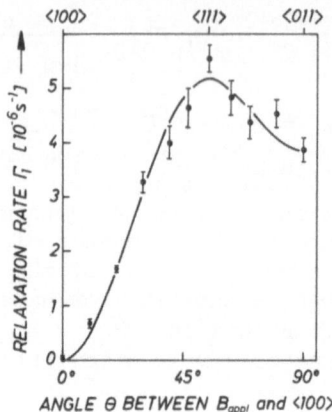

Fig. 8. Muon decay spectra of an α-iron single crystal at 52 K. Applied magnetic field 2 T parallel to the muon polarization and to a <111> crystallographic direction.

Fig. 9. Dependence of the longitudinal relaxation rate Γ_1 of α-Fe on the angle θ between <100> and an applied field in a (001) plane at 52 K [21]. B_{appl} = 2 T. Solid curve: prediction (6) for muon sites of tetragonal symmetry.

to 5 T consists of a pair of coils in Helmholtz geometry (homogeneity $\Delta B_{appl}/B_{appl} < 10^{-4}$ within a sphere of 3cm diameter). The helium-evaporation cryostat with 25μm thick titanium windows allows measurements between 3 K and 300 K. The target, a single crystal of α-iron 9mm in diameter [20], was mounted on a rotatable holder with a <110> direction parallel to the rotation axis in order to measure the dependence of Γ_1 on the crystallographic direction of $\underline{B}_{appl}$.

Timing signals are given by the scintillators M (0.2mm thick) for incoming muons and F1 or B1 (5mm) for "forward" or "backward" decay positrons. The degrades (1cm of Cu) between the positron telescopes F1F2 and B1B2 absorb low energy positrons in order to enhance the analyzing power for the measurement of the forward-backward asymmetry of the μ^+ decay. Fig. 8 shows (on a semilogarithmic scale) spectra measured at 72 K with B_{appl} = 2 T. At times t > $2 \cdot 10^{-6}$s they are entirely determined by the muon lifetime $\tau_\mu = 2.2 \cdot 10^{-6}$s; the deviation from a straight line at shorter times is due to the logitudinal relaxation of the μ^+-spin polarization.

Results

Fig. 9 gives an example of the dependence of Γ_1 on the crystallographic direction of $\underline{B}_{app}$ [21]. The agreement between experiment and theory (cf. (6) and solid line) is good. This was confirmed at different temperatures and different applied fields. From these results we conclude that the muons in α-iron occupy sites of tetragonal symmetry i.e., tetrahedral or/and octahedral interstics.

Fig. 10 shows an Arrhenius plot of $\Gamma_1^{<111>}$ as a function of temperature. The comparison with zero-applied-field relaxation rates $\Gamma_2^{<100>}$ [22] demonstrates the sensitivity of transverse-field μSR to impurities and the superiority of logitudinal field experiments: the deviation of $\Gamma^{<100>}$ from the logitudinal relaxation rate Γ_1 above ~70 K is due to lattice imperfections.

At temperatures above 20 K the longitudinal relaxation rates show no significant dependence on the strength of B_{appl}, indicating that we are in the regime $\omega_\mu\tau_c \ll 1$. Therefore, the maximum in Γ_1 at 40 K cannot be the maximum at $\omega_\mu\tau_c = 1$ predicted by (7) but has to be explained in terms of a change in the intrinsic diffusion mechanism.

Yagi et al. [21] have suggested that at low temperatures (below about 20 K) positive muons in α-Fe occupy preferentially tetrahedral interstices, whereas at higher temperatures there is an equipartition between octahedral and tetrahedral interstices (which both possess tetragonal symmetry and, therefore, give both rise to the orientation dependence (6)). All the experiments carried out since have supported this hypothesis.

Fig. 10 shows that below 20 K there is a definite dependence of the longitudinal relaxation rate $\Gamma_1^{<111>}$ on the strength of the applied field but that not even at the largest applied fields a maximum in the temperature dependence of the relaxation rate has been found down to 4 K, the lowest temperature employed so far. Measurements down to 0.3 K with the help of a ^{3}He-cryostat are being prepared and should clarify the situation.

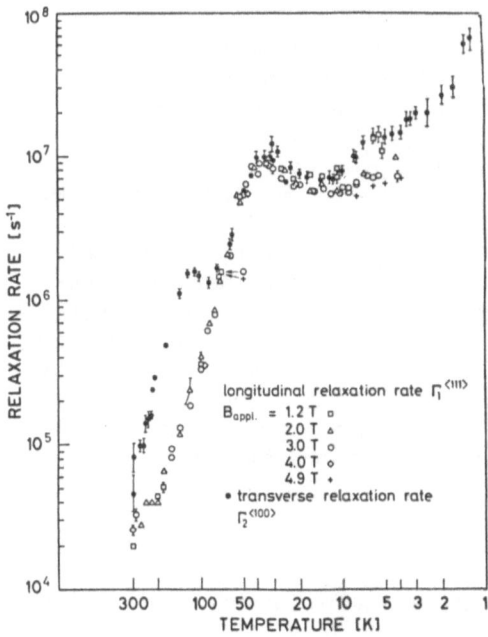

Fig. 10.

Temperatures dependence of the longitudinal relaxation rate $\Gamma_1^{<111>}$ at applied fields B_{appl} ranging from 1.2 to 4.9 T. For comparison, recent zero-applied-field relaxation measurements [22] are included. The deviation of these data from Γ_1 at temperatures > 70 K demonstrates the sensitivity of transverse relaxation rates to impurities.

If more than one type of interstices is occupied the theory outlined in Sect. 3.2. does no longer suffice. Seeger and Monachesi [18]
have recently given a detailed general theory, covering in particular
the case of transitions between octahedral and tetrahedral interstices. The general treatment shows that at high temperatures the
simple theory of Sect. 3.2. remains valid provided $(B_{dip}^{\parallel})^2$ is treated
as an adjustable parameter. For temperatures above $60\,K$ one may fit
the $\Gamma_j^{<111>}$ data of Fig. 10 by assuming that the muon diffusion
coefficient D varies with temperature either according to the classical
law

$$D \sim \exp(-H^M/kT) \tag{10}$$

with the migration enthalpy $H^M = (0.033\pm0.001)$ eV or according to the
multi-phonon incoherent tunnelling law [23]

$$D \sim T^{-1/2}\exp(-H_a/kT) \tag{11}$$

with the lattice-activation enthalpy $H_a = (0.038\pm0.001)$ eV.

According to Teichler and Seeger [24], at low temperatures (11) has
to be replaced by the one-phonon law

$$D \sim T \ . \tag{12}$$

The available experiments are in agreement with (12) (which predicts
$\Gamma_2^{<100>} \sim T^{-1}$) but a final test has to await the outcome of the above-
mentioned experiments in the ^{3}He cryostat.

Acknowledgement
The authors gratefully acknowledge the help of G. Jünemann, G. Flik,
D. Herlach, and H.-D. Carstanjen, who collaborated on the channelling
work. The measurements on Ge were performed on ultra-high-purity crys-
tals kindly provided by E. Haller (Berkeley). The work reported in
Sect. 3 is part of the Stuttgart—Heidelberg μ^+SR collaboration pro-
gramme with the participation of T. Aurenz, H. Bossy, K.P. Döring, M.
Gladisch, N. Haas, D. Herlach, W. Jacobs, M. Krauth, M. Krenke, H.
Matsui, P. Monachesi, H. Orth, G. zu Putlitz, J. Vetter, and E. Yagi.
We are further indebted to G. Wiederoder, R. Henes, P. Keppler, and
W. Maisch for their technical assistance, to Miss L. Holl and Miss H.
Schweyer for preparing the typescript, and to Prof. W. Frank for cri-
tically reading the manuscript. The experiments were made possible by
the financial support of the Bundesministerium für Forschung und Tech-
nologie, Bonn, through the programme "Mittelenergiephysik" and, last
not least, by the excellent experimental conditions at SIN.

REFERENCES

[1] A. Seeger, Hydrogen in Metals I (eds. G. Alefeld and J. Völkl)
      p. 349, Springer, Berlin—Heidelberg—New York 1978

[2] H. Orth, K.P. Döring, M. Gladisch, D. Herlach, W. Maysenhölder,
     H. Metz, G. zu Putlitz, A. Seeger, J. Vetter, W. Wahl, M.
     Wigand, and E. Yagi, Hydrogen in Metals—Münster 1979, p. 631,
     Akademische Verlagsgesellschaft, Wiesbaden 1979; see also
     Z.Phys.Chemie, N.F., 116, 241 (1979)
[3] K. Maier, Nuclear Physics Methods in Materials Research (eds.
     K. Bethge, H. Baumann, H. Jex, and F. Rauch) p. 264, Vieweg,
     Braunschweig 1980
[4] K. Maier, G. Flik, D. Herlach, G. Jünemann, H. Rempp, A. Seeger,
     and H.-D. Carstanjen, Phys. Letters 83A, 341 (1981)
[5] K. Maier, G. Flik, D. Herlach, G. Jünemann, A. Seeger, and
     H.-D. Carstanjen, Phys. Letters 86A, 126 (1981)
[6] A. Seeger, Proceedings of the 5th Meeting of the International
     Collaboration on Advanced Neutron Sources (ICANS-V) (eds. G.
     S. Bauer and D. Filges), Jül-Conf. 45, 1981, p. 113
[7] D.S. Gemmell, Rev. Mod. Phys. 46, 129 (1974)
[8] R. Sizmann and C. Varelas, Festkörperprobleme XVII (ed. J.
     Treusch) p. 261, Vieweg, Braunschweig 1977
[9] J. Lindhard, K. Dan, Vid. Selsk, Mat.-Fys. Medd. 34 (1965) no. 4
[10] K. Maier, G. Flik. A. Seeger, D. Herlach, H. Rempp, G. Jünemann,
     and H.-D. Carstanjen, Nucl. Instr. and Meth., in the press
     (Proc. 9th Intern. Conf. on Atomic Collisions in Solids
     Lyon 1981)
[11] D. Van Vliet, AERE Report 6395, Harwell 1970
[12] H.-D. Carstanjen, phys.stat.sol. (a) 59, 11 (1980)
[13] A. Seeger, Positron Annihilation (eds. R.R. Hasiguti and K.
     Fujiwara) p. 771 (The Japan Institute of Metals, Sendai 1980)
[14] D. Herlach, Recent Developments in Condensed Matter Physics,
     Vol. 1 (ed. J.T. Devreese) p. 93 Plenum Publishing Corp.,
     New York 1981
[15] A. Seeger, Proceedings 5th Intern. Symposium on High-Purity
     Materials in Science and Technology (eds. J. Morgenthal and
     H. Oppermann) Vol. II, p. 253, Akademie der Wissenschaften
     der DDR, Dresden 1980
[16] P.F. Meier, Exotic Atoms '79 (ed. by K. Crowe, J. Duclos, G.
     Fiorentini, and G. Torelli) p. 331, Plenum Press, New York
     and London 1980
[17] B.D. Patterson, Hyperfine Interact. 6. 155 (1979)
[18] A. Seeger and P. Monachesi, Phil. Mag., in the press
[19] T. Aurenz, H. Bossy, M. Gladisch, N. Haas, D. Herlach, W. Jacobs,
     M. Krauth, M. Krenke, H. Matsui, P. Monachesi, H. Orth, G.
     zu Putlitz, A. Seeger, J. Vetter, E. Yagi, and K.P. Döring,
     SIN Newsletter Nr. 14 (1982)
[20] K. Lubitz and G. Göltz, Appl. Phys. 19, 237 (1979)
[21] E. Yagi, H. Bossy, K.P. Döring, M. Gladisch, D. Herlach, H.
     Matsui, H. Orth, G. zu Putlitz, A. Seeger, and J. Vetter,
     Hyperfine Interact. 8, 553 (1981)
[22] H. Graf, G. Balzer, E. Recknagel, A. Weidinger, and R.I.
     Grynszpan, Phys. Rev. Letters 44, 133 (1980)
[23] C.P. Flynn and A.M. Stoneham, Phys. Rev. B 1, 3966 (1970)
[24] H. Teichler and A. Seeger, Phys. Letters 82A, 91 (1981)

BEHAVIOR OF POSITIVE MUONS IMPLANTED IN IRON ALLOYS

C.E. Stronach[†], K.R. Squire[†], A.S. Arrott[‡],
B. Heinrich[‡], W.F. Lankford[ζ], W.J. Kossler[†]
and J.J. Singh[*]

[†]Virginia State University, Petersburg, VA 23803
[‡]Simon Fraser University, Burnaby, BC Canada V5A 1S6
[ζ]George Mason University, Fairfax, VA 22030
[†]College of William and Mary, Williamsburg
VA 23186
[*]NASA Langley Research Center, Hampton
VA 23665

ABSTRACT

Muon spin rotation measurements were made upon Fe alloyed with
small amounts of N, Al, Si, Ge, Ti, V, Cr, Mn, Co, Ni, Nb, Mo and
W. The measurements are described and the results are discussed in
terms of the effect of impurities and associated strain upon B_μ
and B_{hf}.

INTRODUCTION

We employed µSR for a study of iron alloys. The time dependent
angular correlation of the muon's positron decay precesses at
13.552 MHz·B_μ (KG) where B_μ is the average field sensed by a μ^+
implanted into the sample. See ref. 1 for more detail on the
technique.

The experiments were conducted at the TRIUMF cyclotron facility.
As previously mentioned, one obtains the local field B_μ directly
from the precession frequency. With no external field applied, and
with the dipolar fields averaged to zero by the motion of the muon,
B_μ has only two major contributions, the Lorentz cavity field
$B_L^\mu = 4\pi M/3$, and the contact hyperfine field B_{hf}, which arises from
a polarized electron density about the μ^+. The hyperfine field is
then given by

$$B_{hf} = B_{\mu} - B_{L}.$$

For Fe at room temperature B_L is about 7.2 KG and B_μ is about
-3.6 KG, so values for B_{hf} are of the order of -10.8 KG.

We parameterize the effect of the impurity upon the hyperfine
field by calculating the fractional change in the hyperfine field,
normalized to the impurity concentration, $\Delta B_{hf}/c\, B_{hf}$. For most of
the alloys this has been determined only at room temperature, but
for Fe(Mo) and Fe(Al) it has been measured over a range of temperatures.

HYPERFINE FIELDS

The table below summarizes our determinations of $\Delta B_{hf}/c\, B_{hf}$
at room temperature:

Impurity	$\Delta B_{hf}/c\ B_{hf}$
N	-8
Al	-0.23
Si	-0.42
Ge	0
Cr	-0.09
Mo	-0.72
W	-0.79
Ti	-13
V	-1.10
Cr	-0.09
Mn	-1.10
Co	0.2
Ni	-4.4

If an impurity served only to dilute the hyperfine field by
acting as a non-magnetic hole at a substitutional lattice site, with
random site sampling by the μ^+, one has $\Delta B_{hf}/c\, B_{hf} = -1$. We note
that Mo, W, V and Mn approximate such behavior at room temperature,
while Cr, Al and Si decrease the magnitude of B_{hf} considerably less
than predicted by pure dilution. These results for Al and Si may
be at least partially explained by the mechanism described in ref. 2,
in which the temperature dependence of B_{hf} in Fe(Al) was measured.
An increase in $\Delta B_{hf}/c\, B_{hf}$ with increasing temperature suggests that
μ^+ are repelled from solutes with p-wave bonding electrons, and thus
experience smaller solute effects. Even so, the high-temperature
limit (or random sampling limit) of $\Delta B_{hf}/c\, B_{hf}$ for Fe(Al) is still
only $\approx$ -0.35.

Co, which increases both the moment per Fe atom and the Curie temperature (Tc), is the only solute which makes B_{hf} more negative. Ge, which also increases the moment per Fe atom and Tc, has the next most positive effect, zero.

Cr is out of line in both the vertical Cr, Mo, W and horizontal W, Cr, Mn sequences in the periodic table. To the best of our knowledge there is no other characteristic of these alloys for which this is seen. We can only speculate that Cr impurities produce less internal strain in the crystals than the other impurities do.

Ti, Ni and N impurities produce very large reductions in the magnitude of B_{hf}, each being several times the prediction of pure dilution. Again, this may be due in part to non-random site sampling by the μ^+. The temperature dependence of B_{hf} was measured for Fe(Mo) and it showed an effect opposite to that found for Fe(Al): the μ^+ is apparently attracted to Mo impurity sites in Fe.[3] The effect of Ti is about double what one would obtain if B_{hf} were reduced to zero at all sites adjacent to Ti impurities. Studies of Fe(Ti) at higher temperatures show that while $|\Delta B_{hf}/c\, B_{hf}| \gg 1$, it is decreasing with increasing temperature. This suggests that the Fe(Ti) result may arise, in part, from preferential sampling by the μ^+ of sites adjacent to Ti atoms.

A study of the temperature dependence of B_μ in Fe(Ge) showed an hysteresis effect upon annealing (Fig. 1). The form of the $B_\mu(T)$ curve changed upon annealing, apparently because of the release of internal strains in the annealing process.

A large decrease in B_μ was found in the two-phase alloy Fe + Fe_2Nb, compared with the pure Fe from the stock material from which all the alloys were formed. This sample consists of Fe_2Nb inclusions in a pure Fe matrix. Two samples with different Nb concentrations showed that the reduction in B_μ is more pronounced with greater Nb concentration. This is contrary to the naive expectation that, since all μ^+ spin rotation takes place in the pure Fe, the frequency of the signal would be unchanged by the Fe_2Nb inclusions with only the amplitude changing. This reduction in magnitude of B_μ is of the order of magnitude of Fe anisotropy fields around inclusions and may result from weak trapping of the μ^+ around these inclusions.

STRAIN EFFECTS

The observation that in some cases the presence of impurities in Fe decreases the magnitude of B_{hf} more than pure dilution, the shift of B_μ in Fe(Ge) upon annealing, and the decrease in the magnitude of B_μ with addition of Fe_2Nb inclusions all lead us to the

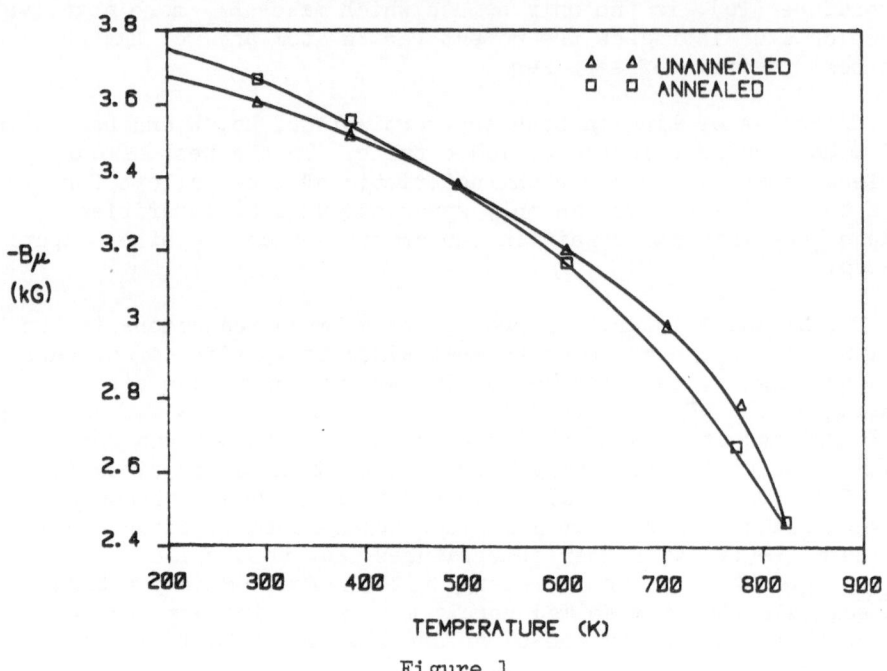

Figure 1

belief that strain plays a major role in determining the field seen by the μ^+.

Shear strains break the symmetry of the dipole lattice in Fe. With complete bcc lattice symmetry the diffusion of the μ^+ averages the dipolar fields to zero. However, shear strains could give a net dipolar field perpendicular to the magnetization. Computer simulations of these fields show that the net effect upon μ^+ is to decrease the precession frequency.[4]

The recent observation that uniaxial tension decreases B_μ in single crystals of Fe supports the motion of strain sensitivity to B_μ.

HIGH TEMPERATURE MEASUREMENTS

The problem of non-random sampling of sites by the μ^+ can presumably be overcome by providing the muons with sufficient thermal energy that kT is much greater than the attractive/repulsive potentials at interstitial sites adjacent to impurity atoms. Measurements at such high temperatures (approaching Tc) are fraught with difficulties. Assuming that the practical problem of constructing ovens

which can maintain stable gradient-free high temperatures is overcome, one must also determine the magnetization of the alloy as a
function of temperature. This varies rapidly near Tc, and the
approximation heretofore used of scaling the magnetization curve of
pure Fe to the M(T = 0K) and Tc values of the alloy probably is not
sufficiently accurate. Therefore both precision μSR experiments and
magnetization measurements will be necessary to fully exploit this
area of study.

ACKNOWLEDGMENTS

These studies were performed at TRIUMF, which is operated by
the University of Alberta, the University of British Columbia,
Simon Fraser University, and the University of Victoria with the
support of the Natural Sciences and Engineering Research Council of
Canada. The experiments were supported in part by the National
Aeronautics and Space Administration (CES and JJS) and the
National Science Foundation DMR8007059 (WJK). We appreciate the
help received from J. Brewer, P. Percival, T. Suzuki and Y. Uemura.
A. T. Fiory and B. D. Patterson participated in the initial studies
of Fe alloys.

REFERENCES

1. W. F. Lankford, K. G. Lynn, W. J. Kossler, A. T. Fiory, R. P.
   Minnich and C. E. Stronach, Nucl. Instr. and Meth. 185 (1981)
   469.
2. C. E. Stronach, W. J. Kossler, J. Lindemuth, K. G. Petzinger,
   A. T. Fiory, R. P. Minnich, W. F. Lankford, J. J. Singh and
   K. G. Lynn, Phys. Rev. B20 (1979) 2315.
3. C. E. Stronach, K. R. Squire, A. S. Arrott, B. D. Patterson,
   B. Heinrich, W. J. Lankford, A. T. Fiory, W. J. Kossler and
   J. J. Singh, Jour. of Mag. and Mag. Nat. 25 (1981) 187.
4. A. S. Arrott, B. Heinrich, C. E. Stronach and W. J. Lankford,
   J. App. Phys., in press.
5. W. J. Kossler, M. Namkung, R. I. Grynszpan, C. E. Stronach,
   B. D. Patterson, W. Kündig and P. F. Meier, private
   communication.

HYDROGEN TRAPS IN COLD-WORKED PALLADIUM

Ted B. Flanagan and S. Kishimoto

Department of Chemistry
University of Vermont
Burlington, VT 05405

INTRODUCTION

The enhancement of α phase H solubility in Pd due to the interaction of dissolved H atoms with the stress fields of dislocations was investigated several years ago by one of the authors and his coworkers (1). This phenomenon has been reinvestigated recently by Kircheim (2,3) using an electrochemical technique which allows the monitoring of much lower H contents than the volumetric technique used in the earlier study (1). Solubility enhancements as great as 10^6 were reported by Kircheim at the lowest H-contents which could be measured. The solubility enhancement is defined as the ratio r'/r where r' and r are the H-to-Pd atom ratios for a sample with a large dislocation density and for a well-annealed sample, respectively, where both values of r are measured at the same chemical potential of H. Such large solubility enhancements were not found in the original study (1) which was limited to solubilities in excess of $r \simeq 10^{-3}$. The solubility enhancements were found to be nearly independent of r for $r > 10^{-3}$; this independence is consistent with the theory of the interaction of the elastic fields of edge dislocations with solute atoms (4). The purpose of the present study is to extend the earlier studies (1) using the volumetric technique to the low H content range where the anomalous solubility was noted by Kircheim (2,3).

EXPERIMENTAL

A large Pd sample, 21 g, was cold-rolled to >78% deformation; 78% appeared from the earlier study (1) to be the approximate level of deformation at which the effects of cold-rolling saturate with respect to solubility enhancements (1). This sample was

623

approximately 10X as massive as the previously employed samples (1) and allowed solubility enhancements to be accurately measured to $r \sim 10^{-4}$. The sample was never heated above 323 K prior to the annealing studies.

It has been noted (5) that if a Pd sample is subjected to the phase transitions $\alpha \rightleftarrows \alpha'$ and a subsequent low temperature evacuation, it exhibits about the same solubility enhancement as does a heavily cold-worked sample. The cold worked sample described above was annealed and then subjected to the $\alpha \rightleftarrows \alpha'$ phase transitions (to r=0.6) and then after low temperature evacuation, it was also investigated in the low, anomalous H content range.

RESULTS AND DISCUSSION

Fig. 1 shows the solubility of H_2 in cold-rolled Pd at 273 K and it is clear that there is an anomalous H solubility, i.e., it is clear that r'/r cannot be constant as r→0. At r' = 10^{-4} (295K) Kircheim (2,3) reports (r'/r) = 2.9 and we find at 295 K, (r'/r) = 2.4 so that these data are in reasonable agreement down to this content and above this, they are in excellent agreement (2). The data in Fig. 1 appear to be rather linear for r'>1.5 x 10^{-3} (dashed line). One interpretation of these data is that a limited number of trapping sites are available which saturate at 273 K at r'≥1.5 x 10^{-3} and thereafter H dissolves as it does in a sample containing dislocations but no traps. Trapping models have been proposed in the literature (6-8); these models are based on the assumption that the chemical potentials of the trapped and untrapped H are equal. The untrapped H is assumed to behave in the present case as in the absence of the stress field. This seems to be a good assumption because the overall interaction of the H atoms and dislocations is weak for Pd, e.g., the normal solubility enhancement in this sample is 1.37 (273 K). This trapping model leads to the result that

$$r_t = \frac{E + r_t(max) + r(1-E) - \{(E + r_t(max) + r(1-E))^2 - 4r(1-E)\}^{\frac{1}{2}}}{2(1-E)} \tag{1}$$

where $E = \exp \{(\Delta H_H^t - \Delta H_H^\circ) - T(\Delta S_H^{\circ,t} - \Delta S_H^\circ)\}/RT$ and ΔH_H^t

and $\Delta S_H^{\circ,t}$ are the partial molar enthalpy and standard entropy of hydrogen solution (relative to $\frac{1}{2}H_2(g,1 \text{ atm})$) in the trapping sites, respectively, and ΔH_H° and ΔS_H° are the same quantities for normal interstices. The three parameters in equation 1 are $r_t(max)$, $(\Delta H_H^t - \Delta H_H^\circ)$ and $(\Delta S_H^{\circ,t} - \Delta S_H^\circ)$. Generally the latter term has been assumed to be zero in analyses of trapping data (7,8) and the other parameters are found by best fit to the experimental data. In this study these parameters were determined from experiment as described below.

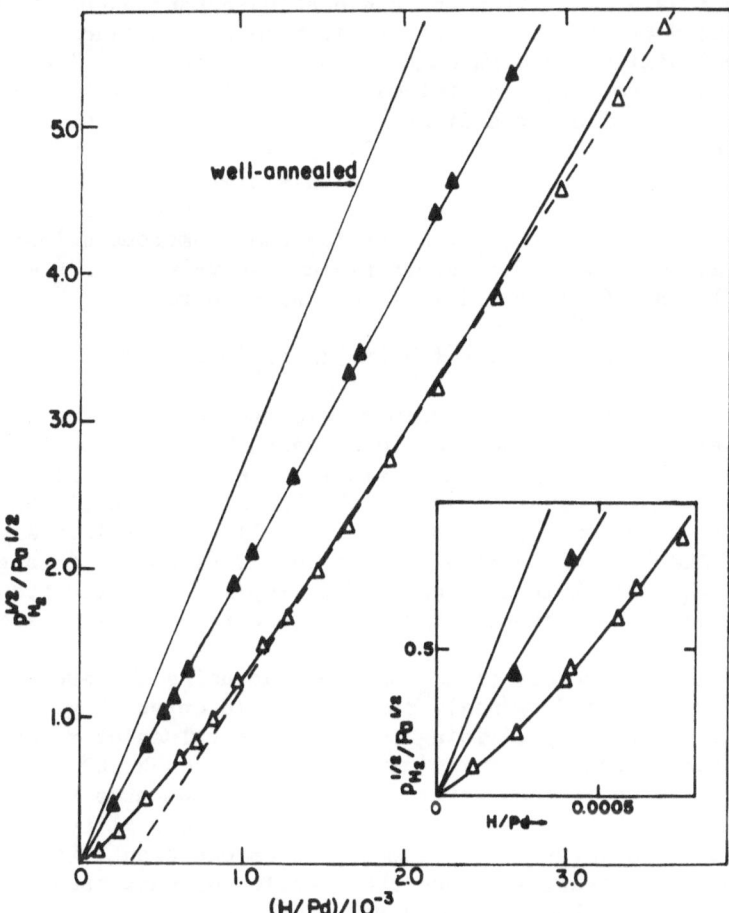

Fig. 1.   Hydrogen Solution in a 21 gram palladium sample
(273 K). Δ, cold-rolled; ▲, cold-rolled sample after
annealing at 447 K. ———, solubility data for well-
annealed sample. ----, straight line drawn through
data points at r' > 0.001 for cold-rolled sample.
Continous curve for cold-rolled sample represents
calculated results for trapping model. Inset shows
the same data for the very dilute region where again
the continuous curve through the data points for the
cold-rolled sample is the calculated curve for the
trapping model.

Solubility measurements were made for the cold-rolled sample at 273, 303 and 323 K. The anomalous solubilities were determined at each $p_{H_2}^{\frac{1}{2}}$ by subtracting from the observed solubilities the contributions arising from the dislocation-enhanced solubilities which were determined from the translation of the linear section of the solubility curve through the origin (Fig. 1). These data were employed to determine values of ΔH_H^t and ΔS_H^t for the anomalous solubility. Similar experiments were performed for the sample which had been subjected to the $\alpha \rightleftharpoons \alpha'$ phase changes. The anomalous solubility was somewhat greater for this sample than for the cold-rolled one and solubility data were also obtained at three different temperatures. Isotherms for the anomalous solubility were obtained as described above giving the values: ΔH_H^t = -23.8 kJ(mol H)$^{-1}$ and $\Delta S_H^{\circ,t}$ = -34.7 J(K mol H)$^{-1}$ where

$$\Delta S_H^{\circ,t} = \Delta S_H^t + R \ln \left[r_t / (r_t (\text{max}) - r_t) \right] \tag{2}$$

Results for the cold-rolled sample were less certain due to the smaller solubility in the anomalous region but the results were the same to within experimental error. The value of $r_t(\text{max})$ was obtained from the intersection of the linear portion of the solubility curve with the abscissa (Fig. 1). The relatively large value of $\Delta S_H^{\circ,t}$ is of interest; it is 20.3 J (K mol H)$^{-1}$ larger than for the normal interstices. This implies that the trapping sites provide the H atom with considerable freedom of motion.

Knowing the values of ΔH_H° and ΔS_H° and using the above values of ΔH_H^t and $\Delta S_H^{\circ,t}$ and $r_t(\text{max})$, the calculated curve (Fig. 1, solid curve) is obtained by dividing the observed values of r into contributions from r_t and r_f (untrapped H) at each $p_{H_2}^{\frac{1}{2}}$. The calculated curve is then obtained by placing computed values of r_f on the straight line extrapolation of the solubility curve and adding to these values of r_t computed from equation 1 for chosen values of r. Essentially then the fraction of untrapped and trapped H is calculated at each value of $r(p_{H_2}^{\frac{1}{2}})$.

The fit of the data to the model shown in Fig. 1 for 273 K is equally good for 303 and 323 K. This does not prove that the model is valid, however, because Kircheim (2,3) has also obtained good fit of his data to a quite different model. He assumes that hydride phase is formed in the tensile-stressed regions about edge dislocations. Ancillary evidence will be cited here, however, which suggests that hydride formation is not likely in the very low H content region.

In all cases where hydride phase is known to form in the Pd-H system it is accompanied by irreversibility, i.e., removal of the hydrogen does not lead to reversible behavior. In the anomalous solubility region reversible behavior appeared to obtain (Fig. 2). Although the absorption and desorption data are not identical,

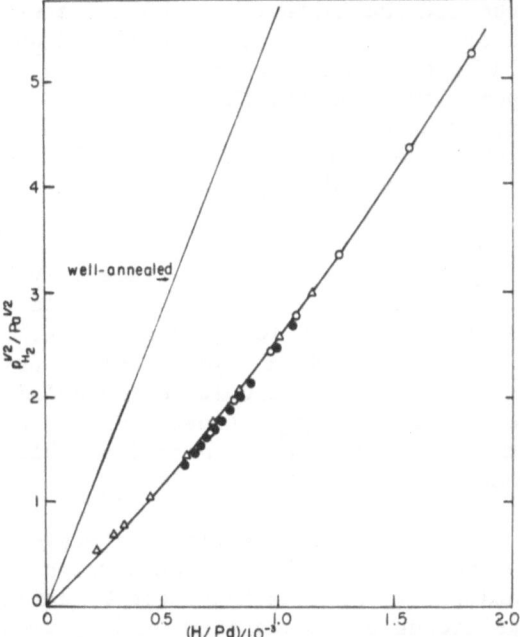

Fig. 2.   Hydrogen solubility data for cold-rolled sample at
           323 K. Δ, absorption data; ○, desorption data; ○,
           absorption data following desorption.

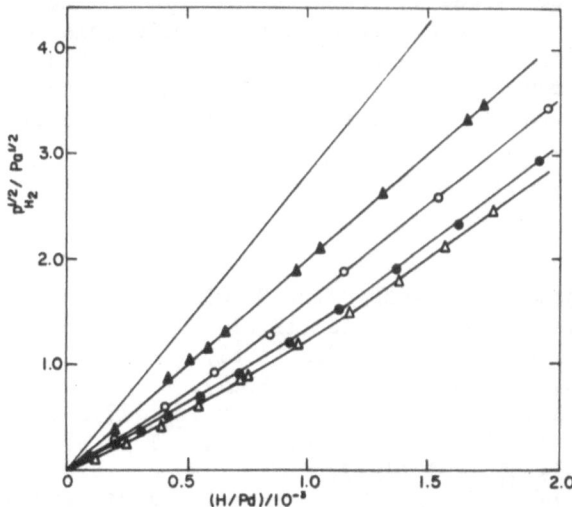

Fig. 3.   Effects of annealing of cold-rolled sample on subsequent
           hydrogen solubility at 273 K. Δ, 323 K; ●, 366 K; ○,
           403 K; ▲, 447 K.——— , represents data for well-
           annealed sample.

they are very close and the latter show the curvature exhibited by the former. If hydride phase were to have formed during hydrogen solution, it would be expected that the desorption data would have a much steeper slope than the absorption data (9). These experiments are somewhat difficult to carry out because the equilibrium pressures are quite low even at 323 K and so only small changes in r can be achieved during desorption.

Annealing results are difficult to reconcile with hydride formation (Fig. 3). It can be seen that relatively low temperature anneals decrease the anomalous solubility and it is completely removed after annealing in vacuo for 12 h at 447 K. Although the anomalous solubility is removed, a significant solubility enhancement remains (Fig. 3). It is difficult to reconcile these results with models which suggest the same defect to be responsible for both the anomalous solubility and the normal solubility enhancement, e.g., edge dislocations.

The results of the present investigation suggest that for cold worked Pd and for Pd which has passed through the phase transitions the solubility of H is increased by interaction with the elastic stress fields of dislocations and by trapping sites. The nature of the latter sites is unknown but they are annihilated following a relatively low temperature annealing treatment

ACKNOWLEDGEMENTS

S. Kishimoto wishes to thank the Ministry of Education (Japan) for a grant under which he was supported while this research was carried out. TBF wishes to thank the N.S.F. for financial support and acknowledgement is made to the Petroleum Research Fund, adminstered by the American Chemical Society for partial support of this research.

REFERENCES

(1) T.B. Flanagan, J.F. Lynch, J.D. Clewley and B. von Turkovich, J. Less Common Metals 49, 13 (1976).
(2) R. Kircheim, Acta Met. 29, 835 (1981).
(3) R. Kircheim, Acta Met. 29, 845 (1981).
(4) J.P. Hirth and B. Carnahan, Acta Met. 26, 1795 (1978).
(5) J.F. Lynch, J.D. Clewley, T. Curran and T.B. Flanagan, J. Less Common Metals 55, 153 (1977).
(6) H.Y. Chang and C.A. Wert, Acta Met. 21, 1233 (1973).
(7) G. Pfeiffer and H. Wipg, J. Phys. F. 6, 167 (1976).
(8) T.B. Flanagan, C.A. Wulff and B.S. Bowerman, J. Solid State Chem. 34, 215 (1980).
(9) T.B. Flanagan, B.S. Bowerman and G.E. Biehl, Scripta Met. 14, 443 (1980).

INFLUENCE OF HYDROGEN ON THE DIRECTION OF

CRACK PROPAGATION IN IRON USING COMPUTER SIMULATION

Mayes Mullins

Polyatomics Research Institute
1101 San Antonio Road, Suite 420
Mountain View, CA 94043

ABSTRACT

The influence of hydrogen on the fracture behavior of α-iron was studied using computer simulation. The results indicate that the relative probability of (101) plane cleavage is enhanced compared with that for the (010) plane, although the lowest load for fracture is not decreased. Some limitations of the interatomic potentials used are discussed.

INTRODUCTION

This paper describes some results obtained with a computer simulation model of fracture in iron containing hydrogen. Several previous simulation studies of this problem have been conducted.[1,2] The work presented here differs from these in the crystal orientation used, and in the use of the finite element method to model the continuum. The latter appears to be a significant improvement since crack propagation to the edge of the discrete region is readily observable with the finite element boundary conditions, while the other continuum treatments which have been used in the past appear to be much more restricting.

MODEL DESCRIPTION

The model used for this study included 624 atoms representing two ($\bar{1}$01) planes of the bcc iron lattice with translational symmetry imposed in the ($\bar{1}$01) direction. This orientation allows all important slip and cleavage planes, (010), (101), and (121), to be

629

active since they are all perpendicular to the two ($\bar{1}$01) planes
modelled. Also, the model appears to most accurately simulate
cleavage in pure iron for this orientation[3] showing definite (010)
fracture to the edge of the discrete region at speeds which are
in good agreement with experiment up to a limiting load level of
$1.8k_G$, where k_G is the Griffith stress intensity factor. At this
load level and beyond, the crack bifurcates and propagates along
(101) planes consistent with the experimental observation of crack
branching at high load levels in many materials. The surrounding
continuum was modelled using the finite element method with 175
nodes and 288 elements.

The interatomic potentials used are shown in figure 1. The
Fe-Fe potential was developed by Johnson[4] (curve 1). It has been
used in a number of simulations of defects in iron and it appears
to offer a reasonable basis for such studies. No H-H interaction
was assumed. For the conditions used in the present study, the
hydrogen concentration in the lattice was not high enough for significant H-H interaction to occur, so this should not be a major
influence on the results.

Previous models of the fracture of iron containing hydrogen
have used three quite different Fe-H potentials. The Walker potential,[5] initially derived for the Fe-H molecule, has been used,[6]
but the applicability of this to the interatomic interactions in
a lattice has been questioned.[1] The Olander potential[7] was formulated specifically for lattice interactions, but the potential

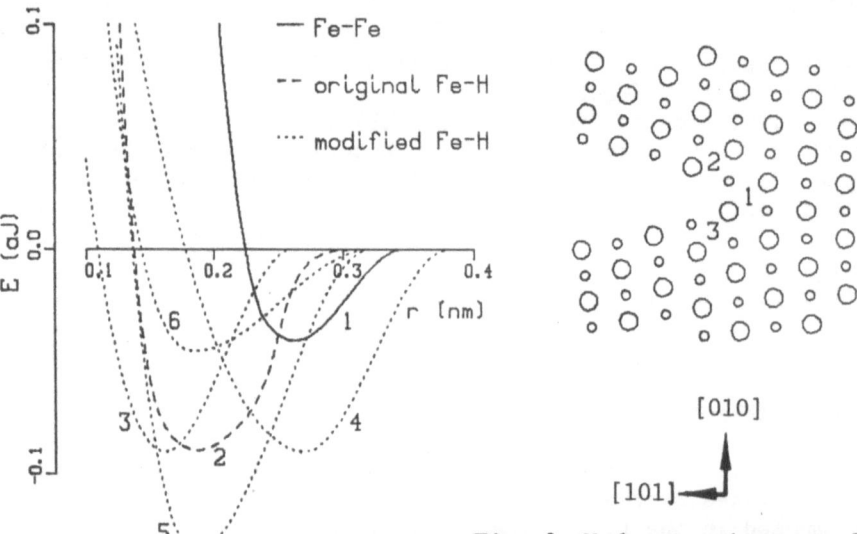

Fig. 1 Potentials used.

Fig. 2 Hydrogen sites at the
crack tip. Circles represent two
planes of iron atoms.

proved so repulsive that the iron lattice rejected the hydrogen.[1]
The potential used by White and Kahn[2] was derived from ab initio
quantum mechanics calculations of the energies of small iron clusters containing hydrogen.[8] It falls between the other two described
above and was the one adopted as the basis for the present work
(curve 2 of figure 1). It is felt that the use of ab initio calculations to provide the basic input for the development of a semiempirical potential can provide as much confidence in the results
as is possible. There are serious limitations associated with the
use of any potential such as these which defines only pair-wise
interactions, however. These limitations are discussed more fully
below.

It was felt that the number of hydrogen distributions which
could be studied would not be sufficient to provide a valid picture
of the statistical distribution of behavior to be expected if a
Monte Carlo scheme was used to place the hydrogens. Because of
this, a more qualitative approach was taken in which a small number
of hydrogen atoms were specifically placed at various sites in order
to assess their influence. It was assumed that hydrogen would
enter the lattice from the environment near the crack tip through
the most highly strained interstitial sites. Points 1, 2 and 3
in figure 2 show the sites considered. The tendency of the hydrogen to diffuse into the lattice was minimized by assigning an
unphysically large mass to the hydrogen atoms. This caused the
hydrogen to remain near the positions assigned and eliminated the
need for a hydrogen source representing the environment within
the crack. The stability of the numerical integrations in the analysis is also enhanced by this modification. The resulting hydrogen distributions are probably physically unlikely but can be considered paradigms which will help to understand more general configurations.

RESULTS

A standard set of initial conditions was assumed for all cases.
These involved placing the atoms at the locations specified by the
linear elastic strain field for the crack tip and setting all
velocities to zero. Figure 3(a) shows that (101) cleavage was
observed when one hydrogen atom was added at every site along line
1 at a load level of $1.5k_G$. Figure 3(b) shows similar results for
the case when hydrogen was added at every site along lines 2 and 3
at the same load. This is significantly lower than the load required to show this behavior in pure iron, which is $1.8k_G$ as described above, and indicates that (101) plane fracture is enhanced
by the presence of hydrogen. There has been growing experimental
evidence in recent years for (101) fracture in hydrogen embrittled
iron.[9] Previous explanations of this effect have concentrated
on the transport of hydrogen away from the crack tip by dislocations

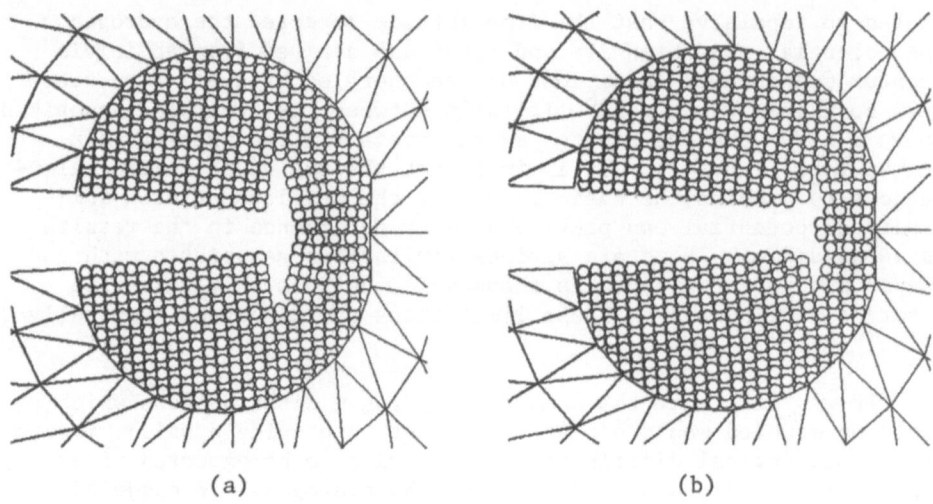

(a)   (b)

Fig. 3 Model configuration after 7 ps at $1.5k_G$ when hydrogen was
added at 1 ps at (a) site 1, (b) sites 2 and 3.

and its subsequent deposition on the dislocation slip planes. The
resulting high hydrogen concentration on these planes could
then lead to decohesion there. The current study indicates that
there may be an additional contribution to this behavior from atomic
scale processes at the crack tip.

When the hydrogen concentration was reduced so that a hydrogen
atom was located at every alternate site along line 1 and along
lines 2 and 3, a load of $1.7k_G$ was required for (101) fracture.
Also, with hydrogen at every site along line 2 only, $1.75k_G$ was
required. These results indicate that the enhancement of (101)
cleavage is probably highly concentration dependent.

Although the presence of hydrogen altered the relative probabilities for fracture on (010) and (101) planes, in no case did it
reduce the load needed to cause fracture on some plane. In fact,
some locations for the hydrogen increased the fracture load. This
result may be interpreted in several ways. It would tend to support hydrogen embrittlement mechanisms which do not require "decohesion" of the iron lattice such as that described in reference
10 for example. Alternatively, it could be an indication of some
inadequacy of the model.

One inaccuracy which does exist is that a negative volume is
predicted for hydrogen in the lattice in contrast to the positive

volume observed experimentally. In order to determine if moderate
changes to the potentials could bring this behavior more into line
with experiment, the Fe-H interaction was varied over a considerable
range with the Fe-Fe potential held constant. It was felt that
changing the Fe-H interaction with respect to the Fe-Fe one should
show the range of behavior which can be expected in such a model.
The Fe-H potentials used in this study are shown as curves 3 to 6
in figure 1. The tetrahedral interstitial site is preferred for
all potentials except number 3. This is the experimentally observed
site. However, all of these potentials gave negative volumes.
This would seem to imply that a positive volume is unlikely to be
achieved without a potential which differs by a large degree from
that derived from the ab initio calculations. The Olander potential[7]
is one example, but as stated above, it is so strongly repulsive
that the iron lattice will not absorb the hydrogen when this potential is assumed. The difficulties in matching the behavior of the
model to that of experiment in this area may be due to the limitations of pair-wise bonding. Multi-body effects may be necessary
to reproduce the correct behavior.

The calculations for hydrogen near the crack tip were repeated
using potentials 3 to 6. The results appear to be qualitatively
the same as those for the original potential in all cases, with
the lowest load required for fracture unchanged or increased, while
(101) plane cleavage is enhanced. This indicates that the results
have some universality, at least when pair-wise bonding is assumed.

CONCLUSIONS

Model calculations indicate that the tendency of a crack in
an iron crystal to branch onto (101) planes off of its original (010)
plane in enhanced when hydrogen is added to the lattice. The lowest
load needed to cause fracture in the crystal was either not changed
or was increased when hydrogen was added to the system. This tends
to support proposed hydrogen embrittlement mechanisms which de-
emphasize the role of "decohesion" of the lattice. There are
deviations in the model predictions for the strain field surrounding
a hydrogen interstitial compared with that observed experimentally,
however. These did not appear to be resolved by making moderate
changes to the potentials. This indicates that either the shape
of the potential curves which were assumed greatly deviate from
those in the real crystal or, more likely that the basic assumption
of two body potentials is inadequate. Work is proceeding on extending the model to include multi-body effects.

ACKNOWLEDGEMENT

This work was supported by National Aeronautics and Space
Administration grant #NCC2-160.

REFERENCES

1. P.C. Gehlen, A.J. Markworth, L.R. Kahn, Atomistic Studies of
    Hydrogen Enhanced Crack Propagation in BCC Iron, in "Computer
    Simulation for Materials Applications", R.J. Arsenault, J.R.
    Beeler, J.A. Simmons, eds., p684 (1976).
2. P.J. White, L.R. Kahn, Computer Model for Hydrogen Effects on
    Crack Propagation in BCC Iron, in "Hydrogen Effects in Metals",
    I.M. Bernstein, A.W. Thompson, eds., p723, AIME, Warrendale,
    Pa. (1981).
3. M. Mullins, Molecular Dynamics Simulation of Propagating Cracks,
    submitted to Scr.Met.
4. R.A. Johnson, Interstitials and Vacancies in α Iron, Phys. Rev.,
    134:A1329 (1964).
5. J.H. Walker, T.E.H. Walker, H.P. Kelly, Ground and Low-Lying
    Excited States of Fe-H, J. Chem. Phys., 57:2094 (1972).
6. A.J. Markworth, M.F. Kanninen, P.C. Gehlen, An Atomic Model of
    an Environmentally Affected Crack in BCC Iron, in "Proc.
    Intern. Conf. on Stress Corrosion Cracking and Hydrogen
    Embrittlement of Iron Base Alloys", p447 (1973).
7. D.R. Olander, Description of the Hydrogen-Metal Interaction by
    a Morse Potential Function, J. Phys. Chem. Solids, 32:2499
    (1971).
8. R.L. Jaffe, personal communication.
9. C.J. McMahon, Effects of Hydrogen on Plastic Flow and Fracture
    in Iron and Steel, in "Hydrogen Effects in Metals", I.M.
    Bernstein, A.W. Thompson, eds., p219, AIME, Warrendale, Pa.
    (1981).
10. S.P. Lynch, Hydrogen Embrittlement and Liquid-Metal Embrittle-
    ment in Nickel Single Crystals, Scr. Met., 13:1051 (1979).

THEORY OF HYDROGEN CHEMISORPTION ON

FERROMAGNETIC TRANSITION METALS

J.L. Morán-Lopez and L.M. Falicov[*]

Departamento de Física, CINVESTAV-IPN, 07000 México F.F.
*Department of Physics, University of California, Berkeley
CA 94720

ABSTRACT

An Anderson-Hubbard model for hydrogen chemisorption on ferro-magnetic 3-d metals is presented. The substrate electronic structure is solved by means of the cluster-Bethe-lattice approximation and the size of the magnetic moment at the adatom and at the metal surface atoms are determined selfconsistently. Results are given for the chemisorption energy and for the change in the substrate magnetization upon chemisorption.

INTRODUCTION

Chemisorption of hydrogen on transition metals has been the subject of extensive research in recent years, from the experimental[1-5] as well as from the theoretical[6-10] points of view. Experiments performed in ferromagnetic 3-d transition metals show that the adsorption of hydrogen changes as the substrate goes through the magnetic transition[3-5]. Thus, in order to understand the low temperature experiments, it is necessary to take explicit account of the substrate magnetization.

Here, we are interested mainly in two questions: i) how does hydrogen chemisorption influence the surface magnetization; and ii) how the energy of chemisorption changes with substrate magnetization. For that purpose, the substrate is described by a model for itinerant magnets, in which local moments are assumed to exist on each lattice site in both the magnetically ordered and in the paramagnetic states[11].

THEORY

The metal-substrate characterization is obtained by means of a Hubbard Hamiltonian in the unrestricted Hartree-Fock (UHF) approximation

$$H_m = \sum_{\alpha\sigma} \left\{ \sum_{ij} t_{ij} c^\dagger_{i\alpha\sigma} c_{j\alpha\sigma} + \sum_i \frac{1}{2} U (n \underline{+} \mu_i) n_{i\alpha\sigma} \right\} - \frac{1}{4} U \sum_i (n^2 - \mu_i^2), \quad (1)$$

where t_{ij} denotes the hopping integral for electronic transitions between lattice sites i and j, σ is the spin index, α is a d-band index ($\alpha = 1, \ldots 5$), U is the intrasite Coulomb integral, and $c^\dagger_{i\alpha\sigma}$, $c_{i\alpha\sigma}$ and $n_{i\alpha\sigma}$ are the usual creation, annihilation and number operators for electrons at site i, spin σ and band α. In addition

$$\mu_i = \sum_\alpha (\langle n_{i\alpha\uparrow} \rangle - \langle n_{i\alpha\downarrow} \rangle) , \quad (2)$$

n is the spatially constant number of electrons at each site

$$n = \sum_{\alpha\sigma} \langle n_{i\alpha\sigma} \rangle . \quad (3)$$

and $\langle n_{i\alpha\sigma} \rangle$ is the average number of electrons of spin σ and band α at site i. In (1) the minus sign holds for spin-up electron and the plus sign for spin-down electrons respectively.

At zero temperature all the magnetic moments μ_i are equal and are aligned. The value for μ is obtained in a selfconsistent manner, by the requirement that

$$\mu \equiv \int_{-\infty}^{\varepsilon_F} (\rho_\uparrow - \rho_\downarrow) d\omega \quad (4)$$

and the value of μ_i used in equation (1) do not differ. In (4) ρ_σ is the density of states for spin σ, obtained from the trace of the Green's function in the Bethe-lattice approximation[12].

To study hydrogen chemisorption we use an Anderson-Hubbard Hamiltonian in the UHF

$$H = H_m + \sum_\sigma \left\{ \left[\varepsilon_a + U_a \langle n_{a\bar{\sigma}} \rangle \right] n_{a\sigma} + t_a c^\dagger_{a\sigma} c_{01\sigma} + t_a^* c^\dagger_{01\sigma} c_{a\sigma} \right\} - U_a \langle n_{a\uparrow} \rangle \langle n_{a\downarrow} \rangle , \quad (5)$$

where ε_a is the energy level of the hydrogen, relative to the substrate, and before chemisorption, t_a is the hopping integral for electronic transition between adatom and substrate, U_a is the Coulomb integral at the adatom, $\bar{\sigma}$ indicates the spin opposite to σ, and $c^\dagger_{a\sigma}$, $c_{a\sigma}$ and $n_{a\sigma}$, which refer to the adatom, have the usual meanings. It should be noted that the electronic states of the adatom ($a\sigma$) are connected in our model only to the electronic states of one atom of the substrate, located at site 0, and only to the d-band of correct symmetry (labelled 1) and of the same spin σ.

At zero temperature the substrate is in the perfect ferromagnetic state with all atomic magnetic moments pointing in a given direction. In this case the local Green's functions at the adatom are given by

$$G_{aa;\sigma}(\omega)=\left[\omega-E_a^\sigma-t_a^2/(\omega-zt\gamma^\sigma)\right]^{-1} \quad , \tag{6}$$

where z is the coordination number of the metal atoms, t is the nearest neighbor hopping integral in the metal,

$$E_a^\sigma=\varepsilon_a+U_a<n_{a\bar\sigma}> \quad , \tag{7}$$

and γ^σ, which do not depend on the adatom properties, are transfer functions which relate two neighboring Green's functions in the clean, ferromagnetic substrate[12].

In general, one might find two, one or no localized states depending on the hopping integral t_a and on the value of E_a^σ. The energy as well as the weight of the localized states can be calculated analytically[12].

In complete analogy with the Anderson model[13] selfconsistency should be achieved by calculating an arbitrarily large number of loops which determine electron occupations and magnetization at the adsorbate and the various metallic atoms, and require that input and output are the same. For the adsorbate-atom orbitals selfconsistency requires that, for $\sigma=\uparrow$ and $\downarrow$, the two equations

$$N\left[N(N_{a\sigma})\right]-n_{a\sigma}=0 \tag{8}$$

are satisfied, where the functions N are defined by

$$n_{a\sigma}=N(n_{a\bar\sigma})=-\frac{1}{\pi}\int_{-\infty}^{\varepsilon_F} Im\ G_{aa,\sigma}(\omega)d\omega+(n_{a\sigma})_L \quad . \tag{9}$$

The last term is the contribution of localized states below the band. If the substrate is in the ferromagnetic state, equation (8) has a variety of solutions. For small t_a there are three solutions: two are ferromagnetic, i.e. with adsorbate magnetization parallel to that of the substrate ($\mu_a>0$) and one, the antiferromagnetic one, with $\mu_a<0$ (magnetization antiparallel to that of the substrate). The antiferromagnetic solution has the lowest energy for the cases we have calculated. For large t_a only one solution, the antiferromagnetic one, exists.

Additional selfconsistency loops must be carried out for all the ferromagnetic-metal atomic orbitals. In our calculations we have only included the $\alpha=1$ orbitals ($\sigma=\uparrow$ and $\downarrow$) of the metal atom immediately below the hydrogen (i=0), and we have frozen the occupancy of all other orbitals. We have tried to mimic the case of hydrogen on ferromagnetic iron, and for that purpose we have chosen a body-centered cubic solid z=8, with a bandwidth of 2W=4eV and such that

$$n=7.7, \quad \mu_{substrate}=1.99, \quad U/W=0.38, \quad t/W=\frac{1}{2}(z-1)^{-\frac{1}{2}}=0.189; \quad (10)$$

for the hydrogen adatom we chose

$$U_a/W=5.6, \quad \varepsilon_a/W=-3.0 \tag{11}$$

and the value of t_a was varied.

Once selfconsistency is achieved, the hydrogen chemisorption energy ΔE can be easily obtained by calculating the total energy of the system, i.e. the expectation value of the hamiltonian $<H>$ given by (5), and subtracting from it the energy obtained by putting $t_a=0$, i.e. by assuming hydrogen and metal infinitely separated.

RESULTS AND DISCUSSION

We have presented a selfconsistent theory of hydrogen chemisorption on ferromagnets. It is a zero-temperature theory where we have assumed that the substrate is in the completely ordered configuration, with all the local moments pointing along the magnetization direction. Finite temperature calculations, which are lengthy but straightforward, can be accomplished by the use of the theory presented in reference 11.

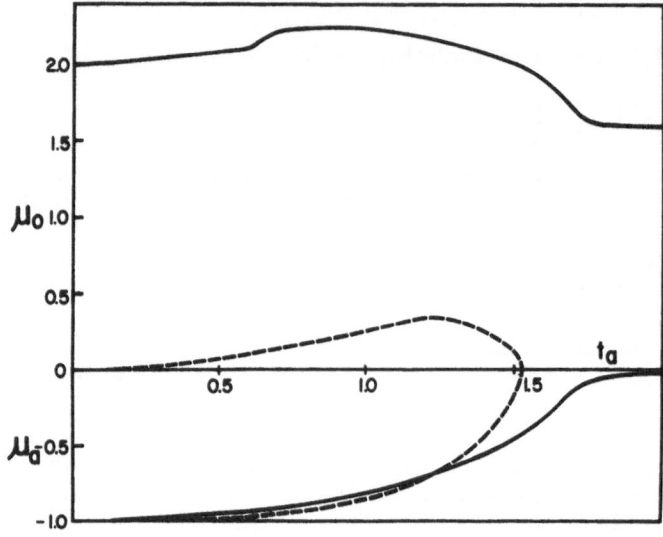

Fig. 1.   The magnetic moment at the adatom μ_a and at the metal atom below μ_0 as a function of t_a for the cases of a paramagnetic (broken line) and ferromagnetic (solid line) substrate.

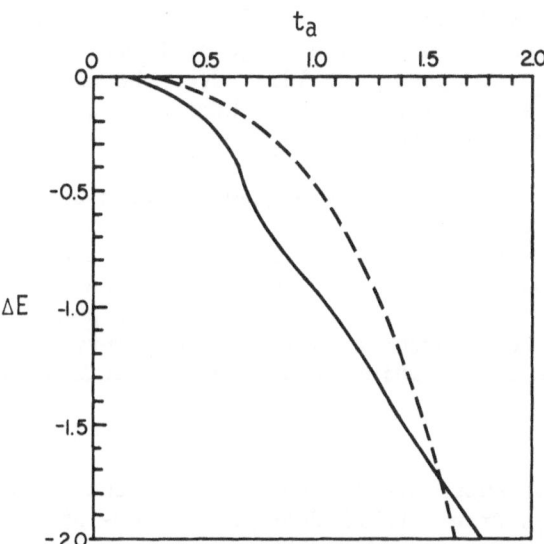

Fig. 2.   The binding energy of hydrogen chemisorbed on a paramagnetic
          (broken line) and a ferromagnetic (solid line) substrate as
          a function of t_a.   The energy is measured in unites of W,
          the d-half-bandwidth ($\sim 2eV$).

We have assumed that the s-hydrogen electrons couple to only
one of the five d degenerate transition-metal bands, and only through
a single atom (on-top configuration).   This approximation has been
justified before[13].

In Figure 1 we show the selfconsistent antiferromagnetic solution as a function of t_a for the paramagnetic as well as for the
ferromagnetic substrate.   In the paramagnetic case, chemisorption
induces a magnetic moment on the metal atom immediately below in the
range of spin-polarized solutions for the hydrogen atom ($0 \leq t_a \leq t_{ac}$).
More interesting is the case of chemisorption on ferromagnetic substrates.   In that case, we see that there is a range of t_a values
where the magnetic moment at the atom below, μ_0, increases upon
chemisorption, and another range where the effect is the opposite.
It is also clear from the results that the range of values of t_a
for which the hydrogen atom is appreciably spin polarized is larger
for the ferromagnetic substrate than for the paramagnetic substrate.

In Figure 2 we show the results for the chemisorption energy
ΔE as a function of the hopping matrix element t_a.   We display the
results for the ferromagnetic substrate (solid line) as well as those
obtained assuming the substrate to be in the paramagnetic state.

We see that ΔE is lower for the ferromagnetic state than for the paramagnetic case over a wide range of t_a values, and that only for very large t_a the opposite holds. Over the range on which the hydrogen atom is spin-polarized, the chemisorption is stronger on the ferromagnetic substrate.

From measurements of hydrogen chemisorption on paramagnetic transition-metal substrates[13], the value for t_a should be between 1.2W and 1.4W for Fe, Co and Ni. Therefore, we expect that the hydrogen atoms would be more tightly bound to the substrate as one goes from the paramagnetic to the ferromagnetic phase. This might explain the observed[5] decrease in the net desorption rate of the H_2/Ni system observed in going across the magnetic phase transition.

ACKNOWLEDGEMENTS

This work was supported in the U.S.A. by the National Science Foundation through Grants DMR81-06494 and INT80-18688, and in Mexico by CONACyT through Grant PCAIEUA-800649.

REFERENCES

1. J. Behm, K. Christmann and G. Ertl, Solid State Commun. 25, 763 (1978).
2. I.A. Toyashima and G.A. Somorjai, Catal. Rev. 19, 105 (1979).
3. R.J.H. Voorhoeve, in Magnetism and Magnetic Materials-1973, ed. C.D. Graham, Jr. and J.J. Rhyne (American Institute of Physics, New York, 1974).
4. M. Landolt and M. Campagna, Phys. Rev. Lett. 39, 568 (1977).
5. M.R. Shanabarger, Phys. Rev. Lett. 43, 1964 (1979).
6. S.R. Schrieffer and R. Gomer, Surface Sci. 25, 315 (1971).
7. D.M. Newns, Phys. Rev. 179, 1123 (1969).
8. H. Kranz, Phys. Rev. B 20, 1617 (1979).
9. J.E. Ure, E.V. Anda and N. Majlis, Surface Sci. 99, 689 (1980).
10. J.L. Morán-López and L.M. Falicov, J. Vac. Sci. and Technol. 20, (to appear March 1982).
11. J.L. Morán-López, K.H. Bennemann and M. Avignon, Phys. Rev. B 23, 5978 (1981).
12. J.L. Morán-López and L.M. Falicov, submitted to Phys. Rev. B.
13. C.M. Varma and A.J. Wilson, Phys. Rev. B 22, 3795 and 3805 (1980).

HYDROGEN CHEMISORPTION ON CLOSE-PACKED SURFACES

F. Mejía-Lira* and J.L. Morán-López[†]

*Instituto de Física, Universidad Autónoma de S.L.P.
78240 San Luis Potosí, México
†Departmento de Física, CINVESTAV-IPN
07000 México D.F.

ABSTRACT

A theory of hydrogen chemisorption on close-packed face-centered-cubic crystal surfaces is presented. The close packing effects are taken into account by solving the local electronic density of states in the Husimi cactus model. The calculation is based on the Anderson-Newns Hamiltonian. Results are given for the local densities of states, at the adatom and at the atoms over which the hydrogen is chemisorbed.

INTRODUCTION

The adsorption of hydrogen on transition metal surfaces has been extensively investigated, mainly because of its catalytic importance. Most of the substrates with catalytic importance, like Pt, Ni, Rh, etc., crystallize in the face-centered-cubic structure and often the surface exposed to hydrogen is the close packed (111)-surface. Experiments[1-3] performed on this surface reveal that hydrogen is chemisorbed at the center of the triangles, forming tetrahedra. It is therefore important to study this chemisorption complex.

The problem has been attacked from the molecular point of view, in which a small cluster of transition-metal atoms is used to model the adsorption site[4]. However, the reduced size of the cluster prevents the description of real surfaces. In opposition to that approach, here we present a solid state theory for hydrogen chemisorption on the (111)-fcc surface. The chemisorption geometry is taken into account in a natural way by describing the substrate by a Husimi cactus[5].

This is a simple model equivalent to a Cayley tree with a complex unit. In order to describe fcc crystals we have taken the unit of the cactus to be a tetrahedron. It has been shown[6] that this approximation gives reasonable results for the electronic structure of fcc crystals. The surface in this model, is a two dimensional net of triangles connected to the bulk Husimi cactus. This is illustrated in Figure 1a.

MODEL AND CALCULATION

The substrate characterization is obtained by means of the tight-binding Hamiltonian

$$H_S = \sum_i \varepsilon \, c_i^\dagger \, c_i + \sum_{ij} t_{ij} \, c_i^\dagger \, c_j + h.c. \tag{1}$$

where ε is the single site energy, t_{ij} denotes the hopping integral for electronic transitions between lattice site i and j, and $c_i^\dagger$ and c_j are the usual creation and annihilation operators for electrons on sites i and j.

The bulk local density of states is given by

$$n(\omega) = -\frac{1}{\pi} \, \text{Im} \, G_{00}(\omega) \tag{2}$$

where $G_{00}(\omega)$ is the local Green function at an arbitrary site zero and it is given by the Dyson equation,

$$G \, (\omega - H) = 1. \tag{3}$$

It can be shown that this equation can be solved by defining a transfer function[6]

$$\gamma \equiv G_{n,0}/G_{n-1,0}. \tag{4}$$

To solve the electronic structure at the surface, we take the geometry shown in Figure 1a, which represents the (111)-fcc surface.

Now, the solution of the surface equation of motion for the Green function is achieved, by defining a surface transfer function

$$\alpha \equiv G^s_{n+1,0}/G^s_{n,0} \, , \tag{5}$$

in which both atoms n and n+1 are on the surface. In units of t=-1/6, ε=1/3 and assuming the values of the parameters at the surface to be the same as those of the bulk, the surface local Green function is

$$G_{00}^{s}(\omega) = \left[\omega - \frac{5}{6} - \frac{1}{2} (1 + 2\alpha) \left(\frac{1}{6D} - 1 \right) \right]^{-1} \tag{6}$$

where

$$\alpha = \frac{1}{\frac{4}{3} \left[\frac{1}{6D} - 1 \right]} \left[\omega - \frac{5}{6} - \frac{2}{3} \left[\frac{1}{6D} - 1 \right] \right.$$

$$\left. + \sqrt{\left[\omega - \frac{5}{6} \right] \left[\omega - \frac{5}{6} - \frac{4}{3} \left[\frac{1}{6D} - 1 \right] \right]} \right], \tag{7}$$

$$D = \omega - \frac{1}{3} + \frac{3}{2} \gamma, \tag{8}$$

and

$$\gamma = -\frac{1}{3} \left[\omega + \sqrt{\omega^2 - 1} \right]. \tag{9}$$

At the (111)-fcc surface, there are two different kinds of triangles. One kind has a substrate atom in the adjacent substrate layer, and the other one does not have these atoms. We tried both configureations, but our model gave us exactly the same results. We present here only the chemisorption of hydrogen on the first kind of triangles (Fig. 1b).

In the calculation of the adatom local electronic density of states we adopt the Hamiltonian

$$H_a = \sum_{\sigma} \left[\varepsilon_a + U \langle n_{a-\sigma} \rangle \right] n_{a\sigma} + t_a c_{a\sigma}^{\dagger} c_{0\sigma} + \text{h.c.} + H_S, \tag{10}$$

where ε_a is the energy level of the adatom before chemisorption, t_a is the hopping integral for electronic transitions between adatom and substrate, U is the adatom Coulomb integral, and $\langle n_{a\sigma} \rangle$ is the average number of electrons with spin σ at the chemisorbed atom.

Using the bulk and surface transfer functions, we obtain for the adatom

$$G_{aa}^{\sigma}(\omega) = \left[\omega - \varepsilon_a^{\sigma} - \frac{3t_a^2}{\omega - \frac{5}{6} - \frac{1}{6}(5+4\alpha) \left[\frac{1}{6D} - 1 \right]} \right]^{-1}, \tag{11}$$

where

$$\varepsilon_a^{\sigma} = \varepsilon_a + U \langle n_{a-\sigma} \rangle. \tag{12}$$

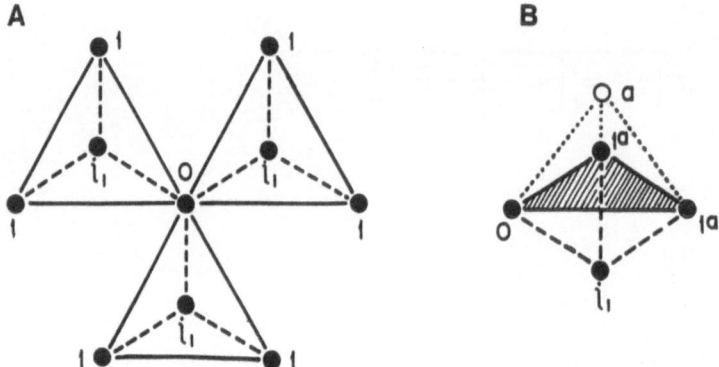

Fig. 1 a). Portion of the surface of a Husimi cactus. b). Chemisorption geometry, <u>a</u> refers to the adatom, $0,1^a$ to the atoms at the surface layer and i_1 to the atoms in the second layer.

Depending on the value for ε_a^σ and t_a, two, one or no localized states may exist. The position as well as its weight can be calculated in an analytical way.

RESULTS

We show in figure 2 results for the local density of states at e the chemisorbed atom (solid line) and at its three nearest neighbors before (long-dashed line) and after chemisorption (short-dashed line). These results correspond to values of the parameters ε_a = -2.56, t_a = 0.4 and U = 0. The localized state at ω = -2.74 belongs to the adatom LDS. Due to the small value of t_a, only a small fraction of

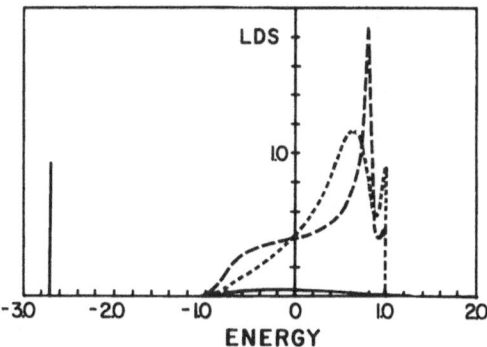

Fig. 2.   Results for the LDS at the adatom (solid line) at its
nearest neighbors before (long-dashed line) and after
chemisorption (short-dashed line). This corresponds to
ε_a = -2.56, t_a = 0.4 and U = 0.

states are available in host band region. We see also that the LDS
at the adatom nearest neighbors is only lightly perturbed. These
are preliminary results, where no self-consistency at the adatom
potential has been carried out. A more detailed account of this
model will be given elsewhere.

ACKNOWLEDGEMENTS

This work was supported in part by Dirección General de
Investigación Científica y Superación Académica de la Secretaría
de Educación Pública (México) under contract 79-04-285 and by
Programa Regional de Desarrollo Científico de la Organización de
Estados Americanos.

REFERENCES

1. R.W. McCabe and L.D. Schmidt, Surface Sci. 60,85 (1977).
2. J. Behm, K. Christmann and G. Ertl, Solid State Commun. 25,763
    (1978).
3. I. Toyoshima and G. A. Somorjai, Catal. Rev. Sci. Eng. 19,105
    (1979).
4. R.P. Messmer, S.K. Knudson, K.H. Johnson, J.B. Diamond and
    C.Y. Yang, Phys. Rev. B13,1396 (1976).
5. C. Domb, Adv. Phys. 9,149 (1960).
6. J.A. Verges and F. Yndurain, J. Phys. F 8,873 (1978).

MAGNETIC SUSCEPTIBILITY, PROTON NMR AND MUON SPIN ROTATION (μSR)

STUDIES OF AN UNSUPPORTED PLATINUM CATALYST WITH ADSORBED H AND O

R.F. Marzke, W.S. Glaunsinger, K.B. Rawlings,
P. Van Rheenen[*], M. McKelvy, J.H. Brewer[†],
D. Harshman[†], and R.F. Kiefl[†]

Arizona State University, Tempe, Arizona 85287,
†University of British Columbia, Vancouver, B.C.
Canada V6T 2A6

ABSTRACT

The effects of adsorbed H and O upon the strong intrinsic magnetism exhibited by small particles of platinum have been studied by magnetic susceptibility measurements, proton NMR and muon spin rotation. All work to date has been on very pure samples of an unsupported Pt catalyst prepared by chemical reduction techniques. Dispersions of the samples are in the range .12-.20. The effect of adsorbed H is slight, but O increases the Curie-Weiss susceptibility of the catalysts substantially at low temperatures. H, however, sharply decreases both the spin-lattice relaxation rate of protons in the samples, whose NMR signals are observable at 4.2K, and the transverse field relaxation rate of muons stopped in the samples.

INTRODUCTION

In order to study the physical properties of small metallic particles it is essential that their surfaces be well-characterized. For this reason we have recently extended our measurements of the magnetism of small platinum particles[1-4] to include: (1) magnetic susceptibility with chemisorbed H and O on the particles surfaces[5], (2) NMR of protons associated with the surfaces[6-8] and (3) muon spin rotation experiments.[9] This paper is a preliminary report of results obtained in these three studies, each of which probes a different aspect of the magnetism of the particles.

---

*Presently at Rohm and Haas Company, Philadelphia, PA 19105

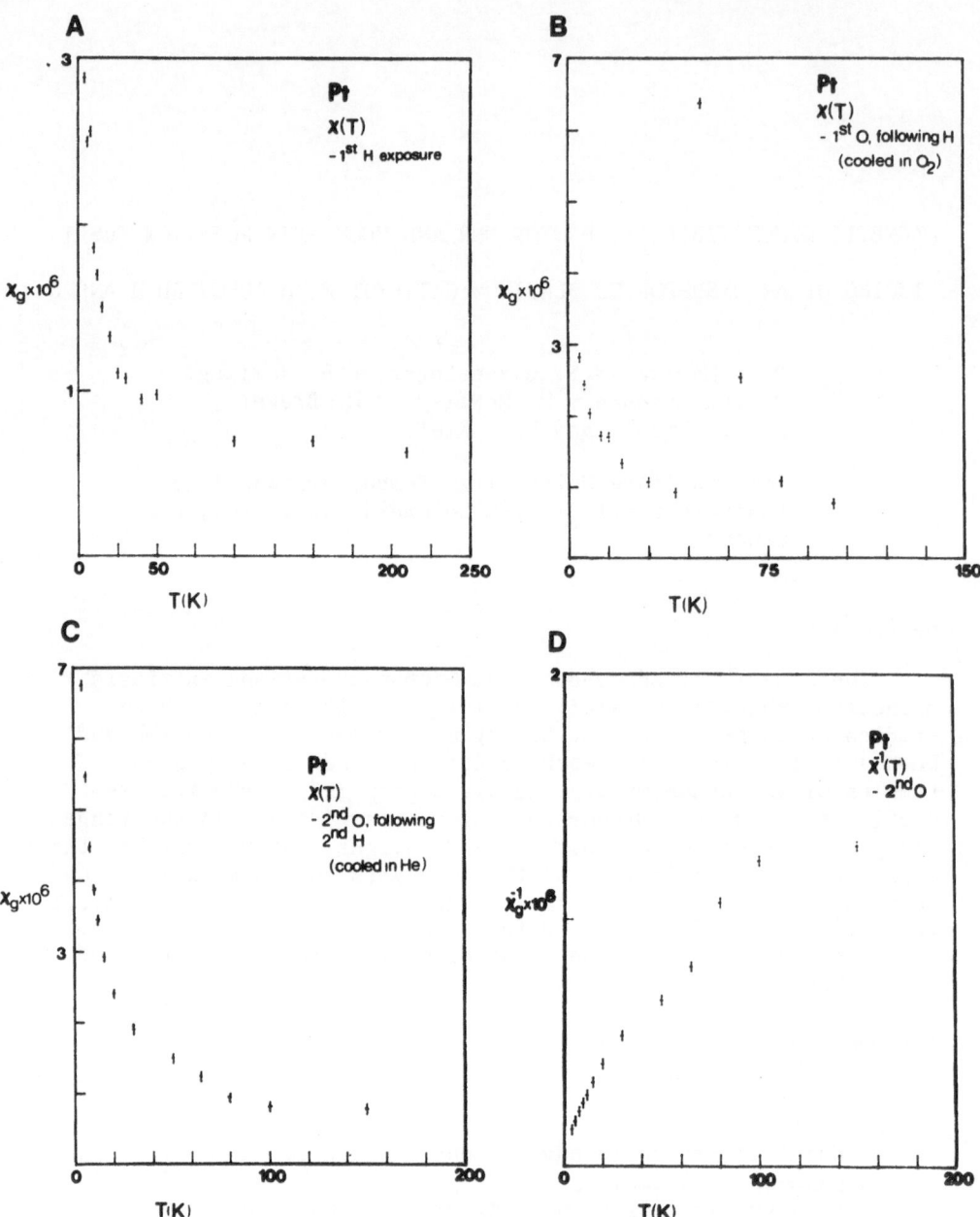

Fig. 1.   (a) Magnetic susceptibility of dispersed Pt after first
exposure to H_2; (b) first exposure to O_2 following H_2
exposure, cooled in O_2; (c) second exposure to O_2 following a second H_2 exposure, cooled in He; (d) Inverse
of magnetic susceptibility shown in (c).

RESULTS

Selected temperature dependences of the particle's magnetic susceptibility χ following various exposures to H_2 and O_2 gases are shown in Figure 1. χ is expressed per gram of Pt. Most of the features can be interpreted easily. In Fig. 1a (single H_2 exposure), χ near 4.2K is almost unchanged from its values before surface treatment (not shown). A small bump is visible near 50K. In Fig. 1b surface H has been replaced by O, and to assure full coverage some residual O_2 remained in the sample chamber upon cooling. He gas was admitted for heat exchange in the temperature range below 50K, where χ is not much changed from Fig. 1a, but now a large peak in χ is observed beginning at 50K, where the solid-liquid phase transition of O_2 is expected at these pressures. At temperatures between 56K and 80K a mass decrease of the sample and a pressure rise were noted. Fig. 1c shows $\chi(T)$ after a second adsorption of O, but this time with only a residual He atmosphere of several hundred millitorr for heat exchange. At 50K there is again a small change, but now at lower temperatures χ is increased over the previous cases by about a factor of two.

A simple picture of what is happening could be as follows: (1) χ is unaffected by the presence of surface H, so there are no dangling carbon bonds. (2) Surface O approximately doubles χ, but (3) frozen O_2 on the surface is sufficiently diamagnetic to reduce the sample's susceptibility in the region below 50K, as shown in Fig. 1b. (4) Liquid O_2 on the surface is strongly paramagnetic, as expected. (5) The persistent 50K feature leads to the conclusion that some O_2 is nearly always present, perhaps because of an imperfect vacuum.

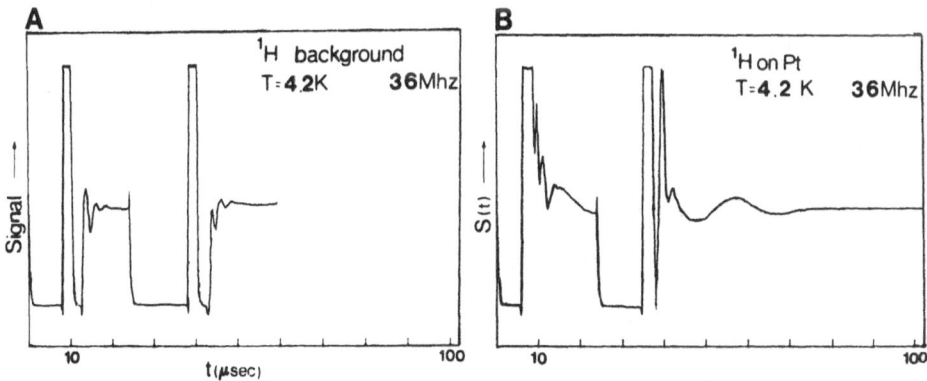

Fig. 2. a. ^{1}H spin echo, empty sample tube.
b. ^{1}H spin echo, untreated evacuated dispersed Pt.

Typical NMR spin echoes for ^{1}H are shown in Figure 2a, b. The large echo in Fig. 2b is unexpected, since it was obtained from an untreated Pt sample that had only been pumped to remove

water. It is generally thought that platinum blacks in air are
covered with an oxygen monolayer, but not with hydrogen to the extent indicated by the size of the proton signals, which are large
enough for an H monolayer to be present as well. Evidence that
the protons being observed in the untreated Pt are associated with
the surface comes from the ∿ 43 msec spin-lattice relaxation time
at 4.2K, which is very short compared to the temperature-independent proton T_1 of 2 seconds measured after surface H-treatment.
This rapid relaxation must result from fluctuations of the particles' electronic magnetism. The linewidth shows a slight decrease, going from 10G to 7G (± 1G) after hydrogen chemisorption.
These values are in the range that would be expected from an average area of 8 square Angstroms per surface Pt atom[10], with one H
per surface Pt. No shifts of the line centers from the diamagnetic
proton resonance field could be detected.

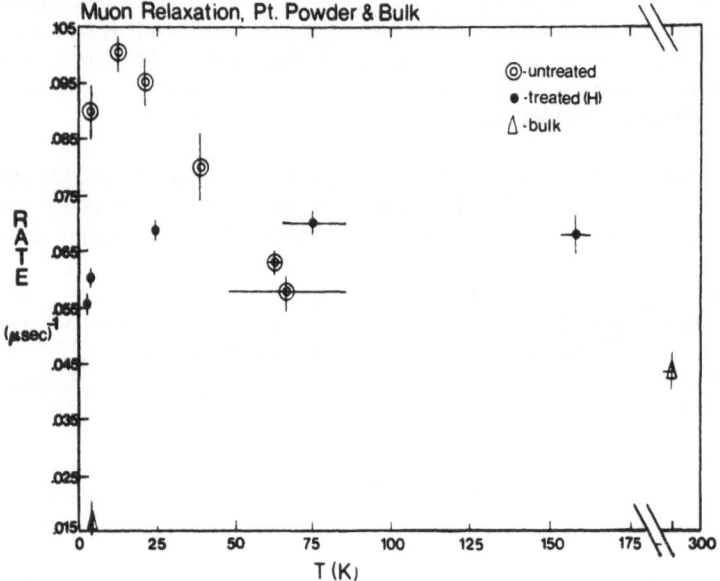

Figure 3.   Transverse field muon relaxation rate as a function of
            temperature.   (Data fitted to Lorentz lineshape).

     Transverse field muon relaxation rates in Pt as a function of
temperature and surface treatment are given in Figure 3. Lorentzian lineshapes were found to give the best fits in all cases. A
bulk foil (4-9's purity) was first run at 4.2K and room temperature and found to exhibit little relaxation at the lower temperature. This is expected if the ∿ 34% abundant nucleus [195]Pt is
responsible for the relaxation.

     Untreated, dispersed Pt samples show a sharply increased muon
relaxation with a strong temperature dependence. Clearly the
surface is having a large effect, but the rates are roughly a fac-

tor of four too low for stationary muons trapped at sites similar to those for the ^{1}H observed by NMR. (This estimate assumes that the same local fields occur at sites for both muons and protons). The observed rates lie in the range of those found in metals doped with paramagnetic impurities[11], at temperatures where the muons diffuse readily toward trapping sites. The temperature dependences of the rates differ significantly from this case, however. In a preliminary test for surface effects a temperature dependence of the relaxation rate was obtained for a sample with surface H, and the results are also given in Fig. 3. There is a strong decrease in both the rate and its temperature dependence. This behavior contrasts with both that of the susceptibility upon H adsorption and that expected from isolated impurities in the particle interiors, but more experiments are needed to determine the actual origins of the relaxation rates observed to date.

DISCUSSION

Fits of the susceptibility data to a Curie-Weiss law indicate that the increase in χ with surface O at low temperatures, shown in Fig. 1c, results from a decrease in the magnitude of the (negative) Curie temperature, rather than from an increase in the Curie constant. The particles' magnetism may thus be due to a molecular species, which is sufficiently dense on their surfaces to have small exchange couplings that are more strongly affected by the larger O atom than by H. The magnetism appears to be too large to be either a simple quantum size effect or to be due to known paramagnetic impurities.[12]

The NMR data indicate that protons are stationary on the surface at low temperatures, and that the surface electronic character is basically that of a non-conductor with appreciable paramagnetism which can relax the protons either rapidly or slowly depending upon surface conditions. The source of this paramagnetism must be the same as that of the large surface susceptibility.

From the shapes of the temperature dependences of the muon relaxation rates, it appears that muons are being trapped at low temperatures near the surfaces of the Pt particles. The rates are not low enough to suggest that the muons decay at any distance outside of the particles, but they would presumably be unable to find vacant hydrogen sites when the surface is treated with H, and thus perhaps are decaying near the surfaces but not on them.

ACKNOWLEDGEMENTS

Part of this work was supported by a grant from the Petroleum Research Fund. The support of Arizona State University is also gratefully acknowledged. We wish to thank M. Norton for helpful discussions, and V. Plachecki for assistance.

REFERENCES

1. R.F. Marzke, W.S. Glaunsinger and M. Bayard, Solid State
    Commun. 18, 1025 (1976).
2. R.M. Wilenzick, D.C. Russel, R.H. Morriss and S.W. Marshall,
    J. Chem. Phys. 47, 533 (1967).
3. P. Van Rheenen, M. McKelvy, W.S. Glaunsinger and R.F. Marzke,
    Bull. Am. Phys. Soc. 26, 415 (1981). See also P. Van
    Rheenen, Ph.D. Thesis, Arizona State University, 1981.
4. M.J. McKelvy, P. Van Rheenen, W.S. Glaunsinger and R.F. Marzke,
    Bull. Am. Phys. Soc. 26, 416 (1981).
5. M.A. Vannice, J.E. Benson and M. Boudart, J. Catal. 16, 348
    (1970). See also reference 3.
6. T. Sheng and I. Gay, J. Catal. 71, 119 (1981).
7. L.C. DeMenorval and J.P. Fraissard, Chem. Phys. Lett. 77, 309
    (1981).
8. H.T. Stokes, H.E. Rhodes, P. Wang, C.P. Slichter and J. H.
    Sinfelt in Nuclear and Electron Spectroscopies Applied to
    Materials Science, edited by E.N. Kaufman and G.K. Shenoy
    (North-Holland, Amsterdam, 1981), p. 253.
9. J.H. Brewer and K.M. Crowe, Ann. Rev. Nucl. Part. Sci. 28, 239
    (1978).
10. J.R. Anderson, Structure of Metallic Catalysts. (Academic
    Press, London, 1975), p. 296.
11. R.H. Heffner, Hyp. Int. 8, 655 (1981).
12. J.A. Mydosh and G.J. Nieuwenhuys in Ferromagnetic Materials,
    edited by E.B. Wohlfarth (North-Holland, Amsterdam, 1980),
    p. 167.

INFLUENCE OF INTERSTITIAL HYDROGEN ON TEMPERATURE DEPENDENCE OF ELECTRICAL RESISTANCE OF DISORDERED Ni-Mn ALLOYS

Istvan Csuzda and Hermann Joh. Bauer

Sektion Physik
University of Munich
Federal Republic of Germany

ABSTRACT

Hydrogenation of disordered Ni-Mn alloys, with manganese contents below 20at%, up to hydrogen concentrations sufficient for hydride formation completely destroys their ferromagnetism and consequently lowers the temperature coefficient of their electrical resistance. Nevertheless known anomalies in the temperature dependence of the electrical resistance near 20K and 50K which have been explained by two possible orientations of the magnetic moments of the Mn atoms are not influenced by the interstitial hydrogen. Furthermore, the addition of hydrogen with its above mentioned neutralization of the matrix ferromagnetism has no significant consequences upon the apparent spin glass character of this range of Ni-Mn alloys. The results are discussed in the context of other papers.

INTRODUCTION

By raising the interstitial hydrogen in nickel to high concentrations[1] it is possible to neutralize the ferromagnetism of the Ni-matrix[2] to the degree that the magnetic moments of dissolved paramagnetic atoms will make themselves felt by anomalous electric and magnetic behavior, as in case of the dilute alloys of nickel with Fe[3]. Such an effect signifies the completion of the hydride phase in the Ni-matrix. Similar behavior is seen in hydrogenated nickel chromium alloys with Cr-contents even of several atomic percent.[4]

This paper deals with the influence of interstitial hydrogen on the temperature dependence of electrical resistance of disordered

653

Ni-Mn alloys with manganese contents up to about 20at% and hydrogen
concentration sufficient to form the non-ferromagnetic hydride
phase. The investigations are performed with regard to anomalies
near 20K and 50K, recently found by Beilin et al[5,6] in alloys with
Mn concentration up to about 18at%.

These anomalies appear as maxima in the temperature dependence
of temperature coefficient dR/dT as well as of thermoelectric power
and of magnetization. The authors infer from the behavior of the
magnetization that both anomalies are connected with the two possible orientations of the magnetic moments of the Mn atoms with
respect to the matrix-magnetization[7], whereby at the temperature
T_1 the Mn($\downarrow$)Mn($\downarrow$)-clusters and at T_2 the Ni($\downarrow$)Mn($\downarrow$)-clusters become
destroyed.

For our investigations it is of special interest that Ni-Mn
alloys, first transformed to hydride by Krukowski et al. under
high pressure conditions[8], lose completely their ferromagnetic
behavior[9,10] in the disordered state when hydrogenated. This is in
contrast to its partial reduction in hydrides with Mn-contents
above 20at%.

RESULTS AND DISCUSSION

We performed measurements of the electrical resistance of cold-
worked Ni-Mn foils of thicknesses between 7.5µm and 20µm. In case
of pure Ni we used a 65µm thick MRC foil with a Mn impurity of
0.03ppm.

With all investigated samples we confirmed the findings of
Beilin et al. (see fig. 1) with respect to the anomalies in the
temperature dependence of the electrical resistance. The temperature coefficient of the electrical resistance calculated from the
data of Fig. 1 further illustrates the suppression of the anomalies
with increasing Mn content.

The hydrogenation was performed electrochemically[1] and the
hydride formation was controlled as earlier[10] by 'in situ' observation of the cancelling of the ferromagnetism by the interstitial
hydrogen (at atomic ratio H/alloy $\geq$ 0.6).

As a consequence of hydrogenation of this range of alloys,
we found a distinct decrease of the temperature coefficient of the
electrical resistance, (explainable by d-band filling), Fig. 2,
but we could observe no influence of the interstitial hydrogen
on the above mentioned anomalies, Fig. 3.

The fact that in spite of the cancelling of ferromagnetism
the anomalies in the temperature dependence of electrical resistance

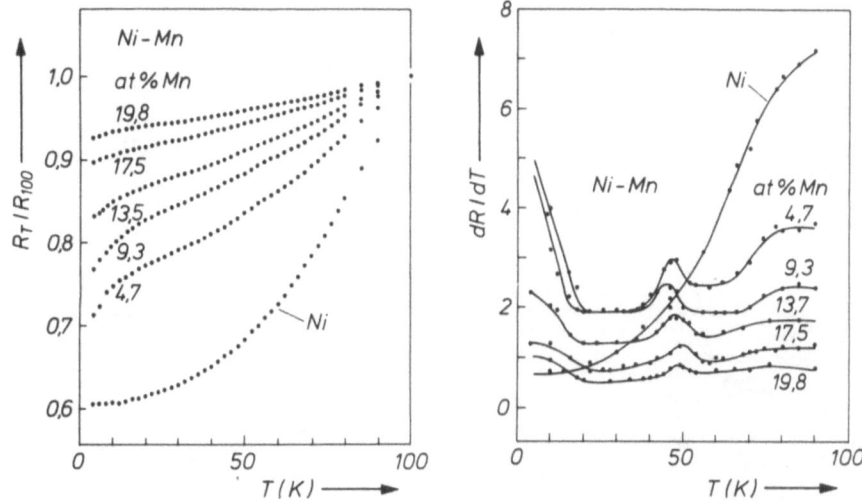

Fig. 1. Temperature dependence of electrical resistance R_T (normalized to R_T at 100K) and its temperature coefficient dR/dT (Ω/K) for Ni and disordered Ni-Mn alloys.

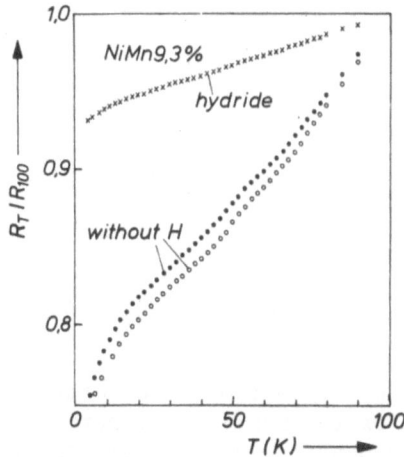

Fig. 2. Influence of hydrogenation of a disordered Ni-Mn alloy up to formation of the non-ferromagnetic hydride phase on the temperature dependence of the electrical resistance R_T (normalized to R_T at 100K). •before hydrogenation, o after desorption.

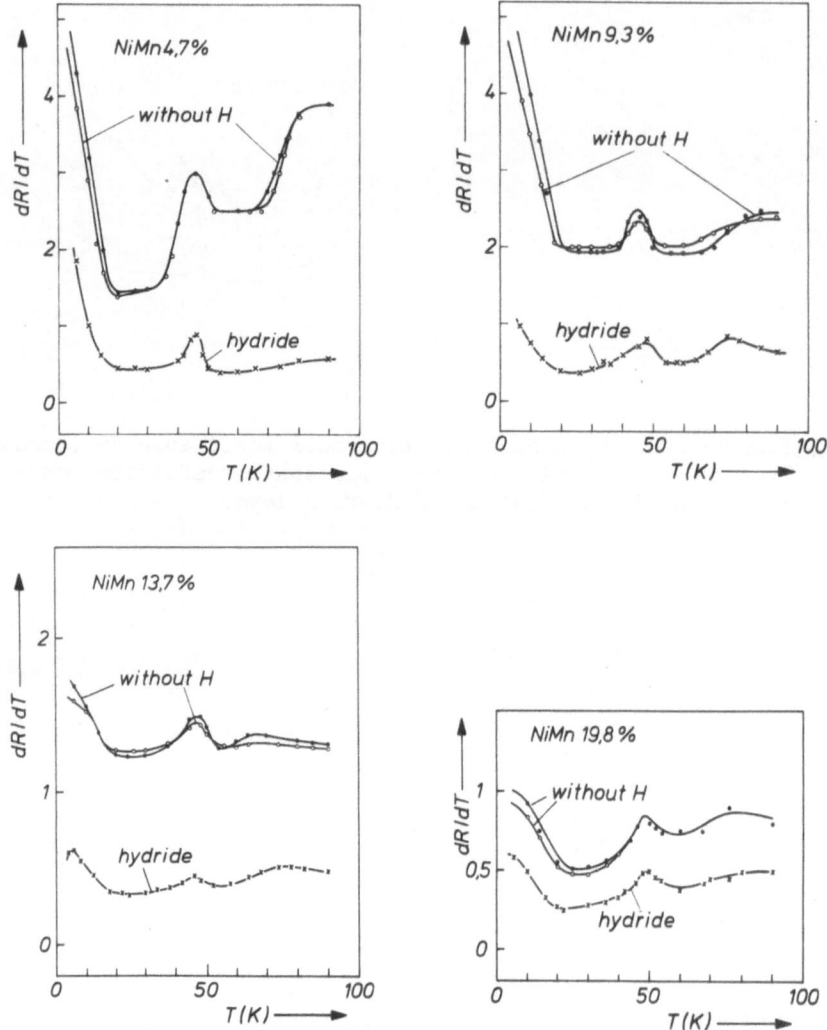

Fig. 3. Influence of hydrogenation of disordered Ni-Mn alloys up
         to formation of the non-ferromagnetic hydride phase on the
         temperature dependence of the temperature coefficient of
         electrical resistance dR/dT (Ω/K). ● before hydrogenation,
         ○ after desorption.

remain is not easy to understand because the anomalies are regarded
as resulting from an interaction between the magnetic moments of
the Mn atoms and the matrix magnetization. This is probably a sign
that a small Ni-area around each impurity has not yet been magnetically neutralized because the hydrogen could not enter the closest
neighborhood of the impurity This assumption is supported by observations of Wagner et al.[11] on the basis of Mossbauer experiments
on substitutional solute metals in the β-phase of the Pd-H system.
The authors found that a repulsive interaction between the solute
impurity and the hydrogen interstitials reduces hydrogen population
in the closest vicinity of the impurity.

   The strong decrease of electrical resistance, Fig. 1, can be
interpreted as that occuring in spin-glass systems.[5,6] Spin-glass
behavior has been discovered recently in a large number of dilute
alloys with transition metals as matrix, as in case of PdCr[12] and
AuCr[13], for example. Characteristically for such systems, the maximum of the impurity resistivity $\Delta\rho(T) = \rho_{alloy} - \rho_{pure}$ metal is at
lower temperatures. This indicates the appearance of magnetic
ordering. For comparison $\Delta R(T) = R_{NiMn}(T) - R_{Ni}(T)$ in our case
of Ni-Mn (see fig. 4) shows a large maximum as observed in the mentioned alloys as well. The analysis of $\Delta R(T)$ with respect to the
$T^{3/2}$ law (not shown) also indicates a spin-glass interpretation.

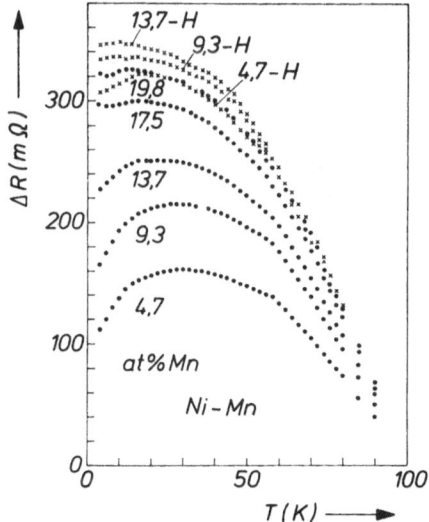

Fig. 4. Temperature dependence of $\Delta R = R_{NiMn(H)}(T) - R_{Ni}(T)$ for
        different Mn-concentrations and after hydride formation.

As to the consequences of hydrogenation, Fig. 4 shows further that there is no considerable influence either. The reason for this may be the same as mentioned above. Moreover it seems that introduction of interstitial hydrogen has the same consequences on the resistance behavior of the Ni-Mn alloys as an increase of the content of (substitutional) Mn atoms in the hydrogen-free alloy.

ACKNOWLEDGEMENT

We are grateful to G. Bacherer and M. Becker for their assistance. The work is supported by the Deutsche Forschungsgemeinschaft.

REFERENCES

1.   B. Baranowski and M. Smialowski, Bull. Acad. Polon. Sci., Ser. sci. chim., geol. geogr. 7, 663 (1959).
2.   H. J. Bauer and E. Schmidbauer, Naturwissenschaften 48, 425 (1961); Z. Phys. 164, 367 (1961).
3.   K. A. Kohler and H. J. Bauer, phys. stat. sol. (a) 14, K27 (1972).
4.   H. J. Bauer and K. A. Kohler, Phys. Lett. 41A, 291 (1972).
5.   V. M. Beilin, I. L. Rogel'berg and V. A. Cherenkov, Sov. J. Low Temp. Phys. 4, 11 (1978).
6.   V. M. Beilin, T. I. Zeynalov, I. L. Rogel'berg and V. A. Cherenkov, Phys. Met. Metall. vol. 46, 5, 163 (1979).
7.   T. Jo, J. Phys. Soc. Japan 40 (3), 715 (1976).
8.   M. Krukowski and B. Baranowski, Roczniki Chem. 49, 1183 (1975).
9.   H. J. Schenk, H. J. Bauer and B. Baranowski, phys. stat. sol. (a) 52, 195 (1979).
10. H. J. Schenk and H. J. Bauer, Z. f. Physikal. Chemie NF 116, 213 (1979).
11. F. E. Wagner, M. Karger, F. Probst and B. Schuttler, paper FR-27 presented at the 'Intern. Symposium on the Electronic Structure and Properties of Hydrogen in Metals', Richmond, Virginia, March 4-6, 1982.
12. R. M. Roshko and G. Williams, Phys. Rev. B16, 1503 (1977).
13. P. J. Ford and J. A. Mydosh, Phys. Rev. B14, 2057 (1976).

HYDROGEN ABSORPTION AND CRITICAL POINT LOWERING IN THIN PdH_x FILMS

H.L.M. Bakker, G.J. de Bruin-Hordijk, R. Feenstra,
R. Griessen, D.G. de Groot and J. Rector

Natuurkundig Laboratorium
Vrije Universiteit, Amsterdam, The Netherlands

INTRODUCTION

The attractive hydrogen-hydrogen interaction which causes phase transitions in metalhydrides is of elastic origin[1,2]. Two hydrogen atoms embedded in a metal will see each other via their respective long-range displacement fields in the lattice. As the displacement field around a dilatation center falls off as $1/r^2$ the range of the perturbation in the lattice is infinite and the H-H interaction depends on the shape and the size of the sample. This has clearly been put in evidence by Zabel and Peisl[3]. These authors showed that the spatial distribution of the segregated phases in cylindrically shaped Nb samples depended crucially on the ratio of the length to the radius of the samples.

As suggested by Alefeld[1] the H-H interaction energy ε should also depend on the boundary conditions imposed on the sample. For two hydrogens in a finite sample with free boundaries the H-H interaction is negative, i.e. attractive. For a sample with perfectly clamped boundaries the H-H interaction is repulsive and for two hydrogens in an infinite medium it vanishes altogether. As a consequence the phase diagram of a metal hydrogen system should depend on the elastic constraints imposed at its boundary.

It is in general difficult to design an experimental arrangement where the clamping of a bulk specimen can be reached without hampering hydrogen absorption. For thin films however an effective clamping is obtained by evaporating them on a massive substrate.

The purpose of this work is to show that both the critical pressure P_c and the critical temperature T_c of thin PdH_x films on quartz substrates are significantly lower than those of bulk PdH_x.

EXPERIMENTAL PROCEDURE

The isotherms of PdH$_x$ films were measured using a quartz crystal microbalance. This type of balance is very similar to film thickness monitors used in vacuum deposition techniques. It consists of an electronic oscillator whose resonance frequency is determined by the mechanical resonance frequency of a piezo electric quartz crystal. The change Δf in resonance frequency of the quartz crystal is directly proportional to the mass of the deposited film. The mass resolution of the balance used in this work is 3.10^{-10} g.

An additional frequency change Δf_H occurs when the deposited metallic film (palladium, in our case) is absorbing hydrogen. From measurements of Δf_H the concentration x = [H]/[Pd] is determined by means of the relation

$$x = \frac{\Delta f_H}{\Delta f_{Pd}} \cdot \frac{M_{Pd}}{M_H} \qquad (1)$$

where M_{Pd} and M_H are the atomic weight of palladium and hydrogen, respectively. This method has originally been used by Bucur et al.[4] for the study of surface effects on the desorption of hydrogen out of palladium films. Recently it was also used by Frazier and Glosser[5,6] for measurements of pressure-composition isotherms of PdH$_x$ films at room temperature.

In order to investigate the influence of the thickness on the critical temperature T_c of PdH$_x$ films it was necessary to extend the range investigated by Frazier and Glosser to higher temperatures (up to 500K) and higher pressures (up to 15 bar). This introduced an extra complication in the determination of the hydrogen concentration x as the resonance frequency of the quartz crystal (coated or uncoated with palladium) depends strongly both on pressure and temperature. In order to correct for these effects, the dependence on hydrogen pressure and temperature of the resonance frequency of each quartz crystal was measured before coating it with palladium.

The palladium films investigated in this work were evaporated by means of a 25 kW Leybold-Heraeus electron beam gun at a pressure of $\sim 10^{-5}$ Torr. The deposition rate was ~ 0.3 nm per second. As shall be discussed in Section III more consistent results for the pressure-composition isotherms were obtained for palladium films evaporated directly on quartz than for films deposited on the gold or silver electrode of standard Balzers or Leybold-Heraeus quartz crystals. Since at the beginning of the deposition the films are not singly connected their thickness could not be determined from the resonance frequency of their quartz substrate. The mass of these films had thus to be determined by means of an auxilliary Inficon XTM thickness monitor.

After having been transfered to a pressure chamber designed to withstand pressures of ~ 150 bar at 800K, the films were annealed in a vacuum of $\sim 10^{-5}$ Torr for ~ 7 hours at 530K. As checked by scanning electron microscopy this resulted in a much smoother surface of the film.

The pressure chamber which fits into a thermostated Heraeus oven was subsequently warmed up to the desired temperature and the freshly evaporated Pd films were activated by means of 10-20 absorption-desorption cycles with 99.999% pure hydrogen at 300-350K. This treatment which caused a roughening of the film surface, improved the hydrogen capacity and led to reproducible pressure composition isotherms.

During a run small amounts of hydrogen gas were successively added to the system. As expected the equilibration times were significantly shorter than for bulk samples. In the α- and α'-phases, equilibrium was normally reached within a few minutes. In the mixed α-α'-phase region the equilibration times ranged from 20 to 180 minutes.

EXPERIMENTAL RESULTS

The pressure-composition isotherms of PdH_x films evaporated on the gold electrode of the quartz crystals used in commercial film thickness monitors were found to be time dependent. Especially at elevated temperatures (T ≃ 400K) the width of the plateaus in the isotherms decreased with time. This behaviour could be explained by assuming that alloying of palladium with the gold electrode is gradually taking place. As gold contributes one extra electron to palladium one expects the critical point of PdAu alloys to be lower than that of Pd (in analogy with the measurements of Brodowski et al.[7] on Pd based PdAg alloys).

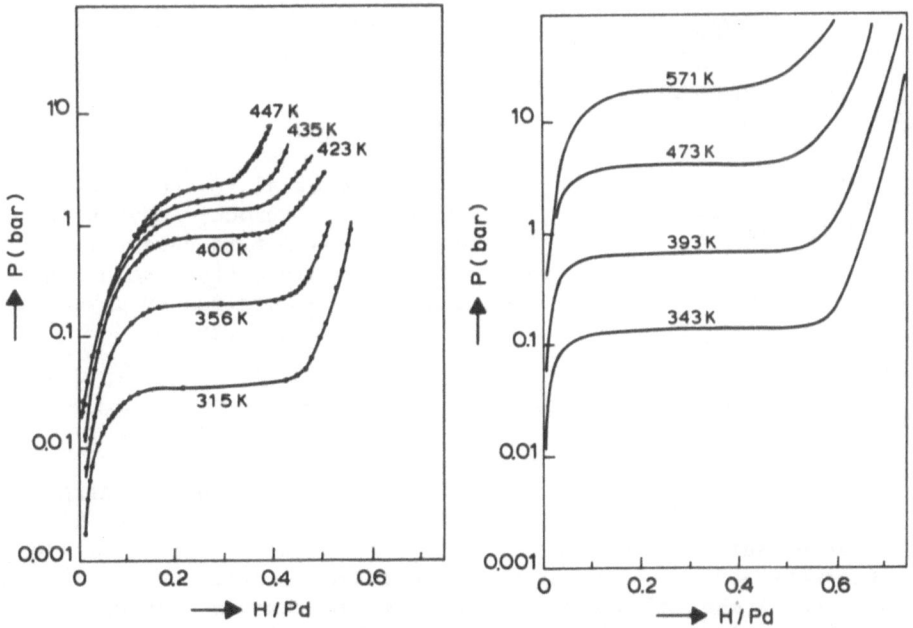

Fig. 1. Hydrogen absorption pressure-composition isotherms of a) a 122 nm PdH_x film and b) bulk PdH_x

In order to eliminate possible alloying effects on T_c we focussed
our attention on palladium samples evaporated *directly* on quartz
crystals with one of their electrodes etched away. Typical absorption
pressure-composition isotherms of a 122 nm thick palladium film are
shown in Fig. 1.a. From a comparison of the isotherms in Fig. 1a
with those of bulk palladium in Fig. 1b one sees that both the
critical temperature (T_c = 470K) and the critical pressure (P_c =
3.1 bar) of the film are considerably lower than the values for bulk
samples (T_c = 565K, P_c = 20.0 bar).

In Fig. 2 the absorption isotherms of the 122 nm thick PdH_x film
are compared to the desorption isotherms of the same film. From the
temperature dependence of the magnitude of the hysteresis loops we
obtained T_c = 450 ± 10K in reasonable agreement with the value
indicated above. From the absorption and desorption isotherms we also
determined the critical concentration to be x_c = 0.26. This value is
comparable to the bulk values x_c = 0.29 and x_c = 0.25 reported by de
Ribaupierre and Manchester[8], and Frieske and Wicke[9], respectively.

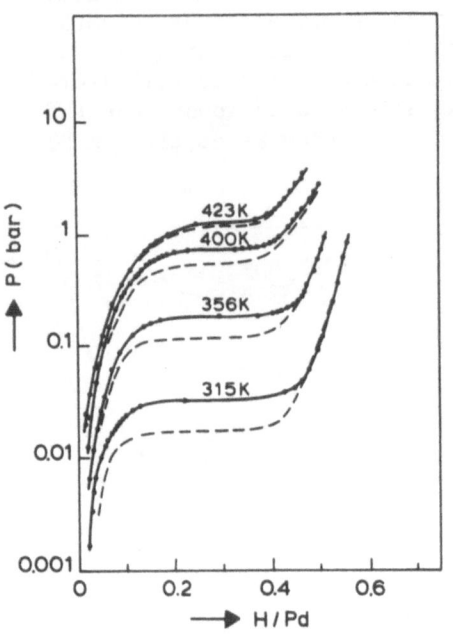

Fig. 2. Hydrogen absorption and
desorption pressure-composition
isotherms of a 122 nm PdH_x film.

As shown in Fig. 3 the absorption and desorption ln p-vs.-1/T
curves of 122 nm and 320 nm PdH_x
films at $x = x_c$ = 0.26 differ
very little from those of bulk
palladium.

DISCUSSION

Picard et al.[11] and recently
Wicke and Blaurock[12] have shown
that a good fit to supercritical
pressure-composition isotherms of
PdH_x can be obtained by means of
the following relation

$$\ln \frac{P}{P_0} = 2 \ln \frac{x}{1-x} + \frac{2\varepsilon nx + 2\varepsilon_0 - \xi_b + 2Ax^3}{kT}$$

(2)

where ξ_b is the binding energy of
the H_2 molecule, ε_0 is the binding
energy of a hydrogen atom on an interstitial site of the Pd-lattice,
ε is the interaction energy between two hydrogen atoms in the lattice
and n is the number of nearest-neighbours. The excess chemical
potential term Ax^3 arises probably from d-band filling[10] and entropy
effects[12]. In the present discussion we shall consider the parameters
εn, ε_0 and A merely as fit parameters.

From the conditions $(\partial\mu/\partial x)_{x=x_c} = 0$ and $(\partial^2\mu/\partial x^2)_{x=x_c} = 0$ which define the temperature and concentration of the critical point we obtain

$$\varepsilon n = -\frac{kT_c(3/2 - 2x_c)}{x_c(1-x_c)^2} \qquad (3)$$

and

$$A = \frac{kT_c(1-2x_c)}{6x_c^3(1-x_c)^2} \qquad (4)$$

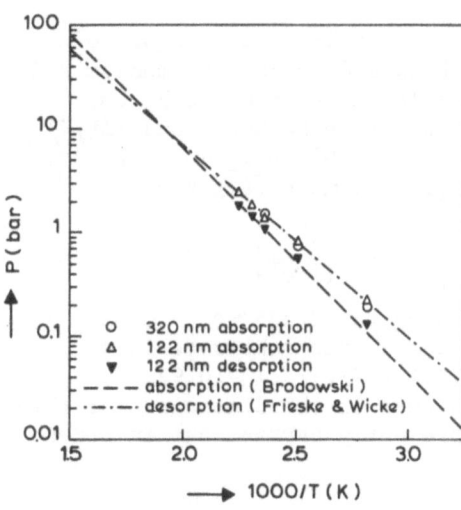

Fig. 3. Temperature dependence of the absorption and desorption dissociation pressures of the 122 nm and 320 nm PdH$_x$ films and of bulk PdH$_x$.[9,10]

From the experimental values T_c = 460K and x_c = 0.26 for the 122 nm thick PdH$_x$ films we derive the following values for εn and A

$$\varepsilon n = -0.273 \text{ eV } (-0.347 \text{ eV}) \qquad (5)$$
$$A = 0.330 \text{ eV } (0.463 \text{ eV}) \qquad (6)$$

In Eqs.5 and 6 the values indicated in parentheses refer to bulk palladium with T_c=565K and x_c = 0.25.

Inserting Eqs.3 and 4 into Eq.2 we find that the critical pressure is given by

$$\ln\left(\frac{P_c}{P_0(T_c)}\right) = 2 \ln \frac{x_c}{1-x_c} + \frac{10x_c-8}{3(1-x_c)^2} + \frac{2\varepsilon_0-\xi_b}{kT_c} \qquad (7)$$

where P_0 is proportional to $T^{7/2}$ (see refs. 13 and 14). For the 122 nm thick PdH$_x$ film P_c = 3.1 ± 0.6 bar and from Eq.7 we obtain

$$\varepsilon_0 - \frac{\xi_b}{2} = -0.121 \text{ eV } (-0.093 \text{ eV}) \qquad (8)$$

The value in brackets corresponds to P_c = 20.0 bar for bulk palladium.

An interesting property of the values given in Eqs.5, 6 and 8, which have been determined from critical point data only, is that the quantity

$$\Delta H^c = \varepsilon nx_c + \varepsilon_0 - \frac{\xi_b}{2} + Ax_c^3 \qquad (9)$$

is independent of the film thickness. In bulk palladium ΔH^c_{bulk} = -0.173 eV while ΔH^c_{film} = -0.186 eV for the PdH$_x$ films. The similarity of the ln p vs. 1/T curves for bulk and films in Fig. 3 indicates that also below T_c $\Delta H^c_{film} \simeq \Delta H^c_{bulk}$. Since the contribution of Ax_c^3

in Eq.9 is negligible $\Delta H^c_{film} \simeq \Delta H^c_{bulk}$ implies that the variation in εn is almost perfectly compensated by the variation in ε_0. This means that in a film the binding energy ε_0 is more negative than in bulk Pd but, at the same time, that the H-H interaction is less attractive in thin samples.

The weakening of the H-H interaction is probably due to the boundary conditions imposed by the quartz substrate on the palladium film. In the case of an isotropic elastic continuous medium[15], for example, the H-H interaction is given by

$$\varepsilon = -\frac{P^2\rho}{B}\left[\frac{(\Delta V/V)}{(\Delta V/V)_{free}} - \frac{1+\nu}{3(1-\nu)}\right] \tag{10}$$

where P is the trace of the dipole moment tensor, ρ the number of hydrogen atoms per unit volume, B the bulk modulus of the medium and ν the Poisson ratio. $(\Delta V/V)_{free}$ is the dilatation produced by one hydrogen atom dissolved in a sample with a free boundary and $(\Delta V/V)$ the dilatation of a partially clamped sample. Since $T_c \sim \varepsilon$, the critical temperature of a partially clamped sample is given by

$$\frac{T_c}{T_{c_{free}}} = \frac{3(1-\nu)}{2(1-2\nu)}\left[\frac{(\Delta V/V)}{(\Delta V/V)_{free}} - \frac{1+\nu}{3(1-\nu)}\right] \tag{11}$$

For the PdH$_x$ films $T_c/T_{c_{free}} = 0.81$. Using $\nu = 0.435$ we calculate[16] then from Eq.11 that the free effect of the quartz substrate is to partially "clamp" the film so that $(\Delta V/V)/(\Delta V/V)_{free} = 0.97$.

Dilatometric measurements of the changes in the dimensions of thin films during hydrogen absorption are presently carried out in order to confirm this conclusion.

REFERENCES

1. G. Alefeld: Ber. Bunsenges. Phys. Chem. 76, 746 (1972)
2. H. Wagner, H. Horner; Adv. Phys. 23, 587 (1974)
3. H. Zabel, H. Peisl; Acta Metall. 28, 589 (1979)
4. R.V. Bucur, V. Mecea, E. Indrea; J. Less Comm. Met. 49, 147 (1976)
5. G.A. Frazier, R. Glosser; J. Phys. D 12, L113 (1979)
6. G.A. Frazier, R. Glosser; J. Less Comm. Met. 74, 89 (1980)
7. H. Brodowski. E. Poeschel; Z. Phys. Chem. N.F. 44, 143 (1965)
8. Y. de Ribaupierre, F.D. Manchester; J. Phys. C 7, 2126 (1974),
   8, 1339 (1975)
9. H. Frieske, E. Wicke; Ber. Bunsenges. Phys. Chem. 77, 50 (1973)
10. H. Brodowski; Z. Phys. Chem. N.F. 44, 129 (1965)
11. C. Picard, O.J. Kleppa, G. Boureau; J. Chem. Phys. 69, 5549 (1978)
12. E. Wicke, J. Blaurock; Ber. Bunsenges. Phys. Chem. 85, 1091 (1982)
13. J.R. Lacher; Proc. Roy. Soc. (London) Ser. A 161, 525 (1937)
14. L.D. Landau, E.M. Lifshitz; Course of Theoretical Physics, vol. 5
    Statistical Physics, Pergamon Press, Oxford (1980)
15. L.D. Landau, E.M. Lifshitz; Course of Theoretical Physics, vol. 7
    Theory of elasticity, Pergamon Press, London (1959)
16. B.M. Geerken, R. Griessen, L.M. Huisman and E. Walker, to be
    published.

HYDROGEN IN METALS 1982

A. M. Stoneham

Theoretical Physics Division, Building 424.4
AERE, Harwell, Didcot
Oxon UK

INTRODUCTION

Any summary of a major conference in an active and diverse field must be a personal view. One can look back on previous meetings to gain an historical perspective, and I shall look back to the Jülich meeting (1) as a guide to the changes over the last decade. It is the future directions, the areas where activity, controversy or promise are most evident, which are both the hardest to identify and the most important to isolate among the ninety or more papers at this meeting.

My summary falls into two main parts. First, there are issues of energetics, i.e. the energies which determine the equilibrium behaviour of dilute or concentrated systems. Secondly, there are questions of kinetics, i.e. those phenomena which are either time-dependent or where lack of equilibrium plays an essential role.

ENERGIES AND ENERGETICS

Energies of various sorts

Experiment gives several distinct types of energy, of which the most important are:

(1) Thermodynamic free energies, typically involving a change in hydrogen concentration (heat of solvation, chemisorption energy, etc.) or a redistribution of hydrogens over sites (as in ordering);

(2) Activation energies, whether describing motion, detrapping, or solid state reaction, defined in terms of a rate $w(T)$ by:

$$E_A^{eff} = kT^2 \frac{d}{dT} \ln[w(T)].$$

For $w(T) \propto T^n \exp(-E_A/kT)$, $E_A^{eff} = E_A + nkT$. Since E_A may be small in hydrogen-metal systems, the (nkT) factor can matter. There is plenty of scope for misinterpreting both the origin of E_A^{eff} (§3) and the magnitude of E_A.

(3) Vibrational energies, whether from neutron, Raman or other methods. It should not be presumed that hydrogen behaves as an Einstein oscillator, and certainly not as a harmonic oscillator; the nature of the excited state is easily misinterpreted.

(4) Electronic excitation energies, as in photoelectron emission, electron energy loss, etc.

Most theories give quite different energies. First, theory (with some exceptions) deals with thermodynamic internal energies ($T = 0$, $\Delta V = 0$ rather than finite T, $\Delta P = 0$;). Secondly, we must distinguish between these distinct expressions related to energies (Table 1). Often so-called total energies are merely sums of one-electron eigenvalues $\Sigma' \varepsilon'_i$ over occupied states i. Such sums double-count electron-electron interactions and omit nuclear-nuclear interactions (possibly core-core terms too); the combined effects are particularly large for interstitial impurities.

Contributions to the total energy

As a framework for discussion, consider an extra hydrogen added to a host (which may be pure metal or a hydride). Before adding hydrogen, the total energy depends on the various atomic displacements. In the harmonic approximation:

$$E_{host} = E_{host}^o + \tfrac{1}{2} \underset{\sim}{u} \cdot \underset{=}{A} \cdot \underset{\sim}{u}, \tag{2}$$

with $\underline{A}$ defining the force constants. The extra hydrogen gives several terms (here we include all vibrational energy corrections too):

$$\varepsilon_H = \varepsilon_{HO} - \underset{\sim}{F}_{HO} \cdot \underset{\sim}{u} + \tfrac{1}{2} \underset{\sim}{u} \cdot \underset{=}{a} \cdot \underset{\sim}{u} + \delta \tag{3}$$

where δ contains higher-order terms, often negligible. We note now (a) that ε_{HO} contains most of the chemical interactions which decide site occupancy; (b) that ε_{HO} depends upon electronic state, and hence upon magnetisation or on electronic excitation; (c) the forces F_{HO} lead to volume change and to elastic interactions; they also incorporate the electron-phonon coupling; (d) the elastic constants and local mode excitation energies are altered by $\underline{a}$ and also by the higher-order terms. We may include the effects of zero-point motion quite neatly by using methods of defect electronic structure (effectively a special case of self-consistent phonon methods).(6)

Table 1

Name	Meaning	Comment
Interaction	When hydrogens are placed in sites i, j the total energy contains terms independent of R_{ij} plus an <u>interaction</u> I_{ij} The <u>sites</u>, not H <u>coordinates</u> are labelled.	Used in empirical analysis of phase diagrams (2) and stability regimes (3). Contains electronic and elastic terms.
Interatomic Potential	Describes the variation of total energy with position of two atoms. Depends on the exact position of the atoms, not just their sites. "Few body" potentials involve more than two atoms.	Used to calculate elastic energies for volume change (4), diffusion (5), etc.
Total Energy	Includes all electron-electron, electron-core and core-core interactions. Can (and should) include zero-point energy.	May be broken down into interatomic potentials (either two- or more-body potentials), possibly not uniquely.
One-electron Energy	Eigenvalue of one-electron Schrodinger equation. For wavevector $\underline{k}$, this gives the $E_n(k)$ of band theory.	Standard in band theory. Optical transitions are often interpreted as differences of one-electron levels; however, this (Koopmans' theorem) should not be trusted without good reason.

Rigid Lattice Energies

At Jülich, Switendick's analysis (7) made it clear the historic anionic and protonic descriptions were at best a partial qualitative description.   Ten years later, a more critical attitude to these extreme pictures (and indeed to simple screened proton models) is apparent, though the improvements in sophistication anticipated do not seem to have happened.   The way in which hydrogen lowers already occupied levels with s-character at the hydrogen site is clearly appreciated.   Charge transfer between metal and hydrogen is now calculated self-consistently as part of many serious calculations. One still needs to stress that different experimental methods measure different apparent charges: lattice dynamics measures dipole moment

Table 2

Tight-binding band model	Semi-Empirical Molecular Cluster model	Phenomenology (e.g. Miedema)
1. ELECTRONEGATIVITY		
Intra-site matrix elements	Electronegativity from ionisation potential or electron affinity	Electronegativity
2. PROPENSITY TO BOND		
Inter-site matrix elements	Bonding parameter (CNDO) or similar	Charge density at edge of Wigner Seitz sphere
3. ORBITAL RADIUS/ATOMIC SIZE		
Atomic spacing; orbital exponents; sometimes core radius	Orbital exponents	Radius of Wigner-Seitz sphere

per unit displacement; electromigration, optical charge transfer and simple band theory all involve different quantities still.

It is useful to begin by stressing similarities of electronic structure methods.   Almost all models contain, effectively (and often disguised) three physical features for each type of orbital: a measure of atomic size, or orbital radius, a measure of electronegativity, and a measure of a tendency to bond for each species (Table 2).   The differences come in questions of flexibility of basis, the quantitative accuracy of different terms, and specific technical aspects like self-consistency and whether it is the charge density in real space or that in reciprocal space near a specific energy (normally the Fermi energy) which is needed.   The good and bad points are summarised in the Table 3, which makes two main points.   First, however good the method, it is always possible to do a bad calculation.   Secondly, virtues depend enormously on what you want to predict.   The best theories go beyond one-electron bands and predict quantities actually observable.   Wavefunctions are as important as energies, and both de Haas-van Alphen (8) and Knight shift work (9) indicate real outstanding problems; the atomic nature of matter, not just electron densities, matter.   Likewise, it is

Table 3

Approach[o]	Good Aspects	Common faults, sometimes avoidable
Jellium	Some exact results	No structural features. Often poor accord with experiment.
Band theory	Good for Fermi surface properties and cases where k selection rules obeyed. No boundary problems	Total energy not given; Distortion hard to include; Not self consistent
Periodically repeated cluster (large unit cell)	Total energy. Self consistent. No surface problem	Only certain stoichiometries; Convergence may be poor; Geometry constrained
Cluster	Self-consistent. Total energy (including free energies) Geometry easily changed. Better theory methods.[+]	Surface problems[*] Where is E_F? One should interpolate on the number of electrons to make E_F match that of the perfect cluster)
Phenomenological (e.g. Miedema)	Extensive comparisons of many systems: trends	Little detail for individual systems
Empirical (from phase diagrams or phonon dispersion)	Systematises data, giving a minimum theory	Not unique; Possibly not transferable

o The reader should be suspicious of any theory called "ab initio". Usually the phrase means merely that the authors found a convenient place to begin.

+ Any model used for a band calculation (e.g. a choice of pseudo-potential) can be used for a cluster calculation, but not vice-versa.

* In our own experience surface problems are not severe if one takes sensible precautions (e.g. (1) avoid problems for which a cluster is clearly inappropriate, (2) take reasonable cluster sizes, (3) check your code is working correctly, (4) make sure surface states are not near E_F).

interesting that hydride stability appears to depend little on the
electronic structure near the Fermi level.   We may also ask what
accuracy we should expect from current experience.

Some of the issues which have arisen at this meeting are con-
tinuing problem areas: the influence of hydrogen on magnetisation
and magnetic factors which affect hydrogen behaviour; the non-
elastic interactions of hydrogen with other hydrogens, other
impurities or with lattice defects; the static lattice contributions
to site preference energies.   Other cases bring new techniques to
bear.   The use of the muon as a probe, as well as a special hydrogen
isotope, is apparent in the studies of Knight shifts and of phase
transitions.   The pion, π^+, made a timely and useful contribution in
understanding site occupancies (10).   The XPS, photoemission and
energy loss experiments illustrate a range of techniques which can be
used to probe excited states.   One of the remaining outstanding
issues concerns non-stoichiometry and the degree to which order
occurs and can be represented simply.   This seems an important
factor in calculations of Dingle temperatures (11).   Clearly a real
test of understanding would be success in predicting the major
qualitative features of phase diagrams: will there be two phases or
one? will there be a metal to non-metal transition?   Obviously it
would be more convincing still if the nature of the ordering and the
way it builds up were predicted too, but this undoubtedly involves
elastic energies too, which should form an important part of any discussions of chemisorption and bulk ordering.

There are two obvious basic questions concerning both the
hydrogen-hydrogen interactions and interatomic potentials.   Are they
pairwise?   Are they transferable?   Several points suggest that any
H-H interaction is sensitive to the presence and arrangement of
neighbouring hydrogens: (i) the asymmetry of phase diagrams with
respect to 50% site occupancy (12), (ii) the apparent long range
of interactions needed to model phase diagrams (2), (iii) the apparent
long range of potentials needed to model phonon dispersion curves
(13), (iv) the complexity of observed phase diagrams (14), and (v)
the various identified mechanisms which could give rise to few-body
terms (Jahn-Teller terms (15), crystal field terms, etc.).

Phonon dispersion curves are a powerful tool for <u>harmonic</u> interactions.   The problems come from effects of disorder in non-
stoichiometric hydrides and from non-uniqueness.   In particular,
one should be suspicious of fitted pairwise interactions of very
long-range; short-range three-body forces may give much the same
effect.   The observed PdD_x (x $\sim$ 0.75) dispersion curves are con-
spicuously similar to those for ZrC_x and HfC_x, where Weber's screened
shell model (16) (with a strong covalent bonding in addition to
metallic and (screened) ionic terms) gives a convincing analysis.
Transferability is harder to assess, though there is some support

from the systematic minimum H-H spacings (3) and also from the fact
that rare-earth/hydrogen potentials from observed defect transitions
in doped CaF_2 give good hydrogen frequencies for stoichiometric metal
hydrides MH_2. (17)

Linear Coupling

We may single out two key areas: (I) The electron-phonon interaction for states near the Fermi level, relating to superconductivity,
electronic transport properties and some aspects of magnetic behaviour;
(II) Elastic energies, appearing in volume changes, in hydrogen-
hydrogen or hydrogen-defect interactions (and so in phase stability
and ordering) and in polaron aspects of diffusion.

Ten years ago, the isotope dependence of hydrogenated Pd was new
and provoking. The challenge is still with us, though clarified by
systematic studies and analysis within MacMillan's theory.   This
meeting showed several useful developments.   The de Haas-van Alphen
method was used to identify where the action occurred at the Fermi
surface (8); likewise, μSR (9) seemed to show an isotope-dependent
Knight shift.

Several other aspects caught my attention. One was the theoretical analysis of trends with d-band occupancy (18). Another (19)
was the large effect of long range order (which reduces T_c by 9-10%
in $Pd\,H_{0.837}$ or $PdD_{0.817}$) and of short range order (which raises T_c
by 7.5% in $PdD_{0.742}$).   The use of ion implantation to identify
rapidly compositions favouring superconductivity is another nice use
of non-equilibrium methods (20).

The enormous importance of elastic interactions has been known
for years, though ignored frequently and persistently, especially in
chemisorption.   At the Jülich meeting, various separate themes - the
Nb:H phase diagram, and diffusion - began to give a consistent picture
of lattice distortion and its consequences.

The major difficulty remaining hinges on the question: How, even
for a pure metal, does one separate volume-dependent and structure-
dependent terms in the total energy? How are isotope effects included? Peisl's review (21) notes sometimes D gives the larger volume
change (Pd; Pd/Ag), sometimes H(Ta); sometimes ΔV is the same for the
two isotopes; in many cases ΔV is independent of <u>host</u>! Most calculations use either continuum elasticity theory (often presumed isotropic) or without a separate volume-dependent term (22). There are now
ic) or one of the many published interatomic potentials, either with
or without a separate volume-dependent term (22). There are now a
number of issues:
(a) If one wants to fix H-metal potentials from the observed
volume change, how does one divide ΔV between short-range and purely

volume terms? Formally, one can separate two terms: ΔV_1 from short-range forces, depending on the virial ($\Sigma F_{\alpha\ell} R_{\alpha\ell}$) and ΔV_2 from the pressure $\frac{\partial^2}{\partial V \partial n} E_{TOT}$ δn from transferring δn electrons to states at the Fermi level.

(b) If one wants to fix H—metal potentials from observed vibrational excitation energies, how does one allow for anharmonic corrections? Indeed, how can one be sure the correct transition is identified, for there are problems even in insulators (23). Are they Franck-Condon or "zero-phonon" transitions? And is the volume constant or not? Anharmonicity is almost certainly important, and one should beware of assuming the same frequency in, e.g. entropies and excitations observed in neutron scatter.

(c) Even if potentials are both pairwise between atoms and transferable (e.g. independent of local fluctuations of H concentration), the elastic interaction energies are not just the sum of component pair energies: explicit calculations (24) show higher-order effects sufficient to influence the evolution of new phases. The separation of volume and structural terms is acute in these cases (25), as in questions of the binding of hydrogen to other hydrogens, to impurities, or to lattice defects.

This problem underlies most analysis of ordering and pairing, and hence many features of concentrated hydrides. The successes of Monte-Carlo methods (2) and of the lattice gas analysis of Horner and Wagner (26) cover an important part of what is seen: it is the remainder and the full quantitative analysis which is missing. Theoretically, for those who believe their electronic structure calculations, the important step would be to give a qualitative form for the short-range potential (including magnetic, crystal field, quadrupole, hybridisation, contributions, etc.) which could be parameterised, and a criterion for separating volume dependent terms.

Second and Higher-order terms

The interatomic interactions just discussed also govern vibrational motion through their higher-order derivatives. Thus smaller, but more rapidly-varying, terms in the total energy are important. Indeed, the observed systematic dependence of frequency on lattice spacing suggests short-range repulsions, not chemical terms, are dominant in the frequencies (but not necessarily in total energies). It was useful to learn that local tunnelling has little effect on optic mode energies. However, effects of H on metal atom motion can be ignored; resonances are seen, and could be important in diffusion so long as the diffusion jump rate does not exceed the resonance frequency.

KINETIC EFFECTS

Diffusion. The Jülich conference came soon after the Flynn-
Stoneham quantum theory of hydrogen diffusion, which combined much of
the agreed physics of hydrogen in metals (e.g. the inertia of lattice
deformation, plus the obvious point that the light hydrogen moves more
rapidly than the heavy host atoms) with the applied mathematics of
small polaron theory (27). The Jülich meeting also included the
first discussions of local tunnelling.

What has happened since? Qualitatively, the theory has
changed little. There have been more extended discussions of
specific points (e.g. the breakdown of the Condon approximation, the
role of excited vibrational states, the origin of isotope effects, all
discussed to some extent in the original paper), and more careful
analyses of data. The real problem continues to be that of potential
energy surfaces and, despite the valiant effort of the Sandia group, a
convincingly complete treatment is still missing. Three examples may
illustrate the present position. First, μSR has now become recog-
nised as an integral part of the hydrogen-metal area, not just an
exotic extra. There are, however, some basic difficulties: already
at the time of the Rorschach meeting it seemed probable that μSR did
not measure simple diffusion rates. The Aℓ:Mn data are a special
challenge. At higher temperatures (T >2K), a conventional diffusion
plus trapping model works well. At lower temperatures two models
have been proposed. Kehr et al (28) argue for rapid self-trapping,
followed with coherent propagation limited by electron scatter.
Browne and Stoneham (29) argued for delayed self-trapping followed by
capture-limited thermalised motion in a delocalised state; impurities
catalyse self-trapping, giving the same higher-T behaviour. Whilst
present data are consistent with both models (the critical aspect is
a muon velocity $\sqrt{(3kT/m_\mu)}$), my personal preference for the model of
delayed self-trapping is based on analogy with other systems plus the
proposal of the role of close octahedral and tetrahedral site
energies, confirmed at this meeting by Kehr et al. Secondly, local
tunnelling states of hydrogen are now respectable. The Nb(O-H) and
Nb(N-H) systems give rather strong tests of the Flynn-Stoneham theory
(30). Direct neutron measurements of tunnel splittings fit in
well too (31). Thirdly, we still have the old problems like the
existence (or otherwise) and interpretation (if needed) of the knee
in the diffusion rate for Nb:H. This leads naturally to questions
of electro and thermotransport, since (thermotransport especially) is
sensitive to mechanism, and no knee is seen in the heat of transport (32).

Phase Transitions. Now that the equilibrium diagrams are better
established and understood (one hopes the complexity will not get
worse!), interest is focussing on mechanisms of phase transitions,
on the dynamical aspects and on new features like incommensurate

states (33) and multiply ordered phases.   There is clearly a whole range of phenomena available for analysis, both for ordering (or dissociation) at constant concentration and for the order which evolves as hydrogen is added.   In particular, metal/non-metal transitions (and their interfacial terms) look promising scientifically.   It is worth noting that μSR may be a good technique for studying the dynamics of such phase transitions in situ, since it will presumably probe the sites most favourable to the next H added.

WHAT ELSE?

Our meeting has covered an enormous variety of topics.   Whilst this meeting stresses the science of hydrogen in metals, its application to technology is of importance as a continual check on our understanding, as a stimulus through the questions applied science poses, and as a potential source of sponsorship in which there are sponsors who actually want to know answers.   I was pleased so see that, as well as the work on superconductivity, there were papers which exploited μSR and hydrogen as probes of catalytic systems and on the role of hydrogen in crack propagation.   However, I would have been happier to detect more awareness of the application of this area of science.   Likewise, I would have expected more parallels with work on other interstitial compounds: carbides, nitrides, borides and oxides.   Looking back again to the Jülich meeting, I was conscious of missing areas.   One is forced diffusion: electromigration and thermomigration, where the whole area seems very poorly understood.   Another concerns hydrogen in liquid metals (34). Why is diffusion in liquid Fe so fast? Is it because of longer jumps of hydrogen excited above the interstitial potential wells (35)?   If so, when do these mechansims take over from small polaron mechanisms?   Do we understand solvation?   For non-hydride formers, hard-sphere models appear to work (36), and differ from the successful solvation cluster models (37) for hydride formers. And why are interstitial impurities more soluble (with rare exceptions like Cu:O) in the liquid than in the solid?   The data are available (38) and show interesting differences between entropy terms (all observed values favouring the liquid) and enthalpy terms, where the liquid is favoured for Nb, Cu, Ag and Aℓ, but the solid for Mg, Fe, Co and Ni.   With the applied interest from liquid-metal coolant systems and from metallurgical processing, I am sure the science should be better understood.

I am indebted to my colleagues for their stimulus and comment, and particularly to Professor W. A. Oates, Dr. D. K. Ross and Dr. H. Schober.

REFERENCES

1.   See the papers in Berichte der Bunsen-Gesellshaft für physikalische Chemie 76 (8) (1972);   contributed papers are given

in reports Jül-Conf-6 (1972).
2. C.K. Hall, this meeting.
3. D.G. Westlake, this meeting.
4. H. Peisl, this meeting.
5. K.W. Kehr, this meeting.
6. A.M. Stoneham and R.H. Bartram, Phys. Rev. B2 3403 (1970).
7. A. Switendick, Ref. 1.
8. R. Griessen and L.M. Huisman, this meeting.
9. A. Schenck, this meeting; see also F.N. Gygax et al, this
      meeting.
10. K. Maier, this meeting.
11. A. Bansil, this meeting.
12. D.K. Ross, private communication; see also ref. 2.
13. J.J. Rush, this meeting; see also C.J. Glinka, J.M. Rowe,
      J.J. Rush, A. Rahman, S.K. Sinha and H. Flotow, Phys.Rev.
      B17 488 (1978).
14. T. Schober, this meeting.
15. G. Abell, this meeting.
16. W. Weber, Phys. Rev. B8 5082 (1973).
17. Anne M. Browne, A.M. Stoneham, Evelyn Wade and J.H. Harding,
      unpublished work (1982).
18. M. Gupta, this meeting.
19. R.W. Standley and C.B. Satterthwaite, this meeting.
20. B. Stritzker, this meeting.
21. H. Peisl, in "Hydrogen in Metals I" edited G. Alefeld and
      J. Völkl, (Springer Verlag 1978).
22. A.M. Stoneham and R.J. Taylor, "Handbook of Interatomic
      Potentials", AERE-R10205 (1981).
23. R.J. Elliott, W. Hayes, G.D. Jones, H.F. MacDonald and
      C.T. Sennett, Proc. Roy. Soc. A289 1 (1965).
24. W.A. Oates, private communication.
25. M.W. Finnis and M. Sachdev, J. Phys. F6 965 (1976).
26. H. Wagner, in "Hydrogen in Metals I" edited G. Alefeld and
      J. Völkl, (Springer Verlag 1978) p.5.
27. C.P. Flynn and A.M. Stoneham, Phys. Rev. B1, 3966-3978 (1970).
      See also Ref. 1.
28. K.W. Kehr, D. Richter, J.M. Welter, O. Hartmann, E. Karlsson,
      L.O. Norlin, T.O. Niinikoski and A. Yaouanc, this meeting.
29. Anne M. Browne and A.M. Stoneham, J.Phys. C15, in press (1982).
30. P.E. Zapp and H.K. Birnbaum, Acta Met. 28 1275 (1980).
31. H. Wipf, this meeting.
32. H. Wipf, in "Hydrogen in Metals II", edited by G. Alefeld and
      J. Völkl, (Springer Verlag 1978).
33. S.C. Moss, this meeting.
34. H.K. Birnbaum and C.A. Wert, Ref. 1, p.806.
35. W.A. Oates, Alison Mainwood and A.M. Stoneham, Phil. Mag.
      A38 607 (1978).
36. T. Emi and R.D. Pehlke, Metall. Trans. 1 2733 (1970).
37. Alison Mainwood and A.M. Stoneham, Phil. Mag. 37B 255 and

263 (1978).

38. E. Fromm and G. Hörz, Int. Met. Rev. $\underline{25}$ 269 (1980).
    For a listing or work on more complex systems (e.g. ternaries)
    see J.H.E. Jeffes and V. Vasantasree, Int. Met.Rev. $\underline{21}$ 128
    (1976) and C.B. Alcock, "Principles of Pyrometallurgy"
    (Academic Press 1976).

PARTICIPANTS

George Abell
Monsanto Research Corporation
P. O. Box 32
Miamisburg, OH 45342

A. Bansil
Physics Department
Northeastern University
Boston, MA 02115

R. L. Armstrong.
Department of Physics
University of Toronto
Toronto, Ontario, Canada
M5S 1A7

Joseph Barak
Soreq Nuclear Research
    Center
Yavne, Israel 70600

A. Attalla
Monsanto Research Corporation
Mound Facility
P.O. Box 32
Miamisburg, OH 45342

Richard G. Barnes
Ames Lab.-USDOE
Department of Physics
Iowa State University
Ames, IA 50011

Hector E. Avram
Department of Physics
University of Toronto
 60 St. George Street
Toronto, Ontario, Canada
M5S 1A7

Hermann Bauer
Sektion Physik der Universitat
Schellingstrasse 4/III
8000 Munich 40, West Germany

D. E. Azofeifa
Department of Physics
Ohio University
Athens, Ohio 45701

M. Belhoul
Physics Department
University of Warwick
Coventry CV4 7AL, England

I. S. Balbaa
Department of Physics
St. George Street, Rm. 074
University of Toronto
Toronto, Ontario, Canada
M4Y 1R6

Raymond Benenson
Department of Physics
Stat Univ. of N.Y. at Albany
1400 Washington Ave.
Albany, N.Y. 12222

Peter Bennett
IE-350 RM
Bell Telephone Labs.
600 Mountain Ave.
Murray Hill, NJ 07974

B. S. Bowerman
University of Birmingham
Physics Department
Birmingham, England

N. F. Berk
National Bureau of Standards
Building 235
Washington, D.C. 20234

R. C. Bowman, Jr.
Monsanto Research Corp.
Mound Facility
P.O. Box 32
Miamisburg, OH 45342

N. F. Bethin
Brookhaven National Lab
Building 480
Upton, NY 11973

D. W. Brown
Department of Physics
The University of Rochester
River Campus Station
Rochester, New York 14627

H. K. Birnbaum
Department of Metallurgical
      Engineering
University of Illinois at
      Urbana-Champaign
Urbana, IL 61801

J. P. Burger
Universite Paris-Sud
Laboratoire de Physique
            des Solides
Bat. 510
91405 Orsay Cedex France

E. B. Boltich
University of Pittsburgh
Department of Chemistry
Pittsburgh, PA 15260

Charles Butler
Department of Physics
Virgina Commonwealth Univ
Richmond, VA 23284

P. Boolchand
Physics Department
University of Cincinnati
Cincinnati, Ohio   45221

Wayne M. Butler
Department of Chemistry
Wheeling College,
Wheeling, W.VA

Russell C. Casella
Reactor Division
National Bureau of Standards
Washington, D.C. 20234

Ted Doiron
Department of Physics
Virgina Commonwealth Univ
Richmond, VA 23284

Miguel Castro
Div. de Estud. de Posgrado
Faculdad de Quimica
Universidad Nacional
Autonoma de Mexico
Mexico 20 D. F. Mexico

J. Eckert
Los Alamos National Lab
MS-805
Los Alamos, NM 87545

G.S. Collins
Department of Physics
Clark University
Worcester, MA 01610

S. Estreicher
Institut fur Theor. Physik
    der Universitat Zurich
Schonberggasse 9
CH-8001 Zurich, Switzerland

Robert M. Cotts
Department of Physics
Clark Hall
Cornell University
Ithaca, NY 14853

Ted Flanagan
University of Vermont
Cook Building, Chem. Dept.
Burlington, VT 05405

Krstyna Dec
Energy Conversion Devices Inc.
1896 Barrett St.
Troy, MI 48084

Carl L. Foiles
Physics Department
Michigan State University
East Lansing, MI 48824

C. Demangeat
L.M.S.E.S.
LA CNRS 306
Universite Louis Pasteur
4, rue Blaise Pascal
67070 Strasbourg Cedex, France

F. Y. Fradin
Materials Science Division
Argonne National Laboratory
Argonne, IL 60439

J. Genossar
Technion-Israel Institute of
        Technology
Faculty of Physics
Technion City, 32000 Haifa
                Israel

M. Gupta
Universite Paris-Sud
Laboratoire de Physique
                des Solides
Bat. 510
91405 Orsay Cedex France

Robert Glosser
University of Texas at Dallas
P.O. Box 688
Physics Programs
Richardson, TX 75080

R. P. Gupta
Commissariat a L'e'nergie
            Atomique
Centre d'etudes Nucle'aires
        de Saclay
Boite Postale No.2
91190 Gif-sur-yvette
        Saclay, France

J. A. Goldstone
Los Alamos National Labs.
MS-805
Los Alamos, New Mexico 87545

Barbara Hall
Materials Science Division
Research & Development Center
Westinghouse
            Electric Corporation
1310 Beulah Road
Pittsburgh, PA   15235

Robert H. Gowdy
Department of Physics
Virgina Commonwealth Univ
Richmond, VA 23284

Carol K. Hall
Dept. of Chemical Engineering
Princeton University
Princeton, NJ   08544

A. V. Granato
Department of Physics
University of Illinois at
        Urbana-Champaign
Urbana, IL 61801

P. A. Hardy
Department of Physics
University of Toronto
Toronto, Ontario, Canada
M5S IA7

Ronald Greissen
Natuurkundig Laboratorium
Vrije Universiteit
                de Boelelaan
    1081
Amsterdam, The Netherlands

O. Hartmann
Institute of Physics
University of Uppsala
P. O. Box 530
751 21 Uppsala
Sweden

J. Haus
Universitat Essen-HGS
D-4300 Essen 1
West Germany

George N. Kamm
Code 6331
Naval Research Laboratory
Washington, D.C. 20375

D. K. Hsu
Physics Department
Colorado State University
Fort Collins, Colorado 80523

T. Kanashiro
Physics Department
Colorado State University
Fort Collins, Colorado 80523

K. Huang
Department of Physics
University of Illinois at
        Urbana-Champaign
Urbana, IL 61801

E. Karlsson
Institute of Physics
Uppsala University
Box 530
S-75121, Uppsala
Sweden

L. M. Huisman
Naturkundig Laboratorium
Vrije Universiteit,
        de Boelelaan
1081
Amsterdam, The Netherlands

Klaus W. Kehr
IFF/KFA Julich
5170 Julich
West Germany

Puru Jena
Department of Physics
Virgina Commonwealth Univ
Richmond, VA 23284

V. M. Kenkre
Department of Physics
The University of Rochester
River Campus Station
Rochester, New York 14627

Bela Joos
Department of Physics
Univ. of British Columbia
2075 Nestbrook Hall
Vancouver, British Columbia
        Canada   V6T-1W5

D. Khatamian
Department of Physics
University of Toronto
Toronto, Ontario, Canada
M5S 1A7

H. A. Kierstead
Solid State Science Division
Argonne National Laboratory
Argonne, Illinois 60439

B. M. Klein
Naval Research Laboratory
Washington, D.C. 20375

J. Kossler
Department of Physics
College of William & Mary
Williamsburg, VA

S. R. Kreitzman
Hebrew University Jerusalem
Racah Institute of Physics
Hebrew Univ, Jerusalem, Israel

Wm. A. Lanford
Department of Physics
SUNY/Albany
1400 Washington Avenue
Albany, NY 12222

William F. Lankford
Department of Physics
George Mason University
Fairfax, Virginia 22030

B. C. Laroy
Phillip Morris Research
    Box 26583
Richmond, VA   23261

Ming-way Lee
University of Texas
    at Dallas
P.O. Box 688
Physics Programs
Richardson, TX 75080

M. Howard Lee
Department of Physics
University of Georgia
Athens, GA 30602

Robert G. Leisure
Physics Department
Colorado State University
Fort Collins, Colorado 80523

M. Leon
MP-3, MS-844
Meson Physics Facility
Los Alamos
        Scientific Laboratory
P. O. Box 1663
Los Alamos, New Mwxico 87545

Irving Lowe
Physics Department
University of Pittsburg
Pittsburg, PA 15260

Harry Lutz
Department of Physics
Virgina Commonwealth Univ
Richmond, VA 23284

R. Marzke
Department of Physics
Arizona State University
Tempe, Arizona 85281

James Lynch
Brookhaven National Laboratory
DEE-815
Upton, NY 11973

L. Mattix
1423-305 Southern Ave.
Oxen Hill, MD 20032

A. Magerl
Institut Max von Laue-
      Paul Langevin
Avenue De Matryrs
156X-38042 Grenoble Cedex
France

F. M. Mazzolai
Department of Metallurgical
      Engineering
University of Illinois at
      Urbana-Champaign
Urbana, IL 61801

K. Maier
Max-Planck Insitut fur
      Metalforschung
Institut fur Physik
Heisenbergstrasse 1
Posfach 80 06 65
7000 Stuttgart 80
            West Germany

P. F. Meier
Physik Institut der
      Universitat Zurich
Schonberggasse 9
CH-8001 Zurich
Switzerland

S. Majorowski
University of Vermont
Cook Building, Chem. Dept.
Burlington, VT 05405

J. L. Moran-Lopez
Departmento de Fisica
Centro de Investigacion y
   de Estudios Avanzados del IPN
Apartado Postal 14-740
07000 Mexico 14 D.F., Mexico

F. D. Manchester
Department of Physics
University of Toronto
Toronto, Ontario, Canada
M5S 1A7

Simon C. Moss
Physics Department
University of Houston
Houston, TX 77004

Mel Mueller
Materials Science Division
Argonne National Laboratory
Argonne, IL 60439

V. Adam Niculescu
Department of Physics
Virgina Commonwealth Univ
Richmond, VA 23284

M. Mullins
Polyatomics Research Institute
1101 San Antonio Road
    Suite 420
Mountain View, CA 94043

M. Numan
Department of Physics
College of William & Mary
Williamsburg, VA

K. Nagamine
Meson Science Lab.
Faculty of Science
University of Tokyo
Hongo 7-3-1, Bunkyo-ku
Tokyo 113, Japan

D.A. Papaconstantopoulos
Naval Research Laboratory
Washington, D.C. 20375

M. Namkung
Department of Physics
College of William & Mary
Williamsburg, VA

Antoni Pedziwiatr
University of Pittsburgh
Department of Chemistry
Pittsburgh, PA 15260

Gary Navrotsky
Reynolds Metals Co.
    Box 27003
Richmond, VA   23225

Hans J. S. Peisl
Geschwister Scholl Pl. I
D 8000 Munchen 22
Germany

K. L. Ngai
Naval Research Lab
Washington D.C.   20375

D. T. Peterson
Ames Lab.-USDOE
Department of Materials Science
Iowa State University
Ames, IA 50011

K. Petzinger
Department of Physics
College of William and Mary
Williamsburg, VA 23185

Peter M. Richards
Sandia National Labs 5132
Albuquerque, NM 87185

Michael Pick
Brookhaven National Lab
76 Cornell St.
Department of Metallurgy
Upton, New York 11973

Wayne Robertson
National Science Foundation
Washington, D.C.   20550

W. E. Pickett
Naval Research Laboratory
Washington, D.C. 20375

D. K. Ross
University of Birmingham
Department of Physics
P.O. Box 363
Birmingham, B15 2TT, England

Faiz Pourarian
University of Pittsburg
Department of Chemistry
Pittsburg, PA 15260

Frank J. Rotella
Argonne National Laboratory
Argonne, IL 60439

A. F. Rex
Department of Physics
University of Virginia
Charlottesville, VA 22901

J. M. Rowe
National Bureau of Standards
A-106 Reactor
Washington, D.C. 20234

James J. Rhyne
Reactor Bldg. 235
National Bureau of Standards
Washington, D.C. 20234

Peter Rudman
INCO/R & D Center, Inc.
Suffern, NY 10901

John J. Rush
Institute of Material Research
  Development
314-0
National Bureau of Standards
Washington, D.C. 20234

T. Schober
IFF
KFA Julich
D-5170 Julich
West Germany

N. Salibi
Department of Physics
Clark Hall
Cornell University
Ithaca, NY 14853

H. Schone
Department of Physics
College of William & Mary
Williamsburg, VA

J. G. Sankar
Allied Corporation Res. & Dev.
P.O. Box 1021R
Morristown, NJ 07960

H. G. Severin
Westfalische
           Wilhelms Universitat
Institut fur Physik. Chemie
Schlossplatz 4
D - 4400 Munster
West Germany

Cam Satterthwaite
Department of Physics
Virgina Commonwealth Univ
Richmond, VA 23284

E. F. W. Seymour
Physics Department
University of Warwick
Coventry, CV4 7AL
England

A. Schenck
Laboratorium fur
  Hochenergiephysik der ETH
CH-5234 Villigen c/o 5IN
Switzerland

Don Shillady
Department of Chemistry
Virgina Commonwealth Univ
Richmond, VA 23284

Mario E. Schillaci
Los Alamos Science Laboratory
MS 844
P.O. Box 1633
Los Alamos, NM 87545

K. S. Singwi
Physics Department
Northwestern University
Evanston, IL 60201

Vijay Sinha
Department of Chemistry
University of Pittsburgh
Pittsburgh, PA 15260

A. M. Stoneham
Building 424.4
Theoretical Physics Division
AERE Harwell, Oxfordshire
Didcot, Oxon OX11 ORA
England

B.W. Sloope
Department of Physics
Virgina Commonwealth Univ
Richmond, VA 23284

B. Stritzker
Institut fur Festkorperforshung
Kernforschungsanlage
D-5170 Julich — Postfach 1913
West Germany

Jim Spivey
Department of Physics
Virgina Commonwealth Univ
Richmond, VA 23284

C. Stronach
Box 358
Virginia State University
Petersburg, VA 23803

K. R. Squire
Philip Morris Inc.
Richmond, VA

W. Studer
Laboratorium
        fur Hochenergiephysik
    der ETH
c/o S.I.N.
CH-5234 Villigen
Switzerland

W. Standish
Department of Physics
SUNY/Albany
1400 Washington Avenue
Albany, NY 12222

Alfred C. Switendick
Division of Materials Sciences
Sandia Laboratories
Albuquerque, NM 87185

Robert W. Standley
Standard Oil Co. (Indianna)
AMOCO Research Center
P.O. Box 400
Naperville, IL 60566

G. Bruce Taggart
Department of Physics
Virgina Commonwealth Univ
Richmond, VA 23284

T. Tanaka
Department of Physics
Ohio University
Athens, Ohio 45701

R. Wappling
University of Uppsala
Institute of Physics
Box 530
S-751 21 Uppsala 1, Sweden

Kiyoyuki Terakura
Institute for
         Solid State Physics
University of Tokyo
Roppongi, Minatoku, Tokyo 106
JAPAN

J. H. Weaver
Synchrotron Radiation Center
University of Wisconsin
Stoughton, Wisconsin 53589

David Loel Tonks
Los Alamos National Laboratory
Group T-11, MS 457
Los Alamos, NM 87545

A. Weidinger
Facultat fur Physik
Universitat Konstanz
Bucklestrasse 13
D-7750 Konstanz, Germany

Eugene L. Venturini
Sandia National Laboratory
Org. 5132
Albuquerque, NM 87185

D. G. Westlake
Materials Science Division
Argonne National Lab
Argonne, IL 60439

Dietrich H. Vincent
Dept. of Nuclear Engineering
University of Michigan
3024 Phoenix Memorial Lab
Ann Arbor, Michigan 48109

S. K. P. Wilson
University of Birmingham
Physics Department
Birmingham, England

F. Wagner
Physics Department E15
Technical University of Munich
D-8046 Garching
West Germany

H. Wipf
Physik-Department der
     Technische
Universitat Munchen
D-8046 Garching b.
Munchen, West Germany

David Witchell
University of Birmingham
Physics Department
Birmingham, England

David Zamir
Soreq Nuclear Research Center
Yavne Israel 70600